普通高等教育"十四五"规划教材

昆虫生理生化

（英汉双语）

主编◎李文楚

Insect Physiology and Biochemistry

(English-Chinese Bilingual)

中国農業大學出版社
China Agricultural University Press
·北京·

内容简介

本书立足中华传统文化，与国际学科前沿研究接轨，本书的出版有利于学习和扩大国际科技交流与合作。本书以英汉双语形式系统介绍昆虫生理生化学研究的基本原理和知识，包括昆虫的体壁、消化系统、呼吸系统、血淋巴循环系统、内分泌系统、排泄系统、感觉系统、神经系统、生殖系统、免疫系统，昆虫信息化合物、滞育和胚胎发育、鳞翅目等昆虫丝腺及丝物质的分泌；重点介绍各大系统的生理解剖组成及其涉及的生理生化过程，从分子水平阐述昆虫中间产物代谢及基因调控机制。本书适用于国内高等农林院校昆虫生理生化课程双语教学，也可供从事昆虫学及相关学科研究的本科生、研究生和科研人员参考。华南农业大学的“昆虫生理生化”双语课程被认定为广东省高等教育“一流课程”。

图书在版编目（CIP）数据

昆虫生理生化：英、汉 / 李文楚主编. --北京：中国农业大学出版社，2023.10

ISBN 978-7-5655-3039-5

Ⅰ.①昆…　Ⅱ.①李…　Ⅲ.①昆虫—生理学—高等学校—教材—英、汉②昆虫—生物化学—高等学校—教材—英、汉　Ⅳ.①Q965

中国国家版本馆CIP数据核字（2023）第150087号

书　名　昆虫生理生化（英汉双语）
Kunchong Shengli Shenghua（Yinghan Shuangyu）

作　者　李文楚　主编

策划编辑　张　程　　**责任编辑**　石　华

封面设计　李尘工作室

出版发行　中国农业大学出版社

社　　址　北京市海淀区圆明园西路2号　　**邮政编码**　100193

电　　话　发行部 010-62818525，8625　　**读者服务部** 010-62732336
编辑部 010-62732617，2618　　**出　版　部** 010-62733440

网　　址　http://www.caupress.cn　　**E-mail**　cbsszs@cau.edu.cn

经　　销　新华书店

印　　刷　运河（唐山）印务有限公司

版　　次　2023年10月第1版　2023年10月第1次印刷

规　　格　210 mm×285 mm　16开本　32.75印张　1 044千字　彩插20

定　　价　98.00元

编审人员

Editing and Reviewing Staffs

主　　编 Chief editor	李文楚 Li Wenchu	华南农业大学 South China Agricultural University
副 主 编 Associate editors	邱宝利 Qiu Baoli	重庆师范大学 Chongqing Normal University
	田　铃 Tian Ling	华南农业大学 South China Agricultural University
	邓惠敏 Deng Huimin	华南师范大学 South China Normal University

编写人员（以姓氏笔画为序）
Editorial staffs (Chinese surname in stroke order)

	邓小娟 Deng Xiaojuan	华南农业大学 South China Agricultural University
	田　铃 Tian Ling	华南农业大学 South China Agricultural University
	邓惠敏 Deng Huimin	华南师范大学 South China Normal University
	李文楚 Li Wenchu	华南农业大学 South China Agricultural University
	邱宝利 Qiu Baoli	重庆师范大学 Chongqing Normal University
	桑　文 Sang Wen	华南农业大学 South China Agricultural University
主　　审 Chief reviewer	韩日畴 Han Richou	广东省科学院动物研究所 Institute of Zoology, Guangdong Academy of Sciences

前言

Foreword

The report of the 20th National Congress of the Communist Party of China pointed out that we will do more to develop basic disciplines, emerging disciplines, and interdisciplinary subjects and speed up the development of world-class universities and strong disciplines with Chinese features. We will nurture a culture of innovation, encourage dedication to science, cultivate fine academic conduct, and foster an enabling environment for innovation. We will expand science and technology exchanges and cooperation with other countries, cultivate an internationalized environment for research, and create an open and globally-competitive innovation ecosystem.

Following the instructions of the 20th National Congress of the Communist Party of China, we have written *Insect Physiology and Biochemistry* (English-Chinese Bilingual). Insect Physiology and Biochemistry is an important course in insect science, and its theory and practice have developed rapidly. After years of development, entomology research in China has stood in the world's advanced ranks. Bilingual teaching has become an important way to cultivate students to broaden their horizons, adjust the structure of professional knowledge, understand the frontiers of discipline development and integrate with international standards.

The book consists of eighteen chapters. Starting from the brief historical development of insect physiology research, according to the knowledge module, systematically introduce integument, digestion and nutrition, fat body and intermediary metabolism, excretory system, respiratory system, hemolymph circulatory system, insect immunity, neuroanatomy and neuro peptides, insect nervephysiology and biochemistry, insect hormones, insect sensory system, light and insect visual receptors, pheromones, insect silk gland and silk protein synthesis, muscle and movement, reproduction and its biochemistry, insect diapause and embryonic development. The content of each chapter is arranged according to insect organizational structure, physiology and biochemical reaction, and molecular process of each organ system. In addition to basic theoretical knowledge, it focuses on the latest progress and development trend of the discipline.

党的二十大报告指出，我们要加强基础学科、新兴学科、交叉学科建设，加快建设中国特色、世界一流的大学和优势学科。培育创新文化，弘扬科学家精神，涵养优良学风，营造创新氛围。扩大国际科技交流合作，加强国际化科研环境建设，形成具有全球竞争力的开放创新生态。

根据党的二十大报告精神，我们编写了《昆虫生理生化》(英汉双语)。昆虫生理生化是昆虫学科的一门重要课程，其理论和实践的发展非常迅速。我国昆虫学研究经过多年发展已站在世界先进行列。双语教学已成为学生开阔视野，调整专业知识结构，了解学科发展前沿以及与国际接轨的重要途径。

本书共 18 章。本书从昆虫生理学研究的简要历史发展入手，按照知识模块，系统介绍体壁、消化与营养、脂肪体和中间代谢、排泄系统、呼吸系统、血淋巴循环系统、昆虫免疫、神经解剖与神经肽、昆虫神经生理生化、昆虫激素、昆虫感觉系统、光和昆虫视觉感受器、信息化合物、昆虫丝腺及丝蛋白质合成、肌肉和运动、生殖及其生物化学、昆虫滞育和胚胎发育。各章节的内容按昆虫各器官系统组织结构、生理和生化反应及分子生物学研究的过程逐一展开。除了基本的理论知识，本书重点介绍学科领域的最新进展和发展趋势。

Experts from several domestic colleges and universities were invited to participate in the writing of some chapters. The names of the staff who participated in the compilation and the newly added content were listed at the end of each chapter of the textbook. The publishing of the textbook has received funding from Guangdong Province's "First Class Curriculum" (Yue Jiao Gao Han〔2022〕No. 10), South China Agricultural University's "14th Five Year Plan" undergraduate textbook construction second phase funding (Huanan Nongjiao〔2023〕No. 43), Plant Protection Discipline Construction in South China Agricultural University's " Double-First Class Project". Hereby I'd like to express my thanks.

The book will be suitable for bilingual education in domestic agricultural and forestry academy. Offering undergraduate and postgraduate entomological students, other researchers and relative technologists, the textbook is an excellent reference.

7th October, 2022

本书的撰写邀请了国内几所院校的专家共同参与。参编和最新增加的编写人员列于每章的末尾。本书的出版得到广东省高等教育“一流课程”（粤教高函〔2022〕10 号）、华南农业大学“十四五”本科教材建设第二期经费（华南农教〔2023〕43 号）、华南农业大学“双一流”植物保护学科建设项目等资助。在此一并致谢。

本书适用于国内高等农林院校开展昆虫生理生化双语教学，可供昆虫学本科生、研究生、其他研究人员和与昆虫学相关行业的专业技术人员参考。

2022 年 10 月 7 日

目录

Contents

绪 论

Introduction

Insects (from Latin *insectum*) are a class of living creatures within the arthropods that have a chitinous exoskeleton, a three-part body (head, thorax, and abdomen), three pairs of jointed legs, compound eyes, and one pair of antennae. They are among the most diverse groups of animals on the planet, including more than a million described species and representing more than half of all known living organisms. The numbers of extant species is estimated at between six and ten million, and potentially represent over 90% of the differing metazoan life forms on Earth (Fig.0-1). So far, fossilized insects of enormous size have been found from the Paleozoic Era, including giant dragonflies with wingspan of 55-70 cm. Insects may be found in nearly all environments, although only a small number of species occur in the oceans, a habitat dominated by another arthropod group—crustaceans. Their morphological diversity and characteristics of life history are highly adaptable to live in various conditions, including extremes of heat and cold, wet and dry, and unpredictable climates. The success of the Insecta is largely related to their high reproductive capacity.

The life cycles of insects vary, but most hatch from eggs. Insect growth of larva is constrained by the inelastic exoskeleton and development involves a series of molts. The immature stages can differ from the adults in structure, habit and habitat, and can include a passive pupal stage in those groups that undergo complete metamorphosis. Insects that undergo incomplete metamorphosis lack a pupal stage and adults develop through a series of nymphal stages.

Insects typically move about by walking, flying or sometimes swimming. As it allows for rapid yet stable movement, many insects adopt a tripodal gait in which they walk with their legs touching the ground in alternating triangles. Insects are the only invertebrates to have evolved flight. Many insects spend at least part of their lives underwater, with larval adaptations that include gills, and some adult insects are aquatic and have adaptations for swimming. Some species, such as water striders, are capable of

昆虫（来自拉丁语 *insectum*）是节肢动物门的一个纲，具有几丁质外壳，由三部分组成的躯体（头、胸、腹部）、三对有关节的腿、复眼和一对触角。它们是地球上最多样化的动物群体之一，囊括超过 100 多万已被描述的物种，相当于所有已知生物的一半以上。现存物种的数量为 600 万到 1 000 万，可能代表着超过地球上 90% 的不同后生动物的生命形式（图 0-1）。截至目前，已经发现古生代规模巨大的昆虫化石，包括翅展为 55 ～ 70 cm 的巨型蜻蜓。昆虫随处可见，只有少部分物种出现在海洋中，由另一节肢动物群——甲壳纲占据的栖息地。它们的形态多样性和生活史特征非常适合在各种条件下生活，包括极端的冷和热、潮湿和干燥以及不可预测的气候。昆虫纲的成功在很大程度上与它们的繁殖能力高有关。

昆虫的生活周期不同，但多数从卵孵化。昆虫幼虫的成长受到无弹性的外骨骼限制，因而发育牵涉一连串的蜕皮。成虫结构、习性和栖息地从幼龄期即可不同，那些进行完全变态的昆虫还包括一个静态的蛹期。不完全变态昆虫没有蛹期，成虫通过一系列若虫期发育。

通常昆虫靠行走、飞行或时而游泳到处活动。考虑到快速而稳定的运动，许多昆虫采用三足步法行走，其腿以交互三角接触地面。昆虫是通过进化飞行的唯一的无脊椎动物。许多昆虫至少在水下度过部分生涯，幼虫的适应性包括气门鳃，有些水生的昆虫成虫能适应游泳。有些物种，如水黾，能在水面行走。

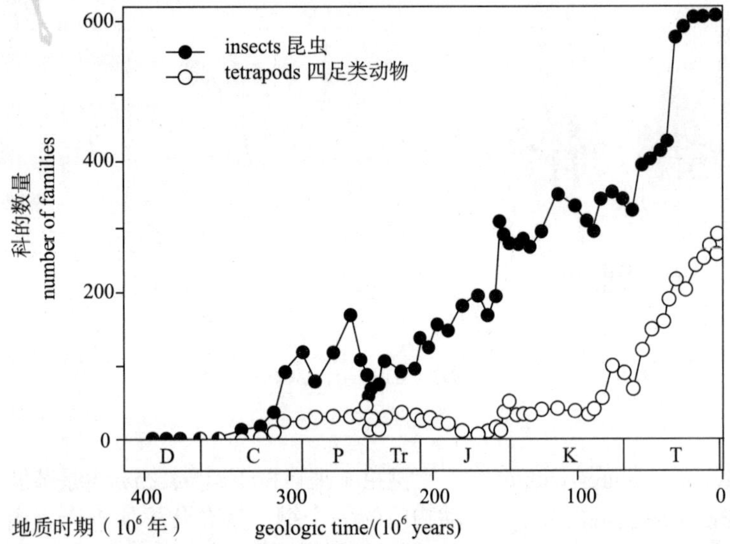

Fig.0-1 Family-level diversity of fossil insects through geologic time. D. Devonian; C. Carboniferous; P. Permian; J. Cretaceous; T. Tertiary. (modified from Labanderia et al., 1993).

图 0-1 地质时期化石昆虫的数量多样性。D. 泥盆纪；C. 石炭纪；P. 二叠纪；J. 侏罗纪；K. 白垩纪；T. 第三纪（仿 Labanderia 等，1993）。

Fig.0-2 Chinese silkworm God, Leizu，who was a legendary empress and wife of Huangdi. According to tradition, she discovered silk and invented the silk loom.

图 0-2 中国蚕神嫘祖，传说的皇后和黄帝的妻子。据说她发现了蚕丝，并发明了丝织机。

walking on the surface of water. Insects are mostly solitary, but some, such as certain bees, ants, and termites, are social and live in large, well-organized colonies. Some insects, such as earwigs, show maternal care, guarding their eggs and young. Insects can communicate with each other in a variety of ways. Male moths can sense the pheromones of female moths over great distances. Other species communicate with sounds: crickets stridulate, or rub their wings together, to attract a mate and repel other males. Lampyridac in the beetle order coleoptera communicate with light.

According to the Chinese legend, the technology of raising silkworms was invented and created by Leizu (Fig.0-2), which indicates the long history of Chinese sericulture, in the middle of Neolithic Period, dating from 3,000 B.C. It witnessed the rising of many variations of this insect in the voltinism, the molting character, markings, body size and shape of the larvae, and the color, shape and silk quality of the cocoons.

Silkworm belongs to the Animal kingdom, Arthropoda, Insecta, Lepidoptera, Silkworm moth family, *Bombyx.mori.*

Insect parasitism was known in China for a long time in the form of parasitic tachinid flies of silkworms (*B. mori* L.). These tachinid flies were first mentioned in the Chinese literature around 300 AD. The developmental cycle of this tachinid (possibly a species of

多数昆虫为独居，有些昆虫是群居的，生活在庞大而有组织性的群体内，例如，蜜蜂、蚂蚁和白蚁。有些昆虫，如蠼螋，呈现出母系呵护，对其卵和幼虫进行监护。昆虫通过各种途径相互联系。雄蛾可以远距离感知雌蛾的信息素。其他种群通过声音联系：蟋蟀的唧唧鸣叫或用它们的翅膀相互摩擦发声来引诱交配，并排斥其他雄虫。甲壳鞘翅目的萤科昆虫即用光线沟通。

根据中国传说，养蚕技术是由嫘祖（图 0-2）发明创造的，表明中国蚕桑的悠久历史可追溯到新石器时期中期，公元前 3 000 年。它见证了这一昆虫在化性、蜕皮特性、斑纹、虫体大小和形状以及茧色、茧形和丝质的许多演化。

家蚕属于动物界节肢动物门昆虫纲鳞翅目蚕蛾科桑蚕属。

昆虫寄生虫在中国很久以前就广为人知，其形式是家蚕的寄生性蝇蛆。这些寄生蝇在大约公元 300 年的中国文献中就被第一次提及。1096 年，明确描

the genus exorista) were clearly described in 1096. This antedates the first descriptions of insect parasitoids from Europe by 600 years. Another parasitic fly, a flesh fly (possibly *Blaesoxipha. lapidosa* Pape) was noted as the main parasitoid of *Locusta migratoria* manilensis Meyen in 1196. A famous Chinese medical book, *Compendium of Materia Medica* recorded a numbers of insects were used as Chinese Medicine, such as cordyceps sinensis and stiff silkworm. The first Chinese record with a correct description of the life cycle of a hymenopteran parasitoid date from 1704 (Cai et al., 2005).

述了寄生蝇的发育周期（可能是家蚕追寄蝇属的一个种）先于欧洲最早的昆虫寄生虫记录600年。1196年，另一种寄生麻蝇（可能是线纹折麻蝇）被记录为飞蝗的主要寄生虫。中医药名著《本草纲目》记载，一些昆虫可作为中药，如冬虫夏草和僵蚕。1704年，中国第一次记录了膜翅目寄生虫准确的生活周期（Cai等，2005）。

Ancient Sumer, the civilization near the river Euphrates, is the cradle of beekeeping, created about 3,000-3,500 BC. In ancient Egyptian poetry honey was used also as a symbol of love. In the few poems written between 1,100 and 2,000 BC and transmitted into our times. Archaeological finds relating to beekeeping have been discovered at Rehov, a Bronze and Iron Age archaeological site in the Jordan Valley, Israel.Thirty intact hives, made of straw and unbaked clay, were discovered in the ruins of the city, dating from about 900 BC. In ancient Greece, aspects of the lives of bees and beekeeping are discussed at length by Aristotle. The art of beekeeping appeared in ancient China for a long time and hardly traceable to its origin. In the book *Golden Rules of Business Success* written by Fan Li (or Tao Zhugong) during the Spring and Autumn Period there are some parts mentioning the art of beekeeping and the importance of the quality of the wooden box for beekeeping that can affect the quality of its honey.

幼发拉底河畔的古苏美尔文明是养蜂业的发源地，创立于公元前3 000至公元前3 500年。在古埃及诗歌里，蜂蜜被当作爱情的符号，少数诗歌创作于公元前1 100至公元前2 000年，并传承至我们的年代。拉霍夫是以色列约旦河谷青铜器和铁器时期遗址，其考古学发现与养蜂业有关。在公元前900年的城市遗址发现了30个用麦秆和未熟黏土制成的完整蜂箱。古希腊，亚里士多德详尽地讨论了蜜蜂生活方式的各个方面及其养蜂业。古代中国的养蜂工艺出现在很久以前，几乎难以追踪其起源。在由春秋时期的范蠡（或陶朱公）所著的《致富奇术》中，有些部分提及了养蜂的工艺以及养蜂木箱质量的重要性，因为它能影响蜂蜜的质量。

It was not until the 18th century that European natural philosophers undertook the scientific study of bee colonies and began to understand the complex and hidden world of bee biology. Preeminent among these scientific pioneers, Jan Swammerdam and Réaumur were the first to use a microscope and dissection to understand the internal biology of honey bees. Réaumur observed queens laying eggs in open cells, but still had no idea of how a queen was fertilized; nobody had ever witnessed the mating of a queen and drone and many theories held that queens were "self-fertile," while others believed that a vapor or "miasma" emanating from the drones fertilized queens without direct physical contact. Huber was the first to prove by observation and experiment that queens are physically inseminated by drones outside the confines of hives, usually at a great distance away. Huber confirmed that a hive consists of one queen who is the mother of all the female workers and male drones in the colony. Together, he and Burnens

直到18世纪，欧洲的自然科学家才着手进行蜂群的科学研究，开始弄懂蜜蜂生物学复杂而神秘的世界。在这些卓越的科学先驱者中，Jan Swammerdam和Réaumur是第一次应用显微镜和切片观察蜜蜂的内部器官生物学。Réaumur观察到蜂王在开放的巢室内产卵，但对蜂王如何受精毫无头绪；没有人见过蜂王和雄蜂交配，许多理论认为蜂王是"自体受精，"而其他人相信雄蜂射出水蒸气或"瘴气"使蜂王受精而无直接体接触。Huber是第一个通过观察和实验证明蜂王是在蜂箱范围之外，通常距离很远的地方，由雄蜂进行体外授精的人。Huber证实蜂群有一个蜂王，是蜂群所有雌工蜂和雄蜂的母亲。他与Burnens dissected bees under the microscope and were among the first to describe the ovaries and spermatheca, or sperm store, of queens as well as the penis of male drones. Huber is universally regarded as "the father of modern bee-science" and his *New Observations on Bees* revealed all the basic scientific truths for the biology and ecology of honeybees (Bricker,2021).

In 1906, Morgan (Fig.0-3; Appendix color graphic Fig.0-3) began his work on *Drosophila melanogaster* and reported his first finding of a *white* (eyed) mutant in 1910 to the academic community. He was in search of a model organism to study genetic heredity and required a species that could randomly acquire genetic mutation that would visibly manifest as morphological changes in the adult. His work on *Drosophila* earned him the 1933 Nobel Prize in Medicine for identifying chromosomes as the vector of inheritance for genes. The white(eyed) mutant and other *Drosophila* species are widely used in studies of genetics, embryogenesis, and other areas.

Malpighi, the founder of comparative physiology, Italian physician and biologist who, in developing experimental methods to study living things, founded the science of microscopic anatomy. He conducted many studies of insect larvae. The most important of which was his investigation in 1669 of the structure and development of the silkworm. The Malpighian tubule system is a type of excretory and osmoregulatory system found in some atelocerata (insects and myriapoda), arachnids and tardigrades. The system consists of branching tubules extending from the alimentary canal that absorbs solutes, water, and wastes from the surrounding hemolymph. The wastes then are released from the organism in the form of solid nitrogenous compounds. The system is named

一起在显微镜下解剖蜜蜂，是所有研究者中第一个描绘蜂王卵巢、受精囊或储精囊以及雄蜂的阴茎的人。Huber 通常被尊称为"现代蜜蜂科学之父，"其《蜜蜂新观察》揭示了蜜蜂生理和生态学的所有基本科学真理（Bricker，2021）。

1906 年，Morgan（图 0-3；附彩图 0-3）着手果蝇的研究，并于 1910 年向学术界报道了第一个白眼突变体的发现。他在寻找一种模式生物研究基因遗传性，需要一种能随机获得基因突变的物种并在成年动物表现出明显的形态变化。他对果蝇的研究使他获得 1933 年的诺贝尔医学奖，以表彰他确认染色体是遗传物质基因的载体。这个白眼突变体和其他果蝇品系已被广泛应用于遗传学、胚胎发生学和其他领域。

意大利生理学家和生物学家 Malpighi 是比较生理学奠基人，发展了生物的研究方法，创立了显微解剖学。他创造了很多研究昆虫幼虫的方法。最重要的发现是 1669 年的家蚕结构和发育的研究。马氏管系统是在一些缺角类动物（昆虫纲和六足类）、蛛形纲动物和多足类动物发现的。该系统由分枝的细管组成，从消化道延伸以便吸收周围血淋巴中的溶质、水分和废料。然后，废料从生物体中以固态氮化合物的形式排放。这个系统以 Malpighi 的名字命名。1762 年，

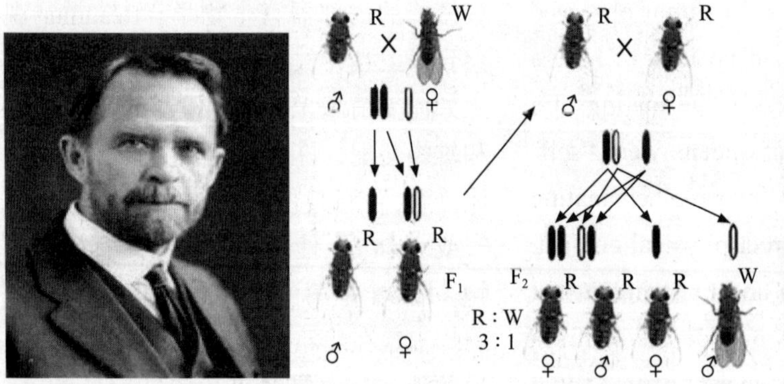

Fig.0-3 Thomas Hunt Morgan and his work on *D.melanogaster* white (eyed) mutant.

图 0-3 摩尔根和他的研究果蝇白眼突变体。

Fig.0-4 "The father of insect physiology" Vincent Brian Wigglesworth.

图 0-4 "昆虫生理学之父" Vincent Brian Wigglesworth。

after Malpighi. In 1762, the French anatomist, Lyonet, described as "granulated vessels", a pair of minute organs located within the thorax of caterpillars. This description was soon forgotten and for 187 years has been buried among the literature pertaining to insect anatomy. Meanwhile, within the twentieth century, the very same organs have been, and apparently continue to be, rediscovered by various investigators (William, 1949). These organs, now known as "prothoracic glands" are among the most important endocrine glands in insects.

A thread that runs a tortuous course from Aristotle's time through the 19th century, visualizing the insect larva and pupa as active embryos was broken by Darwin, who saw the relationship of larva and pupa to the adult insect, with the pupa as a bridge passage. That concept of metamorphosis as sequential polymorphism finally crystallized out to a spare sentence: "Metamorphosis is merely one type of polymorphism" in V.B.Wigglesworth's (Fig.0-4) slim monograph *The Physiology of Insect Metamorphosis* (1954).

Weismann (1864), who first fully recognized the function and significance of the imaginal discs, illustrated 100 years earlier by Lyonet (1762), who proposed the "germplasm" theory of heredity. The germplasm theory states that an organism's cells are divided into somatic cells (the cells that make up the body) and germ cells (cells that produce the gametes). But the mechanisms of molting and postembryonic development remained mysterious even long after the importance of hormones had been recognized in the vertebrates. Kopéc's (1922) experiments, leading to the then heretical proposal that the insect brain secreted a hormone responsible for metamorphosis, opened the book to the final chapter in the endocrine physiology of postembryonic development of insects, and yet it was to take another thirty years before a coherent theory emerged. The subject attracted many skilled experimentalists in the years following that first postulate that nerve cells could secrete and that hormones might regulate postembryonic development, but it was Wigglesworth who brought together his own brilliant experiments with the work of others, from the 1930's to the 1950's. By 1939 he had completed the first edition of *The Principles of Insect Physiology*. With *Rhodnius* he first showed by implantation experiments that protocerebral neurosecretory cells were the source of the hormone that initiates the molting cycle (Wigglesworth, 1939). This finding, building on Kopéc's, was the

法国解剖学家Lyonet描述了位于毛虫胸部的一对极小的器官“粒状小管”。他的描述很快被人遗忘，并一直被埋藏在昆虫解剖学相关的文献中达187年。其间的20世纪，不同的学者陆续重新发现了非常相似的器官（William,1949）。这些器官现在被称为“前胸腺”，是昆虫最重要的内分泌腺体。

从Aristotle年代到19世纪的线索曲折地行进，将昆虫幼虫和蛹当成活跃胚胎的设想被Darwin打断，他考虑了昆虫幼虫和蛹与成虫的关系，其间蛹起着桥梁通道作用。变态的概念就是时序化的多态性，最后凝练成简化的命题：在Wigglesworth（图0-4）单薄的专著《昆虫变态发育生理学》（1954）中“变态只不过是多态性的一种类型。”

Weismann（1864）是第一个完全认识到成虫盘的功能及其意义的，Lyonet（1762）在100年前提出遗传的“种质”理论。种质理论认为，生物细胞被分裂成体细胞（组成躯体的细胞）和生殖细胞（产生胚子的细胞）。但直至认识到脊椎动物激素的重要性之后的很长时间，蜕皮和胚后发育机制仍然保持着神秘。Kopéc（1922）的实验得出“离经叛道”的动议，昆虫脑分泌的激素是变态的原因，这打开了昆虫胚后发育学书籍关于内分泌生理学的最后篇章，此后又用了30年才出现一致的理论。这个学科早前已吸引了许多技能高超的实验学家，跟随着早期的假定是神经细胞能分泌，而激素可调节胚后发育，但Wigglesworth在1930—1950年才将自己的杰出实验与其他人的工作联系在一起。1939年，他完成了成为新学科经典的第一版《昆虫生理学原理》。他第一次用长红猎蝽通过移植实验证明，前脑的神经分泌细胞是启动蜕皮周期的激素来源（Wigglesworth, 1939）。这个发现

first experimental demonstration of an endocrine role for neural cells in any animal. But molting and growing are not the only processes to need explanation: what regulates maturation and metamorphosis in insects? The first indication that the assumption of adult features (metamorphosis) was under control from the head again came from decapitation experiments with *Rhodnius* (Wigglesworth, 1934). Insects that were deprived of endocrine glands attached to the brain (the corpora allata) underwent a precocious molt to the adult state, while implantation of corpora allata from juvenile stages ensured a juvenile molt. Thus the third one of the major developmental hormones–the juvenile hormone was recognized and the triumvirate has since been demonstrated in innumerable insects. The composition and modes of action of the three insect hormones has, of course, become a major effort of insect developmental biology (Nijhout,1994). In the second half of the 20th century, the pioneers who explored a wide variety of invertebrate phyla and developed various preparations appropriate for electrophysiological studies. Their work advanced general knowledge about neuronal properties (dendritic, somatic, and axonal excitability; pre- and postsynaptic transduction mechanisms).

Modern biotechnologies provide advanced facilities to explore every aspect of insects. The morphologies, biochemical processes, and molecular mechanisms of insect development were developed very fast in the recent years. It let us understand more and more knowledge about insect embryonic development, digestive tract, hemolymph and defense system, excretion, respiration, reproduction, muscle, endocrine system, neurophysiology, neuroanatomy, and intermedial metabolism.

The vast amount of literature of insect studies covered all orders and species group of insects. But, indeed, the discoveries of insects which involved in this book just from a part to a whole. Because of space forbids, the book not includes all references.

(Li Wenchu)

是基于Kopéc的实验，是第一个实验证明在任何动物中神经细胞起内分泌作用的。但蜕皮和成长并不是唯一需要说明的过程：什么在调节昆虫的成熟和变态？第一个象征是成虫特征（变态）由头部控制的这一假说再次来自长红猎蝽的头部切除实验（Wigglesworth, 1934）。昆虫被切去连接到脑的内分泌腺（咽侧体），经历早熟蜕皮到成虫状态，而从幼虫期植入咽侧体确保了幼虫期蜕皮。因此，重要发育激素中的第三个激素——保幼激素得以公认，三类昆虫激素从此在无数昆虫中被证实。当然，三类昆虫激素的组成和作用模式也就成为昆虫发育生物学的主要成就（Nijhout, 1994）。20世纪下半叶，先驱者们探索了种类繁多的无脊椎动物门，研制了各种适用于电生理学的制剂。他们的工作使神经特性的普遍知识更为先进（树突、细胞体和轴突兴奋性；突触前膜和后膜转导机制）。

现代生物技术为探讨昆虫的各方面提供了先进的设备。昆虫发育形态学、生化过程和分子机制近年得到快速发展。让我们懂得越来越多关于昆虫胚胎发育、消化道、血淋巴和防御系统、排泄、呼吸、生殖、肌肉、内分泌系统、神经生理学、神经解剖学以及中间代谢的知识。

大量昆虫学研究文献涵盖了昆虫所有种类和种群。但是本书所涉猎的关于昆虫的发现实在是管中窥豹，限于篇幅并没有列出所有参考文献。

（李文楚）

Exercises

1.What do you think about the developmental history and life cycle of insects?

2.Briefly introduce several important tissues and organ systems of insects.

思考题

1. 关于昆虫发展史和生活周期有哪些思考？

2. 简要介绍昆虫的几个重要器官组织系统。

第 1 章 体 壁

Chapter 1 Integument

Key points

1.The insect integument is composed of basenent membrane, epidermis and cuticles. The latter can be divided into more details, such as epi-, exo-, and endocuticle. It serves as the exoskeleton, muscle attachment, first physical line of defense, physical and biochemical behavior, protection from predator, etc.

2.The important constituents of cuticle are chitin and cuticular proteins, however, they are not all present at every part of the cuticle.

3.Molting and formation of new cuticle is the process occurring every stage of nearly all insects as the regular development, and control under a series of hormones of insects.

知识点

1. 昆虫的表皮由基［底］膜、真皮和表皮组成。后者可进一步细分，如上表皮、外表皮和内表皮。它起着外骨骼、肌肉附着、第一道物理防线、物理和生化行为、抵御捕食者等作用。

2. 表皮的重要成分是几丁质和表皮蛋白质，然而，它们并非都存在于表皮的每个部分。

3. 蜕皮和新表皮的形成是几乎所有昆虫正常发育的每个阶段都会发生，并受到一系列昆虫激素控制的过程。

The insect integument comprises of basal lamina, dermal cells, and cuticle. Not only does it provide physical protection for internal organs, but also, because of its rigidity, it primary serves as a skeleton to which muscles can be attached. It also reduces water loss to a very low level in most Insecta, a feature that has been of great significance in the evolution of this predominantly terrestrial class. In addition, the cuticular component of the integument acts as a metabolic reserve, to be used cyclically to construct the next stage, or during periods of great metabolic activity or starvation. It prevents entry of foreign material, both living organisms and nonliving substances, into an insect. In many insects the waxy outer layer serves as a repository for contact sex pheromones. The color of insects is also a function of the integument, especially the cuticular component.

昆虫的体壁由底膜、真皮细胞和表皮组成。它不仅为内脏器官提供物理保护，而且由于其刚性，它主要充当骨骼，肌肉可以附着在上面。在大多数昆虫纲中，它还能将水分的损失降低到非常低的水平，这个特征在一个以陆地为主的种群进化中具有重要意义。此外，体壁的表皮组成可以周期性地用于构建下一阶段，或者在代谢活跃、饥饿时，起着代谢储备的作用。它防止生物和非生物外源物进入昆虫体内。在许多昆虫中，蜡质外层是接触性信息素的贮存库。昆虫的颜色也是体壁的一种功能，尤其是表皮成分。

1.1 体壁的结构

1.1 The Structure of the Integument

The insect cuticle is mainly composed of the polysaccharide chitin, proteins and lipids that interact with each other to form a layered structure. Secreted by the underlying epidermal cells, it covers the surface of the insect body and serves as a muscle attachment site needed for locomotion, maintains the shape of the insect body, prevents water loss and resists microbial penetration. The cuticle is composed of epicuticle, exocuticle and endocuticle (Fig.1-1). The single layer of epidermal cells secretes the cuticle at each molt. The epidermal cells secrete the overlaying cuticle, the lipids (waxes), cement, and many additional chemical components.

The cuticle forms an apical extracellular-matrix that covers exposed organs, such as epidermis, trachea and gut, for organizing morphogenesis and protection of insects, which controls development and structural maintenance of organisms, organ morphogenesis, wound healing, cell signaling, and the local defense barrier against a hostile environment. Numerous events, such as tissue metamorphosis and remodeling, require dynamic alterations and degradations of the extracellular-matrix architecture. Epithelial cells secrete essential components, such as proteins, lipids and polysaccharides, into the extracellular space

昆虫的表皮层主要由多糖、几丁质、蛋白质和脂质组成，它们相互作用形成一个分层结构。它由下面的真皮细胞分泌，覆盖昆虫体表面，作为运动所需的肌肉附着点，保持昆虫身体的形状，防止水分流失，并抵抗微生物穿透。表皮由上表皮、外表皮和内表皮组成（图 1-1）。真皮细胞的单细胞层在每次蜕皮时分泌表皮。真皮细胞分泌覆盖其上的表皮、脂质（蜡质）、黏胶层和许多其他化学成分。

表皮层形成顶端细胞外基质，覆盖暴露的器官，如真皮、气管和肠道，用于组织昆虫的形态发生和保护，控制生物体的发育和结构维持、器官形态发生、伤口愈合、细胞信号传导以及抵抗恶劣环境的局部防御屏障。诸多环节，如组织变态和重塑，需要细胞外基质体系结构的动态改变和降解。上皮细胞将蛋白质、脂类和多糖等重要成分分泌到

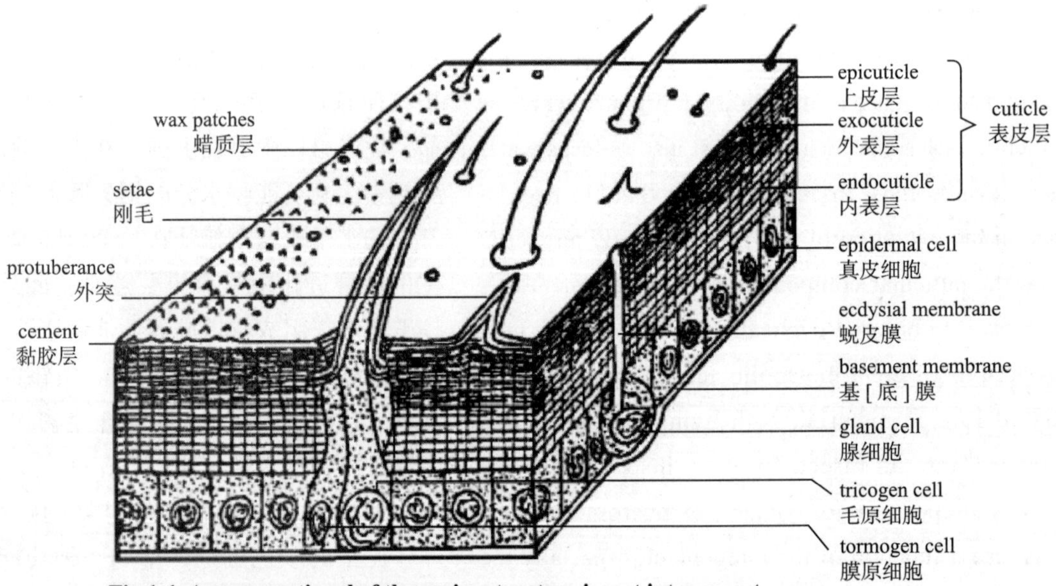

Fig.1-1 A cross-sectional of the major structure insect integument.

图 1-1　昆虫体壁横切面主要结构。

where they form a complex network of proteins and proteoglycans. These extracellular components, together with apical membrane bound proteins, organize the extracellular-matrix as a first barrier at the epithelial cell surface.

细胞外空间，形成蛋白质和蛋白聚糖的复杂网络。这些胞外成分连同顶端膜结合蛋白组成胞外基质，成为上皮细胞表面的第一道屏障。

1.1.1 Epicuticle

1.1.1 上表皮

Epicuticle is the outermost part of the integument in contact with the environment. It is known to contain sclerotized proteins impregnated with lipoproteins, lipids, waxes, cement, and minor amount of various minerals and other chemical components. It does not contain chitin. The proteins and some of the lipids appear to be covalently linked, and the proteins are tanned or sclerotized by phenolic compounds and other oxidized products, quinones. Sclerotization, the cross- linking of molecules, gives the epicuticle strength, hardness, and low water permeability. Lipids and the cement layer on the surfaces also provide reduced.

Permeability to water. Because the epicuticle bears the surface sculpturing of insects, its appearance is highly variable in different species. It does, however, contain one common layer in all insects, the cuticulin layer. Cuticulin is a very thin layer of protein about 7.5 nm thick. It is impregnated with lipids, some of which may be covalently bound to proteins as lipoproteins. There may be additional layers in the epicuticle, such as an inner cuticulin layer. Cement, is a shellac-like substance at the air interface of the cuticle, is secreted by dermal glands and transported to the surface of the epicuticle. The epicuticle represents a limitation on expansion of the cuticle in immature insects, necessitating molting during growth.

上表皮是与环境接触的体壁最外面的部分。已知它包含骨质化的蛋白质，渗有脂蛋白、脂类、蜡质、黏胶层和少量的各种矿物质以及其他的化学成分。它不含几丁质。蛋白质和一些脂类共价结合，蛋白质被酚类化合物和其他氧化产物醌类鞣化和骨化。骨化作用交联了其他分子，赋予上表皮强度、硬度和低水渗透性。表面上的脂类和黏胶层也降低了水的渗透性。昆虫的上表皮承载体表纹饰，不同种的外观变化很大。然而，在所有昆虫中的确有一个共同层，即表皮质层。表皮质层是非常薄的蛋白质层，厚约为 7.5 nm。它渗有脂质，其中一些脂质可能与蛋白质共价结合形成脂蛋白。在上表皮中也许还有另外一层，比如，内表皮质层。黏胶层是表皮与空气界面的虫胶样物质，由真皮腺分泌，并转运到上表皮的表面。上表皮代表了未老熟昆虫表皮扩张的局限性，在成长过程中使蜕皮成为必要。

1.1.2 Exocuticle

1.1.2 外表皮

The exocuticle, containing both chitin and protein, lies just beneath the epicuticle. It is highly sclerotized and is therefore hard and rigid. Lamellae or layers within the exocuticle may refract light in such a way to produce structural colors in some insects. The thickness of the exocuticle is variable and species specific. In particular, the thorax in flying insects has heavily sclerotized exocuticle to support the strong flight muscles. Many larvae have a soft flexible cuticle with little or more exocuticle. Because of sclerotization, little or none of the exocuticle is digested by molting fluid, and it is shed, along with the epicuticle, at molting.

外表皮包含几丁质和蛋白质，位于上表皮的正下方。它是高度骨化的，因而坚硬、有刚性。在某些昆虫中，外表皮内的薄片或薄层以这种方式折射光线，从而产生结构色彩。外表皮的厚度是多变的，而且是物种特有的。特别是飞行中的昆虫，胸部有高度骨质化的外表皮以支撑强壮的飞行肌。许多幼虫具有柔软、有弹性的表皮，几乎没有或有一点外表皮。因为骨化，蜕皮液很少消化或不消化外表皮。在蜕皮时，它与上表皮一起脱落。

1.1.3 Mesocuticle

Some insects have a layer of less highly sclerotized cuticle between the exo- and endo-cuticle called mesocuticle. It also contains chitin and protein.

1.1.4 Endocuticle

The endocuticle is a soft, elastic cuticle containing both chitin and protein. The protein and chitin in the endocuticle are stable through some covalent bonds, hydrogen bonds, and perhaps occasional quinone crosslink. Generally speaking, soft insects have relatively thick endocuticle to keep it flexible and soft, and always an epicuticle layer at the body surface of soft-bodied insects.

1.1.5 Dermal gland and epidermal cells

Each dermal gland consists of three cells: duct cell, saccule cell, and secretory cell. Secretory cells are present throughout larval life and grow in size without any cell divisions. The size of each secretory cell and its vacuoles is closely coupled to each molt cycle and hormonal titers at each molt cells attain their maximum size for the preceding instar. At each molt, the contents are abruptly expelled, and secretory cells shrink dramatically in size. The contents of the vacuoles are expelled into the surrounding hemocoel and not into the contiguous saccule cell. Growth of the secretory cell increases throughout the subsequent intermolt and continues until the next molt. At the larval-pupal molt, dermal glands undergo programmed cell death.

The cells underlying the cuticle are arranged in a single layer, usually called epidermal cells. All epidermal cells probably secrete chitin and proteins, and some may secrete lipids. Specially modified glandular epidermal cells may secrete cement.

Epidermal cells are separated from the circulating hemolymph by basement membrane with pores large enough to permit passage of larger hemolymph proteins and other molecules from hemolymph into the epidermal cells. The origin of the basement membrane is not defined for most insects, some evidence supports secretion by hemocyter, while desmosomal attachments to epidermal cells suggest that the epidermal cells secrete it themselves. Tracheae, tracheoles, and nerve pass through the basement membrane to reach the epidermal cells. Hemidesmosome hold the basement

1.1.3 中表皮

有些昆虫外表皮和内表皮之间有一骨化稍轻的表皮层，称为中表皮。它也含有几丁质和蛋白质。

1.1.4 内表皮

内表皮是柔软、有弹性的表皮，含有几丁质和蛋白质。内表皮的蛋白质和几丁质通过一些共价键、氢键，也许偶尔有醌类交联，呈现稳定性。总体来说，软体昆虫有相对较厚的内表皮，使之保持弹性和柔软，通常在软体昆虫的体表有一层上表皮。

1.1.5 真皮腺和真皮细胞

每个真皮腺由三个细胞组成：管细胞、球囊细胞和分泌细胞。分泌细胞存在于幼虫的整个生命周期，仅有细胞增长而无任何细胞分裂。每个分泌细胞及其液泡的大小与每个蜕皮周期密切相关，每个蜕皮细胞的激素滴度在前一龄达到最大。在每次蜕皮时，内含物会突然排出，分泌细胞的大小会急剧缩小。囊泡的内容物被排出到周围的血腔中，而不是进入相邻的球囊细胞。分泌细胞的生长在随后的蜕皮中间期增加，并持续到下一次蜕皮。在幼虫——蛹蜕皮时，真皮腺经历程序性细胞死亡。

表皮下排列成单层的细胞常被称为真皮细胞。所有真皮细胞也许都分泌几丁质和蛋白质。有些分泌脂质，特别是改型的腺真皮细胞可分泌黏胶层。

表皮细胞与循环血淋巴通过基［底］膜分离。基［底］膜上有足够大的孔，允许大分子血淋巴蛋白质和其他分子从血淋巴进入真皮细胞。多数昆虫基［底］膜的起源未明，有些证据支持由血淋巴分泌，而与真皮细胞的桥粒连接暗示其为真皮细胞自身分泌。气管、微气管和神经通过基［底］膜到达真皮细胞。半

membrane to the epidermal cells. Contiguous epidermal cells, as well as gut and Malpighian tubule epithelium, follicular cells in the ovary, and other cells in insects, are hold together by various types of junctional contact. The junctional contacts not only hold cells together, but depending on the type of contact, may have other functions.

① Septate junctions are close together and numerous, giving a ladder-like appearance between cells. It occurs between the lateral faces of cells, particularly toward the apical surface (cuticle), and plays an important role in preventing inward and outward movement of materials between the cells.

② Gap junctions also occur between the lateral faces of cells. The two membranes are very close together at gap junctions. Gap junctions appear to confer electrical coupling up on cells, can make cells function like a syncytium, and play a role in cells to cell communi cation. It also acts as sieves, allowing certain sizes of molecule to pass through, but excluding others, and may function in controlling the speed of entry of some types of molecules.

③ Tight junctions occur in some insect tissues such as between cells in the compound eyes, testis, rectal pads, and between perineural cells forming part of the blood-brain barrier in the central nervous system. Tight junctions seal the passage way between adjacent cells and present a barrier to passage of substances.

Epidermal cells have extensive rough endoplasmic reticulum where protein synthesis occur, and areas of smooth endoplasmic reticulum for lipid synthesis. Cell nuclei are often polyploidy, with multiple nucleoli. The Golgi complex is prominent in epidermal cells, and probably serves several functions, including the following: ① processing of secretory substances necessary to synthesize cuticle; ② production of material for plasma membrane of the cell; ③ modification of newly synthesized proteins; ④ packaging of cellular components in isolation envelops for later autophagy; ⑤ processing and packaging of lysozyme needed for autophagy and heterophagy.

Epidermal cells are involved in wound repair and can move from an area of undamaged cells into an area of damage or destroyed cells. Cells at the leading edge spread over the wound until they cover it and establish contact with another epidermal cell.

Oenocytes are large, prominent cells scattered among the epidermal cells, and also clustered at spiracles, near the origin of larger trachea, and scattered among fat body cells. Their morphology suggests lipid secretion and lipid metabolism.

桥粒将基［底］膜固定于真皮细胞。邻近的真皮细胞与中肠和马氏管上皮、卵巢的滤泡细胞以及昆虫其他细胞一样，通过不同类型的接触连接维系在一起。接触连接不仅把细胞连在一起，也依赖连接的类型具有其他功能。

①隔壁（隔膜）连接是无间隙连接，数量众多，在细胞间呈梯状外观。出现在细胞的侧面之间，特别是朝向顶部表面（表皮），在防止细胞间物质的内流和外流中起着重要作用。

②间隙连接也出现在细胞侧面之间。间隙连接处两个膜靠得非常近。间隙连接好像赋予细胞电耦合，细胞能起到合胞体功能，在细胞通信中起着重要作用。它也起着滤网作用，允许一定大小的分子通过而防止其他分子进入，在功能上控制某类型分子进入的速率。

③紧密连接出现在一些昆虫的组织中，如复眼、睾丸、直肠垫以及中枢神经系统在外周神经细胞之间构成部分血 - 脑屏障。紧密连接封闭了邻近细胞间通道，物质通道出现屏障。

真皮细胞有大量粗面内质网合成蛋白质，滑面内质网的区域合成脂类。细胞核通常是多倍体，有多种多样的核仁。真皮细胞的高尔基复合体非常突出，可能包括以下几种功能：①加工合成表皮分泌的必需物质；②产生细胞质膜的物质；③修饰新合成的蛋白质；④将细胞成分包装在隔离包膜中，以备后期自噬；⑤加工和装配自噬和异体吞噬所需的溶菌酶。

真皮细胞参与伤口修复，能从未受伤的细胞区域移动到受伤或被破坏的细胞区域。前缘的细胞扩散到伤口正上方，直到它覆盖伤口，并与另一个真皮细胞建立联系。

绛色细胞是分散在真皮细胞之间的大而突出的细胞，也聚集在气门处，靠近大气管的起源处，并分散在脂肪体细胞之间。其形态暗示它的功能是脂质分泌和脂质代谢。

1.2 表皮的化学成分

1.2 Chemical Composition of Cuticle

1.2.1 Chitin

Chitin is the Earth's second most abundant polysaccharide after cellulose, being the main structural component of arthropod's exoskeleton and the cell wall of fungi. It is a polymer of increasing commercial potential. It is used in agriculture for the biological control of pathogens, and in medical tissue regeneration, for wound dressing, as moisturizing agent in cosmetics and beauty clinic, and in industry, for adsorption of dyes and heavy metals from waste water, as additive in paper production. Furthermore, chitin can be converted into chitosan, a polymer with increased solubility and reactivity provided by free amino groups, which can find a wider range of applications (Tomas et al.,2022).

One of the important constituents of cuticle is chitin. A polymer of *N*-acetyl-*β*-*D*-glucosamine residues held together by *β*-(1-4)-glycosidic linkages (Fig.1-2). Enzymatic hydrolysis of chitin with chitinase and chitobiase releases *N*-acetylglucosamine as well as some free glucosamine. Every sixth or seventh residue in chitin may be glucosamine, with the remainder being *N*-acetyl-glucosamine. The inertness and insolubility of chitin in the cuticle serve insects well, but make it difficult to characterize the molecular arrangement of chitin within cuticle. Chitin exhibits a crystalline structure, but unit crystal dimensions and number of chitin chains per unit crystal vary with the source of the cuticle. Three types of chitins, named *α*-, *β*-, and *γ*-chitin, occur in some form in insects, differing in crystal cell size, number of chitin chains per unit cell, and degree of hydration. In *α*-chitin, adjacent chains run antiparallel to each other, which allows them to pack closer together and maximizes the number of within- and between-chain hydrogen bonds. Typically, 10 to 20 *α*-chitin chains are packed together in a roughly circular rod or "crystallite" that is embedded in a protein matrix (Fig. 1-3; Appendix color graphic Fig.1-3). In *β*-chitin,

1.2.1 几丁质

几丁质是地球上第二丰富、仅次于纤维素的多糖，是节肢动物的外骨骼和真菌细胞壁的主要结构成分。它是一种商业潜力越来越大的聚合物。它用于农业病原物的生物控制，医学上的组织再生、伤口敷料，作为保湿剂在化妆品和医美诊所用作添加剂以及工业上用于吸附从废水中提取的染料和重金属，纸张生产中的添加剂。此外，几丁质可以转化为壳聚糖，一种溶解度增大和反应性增加的多聚物，它可以找到更广泛的应用（Tomas 等，2022）。

表皮的重要成分之一是几丁质。它是一种由 *N*- 乙酰 -*β*-*D*- 葡糖胺残基通过 *β*-(1，4)- 糖苷键连接在一起的聚合物（图 1-2）。几丁质酶和壳聚糖酶酶解几丁质释放出 *N*- 乙酰葡糖胺以及一些游离的葡糖胺。几丁链每 6 ～ 7 个残基是葡糖胺，余下的是 *N*- 乙酰葡糖胺。几丁质在表皮中的惰性和难溶性对昆虫很有用，也给几丁质在表皮中的分子排列带来了困难。几丁质呈现一种结晶质结构，但每单位结晶规模以及每单位结晶几丁链的数量随表皮来源而变化。三种类型的几丁质分别被命名为 *α*- 几丁质、*β*- 几丁质和 *γ*- 几丁质，以某种形式出现在昆虫体内，它们在晶体细胞大小、每单位细胞结晶体几丁链的数量和水解程度上有所不同。在 *α*- 几丁质中，毗邻的链相互之间反平行延伸，这使它们密集装配在一起，并使链内和链间氢键的数量最大化。通常，10 ~ 20 根 *α*- 几丁链组装在一起，形成一个大致圆形杆状或“微晶”，嵌入蛋白质基质（图 1-3；附彩图 1-3）。在 *β*- 几丁质中，链

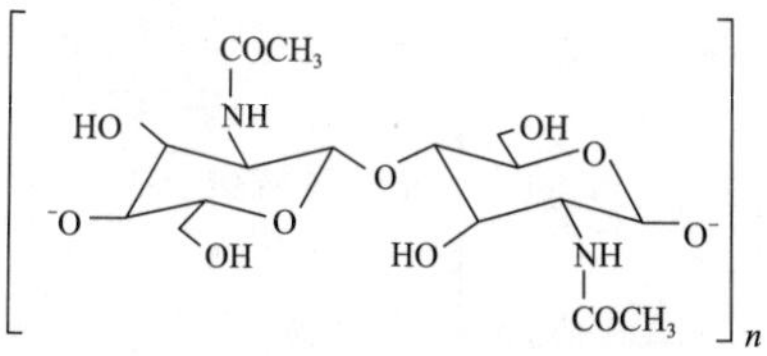

Fig.1-2 Chemical structure of chitin unit.

图 1-2 几丁质单元的化学结构。

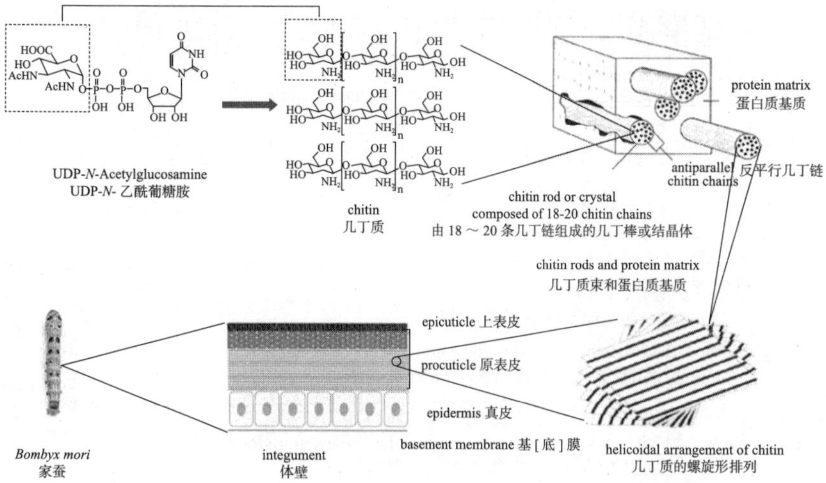

Fig.1-3 Structure of hierarchy of silkworm cuticle. Typically, about 18-20 Chitin chains antiparallel arrangement held together by hydrogen bonds to form chitin rods or crystallites embedded in a protein matrix in cuticle (modified from Liu et al., 2019; Nation,2022).

图 1-3 家蚕表皮的层次结构。通常由 18 ～ 20 根几丁链反平行组成的几丁棒或微晶包埋在表皮的蛋白质基质中（仿 Liu 等，2019；Nation，2022）。

the chains are arranged parallel to each other, whereas in *γ*-chitin, the chains are grouped in sets of three strands where two parallel chains are in one direction and the third one in the opposite direction.

The starting point for synthesis of chitin is *β-D*-glucose, although the *β-D*-glucose may come from a storage form such as trehalose or glycogen (Fig.1-4; Appendix color graphic Fig.1-4). The latter two compounds are formed from *α-D*-glucose, but a rapid and dynamic isomerization of glucose in aqueous solution occurs so that at any given moment there is 37% *α*- and 63% *β-D*-glucose. Any removal of one form, rapidly leads to a new equilibrium. Chitin synthase (CHS), necessary for the final step. CHS-A, which catalyzes the transfer of sugar moieties from activated sugar donors to specific acceptors in all chitin-containing organisms is responsible for chitin synthesis in the cuticle and cuticular lining of the foregut, hindgut, and trachea, whereas CHS-B is dedicated to chitin synthesis in the peritrophic matrix. Multiple forms of CHS may exist as a zymogen that requires activation when chitin synthesis is under way.

Chitin remodeling involves chitin synthases, which make the nascent chitin chains, and chitin deacetylase that partially deacetylate some of the *N*-acetyl-glucosamine residues either

彼此平行排列，而在 *γ*- 几丁质中，链以三股为一组，其中两条平行链在一个方向，第三条在相反方向。

几丁质合成的起点是 *β-D*- 葡萄糖，尽管 *β-D*- 葡萄糖来自其贮存形式，如海藻糖或糖原（图 1-4；附彩图 1-4）。后两种化合物由 *α-D*- 葡萄糖形成，但葡萄糖在水溶液中会发生快速和动态的异构化，因此，在任何给定的时刻都是 37% 的 *α-D*- 葡萄糖和 63% 的 *β-D*- 葡萄糖。移走任何一种形式的葡萄糖都会很快引起新的平衡。几丁质合成酶（CHS）是最后一步必需的。在所有含有几丁质的生物体中，CHS-A 催化糖部分从活性糖供体转移到特定受体，负责表皮和前肠、后肠以及气管的表皮层里衬中的几丁质合成，而 CHS-B 则致力于围食膜中的几丁质合成。多种形式的几丁质合成酶也以酶原形式存在，当几丁质合成时需要被激活。

几丁质重塑涉及生成新生几丁质链的几丁质合成酶和几丁质脱乙酰酶，一些 *N*-乙酰氨基葡萄糖胺残基的部分脱

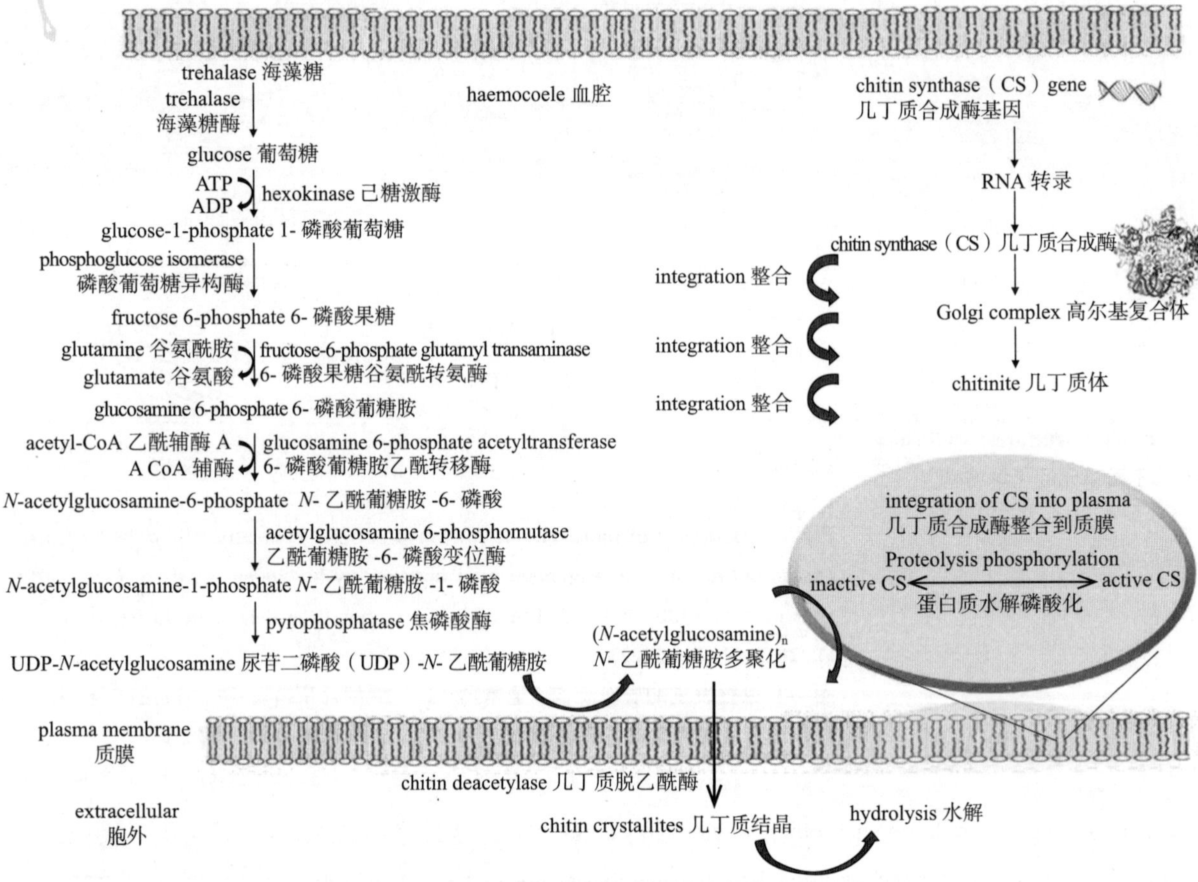

Fig.1-4 Cascade steps in chitin biosynthesis (modified from Nation, 2022)

图 1-4 几丁质合成的级联反应（仿 Nation，2022）

randomly or sequentially to yield local chitosan-like regions. Other proteins secreted into the procuticle or the midgut help in the assembly of single chitin chains into larger crystalline aggregates. They are further embedded in a complex matrix of cuticular proteins or become associated with proteins containing chitin-binding domains to constitute the laminar procuticle or the lattice-like peritrophic matrix. During molting, previously formed laminar cuticle or PM are decrystallized/depolymerized to unmask the chitin chains, which then are degraded by a mixture of chitinolytic enzymes consisting of chitinases and *N*-acetylglucosaminidases present in molting fluid or in gut secretions (Muthu krishnan et al., 2019). Some of the degradation products may be recycled for the synthesis of new matrices. The wide range of properties of individual cuticles at specific anatomical locations and developmental stages is attributable to difference in chitin/protein ratio, protein composition of the matrix, cross-linking, presence of minerals and the degree of dehydration.

乙酰化，随机或顺序地生成局部壳聚糖样区域。分泌到原表皮或中肠的其他蛋白质有助于将单个几丁质链组装成更大的晶体聚集体。它们进一步嵌入表皮蛋白质的复杂基质中或与含有几丁质结合域的蛋白质结合，构成薄片状原表皮或晶格状围食膜基质。在蜕皮过程中，先前形成的薄片状表皮或围食膜被脱结晶/解聚，以暴露几丁质链，然后被蜕皮液或肠道分泌物中存在的几丁质酶和*N*-乙酰氨基葡萄糖苷酶组成的几丁质酶混合物降解（Muthu krishnan 等，2019）。一些降解产物可回收以用于合成新基质。在特定的解剖位置和发育阶段，个体表皮的广泛属性可归因于几丁质/蛋白质比率、基质的蛋白质组成、交联，矿物质的存在和脱水程度的差异。

1.2.2 Cuticular proteins (CPs)

Cuticle proteins are synthesized in the epidermal cells and are secreted from the apical surface of the cells into the cuticle. In the cuticle, they fill in the matrix around chitin rods. In hard cuticles, the proteins are synthesized by phenolic and quinone compounds that form covalent bonds and cross-link to each other, forming a very hard, rigid structure. Even in very soft cuticles, there is some degree of stabilization, but probably relatively few cross-links.

The cuticle matrix consists of many different proteins that confer the physical properties of the exoskeleton. Proteins of soft cuticles are often different from proteins in hard cuticles in ways other than degree of sclerotization. Proteins from soft cuticles frequently have a high content of polar amino acids, while proteins in the harder or more sclerotized cuticles have a higher content of hydrophobic amino acids. Most of the cuticular proteins have the Rebers and Riddiford consensus, is by far the largest cuticular proteins family and is found in every examined species of arthropod, which contains the chitin binding domain and binds chitin. The conserved R&R consensus sequence most likely interacts with chitin through a β-sheet structure, while β-sheets occurred frequently throughout the chitin binding region, which is an optimal location for interaction with chitin. The R&R Consensus refer to G-(x8)-G-x(6)-Y-x(2)-A-x -E-x-G-F-x(7)-P-x-P where x represents any amino acid, the values in parentheses indicate the number of amino acid residues(Rebers and Riddiford, 1988). The strict R&R consensus:G-x8-G-x6-(Y/F) remains almost invariant in cuticle protein R& R (Sun et al., 2013).

Three types of CP R&R(CPRs) have been classified: RR1 is generally found in soft cuticle, RR2 in hard cuticle, and RR3 group comprises a few CPs. Many of these proteins play a fundamental and indispensable role in maintaining the structural integrity of the cuticle in different parts of the insect anatomy. In *B.mori*, the earliest expressed protein group, which is composed of RR2 CP and CPH (a group of hypothetical *Bombyx* CPs), contains high percentage of *His*. It is suggested that these proteins bind *via His* residues to make the tanned exocuticle.RR2 CPs, which are expressed at a later stage, have a lower percentage of *His*, which suggests that they play a minor role in sclerotization of the cuticle. The transcription of RR2 CP

1.2.2 表皮蛋白质（CPs）

表皮蛋白质在真皮细胞中被合成，并从细胞顶端面被分泌到表皮。在表皮中，它们填满几丁质棒状体周围的基质。在硬表皮内，蛋白质由酚和醌类化合物通过共价键并相互交联合成，形成非常坚固、刚硬的结构。即使非常柔软的体表，也有一定程度的稳定性，但可能交联很少。

表皮基质由许多不同的蛋白质组成，赋予外骨骼的物理特性。软表皮蛋白质和硬表皮蛋白质除了硬化程度，通常是不同的。软表皮蛋白质经常有高含量的极性氨基酸，而硬表皮或更骨质化的表皮蛋白质有相对高含量的疏水氨基酸。大多数表皮蛋白质具有 R&R 共有序列，是目前发现的最大分子质量的表皮蛋白质家族，在检测的所有节肢动物中被发现，它包含几丁质结合域和结合几丁质。保守的 R&R 共有序列很可能通过 β-片状结构与几丁质相互作用，而 β-片经常出现在几丁质结合区域，这是与几丁质相互作用的最佳位置。R&R 共有序列是指 -(x8)-G-x(6)-Y-x(2)-A-x-E-x-G-F-x(7)-P-x-P 其中 x 代表任一种氨基酸，括号内的数值指氨基酸残基数量（Rebers 和 Riddiford, 1988）。精确的 R&R 共有序列 G-x8-G-x6-(Y/F) 在表皮蛋白质 R&R 中几乎保持不变（Sun 等，2013）。

已鉴定出三类表皮蛋白质 R&R（CPRs）：RR1 普遍出现在软表皮，RR2 在硬表皮，RR3 组成少数表皮蛋白质。其中许多蛋白质在保持昆虫解剖学不同部位的表皮结构完整性方面起着根本和不可或缺的作用。在家蚕中，最早表达的蛋白质组由 RR2 CP 和 CPH（一组假设的家蚕表皮蛋白质）组成，含有高百分比的组氨酸 *His*。据推测，这些蛋白质通过 *His* 残基结合形成揉化的外表皮。RR2 CPs 在后期表达的 *His* 的百分比较低，这表明它们在表皮层硬化中的

genes ceases before pupation, while that of *RR1 CP* genes continue until after pupation even if the start of transcription was at the same time as that of the *RR2 CP* genes. This suggests that RR1 CPs construct both layers of the exocuticle and endocuticle, while RR2 CPs construct only exocuticle (Kawasaki,2016).

作用较小。*RR2 CP* 基因在化蛹前停止转录，而 *RR1 CP* 基因的转录一直持续到化蛹后，即使转录起始与 *RR2 CP* 基因的转录起始同时。这表明 RR1 CP 构成外表皮和内表皮两层，而 RR2 CP 仅构成外表皮（Kawasaki，2016）。

1.2.3 Resilin

Resilin is an elastomeric protein found in insects, in the tendons and at the wing hinges. Thus, it helps insects for jumping and to move their wings efficiently. Resilin and chitinous cuticles together form a composite structure for energy storage needed in the jumping of insects. The catapult mechanism of rubber-like protein as an elastomer is a notable feature. Resilin is found in between the two cuticle layers of exocuticle to endocuticle and is in contact with the epidermis. Resilin consists of a protein having 620 amino acid molecules long chain included with a signal peptide of 17 residues at N-terminus. This protein is secreted in the extracellular space. Its relative molecular mass is 56,771 and its isoelectric point of 5.0 is composed of three domains. Resilin exhibits elasticity due to the cross links, which are composed of dityrosine and trityrosine. It also possesses fluorescent properties and present pH dependent, and be degraded by proteolytic enzymes as other proteins. It is unaffected by deep freezing and also at temperatures up to 125 ℃.

Resilin is an important structure protein of cuticle that is colorless, transparent, and insoluble in water. It is detected in the tracheal endings, the spermatheca ducts, at the wing articulations, at or near to limb joints and the proboscis and has remarkable properties of elasticity. To be flexible, resilin obviously cannot be very sclerotized nor have many cross-links with other protein chains, but dityrosine and trityrosine residues provide a few internal cross-links, giving resilin chains some stability, yet allowing high elasticity. Resilin occurs in the wing hinges of some insects and in the hinge mechanism of the jumping leg of fleas. The prealar arm connecting the mesonotum to the first basalar sclerite of the thoracic wall in *S. gregarai* wings contains about 50 μg resilin and 15 μg chitin. Some parts of dragonflies, large

1.2.3 节肢弹性蛋白

节肢弹性蛋白是一种存在于昆虫、肌腱和翅膀铰链处发现的弹性蛋白质。因此，它有助于昆虫跳跃和有效地移动翅膀。节肢弹性蛋白和几丁质表皮层形成了一个复合结构，用于贮存昆虫跳跃所需的能量。橡胶样蛋白作为弹性体的弹射机制是一个显著特征。节肢弹性蛋白位于外表皮和内表皮两表皮层之间，与真皮连接。节肢弹性蛋白由一种蛋白质组成，该蛋白质具有 620 个氨基酸分子长链，N- 末端含有 17 个残基的信号肽。这种蛋白质是在胞外空间分泌的。其相对分子质量为 56 771，等电点 5.0，由三个结构域组成。由于二酪氨酸和三酪氨酸组成的交联，节肢弹性蛋白表现出弹性。它还具有荧光特性并呈现 pH 依赖性，也像其他蛋白质一样被蛋白质水解酶降解。它不受深度冷冻，也不受高达 125℃温度的影响。

节肢弹性蛋白是表皮的一种重要结构蛋白质，无色、透明，不溶于水。它在气管末端、受精囊管、翅膀关节处、四肢关节处或附近以及喙中被检测到，有不同寻常的弹性特点。为了柔韧性，节肢弹性蛋白显然不能太骨质化或与其他蛋白质链有太多的交联，但二酪氨酸和三酪氨酸残基提供了少量内部交联，使节肢弹性蛋白链具有一定的稳定性，还允许高伸缩性。有些昆虫翅膀铰链和跳蚤跳跃足的铰链机制中有节肢弹性蛋白。沙漠蝗的翅前臂连接中胸背板与胸壁前上侧片含节肢弹性蛋白约为 50 μg 和几丁质为 15 μg。大蜻蜓和蝗虫的有些部位含有节肢弹性蛋白。然而，节肢

Aeshna spp. and locusts contain resilin. However, resilin-matrices are composite in the way that varying amounts of different elastic proteins and dityrosination define material properties. For the group of proteins that account for the elasticity and resilience of the cuticle, Lerch et al.(2020) coined the term resilome, this does not consist of a single type of protein, Pro-Resilin, but, in *D. melanogaster*, of at least two proteins that are combined in different ratios to accommodate the respective specific function of the cuticle.

弹性蛋白基质是不同数量的不同弹性蛋白和二酪氨酸化限定的材料特性以一定方式组成的复合物。对于引起表皮弹性和弹力的蛋白质组而言，Lerch 等(2020) 创造了节肢弹性蛋白组一词，这不是由一种单一类型弹性蛋白原的蛋白质组成，而是在果蝇中至少由两种蛋白质以不同的比例组合而成，以适应表皮各自的特定功能。

1.2.4 Stage-specific differences in cuticle proteins

Regional-specific, stage-specific, and temporal differences in the proteins of *H. cecropia* have been found, but a cecropia silk moth larva also conserving proteins from some stages to the next, as well as synthesizes new proteins specific to its current stage of development. Stage-specific proteins occur in *D. melanogaster*, *M. sexta*, *A. polyphemus*, *B. mori and T. molitor*. Cuticular proteins from larval, pupal, and adult cuticle of *polyphemus* are specific for each stage. Electrophoretic and immunoblot analyses have been used to demonstrate different cuticular proteins in pupal and larval cuticles of the silkworm *B. mori*.

1.2.4 表皮蛋白质的发育阶段特异性差别

虽然已在惜古比天蚕中发现区域特异性、发育阶段特异性和时差蛋白质，但天蚕蚕蛾幼虫也将蛋白质从某些阶段保存到下一个阶段，并合成适合当前发育阶段的新特异性蛋白质。在果蝇、烟草天蛾、多音天蚕、家蚕和黄粉甲中发现了阶段特异性蛋白质。多音天蚕幼虫、蛹和成虫的表皮蛋白质是每个龄期特异性的。电泳和免疫印迹分析可用于证实家蚕蛹期和幼虫表皮的不同表皮蛋白质。

1.2.5 Protective cuticular proteins

Cuticular proteins in some or all insects may function in protection against invading bacteria or other organisms. When the surface of the epicuticle of silkworm larvae is light abraded and treated with bacteria (*Bacillus lichenformis* or *Enterobacter cloacae*) or bacterial cell wall compounds, mRNA are produced that code for cecropin, a large protein with antibacterial properties. The response is systemic, and cecropin appears in epidermal cells and fat body cells remote from the abrasion.

1.2.5 保护性表皮蛋白质

有些昆虫或全部昆虫的表皮蛋白质在功能上起着抵抗细菌和其他生物体入侵的作用。当家蚕上表皮的表面被轻微损伤，并用细菌（地衣芽孢杆菌或肠内杆菌）或细菌胞壁化合物处理时，就会产生编码天蚕素的 mRNA。天蚕素是一个具有抗菌特性的大蛋白质。这种反应是系统性的，天蚕素出现在远离擦伤处的真皮细胞和脂肪体细胞中。

1.2.6 Cuticular lipids and hydrocarbons

Insects are protected from desiccation and pathogens with a thin layer of lipids covering their cuticle. Cuticular lipids often are involved in inter- and intraspecific communication of insects. The lipids on the cuticle of insects not only promote water conservation, but they often have additional behavioral, pheromonal, ecological

1.2.6 表皮脂质和碳氢化合物

昆虫的表皮上覆盖着一层薄薄的脂类，以防止干燥和病原体。表皮脂质通常参与昆虫的种间和种内通信。昆虫表皮的脂质不仅促进水分的保持，而且它们经常有其他行为、信息素、生态学和

and taxonomic significance. Among the lipids on the cuticle are true chemical waxes, hydrocarbons, alcohols, fatty acids, glycerides, sterols, ketones, aldehydes, and esters.

In very wet environments, insect larva has only small quantities of cuticular hydrocarbons on the cuticle. Pupae, representing a closed system to water, and adults subject to the drying influence of the air have greater quantities of surface lipids. Diapausing pupae have a thicker lipid layer on the cuticle than non-diapausing ones, apparently providing greater protection from desiccation during long diapause. Season, and apparently temperature, are correlated with the type and quantity of hydrocarbons on the cuticle of insects.

Hydrocarbons are often the major components on the cuticle which protect terrestrial insects against desiccation and serve as signaling molecules in a wide variety of chemical communication systems. Many different compounds are also usually present, including straight-chain saturated compounds (alkanes), olefins (unsaturated alkenes), and alkanes with methyl branches. In many insects, cuticular hydrocarbons are species specific, leading to the use of cuticular hydrocarbon analyses in taxonomy and systematics. At ecdysis, new lipids are secreted again just before and just after ecdysis of the old cuticle. Hydrocarbons are synthesized by abdominal oenocytes and epidermal cells in female *B.gervmanica*, and transported by lipophorin in the hemolymph to distribution sites in the body, including the cuticle.

分类学上的重要意义。表皮脂质的化学本质是蜡质、碳氢化合物、醇、脂肪酸、甘油酯、甾醇、酮、醛和酯。

在非常潮湿的环境，昆虫幼虫表皮仅有少量的表皮碳氢化合物。蛹期对水是一种封闭的系统。成虫期受到干燥空气的影响，体表有大量脂质。滞育蛹比非滞育蛹有一层更厚的脂质层。显然，在长滞育期内，脂质提供了最大的保护免于脱水。季节、温度与昆虫表皮碳氢化合物的类型和数量相关。

通常，碳氢化合物是保护陆生昆虫以防脱水的表皮主要成分，并在广泛的化学通信系统中作为信号分子。通常还有很多不同的化合物，包括直链饱和化合物（烷烃）、石蜡（不饱和烷烃）和带有甲基分支的烷烃。在许多昆虫中，表皮碳氢化合物具有种属特异性，导致在分类学和系统学中使用表皮碳氢化合物分析。在蜕皮期，新的脂质在刚好蜕去旧表皮之前和之后重新分泌。雌蜚蠊碳氢化合物由腹部绛色细胞和真皮细胞合成，并通过血淋巴中的载脂蛋白转运到包括表皮在内的身体的各个部位。

1.2.7 Cuticular wax

A tenebrionid beetle, *C.verrucosa*, has a “wax bloom” on the surface. The wax bloom serves as camouflage, reduces water evaporation, and provides some thermal protection by restricting air movement at the cuticular surface in the desert. The beetle change color, depending on how much wax is secreted to the surface. Surface waxes contain many groups of organic compounds. These are mainly hydrocarbons, fatty acids, alcohols, aldehydes, ketones, and esters. Insects use individual compounds from these groups for strictly defined purposes, including protection against water loss, chemical communication, and defense against entomopathogenic microorganisms (Gołębiowski et al., 2022).

1.2.7 表皮蜡质

拟步甲科甲虫表面有蜡质花饰。蜡质花饰用于伪装、减少水分散失，并在沙漠中通过体表面空气运动的扼流提供一些热保护。这种甲虫根据分泌到体表蜡质的多少来变色。表面蜡含有许多组有机化合物。其主要是碳氢化合物、脂肪酸、醇、醛、酮和酯。昆虫将这些组中的独特混合物用于严格明确的目的，包括防止水分流失、化学通信和抵御昆虫病原微生物（Gołębiowski 等，2022）。

1.3 蜕皮和新表皮形成

1.3 Molting and Formation of New Cuticle

The external skeleton gradually becomes too small for the growing body tissue of an immature insect, and it must molt its cuticle. The muscles that move the body must be detached from the old cuticle, but they are detached only immediately before the ecdysis and new attachment are made quickly to the new epicuticle.

对于一个未成熟的昆虫正在成长的体组织来说，它的外骨骼逐渐变得太小，必须脱去旧表皮。移动身体的肌肉必须从旧表皮上分离，但它们仅在蜕皮前快速解离，新的附着迅速连接上新表皮。

Preparation for molting is under endocrine and nervous control. A tobacco hornworm each instar grows until it reaches a certain size and/or weight. Stretch receptors in the body are probably stimulated by the increasing growth of body tissues; and when it attains the critical size, the brain secretes prothoracicotropic hormone (PTTH). PTTH is released into the circulating hemolymph, circulates around the body, and finds specific receptor proteins on the surface of prothoracic gland (PGL) cells to which it binds. Multiple biochemical reactions are initiated that result in synthesis of ecdysone by the PGL. Ecdysone is released into the circulating hemolymph and is converted into 20-hydroxyecdyone by separating from the old cuticle (apolysis, Fig.1-5B) and by mitotic activity that produces new cells to spread over the larger body surface that must be enclosed within the new epidermal cell layer and new cuticle (Fig.1-5C). Apolysis or separation of the epidermal cells from the old cuticle marks the beginning of a molt and of a new instar, and the animal within the loosened, but not yet shed, cuticle is the pharate next instar or stage if the next form is the pupae or adult.

预蜕皮受到内分泌和神经的控制。每一龄的烟草天蛾成长到一定大小和/或体重。体内的张力感受器受到不断增长的体组织刺激，当它的大小达到临界值时，脑分泌促前胸腺激素（PTTH）。PTTH被释放到循环中的血淋巴，在身体周围循环，找到前胸腺细胞表面与其结合的特异性受体蛋白。前胸腺多种生化反应，导致前胸腺（PGL）启动蜕皮激素合成。蜕皮激素被释放到循环的血淋巴中，并被转换成20-羟蜕皮酮，通过分开旧表皮（皮层溶离，图1-5B）和有丝分裂产生新的细胞，展开后覆盖大部分体表，以封闭新的真皮细胞层和新表皮（图1-5C）。皮层溶离或从旧表皮分离真皮细胞标志着蜕皮和新龄期的开始，其内动物松软但旧表皮仍未脱落。表皮蜕裂到下一个龄期或下一个形态是蛹期或成虫期。

1.3.1 The apolysial space

The apolysial space is at first a minute space created by the separation of the epidermal cells from the old cuticle. Soon, molting fluid is secreted into the space and activated; and later, as the molting fluid digests some of the old cuticle, the space widens. The vesicles of secretion, often called apolysial droplets, are secreted by exocytosis from the plasma membrane of epidermal cells.

1.3.1 皮层溶离腔

皮层溶离腔起初是真皮细胞从旧表皮分开产生的瞬时间隙。不久，蜕皮液被分泌进间隙并激活，随后，由于蜕皮液消化了一些旧表皮，间隙变宽。分泌的囊泡常被称为皮层溶离液滴，是真皮细胞的质膜通过胞吐作用分泌出来的。

1.3.2 Molting fluid secretion

Insect molting fluid is a mixture containing various proteins that have a close relationship with physiological functions, such as recycling of old cuticle proteins and chitin, balance of molting fluid pressure, detoxification of toxins, and signal detection and transduction.

Molting fluid contains both proteinases and chitinases that digest the proteins and chitin, respectively, in old endocuticle. Chitins (EC 3.2.1.14) is an endoenzyme and attacks a chitin chain at random by internal hydrolysis. It produces smaller, soluble oligosaccharides that attacked at the ends by *N*-acetyl-*β*-*D*-glucosamini dase, an exoenzyme, yield free *N*-acetyl-glucosamine. Injection of 20-hydoxyecdysone into 5th instar of the silkworm, *B. mori* and *M. sexta* causes secretion of chitinase and *N*-acetyl-*β*-*D*-glucosaminidase, but induction of chitinase requires higher levels of hormone than induction of *N*-acetyl-*β*-*D*-glucosaminidase. Chitinase usually exists in two molecular size (88 kDa and 65kDa), and may exist as a zymogen (215kDa) that is glucosaaminidase. Chitinase converted to the active enzyme at the proper time in *B.mori.* Chitinases and imaginal-disc-growth-factors (IDGFs) are essential for larval and adult molting. Both of their depletion of them results in deformed cuticles, larval and adult molting defects, and insufficient protection against wounding and bacterial infection, which altogether leads to early lethality. Nonenzymatic IDGFs act as structural proteins to maintain the extracellular-matrix scaffold against chitinolytic degradation in *Drosophila* (Pesch et al., 2016), and maintains cuticle structure and controls melanization and the melanin metabolism in spot pattern formation of *B. mori* (Gao et al., 2020).

At least ten proteinases occur in the molting fluid from tobacco hornworm pharate pupae. Both endo- and exo-cleaving proteolytic enzymes occur and are most active in the neutral to alkaline pH range. Some of the proteases have trypsin-like and chymotrypsin-like activity, but they differ from similar gut enzymes in that they are not affected by some inhibitors of gut trypsin and chymotrypsin. At least some of the enzymes are secreted in a zymogen form.

In general, it appears likely that digestion of the old cuticle starts with proteolytic enzymes acting upon the proteins of the cuticle, and thereby exposing chitin rods or crystallites embedded in the protein matrix.

1.3.2 蜕皮液的分泌

昆虫蜕皮液是一种含有多种蛋白质的混合物，与其生理功能密切相关，如旧表皮蛋白质和几丁质的循环利用、蜕皮液的压力平衡、毒素的解毒、信号检测和传导。

蜕皮液包含蛋白酶和几丁质酶，在旧的内表皮内分别消化蛋白质和几丁质。几丁质酶（EC 3.2.1. 14）是一种内酶，通过内部水解随机破坏几丁质链。它产生较小的水溶性低聚糖，通过一种外酶 *N*- 乙酰 -*β*-*D*- 氨基葡糖酶，破坏几丁质末端，产生 *N*- 乙酰葡糖胺。将 20- 羟基蜕皮激素注射到家蚕和烟草天蛾 5 龄幼虫体内可引起几丁质酶和 *N*- 乙酰 -*β*-*D*- 氨基葡糖酶的分泌，但诱导几丁质酶比诱导 *N*- 乙酰 -*β*-*D*- 氨基葡糖酶需要更高水平的激素。几丁质酶常以两种分子存在（88 kDa 和 65kDa），能以酶原（215 kDa），即葡萄糖氨基化酶存在。在家蚕中适时转换成活性酶。几丁质酶和类成虫盘生长因子（IDGFs）对幼虫和成虫蜕皮至关重要。它们的消失会导致表皮层变形，幼虫和成虫蜕皮缺陷以及对伤害和细菌感染的保护不足都导致早期死亡。非酶 IDGFs 起结构蛋白的作用，以维持细胞外基质骨架，防止果蝇中的甲壳素降解（Pesch 等，2016），维持表皮结构，控制桑蚕斑点花样形成中的黑色素化和黑色素代谢（Gao 等，2020）。

烟草天蛾预蛹蜕皮液至少有 10 种蛋白酶。内切和外切两种蛋白质水解酶在 pH 中性和碱性范围内更为活跃。有些蛋白酶具有胰蛋白酶和胰凝乳蛋白酶的活性，但与中肠的相同酶不一样，它们不受中肠胰蛋白和胰凝乳蛋白抑制剂的影响。至少有些酶是以酶原形式分泌的。

总之，对旧表皮的消解似乎起始于对表皮的蛋白质起作用的蛋白质水解酶，因而露出嵌入蛋白质基质中的几丁质杆或几丁质结晶。

1.3.3 New cuticle formation

New cuticle secretion begins soon after the ecdysial space opens. During new cuticle secretion, an epidermal cell has a series of ridges or knobby projections on its apical face where the proteins and fibers of chitin are secreted. At the initial of new cuticle synthesis, cuticulin, the first new cuticular layer secreted, begins to form on these knobby plaques (Fig.1-5D, Fig.1-5E). It is secreted initially as small discontinuous patches over the plasma membrane plaques, but the patches enlarge and eventually form a continuous layer of cuticulin. There also may be long microvilli on the apical face of epidermal cells during the molting process. Molting fluid digests the old endocuticle, and the products of digestion are reabsorbed by epidermal cells and used in the synthesis of new cuticle. The ecdysial membrane that is formed represents a barrier between the molting fluid and the new cuticulin layer being secreted just below the membrane. As old endocuticle is digested, new unsclerotized procuticle is secreted below the cuticulin envelop, thus pushing the cuticulin upward and out-ward. The new cuticle secreted at this stage is unsclerotized, called procuticle (Fig.1-5F, Fig.1-5G). Later, after ecdysis, some or much of the new cuticle, may become heavily sclerotized and very hard.

1.3.3 新表皮形成

新表皮的分泌在蜕皮腔打开之后会很快开始。在新表皮分泌过程中，真皮细胞顶面有一系列的嵴状突起或多节的突起物，是蛋白质和几丁质纤维分泌的场所。在新表皮合成初期，最先分泌的新表皮开始在这些多节状补片上形成（图 1-5 D、图 1-5 E）。它最初分泌的是小而不连续的小片，覆盖在质膜补片上，但小片扩大，最终形成连续的表皮质层。在蜕皮过程中，真皮细胞的顶端面有长长的微绒毛，蜕皮液溶解旧的内表皮，消化产物被真皮细胞再吸收，并用于合成新表皮。形成的蜕皮膜相当于在蜕皮液和膜下正在分泌的新表皮皮质层之间有一道屏障。当旧表皮被消化，未骨化的原表皮在表皮质包被下分泌，这样就把表皮质层向上和向外推。这时分泌的新表皮未骨质化，被称为原表皮（图 1-5 F、图 1-5 G）。随后，有些或多数新表皮蜕皮后会变得高度骨质化，非常坚硬。

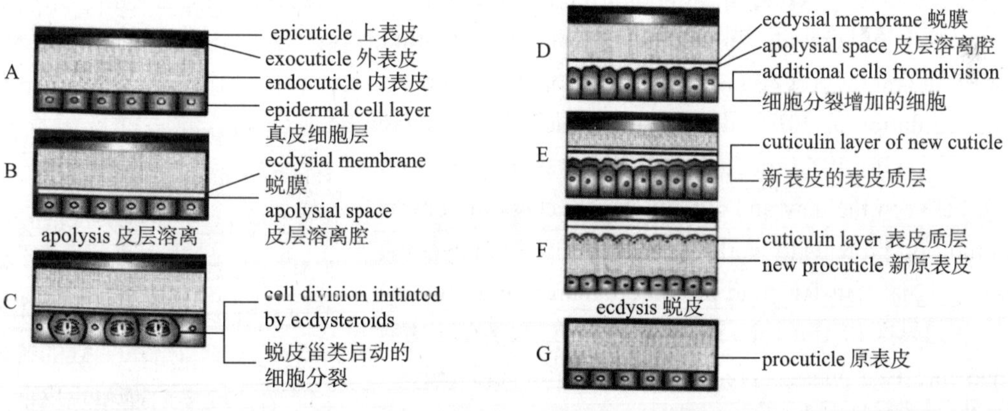

Fig.1-5 Diagrammatic illustration of the process of apolysis, secretion of new cuticle, and ecdysis of the old cuticle. A. Old cuticle just before molting begins; B. Formation of the ecdysial membrane and apolysial space; C. Initiation of cell division in epidermal cell layer in response to molting hormone; D. New epidermal cells; E. New cuticle secretion begins with secretion of cuticulin layer; digestion of the old endocuticle continues; F. New unsclerotized procuticle is formed; G. The old cuticle shell has been ecdysed and the new cuticle will be covered with a wax and cement layer, and some of the procuticle may be sclerotized into exocuticle (modified from Nation, 2022).

图 1-5 皮层溶离、新表皮分泌和脱去旧表皮过程的图解说明。A. 开始蜕皮前的旧表皮；B. 蜕皮膜和皮层溶离间隙的形成；C. 真皮细胞层对蜕皮激素的响应，启动细胞分裂；D. 新真皮细胞；E. 新表皮开始分泌，伴随着表皮质层的分泌；旧的内表皮继续消化；F. 未骨质化的新原表皮形成；G. 旧表皮蜕去，新表皮和蜡质、黏胶层覆盖体表，新的原表皮可骨质化成外表皮（仿 Nation, 2022）。

1.3.4 Reabsorption of molting fluid

The molting fluid that accumulates in the apolysial space disappears shortly before ecdysis. Most of it appears to be reabsorbed by the insect. At this critical stage, preventing loss of fluid volume by terrestrial insects may be vital. Some of the molting fluid may be reabsorbed through the epidermal cells, but in *M.sexta* and *C. ethlius* pharate pupae, it primarily flows beneath the old and new cuticle to the mouth and anal openings and is accumulated in the midgut. The fluid probably contains protein and other potentially useful nutrients, and the water may be useful in helping to flush the gut and/or malpighian tubules of accumulated waste products, particularly in newly eclosed adults, many of which excrete the meconium as the accumulated waste from molting. Excretion of any excess water also lightens the body for flight.

1.3.4 蜕皮液的再吸收

在蜕皮前，皮层溶离腔积累的蜕皮液很快消失。多数被昆虫再吸收；在这个关键时期，防止陆生昆虫体液容量的流失极为重要。一些蜕皮液可通过真皮细胞再吸收，但在烟草天蛾和美人蕉卷叶弄蝶的蜕裂蛹中，蜕皮液主要在新旧表皮下向口器和肛门附近流动，在中肠积聚。蜕皮液可能含有蛋白质和其他潜在的营养物质，水分有助于清洗中肠和/或马氏管积累的废弃产物，特别是在新羽化的成虫中，多数成虫排泄蜕皮时积累的蛹便。排除任何多余的水分也能减轻飞行时的体重。

1.3.5 Ecdysis and hormonal control

To facilitate ecdysis of the old cuticle, some insects swallow air or swallow water if they are aquatic insects to expand the gut and split the old cuticle. The muscular actions involved in ecdysis are controlled by nervous motor programs. Events leading up to and completion of molting in insects illustrate interactions between the endocrine and nervous systems, for although the nervous system controls muscular movements, in moths, and perhaps in other insects as well, the motor program is initiated by hormone action.

Under the regulation of 20E and juvenile hormone (JH), many molting proteins with different functions being expressed, and accumulated between the new and old cuticles. Successful ecdysis requires all molting proteins to work together for nutrition cycling, detoxification, signal transfer, fluid pressure balance, and immune protection. Ecdysis of the tobacco hornworm, *M.sexta*, has been divided into two phases: pre-ecdysis behavior and ecdysis (eclosion). The eclosion hormone (EH) and the pre-ecdysis-triggering hormone (PETH) and ecdysis-triggering hormone (ETH) are involved.

Eclosion hormone, a neurosecretory polypeptide of 62 amino acids, is secreted in *M.sexta* by ventrally located neurosecretory cells in the tritocerebrum in response to facing ecdysteroids. The cells send an axon to the corpora cardiaca (CC). EH is transferred through this axon for storage in the CC and released into the circulating hemolymph for pharate adult eclosion. Hemolymph-born EH cause the release of ETH from the Inka cells. The

1.3.5 蜕皮和激素控制

为了方便蜕去旧表皮，有些昆虫吞入空气或水生昆虫吞入水以促使中肠膨胀并使旧表皮破裂。牵涉蜕皮的肌肉受到运动神经编码指令的控制。引导昆虫蜕皮并完成蜕皮的事件说明了内分泌和神经系统的相互作用，尽管蛾与其他昆虫一样，由神经系统控制肌肉的运动，其运动神经编码指令由激素作用启动。

在 20E 和保幼激素的调节下，许多具有不同功能的蜕皮蛋白质在新旧表皮之间被表达和积累。成功的蜕皮需要所有蜕皮蛋白质协同工作，以实现营养循环、解毒、信号传递、体液压力平衡和免疫保护。烟草天蛾的蜕皮可分为两个时期：蜕皮前期和蜕皮期（羽化期）。涉及羽化激素（EH）和预蜕皮触发激素（PETH）以及蜕皮触发激素（ETH）。

羽化激素是由 62 个氨基酸组成的神经分泌多肽，由位于烟草天蛾后脑腹面的神经分泌细胞对蜕皮甾类反应时分泌的。这些细胞向心侧体（CC）发出轴突。通过这条轴突，EH 被转运到心侧体贮存，并在蜕裂成虫羽化时释放到循环中的血淋巴。血淋巴产生的 EH 引

epitracheal glands consist of several cells, but the most prominent cell is a large opaque cell, named the Inka cell. The opaqueness appears to be due to the protein hormone it is ready to secrete because, after secretion, it is neither so large nor opaque. Further action of ETH and EH on the tritocerebrum and the subesophageal ganglion sets in motion a neural network in abdominal ganglion by elevating cGMP levels. Ecdysis is delayed by inhibitory factors from the cephalic and thoracic ganglion until an independent clock mechanism in each abdominal ganglion as activated by central release of crustacean cardioactive peptide (CCAP). ETH acts directly on the central nervous system, and triggers pre-ecdysis and ecdysis behavior. EH acts to release ETH, and then ETH act directly on the central nervous system to initiate the motor program for pre- ecdysis. Pre-ecdysis behaviors include dorsoventral contractions occurring synchronously in thoracic and abdominal segments. These pre-ecdysis contractions occur every 10 to 12 seconds, with a contraction lasting 5 to 7 seconds. Isolated ventral nerve cords (VNC) of *H.cecreopia* are capable of generating the motor program. Each ganglion along the VNC produces bursts of action potentials alternately from the right and left sides. In an intact animal, these nerve impulses would initiate muscle action in each segment that could cause wiggling and rotatory movements of the abdomen. After an hour—sometimes less, sometimes more—the pre-ecdysis contractions give way to ecdysis contractions, a series of peristaltic waves of contractions that originate in the most posterior segment and pass anteriorly for about 10 minutes until ecdysis is complete. Synchronous bursts of action potentials occur from both sides of each ganglion, but alternating in time between successive ganglion. These action potentials cause the peristaltic muscle contractions necessary to push the insect out of the old cuticle. Some minutes after adult is out of the pupal cuticle, all ganglia simultaneously begin to produce prolonged bursts of action potentials, resulting in a steady tonic contraction of the abdomen that aids the pumping of hemolymph into the wings to inflate them. EH probably exist in most or possibly all insects.

After ecdysis, the air or water filled gut that aided in splitting the old cuticle expands the body size and stretches the new cuticle while it is still pliable and unsclerotized, thus giving the insect some room for growth if it is at an immature stage. Lipids and cement are secreted onto the new epicuticle, muscle reattachments become firmly fixed, and the cuticle begins to tan. Formation

起 ETH 从 Inka 细胞中被释放。上皮腺由几个细胞组成，但最为突出的细胞是一个不透明的大细胞，命名为 Inka 细胞。这种不透明好像是由它准备分泌的蛋白质激素导致的。因为分泌后，它既没有这么大，也没那么不透明。ETH 和 EH 进一步作用于后脑和咽下神经节，以提升 cGMP 水平，开始腹部神经节的神经网络的运行。蜕皮被头、胸部神经节的抑制因子延缓，直到每个腹部神经节内独立的生物钟机制被甲壳纲心激肽（CCAP）激活。ETH 直接作用于中枢神经系统并引发预蜕皮和蜕皮反应。EH 起作用释放 ETH，然后 ETH 直接作用于中枢神经系统以启动预蜕皮的运动编码指令。预蜕皮行为包括胸、腹体节背腹部的同步收缩。这些预蜕皮收缩每 10 ～ 12 s 就发生一次，每次收缩持续 5 ～ 7 s。游离的惜古比天蚕腹部神经索（VNC）能够产生运动编码指令。沿腹部神经索从左到右交替爆发动作电位。在正常的动物中，这些神经脉冲将启动每个体节的肌肉活动，从而导致腹部的摆动和旋转运动。在 1 h 后——有时更短，有时更长——预蜕皮收缩被蜕皮收缩代替，多数来自后部体节的一系列蠕动收缩波传递到前部体节，约 10 min 后蜕皮完成。动作电位的同步爆发出现在每个体节的两边，但经过一段时间后在连续的体节间交替。这些动作电位引起的肌肉蠕动收缩是推动昆虫离开旧表皮所必需的。在成虫离开蛹表皮几分钟后，所有的体节同时开始产生持续的动作电位爆发，引起腹部稳定的强直性收缩，这有助于将血液泵入翅膀使之膨胀。EH 可能存在于多数或所有昆虫中。

在蜕皮后，被空气或充满水的肠道帮助分裂旧表皮，在其还很柔软和未骨质化时扩张身体的大小并伸展新表皮，这样就给未成熟时期的昆虫提供了一些成长的空间。脂质和黏胶层被分泌到新表皮的表面，肌肉连接变得更加牢固，

of layers of endocuticle continues in many insects during internal periods, and even during adult life. In some insects, there are two distinctive layers formed daily: one during the day and another at night. so that growth rings can be observed and counted in a cross section of cuticle.

表皮开始鞣化。在许多昆虫幼虫期间，甚至成年期，仍在继续形成内表皮。一些昆虫有两层独特的表皮：白天形成一层；夜晚形成另一层。因此在表皮切片中可观察计数生长环。

1.3.6 Sclerotization of cuticle

Sclerotization is the process of cross-linking protein to protein chains, chitin to chitin chains and possibly protein to chitin chains. It is also called tanning, sometimes simply described as hardening of the cuticle. Tanning refers to the cross-linking process itself, and not to a color change, although sclerotization is often accompanied by tan, brown, or black colors. The colors are created by a variety of pigments, including melanin. The phenols associated with sclerotization usually leading auto-oxidation and quinones readily polymerize, process usually leading to melanin and tan or brown to black color. Tanning is a chemical process by which certain regions of the insect cuticle undergo an irreversible change from a soft layer into a stiffer and harder structure. Sclerotization mainly occurs in the exocuticle and epicuticle, which are the two outer most layers in insect cuticle. Sclerotized cuticle has decreased deformability, extractability of matrix proteins, and increased resistance towards dissolvants and digestive enzymes (Dongs et al., 2012).

Tanning of the newly formed, soft cuticle in insects includes two separate processes, termed sclerotization (hardening) and melanization (or darkening). Before eclosion or during molting, the newly formed insect cuticle consists of a thin layer of hydrophobic, waxy, chitin free epicuticle and a thick layer of protein- and chitin-rich procuticle. After eclosion or ecdysis, insect cuticular tanning occurs immediately, which hardens and darkens insect cuticle, and provides enough protection from desiccation or from imbibing excessive water in the case of aquatic insects and pathogens.

With tyrosine as the initial substrate, the cuticle tanning reactions include hydroxylation of tyrosine to dihydroxy-phenylalanine (DOPA) and decarboxylation of DOPA to dopamine. For melanization, furthermore, oxidation of DOPA and dopamine to DOPA-quinone and dopamine quinone, conversion of these quinones to dihydroxyindole (DHI) and/or 5,6-dihydroxyindole-2-carboxylic acid (DHICA), oxidation of DHI and DHICA to DHI-chrome and DHICA-chrome (melanochromes) and finally polymerization of melanochromes to

1.3.6 表皮的骨质化

骨质化是指蛋白质和蛋白质链、几丁质和几丁质链以及可能的蛋白质和几丁质交联的过程。它也被称为鞣化，有时简单描述为表皮的硬化。鞣化是指交联过程本身，而不是指颜色的变化，尽管骨质化常伴随着鞣化，变棕色或黑色。其颜色是由各种各样的色素，包括由黑色素产生的。与骨质化相关的酚类物质通常导致自氧化，醌类物质容易聚合，这个过程导致黑化和鞣化或颜色由棕变黑。鞣化是一种化学过程，通过这种化学过程，昆虫表皮的某些区域发生了不可逆转的变化，从柔软的一层变为更坚硬的结构。硬化主要发生在昆虫表皮最外层的外表皮和上表皮。硬化的表皮降低了基质蛋白的可变形性、可提取性，并增加了对溶剂和消化酶的抵抗力（Dongs 等，2012）。

昆虫新形成的柔软表皮的鞣化包括两个独立的过程，称为骨质化（硬化）和黑化（或变黑）。在羽化前或蜕皮过程中，新形成的昆虫表皮由一薄层疏水、蜡质、无几丁质的上表皮和一层厚的富含蛋白质和几丁质的原表皮组成。在羽化或蜕皮后，昆虫表皮立即发生鞣化，使昆虫表皮变硬、变黑，并提供足够的保护，以防止干燥或水生昆虫吸收过多水分和病原体。

以酪氨酸为初始底物，表皮鞣化反应包括酪氨酸羟基化为二羟基苯丙氨酸（DOPA）和DOPA脱羧产生多巴胺。此外，对于黑化而言，多巴和多巴胺氧化为多巴醌和多巴胺醌，这些醌转化为二羟基吲哚（DHI）和 / 或 5,6- 二羟基吲哚 -2- 羧酸（DHICA），DHI 和 DHICA 氧化为 DHI 铬和 DHICA 铬（黑色素铬），最后

form melanins. For pigmentation involving acylated quinones, the reactions include *N*-acetylation of dopamine to *N*-acetyldopamine (NADA) or *N*-β-alanylation to *N*-β-alanyldopamine (NBAD), oxidation of NADA and NBAD to NADA-quinone and NBAD-quinone and their polymerization to form their corresponding pigments. These quinones, in addition, undergo isomerization to quinone methides and cross-linking reactions with CP side chains (most likely histidyl residues) for cuticle sclerotization (Arakane et al., 2016).

The cross-linking or sclerotsing agents are phenols and their products, quinones. A number of phenols and quinones exist in various insect cuticles and all probably participate to some extent as sclerotizing agents, but the chemistry has been best elucidated for production of *N*-acetyldopamine, a common and major sclerosing agent in many cuticles.

N-acetyldopamine is formed from tyrosine by a number of enzymatically controlled steps. *N*-acetyl-dopamine has been implicated as the principal tanning agent in the pupal cuticle of *M. sexta*. It is a major cuticular constituent of a number of insects in various orders, and may be the typical sclerosing agent in the pupal—diphenol oxidase oxidizes it most readily among a variety of potential substrates.

Quinones cross-link protein chains by reacting with free amino groups, such as those of lysine, tryptophan, arginine, histidine, and terminal amino group at one end of a protein. Chitin chains are also linked to each other and possibly to protein chains through the amino group of *N*-acetylglucosamine. The sulfhydryl group of the amino acid cysteine may also participate in cross-linking protein chains through formation of disulfide linkages.

When protein chains are linked to the ring of the phenolic cross-linking agent, the process is called quinone tanning or quinone sclerotization. Proteins may also link to the β-carbon of *N*-acetyldopamine. Quinone also is involved in producing this type of sclerotization but the process is called β-sclerotization to distinguish it from protein attachment to the ring. Both quinone and β-sclerotization can occur in the same small region of cuticle. β-sclerotization can harden cuticle to produce lighter colored or transparent cuticle, which is important in the covering of the compound eyes, for example, and many insects have lightly colored patches of cuticle elsewhere on the body.

1.3.7 Hormone control of sclerotization

Pigmentation development involves activating agents, such as hormones that regulate the process, positioning agents that

黑色素铬聚合形成黑色素。对于涉及酰化醌的色素沉着而言，其反应包括多巴胺*N*-乙酰化为*N*-乙酰多巴胺（NADA）或*N*-β-丙氨酰化为*N*-β-丙氨酰多巴胺（NBAD），NADA和NBAD氧化为NADA-醌和NBAD-醌，并聚合形成相应的色素。此外，这些醌经历异构化，形成醌甲酰胺，并与表皮蛋白侧链（最有可能是组氨酸残基）发生交联反应，以使表皮硬化（Arakane等，2016）。

交联或骨化剂是酚及其产物醌类。许多酚类和醌类存在于各种昆虫表皮，它们在一定程度上作为硬化剂参与其中，但阐释得最清楚的是*N*-乙酰多巴胺产物，它是多数表皮普遍和主要的硬化剂。

N-乙酰多巴胺由酪氨酸通过一些酶控制步骤形成。*N*-乙酰多巴胺是烟草天蛾蛹表皮牵涉的主要鞣化剂。它是若干目昆虫的一些表皮主要成分，可能是蛹期典型的骨化剂——二酚氧化酶在各种潜在的一些底物中最易氧化。

醌类通过蛋白质链的游离氨基基团，如那些赖氨酸、色氨酸、精氨酸、组氨酸和蛋白质某一个末端氨基基团交联。几丁质链也可能通过*N*-乙酰氨基葡糖的氨基交联，并可能与蛋白质链相连。氨基酸半胱氨酸的巯基基团可能也通过形成双硫键参与蛋白质链的交联。

当蛋白质链连接到酚交联剂的苯环时，这个过程被称为醌类鞣化或醌类骨化。蛋白质也与*N*-乙酰多巴胺的β-碳原子连接。醌类也参与这类骨质化的产生，但这过程被称为β-骨质化，以区别于蛋白质与苯环的连接。在表皮同一个小区域，醌类和β-骨质化同时发生。β-骨质化能使表皮变硬，产生较淡颜色或透明表皮，这在复眼的覆盖层很重要，如多数昆虫在身体各处具有淡色表皮斑块。

1.3.7 骨质化的激素控制

色素沉着的发展涉及激活剂，如调节过程的激素、在空间和时间中产生色素

generate pigments in space and time, and the biochemical synthesis of pigments. There are two main classes of insect hormones, the hormones produced by epithelial glands belonging to the ecdysteroids and juvenile hormones (JH) and also the neuropeptide hormones produced by neurosecretory cells. The pigment biochemical module is regulated primarily by ecdysteroids such as 20-hydroxyecdysone (20E) and JH as well as transcription factors expressed in a marking-specific pattern. Neuropeptide hormones such as bursicon also play regulatory roles in pigment metabolism. Bursicon known as "tanning hormone" is a neuropeptide that promotes sclerotization and specifies how much cross-linking of molecules will occur. It is an insect heterodimeric neuropeptide hormone that is secreted from the central nervous system into the hemolymph, which initiates cuticle tanning as well as wing expansion after adult eclosion. Rickets (rk), the receptor for Bursicon, is a member of the G protein-coupled receptor superfamily. RNA interference mediated silencing of the *Bursicon* gene induces defects in wing expansion of silkworm. It has been found in various ganglia of the central nervous system of many insects and is now believed to occur in most, or possibly all, insects.

的定位剂以及色素的生化合成。两类主要的昆虫激素：一类是由上皮腺产生的蜕皮甾类和保幼激素（JH）；另一类是由神经分泌细胞产生的神经肽激素。色素生化模块主要由蜕皮甾类，如20-羟基蜕皮酮（20E）和JH以及斑点特异性模式表达的转录因子调节。神经肽激素也在色素代谢中发挥调节作用，如Bursicon被称为"鞣化激素"，是一种启动骨化并特定多少分子将出现交联的神经肽。这是一种昆虫异二聚体神经肽激素，由中枢神经系统分泌进入血淋巴，在成虫羽化后启动表皮层鞣化和翅膀扩张。Rickets（rk）是Bursicon的受体，属于G蛋白偶联受体超家族。RNA干扰介导*Bursicon*基因沉默导致家蚕翅膀伸展缺陷。在很多昆虫中枢神经系统的各种不同的神经节中都有发现，现在认为它出现在大多数或可能所有昆虫中。

1.4 昆虫体色及其分子机制

1.4 Insect Body Coloration and Molecular Mechanisms

1.4.1 The classification of insect coloration

Most insects exhibit colors *via* absorption or reflection of sunlight using pigments, cuticular surface structures, or their combination. Only some springtails, flies, and beetles possess luciferases, which catalyze light-producing biochemical reactions, were being capable of bioluminescence. Several classes of pigments are involved with insect coloration. Generally, melanins produce shades from black to reddish-brown, and pterins, ommochromes, and carotenoids contribute to red, orange, and yellow colors. Furthermore, flavonoids, which are plant secondary metabolites, also create orange colors in insects. Bile pigments or bilins result in greenish and bluish tints, violet, and golden colors often appear because of structural interference. Interestingly, some insect taxa have specific pigments, like aphins, producing a variety of tints in aphids, and papilochromes, resulting in yellow, orange,

1.4.1 昆虫颜色的分类

大多数昆虫通过色素、表皮结构或它们的结合物吸收或反射阳光来呈现颜色。只有一些跳蚤、蝇类和甲虫拥有荧光素酶，这种酶能催化产生光生化反应而发光。昆虫着色涉及几种色素。一般来说，黑色素会产生从黑色到红棕色的色调，蝶呤类、眼色素和类胡萝卜素会产生红色、橙色和黄色。此外，作为植物次生代谢物的黄酮类化合物也能在昆虫体内产生橙色。胆色素类或后胆色素类会产生绿色和蓝色色调，由于结构干扰，通常会出现紫罗兰色和金色。有趣的是，一些昆虫类群有特定的色素，如蚜虫一样，产生多种蚜虫和乳突色素的颜色，导致蝴蝶会产生黄色、橙色和

and red colors in butterflies. Despite all diversity, melanins and pterins are two prevalent classes of insects' pigments. Sclerotisation (hardening) and melanization (darkening) of insect cuticle can act in conjunction, and the appearance of a color often is the result of both processes.

红色。尽管存在各种多样性，但黑色素和蝶呤类色素是昆虫色素的两种常见类别。昆虫表皮层的骨质化（硬化）和黑化（变黑）可以协同作用，颜色的出现通常是这两个过程的结果。

1.4.2 Molecualr mechanism of insect coloration

Insect colors depend on both pigments and structures. Pigments are deposited in distinct places of the cuticle, which can be venation structures on the butterflies' wings or beetle elytra. Often these places are related to inner structures, such as outgrowths of the cuticle or places of the muscle attachments. Pigmented cuticle patterns are related to muscle bundle topography, where muscles are attached to cuticle, and the cuticle surface possess melanin spots in those places where they are attached. These structures have a morphological basis as described above or can be photonic structures, capable of interfering with visible light scale and possessing layers of the relatively high and low reflective index. Some specific color tints in insects, for example, transparent white, golden-bronze, iridescent blue, or violet have only a structural basis.

The genes and their biochemical transformations within the melanin synthesis pathway in Lepidoptera are highly conserved, and their spatiotemporal expression determines the location and abundance of pigmentation. Insect melanization starts from the hydroxylation of tyrosine to 3,4-dihydroxyphenylalanine (DOPA) encoded by the gene of tyrosine hydroxylase (*TH*). Phenol oxidase (PO) catalyzes the conversion of DOPA to dopamine melanin encoded by *yellow* and *laccase2* genes, producing black melanin. In a parallel branch, DOPA can be transformed to dopamine through decarboxylation pigmentation. However, *TH* and dihydroxyphenylalanine-DOPA decarboxylase gene (*Ddc*) showed higher expression in the albino caterpillars. Moreover, 6-pyruvoyl-tetrahydropterin (*PTS*), a gene outside the melanin pathway that encodes tetrahydrobiopterin (BH4), is a cofactor that impacts the hydroxylation of tyrosine, the precursor of melanin synthesis. Mutations in *PTS* can promote albinism in the silkworm (*B. mori*). This indicates a nondepletion of master genes (*TH*, *Ddc*, *PTS*) in albino caterpillars, suggesting that melanin synthesis may be disrupted during the conversion of DOPA and dopamine to melanin by *yellow* and *laccase2* genes. The observed low expression of these genes in albino caterpillars further supports this notion. Downstream the melanin pathway, the production of dopamine melanin can be suppressed by converting dopamine to

1.4.2 昆虫着色的分子机理

昆虫的颜色取决于色素和结构。色素沉积在表皮层的不同位置，表皮层可以是蝴蝶翅膀或甲虫鞘翅上的脉序结构。这些部位通常和内部结构有关，例如，表皮层的外生长或肌肉附着部位。色素表皮层图案与肌肉束表面形状有关，肌肉附着在表皮层上，表皮层表面附着的地方有黑色素斑点。这些结构具有上述形态基础，也可以是光学结构，能够干扰可见光幅度，具有相对较高和较低反射指数的结构层。昆虫的某些特定颜色，例如，透明白色、金青铜色、彩虹蓝色或紫罗兰色，只有一个结构基础。

鳞翅目昆虫黑色素合成途径中的基因及其生化转化是高度保守的，其时空表达决定了色素沉着的位置和丰度。昆虫黑色素化始于酪氨酸羟化酶（*TH*）基因编码的酪氨酸羟基化为 3,4- 二羟基苯丙氨酸（DOPA）。酚氧化酶（PO）催化 DOPA 转化为 *yellow* 和 *laccase2* 基因编码的多巴胺黑色素，产生黑色黑色素。在平行分支中，DOPA 可以通过脱羧色素沉着转化为多巴胺。然而，*TH* 和二羟基苯丙氨酸 DOPA 脱羧酶基因（*Ddc*）在白化毛虫中表达较高。此外，6- 丙酮酸四氢蝶呤（*PTS*），一种黑色素途径外的基因，编码四氢蝶呤（BH4），是影响黑色素合成前体酪氨酸羟基化的辅因子。*PTS* 突变可促进家蚕白化。这表明主基因（*TH*、*Ddc*、*PTS*）在白化毛虫中，表明在 DOPA 和多巴胺转化为黑色素的过程中，*yellow* 和 *lacase2* 基因可能会破坏黑色素合成。这些基因在白化毛虫中的低表达进一步支持了这一观点。在黑色素途径的下游，多巴胺-黑色素的产生可以通过将多巴胺转化为 *N*-*β*-丙氨酸多巴胺（NBAD）

N-β-alanyldopamine (NBAD). In this branch, *β*-alanine binds to dopamine by the activity of *ebony* forming NBAD, the precursor of yellow sclerotin, resulting in a yellow pigment. This can be reversed by the activity of *tan*, by which dopamine is converted back to dopamine melanin, thus promoting dark pigmentation. Finally, dopamine also can be converted to NADA sclerotin, which is colorless and encoded by arylalkylamine *N*-acetyltransferase gene (*aaNAT*). Previous studies in Lepidoptera and other insects have found an upregulation of *ebony* and *aaNAT* in nonblack body regions, while *tan* is generally expressed in black regions. The melanin-promoting genes *tan* and *yellow* showed a higher expression in the colored caterpillars, whereas the melanin-suppressing genes *ebony* and *aaNAT* were more expressed in the albino caterpillars (Fig.1-6; Appendix color graphic Fig.1-6).

来抑制。在这个分支中，*β*-丙氨酸通过 *ebony* 形成 NBAD 的活性与多巴胺结合，NBAD 是黄色壳硬蛋白的前体，产生黄色色素。这可以通过 *tan* 活性来逆转，通过这种活性，多巴胺被转化回多巴胺黑色素，从而促进黑色素沉着。最后，多巴胺也可以转化为 NADA 壳硬蛋白，它是无色的，由芳烷基胺 *N*-乙酰基转移酶基因（*aaNAT*）编码。鳞翅目和其他昆虫的 *ebony* 和 *aaNAT* 在非黑体区域表达上调，而 *tan* 通常在黑色区域表达。其中促进黑色素的基因 *tan* 和 *yellow* 在有色毛虫中表现出更高的表达，而黑色素抑制基因 *ebony* 和 *aaNAT* 在白化毛虫中的表达更高（图 1-6；附彩图

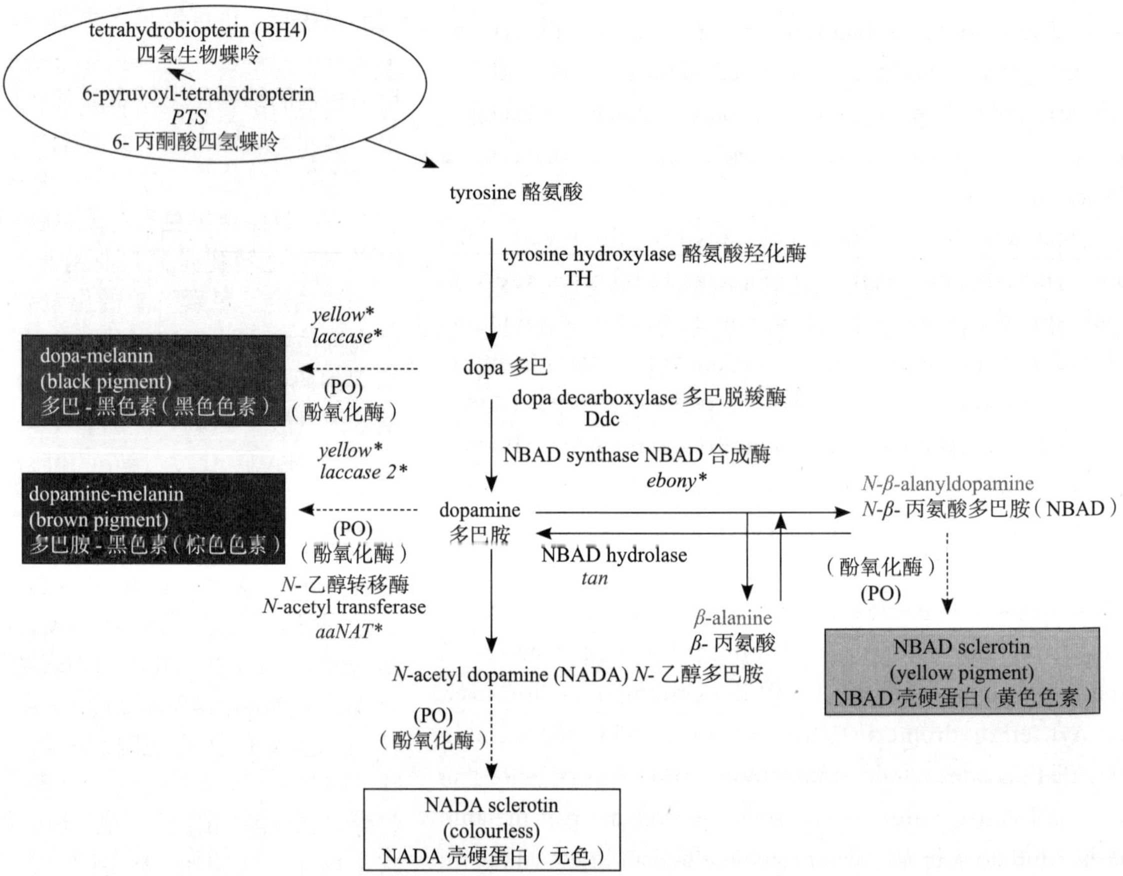

Fig.1-6 Insect body coloration melanin biosynthesis pathway. Green shows the cofactor outside the melanin pathway impacting the hydroxylation of tyrosine, the precursor of melanin biosynthesis. Biochemical phenoloxidase reactions catalyzing pigment conversion are shown in dotted lines, and the final pigments produced are shown in boxes. Asterisk indicates the genes putatively involved in caterpillar albinism (modified from Galarza, 2021).

图 1-6 昆虫体色黑色素生物合成途径。绿色显示黑色素途径外的辅助因子影响酪氨酸的羟基化，酪氨酸是黑色素生物合成的前体。催化色素转化的酚氧化酶生化反应用虚线表示，最终产生的色素用方框表示。星号表明可能与毛虫白化病有关的基因（仿 Galarza，2021）。

Thus, in addition to a defective conversion of melanin from DOPA and dopamine upstream the pathway, melanin may be further suppressed downstream the pathway in albino caterpillars.

1-6)。因此，除了DOPA和多巴胺上游的黑色素转化有缺陷外，在白化毛虫中，黑色素可能会在下游被进一步被抑制。

1.4.3 The functions of insect body colors

1.4.3 昆虫体色的功能

1.4.3.1 Body physical, immunological protection

Darker cuticles are generally thicker, which prevent an insect body from the penetration of pathogens and parasites. Melanin pigments contribute both to insects' cuticle coloration and immunity (encapsulation response). Therefore, melanin-based coloration can be inter-connected with immunological defense.

1.4.3.1 身体物理、免疫保护

颜色较深的表皮层通常较厚，它可防止昆虫身体受到病原体和寄生虫的侵入。黑色素有助于昆虫的表皮层着色和免疫（包囊反应）。因此，基于黑色素的着色可以与免疫防御相互关联。

1.4.3.2 Signaling for camouflage, warning coloration, rival quality, mate choice

Colors help insects to hide from or scare away predators. The main mechanisms for color-concealment are crypsis, disruptive patterning, mimesis, counter shading, and counter illumination. From the evolutionary point of view, insect camouflage is one of the two main adaptations that reduce predation risk. The other is a warning or aposematic coloration, which rely on their coloring and patterning to act as a warning signal advertising unpalatability to predators. The colors that combine with black on insect species to produce aposematism, namely red, orange, yellow, and white.

1.4.3.2 伪装信号、警告色、竞争质量、配偶选择

颜色有助于昆虫躲避或吓跑捕食者。颜色隐藏的主要机制是保护色、扰乱模式、拟态、反荫蔽和反照明。从进化的观点来看，昆虫伪装是减少捕食风险的两种主要适应方式之一。另一种是警告或警戒色，它们依靠颜色和图案来作为警告信号，向捕食者宣告其不适口性。昆虫身上与黑色结合产生警戒的颜色，即红色、橙色、黄色和白色。

1.4.3.3 Physiological condition such as thermo-regulation, UV-resistance, drought-resistance.

Beneficial functions of colors in insects are interrelated with insect biochemical cuticle properties, color types, life strategies, behavior, and physiology. If a color pattern reflects the physiological condition of an individual, it is considered as a true quality signal. Such quality signals are needed in the assessment of potential rivals or for the mate choice. Different insect species use colors for these purposes. For example, blue colors, distinguishing coloration of damselflies, require the development of bile pigments in many cases; these functions are interacting, which highlights the multifunctional and integrative role of coloration. For example, camouflage and warning coloration can be referred to as both body protection and signaling; or the thermoregulatory function of a color can also possess a UV-radiation protective component. Many butterflies focus on wing colors and color elements to select mate partners. For several other beetle species, it was found that closely

1.4.3.3 体温调节、低抗紫外线和抗旱等生理条件

颜色在昆虫中的有益功能与昆虫生化表皮的特性、颜色类型、生活策略、行为和生理学相关。如果一个颜色模式反映了某个体的生理状况，它就被认为是一个真正的特征信号。在评估潜在竞争对手或选择配偶时，需要此类特征信号。不同的昆虫种类使用颜色来达到这些目的。例如，在许多情况下，区分豆娘颜色的蓝色需要进化胆色素；这些功能相互作用突出了色彩的多功能和综合作用。例如，伪装和警告色可以被称为身体保护和信号，或者颜色的温度调节功能也可以具有紫外线辐射保护成分。许多蝴蝶注重翅膀的颜色和色彩元素来选择伴侣。对于其他几种甲虫物种

related species, sharing the same habitat, had noticeable color differences associated with a quick search of a mating partner. The thermal melanism hypothesis states that individuals tend to be darker in colder environments and lighter in the warmer ones. The other environmental driving force, shaping color variations in insects, is UV-light protection. UV-protection hypothesis states that darker individuals are better protected from insolation. The interrelation between drought-resistance and coloration in insects is also conceptualized in the melanism-desiccation hypothesis. It states that the increased melanization favors decreasing cuticle permeability, and therefore darker individuals are better adapted to drier environments.

而言，研究人员发现，共享同一栖息地的近亲物种在快速寻找交配伙伴时有明显的颜色差异。热黑化假说认为，在较冷的环境中，个体往往较暗；在较热的环境中，个体则较亮。形成昆虫颜色变化的另一个环境驱动力是紫外线保护。紫外线防护假说表明，肤色较深的离群个体更容易受到保护。昆虫的抗旱性和颜色之间的相互关系也在黑色素干燥假说中被概念化。研究表明，黑化程度的增加有助于降低表皮层的渗透性，因此肤色较深的个体能更好地适应干燥的环境。

1.4.3.4 Various groups of social insects, including bees, wasps, and ants are identified using color variations on the different body parts

1.4.3.4 根据身体不同部位的颜色差异，可以识别各种群居昆虫，包括蜜蜂、黄蜂和蚂蚁

Entomologists used coloration of ant species in Guyana for species identification. Color of the mandible (whitish), antennae and legs (yellowish), head and gaster (dark-brown or black) as well as waist segments (light brown) were main distinguishing characteristics of *S. royi*. Discovering insect coloration has several research assisting aspects. Those are revealing the range of intra- and within species variability, biochemical properties of pigments and biophysical properties of nanostructures, discovering insects' age or a specific physiological condition, related to coloration (Badejo et al., 2020).

昆虫学家使用圭亚那蚂蚁物种的着色进行物种识别。下颌骨的颜色（白色）、触角和腿的颜色（黄色）、头部和胃间膜的颜色（深棕色或黑色）以及腰部的颜色（浅棕色）是蚂蚁的主要区别特征。发现昆虫的颜色有几个研究方面的辅助作用。这些研究揭示了物种间和物种内的变异范围、色素的生化特性和纳米结构的生物物理特性，发现了昆虫的年龄或与颜色相关的特定生理条件（Badejo 等，2020）。

1.5 昆虫体壁的功能

1.5 The functions of Insect Integument

The integument is the "skin" of the insects. It must provide a barrier to the environment and a shell that supports the body and a diverse set of "tools" in the form of functional appendages for locomotion, foraging and defense such as legs, wings,

体壁是昆虫的“皮肤”。它必须针对环境提供屏障，为身体提供支撑的外壳以及各种各样的“工具”，如腿、翅膀、下颌和爪等功能性附件，用于移

mandibles, and claws. In addition, the cuticle plays an important role in environmental perception and inter- and intraspecific communication with a variety of embedded sensors and by generating color and sound.

The functions might include but not limit: ① serves as the exoskeleton of the insect, keep the main shape of the worm body. ② the site for muscle attachment. ③ the first physical line of defense from bacteria, fungi, predators and parasites, and environmental chemicals, including pesticides. ④ the integument functions in some or all insects in locomotion, breathing and respiration, feeding, excretion, protection from desiccation, behavior, osmoregulation, water control, and as a food reserve. ⑤ the many roles by the integumentary covering of insects are, in part, reflected in the complexity of its structure and chemistry, and in the special ways it is adapted to varied, reflecting species specificity and diversity, such as epidermal cells may be differentiated into sense organs or specialized glandular cells; Oenocytes also synthesize components of the cuticular wax, including some contact sex pheromones; Dermal glands secrete the cement layer of epicuticle. ⑥ a variety of pigments have been found in the cuticle or the epidermal cell which may give an insect its characteristic color and camouflage to protect them from predators. The beauty, color, shape, and intricate sculpturing on the surface of insects attracts amateurs and professionals alike to collect and study insects. ⑦ taxonomists and systematics have traditional used the surface sculpturing, setae, and sutures on the cuticle in classification of insect species.

(Li Wenchu)

动、觅食和防御。此外，表皮在环境感知中起着重要作用以及通过产生颜色和声音，与各种嵌入式传感器进行种间和种内通信。

其功能包括但不限于：①起着昆虫外骨骼的作用，保持虫体主要形状。②是肌肉附着点。③防御细菌、真菌和寄生虫以及环境化学物质包括杀虫剂的第一道物理防线。④体壁在某些或全部昆虫中的功能是运动、呼吸、取食、排泄、脱水保护、行为、渗透压调节、水分控制和作为食物来源。⑤昆虫体壁覆盖层的许多作用部分反映了它的结构和化学成分的复杂性。在特殊情况下，它能适应多变的状况，反映了品种的特异性和多样性。如真皮细胞可分化为感觉器官或特殊的腺细胞；绛色细胞还合成表皮蜡质成分，包括一些接触性信息素；真皮腺体分泌表皮的黏胶层。⑥在昆虫的表皮或真皮细胞中发现了多种色素，它们可能赋予昆虫特有的颜色和伪装色，以保护它们免遭天敌捕食的侵害。昆虫体表的华丽、色彩、外形和错综复杂的纹饰还吸引了业余和专业人员收集和研究昆虫。⑦分类学家和分类学传统上使用表面纹饰、刚毛和表皮上的骨缝来分类昆虫物种。

（李文楚）

Exercises

1.Brief discuss the composition and function of the insect integument.

2.How is the insects' molting process initiated and stimulated?

3.Which enzymes are involved in chitin synthesis?

思考题

1. 简述昆虫体壁的组成及其功能。

2. 昆虫蜕皮过程是如何启动和激发的？

3. 几丁质合成过程有哪些酶参与？

第2章 消化与营养

Chapter 2 Digestion and Nutrition

Key points

1. The digestive tract of insects is generally composed of mouthparts, pharynx, salivary glands, crop, foregut, midgut, hindgut, colon, ileum and rectum. The organizational structure of each part varies with different functions.

2. The morphological characteristics of the ultrastructure of midgut cells enable insects to absorb nutrients and secrete digestive enzymes. The diversified functions of the peritrophic membrane also make the midgut a physical barrier to prevent the invasion of pathogens.

3. The hindgut is an important place for insects to reabsorb nutrients and recover useful substances, thus forming a reverse flow cycle. Therefore, the complex structure of the cryptonephridial duct plays a particularly important role in the adaptation of insects to the environment.

4. Symbiotic microorganisms not only provide important digestive enzymes for host insects, but also change many charact-eristics of insects, such as reproduction, habitus and feeding habits.

知识点

1. 昆虫的消化道普遍由口器、咽喉、唾液腺、嗉囊、前肠、中肠、后肠、结肠、回肠和直肠等组成，各部分组织结构因功能不同而异。

2. 中肠细胞超微结构形态特征使昆虫能吸收营养物质和分泌消化酶，微食膜功能多样化也使得中肠成为防止病原物入侵的物理屏障。

3. 后肠是昆虫再吸收营养物质和回收有用物质的重要场所，从而形成逆向液流循环，复杂的隐肾管结构在昆虫适应环境中具有特别重要的作用。

4. 共生微生物不仅为宿主昆虫提供重要的消化酶，也改变了昆虫的许多特性，如生殖、习性和食性等。

Insects are the most abundant and diverse class of animals on the planet. The evolutionary success and diversity of insects have been driven by their ability to occupy many ecological niches and utilize different food sources. On the contrary, insect's diet, environmental habitat, and developmental stage modified the insect host's-associated gut microbiota, so that insects can quickly adapt to environmental changes. Usually, newly hatched insects must obtain food soon or die. Some newly hatched insects eat egg shell and the small amount of yolk left in the shell after hatching, or begin eating the food on or in which the egg was laid; some have to search for food. With suitable food, a larva may successfully grow, molt, and eventually become an adult.

昆虫是地球上数量和种类最多的动物。昆虫进化的成功和多样性使它们具备占领许多生态小环境并利用不同食物源的能力。相反，昆虫的取食、环境栖息地和发育阶段改变了昆虫宿主相关的肠道微生物群，从而使昆虫很快适应环境的变化。通常，刚孵化的昆虫必须尽快获取食物，否则就会死亡。有些刚孵化的昆虫吞食卵壳和少量孵化后残留的卵黄或开始吞食产卵于其上或其内的食物；有些不得不搜寻食物。有了适宜的食物，幼虫成功成长、蜕皮并最终成为成虫。

According to the nature of insect feeding, insects can be divided into phytophagous, saprophagous, and polyphagous. It also can be divided into polyphagous, omnivorous, oligophagous, and monophagous based on the narrow classification of recipes. In recent years, an increasing amount of literature focuses on insect gut symbiotic microbiota. It is now fully recognized that microorganisms form symbiotic relationships with insects, influencing their survival, fitness, development, mating habits and the immune system and other aspects of the biology and ecology of the insect host.

按照昆虫取食的性质，昆虫可分为植食性、腐食性和肉食性。根据食谱的广狭分类，昆虫也可分为多食性、杂食性、寡食性和单食性。近年来，越来越多的文献关注昆虫肠道共生微生物群。现已充分认识到微生物与昆虫形成共生关系影响着昆虫的生存、健康、发育、交配习惯和免疫系统以及昆虫宿主生物学和生态学的其他方面。

2.1 消化道的结构和功能

2.1 Structure and Function of Alimentary Tract

The alimentary canal of insects runs through the entire body cavity from mouth to anus and is divided into three main regions: foregut, midgut, and hindgut.

昆虫的消化道贯穿全身体腔，从口到肛门又分为三大类区域：前肠、中肠和后肠。

2.1.1 The foregut

During embryonic development, the foregut develops from invaginating ectodermal tissue at the anterior end of the body. The foregut can be divided into a buccal cavity (mouth), pharynx, esophagus, crop, proventriculus, and esophageal invagination (Fig.2-1). At each molt, the old cuticular lining from the foregut is sloughed off into the gut, and any undigested residue is excreted with the feces. Epithelial cells in the foregut are usually flattened squamous cells that do not secrete digestive enzymes.

The mouth or buccal cavity is usually just an enlarged opening that receives the slightly chewed food in mandibulate insects or fluid ingested by insects with piercing and sucking mouthparts. Powerful muscles in the wall of the pharynx pump fluid into the buccal cavity and aid swallowing in blood feeders, and xylem and phloem feeders.

The generalized mouthparts include the anterior labrum, paired mandibles, maxillae bearing a pair of maxillary palpi, and the posterior labium, bearing a pair of labial palpi as well as the hypopharynx in the preoral cavity. The labrum houses many sensory structures, such as setae, pressure receptors, trichoid sensilla, and coeloconic sensilla, and serves a chemosensory

2.1.1 前肠

在胚胎发育时期，前肠由身体前端外胚层组织内陷发育而来。前肠分为口腔、咽喉、食道、嗉囊、前胃和食道内陷（图 2-1）。在每次蜕皮时，前肠的旧表皮膜脱落进入肠道，与任何未消化的残渣随粪排出。前肠的上皮细胞通常是扁平的鳞片状细胞，不分泌消化酶。

嘴或口腔通常是一个真正扩大的开口，咀嚼式昆虫接纳被轻微咀嚼过的食物或用刺吸式口器接纳吞入的汁液。在进食血液、木质部和韧皮部的昆虫中，咽喉壁强而有力的肌肉将汁液泵入口腔并协助吞咽。

口器普遍包括前部上唇、成对的下颌骨、上颌须上的成对上颌骨，带有一对唇须的后部下唇和前口腔中的咽下道。上唇有许多感觉结构，比如，刚毛、压力感受器、毛状感受器和腔锥感

function. Insect mouthparts can be categorized in three principal functional types: ① mandibulate biting and chewing mouthparts; ② haustellate mouthparts forming variously composed proboscises; ③ filter-feeding mouthparts of aquatic immature stages. Additional functions, like defense, brood care, and malemale competition, modified the mouthparts in some insects. Fluid feeders have evolved to exploit a remarkable diversity of food resources including nectar, phloem, and xylem of plants, and blood, carrion, dung, sweat, tears, and urine of animals. The feeding devices of all fluid-feeding insects consist of a proboscis paired with a sucking pump. The proboscis surface consists of hydrophobic chitin and different types of proteins, lipids, and waxes.

受器，并具有化学感受功能。昆虫口器可分为三种主要功能类型：①下颚咬合和咀嚼口器；②组成各种喙的吸器口器；③水生未成熟阶段的滤食口器。其他功能，如防御、产卵护理和雄性竞争改变了某些昆虫的口器。以液体为食的昆虫已经进化到可以利用多种多样的食物资源，包括植物的花蜜、韧皮部和木质部以及动物的血液、腐肉、粪便、汗水、泪液和尿液。所有以液体为食的昆虫的取食装置都由一个喙和一对吸吮泵组成。喙表面由疏水几丁质和不同类型的蛋白质、脂质和蜡组成。

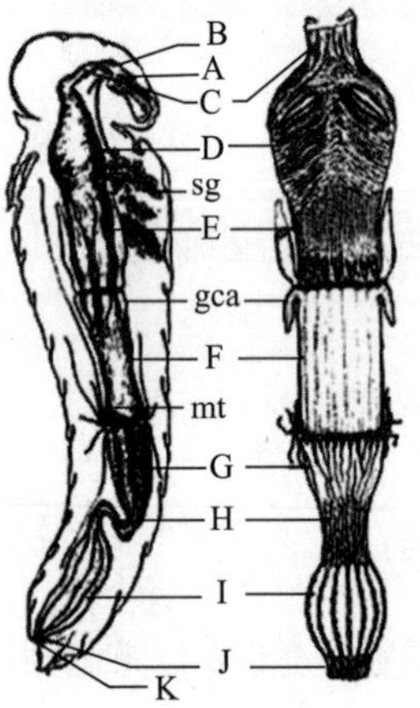

Fig.2-1 Digestive tract of L. *migratoria menilensis* Meyen. A. buccal cavity; B. pharynx; C.esophagus; D.crop. E.proventriculus; F.ventriculus; G. ileum; H.colon; I. rectal pad; J. rectum proper; K. anus; sg. salivary gland; gca. gastric caeca; mt. malpighian tubules(Liu, 1955).

图 2-1 飞蝗消化道结构。A. 口器；B. 咽喉；C. 食道；D. 嗉囊；E. 前胃；F. 胃；G. 回肠；H. 结肠；I. 直肠垫；J. 直肠；K. 肛门；sg. 唾液腺；gca. 胃盲囊；mt. 马氏管（刘，1955）。

The pharynx passes food to the esophagus, which may be a simple tube that continues to the proventriculus at end of the foregut in some insects. Alternatively, the foregut may expand into a much enlarged and dilated crop, or the crop may be a diverticulum from the main part of the foregut, as in some Diptera. The pharynx is rather soft and bears cuticular structures and sensilla; it is suspended by small sclerites and moved by numerous small muscles. Swinging movements in anterior and posterior directions are achieved by a set of muscles, which originate at the fronts of the head, the tentorium, and the labium, and insert at the small suspensory sclerites. In addition, intrinsic hypopharyngeal muscu lature may form part of the salivarium (Krenn et al., 2019).

咽喉将食物传到食道，在有些昆虫中，食道可能是一种简单的管道，延伸到前肠末端的前室。另一种情况是前肠可能扩张成一个更加扩大和膨大的嗉囊或嗉囊可能是前肠主要部分的一个支囊，如某些双翅目昆虫。咽喉相当柔软，有表皮结构和感觉器官；它由小骨片悬吊，通许多小肌肉移动。前后方向的摆动由一组肌肉完成，这些肌肉起源于头部、天幕和下唇的前部，并在小悬骨片处插入。此外，体内的下咽肌可能构成唾窦的一部分（Krenn 等，2019）。

The crop periodically releases some of its contents to pass through the preventriculus and into the midgut. Extra oral digestion occurs in many insects. Some insects reflux enzyme secretions and partially digested products by repeatedly sucking up and reinjecting the liquefied juices into the food. Refluxing mixes the

嗉囊定期释放一些内含物通过前胃进入中肠。许多昆虫进行口外消化。一些昆虫将溶解的汁液反复吸吮和注入食物使酶分泌物和部分消化的产物回流。回流混合了分泌物和汁液，尽可能

secretions and fluids and extends the effective life of the digestive enzymes. Refluxing is particularly effective when the food contains a limiting boundary, such as the shell of a seed or the cuticle of an insect that acts as a container for the liquefying body contents. The preventriculus at the end of the foregut may be very muscular and contain heavily sclerotized teeth, ridges, and spines (Fig.2-2) for further grinding and tearing of the food, or it may be reduced to a simple valve at the entry to the midgut.

延长消化酶的有效活力。回流对限定界线的食物特别有效，如种子的壳或昆虫的表皮可作为液化躯体内容物的容器。前肠末端的前胃可能更加肌肉化，含有高度骨质化的齿、嵴和刺（图 2-2），用于进一步碾磨和撕扯食物，或者在中肠入口处可能简化为一个简单的瓣膜。

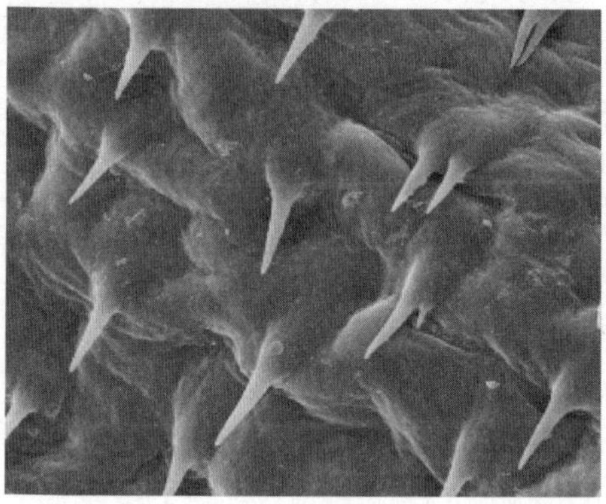

Fig.2-2 The sclerotized thorns in the preventriculus of *Fruhstorferiola Tonkinesis* (unpublished picture).

图 2-2 粤北腹露蝗前胃内部骨质化的刺（未发表图片）。

2.1.2 The hindgut

The hindgut, like the foregut, develops in the embryo from ectodermal tissue and, consequently, hindgut cells have an attached cuticular lining on their surface. The part of the hindgut immediately past the Malpighian tubules has been called the ileum. Sometime the ileum grades into the rectum, but in some insects there is a distinct middle region, called the colon. The terminal part of the hindgut is the rectum.

Both circular and longitudinal muscles lie on the outer or hemolymph side of the hindgut. The entire hindgut has a chitinous lining on the lumen surface of the cells. The cells of the hindgut walls are typically arranged in a single layer of irregularly shaped epithelia cells. Numerous septate desmosomes connect the lateral borders of cells and serve to hold cell together and present a barrier to fluid and molecules that might otherwise pass-through adjacent cells.

The hindgut is the most specialized in those insects that digest cellulose, such as termites. Their hindgut is usually divided into several chambers harboring either bacteria or protozoa that digest cellulose. Glucose is the principal carbohydrate liberated from cellulose digestion, and it is usually fermented by the resident microorganisms, with the end products being short-chain fatty acids that can be absorbed by the termite and used as an energy source.

2.1.2 后肠

与前肠一样，后肠由胚胎的外胚层组织发育而来，所以后肠细胞在表皮上有连接的表皮内衬。直接通过马氏管的后肠被称为回肠。有时，回肠渐入直肠，但有些昆虫有一个明显的区域，称为结肠。后肠的末端部分是直肠。

环肌和纵肌都位于后肠的外侧或血淋巴侧。整个后肠细胞的腔表面有几丁质内衬。通常，后肠壁细胞是排列成单层的无规则形状的上皮细胞。大量分隔的桥粒连接外侧细胞，将细胞连在一起，形成屏障防止液体和分子通过相邻的细胞。

在那些可消化纤维素的多数昆虫中，后肠是最专业化的，如白蚁。白蚁的后肠通常分为几个室，藏匿着有消化纤维素的细菌或原生动物。葡萄糖是纤维素被消化释放后的主要碳水化合物，纤维素常被寄生微生物发酵，终产物短链脂肪酸能被白蚁吸收并可作为贮存能源。

2.1.3 The midgut

The midgut is the principal site for secretion of digestive enzymes and for digestion and absorption in most insects. In many insects, gastric caeca arise at or near the origin of the midgut, but they may be located at various points along the midgut. In some insects, the gastric caeca are a major site of absorption of digestion products, and they also produce digestive enzymes. In some insects at least, the midgut may be derived from ectodermal tissue, as are the fore- and hindguts. The midgut does not attach cuticular lining on the surface of the cells, but midgut cells in the majority of insects can secrete a chitin- and protein- containing membrane, the peritrophic membrane.

The midgut epithelium consists of a single layer of epithelial cells, containing four cell types: the intestinal stem cells (ISC) and the enteroblasts (EBs), which are immature cells that can further differentiate into enteroendocrine cells (EECs) or enterocyte cells (ECs). EECs act as sensors for the internal intestinal environment and are able to secrete digestive enzymes, produce specific gut regulatory hormones and peptides, as well as mediate the communication of the nutritional status to other organs. ECs are the most prevalent cells in the midgut epithelium. They are characterized by a columnar shape and bear microvilli on the apical membrane that significantly increase the surface area of the midgut. The enterocytes of the midgut are responsible for the production of digestive enzymes and the absorption of digested products. In addition, in some insects, enterocytes are also specialized in controlling the midgut luminal pH. In the midgut of lepidopteran insects, specialized ECs, called goblet cells, are responsible for the global alkalization of the midgut lumen (Holtof et al.,2019), while EECs refer to columnar cells, and both ISC and EBs are called regenerative cells.

2.1.3.1 The ultrastructure of the midgut cells and their functions.

1.Columnar cells

Columnar cells, the most numerous cells in the midgut, conduct most of the absorption of digested products and secretion of enzymes. The cells have microvilli on the apical or lumen surface and extensive invaginations of the basal cell membrane (Fig.2-3). The extensive membrane infoldings at the basal face and the microvilli at the apical face of midgut cells are adaptations to present a large membrane surface for metabolic functions such as absorption and secretion. Midgut cells exhibit other characteristics

2.1.3 中肠

中肠是多数昆虫分泌消化酶和消化吸收的主要场所。很多昆虫的胃盲囊从中肠或其附近发出，但它们位于中肠的不同位置。有些昆虫的胃盲囊是消化产物吸收的主要场所，也能产生消化酶。在某些昆虫中，中肠由外胚层衍生而来，这与前肠和后肠一样。中肠不与细胞表面的表皮质内衬相连，但大多数昆虫的中肠细胞能分泌含几丁质和蛋白质的膜，即围食膜。

中肠上皮由单层上皮细胞组成，包含四种细胞类型：肠干细胞（ISC）和成肠细胞（EBs），它们是未成熟细胞，可以进一步分化为肠内分泌细胞（EECs）或肠上皮细胞（ECs）。作为肠道内部环境的传感器，EECs 能分泌消化酶，产生特定的肠道调节激素和肽以及将营养状况传递给其他器官。ECs 是中肠上皮中最常见的细胞。其特征为柱状，顶膜上有微绒毛，能显著增加中肠的表面积。中肠的肠细胞负责消化酶的产生和消化产物的吸收。此外，在一些昆虫中，肠细胞也专门控制中肠腔的 pH。在鳞翅目昆虫的中肠中，被称为杯状细胞的特殊 ECs，负责中肠腔的整体碱化（Holtof 等，2019），而 EECs 是指圆筒形细胞，ISC 和 EBs 都被称为再生细胞。

2.1.3.1 中肠细胞超微结构及其功能

1. 圆筒形细胞

圆筒形细胞是中肠数量最多的细胞，控制多数消化产物的吸收和酶的分泌。细胞顶部或肠腔表面有微绒毛，基部细胞膜有大量的内陷（图 2-3）。基部表面大量的膜内褶和中肠细胞顶端面的微绒毛具有辅助代谢功能，如适应吸收和分泌所必需的大量膜表面。中肠细胞

typical of secretary epithelium, such as a rough endoplasmic reticulum, a large Golgi complex, and secretary vesicles.

2.Regenerative cells

Midgut cells wear out rapidly and are replaced by new cells that grow from small regenerative cells lying randomly near the base of mature cells in larval Diptera and Lepidoptera, or as small cell clusters called nidi (nests). Regenerative cells grow into mature cells gradually and replace cells lost through age, wear, and apocrine and holocrine secretion. Damage to the midgut regenerative cells after irradiation for inducing sterility in insects for population control has been one of the limiting factors in the use of the technique.

3. Goblet cells

Goblet cells are, indeed, somewhat goblet shaped, with a large central cavity lined with microvilli (Fig.2-4). The sides of goblet cell curve around to enclose the cavity, and at the apex, the apical lips have interdigitating microvilli that control fluid exchange between cavity and midgut lumen. The apical membrane facing the goblet cavity houses a vascular-type H^+-ATPase (proton ATPase

呈现出分泌表皮的其他典型特征，如粗面内质网、大高尔基体和分泌泡。

2. 再生细胞

中肠细胞磨损很快，可被小再生细胞成长而来的新细胞取代。再生细胞在双翅目和鳞翅目幼虫，随机位于成熟细胞的基部附近或成为小细胞簇，称之为巢（窝）。再生细胞逐渐成长为成熟细胞并取代那些因为龄期、磨损、顶浆分泌和全浆分泌失去的细胞。利用辐照后的中肠再生细胞诱导昆虫不育以控制种群数量一直是该技术应用的限制性因素。

3. 杯形细胞

杯形细胞的外形在某种程度上确实像杯，有一个内衬微绒毛的大中心腔（图 2-4）。杯形细胞的边缘成弧形弯曲围成一个腔，在顶端边缘有突起的微绒毛控制着肠腔和杯腔之间的液体交换。面向杯腔的顶膜安置有液泡型 H^+-ATP 酶 (ATP 酶质子泵），形成跨越顶膜的

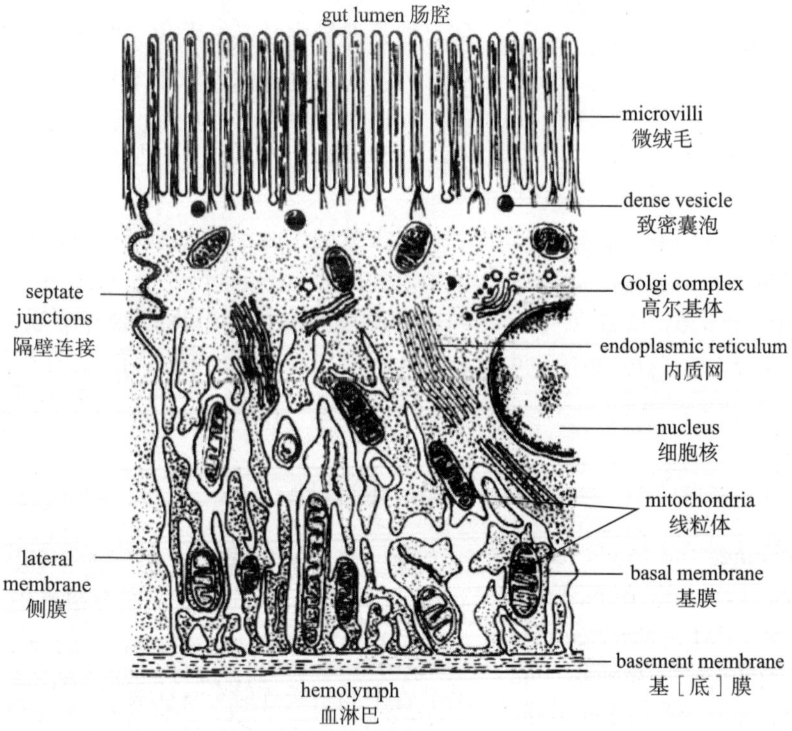

Fig.2-3 The structure of Columnar cell (modified from Berridge,1970).
图 2-3 圆筒形细胞结构（仿 Berridge,1970）。

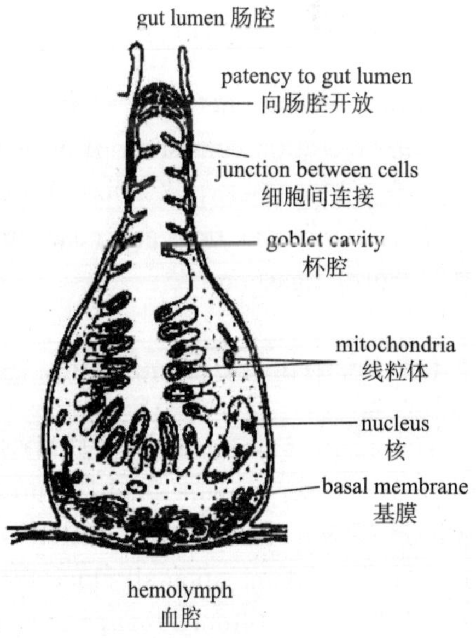

Fig.2-4 The structure of goblet cell (modified from Del Bene et al., 1991).
图2-4 杯形细胞的结构 (仿 Del Bene 等,1991)。

pump) that establishes a voltage across the apical membrane and pumps H^+ into the goblet cavity. In goblet cells, a K^+/nH^+ antiporter mechanism exchanges H^+ for K^+ in the goblet cavity. The pump is strongly electrogenic and creates transmembrane voltages that can exceed 240 mV and transmembrane pH gradients that may exceed 4 pH units. The transmembrane of voltage created by the pump enables a co-transporter mechanism in the apical membrane of columnar cells to reabsorb K^+ and amino acids from protein digestion.

The high midgut pH may help protect against tannins that are common in the plant hosts of Lepidoptera larvae. Tannins can complex with the insect's own enzymes and proteins in the food, and may result in reduced digestion of proteins.

2.1.3.2 Microvilli of midgut cells

Microvilli at the apical surface of midgut cells greatly increase the surface area for enzyme secretion and for absorption of digested products. Center-to-center spacing of microvilli in midgut cells of various insects range from 150 to 200 nm, hence the difficulty of resolving them clearly with light microcopy. The microvilli on the midgut of the mosquito *A.aegypti* are covered by a network of fine strands called the microvilli-associated network. A major function of this dense network may be to protect the midgut microvilli from phagocytes and other cells in the blood meal until digestion begins.

2.1.4 The glycocalyx

Most insects do not have a mucous lining, but often there is a viscous secretion consisting of protein and carbohydrate on the surface of, and between, microvilli called the glycocalyx. The viscous glycocalyx traps and concentrates secreted enzymes and products of digestion.

2.1.5 Peritrophic membranes

The peritrophic membrane (PM) surrounds the foods in the midgut and serves as a shield to protect microvilli from direct physical contact with food particles. Type I PM is secreted as a continuous delamination all along the length of the midgut, and type Ⅱ is secreted from a ring of cells at the anterior margin of the midgut. Larval mosquitoes form a type Ⅱ PM, but adult mosquitoes secrete a type I PM. Some insects secrete a PM only after taking food into the gut. Hemiptera and Homoptera as a group appears not to form a PM.

电压，并将 H^+ 泵入杯腔。在杯形细胞中，一种 K^+/nH^+ 反向转运机制在杯腔内用 H^+ 交换 K^+。这种泵是强大的生物电泵，能产生超过240 mV的跨膜电压，跨膜 pH 梯度可超过 4 个 pH 单位。泵产生的跨膜电压促使圆筒形细胞顶膜的共转运机制再吸收 K^+ 和蛋白质消化所产生的氨基酸。

中肠高pH可能有助于中肠抵抗鞣酸，这在鳞翅目幼虫的植物宿主中很普遍。鞣酸与昆虫本身的酶和食物中的蛋白质形成复合体，从而导致蛋白质消化减少。

2.1.3.2 中肠微绒毛

中肠细胞顶部表面的微绒毛大大增加了酶分泌和消化产物吸收的表面积。不同昆虫中肠细胞微绒毛的中心间距为 150 ～ 200 nm，因此很难用光学显微镜清晰地分辨。埃及伊蚊的中肠微绒毛覆盖有一层纤细的丝网，称为微绒毛关联网。在消化开始前，这种高密度网的一个主要功能可能是保护中肠微绒毛免受吞噬细胞和血粉中的其他细胞的侵害。

2.1.4 糖萼

大多数昆虫的中肠无黏膜层，但在微绒毛表面及微绒毛间有一层由蛋白质和多糖组成的黏性分泌物，称为糖萼。黏性糖萼能诱导和浓缩分泌的酶以及消化产物。

2.1.5 围食膜

围食膜（PM）在中肠包围着食物，并起着屏障作用保护微绒毛，避免食物颗粒与其产生直接的物理接触。I 型 PM 是沿着中肠全长连续分层分泌的，Ⅱ型由中肠前部边缘的环形细胞分泌。幼虫蚊形成Ⅱ型 PM，但成虫蚊分泌 I 型 PM。有些昆虫仅在进食后分泌 PM。半翅目和同翅目昆虫是不形成 PM 的种群。

Peritrophic membranes contain both chitin and protein, with chitin making up from 4% to about 20%, and protein composing up to 40% in various insects. Other components that have been reported are acid mucopolysaccharides, neutral polysaccharides, mucins, hyaluronic acid, hexosamine, glucose, and glucuronic acid. Enzymes must pass through the PM to get at the food, and small molecules resulting from digestion must pass out, so the PM is porous. As a semipermeable membrane, the PM also controls the passage of molecules, hence dividing the midgut lumen into two physiologically separate regions: the endoperitrophic space and the ectoperitrophic space. The pore size at the anterior side of the PM allows the movement of digestive enzymes into the endoperitrophic space. The initial digestion of the food thus takes place inside the endoperitrophic space, where most large macromolecules are degraded. The pore size of the PM gradually decreases towards the posterior end, only allowing smaller sized molecules to pass through the membrane to the ectoperitrophic space. Inside the ectoperitrophic space, a countercurrent flow pushes the food particles retrograde towards the caeca. This flow is created by the excretion of water and ions by the Malpighian tubules and their subsequent uptake by the caeca.Functions of the peritrophic membrane as follows.

① Protection of the delicate microvilli on the surface of midgut cells from contact with rough food particles.

② A barrier against entry of viruses, bacteria, or other parasites that would be too large to pass through the unbroken PM.

③ An aid in preventing the rapid excretion of digestive enzymes.

④ Compartmentalization of digestion within the midgut; The PM separates the midgut cavity into three compartments: the endoperitrophic space (the gut lumen including the food bolus), the ectoperitrophic space (between the PM and the epithelium), and the PM itself.

⑤ Prevention of nonspecific binding of undigested materials or plant allelochemicals to midgut microvilli surfaces and/or binding to transport proteins at the midgut surface.

PM 含有几丁质和蛋白质。在各种昆虫中，几丁质组成从 4% 到 20%，蛋白质组成多达 40%。其他已报道的组分是酸性黏多糖、中性多糖、黏蛋白、透明质酸、氨基已酸、葡萄糖和葡萄糖醛酸。酶必须通过 PM 才能触及食物，消化产生的小分子必须扩散出去，因此 PM 是有孔的。作为一种半透膜，PM 还控制分子的通过，因此将中肠腔分为两个生理上独立的区域：围食膜内空间和外空间。PM 前部的孔径允许消化酶进入围食膜内空间。因此，食物的最初消化发生在围食膜内空间内，多数大分子在那里被降解。PM 的孔径向后端逐渐减小，只允许较小尺寸的分子通过膜进入围食膜外空间。在围食膜外空间内，逆流将食物颗粒推向盲肠。这种流动是由马氏管排泄水和离子并随后被盲肠吸收而产生的。围食膜功能如下。

①保护中肠细胞表面纤细的微绒毛，避免与粗糙食物颗粒接触。

②是抵御病毒、细菌或其他寄生物进入的屏障，由于它们太大而不能通过完整的围食膜。

③防止消化酶的快速分泌。

④中肠内消化的划分；PM 将中肠腔分为三个部分：围食膜内空间（肠腔包括食物团）、围食膜外空间（PM 和上皮之间）和 PM 本身。

⑤防止未消化物质的非特异性结合或植物异种化感物与中肠微绒毛表面的结合和 / 或与中肠表面的转运蛋白结合。

2.1.6 Salivary glands

2.1.6.1 The types of salivary glands

Salivary glands are glands associated with the mouth or oral cavity and produce secretions (saliva) that are mixed with the food during feeding and are ingested along with the food. Salivary glands are named after their associated mouthpart: the mandibular,

2.1.6 唾腺

2.1.6.1 唾腺的类型

唾腺是连接嘴或口腔的腺体，取食期间产生的分泌物（唾液）与食物混合并与食物一起被消化。唾腺根据其相连的口器命名：上颚腺、下颚腺、咽下腺

maxillary, hypopharyngeal, and labial glands.

A common function of mandibular glands in social Hymenoptera is to produce pheromones such as alarm pheromones in ants and honey bees, and the queen substance in queen honey bees. Function can vary with age. Maxillary glands occur in Procure, Collembolan, some Heteroptera, and some larval Neuroptera and Hymenoptera. They are believed to provide secretions to lubricate the mouthparts, and thus serve one of the salivary functions. Hypopharyngeal glands occur in the Hymenoptera, and in honey bee workers they produce secretions that are fed to the larvae (a substance different from royal jelly). Hypopharyngeal glands are vestigial in honey bee queens and absent in males. The hypopharyngeal glands of honey bee workers also produce invertase, a common salivary enzyme that hydrolyzes sucrose, and another enzyme that oxidizes glucose to an acid, which is believed to serve as a preservative in honey. Labial glands occur in the great majority of insect orders (an important exception is the Coleoptera), and most commonly function as salivary glands, but in the larval stage of some groups of silk- producing insects, such as Lepidoptera, Trichoptera, and Hymenoptera, labial glands are the silk-producing organs either throughout the larval stage or at its end, just before pupation. In the Psocoptera, adults have two pairs of labial glands, one pair functioning as silk glands and the other pair functioning as salivary glands. In the primitive orders Collembola and Thysanura, which lack Malpighian tubules (the usual excretory organs of insects), the labial glands function as excretory organs.

2.1.6.2 Structure of salivary gland

The glands occur in pairs, and the ducts from each gland usually join to form a single common duct that opens to the oral cavity at a single orifice (Fig. 2-5). The glands generally have at least two regions: a secretory region and a reabsorptive region.

The secretory region produces the primary saliva. The major component of saliva is water. Water is transported from the hemolymph across cells of the salivary gland and into the lumen of the gland. Movement of water from the blood to the gland lumen is accomplished by active transport of potassium or sodium ions from the hemolymph to the lumen. Cells responsible for water transport generally have deep infoldings of the cell membrane and/or dense microvilli on the side of the cell adjacent to the lumen of the gland. This serves to greatly increase the cell's luminal surface area, and also serves to enclose very narrow extracellular spaces into which ions are pumped. The enclosed nature of the spaces helps contain the ions to keep their

和下唇腺。

社会性膜翅目昆虫的上颚腺的功能是用于产生信息素，如蚂蚁和蜜蜂的警报信息素以及蜜蜂的蜂王物质。功能随龄期而变化。下颚腺在原尾目、弹尾目、一些半翅目以及一些脉翅目和膜翅目中被发现。它们被认为提供分泌物来润滑口器，从而发挥唾液的功能之一。咽下腺存在于膜翅目和工蜂，产生的分泌物用于饲喂幼虫（不同于蜂王浆的一种物质）。咽下腺在蜂后中退化，雄蜂没有咽下腺。工蜂的咽下腺也产生蔗糖酶，一种是常见的催化蔗糖的酶，另一种酶则氧化蔗糖产生酸，因此其被认为在工蜂中起着防腐剂的作用。下唇腺出现在绝大多数昆虫目中（重要的例外是鞘翅目）。通常，多数起着唾腺的作用，但是在一些产丝昆虫的幼虫期，如鳞翅目、毛翅目和膜翅目，下唇腺是整个幼虫期或化蛹前的产丝器官。啮虫目昆虫成虫有两对下唇腺，一对起着丝腺的作用，另一对起着唾腺的作用。原始的弹尾目和缨尾亚目缺少马氏管（昆虫常见的排泄器官），下唇腺起着排泄器官的作用。

2.1.6.2 唾腺的结构

腺体成对存在，由每个腺体而来的导管常汇聚形成一根单独的共通管，其开口以单孔通往口腔（图 2-5）。总体来说，腺体至少有两个区域：一个分泌区和一个再吸收区。

分泌区主要产生唾液。其主要成分是水。唾腺的水从血淋巴跨细胞转运并进入腺腔。水从血液到腺腔的运动是通过钾离子或钠离子的主动运输来完成的。对水分转运起反应的细胞通常有深深的细胞膜内褶和/或细胞与腺腔毗邻的一侧有稠密的微绒毛。这主要起到极大地增加细胞腔表面的作用，也起着封闭离子被泵入非常窄的胞外间隙的作用。间隙的密闭特性有助于其内含的离

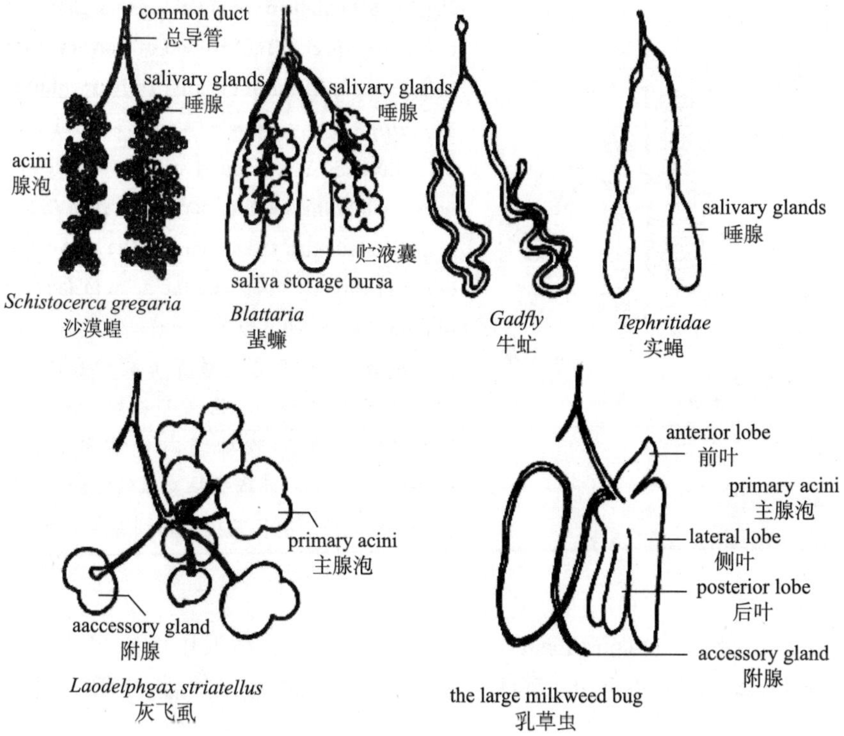

Fig.2-5 Salivary glands of representative insects (modified from Wang Yinchang, 2004).

图 2-5 代表性昆虫的唾腺（仿王荫长，2004）。

concentration high, thus facilitating the osmotic movement of water from the cell into the space. The infoldings and microvilli usually are associated with abundant mitochondria to provide the energy for the ion pumps. The secretory region of the gland synthesizes proteins, such as salivary enzymes and other organic components of the saliva.

The reabsorptive region of the salivary gland reabsorbs potassium or sodium ions from the saliva and transports them back into the hemolymph. As a result, potassium and sodium ions are conserved, and the saliva is usually hypotonic to the hemolymph. Reabsorptive cells often have infoldings, especially on their basal side (hemolymph side), to increase surface area. Reabsorptive cells also have abundant mitochondria to power the active transport of ions from saliva to hemolymph. Insect salivary glands are under the control of neuroactive substances such as serotonin, dopamine and several peptides, including members of the FMRFamide related family of peptides and may be innervated from several sources including the suboesophageal ganglion, the stomatogastric nervous system and the unpaired median nerves (Fig. 2-6).

子保持高浓度，这样，水从间隙进入细胞，使得渗透压的调节更为便利。内褶和微绒毛常与丰富的线粒体有关，以提高离子泵的能量。腺体分泌区合成蛋白质，如唾液酶和唾液的其他有机成分。

唾腺的重吸收区从唾液中再吸收钾离子或钠离子，并把它们运回血液。因此，钾离子和钠离子被保存，唾液对血淋巴一般为低渗。再吸收的细胞常有内褶，特别是基部一侧（血淋巴一侧）来增加表面积。再吸收的细胞也富含线粒体，为腺体到血淋巴活跃的离子转运提供动力。昆虫唾腺受到神经活性物质，如血清素、多巴胺和几种多肽的调控，包括酰胺化 FMRF 相关肽家族的成员，并且可能受到包括食道下神经节、口胃神经系统和不成对的正中神经在内的几个来源的支配（图 2-6）。

2.1.6.3 The functions of saliva

Perhaps, the most fundamental and ubiquitous function of saliva in insects is lubrication of the mouthparts and lubrication of the food bolus to assist its transport through the foregut. Lubrication

2.1.6.3 唾液的功能

也许，昆虫唾液最基本和普遍的功能是润滑口器和润滑食物丸，以助其通过前肠。润滑能通过唾液中最基本的

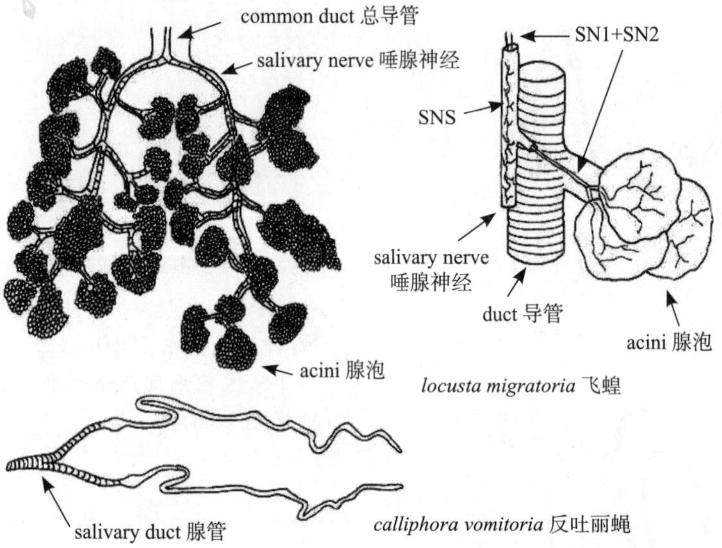

Fig.2-6 Schematic drawing of salivary glands of two insect species. Innervation of salivary acini in *Locusta* and uninnervated salivary glands of *Calliphora vomitoria*. Axons of SN1 and SN2 (SN1+SN2) are found within the salivary nerve, while the satellite nervous system (SNS) is associated with the surface of the nerve and does not directly innervate the acini in *Locusta* (modified from Ali, 1997).

图 2-6 两种不同昆虫唾腺神经分布图解。飞蝗唾腺的神经分布和没有神经分布的反吐丽蝇。飞蝗唾腺神经内有轴突 SN1 和 SN2(SN1+SN2)，而随体神经系统（SNS）与神经表面相连，并不直接分布到腺泡（仿 Ali, 1997 年）。

can be achieved primarily by water, the most abundant constituent in saliva. Water in the saliva also can dissolve components in the food, such as sugars, which then become detectable by chemoreceptors on the mouthparts. Thus, saliva also can aid in food recognition.

成分水来实现。唾液中的水也能溶解食物成分，如蔗糖，随后被口器的化学感受器检测到。因此，唾液能帮助辨别食物。

The most common class of organic constituents of saliva consists of digestive enzymes, such as amylase, invertase, various proteases, and lipases. In many insects with chewing mouthparts, salivary enzymes are mixed with the food during chewing and swallowing, and initiate digestion. However, the midgut usually is the main site of production and secretion of digestive enzymes, and salivary enzymes in these insects generally play only a secondary role in digestion.

唾液最常见的分类包括消化酶的有机成分，如淀粉酶、转化酶、各种蛋白酶和脂肪酶。多数咀嚼式昆虫口器，唾液酶在咀嚼时与食物混合、吞咽，并开始消化。然而，中肠是消化酶产生和分泌的主要场所，这些昆虫中的唾液酶在消化中通常仅起次要作用。

2.2 消化酶

2.2 Digestive Enzymes

Midgut cells secrete enzyme in three ways. In the most common type of enzyme secretion, called merocrine secretion and also called exocytosis, enzymes are produced processed in the Golgi complex of the columnar cells and enclosed in small vesicles. These enzyme containing vesicles fused with the cell plasma membrane, and the enzymes are released to the gut lumen. In another, probably more costly and thus less common form of secretion, the entire midgut cell break down, and the cytoplasmic contents are discharged into the lumen of the gut. This is called holocrine secretion. In a

中肠细胞以三种方式分泌酶。最普遍的一种酶分泌方式称为局部分泌，也称为胞吐，酶是在圆筒形细胞的高尔基复合体中加工产生的，并包被成小囊泡。这些含酶的囊泡与细胞质膜融合，酶被释放到肠腔。另一种分泌方式的代价可能更高，因此分泌方式一般较少，整个中肠细胞破裂，胞质内含物排出进肠道

variation of this called apocrine secretion, only parts of the cell, typically just the microvilli membranes, fragment and disintegrate into the gut lumen. A further variation on apocrine secretion is microapocrine secretion, in which small single- or double-membrane vesicle is pinched-off from the cell microvilli. Apocrine and microapocrine secretions are typical of the anterior part of the midgut, while exocytosis most often occurs in the posterior midgut. The mechanism of enzyme secretion may be related to the region of the gut and its particular function. For example, the anterior part of midgut is often involved with absorption of digested products, and apocrine or microapocrine secretion in this region may be an adaption to promote dispersion of secretory vesicle contents into the midgut lumen in a region undergoing absorption processes.

的肠腔，这称为全浆分泌。在这种称为顶浆分泌的演变中，只有细胞的部分通常只是微绒毛膜、碎片并分解进入中肠。顶浆分泌进一步的演变是微顶浆分泌，单层或双层膜小囊泡从细胞微绒毛捏掐出来。顶浆分泌和微顶浆分泌在中肠的前部具代表性，而胞吐多数发生在中肠后部。酶分泌机制可能与肠道区域有关并具有其特殊功能。比如，中肠前部常涉及消化产物的吸收，这个区域的顶浆分泌或微顶浆分泌可能是吸收过程中促进分泌囊泡内容物分散到中肠管腔的一种适应。

2.2.1 Carbohydrate digesting enzyme

Carbohydrate digesting enzymes are secreted by the salivary glands as well as by the midgut epithelium. *α*-amylase, acting upon starch glycogen, is a common digestive enzyme in insects. It attacks interior glycosidic linkages of starch and glycogen, thus giving rise to a mixture of shorter dextrin. *α*-glucosidase and oligo-1, 6-glucosidase (isomaltase) assist in digesting the smaller dextrin, releasing glucose. Many insects also have one or more *α*- or *β*-glycosidase that digest a broad range of small carbohydrates. *α*-glycosidase hydrolyzes maltose, sucrose, trehalose, melezitose, raffinose, and stachyose; α-galactosidase hydolyzes melibiose, raffinose, and stachyose. An *α*, *α*-trehalase in the gut of some insects digests the trehalose that occurs in the body of other insects preyed upon as food. *β*-glycosidase attacks cellobiose, gentiobiose, and methyl-*β*-glycosides. Lactose is hydrolyzed to glucose and gelatose by *β*-galactosidase, and *β*-fructofuranosidase acts upon sucrose and raffinose to release simple sugars.

Termites feed on and digest cellulose, as do some beetle, a few cockroaches, and wood wasps in the family Siricidae. Cellulose cannot be completely digested by one cellulose enzyme; the crystalline structure and the *β*-linkage of glucose units in cellulose make it difficult to hydrolyze, and complete digestion requires a complement of three enzymes—C_x-cellulase, C_1-cellulose, and *β*-glucosidase, acting in sequential attacks. No insect is able to secrete the complete complement of enzymes from its own cells. C_x-cellulose and *β*-glucosidase enzymes are secreted by many insects, but the C_1- cellulose must be obtained

2.2.1 碳水化合物消化酶

碳水化合物消化酶由唾液腺和中肠上皮分泌。作用于淀粉糖原的 *α*-淀粉酶是昆虫中常见的消化酶。它破坏淀粉和糖原内部的葡萄糖苷连接键，因此产生更短的糊精混合物。*α*-葡萄糖苷酶和低聚 -1,6- 葡糖苷酶（异麦芽糖酶）协助消化较小的糊精，释放葡萄糖。许多昆虫也有一种或多种 *α*-或 *β*-糖苷酶，消化各种小分子碳水化合物。*α*-糖苷酶水解麦芽糖、蔗糖、海藻糖、松三糖、蜜三糖和水苏四塘。*α*-半乳糖苷酶水解蜜二糖、蜜三糖和水苏糖。有些昆虫中肠的 *α*，*α*-海藻糖酶可消化作为食物捕食的其他昆虫体内出现的海藻糖。*β*-糖苷酶破坏纤维二糖、龙胆二糖和甲基-*β*-糖苷。乳糖通过 *β*-半乳糖苷酶水解成葡萄糖和半乳糖，*β*-呋喃果糖苷酶作用于蔗糖和蜜三糖从而释放单糖。

与一些甲虫、少数蜚蠊和树蜂科的树蜂一样，白蚁取食和消化纤维素。纤维素不能用一种纤维素酶完全消化；纤维素的晶体结构和葡萄糖单位的 *β*-连接键使得它难以水解，完全水解需要三种酶的互补—— C_x-纤维素酶、C_1-纤维素酶和 *β*-葡糖苷酶的依次作用。没有任何昆虫能从其自身细胞分泌出完全互补的酶系。很多昆虫分泌 C_x-纤维素酶和 *β*-葡糖苷酶，但 C_1-纤维素酶必须从共生

from protozoa or bacterial symbionts or from fungi ingested with the food. The C_x- cellulose disrupts the crystalline structure of cellulose, and the C_1-cellulose digests long chains of cellulose into shorter cellobiose chains. *β*-glucosidase releases glucose from cellobiose. So, the symbiotic bacteria and protozoa in the hindgut of termite that play an important role in hydrolysis of cellulose and hemicelluloses.

原生动物或细菌或伴随食物消化而来的真菌中获得。C_x- 纤维素酶破坏了纤维素的晶体结构，C_1- 纤维素酶将纤维素的长碳链消化成较短的纤维二糖。*β*- 葡糖苷酶从纤维二糖释放葡萄糖。因此，白蚁后肠中的共生细菌和原生动物在纤维素和半纤维素的水解中起重要作用。

2.2.2 Protein digesting enzymes

Protein digestion in insects is mediated by several classes of protein digesting enzymes, some of which are free in the gut lumen while others are membrane bound. The significance of having different protein digesting enzymes is that some proteases are end proteases that attack large proteins internally at the linkage between certain amino acid, thus breaking the protein into smaller polypeptides, while exopeptidases attack the smaller pieces by cutting off the terminal amino acid. In addition, the presence of different types of proteinase inhibitors in the food eaten, especially plant-derived food, may make it advantageous to have a variety of protein digesting enzymes so that some enzyme molecules will escape inhibition. Chymotrypsin-like endoproteinases have been found in several insects, which can attack at phenylalanine, tryptophan, and tyrosine residues.

Proteinases also classified as serine, cysteine, aspartic acid, and metallo-proteinases, depending on the amino acid or metal at the active site of the enzyme. Some proteinases act optimally at alkaline pH, while others have maximum activity at acidic pH.

The endoproteinases and some exoproteinases must pass through the peritrophic membrane to promote digestion of the large proteins. Smaller proteins and peptides released by the digestive processes may diffuse out the peritrophic membrane through pores of the PM, with final digestion taking place at the surface of the microvilli, where there are both free and membrane bound aminopeptidases. Aminopeptidases, also called exopeptidases, are exoenzymes that remove one amino acid after another from the end of a small peptide.

2.2.2 蛋白质消化酶

昆虫蛋白质的消化由几类蛋白质消化酶促成，其中有些在肠腔中游离而其他的是膜结合型。具有不同蛋白质消化酶的意义在于有些蛋白酶是内切蛋白酶，破坏大蛋白质内部氨基酸之间的连接键，因而将蛋白质降解为较小的多肽，而外切蛋白酶通过切除末端氨基酸破坏较小的片段。此外，吃下的食物中存在不同类型的蛋白酶抑制剂，特别是植源性食物占有各种各样蛋白质消化酶的优势，因此，有些酶分子可避开抑制剂。在几种昆虫中发现了糜蛋白酶样内切蛋白酶，它能破坏丙苯氨基酸、色氨酸和酪氨酸残基。

依据酶活性部位的氨基酸或金属元素，蛋白酶也分为丝氨酸、胱氨酸、天冬氨酸和金属 - 蛋白酶。有些蛋白酶在碱性 pH 下作用最佳，而其他酶在酸性 pH 时具有最大活性。

内切蛋白酶和外切蛋白酶必须通过围食膜才能启动大分子蛋白质的消化。消化过程释放的较小蛋白质和肽可通过围食膜的孔扩散出围食膜，最终的消化发生在微绒毛中，那里有游离和膜结合的氨基肽酶。氨肽酶，也称作外肽酶，是一种从小肽末端逐个水解氨基酸的外切酶。

2.2.3 Lipid digesting enzymes

Most of the fats eaten by an insect consist of triacylglycerols. Lipases secreted from the midgut, and in some insects probably from symbionts as well, release fatty acids and glycerol from

2.2.3 脂类消化酶

昆虫食下的多数脂肪由三酰基甘油组成。从中肠分泌的脂肪酶在一些昆虫中可能也来自共生体，从脂肪中释放脂

triacylglycerols. Components of the glycocalyx layer in the gut may aid in emulsifying fats and in promoting contact between lipase and triacylglycerol.

肪酸和三酰甘油。肠道糖萼层的成分可能有助于乳化脂肪，并有助于促进脂酶和三酰甘油间的接触。

2.2.4 Proteinase inhibitors

There may be a relationship between the type of proteinase inhibitors in food and the evolutionary selection for type and number of proteinase secreted by insects, but only a few insects have been studied. Many plants contain varying levels of naturally occurring proteinase inhibitors, some of which inhibit serine proteinases while others act on cysteine proteinases. Some insects respond to the consumption of a trypsin inhibitor by secreting additional trypsin-like enzyme(s) that allow different susceptibility to the inhibitor and/or hyperproducing enzymes to compensate for inhibition.

2.2.4 蛋白酶抑制剂

食物中的蛋白酶抑制剂的类型可能和昆虫分泌的蛋白酶类型和数量有关，但只对少数昆虫进行了研究。很多植物含有不同水平的天然蛋白酶抑制剂，其中有些抑制丝氨酸蛋白酶，而另一些作用于胱氨酸蛋白酶。有些昆虫对胰蛋白酶消耗的反应是分泌额外的类胰蛋白酶，使之对抑制剂具有不同的敏感性，和 / 或过量产生酶类补偿抑制。

2.2.5 Hormone influence on midgut

Hormonal control of digestive enzyme secretion does not appear to be a widespread mechanism among insect groups, but a well-defined case occurs in *A.aegypti* female mosquitoes. A decapeptide hormone (trypsin modulating oostatic factor, TMOF), synthesized by ovarian follicular epithelium cells, is released 24 to 42 h post feeding and transported by the hemolymph to receptors on the hemolymph side of midgut cells. The hormone, possible through a second massager, signals midget cells to cease producing late trypsin that is necessary to finish digesting the blood meal. Younger and secondary oocytes cease growing, apparently from lack of nutrients, until the inhibition of digestion is released after the female lays the mature eggs.

Two forms of trypsin are secreted in *A.aegypti* and both forms appear to be critical to the ultimate formation of egg. An early trypsin is present in the midgut during the first 4 hours after feeding. Transcription of the early trypsin gene is controlled by JH level, and occurs in the midgut after adult emergence. Late trypsin is the major endoprotease in the midgut and it is necessary to finish the blood meal digestion. Late trypsin synthesis is the form primarily influenced by TMOF. Sufficient amino acids result from early trypsin action to promote maturation of a batch of eggs before TMOF stops late trypsin synthesis. synthesized TMOF with the structure of NH_2-YGPAPPPPPP-COOH, the amino terminal

2.2.5 激素对中肠的影响

激素对消化酶分泌的控制似乎并不是昆虫种群中普遍存在的机制，但埃及伊蚊雌蚊有意义明确的实例。一种十肽激素（胰蛋白酶调控抑卵因子，TMOF）由卵巢滤泡上皮细胞合成，取食后 24 ～ 42 h 释放，通过血淋巴转运到血淋巴一侧的中肠细胞受体上。该激素可能通过第二信使向中肠细胞发出信号，停止产生完全消化血粉所必需的后期胰蛋白酶。未发育的卵和次级卵母细胞显然因为缺乏营养停止发育，直至雌蚊产下成熟的卵后才解除消化的抑制。

埃及伊蚊分泌两种形式的胰蛋白酶，它们对卵的最终形成都起着关键性的作用。早期胰蛋白酶出现于取食后的第一个 4 h 内的中肠。早期胰蛋白酶基因的转录受到保幼激素水平的控制，在成虫羽化后的中肠内出现。后期胰蛋白酶是中肠内一种主要的内切蛋白酶，是完全消化血粉所必需的。后期胰蛋白酶合成是受到 TMOF 影响的主要形式。早期胰蛋白酶的作用产生了足够的氨基酸，在 TMOF 停止合成后期胰蛋白酶之前，促

end of only four amino acid residues (NH_2-YDPA) had 95% as much activity as the decapeptide, and also indicated this is the part of the molecule that binds to the midgut receptor. A similar hormonal mechanism involving TMOF also has been demonstrated in the grey flesh fly *N. bullatta*, except that the neuropeptide is a hexapeptide, called Neb TMOF. Neb-TMOF has been isolated from larvae of the blue blowfly, *C. vicina*, and it seems to have dual functions, acting in larvae as an ecdysiostatin that inhibits the synthesis of ecdysteroids, while inhibiting synthesis of trypsin in the midgut of adult female.

The midgut of *P. regina* releases a hormone targeted for the brain about 4 to 8 hours after the meal. In the brain, it stimulates median neurosecretory cells to initiate the neuroendocrine cascade leading to ovary and egg development.

进一批卵的成熟。合成的 TMOF 结构为 NH_2-YGPAPPP PPP-COOH，氨基端仅有的四个氨基酸残基（NH_2-YDPA）具有 95% 的十肽活性，也表明了分子的这个部分是与中肠受体结合的。灰麻蝇也被证实具有与 TMOF 相似的激素机制，除神经肽是六肽，还被称为 Neb-TMOF。Neb-TMOF 已在红头丽蝇幼虫中分离得到，它似乎具有双重功能，像一种抑蜕皮激素作用于幼虫，抑制蜕皮甾类的合成，而在雌成虫中肠抑制胰蛋白酶的合成。

伏蝇的中肠在取食后 4 ～ 8 h 释放一种针对大脑的激素。在大脑中，它刺激了中间神经分泌细胞，启动神经内分泌级联，引起卵巢和卵的发育。

2.3 反向循环和消化产物的吸收

2.3 Countercurrent Circulation and Absorption of Digested Products

2.3.1 Countercurrent circulation

Insect might have a countercurrent circulation of fluid contents in the midgut, called endo- to ecto- peritrophic countercurrent flow. The process has the following effects: ① serve to increase digestive efficiency; ② conserve nutrients that might be lost by a rapid passage through a short midgut; ③ conserve and reuse enzymes that would otherwise be excreted rapidly with the bulk of food moving through the gut; ④ allow absorption of digested products along the entire length of the midgut and/or absorption in the gastric caeca by permitting fluid containing digested products to flow forward.

In a countercurrent circulation, food and gut contents within the peritrophic matrix move posteriorly after having entered the midgut from the foregut, while fluid containing partially or completely digested food materials outside the peritrophic matrix moves anterior between the midgut cell surfaces and the peritrophic matrix (Fig.2-7). Four criteria for determining whether an endo-ectoperitrophic flow occurred in insects: ① a posterior region of the gut (could include Malpighian tubules) should be specialized

2.3.1 反向循环

昆虫中肠流体内含物具有反向循环，称之为围食膜内—外腔反向循环。这个过程具有如下作用：①有助于增加消化率；②保存快速通过短中肠可能流失的营养；③保存和再利用那些随着大部分食物通过中肠可能快速排出的酶；④允许沿着全中肠长度的消化产物的吸收和 / 或允许胃盲囊吸收向前流动、含有消化产物的液流。

在反向循环中，围食膜基质内的食物和肠道内容物在从前肠进入中肠后向后移动，而围食膜基质外含有部分或全部食物消化物的液流在中肠细胞表面和围食膜基质间向前移动（图 2-7）。判断昆虫是否出现围食膜内—外腔反向循环的四个准则：①肠道后部区域（可包括马

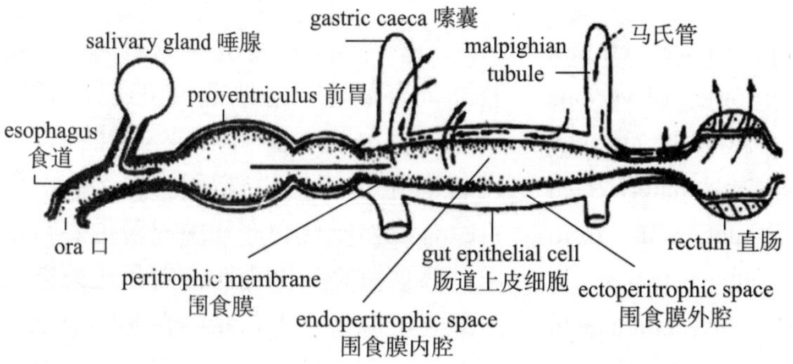

Fig.2-7 Diagrammatic illustration of the counter-current flow that occurs in the midgut of some insects. Fluid may be passed forward from the Malpighian tubules to help create the forward flow that carries small end products of digestion to the gastric caeca, which are very efficient in absorption of fluid and nutrients (modified from Berridge, 1970).

图 2-7 有些昆虫中肠发生的液流循环图解说明。液体可以从马氏管向前部流动，以帮助产生向前的液流，将消化的小最终产物输送到胃盲肠，这在吸收液体和营养方面非常有效（仿 Berridge, 1970）。

for secretion of fluid into the posterior midgut; ② an anterior region of the midgut should be specialized for absorption; ③ there should be a concentration gradient of small molecules between the point of fluid secretion and that of absorption; ④ metabolites from a meal should stay in the gut longer than the bulk of the solid food originally ingested.

氏管）应该专门用于中肠后部分泌液体；②中肠有一个前部区域应该专门用于吸收；③在液流分泌与吸收的节点之间应该有一个小分子的浓度梯度；④从食物中获得的代谢物在肠道内停留的时间应该比当初摄食的大部分固体食物的长。

2.3.2 Absorption of digested products

In several lepidopteran (*M.sexta*, *P. cynthia*, *B. mori*), amino acids are actively absorbed by transport proteins specific to particular amino acids. The driving energy for amino acid absorption from the gut in these lepidopterans is the proton-ATPase pump in goblet cells. The pump creates the high K^+ concentration in the gut lumen and the high transepithelial potential across the gut wall, both shown to be important for amino acid absorption by columnar cells in lepidopte rans. There are at least six transport system, including: ① a transporter of neutral amino acids; ② a specific system for proline; ③ a specific system for glycine; ④ a specific system for *L*-lysine; ⑤ a specific trans- porter for glutamic acid; ⑥ a transporter that is stero specific for *D*-alanine.

Blocking the absorption of amino acids and disruption of their transport systems have been suggested as potential targets for insect control.

2.3.2 消化产物的吸收

在几种鳞翅目昆虫（烟草天蛾、伏蝇和家蚕）中，氨基酸是通过其特异性运输蛋白主动地吸收的。这些鳞翅目昆虫肠道内氨基酸吸收的驱动能是杯形细胞的质子 -ATP 酶泵。泵在肠腔内产生高浓度 K^+ 和横跨肠壁的跨上皮电位，二者在鳞翅目圆筒形细胞氨基酸的吸收中起着重要作用。至少有六种转运系统，包括：①中性氨基酸转运系统；②特异性脯氨酸转运系统；③特异性甘氨酸转运系统；④特异性 *L*- 赖氨酸转运系统；⑤特异性谷氨酸转运系统；⑥ *D*- 丙氨酸立体特异性转运系统。

阻断氨基酸的吸收及破坏其运输系统已被建议为害虫控制的潜在靶标。

2.3.3 The transepithelial and oxidation-reduction potential of the gut and gut pH

There is a transepithelial potential across the gut wall and an oxidation-reduction or redox potential within the lumen of various parts of the gut. Redox potential in the various region of the gut may play a large role in digestion and assimilation of food materials, in detox cation reactions, and in production of toxic metabolites from ingested food materials. Ability to adjust gut redox potentials in plant-feeding insects may be one of the ways that insects adapt in the co-evolutionary race with plants in combating the wide range of allelochemicals common in many plants.

The pH of a gut segment greatly influences the action of any enzymes secreted into or carried with the food into that segment. In addition, gut pH may influence solubility of ingested components, toxicity of some potential toxins, and the population of gut microorganisms. The crop tends to be slightly acidic in most insects, due to organic acids produced when digestion occurs in the crop. Larvae of Lepidoptera and Trichoptera tend to have a very high midgut pH, varying from about 8 to 10. An awerage pH value of 9.42 was tested from 290 silkwom species. The tobacco hornworm larvae have midgut pH of 10 to 12 in different parts of midgut. Very acid conditions prevail in the special hindgut regions of some termites, crickets, and possibly other insects that have hindgut fauna to digest cellulose. The acid conditions are caused by the anaerobic fermentation of glucose from cellulose digestion, resulting in production of short-chain fatty acids, including, acetic, propionic, and butyric acids.

2.3.3 肠道的跨上皮电位和氧化还原电位及其 pH

肠壁有跨上皮电位，肠道的不同肠腔部位有氧化 - 还原或氧化还原电位。氧化还原电位在肠道不同区域的消化、食料同化作用、脱毒反应以及摄入食料的有毒物质代谢产物中起着重要的作用。取食植物的昆虫肠道氧化还原电位的调整可能是昆虫与植物在共同进化竞争中的适应途径之一，许多植物有普遍存在的广谱性异种化感物。

肠道某区段的 pH 对任何分泌进入本区段或随食物携带进入本区段的酶的作用影响很大。此外，肠道 pH 还影响摄入组分的溶解度、某些潜在毒素的毒性和肠道微生物的群落。多数昆虫的嗉囊呈微酸性，源于由嗉囊的消化作用而产生的有机酸。鳞翅目和毛翅目幼虫的中肠 pH 非常高，变化范围为 8 ～ 10。测得 290 个家蚕品系 pH 平均值为 9.42。烟草角虫幼虫中肠 pH 在不同部位不同，为 10 ～ 12。有些白蚁、蜚蠊和其他可能在后肠消化纤维素的昆虫种群，特化的后肠区普遍存在非常酸的环境。酸性环境是由纤维素消化产生的葡萄糖进行无氧发酵而引起的，导致了短链脂肪酸的产生，包括乙酸、丙酸和丁酸。

2.4 昆虫的基本营养需求

2.4 The Basic Nutritional Requirements of Insects

2.4.1 Essential amino acids

The essential amino acids of a number of insects from different orders have been demonstrated to be arginine, histidine, isoleucine, leucine, lysine, methionine, phenylalanine, threonine, tryptophan, and valine (all in the *L*-form). In the absence of one of these essential amino acids, growth and development ceases in the insects. In some cases nonessential amino acids stimulate growth, and this may be related to the optimization of nutrient balance and

2.4.1 必需氨基酸

许多不同目昆虫必需氨基酸被证实为精氨酸、组氨酸、异亮氨酸、亮氨酸、赖氨酸、蛋氨酸、苯丙氨酸、苏氨酸、色氨酸和缬氨酸（均为 *L* 型）。缺少任何一种必需氨基酸，昆虫的生长和发育都会停止，这可能与营养平衡的优

the efficiency of the biochemical pathways involved in synthesis of the nonessential amino acids.

化以及牵涉非必需氨基酸合成的生化途径效率有关。

2.4.2 Carbohydrates

Most insects do not have an absolute growth requirement for a specific carbohydrate in the diet, although carbohydrates are a major energy source for most insects. Generally, insects can synthesize carbohydrates from amino acids and from lipids. Species in the genera *Tenebrio*, *Ephestia*, and *Oryzaephilus* need a carbohydrate source to reach maturity. Other stored grain insects, such as species of *Tribolium*, *Lasioderma*, and *Ptinus*, can be reared to maturity on diets lacking carbohydrates. The adults of the dipterans *C. erythrocephala*, *L. cuprina*, *A.* suspense, some other tephritidae fruit flies, and probably many adult dipterans require carbohydrates for an energy source and continued survival.

2.4.2 碳水化合物

虽然碳水化合物是昆虫的主要能量来源，但多数昆虫对食物中某种碳水化合物没有绝对的生长需求。通常昆虫能利用氨基酸和脂类合成碳水化合物。拟步行虫属、欧洲麦粉蛾和锯谷盗属的一些种群需要碳水化合物源来达到成熟。其他仓储昆虫，如拟谷盗属、烟草甲和蛛甲能用无碳水化合物饲料饲养到成熟。双翅目红头丽蝇、铜绿蝇，一些其他实蝇科果蝇的成虫，可能还有很多双翅目成虫需要碳水化合物作为能源和生存的延续。

2.4.3 Lipids

Typical lipids in biological organisms include free and bound fatty acids; short- and long-chain alcohols; tri-, di-, and mono-acylglycerols; steroids and their esters; phospholipids; and several other groups of compounds. Most insects have the necessary metabolic machinery to convert carbohydrates into lipids, and many insects synthesize lipids and store them in the fat body tissue. A specific lipid requires in the diets is sterol and some insects require polyunsaturated fatty acids. They survive and reproduce by using nutrients accumulated as larvae and stored in the body during pupation.

2.4.3 脂类

生物有机体代表性的脂类包括游离和结合态脂肪酸；短链和长链醇类；三酰甘油、二酰甘油和单酰甘油；甾体及其酯类；磷脂和其他几组化合物。多数昆虫有必需代谢机制将碳水化合物转化为脂类，很多昆虫合成脂类并将它们贮存在脂肪体组织中。日常进食有特别需求的一个脂类是甾醇，有些昆虫需要多不饱和脂肪酸。生存和繁殖靠利用幼虫期积累的营养并在蛹期存于体内。

2.4.4 Sterols

Because their inability to synthesize sterols, insects must obtain sterols from their food and/or from their symbionts. They use sterol as a precursor for synthesis of the ecdysteroids molting hormone, and as a component of all cell membranes. A sparing sterol is one that can be incorporated into cell membranes, but cannot be used to synthesize the molting hormone.

A few species are known to have a requirement for a very specific dietary sterol. Sterol deficiency may be manifest in any of the stages of an insect. Newly hatched larvae that lack dietary sterol usually die in the first or second instar because they exhaust

2.4.4 甾醇

因为它们没有合成甾醇的能力，昆虫必须通过食物和 / 或从它们的共生物中获取甾醇。它们利用甾醇作为合成蜕皮甾类的前体，并作为所有细胞膜的组分。备用甾醇是一种能掺入到细胞膜，但不能用于合成蜕皮激素的甾醇。

已知少数昆虫对食物甾醇有非常特别的需求。昆虫在任一时期都可能缺乏甾醇。食物中缺少甾醇，刚孵化的幼虫通常在一龄期或二龄期死亡，因为它们

the sterol received in the egg from the mother. Lacking of a sterol by adult female houseflies result in 80% reduction in egg hatch, although the number of eggs laid is not affected. Replacement of more than one-half the cholesterol requirement with cholestanol, however, results in eggs that fail to hatch.

耗尽了来自母体的卵内甾醇。家蝇的雌虫缺少甾醇，尽管产卵数没有受到影响，但卵孵化率减少 80%。而用胆甾烷醇替代所需甾醇一半多的量，将引起卵不能孵化。

2.4.5 Polyunsaturated fatty acids

About 50 species of insects from five orders have been shown to require a dietary source of polyunsaturated fatty acids.

The requirement for the polyunsaturated fatty acids was initially discovered in lepidopterans, in which a deficiency is dramatically displayed in failure of pupal or adult ecdysis. Individuals that successfully or partially ecdyse are likely to have misformed wings and lack normal scales on the body. Some hymenopterans show similar difficulty in ecdysis when linoleic and lenolenic acids are absent, or very low in the diet. Acridid grasshopper also tends to produce deformed adults on fatty acid deficient diets. Coleopterans show slowed growth and decreased adult fecundity in response to deficiency in polyunsaturated fatty acids. In general, it appears that dipterans may not have an absolute requirement for polyunsaturated fatty acids, even though they are unable to synthesize them. Some dipterans, however, show improved growth when polyunsaturated fatty acids are added to a diet in a little amount.

2.4.5 多不饱和脂肪酸

约 5 个目的 50 种昆虫表现出对多不饱和脂肪酸的食物源需求。

最初在鳞翅目中被发现昆虫对多不饱和脂肪酸的需求。在缺乏时，其表现出明显的化蛹和成虫蜕皮的失败。即使蜕皮成功或部分蜕皮的个体很可能会出现错误成形的翅膀和缺少正常的体表鳞片。没有亚油酸和亚麻酸或食物中含量很低时，有些膜翅目表现出相似的蜕皮困难。以脂肪酸缺乏的饲料短触角蚱蜢也会产生畸形的成虫。鞘翅目在缺乏多不饱和脂肪酸时对应地表现出生长缓慢和成虫生殖力下降。总之，虽然双翅目不能合成多不饱和脂肪酸，但对其没有绝对的需求。然而，在有些双翅目饲料中加入少量多不饱和脂肪酸，它们的生长情况会得到改善。

2.4.6 Vitamins

Insects need thiamine (vitamine B_1), riboflavin, pyridoxine (vitamine B_6), niacinamide, pantothenic acid, biotin (vitamin H), folic acid, and choline. Carnitine, also called vitamine B_T, is a requirement for *T.molitor*, *T. obscurus*, *T.confusum* and *T.castaneum*. Different strains of these insects show variable requirements. One of the critical roles for carnitine is a participant in the passage of fatty acids across mitochondrial membranes

There is demonstrated requirement for water soluble vitamins in the nutrition of a few insects. Ascorbic acid is required for normal growth and development of some insects, and seems particularly needed by phytophagous insects. Carotene and vitamin A are required by insects for normal pigmentation and eye function. Vitamin B_{12} stimulates the growth of some insects, but a clear-cut requirement for growth has not been shown. Vitamin E is necessary to *C.montrousieri* in order for

2.4.6 维生素

昆虫需要硫胺（维生素 B_1）、核黄素、吡哆醇（维生素 B_6）、烟酰胺、泛酸、生物素（维生素 H）、叶酸和胆碱。肉毒碱也称为维生素 B_T，是黄粉甲、黑粉虫、杂拟谷盗和赤拟谷盗生长发育所必需的。这些昆虫的不同品种表现出不同的需求。肉毒碱的重要作用之一是脂肪酸跨越线粒体膜通道的参与者。

少数昆虫有水溶性维生素的营养需求。抗坏血酸是某些昆虫正常生长和发育必需的，特别是植食性昆虫。胡萝卜素和维生素 A 是昆虫正常色素和视力功能必需的。维生素 B_{12} 刺激有些昆虫的生长，但没有明确的生长需求。维生素 E 是孟氏隐唇瓢虫必需的，为了雌

females to mature and oviposit eggs. The vitamin also is required for spermatogenesis in male house crickets, *Acheta domesticus*.

虫成熟和产卵。维生素也是蟋蟀雄虫精子产生所必需。

2.4.7 Minerals

2.4.7　矿物质

Metal ions are required as enzyme co-factors and as constituents of metalloenzymes. For example, molybdenum is a part of the xanthine dehydrogenase involved in purine metabolism of insects. Many phytophagous insects need relatively large quantities of potassium and only trace amount of sodium. Male transfer sodium acquired by puddling to females at mating, and females incorporate it into eggs. Two aphids, *M.persicae* and *A. fabae*, require trace amounts of Fe, Mn, Zn, and Cu as well as major quantities of K, Mg, and phosphate. The balance of minerals in a salt mixture and the proportion of minerals to other groups of nutrients are important.

金属离子是酶辅因子和金属酶组分所必需的。例如，钼是参与昆虫嘌呤代谢的黄嘌呤脱氢酶的一部分。许多植食性昆虫需要相对大量的钾离子以及痕量的钠离子。雄虫在交配时通过揉捏将其获得的钠给雌虫，雌虫将其结合到卵中。两种蚜虫，桃蚜和黑豆蚜，需要痕量的 Fe、Mn、Zn 和 Cu，也需要大量的 K、Mg 和磷酸盐。盐混合液中的矿物质的平衡以及矿物质比例对其他种群是很重要的。

2.5　食物的摄取

2.5　Food Intake

2.5.1 Phagostimulants

2.5.1　助食素

Phagostimulants are chemicals that induce feeding. Insects are induced to feed by chemosensory stimulation from components in or on their food. Gustatory sensilla are located on the mouthparts, tarsi, and antennae. Adult western corn rootworms show strong phagostimulatory responses to several *L*-amino acids, to *γ*-amino butyric acid, and to as little as 0.1μmol/L of cucurbitacin B, as well as silkworm *B.mori* to sugar, volatile terpene substances such as citral, leafalcohol, leafaldehyde and linalool. In a few cases, an insect's behavior in the presence of phagostimulants has been divided into a biting response and a swallowing response. In *B.mori*, biting responses are stimulated by *β*-sitosterol, isoquercitrin and morin, while swallowing responses are stimulated by cellulose, phosphate, silicate, inosite and vitamin C.

助食素是引诱取食的化学物质。昆虫通过食物内或食物上成分的化学感觉刺激，被引诱取食。味觉感器位于其口器、跗骨和触角上。成年西部玉米根虫表现出对几种 *L*- 氨基酸、*γ*- 氨基丁酸，少至 0.1 μmol/L 的葫芦素 B 以及家蚕对糖分、挥发性萜烯类物质，如柠檬醛、青叶醇、青叶醛和芳樟醇等均有吞噬刺激反应。在少数情况下，昆虫对助食素的行为划分为咬食反应和吞咽反应。家蚕的咬食反应受到 *β*- 谷甾醇、异槲皮苷和桑色素的刺激，而吞咽反应受纤维素、磷酸盐、硅酸盐、肌醇和维生素 C 的刺激。

Among primary nutrients, sugars, and sucrose, especially, are phagostimulants for many insects. Glucosides, which are combinations of glucose with a nonsugar molecule, serve as both phagostimulants and deterrents to feeding in different insects. Combination of nutrients is important in many cases. Whole

在主要营养物质中，糖分和蔗糖是多数昆虫的助食素。糖苷是葡萄糖和非糖分子的组合物，在不同昆虫取食中同时起着助食素和阻碍素的作用。营养组合对许多昆虫是很重要的。全蛋白有时

proteins are sometimes phagostimulants, but more frequently amino acids induce feeding. Leucine, methionine, lysine, and isoleucine in phosphate ions was important to the phagostimulants activity of the amino acids, but in some insects amino acids may act alone or in combination with other amino acids or derivatives of amino acids. Reduced glutathionc, a tripeptide composed of glutamic acid, cysteine, and glycine, is a phagostimulants for some and, perhaps all ticks. Adenosine triphosphate (ATP) in the presence of sodium ions is a phagostimulants for the adult mosquito *A. aegypti.* Lipids frequently are phagostimulants. Phospholipids, triglycerides, sterols, sterol esters, and fatty acids are important phagostimulants for various insects.

Many secondary plant substances induce insect to feed. Combinations of secondary plant substances are frequently more effective stimulants than single compounds.

是助食素，但氨基酸更经常引诱取食。亮氨酸、蛋氨酸、赖氨酸和异亮氨酸的磷酸盐离子对氨基酸的助食素活性是重要的，但有些昆虫的氨基酸单独起作用或与其他氨基酸或者与氨基酸衍生物联合作用。还原性的谷胱甘肽是一种由谷氨酸、半胱氨酸和赖氨酸组成的三肽，是一些，甚至是所有壁虱的助食素。腺苷三磷酸（ATP）钠是埃及伊蚊成虫的助食素。脂类常常是助食素。磷脂、甘油三酯、甾醇、甾醇酯和脂肪酸是不同昆虫重要的助食素。

许多植物的次级代谢产物引诱昆虫取食。次级代谢产物的组合常常比单个化合物更为有效。

2.5.2 Feeding deterrents

Many substances are known that have an inhibitory effect upon insect feeding, and chemoreceptors responsive to specific compounds are involved, as in phagostimulants. For example, ammonium nitrate inhibits feeding of the sweet clover weevil, *S. cylindricollis*. Feeding by the American cockroach, *P.americana*, is inhibited by 1, 4-naphthoquinone. Food intake by the desert locust, *S.gregaria*, is decreased by injection of Lom-sulfakinin, a neuropeptide found in the corpus cardiac of locusts. Although the mechanism has not been elucidated, it may reduce the sensitivity of taste receptors.

2.5.2 取食抑制剂

已知许多物质对昆虫取食具有抑制作用，像助食素一样涉及化学感受器对特殊化合物的反应。例如，硝酸铵能抑制芳香三叶草象鼻虫的取食。1,4- 白花丹素抑制美洲大蠊的取食。通过注射一种在蝗虫心侧体发现的神经肽蝗虫硫激肽时，沙漠蝗食物的摄取会减少。尽管其机制没有被阐明，但它可能会降低味觉感受器的敏感性。

2.6 昆虫的肠道微生物

2.6 Intestinal Microorganisms of Insects

Insect gut symbiotic microbiota plays an essential role in the growth, development, pathogenesis, and environmental adaptation of host insects. A combination of high throughput sequencing, metagenomics, metatranscriptomics and metaprotoeomics with phylogenetic analyses has revealed a high diversity of bacterial communities living in the intestine of a variety of insects.

昆虫肠道共生菌群在寄主昆虫的生长发育、发病机制和环境适应中起着至关重要的作用。高通量测序、宏基因组学、宏转录组学和宏蛋白质组学与系统发育分析相结合揭示了生活在多种昆虫肠道中的细菌群落的高度多样性。

γ-proteobacteria are able to increase insect tolerance to high temperatures and mediate specificity toward certain host plants. Proteobacteria living in the gut of Hymenopteran insects have been shown to improve host resistance to pathogenic parasites. The bumble bee susceptibility to the parasite *Crithidia bombi* was influenced by the composition of the gut microbiota. This result substantiates the role of the microbiota of social bees in gut homeostasis and resistance against parasites and diseases. Some proteobacteria, such as *Wolbachia* sp., can induce male-killing, cytoplasmic incompatibility, parthenogenesis, feminization of genetic males in insect models such as *Asobara japonica*. *Bacillus thuringiensis* has shown variations in its pathogenicity against insects, depending on the intestinal microbiota associated with its host (Mondin et al., 2021), while the bacteria *Stenotrophomonas maltophilia* and *Enterococcus mundtii*, isolated from the intestine of the silkworm *B. mori* were shown to influence resistance to the organophosphate insecticides (Chen et al., 2020).

γ- 变形菌能够提高昆虫对高温的耐受性，并介导对某些寄主植物的特异性。生活在膜翅目昆虫肠道内的变形菌已被证明能提高宿主对致病性寄生虫的抵抗力。大黄蜂对寄生熊蜂短膜虫的易感性受肠道微生物群组成的影响。这一结果证实了群居蜜蜂的微生物群在肠道内环境稳定和抵抗寄生虫和疾病方面的作用。一些变形菌，如沃尔巴克氏菌，可以在日本开臂反颚茧蜂等昆虫中诱导杀雄作用、胞质不亲和、孤雌生殖和雄性个体雌性化。苏云金芽孢杆菌对昆虫的致病性表现的不同取决于与其宿主相关的肠道微生物群（Mondin 等，2021 年），而从家蚕肠道分离出的嗜麦芽窄食单胞菌和蒙氏肠球菌表现出对有机磷杀虫剂的抗性（Chen 等，2020 年）。

Phytophagous insects possess enzymatic tools (e.g., nitrogenases) that can aid in the conversion of toxic compounds and assist in digestion of their many available food sources. Many insect herbivores utilize epithelial tissue present within the midgut as a sensing organ and converting cyanide to ammonia *via* endogenous nitrogenase production. Such enzymatic tools stimulate the insect's digestion of different plants by allowing the metabolization, detoxification, as well as uptake of an array of chemical compounds. This capacity towards rapid metabolization of some chemical compounds may be potentiated if the insect population is kept on an exclusive diet, which is often the case for pest-populations living in monocultured agricultural systems. The presence of specific microbiota in the gut has been shown to be involved in the adaptation of lepidopteran pests to different host diets and metabolites within those foods.

植食性昆虫拥有酶工具（例如，固氮酶），可以帮助有毒化合物的转化，并帮助消化许多可用的食物来源。许多食草昆虫利用中肠内的上皮组织作为感应器官，通过内源性固氮酶的产生将氰化物转化为氨。这种酶工具通过代谢、解毒以及吸收一系列化合物来刺激昆虫对不同植物的消化。如果昆虫种群保持专一的饮食，某些化合物快速代谢的能力就可能会增强，这通常是生活在单一农业系统中的害虫种群的情况。肠道中特定微生物群的存在已被证明与鳞翅目害虫适应不同宿主的取食和这些食物中的代谢物有关。

Influence or interference to these gut microbes due to internal and external factors could potentially affect the silkworm health including the adaptation, biomass degradation, nutrient production, silk production and quality, and compound detoxification; or else they might cause of serious diseases (Ponnusamy et al., 2019). In eri silkworm, *Samia ricini*, maximum symbiotic bacterial cellulose digestion may be occurring in the midgut. The presence of cellulolytic bacteria in foregut and hindgut compartments suggests that preliminary digestion of cellulosic food diets may

由于内部因素和外部因素对这些肠道微生物的影响或干扰可能会影响家蚕的健康，包括适应性、生物量降解、营养产生、丝的产量和质量以及化合物解毒；否则，它们会引起严重疾病（Ponnusamy 等，2019 年）。在蓖麻蚕体内，共生细菌纤维素消化发生最多的可能发生在中肠。前肠和后肠部位纤维素分解菌的存在表明纤维素食物的初

be occurring in the foregut and to some extent in the hindgut and that these isolates are enriched differently in the different gut compartments based on the localized physiological requirements of the host. However, the higher bacterial presence in the midgut compartment may be due to the fact that this is the major sight of food digestion and these bacteria may be aiding the host in the breakdown of cellulosic food diets into simpler sugars (MsangoSoko et al., 2021).

步消化可能发生在前肠，并在一定程度上也发生在后肠，并且根据宿主的局部生理需求，这些分离菌在不同的肠道部位富集的程度也不同。然而，中肠部位的细菌含量较高可能是因为这是食物消化的主要场所，这些细菌可能会帮助宿主将纤维素食物分解成更简单的糖（MsangoSoko 等，2021）。

2.6.1 Diversity of Gut Microflora

Due to their widespread distribution, insects are inevitably associated with an extremely large variety of microscopic life forms, including viruses, bacteria, fungi, protozoa, nematodes, and multicellular parasites. Although some of these microorganisms exhibit a rather wide host range, many associations are highly specialized and involve not only certain insect species but also particular life stages of the insect host. For example, gut microbes especially several bacterial genera and species with respect to seasons and breeds in silkworm.

The insect gut provides a suitable habitat for bacteria, and in many insect species, the gut possesses different types of bacteria, which are transient and do not remain in the gut during all life stages. However, in some insects, a variety of permanent microorganism's habitats and they supply essential nutrients to their host. Various *Escherichia coli*-borne plasmids could be effectively transferred to different gut bacteria of *F. candida* indicating that insect gut is an important environment for horizontal gene transfer. Termite and cockroach gut houses a complex microbiota ranging from protozoan spirochetes, gram-positive and gram-negative bacteria, archaea, and yeast.

Many insect species derive their gut microbiota from the surrounding environment such as the phylloplane of food plants. For instance, aphid's gut microbes share a common ancestry with intracellular symbionts and bacteria ingested from food plants. The plant epiphyte *Erwinia herbicola* in the gut of *B. mori* has been shown that they were able to grow and survive in the gut environment.

2.6.1 肠道菌群多样性

由于其分布广泛，昆虫不可避免地与各种各样的微观生命形式联系在一起，包括病毒、细菌、真菌、原生动物、线虫和多细胞寄生虫。尽管其中一些微生物的宿主范围相当广泛，但许多关联菌都是高度专一化的，不仅涉及某些昆虫物种，还涉及昆虫宿主的特定生命阶段。例如，家蚕的肠道微生物有几个与季节和品种特别有关的细菌属和种。

昆虫的肠道为细菌提供了一个合适的栖息地，在许多昆虫物种中，肠道拥有不同类型、短暂的细菌，它们不会在所有生命阶段都留在肠道中。然而，一些昆虫肠道中有多种永久性的微生物栖息其中，它们为宿主提供必需的营养。不同的大肠杆菌载体质粒可以有效地转移到念珠菌的不同肠道细菌中，表明昆虫肠道是水平基因转移的重要环境。白蚁和蟑螂的肠道有一个复杂的微生物群，包括原生动物螺旋体、革兰氏阳性菌、革兰氏阴性菌和古细菌和酵母。

许多昆虫物种的肠道微生物群来自周围环境，如食用植物的叶面。例如，蚜虫的肠道微生物与从食用植物中摄取的细胞内共生体和细菌有着共同的祖先。植物附生的欧文氏菌已被证明它们能够在桑蚕肠道环境中生长和存活。

2.6.2 Functional Role of Gut Bacteria

Insect gut symbionts play an essential role in the insect

2.6.2 肠道菌的功能

昆虫肠道共生菌在昆虫对各种食物的

adaptation to various food types, and an important role in lignocellulosic degradation, nutrient production, compound detoxification, immunity, and environmental adaptation.

适应中起着至关重要的作用，它们在木质纤维素降解、营养物质生产、化合物解毒、免疫和环境适应中也发挥着重要作用。

2.6.2.1 Nutrient supplement

Nutritional symbioses in microorganisms and insects evolve when a major component of the insects' diet lacks sufficient quantities of specific nutrients or when nutrients present in the diet are inaccessible because the insect lacks the requisite metabolic tools to fully digest their food. In these systems, bacteria or fungi help in nitrogen processing, sulfate assimilation, and fatty acid metabolism and help to contribute deficient sterols, vitamins (especially B-vitamin groups), digestive enzymes, and essential amino acids to their insect hosts. Microorganisms supply essential vitamins and other substances hence changes a poor diet into an adequate nourishing diet. In silkworm, cobalamine content in *B. mori* due to the presence of actinomyces changes according to the stage of development with maximum in larvae, minimum in eggs, and greatest in the Malpighian tubules.

2.6.2.1 营养物的补充

当昆虫摄食的主要成分缺乏足够数量的特定营养，或者由于昆虫缺乏充分消化食物所必需的代谢工具而无法获得饮食中的营养时，微生物和昆虫就会进化出营养共生体。在这些系统中，细菌或真菌有助于氮处理、硫酸盐同化和脂肪酸代谢，并有助于向昆虫宿主提供缺乏的甾醇、维生素（尤其是 B 族维生素）、消化酶和必需氨基酸。微生物提供必需的维生素和其他物质，因此将不良取食转变为充足的营养摄食。在家蚕中，由于放线菌的存在，桑蚕体内的维生素 B_{12} 含量会随着发育阶段的不同而变化，幼虫含量最高，卵含量最低，马氏管含量巨大。

2.6.2.2 Digestion, absorption and feeding halits

The gut microbes rapidly adapt to changes in the insect diet by induction of enzymes and through population changes in microbial community. Ingested microorganisms liberate enzymes that remain active in the gut surroundings and thus expand or extend the digestion and metabolic capabilities of organisms that harbor them. Additionally, microbial products play subtle roles in the life of the insect, being involved in the digestion of refractory food and detoxification of secondary plant compounds. In silkworm, the enzymes amino peptidase, β-glucosidase, alkaline phosphatase, and ATPase in microvilli (brush borders) in midgut cells of *B. mori* and thus contributing in efficient food digestion. Enzymes such as cellulase (β-endoglucanase, cellobiohydrolase/cellulase), xylanase, and pectinase are of exogenous microbial origin, while enzymes including amylase and β-glucosidase are produced endogenously. The bacteria inhabiting the gut of silkworm were found to be elaborating amylase, caseinase, gelatinase, lipase, and urease. The highest percentage of isolates was protease producers followed by lipid and polysaccharide splitters. These findings collectively indicate that the bacterial flora play an important role in digestion of ingested food material.

Gut microbiota promotes weight gain of both whole body and

2.6.2.2 消化、吸收和食性

肠道微生物通过酶的诱导和微生物群落的种群变化迅速适应昆虫取食的变化。被摄入的微生物释放出在肠道环境中保持活性的酶，从而扩大或扩展容纳它们的生物体的消化和代谢能力。此外，微生物产物在昆虫的生命中扮演着微妙的角色，参与难降解食物的消化和次生植物化合物的解毒。在家蚕中，中肠细胞微绒毛（刷状缘）中的氨基肽酶、β-葡萄糖苷酶、碱性磷酸酶和 ATP 酶有助于有效的消化食物。酶，如纤维素酶（β-内葡聚糖酶、纤维二糖水解酶 / 纤维素酶）、木聚糖酶、果胶酶是外源微生物来源，而淀粉酶和 β-葡萄糖苷酶是内源性产生的。家蚕肠道栖息的细菌主要分泌淀粉酶、酪蛋白酶、明胶酶、脂肪酶和脲酶。最高比例的菌株是蛋白酶产生菌，其次是脂质和多糖分解酶。这些发现共同表明菌群在消化摄入的食物材料中起着重要作用。

肠道微生物群促进蜜蜂全身和肠道

the gut in individual honey bees. This effect is likely mediated by changes in host vitellogenin, insulin signaling, and gustatory response. Microbial metabolism markedly reduces gut pH and redox potential through the production of short chain fatty acids and that the bacteria adjacent to the gut wall form an oxygen gradient within the intestine (Zheng et al., 2017). Mondin et al. (2021) demonstrated that the gut microbiota composition of *Trichoplusia ni* is altered by diet and may influence its polyphagous behavior.

的增重。这种效应可能是由宿主卵黄蛋白原、胰岛素信号和味觉反应的变化介导的。微生物代谢通过产生短链脂肪酸显著降低了肠道 pH 和氧化还原电位，肠壁附近的细菌能在肠道内形成氧梯度（Zheng 等，2017）。Mondin 等（2021）证明，拟尺蠖的肠道微生物群组成因取食而改变，并可能影响其多食行为。

2.6.2.3 Immunity and Protection

The production of indole derivatives and siderophores by Pseudomonads in the gut of *Plutella xylostella* was shown to have antagonistic effect toward entomopathogens. Production of extracellular chitinase has been shown to help in maintaining the physical property of peritrophic membrane. Lipase isolated from silkworm larval alimentary canal showed strong antiviral activity against BmNPV, providing evidence that digestive juice might play an important role during per oral infection with BmNPV. The understanding of the diet-derived intestinal bacterial community might yield insight into the relationship between gut bacteria and disease resistance of the silkworm.

The gut bacteria may also resist the colonization of invading pathogens by production of a series of secondary metabolites. Hemolymph has a key role in eliciting innate immune responses, which is triggered when pathogens enter into silkworm. Microbes can also play other roles such as preventing the colonization of pathogenic microbes by either mass action or active involvement in immune reactions. The microbes may synthesize various compounds and small molecules that can be used by the insects which increases fitness in extreme abiotic environments and provides protection from natural enemies such as predators and parasitoids (Ponnusamy et al.,2019).

(Li Wenchu)

2.6.2.3 免疫和保护

小菜蛾肠道假单胞菌产生的吲哚衍生物和铁载体对昆虫病原菌具有拮抗作用。胞外几丁质酶的产生有助于维持围食膜的物理性质。从家蚕幼虫消化道分离得到的脂肪酶对 BmNPV 有较强的抗病毒活性，证明消化液在 BmNPV 经口感染的过程中可能起着重要作用。了解家蚕日常食物衍生的肠道细菌群落可能有助于深入了解肠道细菌与家蚕抗病性之间的关系。

通过产生一系列次级代谢产物，肠道细菌也可以抵抗入侵的病原体定殖。血淋巴在激发先天免疫反应中起着关键作用，当病原体进入家蚕体内时，会触发先天免疫反应。微生物还可以发挥其他作用，例如，通过大规模行动或积极参与免疫反应来防止病原微生物的定殖。这些微生物可以合成各种化合物和小分子可供昆虫使用，从而增强昆虫在极端非生物环境中的适应能力，并保护其免受天敌和寄生虫的侵害（Ponnusamy 等，2019）。

（李文楚）

Exercises

1.How to classify the kinds of insect midgut cells and what their functions?

2.What functional role of gut bacteria in insect?

3.What are the basic nutritional requirements of insects?

4.How about countercurrent circulation and reabsorption.

5.Why silkworms can be fed with mulberry only?

思考题

1. 昆虫中肠细胞如何分类？有什么功能？

2. 昆虫肠道微生物有什么功能？

3. 昆虫的基本营养需求有哪些？

4. 反向循环和再吸收是怎样的？

5. 为什么家蚕只能喂桑叶？

第 3 章 脂肪体和中间代谢

Chapter 3 Fat Body and Intermediary Metabolism

Key points

1. Insect fat body is the analogous organ to vertebrate adipose tissue and liver, and plays key roles in energy metabolism and nutrient storage.

2. Fat body originates from the mesodermal walls of the coelomic cavities of embryo, and are classified into trophocytes, mycetocytes, urate cell, and oenocytes.

3. Hypertrehalosemic hormone (HTH), adipokinetic hormone (AKH), transforming growth factor β (TGF-β), and insulin-like peptides (ILPs) play important roles in metabolism of carbohydrate.

4. Carbohydrate, lipids, and proteins are the three major fuels for energy during insect flight.

知识点

1. 昆虫脂肪体类似于脊椎动物的脂肪组织和肝脏，在能量代谢与营养贮存中发挥重要功能。

2. 脂肪体来源于胚胎发育时期的体壁中胚层，其主要的细胞种类有营养细胞、含菌细胞、尿盐细胞和绛色细胞。

3. 高海藻糖激素（HTH）、脂动激素（AKH）、转化生长因子 β（TGF-β）和胰岛素样肽（ILPs）均在碳水化合物的代谢中发挥重要功能。

4. 碳水化合物、脂质和蛋白质是给昆虫飞行供能的三个主要燃料。

Insect fat body, which is analogous to mammalian liver and adipose tissues, plays potential roles in growth and development. Fat body is the key tissue for storage and usage of nutrients such as proteins and glycogen, and acts as nutrient sensors through nutrient signaling receptors (include Rheb, MTOR, and S6 kinase) and amino acid transporters. In addition, fat body is the center for energy metabolism including lipid and carbohydrate mobilization. Of note, fat body plays key roles in innate immunity in insects. Fat body has a series of receptors sensitive to hormones such as juvenile hormone (JH), 20-hydroxyecdysone (20E), insulin-like peptides (ILPs), and adipokinetic hormone (AKH). Studies have revealed that fat body is the major site for integrating hormonal and nutritional signals to regulate larval growth, body size, circadian

昆虫脂肪体相当于哺乳动物的肝脏和脂肪组织，在昆虫的生长发育中发挥重要作用。脂肪体是蛋白质和糖原等昆虫营养贮存与利用的主要器官，通过营养信号传导受体（包括 Rheb、MTOR 和 S6 kinase）和氨基酸转运蛋白来行驶营养传感器的功能。此外，昆虫脂肪体还是脂质和碳水化合物等能量物质发生动员的中心。同时，脂肪体在昆虫先天免疫中也发挥着重要作用。脂肪体中存在感受保幼激素（JH）、蜕皮激素（20E）、胰岛素样肽（ILPs）以及脂动激素（AKH）的受体分子。研究表明，脂肪体通过整合激素、营养信号并部分反

clock, pupal diapause, longevity, feeding behavior, and courtship behavior, partially by secreting fat body signals to remotely control the release of insulin-like peptides (ILPs) from the insulin producing cells (IPC) in the brain.

馈调控昆虫脑中胰岛素生产细胞（IPC）中胰岛素样生长因子（ILPs）的释放来调控幼虫生长、身体大小、节律、蛹期滞育、寿命、取食以及求偶行为等。

3.1 昆虫脂肪体

3.1 Insect Fat Body

Fat body fills the cavities of head, thorax, and the abdomen, functioning as connective tissue between the organs in insects. Insect fat body cells experience diverse physiological events at different developmental stages. They undergo mitosis during embryogenesis, endoreplication during the larval stages, and remodeling during metamorphosis, and regulate reproduction in adults under the coordinative control of hormonal and nutritional signals. Fat body is the major site for nutrient storage, energy metabolism, innate immunity, and detoxification. Moreover, fat body cells can actively migrate to wounds using a peristaltic-like "swimming" motility in *D. melanogaster.* Once there, they multitask to clear wound cell debris, plug the epithelial gap, and upregulate the expression of antimicrobial peptides (AMPs) to prevent infection. Together, fat body has emerged as a fascinating model for studying metabolic disorders and immune diseases (Franz et al., 2018; Li et al., 2019).

脂肪体充填头部、胸部和腹部的空腔，在器官间起着连接组织的作用。昆虫脂肪体在不同发育阶段经历着不同的生理事件，它们在胚胎发育时期经历着有丝分裂，在幼虫期经历细胞内复制，在变态发育期经历重塑，并在激素和营养信号的协同下参与调控成虫的繁殖。脂肪体是营养贮存、能量代谢、先天免疫以及解毒的主要场所。此外，黑腹果蝇的脂肪体细胞还被发现可以通过蠕动“游泳”的方式主动地迁移至伤口。到达伤口后，脂肪体细胞则会清除受伤的细胞碎片、堵塞上皮间隙，并上调抗菌肽（AMPs）的表达，从而抑制病原的感染。总之，昆虫脂肪体已经被视为研究代谢紊乱和免疫疾病的极好模型（Franz 等，2018; Li 等，2019）。

3.1.1 Origin, localization, and morphology of fat body

Fat body originates from the mesodermal walls of the coelomic cavities during embryogenesis, and is one of the most important tissues for life maintenance and reproduction. Fat body distributes throughout the body of insects, and is usually divided into two types: the perivisceral layer, located around the organs, such as the digestive tube and ovaries, and the peripheral layer, located adjacent to the tegument and the intersegmental muscles (Toprak et al., 2020; chapman, 1998; Fig. 3-1).

In the majority of insects, the cells of fat body are weakly associated with each other, the ramifications of tracheoles that bind one cell to another or the cells to other organs keep the consistency of this tissue. In *D. melanogaster*, a series of molecules including F-actin, DE-cadherin, and integrin are involved in the maintenance

3.1.1 脂肪体的来源、位置和形态

脂肪体来源于胚胎发育时体壁的中胚层，是维持生命和繁殖最重要的组织之一。脂肪体分布于昆虫的整个体躯，通常分为两种形式：围绕器官，如消化道和卵巢等的围脏层以及连接体壁表皮或是节间肌肉的外周层（Toprak 等，2020; Chapman, 1998; 图 3-1）。

多数昆虫脂肪体细胞连接疏松，微气管将细胞与细胞或细胞与组织连接起来以维持脂肪体的稳定性。在黑腹果蝇中，一些蛋白质分子，如 F- 肌动蛋白、DE- 钙黏合素和整合素等被发现与维持脂肪体细胞的形状、细胞 - 细胞连

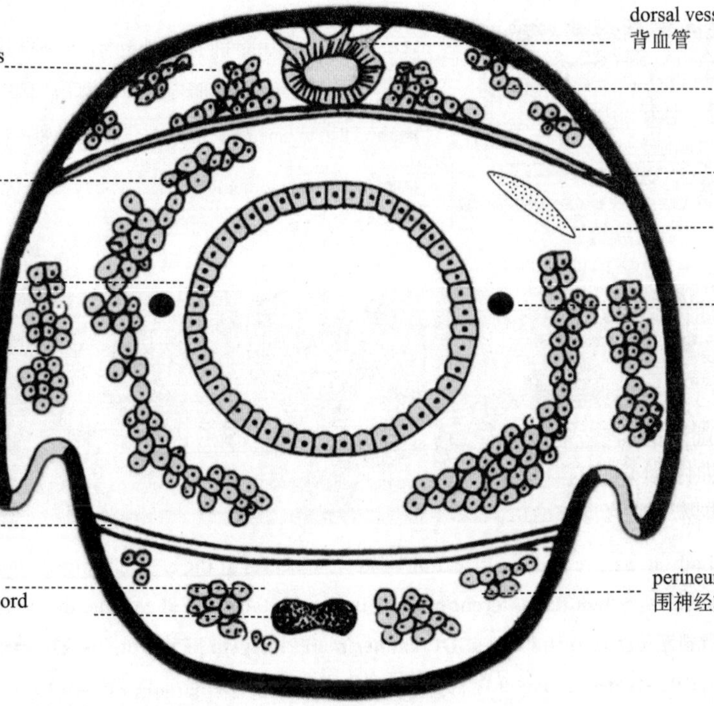

Fig.3-1 Abdominal transection of larvae and distribution of fat body in the larval insects (modified from Chapman, 1998).
图 3-1 幼虫腹部的横切面以及体内脂肪体的分布（仿 Chapman，1998）。

of fat body cell shape, cell-cell junctions, and cell-base membrane junctions (Jia et al., 2014; Fig. 3-2; Appendix color graphic Fig.3-2). In the ants *Attini tribe*, fat body cells are also joined by fusion of the basal lamina in specific regions. Moreover, the strong contact of the fat body cells with the hemolymph would facilitate the communication between both environments, making possible larger metabolic exchanges.

Fat body of the adult insects would develop through the remodeling of the larval tissue or differentiates from pupal stem cells. It was described that the larval fat body would persist in the adult in *Aedes aegypti;* however, in some Dipteran fat body is completely destructed at the end of larval stage. In Hymenoptera, severe autophagy occurs in the larval oenocytes. The autophagy process would not be uniform in different type of fat body cells, and only some regions of fat body would suffer lysis, while others would remain intact, thereby standing in the adult. In silkworm, autophagy and apoptosis are simultaneously induced by molting hormone 20E during larval-pupal metamorphosis by global induction of related gene expressions as well as inhibition of nutrient signals (Tian et al., 2013; Li et al., 2022). Moreover, cholesterol, the precursor of 20E biosynthesis, is also able to upregulate autophagy through dephosphorylating histone deacetylase HDAC1, which results in its nucleo-cytoplasmic translocation, subsequently, the dephosphorylated HDAC1 further

接以及细胞 - 基膜的连接有关（Jia 等，2014; 图 3-2 ；附彩图 3-2）。切叶蚁的脂肪体细胞还可以通过脂肪体细胞基底层特定区域的融合连接在一起。此外，脂肪体细胞和血淋巴紧密的接触有利于二者环境间的通信，使更大的代谢交换就成为可能。

昆虫成虫的脂肪体由幼虫脂肪体组织的重塑或从蛹期的干细胞分化而来。埃及伊蚊幼虫的脂肪体保留到成虫期。然而，在一些双翅目昆虫中，脂肪体在幼虫末期被完全解离。膜翅目昆虫幼虫的绛色细胞则发生剧烈的细胞自噬。在不同类型的脂肪体细胞中，细胞自噬的发生并不完全相同，只有部分脂肪体组织被分解，而另外一部分组织则保持完整直至成虫期。在家蚕中，昆虫蜕皮激素 20E 可以通过诱导大规模相关基因的表达以及抑制营养信号，从而诱导细胞自噬和凋亡发生（Tian 等，2013; Li 等，2022）。此外，20E 合成的前体胆固醇也能够通过诱导组蛋白去乙酰化酶 HDAC1 发生去磷酸化，导致其从细胞质迁移至细胞核，随后，去磷酸化的

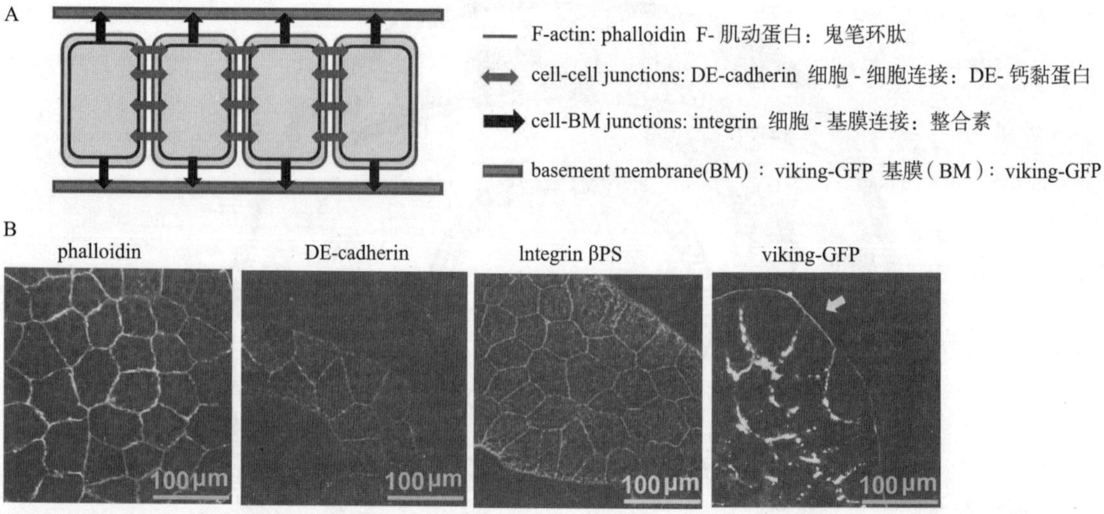

Fig.3-2 A model of the fat body structure and the detection of key molecules of the cell-cell conjugations and basement membrane (BM) junctions. A. Schematic diagram of the different marker molecules localized at cell membrane; B. Integrity of fat body cells stained by phalloidin. Immunofluorescent staining of DE-cadherin and integrin βPS proteins. Fluorescence observation of viking-GFP. viking-GFP: a fluorescently-tagged collagen Ⅳ reporter(permitted by Qiangqiang Jia, 2014).

图 3-2 果蝇脂肪体结构示意图及关键连接分子检测。A. 脂肪体细胞膜上的不同标志性分子示意图；B. 鬼笔环肽染色观察细胞膜的完整性，DE- 钙黏蛋白以及整合素的免疫荧光染色及 viking-GFP 的荧光标记观察。viking-GFP 为荧光标记的胶原蛋白Ⅳ报告蛋白）（经 Qiangqiang Jia 博士允许，2014）。

deacetylates autophagy-related proteins to promote autophagy in both insects and mammals (Wu et al., 2021a, b).

Besides the function of storage of lipids, fat body is also considered as the main source of proteins and carbohydrates, which would be sent to the hemolymph to serve as a precursor of various other substances from metabolism. Moreover, fat body would have a role in the storage and neutralization of non-utilized substances for the insect, being compared to the liver of vertebrates, since both would store excess nutrients in hemolymph, detoxifying it and serving as a biosynthetic source for metabolites that are set free in the circulation system.

Fat body would also act in the process of synthesis and secretion of cuticle proteins, supplying energy to this transformation phase and hydrocarbons for the formation of the new cuticle. Besides, fat body takes part in the protein synthesis and release to the oocytes such as vitellogenins, proteins with 180 kDa molecular weight and connected to the lipids and carbohydrates. In general, the vitellogenin synthesis specifically corresponds to about 70%–90% of proteins synthesized by the fat body of adult females of insects, and approximately 90% of the yolk proteins of insects. The majority of proteins in hemolymph is synthesized in fat body. In silkworm, ~

HDAC1 催化家蚕自噬相关蛋白质的去乙酰化修饰，促进昆虫和哺乳动物的自噬（Wu 等，2021a, b）

除了具有贮存脂质的功能，脂肪体还是蛋白质和碳水化合物的主要来源，它们会被输送到血淋巴以作为各种其他物质的代谢前体。此外，脂肪体在昆虫不可利用物质的贮存和中和上起着重要作用，相当于脊椎动物的肝脏，因为二者都可以贮存血液中过量的营养物质，为其解毒，并作为在循环系统中释放的代谢产物的生物合成来源。

脂肪体也作用于表皮蛋白质合成与释放的过程，为这个转变时期提供能量，为新表皮的形成提供碳氢化合物。除此以外，脂肪体还参与被运送至卵母细胞的蛋白质合成和分泌的过程，如分子质量为 180 kDa 的卵黄蛋白原，其可与脂类和碳水化合物相连接。总体来说，雌性昆虫成虫脂肪体中 70% ～ 90% 的合成的蛋白质均为卵黄蛋白原，其约占昆虫卵黄蛋白的 90%。昆虫血淋巴中的蛋白质主要合成于脂肪体。在家蚕幼虫末龄中期时，血淋巴中 40% 左右的总蛋白质分子质量为

40% of total proteins were those with the molecular weight between 55-100 kDa in hemolymph during the middle stage of the last larval instar, which was gradually increased until the end of larval stages. Whereas, the 55-100 kDa proteins in hemolymph were significantly absorbed by fat body and intensively increased to 60% of the total proteins in fat body at day 1 of prepupal stages. Mass-spectrum analysis showed that the 90% of 55-100 kDa proteins in both fat body and hemolymph were hexamerins (storage proteins), which provided the amino acid source for the biosynthesis of vitellogenins (Wang et al., 2020).

Fat body consists of thin layers or strings, generally one or two cells thick, or small nodules suspended in the hemocoel through connective tissues and trachea. Within a species, the structure of this tissue is more or less constant, but can have considerable differences between insects of different orders. This tissue is formed principally by cells denominated trophocytes, however, other cellular types can be found associated with them, such as mycetocytes, urate cell, oenocytes originated from embryonic ectoderm, chromotocytes (mainly in aquatic insects), and hemoglobin cells (those insects do not have body fluids related to the transport of respiratory gases) (Fig. 3-3; Appendix color graphic Fig.3-3).

55 ～ 100 kDa，并且这些蛋白质的含量在血淋巴中逐渐增加直至幼虫末期。然而，血淋巴中蛋白质分子质量为 55 ～ 100 kDa 的蛋白质在预蛹第一天显著地被脂肪体所吸收，并增加至脂肪体总蛋白质的 60% 左右。蛋白质谱分析显示，在血淋巴和脂肪体中，90% 的 55 ～ 100 kDa 的蛋白质属于血蓝蛋白家族（储藏蛋白），它们为卵黄蛋白原的合成提供了氨基酸来源（Wang 等 , 2020 年）。

脂肪体呈薄层或带状，一般为 1 ～ 2 个细胞的厚度，或者以小结的形式通过结缔组织或气管悬浮于血腔内。在同一物种中，脂肪体的结构或多或少都是恒定的，但是在不同目的昆虫中可能存在较大的差异。脂肪体组织主要由营养细胞组成，其他与之相关的细胞类型有含菌细胞、尿酸盐细胞、来自外胚层的绛色细胞、嗜铬细胞（主要是水生昆虫）和血红素细胞（没有与运输呼吸气体相关的体液的昆虫）（图 3-3；附彩图 3-3）。

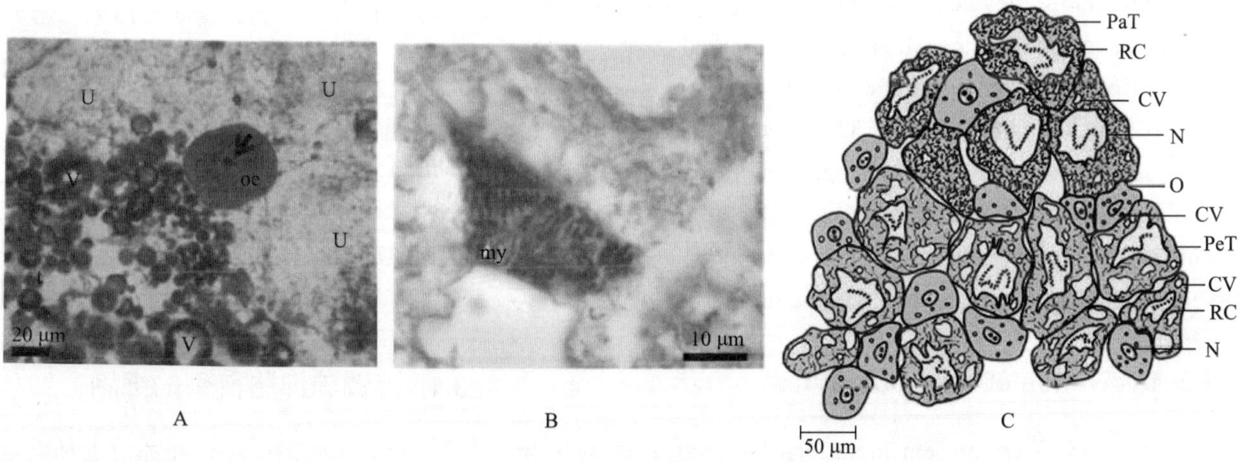

Fig. 3-3 The main types of insect fat body cells: trophocytes, urate cells, oenocytes, and mycetocytes.

A. Histochemical test (mercury bromophenol blue) for protein in the fat body cells of *Melipona quadrifasciata* pupae. Trophocyte (t) with vacuoles (v) positive for protein and oenocyte (oe) with positive cytoplasm and nucleus. u. urocytes (modified from Waléria et al., 2013). B. Detail of a mycetocyte (my) in the fat body of a *Mastotermes darwiniensis* worker. Stain: hematoxylin–eosin (modified from Costa-Leonardo et al., 2013). C. Fat body cells from worker of *A. disciger* (modified from Roma et al., 2010). PaT. present perivisceral trophocyte; RC. reticular cytoplasm; CV. cytoplasmic vacuole; N. nucleus; O. oenocyte; PeT. perivisceral trophocyte (modified from Waléria et al., 2013).

图 3-3　昆虫脂肪体细胞的主要种类：营养细胞、尿酸盐细胞、绛色细胞和含菌细胞。

A. 利用汞溴酚蓝对膜翅目蜜无刺蜂 *Melipona quadrifasciata* 蛹期脂肪组织中的蛋白质进行染色观察。t. 营养细胞；u. 尿酸盐细胞；oe. 绛色细胞，v. 包含蛋白质的囊泡（仿 Waléria 等，2013）。B. 苏木精 - 伊红染色观察达尔文澳白蚁工蚁脂肪体中的含菌细胞（仿 Costa-Leonardo 等，2013）。C. 切叶蚁（*Acromyrmex disciger*）工蚁脂肪体细胞（仿 Roma 等，2010）。PaT. 腔壁营养细胞；RC. 网状细胞质；CV. 胞质液泡；N. 细胞核；O. 绛色细胞；PeT. 围脏营养细胞（仿 Waléria 等，2013）。

3.1.2 Physiology and function of the trophocytes

The main trophocytes of fat body are described as rounded or polygonal cells, possessing a large irregular nucleus, which in general, is dislocated from the periphery of the cell and with many nucleoli, as well as different types of substances stored in their cytoplasm. The main ultrastructural characteristics of the trophocytes in the majority of the insects show that they are supported by a thin basal lamina and possess cytoplasm with large amounts of rough endoplasmic reticulum and Golgi regions, mitochondria, peroxisomes, lysosomes and multivesicular bodies, as well as lipids, carbohydrates, urates and proteins. The presence of these organelles suggests that the trophocytes would be capable of the synthesis, absorption, and storage of materials, especially those coming from the hemolymph. The trophocytes in some species, e.g. Diptera, possess ultra-structural and functional differences when compared to the parietal and perivisceral regions in all the phases of development. In this group of insects, the parietal fat body would mainly be related to the synthesis and the storage of lipids, while the perivisceral would be related to the storage of proteins. The main ultrastructural characteristic of the trophocytes is the developed vacuolar system, which possesses different vesicular structures, that can be classified as pre- or post-transcriptional, according to the content that was pre- or post-processed by the Golgi region. These vesicular structures are classified in two categories: ① digestive vacuoles that can be autophagic, heterophagic or multivesicular bodies; ② storage vacuoles containing urates, proteins or symbionts, associated with glycogen deposits and formations of lipid droplets.

3.1.2 营养细胞的生理学和功能

脂肪体的营养细胞主要为圆形或多边形，有一个大而不规则的核，通常与细胞的边缘脱离，有很多个核仁，细胞质内存储有不同类型的物质。多数昆虫营养细胞超微结构的主要特征显示，它们由薄的基底膜支撑，细胞质含有大量糙面内质网和高尔基体区域、线粒体、过氧化物酶体、溶菌酶和多囊泡小体之外，还有脂类、碳水化合物、尿酸盐和蛋白质等物质。这些细胞器的存在表明营养细胞能合成、吸收和贮存特别是那些来自血淋巴的物质。有些物种，如双翅目昆虫，所有发育阶段中腔壁和围脏区域的营养细胞在超微结构和功能上存在差异。在这些双翅目昆虫中，腔壁脂肪体可能主要参与脂类合成和贮存，而围脏脂肪体与蛋白质贮存有关。营养细胞主要的超微结构特征是拥有发达的不同泡状结构的液泡系统，根据是否经过高尔基体区域加工分为转录前小泡和转录后小泡。这些小泡结构分成两类：①消化型液泡可以是自体吞噬泡、异体吞噬泡或多囊泡小体；②含有尿酸盐、蛋白质或共生体的贮存型液泡，它们与糖原沉积及脂滴的形成相关。

3.1.3 Derivation of the trophocytes

The trophocytes are present in various forms as a result of the processes of functional specialization. The more frequent cell types resulting from its differentiation are urocytes, mycetocytes, chromatocytes, and the hemoglobin cells. The urocytes of Lepidoptera, mainly reported from the silkworm and *Calpodes ethlius*, are specialized in the storage of urates, having reduced rough endoplasmic reticulum, few mitochondria and large vacuoles containing urate granules. The origin of these granules is mainly from the degradation of proteins absorbed from the hemolymph to the interior of a special type of the multivesicular body or from the metabolism of nucleic acids, resulting from autophagy of the rough

3.1.3 营养细胞衍生的细胞

营养细胞因功能特化而以多种形态出现，其分化产生的最为常见的细胞类型是尿酸盐细胞、含菌细胞、嗜铬细胞和血红素细胞。鳞翅目昆虫，尤其是家蚕和巴西白斑弄蝶脂肪体的尿酸盐细胞特化专门用于贮存尿酸盐，伴随减少的粗内质网、极少的线粒体和含有尿酸盐颗粒的巨大囊泡。这些颗粒的来源主要是血淋巴中的蛋白质降解产物被吸收后进入特殊类型的多囊泡小体内部或由蛋白质剧烈合成时产生的废料泄漏导致糙

endoplasmic reticulum, due to the waste let by the intense protein synthesis.

The mycetocytes described in *N. cinerea* cockroaches are cells that possess vacuoles with symbiotic microorganisms that occupy most of the cytoplasm. The microorganisms can synthesize cholesterol for the host cell, which eventually digests its symbionts through the fusing of primary lysosomes to these vacuoles. A rare type of a modified trophocyte is the chromatocyte, a thin cell with a prominent central nucleus. As in the trophocytes, the chromatocytes accumulate lipids during the larval periods, mainly in the aquatic insects of Simullidae and Thaumaleidae, reserves that are intensely used during the development of the adult. The hemoglobin cells frequently are found in insects that do not have body fluids related to the transport of respiratory gases, e.g. larvae of *Chironomus* (Diptera), *Gastrophilus* (Diptera), as well as larvae and adults of *Anisops* and *Buenoa* (Hemiptera).

面内质网发生强烈的自噬而产生的核酸代谢物。

中华蟑螂的含菌细胞具有含共生微生物的液泡，其占据了细胞质的大部分。微生物为宿主细胞合成胆固醇，宿主细胞最终通过初级溶酶体与这些液泡融合而消化共生体。嗜铬细胞是一种罕见的营养细胞特化类型，在其薄的细胞中有一个突出的中心核。在蚋科和奇蚋科水生昆虫中，嗜铬细胞在幼虫期主要积累脂类，用以满足成虫发育对能量的强烈需求。血红素细胞通常存在于体液不参与呼吸气体转运的昆虫中，如摇蚊属（双翅目）、胃蝇属（双翅目）以及仰蝽属、*Buenoa* 属（半翅目）幼虫和成虫。

3.2　昆虫飞行的能量需求

3.2　The Energy Demands for Insect Flight

Metabolism is the sum of all chemical reactions occurring in an organism. For example, insects demand intense and rapid supply of energy during their flighting. The tracheal system of insects is able to supply oxygen to mitochondria even during flight so that insects do not occur an oxygen debt in flight. Mitochondria of flight muscle are highly specialized to support a high rate of metabolism. The cristae in mitochondria of flight muscle are numerous, and surrounded inside by the liquid matrix. The outer membrane of insect mitochondria is permeable to most soluble components, but the inner membrane has high-selective permeability. For example, cytochrome c reductase and hexokinase are located on the outer membrane. The space between the outer and inner membrane contains adenylate kinase and nucleoside diphosphokinase activities. The outer surface of the inner membrane contains glycerol-3-phosphate dehydrogenase, proline dehydrogenase, and trehalase. The inner membrane is located with the enzymes involved in respiratory chain reactions, ATP synthesizing enzymes, and α-ketoglutarate dehydrogenase. The matrix contains enzymes involved in tricarboxylic acid cycle,

代谢是有机体发生的化学反应的总称。例如，大部分昆虫飞行需要快速和持久的能量供应。即使在飞行过程中，昆虫气管系统也能向线粒体供氧，因此，飞行不会引发氧债。昆虫肌肉纤维之间的线粒体高度特化以支持高代谢率。飞行肌的线粒体有很多嵴，嵴的周围充斥着基质。线粒体外膜对大部分可溶性化合物具有可透性，但内膜具有高度选择性。例如，细胞色素 C 还原酶和己糖激酶位于线粒体的外膜。内外膜之间的腔含有腺苷酸激酶和核苷二磷酸激酶活性。内膜的外表面有甘油 -3- 磷酸脱氢酶、脯氨酸脱氢酶和海藻糖酶。内膜上含有电子呼吸链反应需要的酶，ATP 合成酶和 α- 酮戊二酸脱氢酶。基质中含有参与三羧酸循环、氨基酸降解和脂肪酸氧化反应的酶，如柠檬酸合成

degradation of amino acids, and oxidation of fat acid, such as citrate synthetase, aconitase, isocitrate dehydrogenase, fumarase, malate dehydrogenase, alanine, and aspartate amino transferase, as well as acetyl and palmityl transferases etc.

酶、顺乌头酸酶、异柠檬酸脱氢酶、延胡索酸酶、苹果酸脱氢酶、丙氨酸和天冬氨酸氨基转移酶、乙酰基和棕榈基转移酶等。

Some flying insects not only have high oxygen and calorie consumption value, but also can reach these high metabolic rates within a few seconds after taking flight. An oxygen debt does not have to be paid after intense activity because flight metabolism in insects is aerobic. Upon cessation of flight, the metabolic rates return almost instantly to a low "resting" rate. Upon initiation of flight, oxygen consumption rate in many insects jumps within seconds to values as much as 100 times. Insects are able to use several different substrates as fuel during flight. As energy is released, ATP is trapped in the universal metabolic currency. As in all other organism, ATP is present in relatively small amounts in insect cells, and more is synthesized as needed. The ATP concentration in cells is one of the regulators of metabolism, with "large" amounts inhibiting some key enzymes involved in ATP synthesis, while decreased amounts stimulate new synthesis. Carbohydrates, lipids, and dietary proteins are utilized as energy sources. Probably in a typical insect, the level of ATP in the flight musculature is sufficient for only about 1 second of flight; a phosphagen reserve of arginine phosphate, sufficient for an additional 2 to 4 seconds of flight, can be used rapidly to synthesize ATP, as shown in the following (Pereira, 2014):

一些飞行昆虫不仅仅具有高氧和卡路里消耗值，而且它们在开始飞行后的几秒钟内就可以达到高代谢速率。因为昆虫的飞行代谢是有氧呼吸，所以剧烈的飞行活动并没有造成负氧的出现。飞行停止时，代谢速率几乎立即恢复到低"静息"速率。飞行启动时，许多昆虫的氧消耗率在数秒钟内激增至 100 倍。飞行时，昆虫能利用几种不同物质作为燃料。ATP 作为能量释放，在普遍的代谢循环中发挥功能。与其他生物一样，昆虫细胞内 ATP 的数量相对较小，更多的只在需要时合成。细胞内 ATP 浓度是代谢的调节器之一，其量"大"时抑制 ATP 合成的几个关键酶，而数量减少时则刺激新的合成。碳水化合物、脂质和摄食蛋白质都可以作为能量的来源。一般昆虫飞行的肌肉组织内 ATP 水平仅够维持约 1 s 的飞行；磷酸原贮备磷酸精氨酸够额外的 2 ～ 4 s 飞行，如下列方程式所示，能快速合成 ATP（Pereira, 2014）：

$$\text{phosphoarginine} + \text{ADP} + \text{H}^+ \xrightleftharpoons[\text{精氨酸激酶}]{\text{arginine kinase}} \text{ATP} + \text{arginine}$$

磷酸精氨酸　腺苷二磷酸　⟷　腺苷三磷酸　精氨酸

Clearly, metabolism of some additional substrates and synthesis of new ATP must start in the first 1 or 2 seconds if flight is to continue. All insects appear to initially metabolize carbohydrates and a little bit of proline to "prime the Krebs cycle" (the tricarboxylic acid cycle) upon taking flight. In Diptera and Hymenoptera, flight only can be sustained as long as carbohydrates are available for metabolism; while some lepidoptera, orthoptera, and a number of other insect groups rapidly switch to metabolize lipids before their carbohydrates are gone.

显然，要维持继续飞行，必须在最初 1 ～ 2 s 内进行额外的物质代谢和新 ATP 的合成。所有昆虫飞行初期都进行碳水化合物的代谢及使用少量的脯氨酸进行"克雷布斯循环"（三羧酸循环）。因此，像双翅目和膜翅目昆虫只有在碳水化合物可供代谢时才能继续飞行，而一些鳞翅目、直翅目和其他一些昆虫种群在其碳水化合物分解完之前可快速切换至脂质代谢。

3.3 贮存与代谢

3.3 Storage and Metabolism

3.3.1 Trehalose sources

Trehalose is the principal storage sugar of insects, and from 200 mg to as much as 1.5 g per 100 mL hemolymph occur in various insect species. Trehalose is a disaccharide (*α-D*-glucopyranyl-*α-D*-glucopyrano side), with two glucose unit linkage *α*-1,1 after condensation (Fig. 3-4).

3.3.1 海藻糖源

海藻糖是昆虫贮存的主要糖类，在不同昆虫血淋巴中其含量从 200 mg 到 1.5 g/ 100 mL 不等。海藻糖是双糖（*α-D*-吡喃葡糖基 -*α-D*- 吡喃葡糖苷），由两分子葡萄糖缩合后通过 *α*-1,1 糖苷键结合形成（图 3-4）。

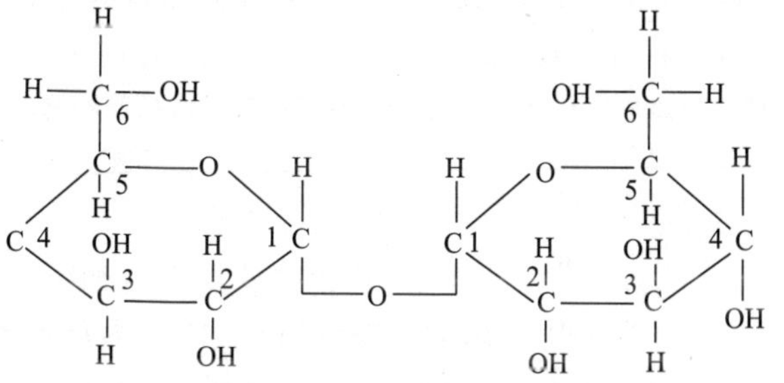

Fig.3-4 The molecular structure of trehalose.
图 3-4 海藻糖的分子结构。

Trehalose is rapidly synthesized from glucose as it is absorbed from the midgut that keeps very low level of glucose in hemolymph, and glucose absorption occurs without an energy-requiring membrane transport mechanism. Trehalose is a costly sugar for insects to synthesize, requiring several enzymes, steps, and input of high-energy phosphate from ATP. There are at least 5 pathways to synthesis trehalose in organisms, while only one exist in eukaryotes. Immediately after absorption from the gut, glucose is converted to glucose-6-phosphate by the enzyme hexokinase, with ATP supplied by the phosphate group and energy of its transferred to glucose. Two molecules of glucose-6-phosphate are needed for trehalose synthesis, which necessitates an investment of 2 ATPs (Fig. 3-5). Phosphoglucomutase isomerizes glucose-6-phosphate to glucose-1-phosphate. Subsequently, the enzyme uridine diphosphate glucose glucose pyro-phosphorylase catalyzes the synthesis of uridine diphosphate glucose (UDPG) from glucose-1-phosphate and uridine triphosphate (UDP).

葡萄糖在中肠中吸收后很快就被合成为海藻糖，以保持血淋巴中低水平的葡萄糖，葡萄糖的吸收是一个不需要能量的跨膜运输机制。海藻糖是昆虫合成代价较大的糖类，需要几种酶、步骤以及 ATP 高能磷酸盐的参与。生物中至少有 5 种海藻糖的合成途径，然而在真核生物中只存在一种合成途径。从肠道吸收后，在磷酸基团供能 ATP 给葡萄糖的情况下，葡萄糖立即被己糖激酶催化转化成 6- 磷酸葡萄糖。合成一分子的海藻糖需要两分子 6- 磷酸葡萄糖，同时需要消耗两分子的 ATPs（图 3-5）。6- 磷酸葡萄糖在葡萄糖磷酸变位酶的作用下异构为 1- 磷酸葡萄糖。随后，尿苷二磷酸葡萄糖 - 葡萄糖焦磷酸化酶催化 1- 磷酸葡萄糖和尿苷三磷酸（UDP）催化合成尿苷二磷酸葡萄糖（UDPG）。

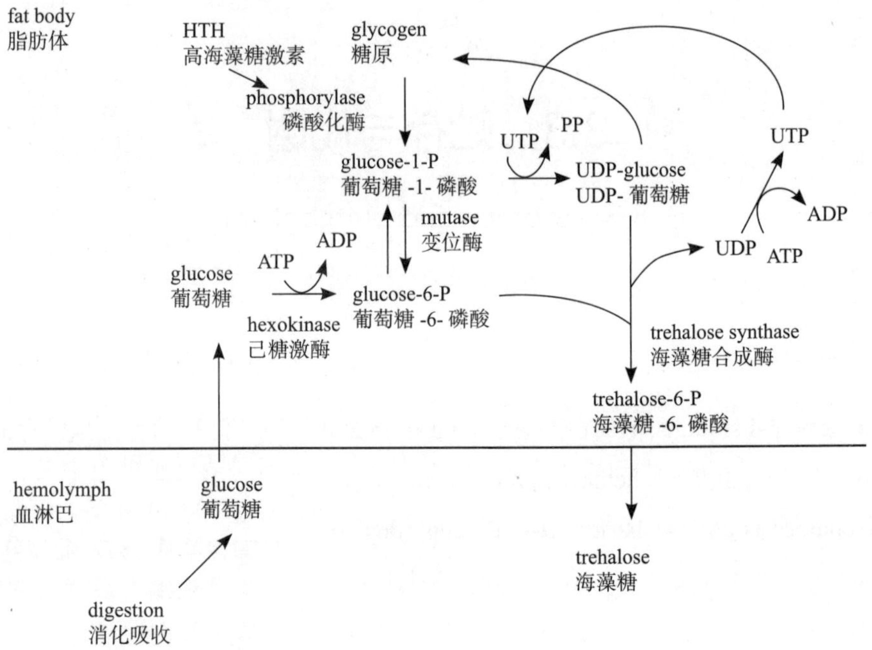

Fig.3-5 Biosynthetic pathway of trehalose and glycogen (The food digested glucose is transferred to fat body by hemolymph circulation, schematic diagram indicates the biosynthetic pathways of glycogen and trehalose); PP. Pyrophosphate.
图 3-5 海藻糖和糖原的生物合成途径（消化吸收的葡萄糖经血淋巴循环运输至脂肪体后经反应合成糖原和海藻糖）；PP. 焦磷酸盐。

Trehalose-6-phosphate synthetase catalyzes the formation of trehalose-6-phosphate from glucose-6-phosphate and UDPG, with release of UDP. Regeneration of UTP (uridine triphosphate) for future reactions requires enzymatically catalyzed phosphate phosphorylation of uridine diphosphate by ATP, so one more ATP must be counted in the cost of synthesis of trehalose. Finally, trehalose-6-phosphate synthetase removes the phosphate group from trehalose-6-phosphate to form trehalose.

Why do insects use a costly storage form for sugar when so many other organisms store sugar as glucose? The answer might be: ① the value of a large reserve of an immediate energy source in the hemolymph that is available to support the energy demand of sustained flight; ② the non-reducing nature of trehalose, in contrast to the reducing nature of glucose, may prevent undesirable reactions in the hemolymph; ③ there may have been evolutionary selection to reduce the osmotic effect of dissolved nutrients in the hemolymph. Most insects have high levels of trehalose in the hemolymph, and in fat body cells, but the enzyme exists as an inactive proenzyme. The mechanism by which an insect converts the proenzyme to the active form under normal physiological processes has not been elucidated.

6- 磷酸海藻糖合成酶催化 6- 磷酸葡萄糖和 UDPG 形成 6- 磷酸海藻糖，并释放 UDP。UTP（尿苷三磷酸）的再生则需要利用 ATP 和酶催化尿苷二磷酸的磷酸化，所以合成海藻糖还要计算多一分子 ATP 的消耗。最终，6- 磷酸海藻糖合成酶催化 6- 磷酸海藻糖去掉磷酸基团形成海藻糖。

为什么昆虫要利用代价高昂的糖类而其他众多的有机体则以葡萄糖作为贮存糖类？答案可能是：①在血淋巴中大量储备即时能源是维持持续飞行对能量的需求；②相对于葡萄糖的还原性而言，海藻糖的非还原性可以避免血淋巴中发生一些不必要的反应；③可能进化选择导致血淋巴中溶解营养的渗透压的降低。多数昆虫血淋巴和脂肪体细胞都含有高水平的海藻糖，但这些催化酶均以非活性的酶原形式存在。在正常生理条件下，昆虫将酶原转化成活性形式的机制尚未阐明。

3.3.2 Metabolism of glycogen

Glycogen is the second form of storage energy. Insect flight muscle contains glycogen, but most muscles are too small to store very much. Glucose can be released from glycogen stored in fat body by glycogen phosphorylase. This enzyme is present in muscle tissue and fat body as inactive phosphorylase b, and it must be activated to phosphorylase a.

Activation is under control of the hypertrehalo semic hormone (HTH), a peptide hormone that belongs to the adipokinetic hormone/red pigment concentrating hormone (AKH/RPCH) family, which is synthesized in neurosecretory cells of corpus cardiacum (CC). Initiation of flight activates CC to secrete the hormone through nervous control. HTH requires the participation of a second messenger, cAMP, at the fat body cell membrane surface to activate phosphorylase b kinase, which converts inactive phosphorylase b to active phophorylase a. Glycogen phosphorylase *a*

3.3.2　糖原的代谢

糖原是昆虫的第二种储能形式。昆虫飞行肌含有糖原，但大多数肌肉太小而不能贮存太多糖原。脂肪体糖原磷酸化酶能够从贮存的糖原中将葡萄糖释放出来。该酶在肌肉组织和脂肪体中以非活化的磷酸化酶 b 存在，并且需要活化为磷酸化酶 a 才能发挥功能。

酶的活化受到肽类激素高海藻糖激素（HTH）的调控，其属于脂动激素/红色素浓缩激素（AKH/RPCH）家族，合成于心侧体（CC）中的神经分泌细胞。飞行通过神经控制活化了心侧体分泌激素。HTH 需要第二信使环磷酸腺苷酸，cAMP，在脂肪体细胞膜表面活化磷酸激酶 b，将非活性磷酸激酶 b 转化成活性的磷酸激酶 a。接着，糖原磷酸化酶 a 从糖原中释放葡萄糖，如图 3-6 所示。

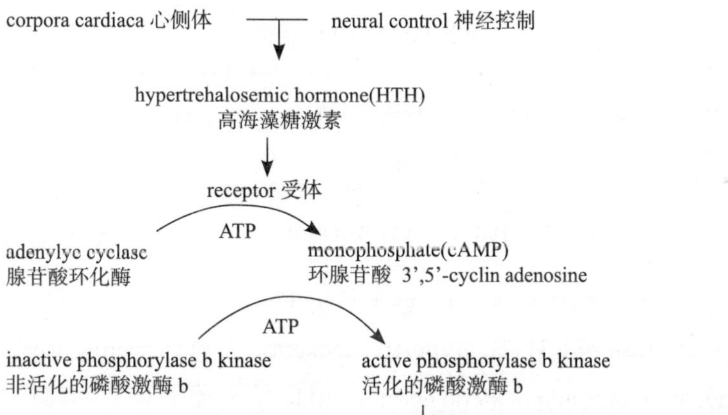

Fig.3-6 The action mechanism of insect hypertrehalosemic hormone (HTH). The neurons control the release of HTH from corpora cardiaca, and then it is transferred to the targeted tissues and organs. Receptor triggers a series of signal transduction and biochemical reactions after binding with HTH, degrading the glycogen to release glucose.

图 3-6 昆虫高海藻糖激素的作用机制。在神经控制咽侧体释放高海藻糖激素后，其经血液循环运送至靶标组织中，与受体结合后启动细胞内一系列的信号转导和生化反应，降解糖原释放葡萄糖。

then release glucose from glycogen as follows in Fig. 3-6. Because glucose-1-phosphate, rather than glucose, is released from glycogen, the investment of 1 ATP is saved in the initial stages of the glycolytic pathway. Free Ca^{2+} at concentration as low as 10^{-8} mol/L and inorganic phosphate stimulate phosphorylase b kinase, and stimulation is near maximum at 10^{-6} mol/L Ca^{2+}. Both free Ca^{2+} and inorganic PO_4 are increased as a result of the initiation of flight. The reverse reaction that inactive glycogen phosphorylase by conversion of phosphorylase a to phosphorylase b is catalyzed by phosphorylase b kinase, while littles is known about how this enzyme functions in insects.

因为从糖原释放的是 1- 磷酸葡萄糖，而不是葡萄糖，所以糖分解途径的起始阶段节约了 1 分子 ATP。10^{-8} mol/L 低浓度的游离 Ca^{2+} 和无机磷酸盐也能激活磷酸化激酶 b，Ca^{2+} 浓度在 10^{-6} mol/L 时对其激活接近最大值。飞行的启动导致游离 Ca^{2+} 和无机磷酸盐增加。将有活性的磷酸化酶 a 逆转化为无活性的磷酸化酶 b 的糖原磷酸化酶失活的反应是由磷酸激酶 b 所催化的，然而这个酶在昆虫中是如何作用的却知之甚少。

3.3.3 Glycogen synthesis

3.3.3 糖原的合成

Storage of glycogen occurs mainly in fat body cells, but some glycogen is also stored in gut epithelial cells, and to a slight extent in muscle cells. Synthesis of glycogen is catalyzed by the enzyme UDP-glucose- glycogen transglucosylase, also known as glycogen synthetase.

糖原的贮存主要发生在脂肪体细胞中，但有些糖原贮存在肠道上皮细胞，肌细胞中也有微量。糖原的合成由 UDP- 葡萄糖 - 葡萄糖转糖基酶催化，其也是众所周知的糖原合成酶。

$$\text{glycogen}_n + \text{UDP–glucose} \xrightarrow{\text{glycogen synthetase}} \text{glycogen}_{n+1} + \text{UDP}$$

糖原n　UDP- 葡萄糖　糖原合成酶　糖原$_{n+1}$　尿苷二磷酸

3.3.4 Hormonal control of carbohydrate metabolism

3.3.4 碳水化合物代谢的激素调控

In some insect species, the key hormone controlling carbohydrate metabolism is the peptide hormone HTH, hypertrehalosemic hormone. A related peptide hormone, adipokinetic hormone (AKH), may supplant the action of HTH in some insects. HTH and AKH are closely and chemically related, and belong to the same family of peptide hormones. HTH is a polypeptide of ten amino acids, and the sequence of amino acids varies slightly from species to species. AKH, also isolated from several different insects, may have from eight to ten amino acids in its structure. AKH is biosynthesized in neural-endocrine cells of CC.

AKH is similar to the mammalian glucagon, and its release into the hemolymph is regulated through membrane depolarization by ATP-sensitive K^+-channels. A rapid decrease in trehalose concentration triggers calcium influx into CC cells, which induces

在一些昆虫物种中，调控碳水化合物代谢的主要激素是肽类高海藻糖激素 HTH。与此相关的肽类激素是脂动激素（AKH），其在有些昆虫中起到替代 HTH 的作用。HTH 和 AKH 在化学上密切相关，被认为属于同一肽类激素家族。HTH 为 10 个氨基酸的多肽，其氨基酸序列在物种间略有不同。AKH 也是从几种不同昆虫中分离到的，其结构有 8 ～ 10 个氨基酸不等。AKH 由心侧体 CC 中的神经内分泌细胞所合成。

AKH 类似于哺乳动物的胰高血糖素，ATP 敏感型的 K^+ 离子通道导致细胞膜去极化后从而释放其进入血淋巴。海藻糖浓度的快速降低启动了 Ca^{2+} 流入 CC 细胞，这诱导了 AKH 向血淋巴的释

the release of AKH into the hemolymph (Mattila et al, 2017; Chatterjee et al, 2021). Of note, AKH secretion is also triggered by high hemolymph trehalose concentration in *Drosophila* larvae fed with a high-sugar diet (Kim et al., 2015; Song et al., 2017). Taken together, the secretion of AKH have biphasic regulation, which promotes by both of low and high hemolymph trehalose concentrations. It can be understood by the necessity to maintain constant hemolymph trehalose concentrations during the rapid larval growth phase, when insulin signaling is high, as well as during the wandering and pupal stages when feeding has ceased.

放（Mattila 等，2017 年；Chatterjee 等，2021 年）。值得注意的是，在饲喂高糖饮食的果蝇幼虫的血淋巴中，AKH 的分泌也较高（Kim 等，2015 年；Song 等，2017 年）。总的来说，AKH 的分泌存在双向调控，血淋巴中的高、低海藻糖浓度均可以提高其释放。可以理解为在幼虫快速生长的时期，需要维持血淋巴内恒定的海藻糖浓度，因此在胰岛素信号比较高的摄食阶段以及游走和蛹等停止摄食的时期，AKH 的释放均较高。

Mammalian insulin/glucagon axis is well-conserved in *Drosophila*. The *Drosophila* genome encodes eight insulin-like peptides (dILPs), which have temporal and spatial expression patterns, suggesting their specialized roles in maintaining growth and metabolic homeostasis at different stages of the life cycle (Mattilaet al., 2017). Adult flies that insulin-producing cells (IPCs) was ablated, or larvae deficient for dILPs 2, 3, and 5 contain high hemolymph glucose concentrations and elevated stored trehalose, glycogen, and lipid (Grönke et al., 2010). Moreover, sugar-inducible transforming growth factor β (TGF-β)/Activin ligand Dawdle (Daw) is involved in the tolerance of sugar in *Drosophila* (Ghosh et al., 2014).

哺乳动物的胰岛素 / 胰高血糖素轴线在果蝇中保守存在。果蝇编码表达 8 个胰岛素样肽（dILPs），它们具有特异的时空表达模式，明示它们在不同生命时期维持生长和代谢稳态中具有特异的功能（Mattila 等，2017 年）。在胰岛素生产细胞（IPCs）被移除的果蝇成虫中或是 dILPs 2、3 和 5 缺陷的幼虫中，血淋巴中含有高浓度的葡萄糖，同时海藻糖、糖原和脂质的含量上升（Grönke 等，2010）。此外，受糖诱导的转化生长因子 β（TGF-β）/ 激活素配体 Dawdle（Daw）与果蝇对糖的耐受性相关（Ghosh 等，2014 年）。

3.4　脂类动员及其激素调控

3.4　Mobilization and Hormonal Control of Lipids

Carbohydrates, lipids, and dietary proteins are the three main resources for supporting energy in insects. Lepidopteran adult, some Orthoptera, and some other insects can use fatty acids as a flight fuel, and well-fed ones generally have sufficient lipids in the body to support flight for much longer periods than can be supported by the available supply of carbohydrates. Virtually, all fatty acids are stored in fat body cells as triacylglycerols and thus before they can be metabolized, they must be released from fat

碳水化合物、脂质和饮食中的蛋白质是三个主要的能量来源。鳞翅目成虫、一些直翅目及其他昆虫能利用脂肪体中的脂肪酸作为飞行燃料，营养好的昆虫体内有足够的脂类。这些脂类比可提供的碳水化合物能支持更长时间的飞行。几乎所有的脂肪酸在脂肪体细胞中都是以甘油三酯贮存，所以在它们被代谢前，必须从脂肪体细胞释放并转运至

body cells and transport to the muscles. Release of lipid from fat body cells is under the control of AKH. The released diacylglycerol is transported through an aqueous medium, the hemolymph, to the thoracic flight muscles.

肌肉。从脂肪体细胞释放脂类受到一种肽类激素的调控，这是一种由心侧体分泌的脂动激素。释放的甘油二酯通过水相介质 - 血淋巴转运到胸部的飞行肌。

3.4.1 Hormone control of lipid mobilization

AKH circulating through the hemolymph binds to a receptor on the surface of fat body cells and activates adenylcyclase to produce cAMP as a second messenger. In turn, cAMP activates a lipase that removes one fatty acid residue from triacylglycerol. The resulting diacylglycerol (DAG) is released from fat body cells to be picked up by high-density (unloaded ApoL-A) lipoprotein particles at the hemolymph fat body cells surface interface. The lipophorin particle also loads a small circulating hemolymph lipoprotein (apoLp-III) to form low density lipophorin particles (LDLp). These LDLp particle are transported to the flight muscles. The process is summarized in Fig. 3-7.

3.4.1 脂质动员的激素调控

AKH 通过血液循环与脂肪体细胞表面受体结合，活化腺苷酸环化酶产生第二信使 cAMP。cAMP 进而活化酯酶从甘油三酯去除一个脂肪酸残基。结果产生的甘油二酯从脂肪体细胞释放出来，被位于血淋巴脂肪体细胞界面的高密度（空载 ApoL-A）脂蛋白颗粒带走。脂蛋白颗粒也装载循环血淋巴的小分子脂蛋白（脱辅基脂蛋白 - Ⅲ）形成低密度脂蛋白颗粒（LDLp）。这些 LDLp 颗粒被转运到飞行肌。这一过程概括如图 3-7 所示。

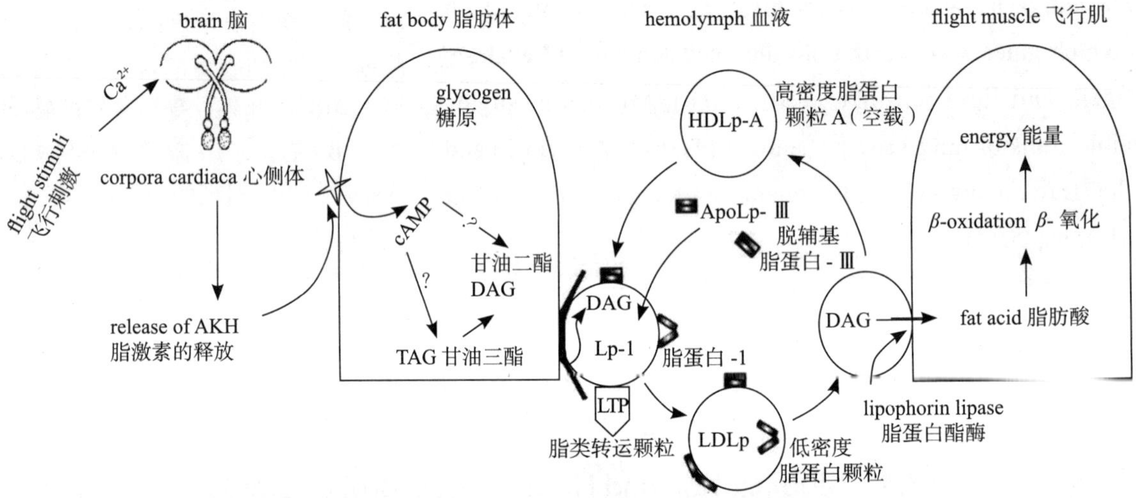

Fig.3-7 A diagram of the role of AKH in promoting DAG release from the fat body, loading of DAG into a lipoprotein transport complex and delivery of it to muscle cells. At the membrane of muscle cell, two fatty acids are removed from DAG by a membrane-bound lipases, subsequently, fat acids released from fat body were transported to muscle cells for metabolism (modified from Nation, 2022).

图 3-7 脂动激素（AKH）在促进 DAG 从脂肪体释放，装载 DAG 进入载脂蛋白转运复合体并释放到肌肉细胞中的作用图解。在肌肉细胞膜上，两个脂肪酸通过膜结合脂酶从 DAG 上切除，随后脂肪酸被转运到肌肉细胞进行代谢（仿 Nation, 2022 年）。

3.4.2 Signal transduction of AKH

AKH realizes its regulation on lipid mobilization via signal transduction. AKH binds to its GPCR (G protein-coupled receptors), the adipokinetic hormone receptor (AKHR), similar

3.4.2 脂动激素（AKH）的信号转导

AKH 通过信号转导来调控脂质的动用。AKH 的脂动激素受体（AKHR）即 GPCR（G protein-coupled receptors）与

to the mammalian gonadotropin-releasing hormone receptor, was first identified from *Drosophila* and the silkworm *B.mori*. AKHR is produced primarily in the fat body (Mattila et al., 2017), but also midgut, muscles, brain, and reproductive organs (Zandawala et al., 2015; Alves-Bezerra et al., 2016; Hou et al., 2017). Binding of AKH to AKHR activates two different second-messenger systems involved in lipid mobilization (Mattila et al., 2017; Fig. 3-8; Appendix color graphic Fig.3-8). In one pathway, their binding results in stimulation of Phospholipase C (PLC), which cleaves membrane lipid phosphatidylinositol 4,5-diphosphate (PIP2) into inositol triphosphate (IP_3) and diacylglycerol (DAG). Subsequently, release of IP_3 activates IP_3 Receptor (IP_3R) in the endoplasmic reticulum, leading to mobilization of the second messenger calcium from the endoplasmic reticulum to cytosol (Toprak, 2020; Fig. 3-8). While, how calcium transmits the AKH

哺乳动物的促性腺激素释放激素受体相似，并在果蝇和家蚕中首先被鉴定。AKHR 主要在脂肪体中产生（Mattila 等，2017），但在中肠、肌肉、脑和生殖系统中也有表达（Zandawala 等，2015；Alves-Bezerra 等，2016；Hou 等，2017）。AKH 和 AKHR 结合后通过激活两条不同的第二信使系统来调节脂质的动用（Mattila 等，2017；图 3-8；附彩图 3-8）。在一条通路中，它们的结合导致磷脂酶 C（PLC）的激活，其切割细胞质膜上的磷脂酰肌醇 4,5- 双磷酸盐（PIP2）生成三磷酸肌醇（IP_3）和甘油二酯。随后，IP_3 的释放进一步激活了内质网中其受体（IP_3R），导致第二信使 Ca^{2+} 从内质网转移至胞质（Toprak, 2020; Fig. 3-8）。然而，Ca^{2+} 是如何精确传递 AKH 信号的尚不

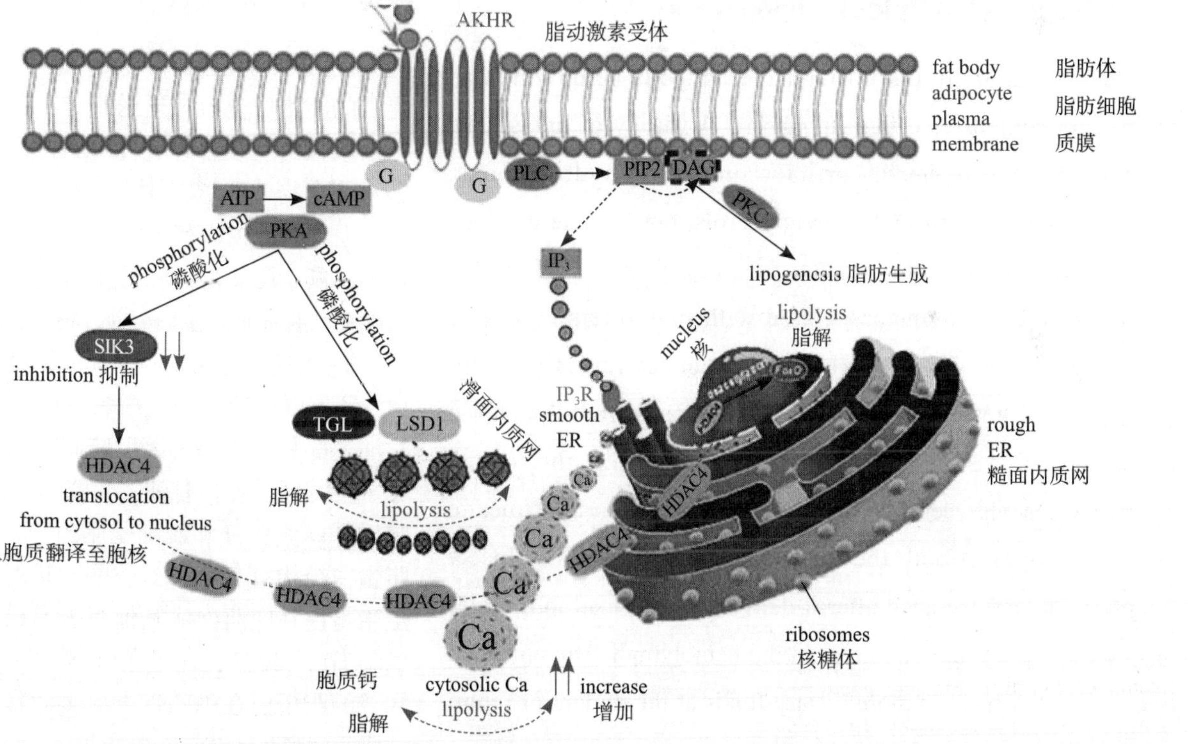

Fig. 3-8 Mode of signaling of adipokinetic hormone (modified from Toprak, 2020). AKH. adipokinetic hormone; AKHR. adipokinetic hormone receptor; Ca. calcium; DAG. diacylglycerol; ER. endoplasmic reticulum; G. g-protein couple; HDAC4. histone deacetylase 4; IP_3. inositol triphosphate; IP_3R. inositol-1,4,5-triphosphate receptor; LD. lipid droplet; LSD1. lipid storage droplet 1 protein; PLC. phospholipase C; PIP2. phosphatidylinositol 4,5-diphosphate; PKA. cAMP-dependent protein kinase; PKC. calcium-dependent protein kinase C; SIK3. salt induced kinase 3; TGL. triglyceride lipase.

图 3-8 脂动激素（AKH）的信号转导（仿 Toprak, 2020）。AKH. 脂动激素；AKHR. 脂动激素受体；Ca. 钙；DAG. 甘油二酯；ER. 内质网；G. G- 蛋白耦联受体；HDAC4. 组蛋白去乙酰化酶 4；IP_3. 三磷酸肌醇；IP_3R. 肌醇 -1，4，5- 三磷酸受体；LD. 脂滴；LSD1. 脂质贮存脂滴 1 蛋白；PLC. 磷脂酶 C；PIP2. 磷脂酰肌醇 4，5- 双磷酸；PKA. 依赖于 cAMP 的蛋白激酶 A；PKC. 钙依赖蛋白激酶 C；SIK3. 盐诱导激酶 3；TGL. 三酰甘油脂肪酶。

signal exactly is unknown. In the second pathway, binding of AKH to AKHR activates adenylate cyclase and mediates a rapid increase of the second messenger cAMP, thereby activation of cAMP-dependent Protein Kinase (PKA), which promotes the phosphorylation of downstream factors, such as the LDs (lipid droplets), TGL and Perilipin 1/Lipid Storage Droplet-1 Protein (PLIN1/LSD1) (Arrese et al., 2010). PLIN1 phosphorylation increases the accessibility of LDs for TGL (triglyceride lipase), thereby allowing lipid mobilization. In addition, PKA inhibits the activity of Salt-Inducible Kinase 3 (SIK3), a member of the AMP-activated Protein Kinase (AMPK) family, by phosphorylation, leading to translocation of Histone Deacetylase 4 (HDAC4) from cytosol into the nucleus, where HDAC4 deacetylates and activates the transcription factor Forkhead Box Class O, FoxO, thereby transcription of *brummer* and lipolysis (Choi et al., 2015; Toprak, 2020).

清楚。在第二条通路中，AKH和AKHR结合激活腺苷酸环化酶，导致第二信使cAMP快速增加，进而激活了依赖cAMP的蛋白酶K（PKA），进而提高了其下游因子，如LDs、TGL和围脂蛋白perilipin1/脂质贮存液滴-1蛋白质（PLIN1/LSD1）的磷酸化（Arrese等，2010）。其中PLIN1的磷酸化促进了甘油三酯脂酶与脂滴的结合，从而提升了脂质的动员。此外，PKA通过磷酸化抑制AMPK家族成员的盐诱导激酶SIK3的活性，导致组蛋白去乙酰化酶（HDAC4）从胞质转运至细胞核，随后HDAC4去乙酰化修饰转录因子叉头蛋白O家族的FoxO，导致中性脂酶（*brummer*）的表达和脂解（Choi等，2015；Toprak, 2020）。

3.4.3 Transport of lipids by lipophorin

Insect transport lipids through the aqueous hemolymph as lipoprotein complexes called lipophorin. A variety of lipids have been found in lipophorin particles, including hydrocarbons, phospholipids, and tri- and diacylglycerols, but the one destined for metabolism in flight muscles is diacylglycerol. There are three identified proteins associated with lipoprotein: apoLp-I, apoLp-Ⅱ, and apoLp-Ⅲ. ApoLp-Ⅲ associates reversibly with diacylglycerol-loaded lipophorin and dissociates from lipophorin in the unloaded state. It is necessary to stabilize the loaded lipophorin-diacylglycerol particle. ApoLp-I and Ⅱ in the structure of lipophorin is in both the loaded and unloaded state. Insect lipophorin is not degraded after it delivers a lipid load to muscles or to other tissues, but circulates in the hemolymph. The unloaded lipophorin may pick up more dietary lipids at the midgut or return to the fat body for a new load of lipids.

3.4.3 载脂蛋白转运的脂类

昆虫通过液态血淋巴以被称为载脂蛋白的脂蛋白复合体运送脂类。在载脂蛋白颗粒内发现各种各样的脂类，包括碳水化合物、磷脂、甘油三酯和甘油二酯，但飞行肌内以代谢为目的的主要是甘油二酯。已鉴定有三种蛋白质与脂蛋白相关：脱辅基脂蛋白 - Ⅲ、脱辅基脂蛋白 - Ⅱ和脱辅基脂蛋白 - Ⅲ。脱辅基脂蛋白 - Ⅲ与可逆的甘油二酯装载脂蛋白和甘油二酯从脂蛋白解离后的卸载状态有关。它为载脂蛋白 - 甘油二酯稳定所必需。脱辅基脂蛋白 -I和脱辅基脂蛋白 - Ⅱ的脂蛋白结构有装载和卸载两种状态。昆虫脂蛋白在向肌肉或其他组织递送装载的脂类后不降解，而是在血淋巴中循环。卸载的脂蛋白在中肠带上更多的食物脂类或回到脂肪体装载新的脂类。

3.4.4 Activation of fatty acids, entry into mitochondria, and *β*-oxidation

At the hemolymph-flight muscle cell interface, DAG is unloaded under the influence of AKH, the apoLp-Ⅲ protein dissociates from the lipophorin, and the remaining high-density lipoprotein particle (HDLp-A) is shuttled back through the hemolymph to transport

3.4.4 脂肪酸的活化、进入线粒体和 *β*- 氧化

在血淋巴 - 飞行肌细胞界面，甘油二酯在脂动激素的影响下是未装载的，脱辅基脂蛋白 - Ⅲ与脂蛋白分离，留下的高密度脂蛋白颗粒（HDLp-A）通过

DAG again, either from the gut to the fat body or from fat body to the muscles. A membrane-bound lipase in flight muscle tissue has high affinity for diacylglycerol, releasing fatty acids and glycerol from diacylglycerol bound to lipoprotein (a), the transport lipoprotein complex, very quickly. Glycerol resulting from the hydrolysis can be phosphorylated and metabolized for energy or returned to the fat body where it can be used again to form triacylglycerols. In the cytoplasm, the fatty acid is activated by reaction with coenzyme A (CoA) and ATP to form a fatty acyl CoA derivative. The total ATPs produced for palmitic acid with 16 carbons, as an example, when it has been completely metabolized to carbon dioxide and water can be calculated from the following equations:

血液循环穿梭回来，再次运送甘油二酯，要么是从中肠到脂肪体，要么是从脂肪体到肌肉。飞行肌组织中的膜结合脂酶对甘油二酯具有高亲和力，可以很快地从二酰基甘油与脂蛋白（a）(转运脂蛋白复合物结合中快速释放脂肪酸和甘油。）水解产生的甘油可被磷酸化、代谢产能或回到脂肪体再次被用于产生甘油三酯。在细胞质中，脂肪酸通过与乙酰辅酶 A（CoA）及 ATP 反应生成脂酰 CoA 衍生物而被激活。如下，以 16 碳的棕榈酸完全代谢成二氧化碳和水产生的总 ATPs 为例，根据下列方程式计算：

$$\text{palmitoyl-CoA} + 7\text{CoA} + 7\text{FAD} + 7\text{NAD}^+ \longrightarrow 8\text{CH}_3\text{COSCoA} + 7\ \text{FADH}_2 + 7\text{NADH}$$

palmitoyl-CoA: 棕榈酰辅酶 A; FAD: 黄素腺嘌呤二核苷酸；$FADH_2$, $FADH_2$: 还原型黄素腺嘌呤二核苷酸；
NAD^+: 烟酰胺腺嘌呤二核苷酸：NADH: 烟酰胺腺嘌呤二核苷酸的还原态；$CH_3COSCoA$: 乙酰辅酶 A。

Each acetyl CoA metabolized through the Krebs cycle will yield 12 ATPs, and a total of 96 ATPs can be produced, as follows:

每个乙酰辅酶 A 通过克雷布斯循环产生 12 分子 ATPs，总共产生 96 分子 ATPs，如下：

$$8\text{CH}_3\text{COSCoA} + 16\text{O}_2 + 96\text{H}_3\text{PO}_4 + 96\ \text{ADP} \longrightarrow 8\text{HsCoA} + 96\ \text{ATP} + 104\ \text{H}_2\text{O} + 16\ \text{CO}_2$$

The total ATPs derived from complete metabolism of one molecule of palmitic acid will then be 35 ATPs + 96 ATPs = 131 ATPs. Net production of ATP will be 129 ATPs due to that 2 ATPs were consumed to activate the fatty acid in the cytoplasm.

1 分子棕榈酸完全代谢产生的总 ATPs 将是 35 分子加上 96 分子等于 131 分子的 ATPs。ATP 的净产量将是 129 分子 ATPs，因为在细胞质中激活脂肪酸时消耗 2 分子的 ATPs。

3.5　支撑肌肉活动的代谢途径

3.5　Pathways of Metabolism Supporting Muscular Activity

3.5.1 Glycolysis

Insects metabolize carbohydrates to support energy first upon taking flight. For some insects, such as dipterans and hymenopterans, carbohydrate is the only fuel they can mobilize fast enough to support flight. Other insects metabolize carbohydrate at the initiation of flight, but if flight continues, they switch to another

3.5.1　糖酵解

昆虫在起飞时首先代谢碳水化合物供能。在一些昆虫如双翅目和膜翅目中，碳水化合物是它们仅有的能快速代谢支持飞行的燃料。其他昆虫在飞行起始时代谢碳水化合物，但是如果继续飞行的

fuel such as proline or fatty acids. Glycolysis, the process by which insects star to metabolize glucose, is similar to the process in vertebrate and other organisms, with the exception that glycolysis in insect flight muscle is always aerobic, never anaerobic as in actively working vertebrate muscle. The tracheal supply to insect flight muscle is extensive, and capable of supplying sufficient oxygen for totally aerobic oxidation during flight.

话，它们就切换代谢其他燃料如脯氨酸或脂肪酸。昆虫开始用糖酵解途径代谢葡萄糖，其过程与脊椎动物和其他有机体一样；例外的是昆虫飞行肌的糖酵解总是有氧的，而不像脊椎动物运动活跃的肌肉进行无氧呼吸。昆虫飞行肌的气管供给系统规模庞大，能为它们在飞行中供应足够的氧气进行完全的有氧氧化。

Glycolysis has another specialization in insect flight muscle that NADH in the cytoplasm is oxidized to NAD^+. The quantity of cytoplasmic NAD^+ is limited in the flight muscles of insects, and maintains glycolysis to continue, in addition, NAD^+ must be constantly regenerated through the glycerol-3- phosphate shuttle. Some slower working skeletal muscles in insects, such as leg muscles, may oxidize NADH to NAD^+, however, by the pyruvate to lactate step. Conversion of fructose -6-phosphate to fructose -1, 6-diphosphate is one of the major control points for carbohydrate metabolism in insects, as it is in other organism.

昆虫飞行肌糖酵解另一个特征是将细胞质中的 NADH 氧化成 NAD^+。昆虫飞行肌细胞质中 NAD^+ 的数量有限，为了维持糖酵解过程的 NAD^+，需要通过三磷酸甘油循环不断再生。昆虫中一些在骨骼工作的慢肌，如腿肌，可通过丙酮酸到乳酸的步骤将 NADH 氧化成 NAD^+。与其他有机体一样，6- 磷酸果糖转换成 1,6- 二磷酸果糖是昆虫碳水化合物代谢的主要控制点之一。

3.5.1.1 The glycerol-3-phosphate (G-3-P) shuttle and regeneration of NAD^+

3.5.1.1 三磷酸甘油（G-3-P）循环和 NAD^+ 再生

Fructose-6-diphosphate is split into two three-carbon products, G-3-P and dihydroxyacetone phosphate. These two compounds are interconvertible catalyzing by the enzyme triosephosphate isomerase. The oxidation of G-3-P to 1,3-diphosphoglycerate is very important step because it is dependent on the availability of inorganic phosphate and the oxidized form of nicotinamide adenine dinucleotide (NAD^+).

6- 磷酸果糖分解成两分子三碳产物，三磷酸甘油和磷酸二羟基丙酮。这两种化合物在磷酸丙糖异构酶的催化下相互转化。三磷酸甘油氧化成 1,3- 二磷酸甘油酸异构酯是非常重要的步骤，因为其依赖于无机磷酸盐和烟酰胺腺嘌呤二核苷酸的氧化形式（NAD^+）的可用性。

This cytoplasmic reaction is the first of two-step shuttle for transferring electrons from the cytoplasm to mitochondria. Regeneration of cytoplasmic NAD^+ allows continued oxidation of G-3-P, with substrate–level production of 1 mol of ATP/3-carbon fragment oxidized to pyruvic acid. Pyruvate rapidly enters mitochondria and leads to further oxidation and production of ATP through the tricarboxylic acid cycle.

细胞质反应的第一步是电子从胞质向线粒体传递的两步穿梭。胞质 NAD^+ 的再生使三磷酸甘油持续氧化，以每三个碳的片段作为底物消耗 1 mol ATP 氧化成丙酮酸。丙酮酸盐快速进入线粒体，并通过三羧酸循环进行下一步的氧化和 ATP 的产生。

$$\underset{\text{磷酸二羟基丙酮}}{\text{dihydroxyacetone phosphate}} + \underset{\text{还原型辅酶}}{\text{NADH}} + H^+ \xrightarrow[\text{胞质三磷酸甘油脱氢酶}]{\text{cytoplasmic glycerol-3-phosphate dehydrogenase}} \underset{\text{三磷酸甘油}}{\text{glycerol-3-phosphate}} + \underset{\text{氧化型辅酶}}{NAD^+}$$

As fast as G-3-P is produced, it crosses the outer membrane of light muscle mitochondria, and fat the outer surface of the inner mitochondrial membrane, it is rapidly oxidized to dihydroxyacetone phosphate by a flavine adenine dinucleotide (FAD)-linked mitochondrial G-3-P dehydrogenase bound to the inner membrane, according to the following reaction.

三磷酸甘油一旦产生，就迅速跨越飞行肌线粒体外膜，使线粒体内膜的外表面增大，通过腺嘌呤黄素二核苷酸（FAD）偶联的线粒体三磷酸甘油脱氢酶与内膜结合，根据下述反应被快速氧化成磷酸二羟基丙酮。

$$\text{glycerol-3-phosphate} + \text{FAD} \xrightarrow{\Delta} \text{dihydroxyacetone phosphate} + \text{FADH}_2$$

三磷酸甘油　　　　　　　　　　磷酸二羟基丙酮

FAD. flavine-adenine dinucleotide 黄素腺嘌呤二核苷酸； $FADH_2$. 还原型黄素腺嘌呤二核苷酸；
Δ. mitochondrial glycerol-3-phosphate dehydrogenase 线粒体三磷酸甘油脱氢酶

The flavoprotein accepts the two electrons and two protons, which ultimately get rapid shuttle through the electron transport system in mitochondria to molecular oxygen as the final acceptor. Several evolutionary adaptations have made this shuttle possible, including: ① high activity of the cytoplasmic G-3-P dehydrogenase; ② location of an active G-3-P dehydrogenase on the outer surface of the inner membrane of flight muscle mitochondria; ③ availability of oxygen and a fully functional electron transport system in mitochondria of working muscle. Dihydroxyacetone phosphate at the inner membrane surface rapidly diffuses out of mitochondria into cytoplasm.

黄素蛋白接受的两个电子和两个质子，在线粒体内以最快的速率通过电子传递系统传递给最终受体氧分子。这种穿梭有几种进化的可能：①细胞质三磷酸甘油脱氢酶的高活性；②飞行肌线粒体内膜外表面上活跃的三磷酸甘油脱氢酶的场所；③工作时的肌肉线粒体内有可供利用的氧气和功能完整的电子传递系统。内膜表面的二羟丙酮磷酸快速从线粒体内外散进入细胞质。

3.5.1.2 Significance and control of the G-3-P shuttle

3.5.1.2 三磷酸甘油穿梭的调节及其重要性

The significance of G-3-P shuttle hinges on the assumption that the shuttle is self-generating when catalytic amounts of dihydroxyaceteone phosphate (DHAP) are introduced. A small number of DHAP molecules converted to G-3-P in the cytoplasm, with subsequent conversion of G-3-P to DHAP in the membranes of mitochondria, may cycle over and over during flight and keep high level of NAD^+ in the cytoplasm. This would allow almost all the DHAP produced from the splitting of the fructose-1,6-diphohate to be converted to G-3-P, and ultimately converted to pyruvate. In this scenario, all the initial glucose can be converted to two pyruvate molecules that enter the tricarboxylic acid cycle; thus, ATP production in glycolysis and in the Krebs cycle is maximized.

三磷酸甘油穿梭的意义在于当一定催化数量的磷酸二羟基丙酮（DHAP）被引入时对其自生的影响。少量的磷酸二羟基丙酮分子在胞质中转化成三磷酸甘油，随后在线粒体膜内将三磷酸甘油转化成磷酸二羟基丙酮，在飞行期间不断循环，从而保持细胞质中 NAD^+ 的高水平。这将使几乎所有由 1,6- 磷酸果糖分解而来的磷酸二羟基丙酮转化成三磷酸甘油，并最终转化成丙酮酸盐。在这种情况下，所有最初的葡萄糖都转化成两分子丙酮酸盐分子进入三羧酸循环。因此，在糖酵解和克雷布斯循环中最大化的产生 ATP。

3.5.1.3 Control of Krebs cycle metabolism and regulation of carbohydrate metabolism in flight muscles

3.5.1.3 飞行肌中克雷布斯循环代谢和碳水化合物代谢的调节

Control points in the Krebs cycle enable insects to rapidly

克雷布斯循环的控制点使昆虫在飞

increase metabolic activity upon taking flight. Isocitrate dehydrogenase, an NAD-linked enzyme, is a major control point. It is inhibited by high levels of ATP and stimulated by isocitrate, ADP, and inorganic PO_4. The latter two compounds accumulate from the use of ATP to initiate and support flight. The concentrations and relative ratios of ATP, ADP, and adenosine monophosphate (AMP) play an important role in regulating metabolism in insects, as they do in other organisms. Upon initiation of flight, the level of ATP falls rapidly, while concentrations of ADP, AMP, and inorganic PO_4 increase.

行时可以迅速地增加代谢活力。异柠檬酸脱氢酶作为一种 NAD 连接酶，是一个主要的控制点。异柠檬酸脱氢酶活性可被高水平 ATP 抑制，而被异柠檬酸、ADP 和无机磷盐促进。后两种化合物在利用 ATP 启动和支持飞行时进行累积。ATP、ADP 和单磷酸腺苷（AMP）的浓度和相对比例在调节昆虫代谢中发挥重要作用，这和其他有机体一样。在飞行起始阶段，ATP 水平快速下降，而 ADP、单磷酸腺苷和无机磷酸盐的浓度增加。

3.5.2 The electron transport system

The electron transport system in insect mitochondria is similar to that in other animals. The components if the respiratory chain are arranged in a sequence on the inner mitochondrial membrane so that electrons flow down the chain (Fig.3-9). Flight muscle mitochondria are not permeable to NAD^+ and NADH, so cytoplasmic nucleotides do not readily enter mitochondria, nor do those inside the mitochondria pass outside. The insect mitochondria have nonheme iron containing coenzyme Q between the flavoprotein and cytochrome b in the chain sequence. Coenzyme Q has an isoprene side chain of varying length in different insects. Water formed through transfer of substrate electrons and protons to oxygen is called metabolic water. It is a very important source of water for all insects, and especially those that live in very dry environments.

The chemiosmotic hypothesis was proposed for the details of ATP production in insects. In the chemiosmotic process, energy

3.5.2 电子传递系统

昆虫线粒体的电子传递系统与其他动物相似。当呼吸链的化合物在线粒体内膜上依次排列时，电子就沿着该链传递（图 3-9）。飞行肌线粒体对 NAD^+ 和 NADH 是不具有渗透性的，所以细胞质核苷酸就不容易进入线粒体，线粒体内的也无法渗透出去。昆虫线粒体中含辅酶 Q 的非血红素铁，其存在于顺序链的黄素蛋白和细胞色素 b 之间。不同昆虫的辅酶 Q 有一个不同长度的异戊二烯侧链。通过底物电子和质子传递给氧产生的水称为代谢水。代谢水对所有昆虫来说是一个重要的水源，特别是那些生活在特别干燥环境中的昆虫。

关于昆虫 ATP 产生的细节，学者提出了化学渗透学说。在化学渗透过程

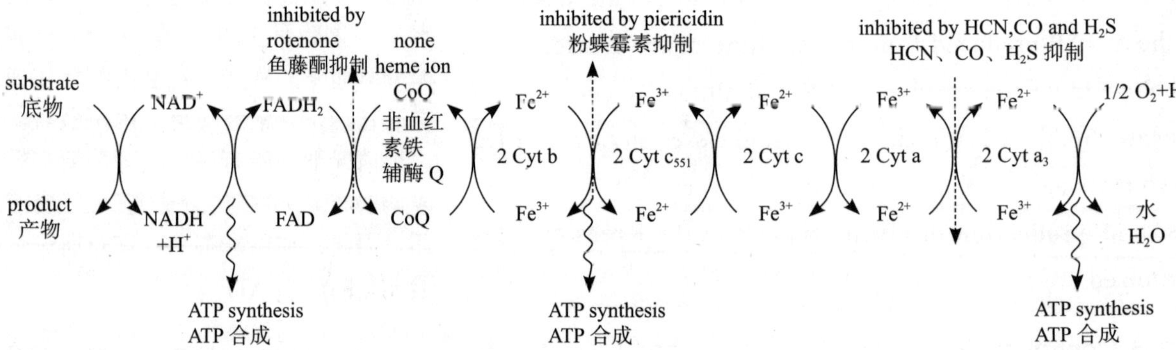

Fig.3-9 The electron transport system in insect mitochondria (modified from Wang Yingchang, 2006； Nation, 2022).
图 3-9 昆虫线粒体的电子传递系统（仿王荫长，2006; Nation, 2022）。

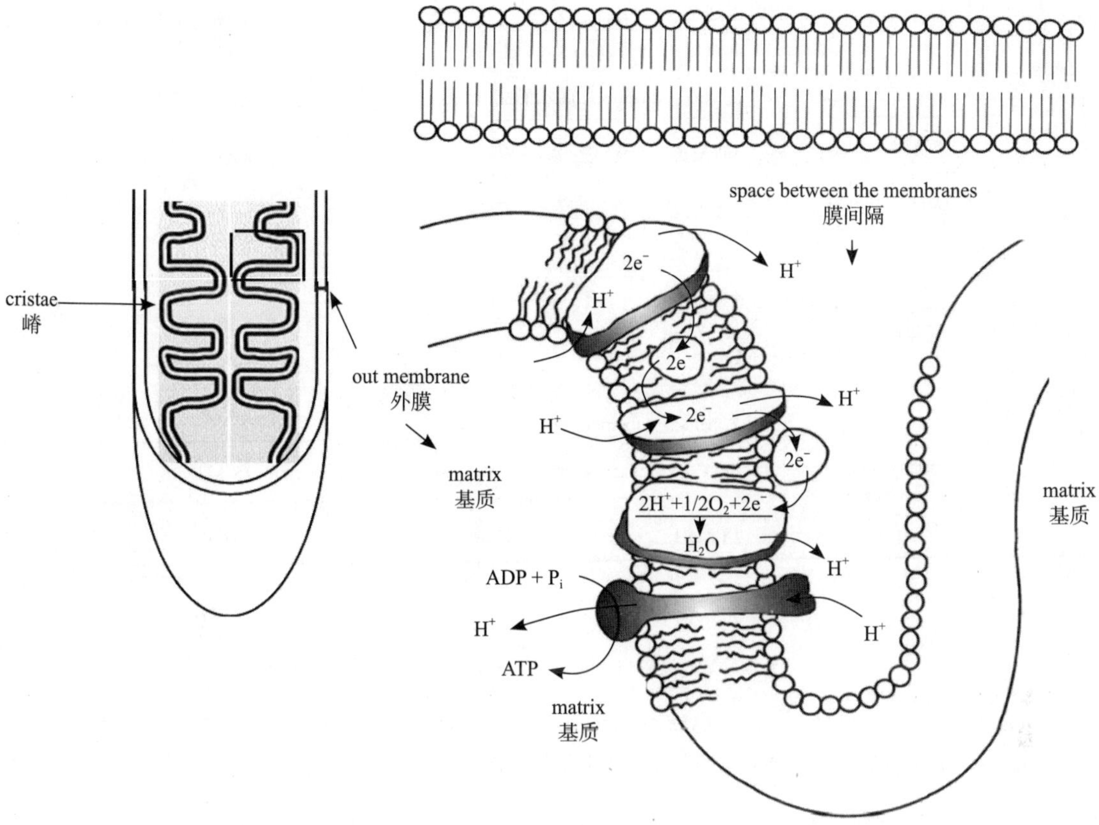

Fig.3-10 An illustration of a typical cristae of flight muscle mitochondria and the steps in transfer of electrons down the electron transfer chain. The pumping of protons into the space between the membranes, and the passage of protons back to the matrix through the ATP synthetase complex causing the formation of ATP (modified from Nation, 2022, integrated).
图 3-10 飞行肌典型的线粒体嵴和电子沿传递链传递的步骤。质子泵将质于泵入膜间隙，通过负责产生 ATP 的 ATP 合成酶复合体回传到基质（仿 Nation, 2022，综合图）。

from the transfer of electrons is used to pump protons into the space between the inner and outer membranes of mitochondria (Fig. 3-10). Mitochondria utilize the potential difference between the membranes to supply the energy for ATP synthesis from ADP as the protons pass down the chemiosmotic gradient by returning to the inner compartment through F1 membrane complexes, also called F1-ATP synthase complex. The complex is a membrane ion channel composed of proteins that allow the passage of protons, and ATP synthesizing enzymes are part of the complex. The energy yield from metabolism of 1 mol of glucose is as much as 36 moles of ATP. If glycogen is the source of metabolism, the number of ATP could be higher due to that one less ATP is needed to phosphorylate the glucose-1-phosphate initially resulting from the cleavage of one glucose unit from a glycogen molecule (Fig. 3-11).

中，电子传递的能量被用于将质子泵入线粒体内外膜之间的间隙（图 3-10）。线粒体利用膜间电势差，在质子沿化学梯度下传回到膜内并通过 F1 膜复合体，也称为 F1-ATP 合酶复合体，利用 ADP 合成 ATP 时供给能量。复合体是蛋白质组成的膜离子通道，使质子经过，ATP 合成酶是复合体的一部分。1 mol 葡萄糖代谢产生的能量多至 36 mol ATP。如果代谢源是糖原，ATP 的数量更高，因为从糖原分子上切出 1 分子葡萄糖（图 3-11）磷酸化形成葡萄糖 -1- 磷酸少用 1 个 ATP。

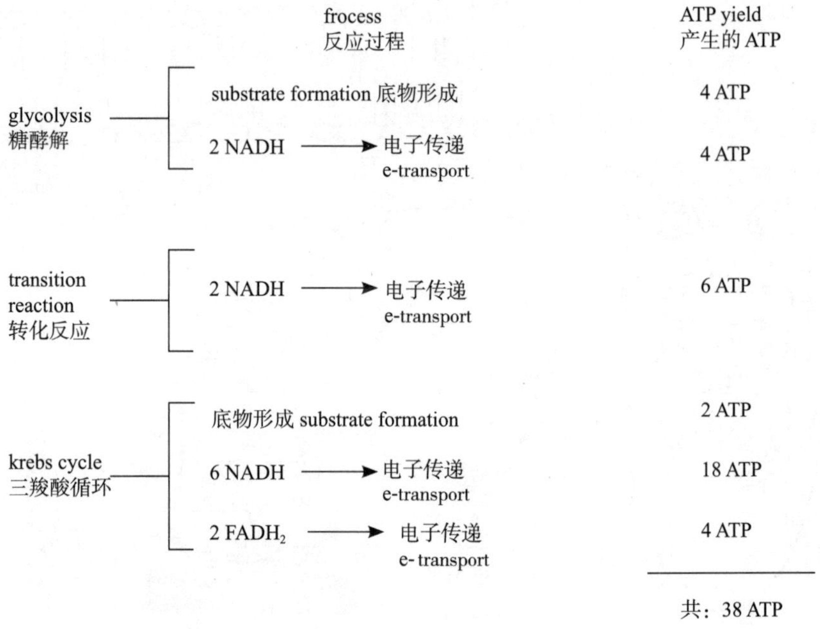

减去糖酵解起始所消耗的 2 分子 ATPs，净得 36 分子的 ATPs

Fig.3-11 Summary of the energy yield in the several stages of glucose metabolism in flight muscle

图 3-11 昆虫飞行肌糖代谢几个步骤中产生能量的概括。

3.5.3 Proline as an energy fuel of flight

Some blood-feeding insects such as tsetse fly, adults of the Colorado potato beetle, *L. decemlineata*, and some beetles in the family Scarabaeidae, use the amino acid proline as their main fuel even though most animals avoid relying on proteins (Teulier et al., 2016). Alanine is increased when proline is metabolized for energy in insects including *M. melolontha*, *H. dilloni*, and *Popillia japonica*, and some beetles in the family Cerambycidae. In *Phryneta spinator*, about 50% of the carbohydrates and 40% of the proline in the hemolymph were metabolized to support 5 minutes of the transamination reaction in which the amino group from glutamic acid is transferred to pyruvic acid, creating alanine and α-ketoglutarate as follows:

3.5.3 脯氨酸作为飞行的能量燃料

尽管大多数动物尽量避免依赖蛋白质，但一些吸血性昆虫如采采蝇、科罗拉多马铃薯甲虫成虫和金龟科的一些甲虫则利用脯氨酸作为主要的能量燃料（Teulier 等，2016 年）。在欧洲蔬菜大金龟子、巨型蜣螂、日本金龟子和天牛科的一些甲虫中，当脯氨酸被代谢用以供能时丙氨酸会相应地增加。在蛀干害虫 *P. spinator* 成虫的血液中，50% 的碳水化合物和 40% 脯氨酸被代谢用于支持持续 5 min 的转氨作用，在此期间，谷氨酸的氨基被转移至丙酮酸上，产生丙氨酸和 α- 酮戊二酸。反应式如下：

$$\underset{\text{谷氨酸}}{\text{glutamic acid}} + \underset{\text{丙酮酸}}{\text{pyruvic acid}} \longrightarrow \underset{\alpha\text{-酮戊二酸}}{\alpha\text{-ketoglutarate}} + \underset{\text{丙氨酸}}{\text{alanine}}$$

Complete metabolism of 1 mol proline releases 14 moles ATP in mitochondria. Thus, flight and proline metabolism are linked to and depend on a rich supply of oxygen to flight muscles. One of the products from proline metabolism, alanine, is rapidly removed

在线粒体中，每摩尔脯氨酸完全代谢释放 14 mol 的 ATP。因此，飞行和脯氨酸代谢是相互联系的，并依赖于飞行肌的充分供氧。脯氨酸代谢产物之一

from the muscle and transported to fat body, where it is converted into proline again by additional of a 2-carbon unit derived from fatty acids. It can then be transported to the muscles to repeat the proline cycle reaction. Thus, the proline pathway represents a shuttle to transfer 2-carbon units from the fat body to the muscles for metabolism. The sum of the reactions is as follows:

的丙氨酸被快速从肌肉中移除并转运到脂肪体，在此加上从脂肪酸衍生而来的另外一个 2- 碳单位，再转换成脯氨酸。随后，脯氨酸被转运到肌肉，重复循环之前的反应。由此，脯氨酸途径代表了 2-碳单位从脂肪体到肌肉的穿梭代谢。总反应如下：

$$\text{proline} \longrightarrow \text{alanine} + 2\ CO_2 + 3\ NADH + 2\ FPH_2 + 1\ GTP$$

脯氨酸

In addition, under extreme energy-deficient conditions, such as prolonged starvation, proteins serve as an important source of energy. During these conditions, amino acids are converted into pyruvate, acetyl-CoA, or TCA cycle intermediates and then are further metabolized to generate ATP (Chatterjee1 et al., 2021). Proline metabolism is under the control of the cellular level of ADP. An increase in ADP stimulates oxidation of proline, while glutamate resulting from oxidation inhibits through a feedback mechanism.

(Tian ling)

此外，在能量匮乏的情况下，如长期的饥饿等，蛋白质可以作为能量的重要来源发挥作用。在这些情况下，氨基酸被转换成丙酮酸盐、乙酰辅酶 A 或者三羧酸循环的中间产物，从而进一步被代谢产生 ATP（Chatterjee1 等，2021）。脯氨酸的代谢受到细胞中 ADP 浓度的调控。ADP 浓度的增加促进脯氨酸的氧化反应，而氧化反应产生的谷氨酸盐则反馈抑制其氧化反应。

（田 铃）

Exercises

1. What are the function and the main cell types of fat body?
2. Which hormones are involved in regulating metabolism in insects? What are their characteristics?
3. What are the main sources of energy and the traits of their metabolism in insects?

思考题

1. 昆虫脂肪体的主要功能及其主要细胞种类有哪些？
2. 调控昆虫代谢发生的激素都有哪些？它们的功能特点是什么？
3. 昆虫能量来源的主要物质都有哪些？它们的代谢有哪些特点？

第 4 章 排泄系统

Chapter 4 Excretory System

Key points

1.In the metabolic process of insects, the two major systems that complete excretion are Malpighian tubules and hindgut. Malpighian tubules are classified into basal segment, middle segment and distal segment according to their functions, including principal cells and stellate cells. The hindgut includes the ileum, rectum and cryptonephridial system.

2.Excretion can be either storage or deposite that involves reabsorption of nutrition and removing harmful substances ammonia, uric acid, and toxic metabolites. It required homoeostasis in electrolytes, water, acid-base, nitrogen, and even small molecular inorganic ions.

3.Excretion is under the control of hormones and some related peptides.

知识点

1. 在昆虫的新陈代谢过程中，完成排泄的两个主要系统是马氏管和后肠。根据功能的不同，马氏管分为基段、中段和端段，包括主细胞和星状细胞。后肠包括回肠、直肠和隐肾系统。

2. 排泄可以是贮存或沉积，参与营养的再吸收，去除有害物质氨、尿酸和有毒代谢物。它要求电解质、水、酸碱、氮，甚至小分子无机离子中都是处于稳态的。

3. 排泄受激素和一些相关肽的控制。

Excretion can be defined broadly as any process that eliminates the interaction of harmful substances with cells and tissues. Even useful substances, such as glucose, amino acids, and certain ions, can be harmful if present in excess amounts. Nitrogenous metabolites, ions, water, and ingested chemicals are substances that an insect may need to excrete. Allelochemicals from eaten plant tissues may be excreted from the body or stored in some inert location in the body. Storage excretion includes materials with future potential use, such as glycogen as a storage form of glucose or amino acids stored as proteins. Deposit excretion as waste material of no further use that needed to be removed from harmful interaction with the tissues.

In insects, both the Malpighian tubules and the hindgut function together as excretory organs. The Malpighian tubules

排泄广义定义为排除有毒物质与细胞和组织相互作用的任一过程。即使有用的物质，如葡萄糖、氨基酸以及某些离子如果过量也是有害的。含氮代谢物、离子、水和吃下的化学物质是昆虫必须排出的物质。食下的植物组织而来的异种化感物可由躯体排泄或贮存在体内某个惰性场所。贮存排泄包括将来有潜在用途的物质，如糖原作为葡萄糖的贮存形式或氨基酸以蛋白质形式贮存。沉积物排泄是再也没有用途的废弃物质，必须从有毒物质与组织的互作中移除。

昆虫马氏管和后肠都是排泄功能器

collect a filtrate from the hemolymph, and pass this primary urine to the hindgut. Additional components are secreted into the excreta by the hindgut, and some substances are reabsorbed into the hemolymph. Excreta describes material actually eliminated from the anus by insects. The hindgut consists of the pylorus, ileum, colon and rectum, which then opens to the anus.

官。马氏管收集血淋巴滤出液，将此原尿向后肠输送。后肠释放其他成分进入排泄物，有些物质被再吸收进血淋巴。排泄物指从昆虫肛门实际清除的物质。后肠由幽门、回肠、结肠和直肠组成，然后开放到肛门。

4.1 排泄系统的结构

4.1 The Structure of the Excretory System

4.1.1 Malpighian tubules

4.1.1.1 The morphology of Malpighian tubules

Malpighian tubules, the first of the two systems involved in excretion, are long, tubular structures which derived from endoblast cells, usually arising at the junction of the mid- and hindgut and terminating blindly in the hemocoel.

The tubules vary in number from 2 to more than 100 in various insect species. Collembola, Aphidae (Hemiptera), and some Thysanura lack Malpighian tubules altogether, and other cells and glands take over the functions of excretion. In some insects of the Lepidoptera and Coleoptera, the distal ends of the tubules are embedded in the wall of the rectum, called cryptonephridial tubules, appear to be a modification that aids water conservation. Tracheal connection to Malpigihian tubules is numerous, and are indicative of a high metabolic demand for oxygen. A small spiral muscle frequently runs along the surface of a tubule, promoting coiling movements that assist proximal flow of fluid and increase hemolymph in contact with the tubules, but in Thysanura, Dermaptera and Thysanoptera, there are no muscles, and the tubules are quiet.

A single layer of epithelial cells surrounds the lumen of a tubule. Several different cell types have been identified. Some tubule cells have a brush border of microvilli on the apical surface have been called principal tubule cells. They are involved in formation of the primary urine. It is usually called basal segment which arising from the junction. Cells in the proximal half of the tubules have a

4.1.1 马氏管

4.1.1.1 马氏管的形态

马氏管是涉及排泄的两个系统之一，为长管状结构，由内胚层衍生而来，通常起源于中、后肠接界处，以盲管终于血腔。

马氏管的数量因昆虫种的不同而有变化，从 2 条到 100 多条。弹尾目、蚜虫（半翅目）和一些缨尾亚目都缺乏马氏管，由其他细胞和腺体承担排泄功能。鳞翅目和鞘翅目的一些昆虫，马氏管的末梢掩埋在直肠壁，称为隐肾管，想必是有助于保持水分的进化类型。气管与马氏管的连接数量众多，表明对氧气的高代谢需求。常有一条螺旋形肌肉沿管表面蔓生，促进螺旋形蠕动有助于最快的液流，并增加血淋巴与马氏管的接触，但是缨尾亚目、革翅目和缨尾目的马氏管上因没有肌肉而相对静止。

单层上皮细胞围绕着管腔。目前，已经鉴定出几种不同的细胞类型。有些管壁细胞顶端有刷状微绒毛，称为马氏管主细胞，参与原尿的形成。起始于连接处的常称为基段。近末端一半的管壁

relatively smooth apical surface over which reabsorption occurs, probably by energy-requiring mechanisms. The free segment suspended in the hemolymph is usually called distal segment. The transitional segment is the region between basal and distal segment.

细胞顶端相对平滑，也许在这些细胞表面通过能量需求机制进行再吸收。悬浮在血淋巴中的游离段常称为端段。基段和端段之间的区域是过渡段。

4.1.1.2 Ultrastructure of Malpighian tubules

Tubule cells are thin, sheet-like cells that wrap around the tubule lumen. Water and hemolymph components entering at the basal side of the cells have only a short distance to traverse to reach the apical surface where they may be secreted, or diffuse, into the tubule lumen.

Malpighian tubule cells are characterized by extensive infoldings of the basal membrane, creating many twisting channels that reach 5 to 10μm or more into the cell (Fig.4-1). Potassium ions in the hemolymph enter Malpighian tubule cells through potassium ion gates in these enfolded channels in the basal membrane, and water and dissolved substances in the aqueous medium of the hemolymph follow the osmotic gradient. Extensive membrane surfaces presented by both the basal and apical faces of Malpighian tubule cells and large mitochondria are indicative of specialization for active transport as well as passive diffusion.

Malpighian tubules are not only important in excretion. They have many functions in insects, including detoxication, metabolic function, a role in immunity, lubrication for ecdysis, and even serve for organotypic study of humengenes.

4.1.1.2 马氏管的超微结构

管壁细胞是包裹围绕管腔的单层薄细胞。水分和血淋巴成分从细胞底部进入，仅穿越一段很短的距离就到达顶部，从此处它们被分泌、扩散进管腔。

马氏管细胞的特征是底膜上有大量内褶，向细胞内产生许多弯曲的通道，达 5 ～ 10 μm 或更大（图 4-1）。血淋巴中的钾离子通过底膜上内这些钾离子门控内褶通道进入马氏管细胞，水分和血淋巴中溶于水相介质中的可溶性物质沿渗透梯度进入。马氏管细胞底部和顶部大量膜表面及大型线粒体象征着主动运输和被动扩散一样活跃。

马氏管不仅在排泄过程中很重要。在昆虫中，它们有许多功能，包括解毒、代谢作用、免疫、润滑蜕皮，甚至用于人类基因组的器官型研究。

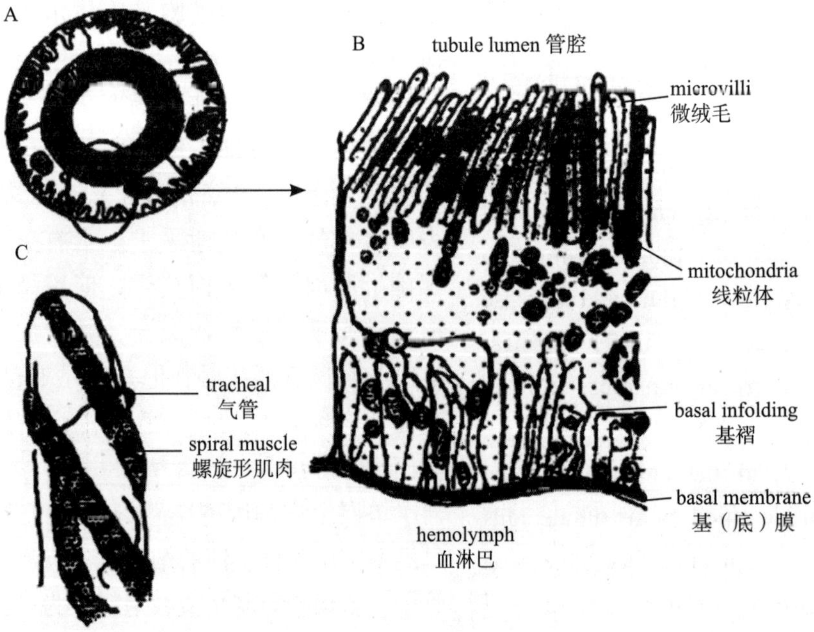

Fig. 4-1 The ultrastructure of Malpighian tubule (modified from Wigglesworth, 1965).
A.cross-section; B.ultrastructure of tubule cell wall; C.spiral muscle and tracheal around tubule.

图 4-1 马氏管的超微结构（仿 Wiggles-worth, 1965）。
A. 横切面；B. 管壁细胞超微结构；C. 管外围螺旋形肌肉和气管。

4.1.1.3 The stellate cells

The Malpighian tubules of *D.hydei* and *D. melanogaster* larvae are composed of two types of cells, principal cells and stellate cells. In the anterior larval Malpighian tubules approximately 26% and 18%, respectively, of all cells are stellate cells. In the larvae of *D. melanogaster*, the stellate cells are fenestrated and the hemolymph space and tubule lumen are separated only by the basal lamina. The principal cells of the distal segment are carbonic anhydrase positive indicating transport activity, whereas the stellate cells lack this enzyme. In the stellate cells of the transitional segment, the sodium content is strikingly high in comparison to the neighboring principal cells and lumen where no sodium was detected. This finding indicates that stellate cells reabsorb sodium. The Malpighian tubules of several insect orders contain not just a mitochondrion-rich principal cell expressing high levels of V-ATPase, but a smaller, intercalated stellate cell. Recent data show that this cell type plays a key role in control of chloride and water flux across the tissue, but also may play other, still unsuspected dynamic roles. In the fruit fly *D. melanogaster*, the main route for chloride transport is transcellular through stellate cells. In both the fruit fly and the mosquito *A. stephensi* the kinin receptor has been immunolocalized in stellate cells, where it regulates transepithelial chloride permeability. Unlike the dipteran Malpighian tubules, which have both primary and stellate cells, each region of the *A.domesticus* Malpighian tubule consists of a morphologically uniform cell type. The mid and distal regions are both secretory in function and increase secretion rate in response to dibutyryl cAMP.

4.1.1.3 星细胞

海德氏果蝇和黑腹果蝇的幼虫马氏管由两类细胞组成，主细胞和星细胞。其幼虫的马氏管前段，星细胞约分别占 26% 和 18%。黑腹果蝇幼虫的星细胞呈网状，血腔和管腔仅由基部薄层分开。端段主细胞对碳酸酐酶呈阳性，标志着运送活性，而星细胞缺少这种酶。在过渡段星细胞中，钠含量与检测不到钠的邻近主细胞和管腔相比高得惊人。这个发现表明星细胞会重新吸收钠。几个目的昆虫马氏管不仅仅是含有线粒体丰富的主细胞，表达高水平的 V-ATPase，而且也夹杂有星细胞。最近的研究证实这类细胞不仅在控制跨越组织的氯和水分流量中起着重要作用，而且有其他静态动力学的作用。黑腹果蝇的氯运送主要途径是经过星细胞的跨细胞转运。果蝇和蚊子的星细胞激肽受体免疫定位于星细胞，是其调节经上皮氯通透性的场所。与有主细胞和星细胞的双翅目马氏管不同，家蟋蟀马氏管每个区域都有形态一致的细胞类型组成。中段和端段都有分泌功能，对双丁酰环磷酸腺苷的反应是增加分泌速率。

4.1.2 The hindgut

The hindgut of many insects consists of three major regions, termed the pylorus, ileum, and rectum. Each region contains a single layer of distinctly different epithelial cell types that contact the intestinal lumen, which are surrounded by circular muscle fibers. The hindgut is the second system that completes the excretion process by selectively reabsorbing some substances into the hemolymph, leaving others in the lumen, and actively secreting some substances into the hindgut lumen. The rectal cuticular lining has greater permeability than the cuticular lining on foregut cells, and the epithelial cells of the hindgut are specialized for both active secretion and active reabsorption.

In *B.mori*, when staining the hindgut with the mixture of dopamine or *L*-DOPA and ethanol, a pair of tubes extended from

4.1.2 后肠

许多昆虫的后肠由三个主要区域组成，称为幽门、回肠和直肠。每个区域包含一层与肠腔接触的明显不同的上皮细胞类型，肠腔被圆形肌纤维包围。后肠是完成代谢过程的第二个系统，通过选择性再吸收一些物质进入血液，留下其他物质在肠腔，并活跃地分泌一些物质进入后肠腔。直肠表皮膜比前肠细胞上的表皮膜渗透性更大，后肠的上皮细胞是专门主动分泌和主动吸收的细胞。

当用多巴胺或左旋多巴和乙醇的混合物对家蚕后肠进行染色时，出现了一

the hindgut connected to the Malpighian tubule was presented. These two tubes became melanized together with the hindgut but not the neighboring Malpighian tubule after stained. So called the PPO-positive tubes show resemblance to a "T" like tissue with the neighboring Malpighian tubule when separated from the hindgut. Inside the PPO-positive tubes, there is a thick lay of peritrophic membrane. Moreover, the PPO-positive tubes may act as a "doorkeeper" to detoxify the metabolic wastes collected by the Malpighian tubules from the hemolymph.

对从后肠延伸到马氏管的管。这两个管染色后与后肠一起变黑，但与之相邻的马氏管没有变黑。当与后肠分离时，所谓的PPO阳性管显示出与邻近的马氏管外观形成类似“T”的组织。在PPO阳性管内，有一层厚厚的围食膜。此外，PPO阳性管可能充当“守门人”，对马氏管从血淋巴收集的代谢废物进行解毒。

4.1.2.1 The pylorus

The pylorus is a gut "gatekeeper" and immune signaling hub. It functions as a contractile sphincter that connects the midgut and hindgut. Contraction of the pylorus is controlled by the hindgut-expressed neuropeptide proctolin. Both muscle and epithelial cells of the pylorus are heavily innervated by sensory and efferent neurons from both the peripheral and central nervous system. This innervation may enable the pylorus to function as an intestinal checkpoint for further passage of gut contents. In addition to functioning as a gut valve, the pylorus is also an important zone of interaction between the *Drosophila* host environment and its microbiota, both symbiotic and pathogenic. Following feeding of silkworms with pathogenic bacteria, prophenoloxidase, a component of the melanization process, is activated specifically in the feces when passing through the hindgut pylorus. Further, feeding *Drosophila*, silkworms, or cotton bollworms with toxic plant phenolic compounds activates a melanization response in the hindgut. This pyloric melanization response is thought to be a last chance for the infected host to clear bacteria or toxic substances before excretion.

4.1.2.1 幽门

幽门是肠道的“门卫”和免疫信号中枢。幽门起着连接中肠和后肠的收缩括约肌的作用。幽门的收缩由后肠表达的神经肽直肠肽控制。幽门的肌肉和上皮细胞都受到周围神经系统和中枢神经系统的感觉和传出神经元的强烈支配。这种神经支配可能使幽门成为肠道内容物进一步通过的肠道关卡。除了作为肠道瓣膜的功能外，幽门也是果蝇宿主环境与其共生和致病微生物群之间相互作用的重要区域。用致病菌喂养蚕后，黑色素化过程中的一个组成部分——酚氧化酶，在通过后肠幽门时在粪便中被特异性激活。此外，用有毒植物酚类化合物喂养果蝇、蚕或棉铃虫激活了后肠的黑化反应。这种幽门黑化反应被认为是感染宿主在排泄前清除细菌或有毒物质的最后机会。

4.1.2.2 The ileum

The ileum is the most anterior part of the hindgut, occurring just posterior to the origin of the Malpighian tubules in most insects. In the desert locust, the ileum is a major site for isosmotic fluid reabsorption, for active Na^+ and Cl^- reabsorption, and for active secretion of proline as an energy source to support metabolic processes. In many wood-eating insects, the ileum is dilated to form a "fermentation pouch"housing bacteria or protozoa which digest wood particles. The productions of digestion, when liberated by microorganism, are absorbed across the wall of the ileum.

The ileum epithelium in *A. gemmatalis* has acidophilic flattened polygonal epithelial cells with elongated nuclei and apical

4.1.2.2 回肠

回肠是后肠最前面的部分，多数昆虫出现在马氏管源头的正后方。沙漠蝗的回肠是等渗体液再吸收的主要部位，活跃地再吸收 Na^+ 和 Cl^-，并活跃地释放脯氨酸作为能源支持代谢过程。许多取食木材的昆虫在回肠膨大形成“发酵袋，”容纳消化木料颗粒的细菌和原生动物。消化产物被微生物释放后，通过回肠壁再吸收。

大豆夜蛾的回肠上皮有嗜酸的扁平多角形细胞，核伸长，胞质顶端有刷状

cytoplasm exhibit brush border-like structure lined by cuticular intima. The brush border structure represents the apical plasma membrane invaginations associated with mitochondria described under transmission electron microscopy in the ileum cell of *S. gregaria*, which allows the water reabsorption that occurs in this region. This absorption is necessary to maintain the insect metabolism as well to promote the feces dehydration that starts into the ileum. The outer circular muscles are more developed than the inner longitudinal fibers; similar aspects were observed in Lepidoptera species and in some Coleoptera.

The epithelium of the ileum is a single layer of large polyploid enterocytes. Underneath an apical cuticle, these enterocytes contain long, microvillar-like, apical plasma membrane infoldings that are closely associated with mitochondria. These infoldings are important for increasing surface area available for reabsorption. In the ileum and rectum, selective reabsorption or secretion occurs to maintain ion and water homeostasis. Major resorbed ions include Na^+, Cl^-, and K^+. Ion reabsorption in the ileum is under antidiuretic hormonal control, principally by the Cl^- transporting neuropeptide and ion transport peptide.

结构，内衬表皮内膜。沙漠蝗的回肠细胞在透射电镜下，其刷状结构表示质膜顶部的内陷与线粒体相连，允许这个区域进行水分的再吸收。这种吸收不仅对维持水分很有必要，而且可促使进入回肠的粪便进行脱水干燥。外层的环肌比内层纵肌纤维更发达；这在鳞翅目和一些鞘翅目昆虫中可观察到相同的样貌。

回肠上皮是单层大型多倍体肠上皮细胞。在顶端表皮下面，这些肠上皮细胞含有与线粒体密切相关的、长的微绒毛状顶端质膜内褶。这些内褶对于增加可用于重吸收的表面积非常重要。在回肠和直肠中，发生选择性重吸收或分泌以维持离子和水的动态平衡。再吸收的主要离子包括 Na^+、Cl^- 和 K^+。回肠中的离子再吸收受抗利尿激素控制，主要由 Cl^- 转运神经肽和离子转运肽控制。

4.1.2.3 The rectum

In the rectum, small groups of cells are variously called rectal cells, rectal pad cells, or rectal papillae cells in different cells. These groups of cells have special modifications for reabsorption. In Diptera, four to six finger-like papillae are attached to the wall of the rectum and project into the rectal lumen. The chitinous lining on the luminal surface of the papillae is continuous with the lining on the inner wall of the rectum. The cells of a rectal papillae are large, usually cuboidal cells that surround a central channel in the papillae that opens into the hemolymph space through a valve. Fluid that crosses the rectal papillae cells and enters the central channel is returned to the hemolymph. A small trachea trunk and a nerve pass into the central cavity, and the tracheal trunk branches into many finer tracheae and tracheoles, suggesting a high demand for oxygen for performance of metabolic and secretory work. The apical cell membrane of rectal papillae cells also is greatly enfolded, and large mitochondria are usually associated closely with the intracellular channels created by the extensive membrane folding (Fig. 4-2).

Rectal pad cells are common in many insects, and typically are enlarged, columnar to cuboidal cells arranged in six clusters separately by smaller, squamous cells between the "pads"of absorptive cells. The rectal pad cells in the cockroach, *P.ameicana*,

4.1.2.3　直肠

在直肠中，不同细胞中的一群细胞被称为直肠细胞、直肠垫细胞或直肠乳突细胞。这些细胞群对重吸收有特殊的改进。双翅目有 4～6 个指形乳突与直肠壁相连，伸入直肠腔。乳突腔表面的几丁质膜与直肠内壁的膜连续。直肠乳突的细胞大，呈立方形，在乳突中围成一个中心通道，通过瓣膜开口于血腔。穿越直肠乳突细胞并进入中心通道的液流被返回到血淋巴。一条小气管干和神经进入中心腔，气管干再分枝成许多更细的气管和微气管，这表明进行代谢和分泌活动需氧量很高。直肠乳突细胞的顶端细胞膜也折叠很大，大型线粒体常与胞间通道密切相连；这些通道由大量膜褶形成（图 4-2）。

直肠垫细胞在许多昆虫中是普遍存在的，一般是扩大的、圆柱形到立方形细胞排列成六簇，被有吸收能力的细胞“垫”之间的小鳞片状细胞隔开。美

have highly folded cell membranes at the surface of the rectal lumen that present 10 to 20 times the surface area of smooth membranes. Mitochondria are located near and within the apical folds, and often occur in compact stacks in conjunction with the highly enfolded lateral membranes. The extensive infoldings of membranes create many intracellular channels that collect fluid from the rectum and direct it toward channels between adjacent cells (intracellular channels). The intracellular channels lead to the basal membrane of pad cells where water and useful solutes reenter the hemolymph. The membranes of the intracellular channels are straight and smooth, thus, they present relatively little surface for back diffusion into the rectal pad cells (Fig. 4-3).

洲大蠊的直肠垫细胞，在直肠腔表面有高度折叠的细胞膜，比平滑的膜表面积多 10 ～ 20 倍。线粒体位于顶部的折叠附近和折叠内，经常出现在密堆积与高度折叠的侧膜结合部。大量的膜折叠产生许多胞间通道，可收集直肠液并引导其流向邻近细胞间的通道（胞间通道）。胞间通道通向垫细胞的基膜，水分和可用溶质再进入血淋巴。胞间通道的膜笔直平滑，因此，回流扩散进入直肠垫细胞的面积相对较小（图 4-3）。

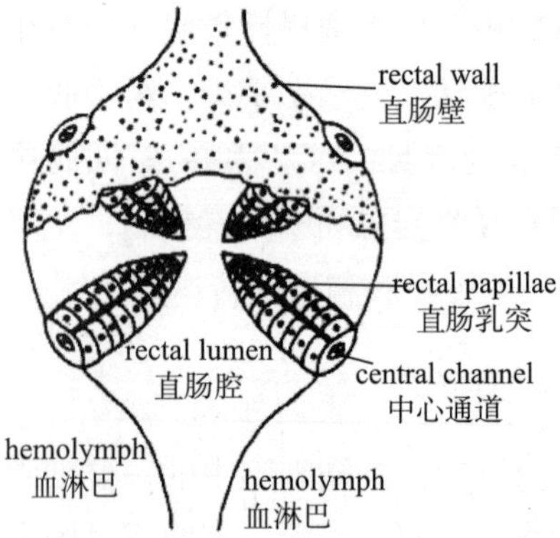

Fig.4-2 The structure of Dipteran larval rectal paillae (modified from Nation, 2022).

图 4-2 双翅目幼虫直肠乳突结构（仿 Nation, 2022）。

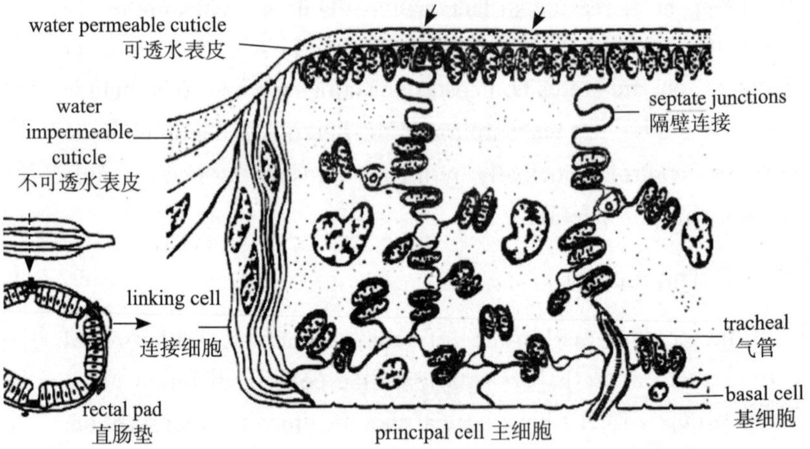

Fig.4-3 The structure of rectal pad of cockroach (arrows indicate the water permeable site; modified from Noirot et al., 1976).

图 4-3 蜚蠊直肠垫的结构（箭头示水透入处；仿 Noirot 等，1976）。

The rectum is the final site of reabsorption, and the site of some of the most elaborate cell membrane networks documented anywhere in nature. Insect papillar enterocytes display heavily folded regions of lateral membrane stacks with tightly associated mitochondria. These stacks greatly increase basolateral membrane surface area available for ion transporter localization and function, with the neighboring mitochondria providing energy for active ion transport. Ions destined for reabsorption into the hemolymph would be absorbed from the gut lumen, and then transported through the papillar membrane stacks into an intermembrane space that ultimately leads to the central canal and hemolymph. The Na^+/K^+ ATPase (P-ATPase) and V-ATPase are required for K^+ transport in

直肠是再吸收的最终场所，是自然界中被记录的一些最复杂的细胞膜网络的场所。昆虫直肠乳突细胞显示出侧膜叠层的高度折叠区域，这与线粒体紧密相关。这些叠层极大地增加了可用于离子转运蛋白定位和功能的基底外侧膜的表面积，相邻的线粒体为活跃离子转运提供能量。用于再吸收到血淋巴中的离子将从肠腔中吸收，然后通过乳突膜叠层输送到膜间空间，最终通向中心管和血淋巴。直肠中的 K^+ 转运需要 Na^+/K^+-

the rectum. The larval rectum plays a critical role in adult hindgut development and is a source of chromosomally unstable cell divisions.

ATP 酶（P-ATP 酶）和 V-ATP 酶。幼虫直肠在成虫后肠发育中起着关键作用，是染色体不稳定细胞分裂的来源。

4.1.2.4 Cryptonephridial systems

Many families of Coleoptera, Lepidoptera, and some sawfly larvae (Hymenoptera) have an arrangement of Malpighian tubules in which the distal ends of the tubules are enveloped within a membrane and held close to the surface of the rectum. The distal ends of the tubules may be simply closed with a terminal cell, or the ends may be attached to and embedded in the tissue that surround the rectum on the hindgut. This arrangement is known as a cryptonephridic tubules system (Fig.4-4). Insects living in the desert habits and eating very dry food have the most extensive development and network of cryptonephridial tubules.

Cryptonephridic tubules do not penetrate the lumen of the rectum, but lie on the outer surface of the rectum, encased within a perinephric chamber bounded by the perinephric membrane (Fig.4-5). The perinephric membrane is composed of thin, elongated cells

4.1.2.4 隐肾系统

鞘翅目、鳞翅目和一些锯蝇幼虫（膜翅目）的许多科都有马氏管的排列，其中管的末端被膜包裹并紧贴直肠表面。端段管的末梢可能是简单地密封在一个端细胞内，或者末端与后肠上的直肠组织相连并被隐埋。这种排列被认为是隐肾管系统（图 4-4）。生活在沙漠中的昆虫，习惯取食非常干燥的食物，拥有最广泛发育和网状系统的隐肾管。

隐肾管并不穿透直肠腔，而只是位于直肠外表面，包入肾周室内以肾周膜为界（图 4-5）。肾周膜由薄而延长的细胞组成，于肠道接触的起点将管从血

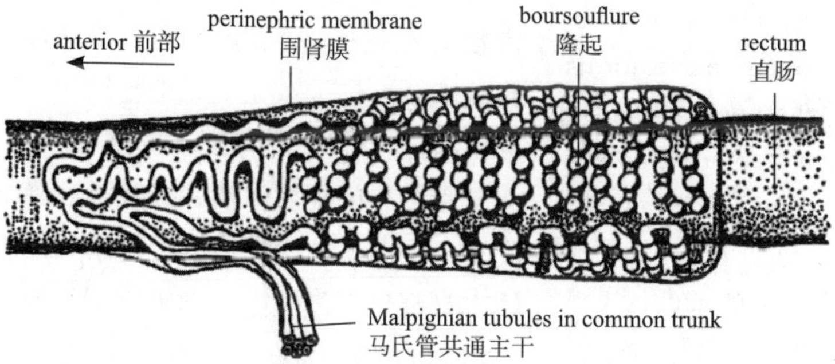

Fig.4-4 The cryptonephridic tubule system characteristic of most Lepidoptera and Coleoptera (modified from Grimstone et al., 1968).

图 4-4 多数鳞翅目和鞘翅目隐肾管系统特征（仿 Grimston 等，1968）。

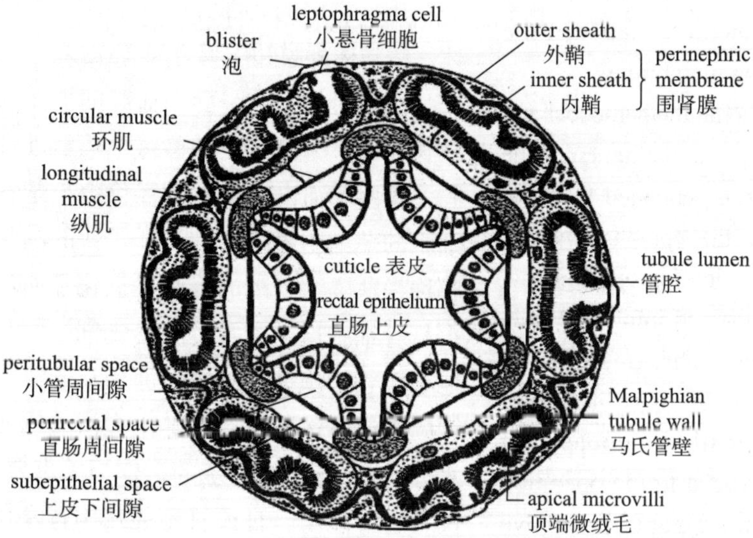

Fig.4-5 Cross-section view of the rectum with cryptongphridic tubules from the yellow mealworm, *T.molitor* (modified from Grimstone et al., 1968).

图 4-5 黄粉甲含隐肾管的直肠横切面（仿 Grimstone 等，1968）。

that seal the tubules from the hemocoel and hemolymph at the initial point of contact with the gut. The tubules do not terminate immediately after contracting the rectum, but typically are thrown into many loops and convolutions, with segments running radially around the rectum as well as looping anteriorly and posteriorly along the length of the rectum. The perinephric membrane follows the various convolutions and turns, always enclosing the tubules like a blanket. Several layers of tubules may lie on the rectum in those insects that live in the driest environments. A small perirectal space occurs between the epithelial cell layer of the rectum and the inner most layer of tubules within the perinephric chamber.

An earlier description identified three spaces within the rectal complex: the perirectal, subepithelial and peritubular spaces. The first two are true intercellular spaces, bounded by basement membranes, but the so-called peritubular space is occupied by necrotic cells. Just at the point where the perinephric membrane is lacking, the Malpighian tubules become extremely thin. The inner sheath of the perinephric membrane is interrupted by a specialized cell, the leptophragma cell, which comprises of the wall in this area and it is attached laterally to both the perinephric membrane and the surrounding cell of the Malpighian tubule wall. Each leptophragma is bounded by a prominent electrondense ring into which the laminae of the inner sheath are inserted. The outer sheath forms a blister over the leptophragma and is completely noncellular in this region. At the base of the blister a basement membrane covers the leptophragma itself, and the body of the leptophragma cell projects into the lumen of the tubule, with a thin layer of cytoplasm lying beneath the basement membrane. Both this layer and the cell body itself bear microvilli. The cell has a normal complement of mitochondria, but these do not invade the microvilli. The ordinary tubule cells differ from the leptophragma cells in that most of their microvilli contain mitochondria, with connexions between the outer mitochondrial membrane and the plasma membrane.

The leptophragma cell has a plane surface facing the hemocoel and long, thin, widely-separated microvilli radiating into the lumen of the tubule. These points of contact when a tubule is separated from the hemolymph only by the thin cell membrane of the leptophragma cell and the very thin perinephric membrane are called leptophragmata. It is reasonable to suppose that KCl is transported into the tubule by leptophragma cells. What makes the leptophragma cells so differently from other Malpighian cells is that the inward transport of KCl is not accompanied by an osmotically compensating amount of water. It is this ability which is crucial to the operation of the system, without it the tubule would not be able to remove

腔和血淋巴中密封起来。在直肠收缩后，隐肾管并不立即终结，但通常被推入很多的环和盘绕管内，不但按体节辐射状环绕直肠，而且沿直肠段前后形成回环。围肾膜紧随着各种盘绕，一直像毛毯一样将管围起来。那些生活在最干燥环境中的昆虫，直肠上有几层隐肾管。直肠上皮细胞层和肾周室的最内层管之间有直肠周小间隙。

早期直肠复合体的研究确定出三个间隙：直肠周间隙、上皮下间隙和小管周间隙。前两个是真正的胞间空隙，由底膜界定，但所谓的小管周间隙被坏死细胞占用。正好是围肾膜缺乏的位置，马氏管变得极薄。围肾膜的内鞘被一种专门的小悬骨细胞中断，这种细胞由这个区域的管壁组成，侧面附着在围肾膜和包围细胞的马氏管壁上。每个小悬骨细胞由凸出的电子致密环界定，着生在内鞘的薄层上。外鞘形成覆盖小悬骨细胞的泡，这个区域完全无细胞结构。泡的底部，一层底膜覆盖小悬骨细胞本身，小悬骨细胞的细胞体伸入管腔，底膜下面有一层薄薄的细胞质层。这层和细胞体本身都具有微绒毛。细胞有正常的线粒体，但这些线粒体并不进入微绒毛。普通的管壁细胞与小悬骨细胞的不同之处在于，它们的大部分微绒毛含有线粒体，线粒体外膜和质膜之间有连接。

小悬骨细胞有一面向血腔的平面以及长而薄、分散广泛的微绒毛辐射进管腔。这些与血淋巴接触的点仅由小悬骨细胞的薄细胞膜隔开，其非常薄的围肾膜称为薄膈。合理的推测是氯化钾通过小悬骨细胞运送进管。小悬骨细胞与其他马氏管细胞如此不同的原因在于氯化钾的内流转运并没有伴随平衡渗透压的水分。这种能力对系统的运作至关重要，没有这种能力，马氏管不能从围肾室排

water from the perinephric chamber. Only two families of beetles, Ptinidae and Anobiidae, do not have leptophragmata.

除水分。甲虫只有蛛甲科和窃蠹科两个科没有薄膈。

In *T.molitor* and some other beetles, the thin perinephic membrane is expanded into a boursouflure above each leptophragma cell. The function of boursouflure is that it may be a site of active secretion of ions from the hemolymph into the perinephric tubules, thus creating high osmotic gradient from rectal lumen to the tubule lumen. A high molecular weight compound is secreted into the perirectal space of *T.molitor,* and that it first absorbs water from the rectum, with the water passed on to the tubules.

黄粉甲和其他甲虫，在每个小悬骨细胞上的薄围肾膜膨胀成隆起。隆起的功能可能是从血淋巴分泌离子进入围肾管的活跃部位，因此，产生从直肠腔到管腔的高渗梯度。黄粉甲分泌一种高分子质量化合物进入直肠周间隙，它首先从直肠吸收，然后将水分传递到管腔。

Cryptonephric tubules of Lepidoptera have neither leptophragmata nor leptophragma cells. Thus, regardless of the function of leptophragma cells in Coleoptera, they are not essential to the system in Lepidoptera. In those larvae living in very dry conditions and eating dry food, the convoluted tubules are closely packed together in both the inner and outer layers and extend the entire length of the anterior rectum. There is no cryptonephridic system of tubules in the aquatic Lepidoptera, *P. stratiotata* and *C. lemnata* (Pyraustidae).

鳞翅目幼虫隐肾管既没有薄膈，也没有小悬骨细胞。因此，不管鞘翅目昆虫的小悬骨细胞功能如何，它们都不是鳞翅目昆虫排泄系统所必需的。那些生活在非常干燥条件下并取食干燥食物的幼虫，盘绕的管把内层和外层紧紧包裹在一起，并延伸到整个直肠前部。水生鳞翅目重筒水螟和水螟（野螟科）没有隐肾管系统。

4.2 原尿的形成

4.2 Formation of Primary Urine

The primary urine formed in the lumen of the Malpighian tubules is a filtrate of the hemolymph, and it contains most of the small ions and molecules that occur in the hemolymph. The urine to hemolymph concentration ratio for many of the filtered substances approaches unity, indicating passive movement across the tubule cell membranes, but some components are actively secreted and their urine: hemolymph ratio is always greater than one.

马氏管腔形成的原尿是血淋巴的滤出液，含有血液中存在的大多数小离子和小分子。许多滤过物的原尿与血淋巴浓度比趋于 1，标志着跨越管壁细胞膜的被动运动，但有些成分是主动分泌的，其原尿：血淋巴比通常大于 1。

4.2.1 Proton pump

4.2.1 质子泵

Urine formation in Malpighian tubules relies upon a proton pump, which has been found not only in the Malpighian tubules, but also in the midgut and rectum. The pump is located in the apical membrane of Malpighian tubule cells and actively secretes protons, H^+, into the tubule lumen against an electrochemical

马氏管原尿的形成依赖于质子泵，不仅在马氏管中发现，也在中肠和直肠中发现。质子泵位于马氏管细胞的顶膜上，逆电化学梯度主动向管腔分泌质子 H^+（图 4-6）。在果蝇的主细胞中鉴定出

gradient (Fig.4-6). Three genes encoding the inward rectifier channels in *Drosophila, ir, irk*2, and *irk*3, have been identified in principal cells. In some insects, such as mosquito adults, *R.prolixus* and test fly adults, that take blood meals rich in Na^+ is actively transported by the pump mechanism. The pump is probably regulated by dissociation/reabsorption and likely is under genetic regulation at the transcriptional and posttranscriptional levels.

三个基因，*ir*、*irk*2 和 *irk*3 编码内向整流通道。有些昆虫，如蚊子成虫、吸血蝽和测试的苍蝇成虫，取食富含 Na^+ 的血粉，Na^+ 通过泵机制主动运输。质子泵也许通过解离 / 再吸收调节，可能在转录和转录后水平上受基因调控。

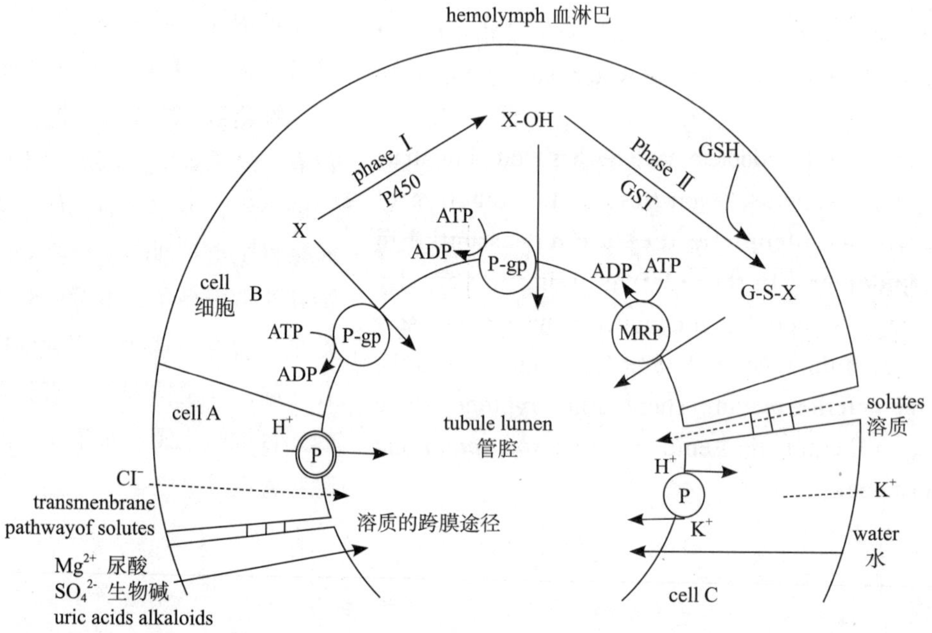

Fig.4-6 An illustration of formation of primary urine in Malpighian tubule process (in cell A and C) and speculative model of xenobiotic resistance and detoxin based on active transport through apical P-glycoprotein (P-gp) and multidrug-resistance-associated protein (MRP), the phase I detoxification enzymes, cytochromes P450 and the phase Ⅱ detoxification enzymes glutathione-S-transferase (GST). X. hydrophobic toxin (Redrawn and modified from Barg, 2000 and O'Donnell, 2008). One single cell may have multifunction, illustration only.

图 4-6 马氏管原尿形成过程（细胞 A、细胞 C）图解及推测的异型生物质抗性解毒模型（细胞 B），基于顶端 P- 糖蛋白（P-gp）和多药抗性相关蛋白（MRP）的主动运输，Ⅰ期脱毒酶细胞色素 P450，Ⅱ期脱毒酶谷胱甘肽 -S- 转移酶（GST）。X. 疏水性毒素（仿 Barg, 2000 及 O'Donnell ,2008 年，重绘）。一个单细胞可有多功能，仅图解方便。

4.2.2 Ion transportation

Cations, such as K^+ and/or Na^+ and probably anions, must enter the Malpighian tubule cells at the basolateral surface from the hemolymph in order for secretion into the lumen at the apical face of cells to continue. Entry of ions at the basolateral membrane appears to be electrically coupled with Cl^- transport in the basolateral membrane. Chloride may be transported by a paracellular pathway in some insects, such as *A.aegypti* or by a transcellular pathway in others. The routine for Cl^- transport

4.2.2 离子的转运

阳离子，如 K^+ 和 / 或 Na^+，也许还有阴离子，必须从血淋巴中进入到马氏管的基底外侧面，以便分泌物持续不断地从细胞顶端面进入管腔。阳离子从基底侧膜的进入可能与 Cl^- 转运进基底侧膜电偶联。有些昆虫，如伊蚊，Cl^- 是通过旁路途径运送的，或者在其他昆虫通过跨细胞途径运送。相同的生物，

may vary within the same organism in response to variable physiological conditions and/or hormonal stimulation. Based on measurement of transepithelial potentials across Malpighian tubules cells and lumen, it is suggested that a stoichiometry of Na^+ : K^+ : $2Cl^-$ cotransport across the basolateral membrane of tubule cells in *R.prolixus*.

Cl^- is the most abundant anion in insect hemolymph and its transport is linked to the bulk of fluid Malpighian tubules transport in ion-transporting epithelia of many groups of insects. In general, Cl^- transport in insect Malpighian tubules is mediated by cation/Cl^- co transporters, Cl^-/bicarbonate exchangers and Cl^- channels. In many cases, Cl^- transport in the Malpighian tubules of dipterans is linked to the transport of HCO_3^- by Cl^-/HCO_3^- exchangers. Four major groups of Cl^- transporters in insect Malpighian tubules: Cl^- channels, Cl^-/HCO_3^- exchangers, ligand-gated Cl^-channels, and Ca^{2+}-gated Cl^- channels known as anoctamins. These transporters may be present in the distal portions of the ileac plexus and their transcript abundance changes when larvae are fed ion-rich diets. The transport of Cl^- in the Malpighian tubules of larval lepidopterans may be correlated with the reabsorption of base, or sensitive to Cl^- concentration in the hemolymph, and could be regulated by helicokinin and glycine.

The formation of urine volume is highly dependent upon K^+ concentration in the bathing hemolymph or saline, but fluid formation stops even when K^+ concentration in the bathing saline is high if the H^+ pump is inhibited. Molecules, such as sugars, amino acids, and some allelochemicals in the hemolymph, follow the osmotic gradient created by the transport of K^+ and other ions. In addition to secreting K^+, Malpighian tubule cells in some insects also secrete Na^+, some other ions, and organic molecules (Fig.4-6 cell A). One organic molecule, the amino acid proline, is actively secreted into the urine by tubules of the desert locusts, *S.gregaria,* and, after passage into the hindgut, it is used as an energy source for adenosine triphosphate (ATP) production to fuel ion pumps in the hindgut epithelium.

The primary urine formed by the Malpighian tubules is isomotic or sometimes slightly hyposmotic to the hemolymph. Malpighian tubules are not capable of producing primary urine that is appreciably hyperosmotic to the hemolymph. The proximal tubules may modify urine by reabsorption of some substances, but many insects transfer the tubule fluid to the hindgut with few or no

Cl^- 的运输途径可能随着生理条件和/或激素的刺激发生变化。根据横穿马氏管细胞和管腔的跨上皮电势的测定，表明吸血蝽管壁细胞有横跨基底侧外膜的 Na^+ : K^+ : $2Cl^-$ 的协同转运化学计量。

Cl^- 是昆虫血淋巴中含量最丰富的阴离子，其转运与许多昆虫马氏管上皮中的大量液体离子转运有关。一般来说，昆虫马氏管中的 Cl^- 转运是经由阳离子/Cl^- 共转运蛋白、Cl^-/碳酸氢盐交换泵和 Cl^- 通道进行的。在许多情况下，双翅目昆虫马氏管中的 Cl^- 转运与 HCO_3^- 转运的 Cl^-/HCO_3^- 交换泵有关。参与昆虫马氏管中 Cl^- 转运的主要转运体有四类：Cl^- 通道、Cl^-/HCO_3^- 交换泵、配体门控 Cl^- 通道和众所周知的 Ca^{2+} 门控 Cl^- 通道 anoctamins。这些转运蛋白可能存在于回肠丛远端中，当幼虫食用富含离子的食物时，其转录物丰度发生变化。鳞翅目幼虫马氏管中 Cl^- 的转运可能与碱的再吸收有关或可能对血淋巴中 Cl^- 浓度敏感以及可由螺旋激酶和甘氨酸调节。

形成的原尿量高度依赖于浸渍血液中 K^+ 或盐的浓度，但如果 H^+ 泵被抑制，即使浸渍盐溶液中 K^+ 浓度很高，原尿的形成也会停止。血液中的糖、氨基酸和一些异种化感物随着 K^+ 和其他离子的运输产生渗透梯度。除了释放 K^+，有些昆虫的马氏管也释放 Na^+、其他离子和有机分子（图 4-6 细胞 A）。沙漠蝗向管腔活跃释放一种有机分子，氨基酸脯氨酸，进入后肠之后，它被用于 ATP 生产的能量来源，为后肠上皮中的离子泵提供能量。

通过马氏管形成的原尿是等渗的或有时对血淋巴略为低渗。马氏管不能产生原尿的原因在于相对血淋巴略微低渗。最接近的马氏管可能通过再吸收一些物质调整原尿，但许多昆虫运送到后

change in its chemical composition or volume. The hindgut then proceeds to concentrate waste products by reabsorbing water and useful substances.

肠的马氏管液，很少或没有改变化学成分或体积。接着，后肠通过再吸收水分和有用物质而浓缩废弃物。

4.3 后肠的选择性再吸收和排泄

4.3 Selective Reabsorption of the Hindgut and Excretion

4.3.1 Reabsorption in the rectum

The rectum is the final and major site for reabsorption of ions, water, and nutrients, and it is capable of reabsorbing fluid against strong osmotic gradients, ultimately producing in any insects a very concentrated, hyperosmotic excreta.

Electrogenic Cl^- pump provides the energy for K^+ reabsorption. The exact route followed by K^+ as it crosses the rectal epithelial cells to reenter the hemolymph or Malpighian tubules varies in different insects depending on the anatomy of the rectum and the cells involved in the process. Water from the rectal lumen and dissolved solutes follows the osmotic gradient created by ion absorption. The result is that water is reabsorbed into the epithelial cells against an increasingly strong concentration gradient in the rectal lumen. The excreta in the rectum become very pasty or even dry in many insects as water is reabsorbed. The rectal epithelia cell actively reabsorbs amino acids from the lumen, and metabolize them to produce ATP needed to energize the pump. Thus, active ion secretion and electrogenic pumps play a major role in the formation of primary urine in the Malpighian tubules and then in reclaiming water from the hindgut. Several different hormones are responsible for regulating the different functions.

4.3.1 直肠的再吸收

直肠是离子、水分和营养物质再吸收的最后和最主要的部位，能够在强渗透梯度下重吸收液体，最终在任何昆虫体内产生非常浓缩的高渗排泄物。

Cl^-生电泵提供K^+再吸收能量。K^+跨过直肠上皮细胞再进入血淋巴或马氏管的确切路径随着不同昆虫直肠解剖结构及其过程中涉及的细胞而变化。来自直肠腔的水分和溶解的溶质遵循离子吸收产生的渗透梯度。结果是水分逆着直肠腔逐渐增强的浓度梯度被再吸收到上皮细胞。许多昆虫直肠的排泄物因水分被再吸收而变得非常黏稠，甚至干燥。直肠上皮细胞从肠腔主动再吸收氨基酸，代谢产生的 ATP 给泵供能。因此，马氏管内原尿的形成过程中，主动的离子释放和生电泵起着主要作用，并随之在后肠回收水分。几种不同的激素负责调节不同的功能。

4.3.2 Reabsorption in the ileum

The ileum is a major site for isosmotic fluid reabsorption, for active Na^+ and Cl^- reabsorption, and for active secretion of proline as an energy source to support metabolic processes. The driving mechanism for ion and water reabsorption in the ileum is an electrogenic Cl^- pump. The ileum plays major role in acid-base balance by secretion of H^+ into the lumen, formation of NH_4^+, and reabsorbption of HCO_3^-. Metabolism of reabsorbed nonessential

4.3.2 回肠的再吸收

回肠是等渗流体再吸收的主要部位，主动再吸收Na^+和Cl^-，并主动释放脯氨酸作为支持代谢过程的能源。回肠的离子和水分再吸收驱动机制是Cl^-生电泵。通过向肠腔释放H^+，形成NH_4^+，并再吸收HCO_3^-，回肠在酸碱平衡中起主要作用。尿中再吸收的非必需

amino acids present in the urine releases energy for ATP synthesis, an essential power source needed to drive the active reabsorption of Na^+ and Cl^-. About 80% of the ammonia produced in the epithelial cells from metabolism of these amino acids is transported into the lumen where it is extracted as NH_4^+.

氨基酸释放的能量用于 ATP 合成，这是驱动主动再吸收 Na^+ 和 Cl^- 必需的基本能源。上皮细胞内这些氨基酸代谢产生的约 80% 的氨被运送到回肠腔，在那里转化成 NH_4^+。

4.3.3 Ammonia Excretion

4.3.3 氨的排泄

Ammonia is a product of protein and amino acid metabolism. Free ammonia cannot be stored in tissues or cells because it is a very strong base influencing pH and, in its free form, it is very toxic to all cells. It must be rapidly excreted or transformed into a less toxic compound. If water is readily available for dilution, ammonia can be excreted as the free base or as an ammonia salt. Animals that excrete ammonia as their primary nitrogenous waste product are described as ammoneoteleic. Ammonia is a major excretory product for larval stages of some Diptera that live in very wet environments. Many insects excrete ammonia into wet surroundings that dilute it to nontoxic levels.

Some terrestrial insects excrete the majority of their excretory nitrogen as ammonia or ammonium salts. Ammonia and ammonium nitrogen can account for 10% to 46% of the total nitrogen excreted by the desert locust, *S.gregaria*. Ammonia nitrogen is rapidly lost from fecal pellets, especially if they are dried prior to analysis. In some cases, ammonium urate may be lost because it is poorly soluble unless excretory material is extracted with a large volume of aqueous solvent.

氨是蛋白质和氨基酸代谢的产物。游离氨不能在组织和细胞中贮存，因为它是影响 pH 的一种非常强的碱，其游离形式对所有细胞毒性都很强。必须快速排泄或转化成低毒化合物。如果有水便于稀释，氨能以游离碱或氨盐形式排泄。以氨作为其主要含氮废物排泄的动物被称为排氨代谢动物。氨是生活在非常潮湿环境中的一些双翅目幼虫期主要排泄产物。许多昆虫将氨排泄到潮湿环境，使之稀释到无毒水平。

一些陆栖昆虫大多数以氨或铵盐形式排泄其主要含氮排泄物。沙漠蝗排泄的氨和铵占排出总氮的 10% ～ 46%。氨态氮从粪便颗粒中快速散失，特别是如果在分析前先干燥。有时尿酸铵可能散失，因为溶解度差，除非用大量水溶剂提取排泄物。

4.3.4 Uric acid synthesis and excretion

4.3.4 尿酸的合成和排泄

Uric acid is synthesized in insects from protein nitrogen as well as from nucleic acid nitrogen. The major portion is synthesized from protein nitrogen simply because insects digest relatively much more protein nitrogen in their diet than nucleic acid nitrogen. Excretion of uric acid as the main product of protein metabolism is energetically costly. Its advantages are that it rids the body of four nitrogen atoms, which make up to 33.3% of the molecular weight of uric acid, and it is very insoluble in water. It often reaches concentrations that promote precipitation from solution in the Malpighian tubules and hindgut and, as a precipitate, it does not contribute to osmotic values across cells that line the tubules or hindgut.

The fat body is the primary site for uric acid synthesis. The final steps in the synthesis of uric acid involve the conversion of

昆虫尿酸的合成是利用蛋白质氮，也用核酸氮。主要蛋白质是用蛋白质氮合成的，仅仅是因为昆虫消化食物中的蛋白质氮比核酸氮相对更多。尿酸的排泄作为蛋白质代谢的主要产物耗能代价高昂。其优势在于它从虫体清除了四个氮原子，接近尿酸分子质量的 33.3%，极难溶于水。它通常能促进马氏管和后肠中沉淀的浓度，并作为沉淀，它对排列在马氏管或后肠中的细胞的渗透值没有贡献。

脂肪体是尿酸合成的主要部位。尿酸合成的最后一步包括将次黄嘌呤和黄

hypoxanthine and xanthine to uric acid. The enzyme that catalyzes the two-step conversion is xanthine dehydrogenase. The insect enzyme does not work without NAD^+ or FAD^+ (or a synthetic acceptor, such as methylene blue) as the proton acceptor, so it is a true dehydrogenase. Evolution of xanthine dehydrogenase in insects, instead of xanthine oxidase, may have occurred as a potential way to recover some of the costs of urate synthesis.

嘌呤转化成尿酸。催化这两步转化的酶是黄嘌呤脱氢酶。昆虫的酶若没有氧化态烟酰胺腺嘌呤二核苷酸（辅酶Ⅰ）或黄素腺嘌呤二核苷酸（或合成受体，如亚甲蓝）作为质子受体就不能运行，因此，它是真正的脱氢酶。昆虫黄嘌呤脱氢酶的演变，替代了黄嘌呤氧化酶；可能作为一条潜在的可弥补尿酸盐合成潜在的途径存在。

In lepidopteran larvae, a large amount of uric acid is stored in the integument *via* storage excretion. Mosaic larvae in which the integument comprised both uric acid containing white and uric acid lacking translucent areas. Significant quantities of uric acid can be synthesized in the larval epidermis in *B. mori* and *S. ricini*. In silkworm larvae, uric acid can be incorporated into the integument after synthesis in the epidermal cells. Uric acid is a major component of the larval integument and accounts for approximately 11% of its dry weight. Epidermal cells of normal silkworms accumulate uric acid within urate granules, which make the skin white and opaque. The urate granules in the epidermis cells provide an effective barrier against damage from ultraviolet radiation. White regions of adult wings also contain high levels uric acid, likely synthesized in the fat body, that act as a pigment.

在鳞翅目幼虫中，大量尿酸通过贮存排泄并将其贮存在体壁中。斑点蚕幼虫，其体壁由含尿酸的白色区域和不含尿酸的半透明区域组成。家蚕和蓖麻蚕幼虫真皮可合成大量尿酸。在家蚕幼虫中，尿酸在真皮细胞中合成后掺入到体壁。尿酸是幼虫体壁的主要成分，约占其干重的 11%。正常家蚕的真皮细胞在尿酸盐颗粒中积累尿酸，使表皮变白但不透明。真皮细胞中的尿酸盐颗粒为幼虫提供了有效的屏障，防止紫外线辐射造成的损伤。成虫翅膀的白色区域也含有高水平的尿酸，可能是在脂肪体内合成的，起着一种色素的作用。

In the silkworm, *B. mori*, *Bmwh3*, a homolog of the fruit fly white gene, is responsible both for transport of ommochrome precursors into pigment granules in the eyes and for transport of uric acid into urate granules in the epidermis. Two half-type ATP Binding Cassette (ABC) transporters, Bmwh3 and Bm-ok, form a heterodimer to transport uric acid in the silkworm epidermis. A member of solute-linked carrier (SLC) family, *B. mori* monocarboxylate transporter 9 (BmMCT9) is also involved in the transport of uric acid in silkworm integument. Whereas the ABC transporters are proposed to transport urate from hemolymph to epidermal cell cytoplasm, BmMCT9 is proposed to function intracellularly, transporting urate from cytoplasm to the interior of the urate granules.

在家蚕中，与果蝇白眼基因同源的 *Bmwh3* 负责将眼色素前体转运到眼睛的色素颗粒中，并负责将尿酸转运到真皮的尿酸盐颗粒中。两个半型 ATP 结合盒（ABC）转运蛋白 Bmwh3 和 Bm ok 形成异二聚体，在家蚕真皮中转运尿酸。作为溶质连接载体（SLC）家族的一员，家蚕单羧酸转运蛋白 9（BmMCT9）也参与了尿酸在家蚕体壁中的转运。而 ABC 转运蛋白被认为可将尿酸盐从血淋巴转运到真皮细胞质，BmMCT9 则被认为是在细胞内发挥作用，将尿酸盐由细胞质转运到尿酸盐颗粒内部。

Uric acid is the least soluble of excretory products, followed by allantoin, hypoxanthine, and xanthine. Uric acid may crystallize as free uric acid and/or as a salt. Ammonia urate is much less soluble than either sodium or potassium urate, and it has the advantage that the NH_4^+ rids the body of additional nitrogen and acid equivalents.

尿酸是排泄化合物中溶解度最小的，其次为尿囊素、次黄嘌呤和黄嘌呤。尿酸可作为游离尿酸和 / 或尿酸盐结晶。尿酸氨比尿酸钠或尿酸钾溶解度小很多，其优势是 NH_4^+ 可从体内清除额外的氮和等量的酸。

In the lepidopteran larva, uric acid has multiple roles: ① it shields

在鳞翅目昆虫幼虫中，尿酸具有多

the larval body from ultraviolet radiation from the sun; ② it acts as a scavenger of photo-induced reactive oxygen species; ③ it is used as a white pigment to form the larval pigmentation pattern.

种作用：①它保护幼虫身体免受太阳紫外线辐射；②它充当光诱导活性氧化物的清除剂；③它用作白色色素以形成幼虫色素沉着模式。

A few insects excrete some allantoic acid, an oxidation product of allantoin, but the enzyme allantoicase is not widespread in insects. Allantoic acid is a very minor excretory product.

少数昆虫排泄一种尿囊素的氧化产物——尿囊酸，但尿囊酸酶在昆虫中分布并不广泛。尿囊酸是一种非常少的代谢产物。

4.4 排泄系统在保持动态平衡中的作用

4.4 Role of the Excretory System in Maintaining Homeostasis

Dynamic changes in salt, water, acid-base, and nitrogen amounts occur from time to time in all organisms as a result of food ingested, environmental conditions, and metabolism. Regulatory mechanisms that respond rapidly to these changes are necessary to preserve the integrity of cells and tissues.

由于摄入的食物、环境条件和新陈代谢，所有生物体中的盐、水、酸、碱和氮含量都会不时地发生动态变化。对这些变化快速的代谢调节反应是维持完整细胞和组织所必需的。

When the diet is supplemented with ions and the gut is full, the rectal complex provides the source of ions, while the entire length of the free Malpighian tubules reabsorbs ions. Reductions in water permeability and paracellular solute permeability in this region in response to the hormone. Helicokinin are thought to be aimed at preventing water and toxin backflow out of the tubule into the hemolymph during the switch from ion secretion to ion reabsorption. In contrast, when the gut is empty, when the diet is poor in ions, or within a few minutes of disconnecting the free Malpighian tubules from the rectal complex, the distal portions of the ileac plexus secrete ions. Gap junction coupling between principal and stellate cells of Malpighian tubules of lepidopteran larvae enables these cell types to transport ions (and presumably water) in opposite directions, allowing larvae to recycle some of the ions and water needed to drive the excretory function of their Malpighian tubules.

当食物中补充了离子并且肠道充满时，直肠复合体提供了离子源，而整个长度的游离马氏管重新吸收离子。该区域的水渗透性和细胞旁溶质渗透性响应于激素而降低。人们认为，在从离子分泌转变为离子再吸收的过程中，螺旋激酶的作用是防止水和毒素从马氏管回流到血淋巴。相反，当肠道排空，食物中离子含量不足时，或在将游离马氏管与直肠复合体断开后几分钟内，回肠丛的远端分泌离子。间隔连接可渗透生理相关离子和水以及各种第二信使，但上皮细胞矢量离子传输模型通常不考虑这种结构。鳞翅目幼虫马氏管的主细胞和星细胞之间的间隔连接耦合使这些细胞类型能够以相反的方向运输离子（可能还有水），使幼虫能够回收驱动其马氏管排泄功能所需的一些离子和水。

4.4.1 Electrolyte homeostasis

Larval and adult *A.aegypti* live in different habitats, have different food habits, and control Malpighian tubule function by different hormones. Adult female mosquitoes need a blood meal in order to mature each batch of eggs, but with the blood comes a large salt load that must be excreted. Sodium excretion is an active process and occurs in Malpighian tubule cells of the adult mosquito in response to stimulation from the mosquito natriuretic peptide (MNP) releases from the corpora cardiaca. A proton pump coupled with a H^+-Na^+ exchanges mechanism secretes sodium into the tubule lumen. The pump appears to work as previously described for Malpighian tubules in general, except that Na^+, rather than K^+, is the principal ion exchanged for protons pumped into the tubule lumen.

Feeding on a blood meal in *A.aegypti* stimulates the release of MNP, and cAMP is produced as a second Messenger at the inner surface of the basolateral membrane of tubule cells. cAMP acts selectively to open Na^+ channels in the basolateral membrane. As Na^+ enters the tubule cell from the hemolymph, the transcellular resistance falls to about 40% of its prefeeding value. The ion flux generated by MNP and cAMP is specifically an increase in secretion of sodium. Potassium movement is not influenced. The large chloride load from the blood meal also must be excreted, and chloride moves from hemolymph to tubule lumen is a passive transport pathway between the cells (called paracellular pathway). The permeability of the paracellular pathway is increased by leucokinin-Ⅷ, a neuropeptide, although it is not known whether this or a similar hormone is secreted by the mosquito.

Three potential physiological processes through which *A. aegypti* may regulate (control) rates of ion and fluid excretion. These processes are: ① the proton pump that supplies energy for Na^+ and K^+ secretion to the tubule lumen; ② the transcellular resistance across the tubule cells that control ion channels in the basolateral membrane; ③ the resistances of the passive transport pathway for chloride movement.

Fly electrolyte homeostasis was shown to be remarkably similar to that in higher species. When absorption of Pi is blocked by the phosphate binder sevelamer or cellular uptake is inhibited by phosphornoformic acid, larval development is delayed. In contrast, adult flies exposed to high dietary Pi die prematurely. Consistent with its role in Pi excretion, dietary Pi induces *MFS2*

4.4.1 电解质稳态

埃及伊蚊的幼虫和成虫，栖息之处不同，食性不一样，不同的激素控制着马氏管的功能。雌蚊成虫为了使每批卵成熟必须进食血粉。钠排泄是发生在蚊成虫马氏管细胞中的一个主动过程，是对心侧体释放的蚊尿钠肽（MNP）刺激的反应。质子泵偶联的 H^+-Na^+ 交换机制向管腔释放钠。总体上，这种泵与之前描述的马氏管的泵运作相同，但泵进管腔交换的主要离子不是 Na^+，而是 K^+。

以血粉为食的埃及伊蚊刺激了 MNP 的释放，作为第二信使的 cAMP 产生于管壁细胞侧膜的内表面，选择性地打开侧膜内的 Na^+ 通道。因为 Na^+ 从血淋巴中进入管壁细胞，穿越细胞的阻力值下降到饲喂前的大约 40%。MNP 和 cAMP 产生的离子流在钠释放时显著增加，而钾的运动不受影响。从进食血粉得到的大量氯负载也必须排出；氯从血淋巴向管腔的运动是细胞间的被动运输途径（称为细胞旁路途径）。细胞旁路途径的渗透性通过一种神经肽，蜚蠊激肽 - Ⅷ而增加，然而，蚊子是否分泌这种激素或相似的激素还尚不明确。

埃及伊蚊调节（控制）离子和流体排泄速率可能通过三个潜在的生理学过程。这些过程是：①质子泵提供 Na^+ 和 K^+ 释放到管腔的能量；②跨越管壁细胞的阻力控制侧膜内的离子通道；③ Cl^- 被动运输途径的阻力。

果蝇电解质的动态平衡被证明与高等物种非常相似。当磷黏合剂司维拉姆阻止磷的吸收或膦甲酸抑制细胞吸收时，幼虫发育延迟。相比之下，成年果蝇暴露在高膳食磷下会过早死亡。与磷排泄中的作用一致，食物中的磷诱导 *MFS2* 表达。这导致在高磷培养基上培养的果蝇中形成马氏管磷酸钙结石，

expression. This results in the formation of Malpighian calcium-Pi stones, while RNAi-mediated knockdown of *MFS2* increases blood (hemolymph) Pi and decreases formation of Malpighian tubule stones in flies cultured on high Pi medium. Conversely, microinjection of adults with the phosphaturic human hormone *fibroblast growth factor 23* (*FGF23*) induces tubule expression of *MFS2* and decreases blood Pi. This action of *FGF23* is blocked by genetic ablation of *MFS2*. Furthermore, genetic overexpression of the fly *FGF branchless* (*bnl*) in the tubules induces expression of *MFS2* and increases Malpighian tubule stones suggesting that *bnl* is the endogenous phosphaturic hormone in adult flies. Genetic ablation of *MFS2* increased fly life span, suggesting that Malpighian tubule stones are a key element whereby high Pi diet reduces fly longevity.

而 RNA 干扰介导的 *MFS2* 敲除则会增加血液（血淋巴）磷，并减少马氏管结石的形成。相反，微量注射人类磷酸尿激素成纤维细胞生长因子 23（*FGF23*）可诱导 *MFS2* 的马氏管表达，并降低血液磷。*FGF23* 的这种作用被 *MFS2* 的基因切除所阻断。此外，果蝇 *FGF*（*bnl*）在马氏管中的基因过表达诱导了 *MFS2* 的表达，增加了马氏管结石，表明 *bnl* 是成年果蝇体内的内源性磷酸尿激素。敲除 *MFS2* 基因增加了苍蝇的寿命，这表明马氏管结石是高磷取食和减少果蝇寿命的关键因素。

4.4.2 Water homeostasis

Water homeostasis is very important to insects because they have a high surface-to-volume ratio and their food often has variable water content. Although many physiological system and behavior are related to water conserva tion, ridding the body of excess water is primarily a function of the excretory system. Water excretion and retention are regulated by hormones. Diuretic hormones promote fluid formation and rapid excretion by the Malpighian tubules, while the currently known anti diuretic hormones act upon the hindgut and promote water reabsorption. The exception is an antidiuretic hormone demonstrated from water-deprived, dehydrated house crickets, *A.domesticus,* that inhibits fluid formation by the Malpighian tubules without action upon the hindgut.

In *D.melanogaster*, 4 genes of the major intrinsic protein family are expressed at a very high level in the fly renal tissue (Malpighia tubules): the aquaporins (AQPs) *Drip* and *Prip*, and the aquaglyceroporins *Eglp2* and *Eglp4*. *Drip*, *Prip*, and *Eglp2* show significant and specific water permeability, whereas *Eglp2* and *Eglp4* show very high permeability to glycerol and urea. Knockdowns of any of these genes result in impaired hormone-induced fluid secretion. The *Drosophila* tubule has 2 main secretory cell types: active cation-transporting principal cells, wherein the aquaglyceroporins localize to opposite plasma membranes, and small stellate cells, the site of the chloride shunt conductance, with these AQPs localizing to opposite plasma membranes. The spatial

4.4.2 水分稳态

对昆虫来说，水分的动态平衡非常重要，因为它们的表面积与体积比高，其食物通常有不同的水分含量。许多生理系统和行为都与水分的保持有关，将多余的水分从体内清除是排泄系统的一个主要功能。水分的排泄和保留受激素调节。利尿激素促进尿液形成，并通过马氏管快速排出，而目前已知的抗利尿激素则作用于后肠，促使水分再吸收。例外情况是缺水的家蟋蟀，已证实其抗利尿激素通过马氏管而不作用于后肠，阻止尿的形成。

在果蝇中，主要内在蛋白质家族的 4 个基因在果蝇肾组织（马氏管）中以非常高的水平表达：水通道蛋白（AQPs）基因 *Drip* 和 *Prip* 以及水甘油通道蛋白基因 *Eglp2* 和 *Eglp4*。*Drip*、*Prip* 和 *Eglp2* 显示出水特异性的渗透性，而 *Eglp2* 与 *Eglp4* 显示出非常高的甘油和尿素渗透性。任何这些基因的敲除都会导致激素诱导的体液分泌障碍。果蝇马氏管有两种主要的分泌细胞类型：活性阳离子转运主细胞，其中甘油水通道蛋白定位于相对的质膜上；小星状细胞是氯分流电

segregation of these components of epithelial water transport may help to explain the unique success of the higher insects in regulating their internal environments. A putative Na^+/Cl^--dependent neurotransmitter/osmotic transporter in the *Drosophila* hindgut is also essential for the maintenance of systemic water homeostasis.

导的位点，这些 AQPs 定位于相对的质膜上。这些上皮水转运成分的空间分离可能有助于解释高等昆虫在调节其内部环境方面的独特优势。假定的 Na^+/Cl^- 依赖性神经递质 / 渗透压转运蛋白对于维持果蝇后肠系统水稳态也是至关重要的。

4.4.3 Acid-base homeostasis

Acidosis or alkalosis maybe experienced by an insect depending on various foods, the presence of certain types of chemical compounds in plants eaten, type of proteins metabolized (whether proteins yield a high proportion of acidic, basic, or neutral amino acids), and metabolic conditions, such as exercise (e.g., flight) that produce acids in the tissues. The western lubber grasshopper, *Taeniopoda equea*, exhibits flexibility in shifting between excretions of excess acid or excess base equivalents depending on need. This flexibility and regulatory ability in a highly polyphagous insect seems adaptive, and many other insects may show similar ability.

Acid-base regulation has been most thoroughly studied in the desert locust, *S.gregaria*. The Malpighian tubules participate only marginally. Secretion of H^+ and formation of ammonium ions NH_4^+ in the ileum is a principal mechanism for excreting excess acid equivalents. The ileum is a major site of ammoniagenesis, the formation of ammonia from precursors, in locusts in which hindgut cells specifically metabolize amino acids and glucose for energy. The excess nitrogen from amino groups is incorporated into formation of ammonia. These metabolites are present in the lumen of the ileum, having come with urine formed in the Malpighian tubules, but because of ileal and rectal secretion of ammonia/ammonium ions, about half of the total nitrogen secretion excreted by desert locusts is ammonia nitrogen. Some of the ammonia reacts with uric acid in the hindgut to form ammonium urate, and spares sodium and potassium, which also react with uric acid to form sodium and potassium urate.

Thus, excretion of total ammonia nitrogen serves several functions in locusts, including: ① ammonium urate allows the insects to conserve Na^+, an ion that is not high in the food of locusts; ② conservation of ammonia (NH_3) to ammonium (NH_4^+) in the ileal cells is equivalent to removal of protons (H^+), and excretion of ammonia is more than sufficient to explain the recovery of hemolymph pH after a load of HCl is injected into the hemocoel; ③ excretion of ammonia by locusts conserves water because of precipitation of not very soluble ammonium urate salt;

4.4.3 酸碱稳态

昆虫随食物的不同可能经历酸毒症或碱毒症，食下的植物中存在某些类型的化学物质、某种蛋白质代谢（蛋白质是否产生高含量的酸性、碱性或中性氨基酸）以及代谢状态，如在组织中产生酸的运动（例如飞行）。西部小翅笨蝗根据需要，在等量排泄过多的酸或过多的碱之间体现出易变的灵活性。高等杂食性昆虫这种灵活性和调节能力似乎是适应性的，许多其他昆虫也有相同的能力。

对沙漠蝗的酸碱平衡调节的研究最为彻底。马氏管不仅仅是少量参与。回肠 H^+ 的分泌和 NH_4^+ 的形成是排泄过量酸的主要机制。回肠是生氨的主要部位，氨从其前体物形成，蝗虫的后肠细胞专门代谢氨基酸和葡萄糖以供能。氨基基团中过量的氮参与氨的形成。这些代谢物伴随马氏管氨的形成出现在回肠腔，但由于回肠和直肠氨 / 铵离子的分泌，沙漠蝗分泌总氮的约一半是通过氨态氮排泄的。有些氨在后肠与尿酸反应形成尿酸铵，闲置的钠和钾也与尿酸反应生成尿酸钠和尿酸钾。

因此，总氨态氮在蝗虫中起着几个作用，包括：① 尿酸铵使得昆虫保存了 Na^+，是蝗虫食物中含量不很高的离子；②在回肠，细胞将氨（NH_3）保存为铵（NH_4^+），等于脱掉质子（H^+）。排泄氨是 HCl 注入血腔后，其血淋巴 pH 复原的最好解释；③蝗虫氨的排泄保存了水分，因为尿酸铵盐溶解性较差；

④ increasing nitrogen excretion by 25% more than excretion of only sodium or potassium urate.

④氮的排泄与仅仅排泄尿酸钠和钾比较，增加量多于 25%。

4.4.4 Nitrogen homeostasis

4.4.4 氮的稳态

The nitrogen, although known to be a growth-limiting nutrient for some insects, is nevertheless excreted in several forms by insects. Excess nitrogen that must be excreted may come from ingestion of proteins leading to an imbalance of amino acids; they save essential amino acids and may metabolize the nonessential ones as energy sources. Insects also ingest nitrogen in the nucleic acids of their food. Excess protein nitrogen is excreted as uric acid, as purines related to uric acid, as ammonia or ammonium salts, and in several other forms. Nitrogen from nucleic acids also is excreted as uric acid or as related purine metabolites or metabolites of uric acid. Typically, only trace or small amounts of urea are excreted. Although uric acid is the primary excretory product, small amounts of urea are excreted.

虽然知道氮也是一些昆虫限制生长的营养元素，但仍有几种形式从未被昆虫排泄。过量而必须排出的氮可能来自蛋白质消化引起的氨基酸不平衡；它们保存必需氨基酸，代谢非必需氨基酸作为能源。昆虫也消化其食物中的核酸氮。多余的蛋白质氮以尿酸、与嘌呤相关的尿酸、氨或铵盐以及其他几种形式排出。核酸中的氮也以尿酸或有关嘌呤代谢物或尿酸代谢物排出。一般情况下，只排泄微量或少量脲。虽然尿酸是主要代谢产物，但也会排泄少量的脲。

Most animals, including insects, synthesize ammonia into less toxic compounds, such as uric acid. Enzymes involved in amino acid metabolism and ammonia production include amino transaminases or transferases, glutamic acid and alanine dehydrogenases, *L*- and *D*- amino acid oxidases, adenosine deaminase, and mono amine oxidase. All of these enzymes have been detected in a number of insects. The amino transaminases are widely distributed in insect tissues, and enable the amino group of one amino acid to be transferred to a ketoacid, thereby forming a new amino acid. For example:

包括昆虫在内的多数动物将氨合成为低毒化合物，如尿酸。参与氨基酸代谢和产物氨的酶包括氨基转氨酶或转移酶、谷氨酸和丙氨酸脱氢酶、*L*-和 *D*- 型氨基酸氧化酶、腺苷脱氨酶和单胺氧化酶。氨基转移酶在昆虫组织中分布广泛，能将一个氨基酸的氨基转移到酮酸，因此形成一个新的氨基酸。比如：

$$L\text{-aspartic acid} + \alpha\text{-ketoglutaric acid} \longleftrightarrow L\text{-glutamic acid} + \text{oxaloacetate}$$

L- 天冬氨酸　α- 酮戊二酸　*L*- 谷氨酸　草酰乙酸

Although ammonia is not directly released in transaminase reactions, the reactions provide a way to interconvert nonessential amino acids and to make amino acids available to enzymes that deaminate with release of ammonia, such as glutamic dehydrogenase and alanine dehydrogenase in the following reactions:

虽然氨不是直接从转氨酶反应中释放的，但是反应提供了一种将非必需氨基酸互换的方法，提供可用的氨基酸给酶，使之脱去氨基释放氨，如下列反应中的谷氨酸脱氢酶和丙氨酸脱氢酶：

$$L\text{-glutamic acid} + NAD^+ + H_2O \longrightarrow \alpha\text{-ketoglutarate} + NADH + NH_3$$

L- 谷氨酸　α- 酮戊二酸

$$\text{alanine} + NAD^+ + H_2O \longleftarrow \text{pyruvate} + NADH + NH_3$$

丙氨酸　丙酮酸

Ammonia formed in these reactions is rapidly excreted or converted into less toxic compounds. The α- ketoglutaric aicd and pyruvate formed can be metabolized through the Krebs cycle as in energy source. Ammonia production also may come from turnover and replacement of an insect's own nucleic acids as well as from the metabolism of nucleic acids ingested with food. The enzymes adenosine deaminase, guanine deaminase, and adenine deaminase, all of which give rise to ammonia from metabolism of nuclei acids or their derived products, have been reported from various insects.

这些反应中形成的氨被快速排泄或转变为低毒化合物。形成的 α- 酮戊二酸和丙酮酸作为能源，通过克雷布斯循环进行代谢。氨产物也可能与来自食物摄取的核酸代谢一样，也来自昆虫本身的核酸翻转和取代。在已报道的各种昆虫中，腺苷脱氨酶、鸟嘌呤脱氨酶和腺嘌呤脱氨酶都能从核酸代谢或其衍生产物中产生氨。

4.4.5 Calcium homeostasis

Malpighian tubules play a major role in excreting excess Ca^{2+} from the hemolymph, either by secretion or sequestration. Secretion refers to the transport of Ca^{2+} in soluble form into the primary urine, whereas sequestration refers to the transport of Ca^{2+} into tubule calcium stores. Malpighian tubules of many insect species exhibit distinct morphological segmentation along their length and in several cases Ca^{2+}-rich granules appear within distinct "storage" segments of the tubules.

Hemolymph calcium homeostasis in insects is achieved by the Malpighian tubules, primarily by sequestering excess Ca^{2+} within internal calcium stores (Ca^{2+}-rich granules) most often located within principal tubule cells. Ca^{2+} transport was specific to mid-tubule segments, where 97% of the Ca^{2+} entering the tubule is sequestered within intracellular calcium stores and the remaining 3% is secreted into the lumen. Antagonists of voltage-gated (*L*-type) calcium channels decreased Ca^{2+} influx ≥fivefold in adenosine 3′,5′-cyclic monophosphate (cAMP)-stimulated tubules, suggesting basolateral Ca^{2+} influx is facilitated by voltage-gated Ca^{2+} channels. Increasing fluid secretion through manipulation of intracellular levels of cAMP or Ca^{2+} had opposite effects on tubule Ca^{2+} transport. The adenylyl cyclase- cAMP -PKA pathway promotes Ca^{2+} sequestration whereas both 5-hydroxytryptamine and thapsigargin inhibited sequestration. The mid-tubules of *A. domesticus* are dynamic calcium stores, which maintain hemolymph calcium concentration by manipulating rates of Ca^{2+} sequestration through stimulatory cAMP and inhibitory Ca^{2+} regulatory pathways.

4.4.5 钙的稳态

马氏管在通过分泌或螯合从血淋巴排出过量 Ca^{2+} 方面起主要作用。分泌是指将可溶形式的 Ca^{2+} 转运到原尿中，而螯合是指将 Ca^{2+} 运输到马氏管钙库中。许多昆虫的马氏管沿其长度表现出明显的形态分段，在一些情况下，富含钙的颗粒出现在马氏管的不同"贮存"段内。

昆虫血淋巴中的钙稳态是通过马氏管实现的，主要是通过将过量的 Ca^{2+} 螯合在主细胞内的胞内钙库（富含 Ca^{2+} 的颗粒）中。Ca^{2+} 转运是马氏管中段特异性的，其中进入马氏管的 97% 的 Ca^{2+} 被螯合在细胞内钙库中，其余 3% 被分泌到管腔中。电压门控（*L* 型）钙通道拮抗剂降低钙内流大于或等于腺苷 3′，5′ - 环磷酸（cAMP）刺激的马氏管的 5 倍，表明电压门控 Ca^{2+} 通道促进了基底外侧 Ca^{2+} 内流。通过控制细胞内 cAMP 或 Ca^{2+} 水平增加液体分泌对马氏管 Ca^{2+} 转运有相反的影响。腺苷酸环化酶 -cAMP-PKA 途径促进 Ca^{2+} 螯合作用，而 5- 羟色胺和毒胡萝卜素均抑制螯合。家蟋蟀的马氏管中段是动态的钙库，通过刺激 cAMP 和抑制 Ca^{2+} 调节途径控制 Ca^{2+} 螯合率来维持血淋巴钙浓度。

4.5 激素对排泄和水分盐分平衡的调控作用

4.5 Hormonal Control of Excretion and Homeostasis

The hormonal control of the Malpighian tubule secretion mainly by neuropeptides Capa, kinin, diuretic hormones (DH_{31}, DH_{44} and Nplp1-4). Given that the diuretic neuropeptides Capa, DH_{31} and DH_{34} are all thought to act to stimulate fluid secretion by activating principal cell V-ATPase, and kinin by activating chloride flux in stellate cells.

马氏管分泌的激素控制主要通过神经肽 Capa、激肽和利尿激素（DH_{31}、DH_{44} 和 Nplp1-4）进行。鉴于利尿神经肽 Capa、DH_{31} 和 DH_{34} 都被认为是通过激活主细胞 V-ATPase 来刺激液体分泌的，激肽即通过激活星细胞中的氯流来激活的。

4.5.1 Diuretic hormones

Diuresis means that fluid secretion into the Malpighian tubules increased, the fluid may not always be excreted from the body, as in certain beetles, which is noted below. Currently more than 20 insect diuretic hormones are known. All are neuropeptides. They increase the formation of fluid by Malpighian tubules in insects in the orders Orthoptera, Lepidoptera, Diptera, Dictyoptera, and Coleoptera. Serotonin (5-hydroxytryptamine, 5-HT), which is not a neuropeptide, also stimulates urine formation. Diuresis reduces the flight load in thosc insects that take large blood meals at one time or ingest plant sap in large quantities, clearly an adaptive physiological mechanism because the excess water is not needed and critical nutrients in the liquid food need to be concentrated and conserved. Diuresis in beetles that live in very dry environments is not so intuitively obvious, but may have important functions.

Rhodnius takes one large blood meal in each instar, and within hours rapidly excretes a large volume of urine. This action leaves concentrated proteins from the blood meal in the midgut, and rids the body of excess water and Na^+. Within minutes after feeding starts, the large volume of blood ingested greatly distends the abdomen and activates stretch receptors located in the tergosternal muscles near the lateral edge of abdominal segments two to seven. Rapid circulation of hemolymph is induced, caused at least in part by vigorous peristaltic movements of the alimentary canal. Diuresis starts within 3 minutes after feeding, initiated by secretion

4.5.1 利尿激素

多尿就是分泌到马氏管的尿液增加，尿液并不是一直从体内排泄的，像一些甲虫以尿液少著称。目前已知的 20 多种昆虫利尿激素都是神经肽类。通过马氏管，它们增加了直翅目、鳞翅目、双翅日、网翅目和鞘翅目昆虫尿液的形成。血清素（5- 羟色胺）不是神经肽，也能刺激尿液的形成。多尿减少了那些一次进食大量血液或大量摄入植物汁液的昆虫飞行负荷，明显是一种适应性的生理机制，因为过量的水分并非必需，液体食物中的重要营养物质必须浓缩和保存。生活在非常干燥环境中的甲虫的多尿没有这么直观明显，但可能具有重要的功能。

长红锥蝽每个龄期都会进食大量血液，数小时内排泄大量尿。这种行为从中肠进食的血液中留存了浓缩的蛋白质，清除体内多余水分和 Na^+。进食开始后的数分钟内，摄取的大量血液使腹部大幅度膨胀，并触发位于第 2 ～ 7 体节侧面附近、背腹板肌肉内的牵引感受器。诱发血淋巴快速循环，部分原因是消化道剧烈蠕动。多尿始于进食后的

of 5-HT and a neuropeptide diuretic hormone synthesized in large neurosecretory cells in the fused mesothoracic ganglion. The neuropeptide is released from a series of enlarged axonal endings of abdominal nerves originating from the mesothoracic ganglion. The hormone is transported to the Malpighian tubules, the target tissue, by rapid circulation of hemolymph. By the time diuresis is over, *Rhodnius* has lost 40% of its freshly fed weight. The osmotic concentration of its hemolymph first falls due to the dilution effect of absorbing so much water from the midgut, but after diuresis the osmotic concentration is about the same as before feeding. Malpighian tubules and possibly other tissues rapidly destroy the diuretic hormone, and sustained release of the hormone is necessary to maintain the excretion of a large volume of fluid. *T.infestans* produce an allatotropin-like peptide that has a diuretic effect and aids the excretion of excess water. Diuresis is also under similar hormone control in the cotton stainer, *D. fasciatus*, which takes plant sap. Medial neurosecretory cells of the brain are the principal source of the hormone in *D.fasciatus*, but some activity is present also in corpora cardiaca and in the mesothoracic ganglion.

A diuretic peptide is defined as any peptide that increases water loss in a whole insect either by fluid reabsorption and/or by inhibiting from the hindgut. In contrast, antidiuretic peptides are those that increasing tubule secretion decreases the loss of water in the whole insect either by inhibiting tubule secretion and/or by promoting the reabsorption of ions and water in the hindgut.

Primary urine production in insect Malpighian tubules is stimulated by two classes of neuropeptides, cortico trophin releasing factor (CRF)-related diuretic peptides and insect kinins. The CRF-related peptide of the locust, *L.migratoria*, has a hormonal role in the control of post feeding diuresis, but the functional role of the kinins is integrated with its receptor in the secondary stellate cells of the Malpighian tubules. The two classes of peptide act synergistically to stimulate tubule secretion, and the kinins may therefore have a modulatory action in the control of diuresis. The peptides differ in their effects on Malpighian tubule ion transport, and this could be important for the regulation of hemolymph volume and composition. A group of eight neurons in the brain of the red flour beetle *T. castaneum* responds bidirectionally to changes in extracellular fluid osmolality by releasing DH37 and DH47 hormones into circulation to remotely control tubule secretion and systemic water balance. CRF-

3 min，由 5- 羟色胺和融合的中胸神经节内大型神经分泌细胞合成的神经肽利尿激素的分泌而启动。神经肽从源自于中胸神经节的一系列腹部神经膨大的轴突末端被释放出来。激素通过血液快速循环被运送到其靶组织，马氏管。当多尿时间结束时，长红锥蝽失去饲喂后鲜重的 40%。血淋巴的渗透浓度因为从中肠吸收太多水分的稀释效应而首次下降，但多尿后渗透浓度与饲喂前大致相同。马氏管或其他组织很快破坏利尿激素，激素的持续释放是维持排泄大量尿液所必需的。锥蝽属害虫产生一种促咽侧体神经肽并有利尿作用，协助排泄多余水分。取食植物汁液的棉椿象，多尿也受到相同激素的控制。脑的中部神经分泌细胞是其主要的激素源，但有些活性也出现在心侧体和中胸神经节内。

利尿肽定义为任何一种通过液流再吸收和 / 或通过抑制后肠吸收而增加整个虫体水分散失的肽。相反，抗利尿肽是那些通过抑制管壁的分泌和 / 或通过启动后肠离子和水分的再吸收，增加管壁分泌而减少整个虫体水分散失的肽。

昆虫马氏管内的原尿的产生受到两类激素刺激：促肾上腺皮质激素释放因子（CRF）– 相关利尿肽和昆虫激肽。飞蝗的 CRF– 相关肽在控制取食后多尿中起着激素的作用，但激肽的功能性作用是与马氏管二级星细胞内的受体结合。两类肽相互作用刺激管壁分泌，因此，激肽可能在控制利尿方面具有调节作用。这些肽类对马氏管离子运输的效果不同，这可能对调节血液体积和成分是重要的。拟谷盗脑中的一组八个神经元通过将利尿激素 DH37 和 DH47 释放到循环中，以远程控制马氏管分泌和系统中水的平衡，对胞外液渗透压的变化作出双向反应。CRF 样受体，称为排尿受体（Urn8R），它独特地定位于一种散

like receptor, named Urinate receptor (Urn8R), which uniquely localizes to a type of SC interspersed along the MTs as underlying this response. Activation of Urn8R increases the luminal directed K^+ flux most likely through SCs, which creates a lumen-positive transepithelial potential that drives fluid secretion *via* a cyclic AMP (cAMP)-dependent mechanism.

A diuretic neuropeptide isolated from *L. migratoria* consists of two antiparalled 9-amino acid peptides jointed by two disulfide bonds, as follows:

布在马氏管上的星形细胞，作为这种反应的基础。Urn8R 的激活增加了管腔定向的 K^+ 流量，很可能是通过星形细胞，这产生了管腔正向跨上皮电位，通过环腺苷酸（cAMP）依赖机制驱动液体分泌。

从飞蝗中分离得到的利尿神经肽由两条 9 个氨基酸的肽通过两对双硫键反平行连接组成，如下：

```
                    Cys-Leu-Ile-Thr-Asn-Cys-Pro-Arg-Gly-NH2
                     |                   |
NH2-Gly-Arg-Pro-Cys-Asn-Thr-Ile-Leu-Cys
```

Malpighian tubules from the cricket *A.domesticus* are stimulated to secrete fluid by tissue extracts of corpora cardiaca, corpora allata, and some other parts of the central nervous system. Tubules in this cricket have three physiologically and morphologically different segments, and different extracts may act on different segment by different mechanisms. The beetle, *T.molitor*, lives in dry grain and grain products and needs no water other than that already present in the food and that derived from metabolism. Surprisingly, it also produces a diuretic hormone. Its function is to act as a "clearance hormone" to flush the Malpighian tubules and hindgut. The tubule may be passed into the hindgut to help move the very dry food residue through the gut, and may help promote a countercurrent flow in the midgut. Fluid reaching the rectal region is nearly all reabsorbed by the cryptonephridic tubules and returned to the hemolymph.

家蟋蟀马氏管受到心侧体、咽侧体和中枢神经系统一些其他部分组织提取物刺激分泌液体。这种蟋蟀的马氏管有三个生理和形态上不同的部分，不同的提取物以不同的机制作用于不同的区段。黄粉甲生活在谷物和谷类产品中，除食物中已经存在的水和新陈代谢产生的水外，不需要其他水。令人惊讶的是，它也产生利尿激素。其功能是起着“清理激素”的作用，冲刷马氏管和后肠。马氏管可能进入后肠，协助移动非常干燥的食物残渣通过肠道，在中肠促进液流的循环。到达直肠区域的流体几乎被隐肾管全部再吸收，并回流至血淋巴。

4.5.2 Antidiuretic hormones

Functionally, there are two types of antidiuretic peptides: those that inhibit Malpighian tubules secretion and those that stimulate fluid reabsorption by the ileum or rectum. Two antidiuretic peptides have been identified from pupal heads of *T. molitor* on the basis of their ability to increase the synthesis of cyclic GMP (cGMP) by isolated tubules. Structurally, they are nonblocked peptides; both peptides use cGMP as a second messenger. Tenmo-DHs act *via* cAMP to increase tubule secretion, whereas Tenmo-ADFs act through cGMP to reduce it.

In desert locust, *S.gregaria,* two fully sequenced poly peptides are known: neuroparsins, which influence the rectum (as does the partially

4.5.2 抗利尿激素

从功能上看，有两类抗利尿激素：抑制马氏管分泌的和刺激回肠或直肠再吸收的激素。基于其增加游离马氏管合成环鸟苷酸（cGMP）的能力，从黄粉甲蛹的头部鉴定出两种抗利尿激素。从结构上，它们都是非封闭肽，二者都利用 cGMP 作为第二信使。黄粉甲利尿肽通过 cAMP 增加马氏管分泌，而抗利尿肽通过 cGMP 降低它。

已知沙漠蝗两个抗利尿肽全序列，影响直肠（像初步纯化的氯化物转运刺

purified chloride transport- stimulating hormone, CTSH), and the ion-transport peptide (ITP), which stimulates ion and water reabsorption from the ileum. The neuroparsins, isolated from the neurosecretory parts of the migratory locust's corpus cardiacum, are apparently three monomers containing intramolecular disulfide bridges. However, neuroparsins have no effect on ion transport across the rectal epithelium, which is required for fluid uptake, and there must therefore be some doubt as to their role as antidiuretic peptides (Gäde, 2004).

激激素，CTSH）和离子运送肽（ITP），激发回肠再吸收离子和水分。从飞蝗心侧体的神经分泌区分离到的蝗抗利尿肽，显然有三个含分子内双硫键的单体。但是蝗抗利尿肽对离子跨越直肠上皮的运输无效，这需要摄取流体，由此得出，它们作为蝗抗利尿肽是存在一些疑问的（Gäde, 2004 年）。

4.5.3 Capa nueropeptide

Capa peptides have been shown to have either diuretic or anti-diuretic effects on tubule function, depending on the specific assay conditions and the species under study. In *D. melanogaster* in particular, Capa peptides are generally thought to be diuretic, and act through increased nitric oxide, cGMP and Ca^{2+} levels in the principal cells of the tubules. It has also been demonstrated, however, that Capa can instead have anti-diuretic and anti-kaluretic effects on the tubules of wild-type *D. melanogaster*. Whether Capa peptides are diuretic or anti-diuretic is of great importance to understanding neuropeptide control of salt balance.

For insects, chilling injuries that occur in the absence of freezing are often related to a systemic loss of ion and water balance that leads to extracellular hyperkalemia, cell depolarization and the triggering of apoptotic signaling cascades. The ability of insect ionoregulatory organs (e.g. the Malpighian tubules) to maintain ion balance in the cold has been linked to improved chill tolerance, and many neuroendocrine factors are known to influence ion transport rates of these organs. Injection of micromolar doses of Capa improve *Drosophila* cold tolerance, and low doses of Capa cause anti-diuresis in insects, including dipterans. low (femtomolar) and high (micromolar) doses of Capa impair and improve chill tolerance, respectively, *via* two different effects on Malpighian tubule ion and water transport. While low doses of Capa are anti-diuretic, reduce tubule K^+ clearance rates and reduce chill tolerance, high doses facilitate K^+ clearance from the haemolymph and increase chill tolerance. Capa peptide can negatively affect cold tolerance in an insect and function as an anti-diuretic peptide in this ubiquitous insect model (Fig.4-7; Appendix color graphic Fig.4-7).

4.5.3 Capa 神经肽

Capa 肽已被证明对马氏管功能具有利尿或抗利尿作用，这取决于特定的测定条件和研究的物种。特别是在果蝇中，Capa 神经肽通常被认为是利尿剂，并通过增加马氏管主细胞中的一氧化氮、cGMP 和 Ca^{2+} 水平发挥作用。然而，研究发现，Capa 可以对野生型黑腹果蝇的马氏管起到抗利尿和抗促尿钾排泄的作用。Capa 肽是利尿剂还是抗利尿剂，对于理解神经肽对盐平衡的控制非常重要。

对于昆虫来说，通常在没有冷冻的情况下发生的冻害与离子和水平衡的系统性失衡有关，导致细胞外高钾血症、细胞去极化和触发凋亡信号级联。昆虫离子调节器官（如马氏管）在寒冷中维持离子平衡的能力与提高耐寒性有关，已知许多神经内分泌因子会影响这些器官的离子转运速率。注射微摩尔剂量的 Capa 可提高果蝇的耐寒性。低剂量的 Capa 可引起昆虫（包括双翅目昆虫）的抗多尿作用。低剂量（fmol）和高剂量（μmol）的 Capa 通过对马氏管离子和水转运不同的作用而引起两种不同影响，分别削弱和提高其耐寒性。虽然低剂量的 Capa 具有抗利尿作用，可降低马氏管 K^+ 清除率并降低寒冷耐受性，但高剂量有助于从血淋巴清除 K^+ 并提高寒冷耐受性。Capa 神经肽会对昆虫的耐寒性产生负面影响，并在这种普遍存在的昆虫模型中起到了抗利尿肽的作用（图 4-7；附彩图 4-7）。

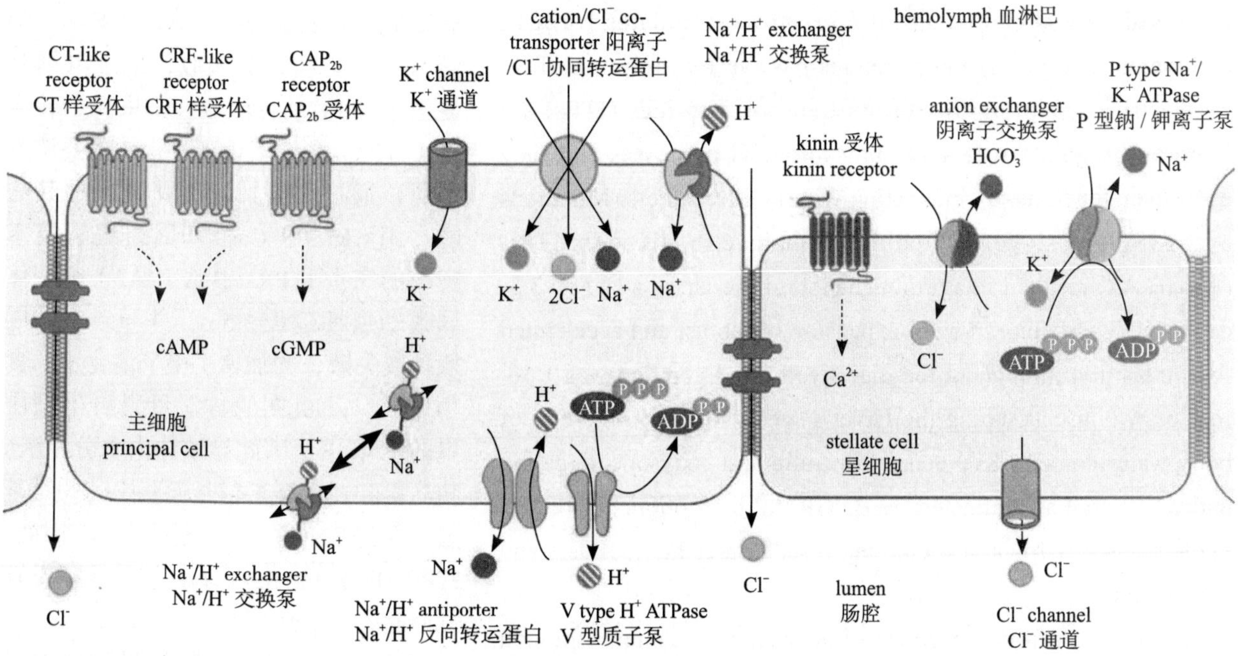

Fig.4-7 Ion transport mechanism and hormonal regulation in the principal cells and stellate cells of Malpighian tubule in insects (modified from Lu er al., 2011).

图 4-7 昆虫马氏管主细胞和星细胞离子转运机制及激素调控（仿 Lu 等，2011）。

4.5.4 The (myo)kinins

Another large family of diuretic peptides are the insect (myo) kinins. The kinin enhances the production of inositol triphosphate (IP_3), which in turn releases Ca^{2+} from IP_3-sensitive stores. The precise location of the Cl^- shunt is disputed; there is evidence for it being paracellular in the mosquito, but transcellular through apical and basal Cl^- channels in *D. melanogaster.* Most structure-activity studies have tested kinin analogs for their diuretic effects on cricket Malpighian tubules *in vitro*. Degradation of kinins is relatively well- researched in *H. zea*. Apparently, a membrane-bound enzyme in Malpighian tubules that cleaves Helze-K-II primarily between Pro5 and Trp6, hence destroying the C-terminal pentapeptide that is so important for biological activity. There is experimental evidence that they can be released independently of one another, which gives great subtlety to the control of tubule secretion.

4.5.4 （肌源性）激肽

另外一个利尿肽大家族是昆虫（肌源性）激肽。激肽提高了三磷酸肌醇（IP_3）的产生，进而从 IP_3- 敏感库存中释放出 Ca^{2+}。精确的 Cl^- 支路位置存有争议；迹象表明，在蚊子中属于旁路，但在果蝇中，跨细胞路径是通过顶部和底部的 Cl^- 通道。多个结构 - 活性的研究测试了激肽类似物对蟋蟀离体马氏管的利尿活性。激肽的降解在棉红铃虫中研究得比较透彻。显然，马氏管膜结合酶主要在第 5 位脯氨酸和第 6 位色氨酸之间切开 Helze-K- II，因此破坏了 C-末端的对生物学活性非常重要的五肽。实验表明，它们能彼此独立地释放，对马氏管分泌的控制相当巧妙。

4.5.5 Ion transport peptide and chloride transport stimulating hormone

A neuropeptide, the ion transport peptide (ITP), isolated from

4.5.5 离子转运肽和 Cl^- 转运激活激素

离子转运肽（ITP）是一种神经

the fused corpus cardiacum of *S.gregaria* stimulates Na^+,Cl^-, and water absorption, and promotes passive reabsorption of K^+ by electrical coupling. Insect ion transport peptides (ITPs) are important regulators of many physiological processes and they exert their functions by interacting with their receptors. Mutations by CRISPR-Cas9 genome editing technique in silkworm ITPR (BNGR-A2) resulted in a prolongnation of the larval stage by 3.5-day as well as failure in wing expansion of moths, and accelerated food transition throughout the digestive tract. Excretion was 1.56-fold of WT insects during the larval stage, resulting in the loss of body water content. Key genes in insulin and ecdysone signaling pathways were also affected by BNGR-A2 disruption. Therefore, ITPR plays key roles in regulating insect water homeostasis and development.

The net result from movement of the excretory contents through the ileum of the desert locust is that the volume of fluid reduced as Na^+, Cl^-, K^+, and fluid are reabsorbed (the ions by active mechanism and fluid following the osmotic gradients).

The driving mechanism for cation and water reabsorption, as in the ileum, is an electrogenic Cl^- pump under the influence of a neuropeptide hormone, chloride transport stimulating hormone (CTSH), from the corpora cardiaca. It acts on the rectal epithelium to promote active Cl^- absorption. The pump provides the energy for K^+ reabsorption.

(Li Wenchu)

肽，从飞蝗融合的心侧体分离而来，刺激 Na^+、Cl^- 和水分吸收，通过电耦合促进 K^+ 的被动再吸收。昆虫离子转运肽（ITPs）是许多生理过程的重要调节因子，通过与受体相互作用发挥其功能。用 CRISPR-Cas9 基因组编辑技术突变的家蚕 ITPR（BNGR-A2）突变体导致幼虫期延长 3.5 d，并导致蚕蛾展翅扩展失败，并加速了整个消化道的食物转换。幼虫期的排泄量是野生型昆虫的 1.56 倍，因此导致体内水分损失。胰岛素和蜕皮激素信号通路中的关键基因也受到 BNGR-A2 中断的影响。因此，ITPR 在调节昆虫水分动态平衡和发育方面起着关键作用。

沙漠蝗排泄内容物通过回肠的最终结果是流体体积减少了，因为 Na^+、Cl^-、K^+ 和液流的再吸收（离子通过主动吸收，液流随渗透梯度）。

阳离子和水分的再吸收驱动机制，就像在回肠内是一种 Cl^- 生电泵，受到心侧体分泌的神经肽激素，Cl^- 转运激活激素（CTSH）的影响。它作用于直肠上皮，促进 Cl^- 的主动吸收。由泵提供能量进行 K^+ 的再吸收。

（李文楚）

Exercises

1.What are the main compositions of excretion system?

2.Try to elucidate the structure and function of cryptonephridial systems.

3.What's the role of the excretion system in maintaining homeostasis?

思考题

1. 排泄系统的主要成分是什么？

2. 试阐明隐肾系统的结构和功能。

3. 排泄系统在维持体内平衡方面的作用是什么？

第5章 呼吸系统
Chapter 5 Respiratory System

Keypoints

1.The respiratory system of insects is mainly composed of spiracles, trunk, branches trachae and tracheloes, forming a complex network in structure, providing sufficient oxygen for aerobic metabolism, so that most insects will not produce oxygen debat during long-term migration and violent movement.

2.Tracheal system in insects is the base of physiological functions of repiration that supply oxygen to all tissues and cells.

3.The types and forms of insect respiration vary greatly in different species, such as wasps, aquatic insects and parasitic insects. However, the common feature is that mitochondria oxidize nutrients to provide energy for insect physiological activities.

知识点

1. 昆虫的呼吸系统主要由气门、主气管、支气管和微气管组成，在结构上形成复杂网络，提供充足的氧气进行有氧代谢，使大多数昆虫长期迁徙、剧烈运动不致产生负氧。

2. 昆虫的气管系统是呼吸生理功能的基础，为所有组织和细胞提供氧气。

3. 昆虫呼吸作用的类型和形式在不同物种中的变化很大，如黄蜂、水生昆虫和寄生昆虫等。其共同特征都是在线粒体氧化营养物质，为昆虫生理活动提供能量。

Respiration is a function of the tracheal system, a tubular network that originates at spiracular openings on the body surface and radiates to all parts of the insect body. The respiratory organs of terrestrial insects consist of tracheal tubes with external spiracular valves that control gas exchange. The tracheal system is very efficient for insects and delivers oxygen to flight muscles in sufficient amounts so that they conduct aerobic metabolism even during flight. Despite their relatively high metabolic rate, many insects have highly discontinuous patterns of gas exchange, including long periods when the spiracles are fully closed.

In some very small insects, simple diffusion of gases through the system of tracheal tubules may suffice, but most insects actively ventilate the system by muscular pumping motion. The cuticular linings of trachea are shed at each molt, but the lining of tracheoles is not molted. Aquatic insects have a tracheal system

呼吸作用是气管系统的一种功能。陆生昆虫的呼吸器官由气管和控制气体交换的外部气门组成。该系统是一种管状网络，起源于体表气门的开口，并辐射到昆虫体的所有部位。气管系统对昆虫非常有效，即使在飞行过程中也能向飞行肌输送足够的氧气供其进行有氧代谢。尽管它们的代谢率相对较高，但许多昆虫的气体交换模式极不连续，包括长时间的气门完全关闭。

在体积较小的昆虫中，气体通过气管系统简单的扩散可满足需要，但多数昆虫通过肌源性气泵的运动主动地使系统通风。气管的表皮膜每次蜕皮都会脱

that is essentially the same in structure as that of terrestrial insects. Many adaptations to an aquatic life occur, however, in the tracheal system, including compressible gas bubbles (gas gills), incompressible gas films (plastrons), and thin cuticular flaps (gills) that take oxygen from water by cutaneous diffusion. Insect eggs often have plastrons as a part of the egg shell that aid gas exchange for the developing embryo. In addition to gas exchange, the tracheal system has been adapted in some insects to serve non-respiration functions, such as an attachment site for endocrine cells, participation in sound reception, and sound production as a source of hissing sounds, and delivery of a distasteful froth.

The increasing volume of the developing eggs can compress the tracheal system of gravid females leading to reduced capacity for active ventilation and therefore a limitation on flight performance. The tradeoff between migratory flight and reproduction in female insects, also known as the 'Oogenesis-Flight Syndrome', results from the high amount of energy and substrates needed to concurrently maintain flight muscles and developing eggs (Urca et al., 2021).

落，但微气管并不脱落。水生昆虫具有与陆生昆虫结构一样的气管系统。水生生活出现很多适应性，但其气管系统，包括压缩气泡（气鳃）、非压缩气膜（气盾）以及薄表皮气门片（鳃），通过皮肤扩散从水中摄取氧。通常昆虫卵有气盾，是卵壳的一部分，有助于发育中胚胎的气体交换。除了气体交换，气管系统在某些昆虫中已适应为非呼吸作用的功能，如内分泌细胞的附着点、参与声音的反应以及作为嘶嘶声源和释放令人讨厌的泡沫。

不断增加的发育中的卵体积会压缩抱卵雌虫的气管系统，导致主动通气能力降低，从而限制飞行性能。雌性昆虫迁徙飞行和繁殖之间的权衡，也称为“卵发生—飞行综合征”，是由同时维持飞行肌肉和发育卵子所需的高能量和底物所致（Urca 等 , 2021）。

5.1 气管系统的结构

5.1 Structure of the Tracheal System

Insect breathe by delivering air through a network of small tubules, the tracheal, arise at opening on the sides of the body, the spiracles. In most insects there are interconnecting longitudinal and transverse tracheal trunks, and sometimes large air sacs that tie together the entire system. Larger tracheal lubes send off branches that become smaller in diameter as they ramify to all tissues, with the smallest diameter tracheoles (Fig.5-1) touching most cells and even indenting some cells. The airflow may be tidal or directed, depending on the insect and its physiological state. The interconnections make directed air flow possible, in which air enters through one or more anterior spiracles, get pumped through the body by muscular ventilatory movements, and is expelled through one or more posterior spiracles. Such a directed flow is more efficient than tidal inflow and outflow from the same spiracles

昆虫呼吸通过小管网络气管传送空气，气管出现在体侧的气门开口，有时大气囊将整个系统连接在一起。较大的气管润滑剂被传送到直径较小的分支，因为它们在所有组织中成枝杈状，以更小的微气管接触多数细胞，甚至缩进某些细胞（图 5-1）。气流可以是潮汐式的或定向的，取决于昆虫和它的生理状态。气管的相互连接使定向空气流动成为可能，其中的空气从一个或多个前部气门涌入，由肌源性通风运动将空气泵入体内，并通过一个或多个后部气门排出。这种定向气流比同一个气门进出气流的潮汐式更为有效，因为系统不断

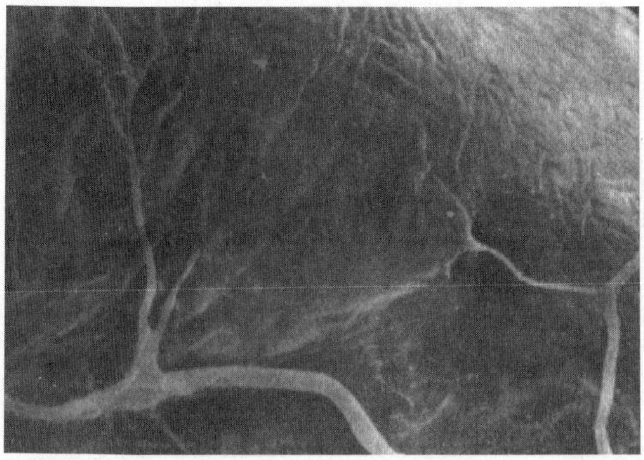

Fig.5-1 A scanning electron microscope view of a tracheal tube branching into many fine tracheoles that disappear beneath the surface of cells.
图 5-1 气管分支为许多微气管，消失在细胞表面下方的扫描电镜图像。

because the system is constantly flushed and incoming air is not mixed with used air.

被冲刷，进入的空气不会和用过的空气混合。

5.1.1 Structure of the tracheal system

5.1.1 气管的结构

5.1.1.1 Trachea and tracheoles structure

5.1.1.1 气管和微气管的结构

The tracheal system develops from embryonic ecto dermal tissue. Trachea and tracheoles have an epi cuticular lining, comprised primarily of a cuticulin layer that is continuous with the external cuticle. Larger tracheal trunks have an endocuticle layer that gives more strength to the tubular structure. A hydrophobic substance is secreted on the lumen surface of tracheae that helps prevent water from entering the tracheae and reduces evaporative water loss from the humid, extensive internal tracheal surfaces. The major distinction between tracheae and tracheoles is one of size. Tubes down to about 1μm in size are called tracheae (trachea) while those smaller than 1μm are tracheoles. Tracheoles typically have fluid in their terminal endings, and the fluid level seems to be related to metabolic demand for oxygen delivery to cells near the endings. The change in fluid level and, thus, the open air path when the fluid recedes or is absorbed into the tissue is probably caused by changes in the osmotic pressure between tracheolar fluid and the intracellular medium. Thickened, tight spirals of the cuticular intima, the taenidial, strengthen tracheae and tracheoles, provide elasticity, and help the tubes resist compression and collapse (Fig.5-2). In large tracheae, the taenidia are up to 450nm in width and are spaced about 300 nm apart, but, in tracheoles, the taenidia are smaller (50 to 80 nm in width)

气管系统从胚胎外胚层组织发育而来。气管和微气管有一外表皮膜，主要由一连续的表皮质层和外部表皮组成。较大的气管主干有一外表皮层，使管状结构更为有力。气管的管腔表面分泌有疏水物质，这有助于防止水分进入气管，并减少潮湿、大量气管内表面的水分蒸发损失。气管和微气管的主要区别是其大小不同。大小下行至约 1μm 的管，称为气管，而那些小于 1μm 的是微气管。微气管终端通常有液体，而液体的水平可能与终端附近的细胞代谢氧的输送需求相关。当液体后退，空气通道打开，或微气管液体和细胞基质之间的渗透压改变引起液体被组织吸收，液体水平都会发生变化。厚化、表皮内膜紧凑的螺旋、螺旋丝、强化的气管和微气管，提供了弹性并有助于气管抵抗压缩塌陷（图 5-2）。大气管中，螺旋丝宽度高达 450 nm，间距相隔约 300 nm，但微气管中的螺旋丝较小（宽度 50 ～ 80 nm），间

Fig.5-2 A scanning electron microscope cut-away view of the inside of a large tracheae of a mole cricket, showing the origin of two smaller tracheae and taenidial windings that strengthen all tracheae and keep them from collapsing(modified from Nation, 2022).

图 5-2 大气管内部截面扫描电镜图，显示两个小气管及支持气管不致崩坍的螺旋丝圈（仿 Nation, 2022）。

and are spaced farther apart than their width. Within the taenidial folds there is a component, probably similar to procuticle, that adds strength to the taenidia. The micelles of the cuticulin layer are oriential so that their long axis is parallel to the long axis of a trachea or tracheole in intertaenidial areas, but perpendicular to the axis within the taenidial thickenings.

距比它们的宽度更大。螺旋丝皱褶内有一种可能和原表皮类似的组成部分，加强了螺旋丝的强度。表皮质层的微胞是定向的，因此，其纵轴与气管和微气管纵轴在螺旋丝内部区域平行，但与螺旋丝厚度轴线垂直。

The ability to respond developmentally to oxygen availability would impart selective advantages on insects. In terms of oxygen supply, hypoxia increased investment in the mean tracheolar diameter, tracheolar tip diameter, number of tracheal branches and number of tracheoles per main branch. Hyperoxia led to a decrease all of that, but only branch points and mean tracheolar diameter were significantly affected. In terms of oxygen demand, with increased oxygen availability during exposure to hyperoxia, the mitochondrial networks varied in the same direction as the tracheal networks (VandenBrooks et al., 2018).

对氧气供应做出发育反应的能力将给昆虫带来选择性优势。在供氧方面，缺氧增加了平均微气管直径、微气管尖端直径、支气管数量和每个主分支的微气管数量的投入。高氧导致所有这些降低，但只有分支点和平均微气管直径受到显著影响。就需氧量而言，随着暴露于高氧环境中氧气可用性的增加，线粒体网络的变化方向与气管网络的变化方向相同（VandenBrooks 等，2018）。

5.1.1.2 Spiracle structure and function

The openings of the tracheal system at the body surfaces are called spiracles. Spiracles usually occur on the pleural surfaces of the body, typically one on each side segment, but numerous variations have evolved (Fig.5-3). Most terrestrial insects can close the spiracles as an adaptation for water conservation. The simplest closing mechanism consists of folds of the integument, which can be pulled together over the spiracular opening by a closer muscle. Many insects have valves or flaps of cuticle that rotate and close over the spiracle. Sometimes an opener muscle is present, but in most cases opening occurs because of elasticity of the cuticle when the closer muscle relaxes.

5.1.1.2 气门的结构和功能

体表面气管系统的开口被称为气门。气门通常出现在体表胸膜，一般每体节一个，但也有许多变化（图 5-3）。大多数陆生昆虫能关闭气门以保存水分。最简单的关闭机制由体壁的折叠组成，通过闭肌拉伸覆盖气门开口。许多昆虫有瓣膜或表皮气门片可以旋转并在气门上关闭。有时有开肌，在大多数情况下，当闭肌松弛时，其表皮的弹性会引起气门的打开。

The opening may be simple unguarded pores, but frequently there are additional structural details associated with the openings. Typically, a ring of sclerotized cuticle, the peritreme, around the spiracular opening provides reinforcement. In many insects there is a slightly enlarged chamber or atrium just inside the spiracular opening from which tracheae branch in a variety of directions. The atrium may contain various structural adaptations to filter the air, such as dust-catching setae or hairs in the atrium space. Many terrestrial Diptera, Coleoptera, Lepidoptera, and some aquatic insects have thin perforated partitions, called sieve plates (Fig.5-4), in the atrium that act as a filter to keep dust particles out of the tracheae. In aquatic insects, it aids in excluding water.

The muscles associated with the spiracles are innervated by a branch of the median nerve from the ganglion in the same segment or from the ganglion in the anterior segment. Repetitive action potentials from the median nerve cause contracture of the closer muscle and close the spiracle. The closer muscle of *H.cecropia* has a myogenic rhythm leading to slow pacemaker potentials that give rise to spike discharges. Experimental increments of hyper-polarization of the muscle membrane potential slows the rate by which the pacemaker potentials depolarize the membrane, and

气门开口可以是简单而无防护的孔道，但通常有与开口相连、额外的细微结构。典型的是硬化的环状表皮、气门板，围绕在气门开口周围提供强化。很多昆虫有一略为扩大的气门腔或前庭，位于气门开口之中，从此分岔出往不同方向的气管。气门前庭含有各种不同的适应性结构滤过空气，如前庭腔内捕获尘埃的刚毛或茸毛。许多陆生双翅目、鞘翅目、鳞翅目以及有些水生昆虫的前庭具有穿孔的薄分隔层，称为筛板（图 5-4），起着隔离尘埃颗粒于气管外的过滤器作用。水生昆虫的筛板有助于排水。

与气门相连的肌肉受到同体节或者前一体节的神经节发出的中神经分支的支配。中神经的重复动作电位引起闭肌收缩并关闭气门。惜古比天蚕的闭肌有肌源性节律导致慢起搏器电位释放峰电位。实验性增加肌膜电位极化使起搏器电位对膜的极化速率变慢，强烈的超极

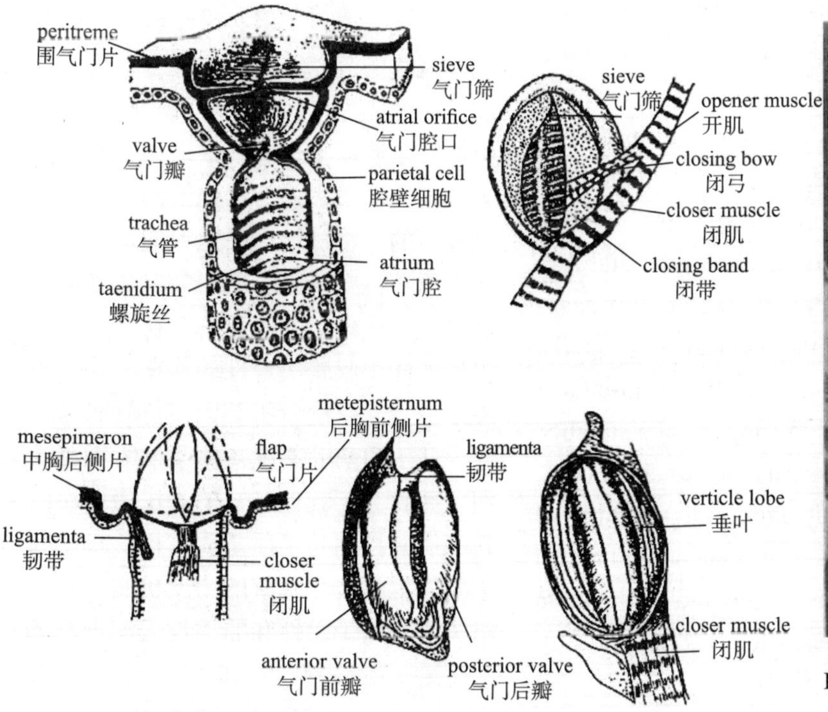

Fig.5-3 The types of the spiracles and their composition.

图 5-3 气门的类型及其组成。

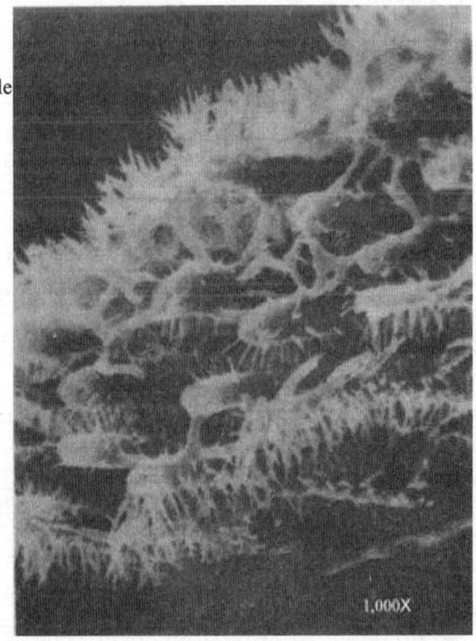

Fig.5-4 Sieve plates in the 5th instar of silkworm, *B. mori*(×1 000) (Wu Zaidc, 1993).

图 5-4 家蚕五龄期筛板（×1 000）(吴载德，1993)。

strong hyperpolarization stops the ability of the graded potentials to exceed the threshold for action potentials. Prolonged exposure of the closer muscle to CO_2 causes sufficient hyperpolarization that the pacemaker potentials are ineffective and the muscle relaxes and the spiracle opens.

A bilateral pair of spiracles occurred on each thoracic segment, and one pair on each abdominal segment in the evolutionary primitive arrangement. Some insects have only one or two functional spiracles, and a few, particular ly some aquatic ones, have no functional spiracles at all, and tracheae do not open to the outside. Gas exchange between tracheae and the environment in this latter type of system occurs by diffusion through a thin cuticle.

5.1.1.3 The tracheal epithelium

Every part of the tracheal system contains living cells. Tracheal epithelial cells, derived like cuticular epidermal cells from embryonic ectoderm, surround tracheal tubes and tracheoles throughout all parts of the system. The epithelial cells secret the cuticulin lining and the hydrophobic compounds at the lumen surface of the tubules. In larger tracheae, the cells secrete a new cuticle lining on their apical side prior to each molt and the old lining of tracheae is shed at each molt.

Insect respiratory cells establish barriers through their extracellular matrices. These chitinous-matrices must be soft and flexible to provide ventilation, but also tight enough to allow oxygen flow and protection against dehydration, infections, and environmental stresses. Five chitinases and three chitinase-like genes organize the tracheal chitin-cuticles. Most of the chitinases degrade chitin from airway lumina to enable oxygen delivery. They further improve chitin-cuticles to enhance tube stability and integrity against stresses. Unexpectedly, some chitinases also support chitin assembly to expand the tube lumen properly. Moreover, chitinase Ⅱ plays a decisive role in the chitin-cuticle formation that establishes taenidial folds to support tube stability. Chitinase Ⅱ is apically enriched on the surface of tracheal cells, where it controls the chitin-matrix architecture independently of other known cuticular proteins or chitinases. It is supposed that the principal mechanisms of chitin-cuticle assembly and degradation require a set of critical glycosyl hydrolases for flexible and not-flexible cuticles (Behr et al., 2020).

5.1.2 Development of new tracheoles

The formation of new tracheae and tracheoles varies with

化阻止了分级电位跨越动作电位阈值的能力。将闭肌延时暴露于 CO_2 导致足够的超极化，即起搏器电位失效，肌肉松弛，气门打开。

在进化的原始排列中，每个胸节上有一对双侧气门，每个腹节上有一对气门。有些昆虫仅有一对或两对功能性气门，极少数昆虫，特别是水生昆虫，根本没有功能性气门，且气管对外不开放。在后一种类型的系统中气管和环境之间的气体交换通过薄表皮融氧完成。

5.1.1.3 气管上皮细胞

气管系统的每一个部分都含有活细胞。气管上皮细胞，像表皮的真皮细胞一样，从外胚层衍生而来，围绕整个系统所有的气管和微气管。上皮细胞在管腔表面分泌表皮质膜和疏水化合物。较大的气管，细胞在每次蜕皮前顶端分泌新的表皮膜，气管的旧表皮膜即脱落。

昆虫呼吸细胞通过其细胞外基质建立屏障。这些几丁质基质必须柔软和灵活，以保证通风，但必须足够紧密，以允许氧气的流动，并防止脱水、感染和环境压力。大多数几丁质酶降解气管腔中的几丁质以实现氧气输送。它们进一步改善几丁质表皮，增强管的稳定性和抗应力的完整性。出乎意料的是，一些几丁质酶也支持几丁质组装以适当地扩展管腔。此外，几丁质酶Ⅱ在几丁质表皮的形成中起着决定性的作用，几丁质表皮建立起螺旋丝折叠以支持管的稳定性。几丁质酶Ⅱ高度富集于气管细胞表面，并独立于其他已知的表皮蛋白质或几丁质酶控制几丁质基质结构。推测几丁质表皮组装和降解的主要机制需要一组关键的用于柔性和非柔性表皮糖基的水解酶（Behr 等，2020）。

5.1.2 新微气管的发育

新气管和微气管的形成随着组织呼

respiratory demand of tissues. The initial event in development of a new tracheole begins when a tracheal epithelial cell begins to grow out, usually in a triangular shape, toward a group of cells that apparently need better gas exchange. It may be that metabolites from tissue cells induce the tracheal epithelial cell to grow a process out toward the tissue. A tracheal epithelial cell that initiates such growth has been called a tracheoblast, stellate cell, transition cell and tracheal end cell. As it grows, the tracheoblast tends to become spindle-shaped, and a single unbranched tracheole may form within it at this stage. More often, however, the cell develops multiple finger-like projections before a tracheole forms, and then tracheole branches may form in the several fingers. A fluid- filled, linear channel begins to appear in the cytoplasm of the tracheoblast and gradually becomes longer as the tracheole grows. Developmental cues drive terminal tracheal cells formation, and oxygen regulates tracheoles remodeling such that low oxygen (hypoxia) induces tracheole formation, whereas increased oxygen (hyperoxia) suppresses it (Tamamouna et al., 2021).

吸的需求而改变。新微气管发育的最初事件起始于气管上皮细胞开始生长之时，通常呈三角形，朝向一组显然需要更好气体交换的细胞群。启动这种生长的气管上皮细胞称为成气管细胞，星形细胞，过渡细胞和气管终端细胞。当它生长时，成气管细胞变成梭形，在这个时期，单条而未分支的微气管可在其内形成。但通常细胞在微气管形成前发育成多指形突起，随后微气管分支在几个指形中形成。液体充填的线形通道开始在微气管的胞质中出现，随着微气管的生长逐渐变长。发育信号驱动气管端细胞的形成，而氧气调节微气管的重塑，使低氧（缺氧）诱导气管形成，而增加氧气（高氧）则会抑制气管形成（Tamamouna 等，2021）。

5.1.3 Air sacs

Dilations of both primary and secondary tracheae occur in many insects and are called air sacs. They are variable in size, but frequently are large in flying insects, such as honeybees, cicadas, many adult Diptera, and some scarab and buprestid beetles. The intima of air sacs may contain typical taenidium, but these may be reduced or irregular. Air sacs provide a large surface in contact with flight muscles for exchange of gases. The rhythmical squeezing action of working flight muscles pumps the air sacs like a bellows and increases the flow of air through the system. Air sacs serve a hydrostatic function in some aquatic insects, and allow more freedom in vibration of the tympanic membrane in some sound-producing insects. They may increase hemolymph concentration of solutes without necessarily increasing total solute and reduce hemolymph volume by restricting space that the volume of circulating hemolymph must serve. Quantitative analysis of three-dimensional image series reveals that the compression of the air sac during respiration in bell crickets exhibits obvious anisotropic characteristics both longitudinally and transversely (Xu et al., 2016).

5.1.3 气囊

许多昆虫的初级和二级气管都会膨胀并称之为气囊。其大小可变，在飞行昆虫中通常很大，如蜜蜂、蝉、很多双翅目成虫以及一些圣甲虫和柑橘丁吉虫。气囊的内膜含有典型的螺旋丝，但这些可能减少或不规则。气囊为飞行肌提供大表面积进行气体交换。飞行肌有节奏的挤压动作像风箱一样给气囊泵气，增加了系统通过的气流。气囊在有些水生昆虫中起着流体静力学的功能，在一些发声昆虫中允许鼓膜自由共鸣。它们能增加溶质在血淋巴中的浓度而无需增加总的溶质，通过限定循环血淋巴必需量占据的空间来减少血淋巴量。对三维图像序列的定量分析表明，钟蟋蟀呼吸过程中气囊的压缩在纵向和横向上都表现出明显的各向异性特征（Xu 等，2016）。

5.1.4 Molting of trachea

Prior to molt, tracheal epithelial cells secrete a new cuticulin layer on their apical surface. Often a fluid layer then separates the two cuticulin layers, the old and the new. The ecdysis of the external cuticle pulls the old, air-filled tubes out, leaving behind the newly formed, fluid-filled ones. The new tracheal tube fills with air rapidly, it seems likely that the fluid in the new system is actively reabsorbed into the surrounding tissue, thus making way for the air. Furthermore, Perochon et al.(2021) demonstrated the crosstalk between adult intestinal stem cells (ISC) in *Drosophila* and the vasculature-like tracheal system, which is essential for intestinal regeneration. Following damage to the intestinal epithelium, gut-derived reactive oxygen species activate tracheal and bidirectional signaling, leading to reversible remodeling of gut-associated terminal tracheal cells and intestinal stem cell proliferation following damage. They identified a mechanism of tracheal-intestinal tissue communication.

5. 1.4 气管的蜕皮

蜕皮前，气管上皮细胞在其顶端面分泌一个新表皮质层。常有一液体层将新旧两个表皮质层分开。外表皮的蜕皮推动旧的、充满气体的气管往外，留下新形成的、充满液体的气管。新气管很快被空气充填，好像是新系统的液体被周围组织主动再吸收，因而为空气清理了路径。此外，Perochon 等（2021）证明了果蝇成虫肠道干细胞（ISC）与对肠道再生至关重要的脉管样气管系统之间的相互作用。肠上皮损伤后，肠源性活性氧激活气管和双向信号，导致肠受损时相关的终末气管细胞可逆重塑和肠干细胞的增殖。它们确定了气管—肠组织沟通的一种机制。

5.2 组织器官的氧供应

5.2 Oxygen Supply to Tissues and Organs

The heart and dorsal musculature usually are aerated by tracheal branches in each segment from the dorsal longitudinal trunks. The visceral and internal reproductive organs receive tracheal branches from the lateral trunked. The ventral nerve cord and ventral musculature are supplied by branches from the ventral trunks or by branches off ventral transverse connective. The legs and wings are supplied by tracheae from the thoracic spiracles, while the head and associated structures are generally supplied with branches from the first spiracle and the dorsal longitudinal trunk. The only cells or tissues not having direct connections with tracheoles are the hemocytes, the cells that circulate in the hemolymph. Working muscles, especially flight muscles, have a very high demand for oxygen and there is an extensive tracheal supply to flight muscles enabling them to avoid an oxygen debt with prolonged flight.

Primary tracheae that originate at a thoracic spiracle may: ① pass through the core of a muscle, giving rise to numerous

心脏和背部肌肉组织常常由背部纵向主干到每个体节的气管分支而供氧充盈。内脏和内生殖器官接纳横走主干而来的气管分支。腹神经索和腹部肌肉组织，通过腹部气管的分支或腹部横肌连接的盲管提供氧气。腿和翅由胸部气门的气管供氧，而头部及其相连结构通常由第一对气门的气管和背纵主干供氧。唯一与微气管没有直接联系的细胞或组织是血细胞，其细胞在血淋巴中循环。运动中的肌肉，特别是飞行肌，对氧的需求非常高，飞行肌有延伸的气管使之在长距离的飞行中避免负氧。

初级气管起源于胸部气门：①通过肌髓，引发大量分支（中心径向系统）；

branches (centro-radial system); ② run along its surface with branches penetrating deep into the interior of the muscle (latero-linear system); ③ the tracheae may expand into air sacs that lie on top of the muscle (latero-linear system). The latter arrangement is common in small insects.

Secondary tracheae branch from the primary tracheae and radiate into the spaces between muscle fibers, and tertiary tracheae branch from the secondary. In the tergosternal muscle of *Aeshna* spp. dragonflies, tertiary tracheae branch from the secondary tracheae at about 20μm intervals, and eventually narrow and each branches into 20 to 30 tracheoles. In metabolically active tissues and, especially in flight muscles, tracheoblast push again the sarcolemma of a muscle fiber and indent the fiber like a finger pushed into a balloon to become intracellular tracheoles. Despite the term "intracellular", the tracheoles are not inside the cells they penetrate. Because they still maintain their own cellular epithelium, and the host-cell membrane, though indented, remains intact. Functionally, intracellular tracheoles bring gas exchange capability within a few tenths of a micrometer of mitochondria.

The tracheal system is very efficient, delivering O_2 through an air path branching into smaller and smaller tracheoles that often end within a few micrometers of mitochondria.

②沿肌肉表面伸出分支，深深渗入肌肉内部（侧向线性系统）；③气管伸展进肌肉上方的气囊（侧向线性系统）。后一种排布在小昆虫中较为普遍。

二级气管从初级气管分支辐射到肌纤维之间的空隙，三级气管从二级气管分支。在蜻蜓背腹肌，从二级气管分发出的三级气管的间隔约 20μm，最终变窄，每根分支分岔成20～30根微气管。在代谢活跃的组织，特别是飞行肌中，成气管细胞按压在肌纤维的肌膜上，肌纤维缩进，就像手指压在气球上成为细胞内微气管。尽管其术语为“胞内”，但微气管并不穿透进细胞内。因为它们仍保持其细胞上皮和宿主细胞膜，虽然缩进去，但保持完整。在功能上，胞内微气管携带交换气体的能力在零点几微米线粒体内。

气管系统非常有效，通过气体路径输送 O_2，再分支到越来越小的微气管，通常在几微米的线粒体内终止。

5.3 通风作用和系统气体扩散

5.3 Ventilation and Diffusion of Gases within System

Movement of O_2 and CO_2 through tracheal tubes is promoted by active ventilation in most insects. Simple diffusion may suffice in very small insects, but in all insects, body movements, and muscle and gut movements serve to pump the tracheal system and move gases within it, and aid gas exchange within the system.

Ventilation is the bulk movement of gases within the tracheal system and their exchange with the atmosphere *via* the spiracles. Muscular compression increases tracheal gas pressures and may drive air out through opened spiracles (expiration) expansion reduces tracheal pressure and may suck air in through spiracles (inspiration). Ventilation occurs in one of three principal forms:

在大多数昆虫中，O_2 和 CO_2 在气管中的运动通过主动通风作用推动。非常小的昆虫简单扩散就已足够，但对所有昆虫来说，躯体移动、肌肉和肠道运动起着气管系统泵的作用，使气体在其内部运行并有助于气体交换。

通气是气体在气管系统内的大量运动，并通过呼吸孔与大气进行交换。肌肉压缩会增加气管气体压力，并可能通过打开的呼吸孔排出空气（呼气）。扩张会降低气管压力，并可能通过呼吸孔

① rhythmical muscular pumping, usually by the abdomen, in which tracheae are alternately compressed and expanded; ② autoventilation, which results as a side effect of other types of movement principally locomotory; ③ passive suction ventilation, dependent on precise spiracle control which allows a negative pressure to be maintained in the tracheae with small amounts of air being periodically inspired.

吸入空气（吸气）。通气有三种主要形式：①有节奏的肌肉泵送，常通过腹部，其气管交替压缩和扩张；②自动通气，这是其他类型运动，主要是移动的连带作用；③被动吸入通气，依靠精确的呼吸控制，使气管保持负压，并定期吸入少量空气。

5.3.1 Simple diffusion

The simple formula for calculating the pressure necessary to cause oxygen to diffuse from the spiracles to the site of use applies to those systems in which the cross-sectional area of all tracheae is constant from the last point of branching, and can be calculated as:

5.3.1 简单扩散

计算氧气从气门到利用位点扩散所需压力的简单公式，适用于所有从最后分支点气管的横截面积是恒定的那些系统，计算如下：

$$\Delta p = m\mathrm{L}^2/2aP$$

where Δp is the pressure necessary to cause oxygen to diffuse over the distance L(cm)to satisfy the metabolic rate; *m*, in mL O_2/g tissue/min that is used; *P* is *a* permeability constant for O_2 diffusion in an air path; the hole fraction is represented by a. “Hole fraction” value, ranging from 10^{-1} to 10^{-2} for the secondary tracheal supply to wing muscles, and 10^{-2} to 10^{-3} for the tertiary tracheae and tracheoles to wing muscles.

Δp 是引起氧气扩散到距离 L（cm）满足代谢率所需的压力；*m* 为代谢率[mL O_2/（g 组织 · min）]；*P* 是 O_2 在空气路径内恒定的渗透率；孔洞分数用 *a* 表示。“孔洞分数”值在二级气管向翅肌供氧为 10^{-1} ～ 10^{-2}，三级气管和微气管向翅肌供氧为 10^{-2} ～ 10^{-3}。

5.3.2 Active ventilation of tracheae

Abdominal pumping may occur by dorsoventral compression of the abdomen or by longitudinal telescoping of abdominal segments, and pumping of large tracheae and air sacs is created by action of the large flight musculature in the thorax when an insect is flying. Moreover, simply the change in the shape of the thorax during flight acts as a pumping mechanism. Finally, some insects seem to get ventilation from suction-ventilation pressure differences in the large tracheae in the thorax with differential opening and closing of thoracic spiracles. Abdominal pumping is common in large insects at rest and continues during flight. In some fast-flying insects, abdominal pumping alone is sufficient to supply the wing muscles during flight.

Used air and carbon dioxide are released from the posterior thoracic spiracles in *M.sexta*. The downstroke of the wings in

5.3.2 气管的主动通风作用

腹部背腹肌挤压或腹部体节的纵肌压缩可形成腹泵，当昆虫飞行时，胸部的大飞行肌组织会产生大型气管和气囊的抽吸。此外，飞行时简单改变胸部的形状可以起到气泵机制的作用。最后，一些昆虫的胸部大气管通过分别打开和关闭胸部气门，用抽吸 - 通风压力差进行通风。腹泵在大昆虫休息时常见的，并在飞行期间持续。有些快速飞行的昆虫，飞行时只需要腹泵就足以向翅肌供氧。

烟草天蛾使用过的空气和二氧化碳通过后胸部气门释放。飞行中翅的

flight slightly increases the volume of the thorax and increases the volume of the thoracic air sacs, creating a slight negative pressure that sucks air in at the mesothoracic spiracles. The openings to the posterior thoracic spiracles are compressed by the downward movement of the wing hinge, but during the upstroke the thoracic air sacs are compressed and the posterior spiracles are open so that expiration occurs. Coelopulses enable insects to breathe through selected spiracles and promote a unidirectional ventilation stream of air through the body. In dragonflies and locusts, thoracic muscular pumping is important for ventilation of wing muscles, and experimentally can be demonstrated to be sufficient in the absence of abdominal mechanisms.

Thoracic pumping tends to create a tidal flow that moves air in and out of the same spiracles, while abdominal pumping promotes a directed flow in through thoracic spiracles and out through abdominal spiracles. Abdominal pumping also can create a tidal flow in some insects. Each abdominal ganglion of *Schistocerca* is capable of initiating ventilatory movements for its segment, but the rhythm that synchronizes the overall movements originates in the metathoracic ganglion. Humoral mechanism, nerve activation, neuromodulators, and ionic or metabolic products from working muscle as possible mechanisms that could be involved in controlling activity-related increase in ventilation rate.

5.3.3 Diffusion from tracheoles to mitochondria

There is little or no difference in the permeability of large tracheae and tracheoles to O_2, but only tracheoles usually will be close enough to mitochondria for O_2 to diffuse across the aqueous path to the mitochondria in quantities needed. In its final path from tracheoles to mitochondria, O_2 must move by diffusion, crossing tracheoles wall, cell membranes, cell cytoplasm, and mitochondrial membranes. This is the slowest part of the pathway for the final delivery of oxygen for metabolism. The diffusion of O_2 in a water-filled path is about one million times less rapid than its movement in an air path, thus the closer the tracheoles can get to the site of mitochondria, the higher the rate of O_2 consumption that can be supported.

Even in large active insects that ventilate, diffusion is a significant process, because the ventilation movements serve only to move the air in the larger tracheae. Diffusion is important in moving gases between the tracheoles and mitochondria of the

下行冲程轻微增加了胸部容积和胸部气囊的容积，造成轻微的负压，从中胸吸入空气。后胸气门的开口在后翅向下运动时被压缩，但上行冲程时胸部气囊被压缩，后部气门被打开，因而出现吐气。腹腔脉动使昆虫通过选择性气门呼吸，推动单向通风气流通过体躯。蜻蜓和蝗虫，胸肌气泵作用对翅肌的通风作用是重要的，实验能证明没有腹泵机制也足够。

胸肌泵趋于形成潮汐式气流，从同一个气门进出气体。而腹泵推动定向气流，从胸部气门进气，腹部气门排气。有些昆虫的腹泵也能产生潮汐式气流。沙漠蝗的每个腹神经节能启动其所在体节的通风运动，但节律同步于中胸神经节引起的整体运动。体液机制、神经活动、神经调质以及肌肉产生的离子或代谢产物可能参与调控与活动相关的通风速率增加的机制。

5.3.3 从微气管到线粒体的扩散

大气管和微气管对 O_2 的渗透性有细微或没有差别，但通常只有微气管在线粒体的 O_2 充足时被关闭，以需求量跨越水相路径扩散到线粒体。在微气管到线粒体的最后途径，O_2 必须通过扩散移动，跨越气管壁、细胞膜、细胞质和线粒体膜。这是代谢终端输送氧气途径的最慢部分。O_2 的扩散在充满液体的途径中，大约小于空气中移动速度的百万分之一，因此，气管离线粒体越近，能支持的氧气消耗率就越高。

即使在通风的大型活动昆虫中，扩散也是一个重要的过程，因为通风运动只会带走较大气管中的空气。扩散在微气管和组织细胞的线粒体之间的气体移

tissue and cells (Fig.5-5).

In many insects, distal parts of tracheoles are not filled with air but interstitial fluid under normal resting conditions. During activity, however, the tracheoles become completely air-filled; that is, fluid is withdrawn from them only to return when activity ceases. The level of fluid in tracheoles depends on the relative strengths of the capillary force drawing fluid along the tube and the osmotic pressure of the hemolymph. During metabolic activity, the osmotic pressure increases as organic respiratory substrates are degraded to smaller metabolites, causing fluid to be withdrawn from the tracheoles and, therefore, bringing gaseous oxygen closer to the tissue cells. As the metabolites are fully oxidized and removed, the osmotic pressure will fall, and once again the capillary force will draw fluid along the tracheoles (Fig.5-6). The tracheoles are the main site of gas exchange with tissues; they closely surround and might even penetrate the tissues to become intimately associated with individual cells and to lie close to mitochondria where cell respiration and energy metabolism occur. Carbon dioxide moves in the opposite direction. Increased ionic strength or acidification of the interstitial fluid withdraws fluid from the tracheoles, facilitating gas exchange during activity or varying oxygen partial pressures.

Menail et al. (2022) demonstrated that mitochondrial functions are more resilient to high temperatures in honeybees compared to fruit flies. The data indicated an important but species-specific mitochondrial flexibility for substrate oxidation to sustain high oxygen consumption levels at high temperatures. Mitochondrial respiration increased with increasing temperature and this respiration is

动中很重要（图 5-5）。

许多昆虫在正常的静息条件下气管的远端不是充满空气，而是充满组织间液。然而，在活动期间，微气管完全充满空气，也就是说，只有当活动停止，流体才会从中抽出并返回。微气管内液体的水平取决于沿毛细管抽吸液体的力量和血淋巴的渗透压。在代谢活动期间，随着有机体呼吸基质降解为较小的代谢物，渗透压增加导致液体从气管中排出，从而使气态氧更接近组织细胞。当代谢物被完全氧化和去除时，渗透压将下降，毛细管压力将再次沿着微气管吸入液体（图 5-6）。微气管是与组织进行气体交换的主要场所；它们紧密围绕，甚至可能穿透组织，与单个细胞密切相关，并靠近线粒体，在线粒体中发生细胞呼吸和能量代谢。二氧化碳的运动方向相反。增加的离子强度或组织间液酸化可将液体从气管中抽出，促进活动期间的气体交换或改变氧分压。

Menail 等 (2022) 证实，与果蝇相比，蜜蜂的线粒体功能更能适应高温。数据表明，线粒体具有重要但又有物种特异的灵活性，以维持在高温下底物氧化的高耗氧水平。线粒体呼吸随温度升高而增加，甲虫的这种呼吸由脯氨酸维

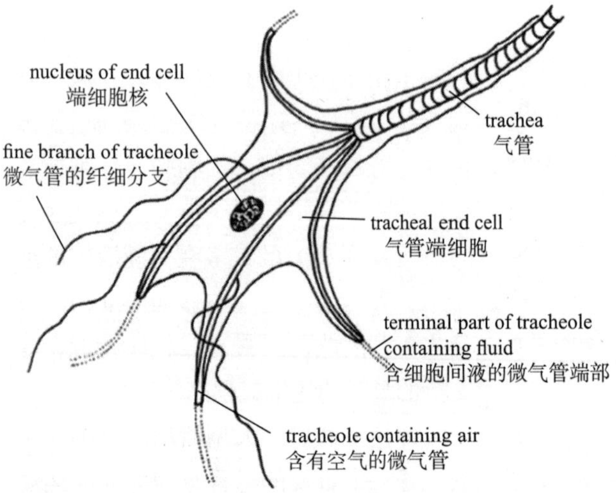

Fig.5-5 The structure of tracheal end cell and tracheoles.
图 5-5 气管端细胞处的结构。

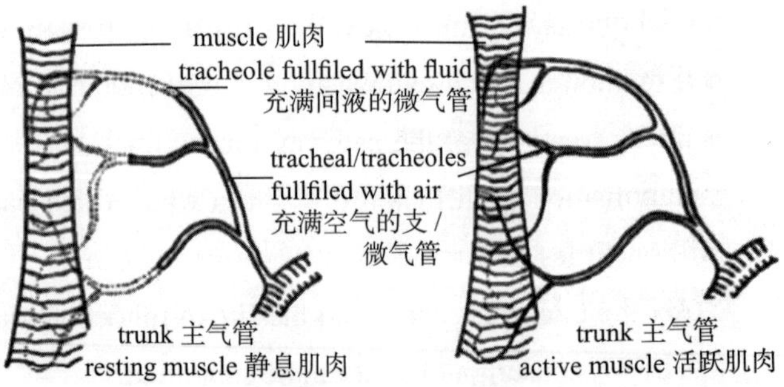

Fig. 5-6 Diffusion mechanism of air between tracheoles and tissues under active and rest condition.
图 5-6 静息和活跃状态下空气在微气管和肌肉组织间的扩散机制。

sustained by proline in beetles (*L. decemlineata*), mainly by glycerol-3-phosphate in fruit flies (*D. melanogaster*), and by succinate and glycerol-3-phosphate in honeybees (*A. mellifera carnica*). Moreover, oxygen consumption rates when fueled with all substrates were still intact at 45 ℃ in honeybees and beetles, but drastically declined in fruit flies. This highlights the metabolic flexibility of insects' mitochondria that can rely on the oxidation of a large set of substrates fueling mitochondrial complexes with different thermal tolerances to withstand increasing temperatures.

持，果蝇主要由甘油 -3- 磷酸维持，蜜蜂主要由琥珀酸和甘油 -3- 磷酸维持。此外，在 45℃条件下，蜜蜂和甲虫在添加所有底物后的耗氧率仍保持不变，但果蝇的耗氧率急剧下降。这突出了昆虫线粒体的代谢灵活性，它可以依赖于大量底物的氧化，为具有不同热耐受性的线粒体复合体提供燃料，以承受不断升高的温度。

5.3.4 Discontinuous gas exchange

Some insects are able to keep the spiracle tightly closed and/or "apparently closed" for a high percentage of the time. Gas exchange occurs in three periods named the open, flutter and closed periods because of the action of spiracles over the duration of a cycle. This functional pattern, variously known as chthonic hypothesis (Lighton, 1996): discontinuous release of CO_2, passive suction ventilation, discontinuous ventilation cycle, and as the discontinuous gas exchange cycle (DGC). During DGC, accumulated CO_2 is discharged periodically in episodic bursts during brief intervals when spiracles are open. After a burst, the spiracles are closed for some period of time that varies from species to species. During the closed interval, oxygen is used by tissues and intratracheal oxygen tension falls. Most insects that exhibit DGC allow the spiracular valve to flutter with an amplitude often imperceptible to the unaided eye during a portion of a cycle. The fluttering (F) phase allows small amounts of O_2 to be sucked into the tracheal system by the slight negative pressure arising from O_2 consumption by the tissues. The F phase is usually considered to involve convective transfer O_2, so named "passive suction ventilation" that sometimes is applied to the process. Spiracular fluttering increases oxygen uptake, further, variation in flutter rate is an effective mechanism for controlling oxygen uptake, and therefore is an important part of insects' solutions to the anatomical, physiological, and environmental challenges (Lawley et al., 2020). The slight internal vacuum retards the outward loss of water vapor and CO_2 during the F phase, and the low influx of O_2 lengthens the time to full opening or "burst"

5.3.4 不连续的气体交换

有些昆虫能在高百分比时间内保持气门紧闭和 / 或"明显关闭"。气体交换发生在三个时段内，由气门在循环周期内的活动而命名为开启、颤动和关闭期。这种不同实用模型被称为索尼克假说（Lighton，1996）：CO_2 的不连续释放，被动抽吸通风，不连续的通风换气循环和不连续的气体交换循环（DGC）。在 DGC 期间，当气门开启时，积聚的 CO_2 在短暂的时间间隔内瞬间爆发式地被定期排出。爆发式排出之后，气门关闭一段时间，其时间的长短因品种而异。关闭间歇期内，氧气被组织利用，气管内氧气分压下降。多数昆虫显示 DGC 允许气门片在循环的某部分通常有细微到肉眼察觉不到的小幅度颤动。颤动期（F）由组织消耗 O_2 引起的轻微负压，允许吸入少量的 O_2 到气管系统。通常认为 F 期牵涉到 O_2 的对流传输，有时适用于这个过程，因此被命名为"被动吸入通风"。气门颤动增加氧气的吸收，而且颤振率的变化是控制吸氧量的有效机制，因此是部分昆虫应对解剖学、生理学和环境挑战的重要组成部分（Lawley 等，2020）。轻微的内部真空延缓了 F 期内水蒸气和 CO_2 的向外损失，O_2 的流入量低，使时间延长以便

of spiracles. CO_2 is produced by tissues even when the spiracles are closed, but the high solubility of CO_2 in aqueous solutions enables insects to accumulate bicarbonate ion, HCO_3^-:

Buffering capacity of hemolymph aids in solubilizing CO_2 as

$$CO_2 + H_2O \longrightarrow H_2CO_3 \longrightarrow H^+ + HCO_3^-$$

HCO_3^-, and this keeps gaseous CO_2 from build ing up rapidly in the tracheal system. At some point, probably different for different insects, the relationship between gaseous O_2 and CO_2, and HCO_3^- in solution reaches an equilibrium at which tracheal tension of CO_2 and O_2, and/or pH change in the hemolymph, trigger spiracle opening and release of CO_2 from the hemo lymph as a gas. However, limits to oxygen diffusion to the tissues during spiracle closure trigger ventilatory movements, which in turn support tissue demands.

Low oxygen tension acts centrally on the ganglion controlling a spiracle to trigger fluttering of spiracular valves, while CO_2 acts directly on the closer muscle of *H. cecropia*. Diapausing cecropia pupae can tolerate very low oxygen tensions in the tracheae before fluttering is triggered. Discontinuous respiration is highly functional and adaptive for cecropia pupae, which spend the winter buried beneath litter and soil.

This pattern might serve to reduce respiratory water loss; may also have initially evolved in underground insects as a way of dealing with hypoxic or hypercapnic conditions; and a necessary consequence of the need to rid the respiratory system of accumulated CO_2, and reduce dangerously high levels of oxygen toxicity. The insects might show variation in their metabolism and respiratory patterns among ontogeny and species. For example, all life stages of *L. migratoria* exhibited discontinuous gas exchange cycles, but gregarious brown locusts, *Locustana pardalina* showed the 1st instars used a form of continuous gas exchange while all the other stages showed discontinuous gas exchange cycles (Duncan et al., 2018).

However, the evolution of DGC in insects is highly controversial. Adaptive hypotheses which have obtained experimental support include a water savings mechanism for living in dry environments (hygric hypothesis); a reduction in oxidative damage due to a high-performance oxygen delivery system (oxidative damage hypothesis); the need for steep intratracheal partial pressure

完全打开气门或使气门“爆开”。即使气门关闭，组织也会产生CO_2，但CO_2在水溶液中的高溶解性使昆虫能够积累碳酸盐离子HCO_3^-：

血淋巴的缓冲能力有助于溶解CO_2和HCO_3^-，从而防止CO_2在气管内的迅速积累。在某个点上，也许不同的昆虫不同，气态O_2和CO_2与溶液中的HCO_3^-间达到平衡，O_2和CO_2分压以及/或者血液pH的变化，触发气门的开放和血淋巴中CO_2以气态释放。然而，在气门关闭期间，限制氧气向组织扩散会触发通气运动，转而支持组织需求。

低氧气分压主要作用于神经节控制气门，促使气门片摆动，而CO_2直接作用于天蚕蛾的闭肌。滞育的天蚕蛹在颤动前能耐受气管内非常低的氧气分压。不连续的呼吸对在杂物和土壤中越冬的天蚕蛹是高度功能性和适应性的。

这种模式可能有助于减少呼吸失水；也可能最初在地下昆虫中进化，作为处理缺氧或高碳酸状态的一种方式；需要清除呼吸系统中累积的二氧化碳，并减少危险的高水平氧毒性的必然结果。昆虫在个体发育和物种间可能表现出代谢和呼吸模式的差异。如东亚飞蝗的所有生命阶段都表现出不连续的气体交换循环，但群居的棕褐色拟飞蝗1龄蝗虫使用了一种连续的气体交换形式，而所有其他阶段都表现出不连续的气体交换循环（Duncan等，2018）。

然而，昆虫体内不连续气体交换的进化过程备受争议。已获得实验支持的适应性假设包括生活在干燥环境中的节水机制（湿度假设）；由于高性能氧气输送系统而减少的氧化损伤（氧化损伤假设）；在地下环境中可能遇到的

gradients to exchange gases under the hypercapnic and/or hypoxic conditions potentially encountered in subterranean environments (chthonic hypothesis) (Terblanche et al., 2008).

高二氧化碳和 / 或缺氧条件下，需要急剧升降的气管内分压梯度来交换气体（chthonic 假说）(Terblanche 等 , 2008)。

5.4 水生昆虫的气体交换

5.4 Gas Exchange in Aquatic Insects

The tracheal system of most aquatic insects is structurally the same as that of terrestrial insects, i.e., open spiracles and an extensive network of tracheae and tracheoles. These aquatic insects breathe air by frequen tly coming to the surface. Water is presented from entering the system by the hydrophobic surface of the tracheae and, in some cases, by closed spiracles. Many aquatic larvae have a metaneustic system, with only one functional pair of spiracles near the tip of the abdomen. When the insect comes to the surface, it does so with the posterior end upper most and only the tip of the abdo men bearing the spiracles is held above the surface. Some have a siphon at the tip of the abdomen that is pushed above the surface of the water for gas exchange. During emergence, the spiracles are kept closed. Some aquatic insects living under water must be able to break the surface film of water to get air from the surface. Typically they have structural adaptations of the body (hydrofuges) that are difficult to wet and these facilitate surfacing and hanging at the surface. The hydrofuge repels water, and tends to buoy the insect at the surface of the water.

Water's high capacity for CO_2 but low oxygen solubility means that ventilation in aquatic animals tends to be driven more by lack of oxygen than by the need to excrete CO_2. Ventilatory responses to hypoxia and hypercapnia indeed differed between the aquatic nymphs and the terrestrial adults and matched the expectation that water-breathing nymphs would not respond to hypercapnia.

多数水生昆虫的气管系统在结构上和陆生昆虫相同，即具有开放的气门和大量气管和微气管网络。这些水生昆虫通过频繁浮出水面呼吸空气。水从疏水性气管表面进入系统，在有些情况下，通过关闭气门。许多陆生幼虫具后气门系统，仅在靠近腹部尖端有一对功能性的气门。当昆虫到达水面，多数用后端向上，只有腹部尖端气门部位露出水面。有的在腹末具虹管，可推出水面进行气体交换，而浮出水面时的气门是关闭的。有些生活在水下的水生昆虫必须能够打破水面薄膜，以便从水面获取空气。通常，它们的虫体有结构上的适应性（防水），很难被弄湿，这些结构使昆虫浮出和悬停于水面非常方便。防水结构排出水分，使昆虫浮在水面上。

二氧化碳在水中的溶解度很高，但氧的溶解度很低，意味着水生动物的通风往往更多地是由缺氧而非排泄二氧化碳的需要驱动的。水生若虫和陆生成虫对低氧和高碳酸血症的通风反应确实不同，这与水栖若虫不会对血液高碳酸症作出反应的预期相符。

5.4.1 Compressible gas gills

A large number of aquatic insects submerge with a bubble or film of air enclosing one or more spiracles. Compressible gas gills, a bubble or film of air somewhere on the body, are wide

5.4.1 压缩性气鳃

大量水生昆虫用气泡或气膜封闭一个或多个气门潜入水中。压缩性气鳃是体表某处的气泡或气膜，在鞘翅目、膜

spread among aquatic members of the Coleoptera, Hemiptera, and Lepidoptera. Compressible gills slowly collapse as the oxygen is used, although additional oxygen is usually extracted from the water before the bubble must be renewed at the surface. Air stores in compressible gas gills may be carried in a fine network of hydrofuge hairs. Sometimes there is a fine, dense set of hairs nearest the body surface from which the gas volume is very slowly used, and a set of longer, larger hairs over these, from which the gas is rapidly used. Alternatively, a gas gill may be carried beneath the elytra as in dytiscid diving beetles, or as a gas bubble on the posterior part of the abdomen. In addition to serving a respiratory function, gas gills also provide buoyancy and a hydrostatic function. When the insect comes to the surface to renew the air store, the buoyancy of the remaining bubble allows the part of the body carrying the bubble to come to the top first. Thus, the air in the bubble is restored with a minimum of exposure of the insect at the surface.

Immediately after an insect has surfaced, any air bubble on its body should contain approximately 21% O_2 and 79% N_2, in equilibrium with the atmosphere. As the insect submerges, O_2 will be used for metabolic processes and O_2 tension in the air bubble will fall, while the partial pressure of the N_2 will rise. The N_2 in the bubble play a crucial role because its relatively low solubility in water causes it to move out of the lapsing, giving oxygen from the water a chance to diffuse in. However, Jones et al (2021) demonstrated that the small gas gill of *P. decempunctatus* contributes less than 10% of the total O_2 used during the dive, while up to 10% is supplied by cutaneous uptake.

5.4.2 A plastron

Incompressible gills do not collapse and oxygen can continue to be extracted indefinitely from the water into the gill, allowing the insect to live underwater. It also called plastron. A plastron consist of any extensive physical meshwork, either of fine hairs or setae, or a meshwork of small pores and channels in the cuticular surface of some insects and eggs that can hold a volume of air and can present a large water-air interface. Plastrons are common adaptations of insects living in aquatic arrangements and can take many physical forms. A plastron contains a film of air so tightly guarded by a dense network of non-wetable hairs or meshwork of pores that even though the gas equilibrium may change due to O_2 use, water

翅目和鳞翅目水生昆虫的成员中广为分布。压缩性气鳃在氧气耗尽时缓缓塌瘪，虽然额外的氧气常常从水中提取，但塌瘪前，气泡必须在水面更新。贮存在压缩性气鳃中的空气可通过纤细的防水毛携带。有时，近体表处有纤细而浓密的毛，气体在此处被非常缓慢地利用；而覆盖此处的毛如果较长而大，空气即被迅速利用，或者气鳃可被像龙虱一样的潜水甲虫在翅基下携带，或在腹后部携带气泡。除了起到呼吸功能，气鳃也提供浮力和流体静力作用。当昆虫浮出水面更新贮存的空气时，气泡剩余的浮力使携带气泡的那部分虫体率先到达。因此，气泡中的空气在昆虫曝于水表面最少时得到恢复。

昆虫一旦浮到水面后，其体表上的任何气泡都将含有大约 21% 的 O_2 和 79% 的 N_2，与大气平衡。当昆虫潜入水中，O_2 将在代谢过程中被利用，气泡内 O_2 分压将下降，而 N_2 的分压将上升。气泡内的 N_2 起着关键作用，因为其在水中的相对低溶解度使之在失效时逸出，给水中的氧气以扩散进入气泡的机会。然而，Jones 等（2021）证实潜水甲虫的小气鳃贡献的氧气不到潜水时使用的总氧的 10%，而高达 10% 的氧气是由皮肤吸收提供的。

5.4.2 气盾

非压缩鳃不坍塌，氧气能无限期地连续从水中抽出进入鳃，使昆虫生活在水下，也称为气盾。气盾由任何广泛的物理网组成，要么是纤毛或刚毛，要么是有些昆虫体表面的小气孔和通道形成的网络，能容纳一定体积的空气并有很大的水 - 气界面。气盾是准备生活在水下的昆虫的普遍适应性，可以有多种物理形式。气盾包含的气膜由密集的不湿毛网络或小气孔网紧紧防护着，即使发

cannot invade the air space.

5.4.3 Use of aquatic plants as air source

Some insects with a hydrofuge are able to capture and utilize the gas bubbles released by aquatic plants, either by inserting a part of the body bearing a spiracle into the plant tissue or by biting into the air spaces of the plant. Some species of Diptera, Coleoptera, and Lepidoptera independently evolved modifications for piercing aquatic plants for air. The larvae mine the leaves, which usually are underwater, and the larvae get their air either from the plant or by cutaneous diffusion through the cuticle. Before pupariation, a larvae leaves the leaf and anchors itself to the stem of the weed by inserting its paired anal spines into the stem, and the pupa gets its air from the plant through the terminal spiracles.

5.4.4 Closed tracheal system

A closed tracheal system without functional spiracles is present in some aquatic insects. The lack of functional spiracles eliminates any chance that water will enter the system, but oxygen also must diffuse through the cuticle, a breathing mechanism called cutaneous respiration. Internally these insects still have an extensive tracheal system like terrestrial insects. Often in larger, more active insects there are tracheal gills that greatly increase the cuticular surface for gas exchange with the water. Relatively large larvae of Trichoptera, Plechoptera, Odonata, and some Lepidoptera utilize cutaneous respiration that is facilitated by extensive elaborations of thin hair-like or flap-like cuticular extensions from the body surface called tracheal gills.

The tracheal gills are highly variable in structure and in location on the body. The cuticle of insects with tracheal gills is very thin, and usually large numbers of small tracheae and tracheoles lie just beneath the cuticle. The highly structured arrangement of tracheoles that are uniform in size and spacing in the gills as an adaptation to utilize optimal functional efficiency with a minimum of tracheoles. The tracheoles run parallel to the gill surface and are not uniform in size, but vary in diameter from 0.2 to 1.0μm.

Pupae of the simuliids (dipterans) develop plastrons on the gill surface at the last larval molt. The plastron fills with air shortly after the larval-pupal ecdysis and, after ecdysis, the plastron expands into its functional appearance before the cuticle hardens.

生改变，水也不能侵入空气间隙。

5.4.3 利用水生植物作为空气来源

有些防水昆虫能捕获和利用水生植物释放的气泡，通过将虫体气门部位插入植物组织或刺进植物的空气间隙来实现。双翅目、鞘翅目和鳞翅目的一些品种，自主进化演变，通过穿透水生植物获取空气。幼虫常开采水下的叶片，幼虫从植物获取氧气或通过表皮进行皮肤扩散。化蛹前，幼虫离开叶片，用成对的肛门棘刺进茎部将其固定在杂草的茎上，蛹通过末端气门从植物中获取空气。

5.4.4 封闭的气管系统

有些水生昆虫存在没有功能性气门的封闭的气管系统。功能性气门的缺乏排除了水进入系统的任何机会，但氧气也必须通过表皮扩散，这是一个被称为皮肤呼吸的呼吸机制。这些昆虫的内部仍然具有和陆生昆虫一样的广泛气管系统。较大、更为活跃的昆虫常有气管鳃，这大大增加了与水进行气体交换的体表面积。相对而言，毛翅目、襀翅目、蜻蜓目和某些鳞翅目大幼虫利用皮肤呼吸，大量精致的细毛类或气门片类的表皮从体表伸展开来称之为气管鳃。

气管鳃在体表上的结构和位置高度变异。有气管鳃的昆虫表皮非常薄，在表皮正下方常有大量小气管和微气管。微气管高度结构化的排列，其大小和鳃的空隙均匀，用于适应以最少的气管利用最佳的功能效率。微气管与鳃表面平行，大小不一，其直径为 0.2 ～ 1.0 μm。

末龄幼虫蜕皮时，蚋（双翅目）的蛹在鳃表面长有气盾。充气的气盾在幼虫到蛹的蜕皮后很短暂，蜕皮后，气盾在表皮硬化前伸展成功能性外貌。当

After O_2 diffuses across the thin cuticle of the typical gill surface and enters the tracheae, it probably is distributed to different body tissues much as in terrestrial insects by diffusion and by body and muscular movements. Movement of water over the gill and body surface of aquatic insects is important in maintaining a fresh supply of oxygenated water in contact with the body, and most use undulations of the body and/or movements of the gills themselves to create ventilatory currents of water. Larvae of some dragonflies draw water into the rectum by elastic expansion of the body as dorsoventral compressor muscles relax.

O_2跨过常见鳃表面的薄表皮扩散进气管，它可能通过扩散身体和肌肉运动分布在不同的具体组织中，就像在陆地昆虫中一样。水流过鳃和水生昆虫体表面的运动对维持与虫体接触的含氧水的新鲜供应是重要的，多数用躯体波动和/或鳃本身的运动造成水的对流。有些蜻蜓的幼虫通过躯体伸缩扩展，如背腹压缩肌松弛将水吸进直肠。

5.5 其他呼吸形式

5.5 The Other Forms of Respiration

5.5.1 Respiration in endoparasitic insects

Many hymenopterans and dipterans are parasitic on other insects. The first instar often has a fluid-filled tracheal system, necessitating cutaneous respiration. Although air displaces the fluid at the molt into the second instar, the mechanism for clearing the fluid from the system is not known. Possibly the fluid is absorbed into the tissues of the larva. The spiracles become functional only just before the larva is ready to leave the host to pupate. Many chalcid wasp and tachinid flies that hatch from eggs laid on the surface of the host insect eat their way into the host. They have a metaneustic system and orient the posterior pair of spiracles at the body surface of the host so that they breathe air directly. Chalcid wasp remain in contact with the hollow pedicel of the egg from which they hatched at the host's integumental surface. Some tachnids stimulate the host's integument to become invaginated into a sheath around the parasite, leaving it with an air opening at the host's surface. The larvae of hot flies, *Hypoderma* spp. migrate to the skin of the vertebrate host where it bores a tiny opening to the surface through which gas exchange occurs. In earlier instars that are migrating, respiration is presumably cutaneous.

5.5.1 内寄生昆虫的呼吸作用

很多膜翅目和双翅目寄生在其他昆虫上。一龄期常具充液气管系统，需要皮肤呼吸。虽然在蜕皮进入二龄期，空气代替了液体，但从系统清除液体的机制尚不清楚；液体可能被幼虫组织吸收了。气门只有在幼虫离开宿主准备化蛹前才有功能。很多小黄蜂和寄蝇从产于其上的寄主昆虫表面的卵孵化，以它们的方式进入宿主体内。它们有后气门系统，后部的成对气门朝向宿主体表以便直接呼吸。寄生黄蜂保留与卵的中空蒂，它们是从宿主体壁表面的卵孵化的。有些寄蝇刺激宿主的体壁变成包围寄生物的凹陷鞘，给其在宿主表面留下一个空气开口。牛皮蝇蛆的幼虫在脊椎动物宿主皮下移动，通过挖掘通往体表小孔进行气体交换。早期移动时，推测其呼吸的类型是皮肤呼吸。

5.5.2 Respiration in eggs and developing embryos

The embryo developing inside the egg must obtain sufficient oxygen for development. The majority of aquatic and semiaquatic insects lay eggs with no special respiratory structures incorporated

5.5.2 卵和发育中胚胎的呼吸

卵内发育中的胚胎必须获得足够的氧气进行发育。多数水生和半水生昆虫产下的卵没有专门的与卵壳合为一体的呼

into the shell, while eggs of a majority of terrestrial insects contain special structures for respiration, including an extensive, inner chorionic meshwork that can function as a plastron when the egg is submerged in well-aerated water. Plastrons on eggs laid in a terrestrial environment may help prevent an O_2 deficiency if the eggs are subjected to only short periods of wetting from dew, rain, or temporary flooding.

吸结构，而多数陆生昆虫的卵有专门用于呼吸的结构，包括广泛的壳内网，当卵浸泡在气体充足的水中时，其功能上类似于气盾。产于陆地环境的卵，在受到短暂的露水、雨水或暂时的水浸湿润时，其气盾有助于防止氧的缺乏。

Gas exchange in eggs with no special respiratory structure occurs by simple diffusion through interstices in the egg shell. Developing eggs of *M.sexta* experienced O_2 limitation even at normoxic conditions, and metabolic rates of eggs were depressed at higher temperatures under hyperoxic conditions. The eggs of many terrestrial insects frequently are laid in wet environments, including decaying organic matter, animal manure, fruits, leaves, and stems of plants. To be effective, the plastron must resist the action of naturally occurring surfactants that often are present in animal dung and decaying flesh. Clearly it is an adaptation to the egg environment. Some eggs have the plastron elevated on a respiratory horn, with the plastron covering most of the surface of the horn. When the egg is submerged completely in water, the length of the longer horns makes the gradient for gas exchange favorable only in the proximal part of the horn. It may be that long horns are useful as a conduit to atmospheric air when the egg is not too deep in water.

没有专门呼吸结构的卵，通过卵壳上的缝隙以简单扩散的方式进行气体交换。烟草天蛾发育中的卵即使在含氧量正常的条件下也会经历氧限制；在含氧量高的条件下，高温使卵的代谢率受到抑制。许多陆生昆虫的卵常常产于潮湿的环境，包括腐烂的有机质、动物粪便、水果、叶片和植物的茎。为了保持其有效性，气盾必须经得住动物粪便和腐肉表面，常常是自然出现的活性剂的作用。显然，这是对卵所处环境的一种适应性。有些卵的气盾升高到呼吸角质物上，角质物的大部分表面都被气盾覆盖。当卵完全浸泡在水中时，比较长的角质物的长度产生斜度，使气体交换仅在角质物顶部顺利进行。也许，在卵入水不太深时，长角质物是通向大气的有用管道。

5.5.3 Respiration and oxygen uptake in pupae

5.5.3 蛹的呼吸和耗氧

For diapausing pupae and certain other postembryonic stages of insects, the slight net inflow of air during the constricted-spiracle phase of the DGC may serve as a means of conserving moisture that would otherwise be lost as a result of gas exchange. Mosquito pupae evolved to adapt for living in the water by using oxygen for respiration. The respiratory system of mosquito pupae, as aquatic insects, has limitations in accessing atmospheric oxygen. Mosquito pupae remain most of the time on the water surface with the top of its thorax and abdomen hanging down. The head and thorax of a pupa are combined into a cephalothorax. The cephalothorax bears two respiratory trumpets on its dorsal surface. Pupae do not use the terminal abdominal spiracles for respiration, unlike the larvae. The respiratory trumpets are connected with mesothoracic spiracles coupled with short tracheal trunks and dorsal longitudinal tracheal trunks. These trumpets are moved posteriorly when a pupa dives to minimize the energy spent for overcoming the interfacial tension (Ha et al., 2017).

对于滞育蛹和昆虫的某些其他胚后阶段而言，在 DGC 缩窄的气门期，空气的轻微净流入可作为保存水分的一种手段，否则水分会因气体交换而流失。蚊子蛹通过利用氧气进行呼吸来适应水中的生活。作为水生昆虫，蚊子蛹的呼吸系统在获取大气氧方面有局限性。蚊子蛹大部分时间用胸部顶端和腹部下垂停留在水面上。蛹的头部和胸部合并成头胸。头胸部背表面有两个呼吸喇叭。蛹不像幼虫那样使用腹部末端的呼吸孔进行呼吸。呼吸喇叭与中胸段的呼吸孔相连，并与短气管干和背侧纵向气管干相连。当蛹潜入水中时，这些喇叭向后移动，以尽量减少克服界面张力所消耗的能量（Ha 等，2017）。

Endogenous variation of metabolism during pupal development offers a great opportunity to study the regulation of respiratory patterns in insects. The pupae of the mealworm *T. molitor* show extremely high metabolic rates after metamorphosis from the larvae and shortly before metamorphosis to the adult, while the metabolic rates in the middle of pupal development were generally lower. In fact, diapausing insect pupae primarily use anaerobic glycolysis during metabolic depression but engage in aerobic respiration through the tricarboxylic acid cycle during periodic arousal. Diapausing pupae also clear anaerobic by-products and regenerate many metabolic intermediates depleted in metabolic depression during arousal.

U-shaped curves of oxygen uptake rate were also found in pupae of other insect species. These changes closely correlated to the activity of a set of enzymes participating in energy production and histogenesis. The endogenously increased metabolic rate in the beginning and the end of the pupal stage requires increased oxygen uptake and carbon dioxide release, which should alter spiracular activity and gas exchange patterns. Pre-pupae and pupae of the silk moth *S. cynthia* had similar metabolic rates, but the pre-pupae had longer cycles with more CO_2 released during longer opening phases than pupae. An occasional occurrence of continuous CO_2 release patterns at low metabolic rates was very likely caused by single defective non-occluding spiracles (Kaiser et al.,2010).

A variety of discontinuous and cyclic gas exchange patterns, such as DGC, irregular patterns, cyclic, and continuous CO_2 release, which were not correlated with any periods of pupal development. However, *Zophobas rugipes* displays DGC during the pupal stage suggesting a high among individual variability. The diversity of gas exchange cycle adjustments demonstrates that different groups of insects most likely have specific regulatory responses to variations in metabolic rate (Kaiser et al., 2010). Metabolic rate, however, varied with gas-exchange pattern, and was lowest for individuals that used discontinuous gas exchange, intermediate for individuals using cyclic gas exchange, and highest for individuals using continuous gas exchange (Gibbs et al., 2004).

5.5.4 Hemoglobin and hemocyanin

Many invertebrates harbor respiratory proteins that store or transport O_2, thereby enhancing their metabolic performance under low oxygen conditions. Because of the highly efficient O_2 diffusion along the tracheal system, it has long been assumed that most insects do not need respiratory proteins. Known exceptions

蛹发育过程中代谢的内源性变化为研究昆虫呼吸模式的调节提供了极好的机会。粉虱黄粉虫的蛹在从幼虫蜕变到成虫前不久表现出极高的代谢率，而代谢率在蛹的发育中期普遍较低。事实上，滞育的昆虫蛹主要利用厌氧代谢进行糖酵解，但在周期性觉醒前期通过三羧酸循环进行有氧呼吸。滞育蛹还清除厌氧副产物，并在觉醒过程中再生代谢抑制中耗尽的代谢中间产物。

其他昆虫蛹的耗氧速率也呈U形曲线。这些变化和参与能量生产和组织发生的一组酶的活性密切相关。在蛹期开始和结束时，内源性代谢率的增加需要增加耗氧量和二氧化碳释放量，这将改变气门活动和气体交换的模式。蓖麻蚕的前蛹和蛹的代谢率相似，但蛹前期比蛹有更长的周期，在更长的开放期释放更多的 CO_2。偶尔出现低代谢率时的连续 CO_2 释放模式很可能是由单一缺陷的非阻塞性气门引起的（Kaiser 等，2010）。

各种不连续和循环的气体交换模式，如不连续的气体交换、无规则模式、周期性的和连续的 CO_2 排放，与蛹发育的任何时期都不相关。然而，褐飞虱在蛹期显示 DGC，表明个体间的高度变异性。气体交换循环调节的多样性表明，不同种类的昆虫很可能对代谢率的变化有特定的调节反应（Kaiser 等，2010）。然而，代谢率随气体交换模式的不同而变化，使用不连续气体交换的个体代谢率最低，使用循环气体交换的居中，使用连续气体交换的最高（Gibbs 等，2004）。

5.5.4 血红素和血蓝蛋白

许多无脊椎动物含有贮存和转运 O_2 的呼吸蛋白，在低氧条件下由此提高其代谢性能。由于沿着气管系统有高效率的 O_2 扩散，很长一段时间被推测为多数昆虫不需要呼吸蛋白。已知的例

were the aquatic larvae of the chironomids, aquatic backswimmers and the parasitic larvae of *G. intestinalis*. In addition, some basal insects have hemocyanin in their hemolymph, a copper-based respiratory protein which they apparently inherited from their crustacean ancestor. Recently, a number of hemoglobin genes were cloned from *D. melangaster*, *H. cecropia*, *M. sexta*, mosquito and the nymphs and adults of the stonefly.

外是摇蚊的水生幼虫、仰泳蝽和肠胃蝇的寄生幼虫。此外，一些基底昆虫在其血液中含有血蓝蛋白，一种明显从其甲壳动物祖先遗传而来的铜基呼吸蛋白。最近，从果蝇、天蚕和烟草天蛾，蚊以及石蝇的若虫和成虫中克隆得到大量血红素基因。

Hemocyanins are copper-containing respiratory proteins that occur freely dissolved in the hemolymph of many arthropod species throughout their development. In the periphery the oxygen is then released to diffuse to target cells, while carbon dioxide is loaded up for return transportation to the primary respiratory surface and elimination to the environment. Hemocyanins are built by six identical or similar subunits of about 75 kDa that assemble into hexamers or oligo-hexamers with masses of up to 3,600 kDa. Each subunit can transport one oxygen molecule by the virtue of two Cu^{2+} ions, which are coordinated by six histidines in the polypeptide chain. Oxygen affinity in hemocyanins is usually low or moderate, but cooperativity can be very high. Hemocyanins occur in the hemolymph of many ametabolous and hemimetabolous insect taxa, but not in the Eumetabola, which include Hemiptera and Holometabola. The expression of hemocyanin is restricted to late embryos, thus being detectable also in whole female extracts and oothecae. Hemocyanin protein is also present in 1st instar nymphs, but not in later developmental stages. The ontogeny of hemocyanin in cockroaches is distinct from that known from Zygentoma and Plecoptera, in which hemocyanin occurs in both nymphal and adult stages. It is suggested a specific role of hemocyanin in embryonic cockroaches, which may be related to an enhanced oxygen supply in the oothecae. For some reason, the fundamental physiological changes associated to the evolution of holometaboly have made hemocyanin unnecessary (Pick et al., 2010).

血蓝蛋白是一种含铜的呼吸蛋白，在许多节肢动物物种的整个发育过程中完全溶解在血淋巴中。在外围，氧气被释放并扩散到靶细胞，而二氧化碳则被装载起来，返回到初始呼吸的表面并排放到环境中。血蓝蛋白由 6 个约 75 kDa 的相同或类似亚单位组成，这些亚单位组装成质量高达 3 600 kDa 的六聚体或低聚六聚体。每个亚单位可以通过两个铜离子转运一个氧分子，这两个铜离子由多肽链中的六个组氨酸配位。血蓝蛋白中的氧亲和力通常较低或中等，但可能具有很高的协同性。血蓝蛋白存在于许多不变态昆虫和半变态昆虫类群的血淋巴中，但不存在于包括半翅目和全变态昆虫的真变态昆虫中。血蓝蛋白的表达仅限于后期胚胎，因此也可在整个雌性提取物和卵鞘中检测到。血蓝蛋白也存在于 1 龄若虫中，但不存在于发育后期。蟑螂血蓝蛋白的个体发育不同于已知的血蓝蛋白发生在若虫和成虫阶段的衣鱼目和襀翅目昆虫的个体发育。这表明血蓝蛋白在蟑螂胚胎发育中起着特殊的作用，这可能与卵鞘中氧气供应增强有关。某些原因所致，与完全变态进化相关的基本生理变化使血蓝蛋白变得不必要（Pick 等，2010）。

5.5.5 Nonrespiratory functions of the tracheal system

Tracheal serve the functions of connective tissue, typically tying cells and tissues together. Tracheae are important as a structural base for at least two important endocrine tissues: ① the diffuse cluster of cells making up the prothoracic glands that are attached to the prothoracic tracheae near the prothoracic spiracles in larvae

5.5.5 气管系统的非呼吸功能

气管起着结缔组织的作用，通常把细胞和组织连在一起。气管至少是两种重要内分泌组织的重要结构基础：①鳞翅目幼虫松散的细胞束组成的前胸腺，附属于前胸气门附近的前胸气管；②鳞

of Lepidoptera; ② the epitracheal glands attracted to the ventral surface of the major ventro-lateral tracheal trunk connection to each spiracle in lepidopterous larvae.

Air sacs that back sound-producing organs in insects, particularly in cicadas that produce very loud sounds, are important to loudness or sound production and its modulation. A large air sac backs up to the tymbal located on each side of the first abdominal segment of male cicadas. The size of the air sac with different species, and its size and tuning are partly responsible for the species-specific quality of the sound produced by male cicadas. The interconnected prothoracic tracheae extending across the prothorax and into the prothoracic legs of cricket act as a resonator of sounds.

Respiratory patterns in insects are not distinct but are instead a continuum of respiratory responses driven by the metabolic demand experienced by the insect. At low temperatures insects may exhibit lengthy periods of spiracular closure reflecting a discontinuous respiratory pattern. As metabolic rate increased, periods of spiracular closure decreased and insects displayed a more cyclical pattern of respiration. As metabolic rates increased even further under the highest experimental temperatures, periods of spiracular closure decreased even more and a continuous respiratory pattern was employed (Contreras et al., 2010). However, DGCs can arise due to a decrease in brain activity and a change in the underlying regulation of gas exchange, and are not necessarily a simple consequence of low respiratory demand.

(Li Wenchu)

翅目幼虫气管外腺附属于较大的腹侧气管主干的腹面，与各气门相连。

昆虫，特别是蝉的背部发声器官产生非常大的声音，气囊对音量或发声及其调节是重要的。大气囊对应位于雄蝉第一腹节每边的鼓室。不同品种蝉的气囊大小不同，其大小和调音在一定程度上决定了雄蝉品种发出声音的种属特异性特征。蟋蟀延伸的前胸气管横跨前胸和进入前胸腿的部分相互连接，起着声音共振器的作用。

昆虫的呼吸模式是不确切的，而是由昆虫经历的代谢需求驱动的一系列呼吸反应。在低温下，昆虫可能表现出长时间的呼吸道闭合，反映出不连续的呼吸模式。随着代谢率的增加，气孔闭合期缩短，昆虫的呼吸呈现出更为周期性的模式。随着代谢率在最高实验温度下进一步增加，气门闭合的时间减少得更多，并采用了连续的呼吸模式（Contreras 等，2010）。然而，DGCs 的出现可能是由于大脑活动的减少和气体交换潜在调节的改变而产生的，不一定是低呼吸需求的简单结果。

（李文楚）

Exercises

1.Why the small tracheal tube couldn't be compressed and collapsed in insects?

2.The types and functions of insect tracheal systems.

3.Briefly introduce discontinuous gas exchange cycle and its mechanism.

4.How aquatic insects exchange their gas under the surface of water?

思考题

1. 为什么昆虫微小的气管不会被压缩和崩塌？

2. 简要描述昆虫气管系统的类型及其功能。

3. 简要描述不连续气体交换周期及其机制。

4. 水生昆虫在水下如何交换气体？

第 6 章 血淋巴循环系统

Chapter 6 Hemolymph Circulatory System

Key points

1. Insects have an open circulatory system, which is composed of the hemocoel, diaphragms, circulatory organs or muscular pumps, including the dorsal vessel and the autonomous accessory pulsatile organs, hematopoietic organs as well as pericardial cells.

2. Hemolymph play critical roles in transport of nutrient, energy, hormone, enzymes, and other components to cells and tissues.

3. Seven most common types of hemocytes are found in insects are prohemocytes, plasmatocytes, granulocytes, spherulocytes, adipohemocytes, oenocytoids, and coagulocytes.

4. Components of hemolymph are predominantly composed of inorganic ions, free amino acids, proteins, and other organic constituents.

知识点

1. 昆虫有开管式循环系统，循环系统内由充满血淋巴的血腔（体腔）、隔膜、循环器官和肌肉泵组成，包括背血管和自主辅搏器、造血器官和围心细胞。

2. 血淋巴在细胞和组织的营养、能量、激素、酶等成分的运输中发挥关键作用。

3. 血细胞有 7 种主要的类型：原血细胞、浆血细胞、粒血细胞、珠血细胞、脂血细胞、类绛色细胞和凝血细胞。

4. 血淋巴的成分主要由无机离子、游离氨基酸、蛋白质以及其他有机物质组成。

Insect circulatory system is at the center of most physiological processes. It delivers nutrients and hormones to cells and removes waste. In general, the circulatory system is composed of a fluid medium called hemolymph, a body cavity called the hemocoel, and a series of muscular pumps(Chapman et al., 2013). Hemolymph is the blood of insect, which typically enters the abdominal portion. Dorsal vessel, located in the central body cavity, is the main driver of hemolymph circulation, which is usually divided into an aorta in the thorax and a heart in the abdomen. Peripheral circulation in the appendages, however, is driven by autonomous pumps known as accessory pulsatile organs, or auxiliary hearts (Hillyer and Pass, 2019). Although insects have an open circulatory system,

昆虫的循环系统处于许多生理过程的中心。它向细胞输送营养物质和激素并清除废弃物。一般而言，循环系统由称为血淋巴的流体介质、被叫作血腔的体腔和一系列的肌肉质的泵组成（Chapman 等，2013）。位于体腔中心的背血管是血淋巴循环的主要驱动器，其可以分为分布在胸部的主动脉和分布在腹部的心脏。身体附肢中的围周循环是由自动泵也就是辅助搏动器官或辅助心脏所驱动的（Hillyer 和 Pass, 2019）。虽然昆虫有开管性循环系统，但是血淋

hemolymph does not diffuse freely, and instead flows along distinct channel-like routes in the hemocoel that are created by the structural organization of the internal organs and by fibromuscular septa or diaphragms.

巴并不能自由扩散，而是在血腔内沿着一定的通道类路线进行流动，这些通道类路线是由内部器官组织结构、纤维肌的隔或横膈膜所形成。

6.1 血淋巴的生理功能

6.1 Physiological Function of the Hemolymph

The principal organ regulating whole-body circulation in insects is a tubular vessel lying beneath the dorsal body wall that generally extends the entire length of insect. Dorsal vessel is organized by serial repetition of building blocks. Each module is comprised of cardiomyocytes, a pair of alary muscles, and connective tissue. Hemolymph typically enters the abdominal portion, the heart, through paired lips (ostial openings), and is pumped anteriorly through the thoracic portion, the aorta. The aorta sends out branches in a few insects, but generally, it is a simple tube with an open end in the head. It is not uncommon for the heartbeat to reverse in some insects, with the contraction wave beginning at the anterior and passing toward the posterior. The heartbeat is myogenic, but modified by nervous input and neurosecretions. Accessory pulsatile organs or hearts often exist at the base of appendages, such as antennae, legs, and wings, and promote circulation into the appendages. Although hemolymph does not transport oxygen, it is very important for transporting nutrients and hormones and removal of waste. The pH of the hemolymph usually is slightly acid but varies over a small range in different insects. Hemocytes, free cells in hemolymph, show variability in structure and function in different insects. Hemolymph contains relatively high concentrations of free amino acids, the disaccharide trehalose, and numerous other chemical substances that are often in transport from the site of synthesis to a site of utilization. The circulatory system also coordinates defense mechanisms, modulates heat transfer, assists in gas exchange, facilitates ecdysis, maintains homeostasis, and more (Fig. 6-1; Appendix color graphic Fig.6-1). The physiological functions of hemolymph include the following:

位于身体背壁下方的管状导管是调控昆虫全身循环的主要器官，它通常贯穿整个昆虫。背血管由一系列连续重复的构建模块组成，每个模块都由心肌细胞、一对翼肌和结缔组织组成。血淋巴一般通过成对的唇瓣（心门开口）进入腹部和头部，并从胸部大动脉泵向前端。有些昆虫的大动脉有分支，但总体上是一条在头部有末端开口的简单血管。较为常见的是，有些昆虫心跳存在逆转，即收缩波从前端开始向后部传递。心跳是肌源性的，但受到神经传递和神经分泌物的调节。辅搏器或辅助心脏常位于附肢，如触角、腿和翅的基部，并促进附肢的血液循环。虽然血淋巴不运输氧气，但对营养和激素的转运以及废物的清除很重要。血淋巴的 pH 一般都是呈弱酸性的，但在不同昆虫中会有小范围的变化。血细胞是血淋巴中的游离细胞，在不同的昆虫中，其结构和功能上可能存在差异。通常，血淋巴含有相对高浓度的游离氨基酸、双糖海藻糖和大量其他化学物质，这些物质通常从合成部位转运到利用部位。循环系统还协调防御机制，调节传热，协助气体交换，促进蜕皮和维持稳态等（图 6-1；附彩图 6-1）。血淋巴的生理功能包括以下内容。

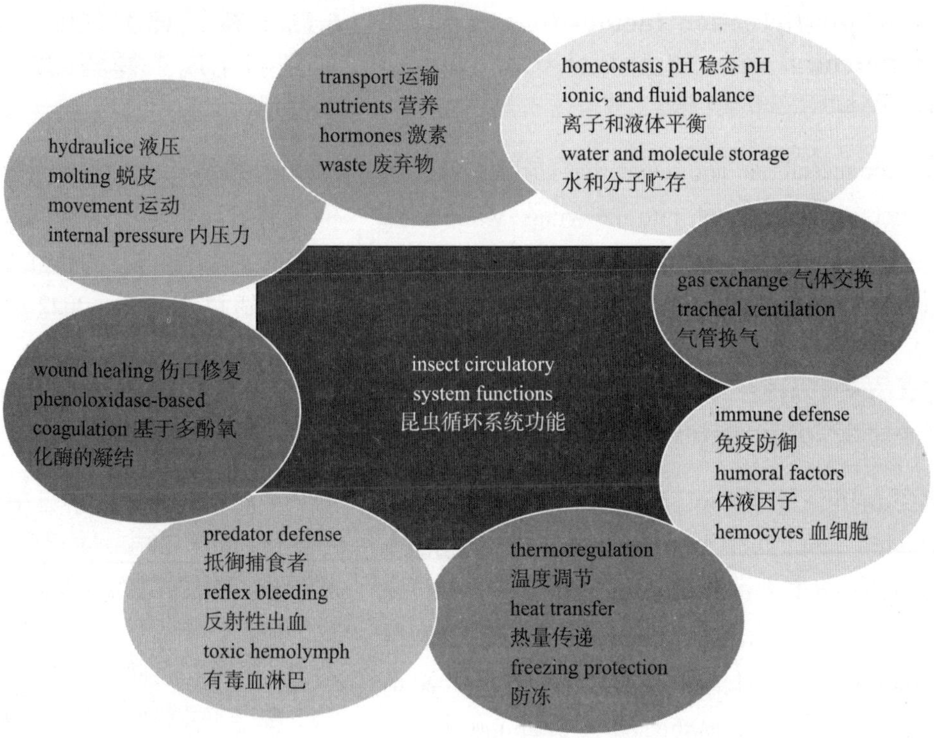

Fig. 6-1 The many functions of the circulatory system of insects (modified from Hillyer and Pass, 2019).

图 6-1 昆虫循环系统的主要功能（仿 Hillyer 和 Pass，2019）。

6.1.1 Insect metamorphosis requires an active circulatory system

During imaginal ecdysis, dorsal vessel and the wing hearts trigger hydraulic effects that are critical for eclosion and the spreading of the wings (Hillyer et al., 2020; Nation, 2022). Prior to eclosion, the heart immediately beats anterograde at an accelerated rate. In addition to peristaltic contractions of the abdomen, this increases hemolymph pressure in the anterior of the body, which finally facilitates the rupturing of the pupal cuticle, resulting in adult emergence and the inflation of the folded wings. After inflation, the epidermal layer of the wing lamellae undergoes apolysis, and then the cell fragments are removed through absorption activated by the wing hearts, and the dorsal and ventral cuticles are glued together with a matrix produced by special hemocytes in a manner that leaves spaces that become the wing veins (Sujkowski et al., 2015). After completion of eclosion, heartbeat reversal sets in, and thereby balancing hemolymph pressure throughout the body.

6.1.1 昆虫的变态需要一个活跃的循环系统

在成虫蜕皮的过程中，背血管和翅辅助心脏触发液压效应，这对羽化和翅膀的展开至关重要（Hillyer 等，2020; Nation，2022）。羽化前，心脏立即顺行加速跳动，再加上腹部的蠕动收缩，增加了身体前部的血淋巴压力，促进了蛹表皮的破裂，最终导致成虫羽化，并使最初折叠的翅膀膨胀。膨胀后，翅膀骨板的真皮层发生溶离，产生的细胞碎片通过翅膀心脏激活的吸收作用被去除，背部和腹侧的表皮通过与特殊血细胞产生的基质黏合在一起，留下的空间成为翅脉（Sujkowski 等，2015）。羽化完成后，心跳发生反向逆转，从而平衡全身血淋巴压力。

6.1.2 secretion of plasticization factors by neurocyte and more play potential roles in proper expansion of the wings in newly emerged adults

In some species, the mesotergal and metatergal accessory hearts are important in directing hemolymph into the wings, while some insects have other accessory hearts pumping hemolymph into the wings (Nation, 2022).

6.1.2 神经细胞分泌的塑化因子对刚羽化成虫翅膀的适当伸展至关重要

在某些种内，中胸和后胸背板的辅助心脏对血淋巴导入翅膀起着重要作用，但有些昆虫另有辅助心脏，可将血液泵入翅膀（Nation, 2022）。

6.1.3 Circulatory system is a reservoir of fluid, nutrients, and a series of proteins

Some proteins function in hemolymph, and some proteins are reserved in hemolymph, such as storage proteins, which is synthesized and released from fat body during larval stages. Hemolymph serves as the storehouse for storage proteins. During larval-pupal metamorphic stage, storage proteins are reabsorbed by fat body from hemolymph, and act as the source of amino acid for the synthesis of vitellogenin (Wang et al., 2020).

6.1.3 循环系统是体液、营养和一系列蛋白质的贮存库

一些蛋白质在血淋巴中发挥功能，一些蛋白质贮存在血淋巴中。例如，储藏蛋白在幼虫期间合成并从脂肪体释放。血淋巴作为贮存蛋白质的仓库。在幼虫 - 蛹变态时期，储藏蛋白被脂肪体从血淋巴重新吸收，作为卵黄原蛋白合成的氨基酸来源（Wang 等，2020）。

6.1.4 Hemolymph and hemocytes provide immunity-type reactions in response to invading bacteria, eggs of parasitoids, and other foreign substances

Antimicrobial peptides with antibacterial, antiviral, or antifungal activity are induced and transported into hemolymph after insect is exposed to bacteria, virus, fungi or some of their chemical by products. Lysozymes and phenoloxidases, act as protective agents that can chemically modify potential toxicants and foreign invaders. Moreover, phagocytosis and encapsulation by hemocytes, and wound healing by coagulation in some insects, are also important immune responses.

6.1.4 血淋巴和血细胞提供免疫反应，以应对入侵的细菌、寄生卵和其他外源性物质

当昆虫暴露于细菌、病毒、真菌或其化学副产物时，抗菌肽被诱导表达并转运至血淋巴中发挥抗细菌、病毒或抗真菌及其一些化学副产物的活性。溶菌酶和酚氧化酶起着保护剂的作用，能对潜在的有毒物质和外源入侵物进行化学修饰。此外，血细胞的吞噬作用和包囊作用、血淋巴凝结愈合伤口都是昆虫中重要的免疫反应。

6.1.5 Circulatory system transports nutrients to cells and tissues

Hemolymph is a main site for storage of trehalose, the major insect sugar used to generate energy, particularly for flight. Fat body is the main store of lipids utilized by insects for flight, meanwhile, rapid circulation of hemolymph is required to bring the fat body-released lipids to the flight muscles.

6.1.5 血液循环向细胞和组织输送营养物质

血淋巴是海藻糖的贮存场所，这是昆虫产生能量的主要糖类，特别是飞行所需。昆虫飞行利用的脂质主要贮存在脂肪体中，而释放的脂质需要血淋巴快速循环将其携带至飞行肌。

6.1.6 Hemolymph transports hormones from neurohemal organs to target tissues

Many neuropeptides, which have good water solubility, are transported in the hemolymph from their release points in the nervous system to target tissues. Circulatory system delivers waste products, excess water absorbed from food, and ingested allelochemicals or metabolites to excretory organs, or in some cases, to storage structures. Hemolymph is a lubricant and hydraulic support to maintain body shape and movement, especially of soft-bodied larvae such as caterpillars.

6.1.6 血淋巴将激素从神经血液器官运送到靶组织

神经肽大都具有很好的水溶性，从其神经系统的释放部位释放进入血液中并被运送到靶组织。循环系统传送废弃物、食物中吸收的过量水分和摄食的异种化感物或代谢物到排泄器官或在有些情况下，运载至贮存结构中。血淋巴是保持体形和运动的润滑剂和液压支撑，特别是软体幼虫如毛虫。

6.1.7 Hemolymph is important as a pool for carbon dioxide

The total carbon dioxide content of the hemolymph varies with an average of near 72 vol. %. The carbon dioxide content of the tissues minus the cuticle is very close to, and follows the variations in, the carbon dioxide content of the hemolymph. Keeping the spiracles closed as much as possible prevents excessive water loss when the insects take respiration, and it is also a critical factor for the closed system in pupal stage.

6.1.7 血淋巴是二氧化碳的重要吸收池

血液中的总二氧化碳含量从变化不等，平均为 72 左右的体积百分比。组织减去角质层的二氧化碳含量后非常接近血液中的二氧化碳含量，并随血液中二氧化碳含量的变化而变化。昆虫呼吸时尽可能保持气门关闭，可防止水分散失，它也是蛹期封闭系统的一个关键性因素。

6.1.8 Circulatory system is an important mean of heat transfer

It can prevent excessively high temperatures in the thorax from activity of flight muscle, or to hold heat in the thorax to warm above the ambient. Many insects do not initiate flight until the temperature in the thorax reaches some critical temperature. They warm the thorax by wing whirring before attempting to flight. The high temperature in the thorax is the result of the intense work by the flight muscles. In comparison, circulation of hemolymph from the hot thorax into the abdomen is initiated to cool the thorax, where little muscle activity is occurring during flight. In moths, the abdomen is covered with a much thinner layer of scales, and heat can escape by convection to the atmosphere more readily. The advantage of the physiological mechanisms that allow heat in the thorax to either accumulate or to be dissipated gives the moth the opportunity to fly at a wide range of ambient temperatures.

6.1.8 昆虫的循环系统是热量转移的重要途径

它可以防止胸部因飞行肌的活动产生过高的温度，或在胸部保温，让胸部温度高于周围环境。有些昆虫只有在胸部温度达到某个临界温才启动飞行。它们在起飞前通过振翅使胸部变暖。胸部的高温是飞行肌剧烈运作的结果。相比之下，当血液从热的胸部向飞行时肌肉活动很少的腹部循环时，则启动了胸部的降温。在蛾类中，腹部有更薄的鳞片层覆盖，热量很容易通过与大气的对流而散发。这种胸部热量积聚或散发的生理机制的优势使得蛾类可以在较大幅度的环境温度范围内飞行。

6.1.9 Contributes to homeostasis of osmoregulation and pH

Probably most components contribute to the osmotic pressure

6.1.9 有助于维持渗透压和 pH 的稳态

也许血淋巴的多数组分对渗透压和

and specific gravity of hemolymph. Osmotic pressure in a number of species, and in different developmental stages of the same species, is not correlated with age, sex, developmental stage, diet, or systematic position among orders. pH of hemolymph is weakly acidic in most insects, but varies somewhat within individuals owing to effects of temperature, activity, discontinuous ventilation, and diet. To maintain a constant blood pH in the face of elevated partial pressure CO_2, air-breathers increase their strong-ion difference, causing an associated increase in bicarbonate concentration [HCO_3^-].

An additional role of the hemolymph is predatory defense in some orders. It can contain unpalatable and malodourous chemicals that will act as a deterrent to predators.

特定的血淋巴密度都有所贡献。一些种类的昆虫或同种昆虫特定发育阶段的渗透压与虫龄、性别、发育时期、食物或其在目内的系统位置无关。大部分昆虫血淋巴的 pH 是弱酸性的，但常受到温度、活动、不连续的通风和饮食等的影响，使其在个体内有所不同。当血淋巴内 CO_2 分压升高时，为了维持其表面恒定的 pH，气息者就会增加强离子差，从而导致碳酸氢根浓度（HCO_3^-）的增加。

在有些目昆虫中，血淋巴的另一个作用是捕食性防御。它含有的不可口和臭味的化学物质是对捕食者的威慑。

6.2 循环系统的结构

6.2 The Structure of Circulatory System

6.2.1 Generation of the circulatory system during embryogenesis

The heart is derived from mesodermal tissue in during embryogenesis. Paired cardioblasts become polarized, and move to the dorsal midline to form two regular, opposing rows, and then extend and meet their contralateral counterparts in the midline, thus forming a cylindrical, spiral arrangement that constitutes the lumen of dorsal vessel in each segment (Hillyer et al., 2019; Nation, 2022; Fig.6-2; Appendix color graphic Fig.6-2). Cardioblasts on each side of the body differentiate into specialized cells that form inflow openings called ostia. Usually, ostial cardioblast develops a flap-like extension at contact points that protrudes into the lumen of the dorsal vessel, which result in a spiral-like arrangement in the dorsal and ventral median lines (Hillyer et al., 2019). Dorsal vessel in abdomen is derived from cardioblasts, cells that fuse with each other, modify their shapes and form the dorsal tubular vessel. Eventually fusion with the cephalic portion of the vessel occurs. In some insects, the heart begins to beat only after dorsal closure.

6.2.1 循环系统的胚胎发育

昆虫的心脏起源于胚胎发育的中胚层。成对的成心肌细胞去极化，并移动到身体背部的中线，对立排列，随后延伸并与对侧的细胞在反方向的中线相遇，从而形成圆柱和螺旋形的排列，在每一个体节闭合形成中空的背血管腔（Hillyer 等，2019; Nation, 2022; 图 6-2；附彩图 6-2）。身体两侧每一侧的成心肌细胞分化成特异的细胞以形成流入的开口，被叫作心门。开口的成心肌细胞通常在它们的接触点上发育为一个片状的延伸，并伸入背血管的腔内，结果在腹部和背部中线位置形成螺旋状的排列（Hillyer 等，2019）。腹部背血管部分来自胚胎发育时相互融合的成心肌细胞，其改变细胞形态，并在背部形成管状，最后和头部血管相连。有些昆虫的背血管封闭后

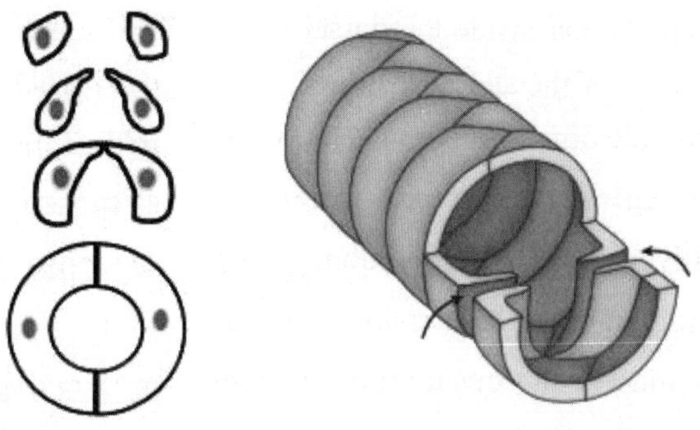

Fig. 6-2 Development and structure of the dorsal vessel. A.Consecutive stages of morphogenesis, from top to bottom; B.The mature dorsal vessel consists of two rows of semicircular cardiomyocytes. The anterior front cut shows the organization of the ostial cells; the two inward directed flaps act together and allows hemolymph to enter the cavity, but hinders outward flow (modified from Hilly and Pass, 2019).

图 6-2 背血管的发育与结构。A. 持续发育过程中背血管的形态，从上至下；B. 成熟的背血管由两排半圆形心肌细胞组成。它们在背侧和腹侧中线的接触点形成螺旋状排列。前部切面显示心门细胞；两个向内的瓣膜共同作用，就像一个阀门一样，允许血淋巴进入管腔，但防止其外流（仿 Hilly 和 Pass, 2019）。

In other insects, the heartbeat may begin before the posterior part of the heart fuses with the anterior portion in the head. During the embryonic development of water strider, *Gerris paludum insularis*, hemocytes circulate within the cavity of the embryonic heart and travel anteriorly, and then are discharged from the cephalic open end.

心脏才开始搏动；在另一些昆虫中，心脏的搏动在心脏的后部和头内的前部融合之前就开始了。水黾胚胎发育时期的血细胞循环于胚胎心脏腔内，并向前部行进，然后从头部开口端流出。

6.2.2 Characters of hemolymph circulatory system in insects

6.2.2　昆虫血淋巴循环系统的特征

Insect has an open circulatory system, and the cavity of whole-body becomes hemocoel. Hemolymph are hypotension, large volume, and variable with feeding and physical conditions. The main functions of hemolymph are transporting nutrients, energy, hormones, signaling molecules, and removal of waste products. Circulatory system is important in maintenance of osmotic pressure and ion balance. Moreover, hemolymph participates in median metabolism, removal of dissociated tissue and debris, wound repair, innate immunity as well as thermoregulation during flight.

昆虫有一个开管式的循环系统，整个体腔即为血腔。血淋巴的血压低、血量大，并随取食和身体状态而发生变化。血淋巴的主要功能是运输营养物质、能量、激素、信号分子和移除代谢废物。循环系统对维持渗透压和离子平衡非常重要。此外，血淋巴参与中间代谢，清除解离的组织和碎片、伤口修复、先天免疫和飞行时体温的调节。

6.2.3 Functional morphology of circulatory system

6.2.3　循环系统的功能形态学

The circulatory system is composed of the hemocoel (body cavity) filled with hemolymph, diaphragms, circulatory organs or muscular pumps: the dorsal vessel (Fig. 6-3) and the autonomous accessory pulsatile organs (accessory hearts) that supply hemolymph to the appendages, hematopoietic organs, and

循环系统内是由充满血淋巴的血腔（体腔）、隔膜、循环器官和肌肉泵，即背血管（图 6-3）和为附肢提供血液的自主辅搏器（辅助心脏）、造血器官

pericardial cells (renal cells). Hemolymph inside the dorsal vessel is pumped toward the head or the tip of the abdomen. Insects also utilize autonomous accessory pulsatile organs to pump hemolymph to the appendages and other regions that would otherwise experience low flow rates (Hillyer et al., 2019; Nation, 2022). Insects have evolved over 400 million years, so it is not surprising that there are numerous variations in the circulatory systems between different species, especially that from water-breathers and air-breathers. Variations include the presence or absence of segmental blood vessels originating from the heart (Salcedo et al., 2020), various forms of ostia, a closed versus open posterior heart (Nation, 2022; Fig. 6-3; Appendix color graphic Fig.6-3), differing populations of circulating hemocytes, and differing ratios of forward to backward hemolymph flow (Klowden, 2013).

和围心细胞（肾细胞）组成的。背血管中的血淋巴被泵向头部或腹部末端。昆虫也利用一系列的辅搏器将血液泵往附肢和其他血液流速较低的区域（Hillyer 等，2019; Nation, 2022）。昆虫经历超过 40 亿年的进化，因此，在不同种类的昆虫中循环系统出现大量变异也就不足为奇，尤其是那些水息者和气息者。变异包括起源于心脏的体节血管的有无（Salcedo 等，2020）、心门的不同形式、封闭的逆向开放式后部心脏（Nation, 2022；图 6-3；附彩图 6-3）、循环中的不同血细胞总数以及血液向前和向后流动比率的不同（Klowden, 2013）。

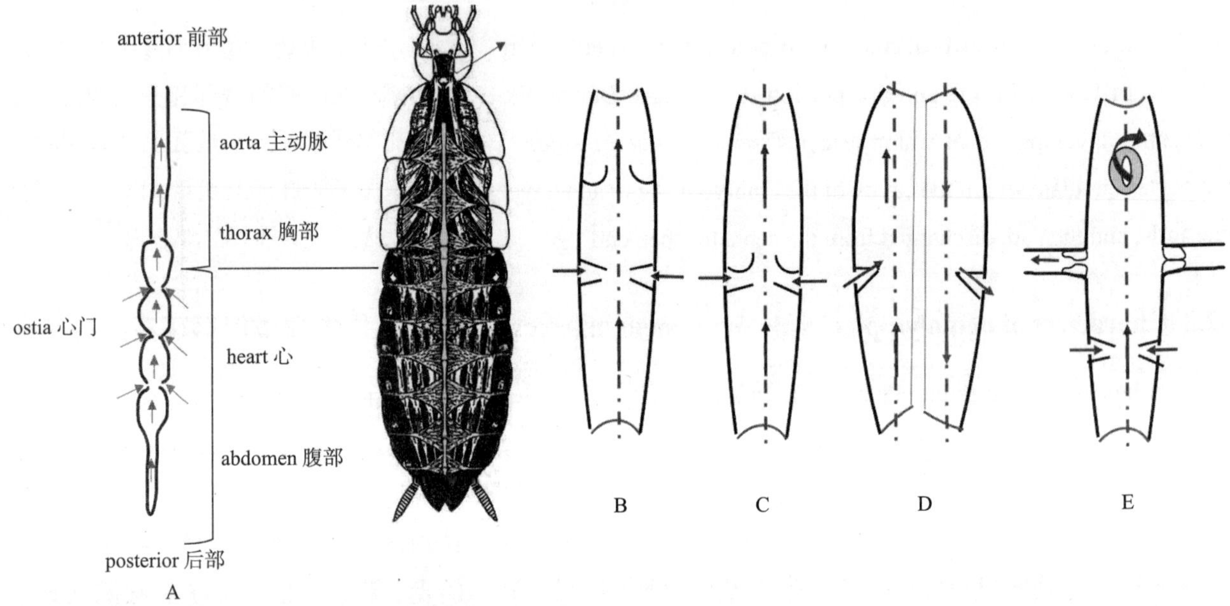

Fig. 6-3 Morphology of the dorsal vessel (heart) and various kinds of ostia in insects. A. Constructure of dorsal vessel; B. Incurrent ostia with paired lips and an intracardiac valve; C. Incurrent ostia with elongate lips that function as a pouch valve; D. Two-way ostia with a single lip. (left) Incurrent flow during the anterograde phase. (right) Excurrent flow during the retrograde phase. E. Incurrent ostia (bottom), excurrent ostia with sphincter-like valves at the base of lateral vessels (middle), and an unpaired excurrent ostium (top). The sphincter-like valve on the left is shown in the open position, whereas the valve on the right is shown in the closed position (modified from Hillyer and Pass, 2019).

图 6-3 昆虫背血管（心脏）的形态及各种心门图。**A.** 背部血管形态与结构；**B.** 长有一对瓣膜结构的流入心门，同时伴有心脏内阀门；**C.** 长有拉长的瓣膜的流入心门，发挥类似于口袋的阀门作用；**D.** 双口单瓣膜的心门，（左）顺行阶段向内的流动，（右）逆行阶段向外流出的流动；**E.** 流入心门（底部），左侧血管的基部具有类似括约肌阀门的流出心门（中央）以及不成对的流出心门（上部）。左侧类似括约肌的阀门显示的是打开的位置，右侧类似括约肌的阀门显示的是关闭的位置（仿 **Hillyer** 和 **Pass, 2019**）。

6.2.3.1 The dorsal vessel: heart and aorta

The dorsal vessel consists of two parts, the heart and the aorta, and the entire length of the dorsal vessel carries a wave of contraction. The heart is the abdominal portion of the vessel, but it may extend into the thorax in some insects. The presence of alary muscles and incurrent ostia in the heart portion is the major recognition feature for deciding where the heart ends and the aorta begins. In comparison, the aorta does not have alary muscles, and is lack of incurrent ostia, but it may have excurrent ostia. Usually, the entire dorsal vessel is a simple tube in most insects. In many Orthoptera, heart containing both alary muscles and incurrent ostia extend into the thoracic portion of the dorsal vessel, and thus there is only a short aorta leading into the head. Of note, the heart has several pairs of long diverticula passing laterally to tergosternal muscles and fat body in Orthoptera and Dictyoptera (Fig. 6-3 and Fig. 6-4; Appendix color graphic Fig.6-4).

6.2.3.1 背血管：心脏和大动脉

背血管由两部分组成：心脏和大动脉，并沿整个背血管传送收缩波。心脏是指背血管的腹部部分，但有些昆虫可延伸至胸部。判断心脏末端和动脉起始的主要特征是心脏存在心翼肌和流入式心门。相对而言，主动脉没有心翼肌，没有流入式心门，但可能有流出式心门。通常，多数昆虫的背血管在整体上是一条简单的管。在许多直翅目昆虫中，包含心翼肌和流入式心门的心脏延伸进入背血管的胸腔部分，因此主动脉只是较短的、深入脑的一段。值得注意的是，直翅目和网翅总目昆虫的心脏有成对的长憩室，并向体侧延伸至后背腹肌和脂肪体（图 6-3、图 6-4；附彩图 6-4）。

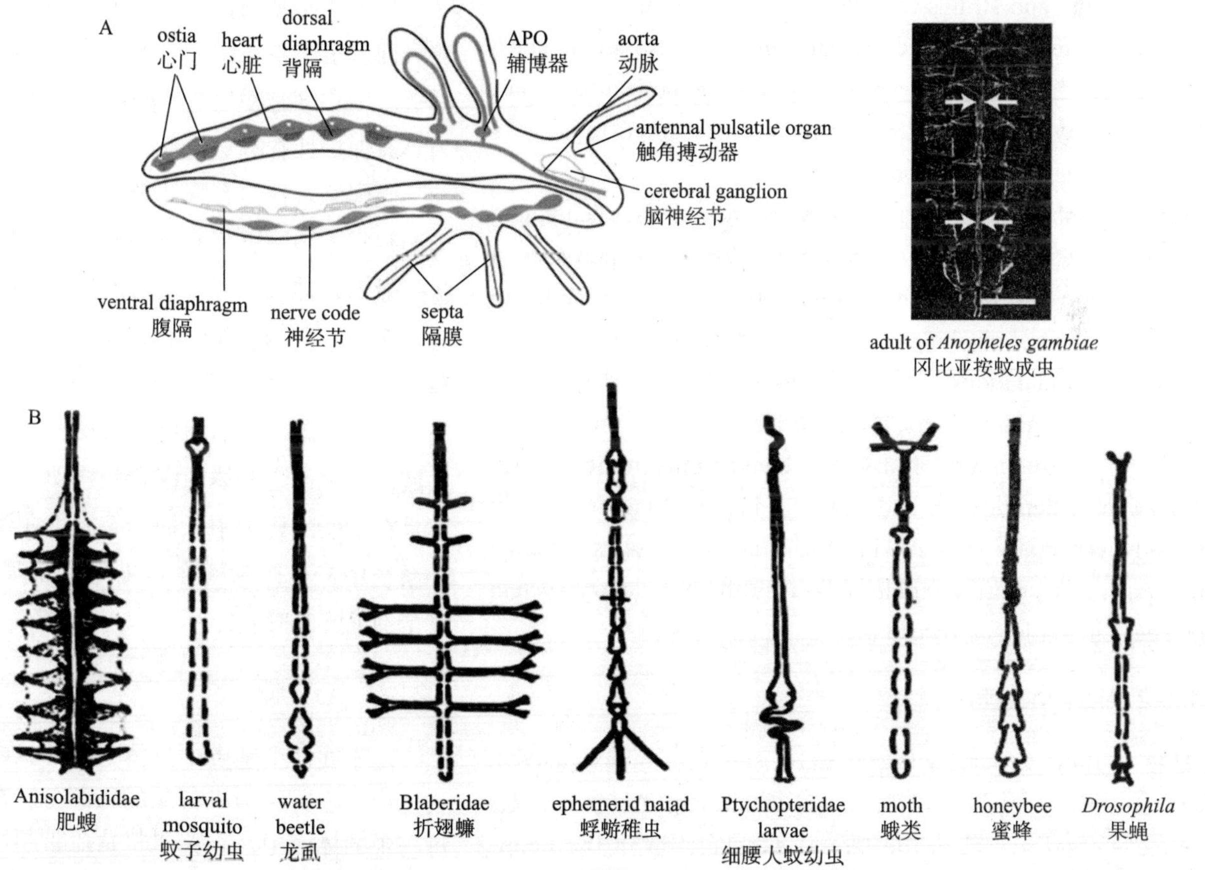

Fig. 6-4 The structure of insect circulatory system. A.The composition of insect hemolymph circulatory system and the routine for circulation as well as the dorsal vessel of adult of *Anopheles gambiae*. B.Variation of the heart and aorta in different insects (modified from other authors).

图 6-4 昆虫血液循环系统结构。A. 昆虫血液循环系统组成和血液循环路线以及冈比亚按蚊成虫的背血管图；B. 不同昆虫的心脏和动脉的变化（仿其他作者）。

Hemolymph pressure is low and enters the dorsal vessel through incurrent opening called ostia, which may be guarded by valvular flaps in the abdominal portion of the vessel. In general, incurrent ostia are localized at the abdomen, but sometimes there are incurrent ostia in the thoracic portion. Occasionally, there are excurrent ostia in the abdomen and thorax, but most of the hemolymph exits from the open anterior end and flows posteriorly through the hemocoel. The continuous outflow at the head pushes hemolymph toward the posterior of the insect, where it flows around tissues and organs. It is aspirated into posterior incurerent ostia by diastole of the heart.

血淋巴血压低，通过被称为心门的流入式开口进入背血管，在腹部的背血管可能被瓣膜片保护。流入式心门常位于腹部，但有时胸部也有流入式心门。胸腹部偶尔有流出式心门，但多数血淋巴从前端的开口流出，再通过血腔向后流动。头部持续的外流，推动围绕昆虫组织器官的血淋巴向后流动。它被心脏的舒张抽吸进入后部流入式心门。

In most but not all insects, the abdomen is divided by the dorsal and the ventral diaphragm into three partitioned regions: the pericardial sinus, the perivisceral cavity, and the perineural sinus. The dorsal diaphragm consists of multiple layers of thin sheets of connective tissue, with small dorsal transverse muscle fibers, the alary muscles, enclosed within the sheets. Hemolymph readily passes between the pericardial sinus and the perivisceral sinus. Loose clusters and strings of cells, called pericardial cells, often occur on the external surface of heart, and they also are attached to the dorsal diaphragm at various places in the pericardial sinus. They phagocytize injected particles, are presumed to serve a protective phagocytic function.

在多数但不是全部昆虫中，腹部被背膈和腹膈划分为三个区域：围心窦、围脏腔和围神经窦。背膈由多层结缔组织薄片组成，薄片内包含有背部小横肌纤维和翼肌。血淋巴很容易在围心窦和围脏窦之间流动。松散的细胞簇和细胞带，称为围心细胞，常常存在于心脏的外表面，也附着在围心窦不同部位的背膈上。围心细胞吞噬注入的颗粒，被认为具有保护性吞噬功能。

Ventral diaphragm often separates the perineural sinus containing the ventral nerve cord and ganglia from organs in the perivisceral sinus, but the ventral diaphragm usually does not extend into thorax. The ventral diaphragm is fenestrated, allowing hemolymph to circulate. Undulations of both the dorsal and ventral diaphragms due to intrinsic muscle fibers, alary muscle action, gut movements, ventilatory movements, and general body muscle action aid circulation of hemolymph and keep it mixed and moving within the body, especially in active insects. Quiescent insects generate micropulses of pressure (coelo-pulses) in the circulatory system that keep circulation and breathing.

腹膈常常把含有腹神经索和神经节的围神经窦与围脏窦的器官分开，但腹膈一般不延伸入胸部，其为网状，充许血液循环。背膈和腹膈的波动归结于内源性肌纤维、心翼肌的作用、肠道蠕动、通风运动和整体肌肉的活动，协助血液循环并使之在体内混合和移动，尤其是在活跃的昆虫中。而静态昆虫在循环系统中产生微脉搏血压（鞘翅脉搏）以帮助循环和呼吸。

6.2.3.2 Alary muscles

6.2.3.2 心翼肌

Due to their general wing or delta shape in many insects, alary muscles form part of the dorsal diaphragm. Alary muscles connect dorsal vessel to the lateral exoskeleton in the hemi-segment (Bataillé et al., 2015). Early reports on alary muscles in adults from different insect species led to proposals of their role in maintaining the heart in its dorsal position or/and controlling ostia opening and heart beating. The muscle fibers typically fan out from a small point of origin on the lateral wall of the dorsum

由于在许多昆虫中总体上呈翼状或三角形，心翼肌构成背膈的一部分。在每一半的体节中，心翼肌将背血管与体侧的外骨骼连接起来（Bataillé 等，2015）。早期关于不同种类成虫心翼肌的报道显示，它们可能对于维持心脏在背部的位置或 / 和调控心门开放和心搏中具有功能。许多昆虫肌纤维一般从背侧体壁的

to a broad insertion on the heart in many insects, presenting the typical delta appearance. In grasshoppers, the origin and insertion are wide and the delta shape is not particularly evident. Some alary muscle fibers pass beneath the heart and extend from lateral side to side. In places, the fibers may also run parallel to the long axis of the heart for a short distance. The pairs of alary muscles tend to agree with the number of pairs of ostia. In addition to support, the alary muscle may assist in the expansion (diastole) of the heart after a contraction wave and, thus, aid in pulling hemolymph into the incurrent ostia. In *Drosophila*, selective elimination of alary muscles by targeted apoptosis revealed that they are required for suspending the heart in proper intra-haemocelic position and for opening of the heart lumen, and that alary muscles constrain the curvature of the respiratory tracheal system during crawling. Cardiac versus visceral attachment of alary muscles is under the control of Hox protein (Bataillé et al., 2020).

6.2.3.3 Ostia

Ostia are small, slit-like, pairs openings in the dorsal vessel that allow hemolymph to enter or leave the vessel. Incurrent ostia allow hemolymph to enter during diastole and excurrent ones permit hemolymph to exit. Most insects commonly have 2, 3, or 5 pairs of ostia, but some Orthoptera have 12 pairs of incurrent ostia, 9 in the abdomen and 3 in the thorax. Pairs of ostia are usually located laterally, with one on each side of the heart, but some are ventrally and dorsally located on the aorta. Most incurrent ostia open with diastole, allowing hemolymph to be forced into the dorsal vessel by general body pressure and/or perhaps slight negative pressure inside the dorsal vessel. Some insects have flap valves in incurrent ostia that open inward, and allow hemolymph to pass the valves into the lumen of the heart.

Reports show that hemocytes have been shown to aggregate around the incurrent ostia in the abdomen, and thus to facilitate the phagocytosis of invading pathogens in the hemolymph (Sigle et al., 2016). The genes *eater*, *draper* as well as complement-like proteins TEP1, TEP2, and TEP3 are involved in the recruitment of hemocytes to the periosteal area in the African malaria mosquito *A. gambiae* challenged with *E. coli* (Sigle et al., 2018a, b; Yan et al., 2019). Using the Gal4-UAS system, the activity of *Drosophila*

一个小起点以扇形展开并广泛连接到心脏，外观呈现典型的三角形。在蚱蜢中，心翼肌的起点和插入很宽，三角形不是特别明显。有些心翼肌纤维通过心脏下方，从一侧延伸到另一侧。在几处，肌纤维也沿着心脏纵轴平行延伸一段短的距离。成对的心翼肌与成对的心门在数量上一致。除了支撑心脏外，心翼肌在收缩波过后有助于心脏扩展（舒张），因而助力推动血淋进入流入式心门。在果蝇中，通过靶向凋亡选择性的消除心翼肌研究发现它们是心脏悬浮在血液腔内适当位置和心腔开放所必需的，而且心翼肌在果蝇爬行过程中抑制了呼吸气管系统的弯曲。心翼肌对心脏与内脏的附着受到了Hox 蛋白的调控（Bataillé 等，2020）。

6.2.3.3 心门

心门是背血管内小裂缝状的成对开口，允许血淋巴进出血管。流入式心门在心脏舒张时让血淋巴流入，而流出式心门允许血淋巴流出。多数昆虫常有 2 对、3 对或 5 对心门，但有些直翅目有 12 对流入式心门，9 对在腹部，3 对在胸部。成对的心门常位于心脏侧面，每侧一个，但有些位于背部和腹部大动脉。多数流入式心门随心脏舒张打开，由整体血压和 / 或背血管内微弱的负压迫使血淋巴流入背血管。有些昆虫的流入式心门内，有向内开放的瓣膜片，允许血淋巴通过瓣膜进入心脏的内腔。

研究表明，血细胞向腹部的流入式心门周围聚集以利于对病原的吞噬（Sigle 等，2016）。当非洲疟蚊冈比亚按蚊在遭受大肠埃希菌感染时，基因 *eater, draper* 以及补体样蛋白 TEP1、TEP2 和 TEP3 参与补充血细胞至骨膜位置（Sigle 等，2018a, b; Yan 等，2019）。利用果蝇的 Gal4-UAS 系统，检测果蝇血细胞特异的血凝素启动子在

hemocyte-specific hemolectin promoter in the adults of *A. gambiae* is characterized in differentiating prohemocytes as well as in differentiated granulocytes after a blood meal (Ponderville et al, 2020).

6.2.3.4 Heartbeat

The heartbeat is a wave of contractions (systole) generally originating at the posterior end of the heart and traveling anterior. The rate of contractions or beats is highly variable in different insects, and varies with physiological conditions, temperature, species, stage of development, nervous activity, and neurosecretions. The beat is considered to be myogenic, originating in the muscle itself, although the heartbeat of many other invertebrate is neurogenic. The rate may be as slow as 15 beats/min in the larva of *Lucanus cervus* (Coleoptera), while near or higher than 100 beats/min in several insects. In ants, an apparatus has been designed to monitor cardiac contractions without harming them (Davis et al., 2021).

The heart of the American cockroach, *P. americana*, has one of the most complex systems of innervation, which continues to express a beat after cutting into numerous pieces, indicating that various parts of the dorsal vessel are capable of acting as a pacemaker. The pacemaker that usually dominates, however, is located at the posterior of the heart, and the contraction wave (systole) usually originates near the posterior of the heart and travels anteriorly. The contraction wave may move very slow in some species, while may be so rapid in other insects that it appears as if the entire vessel contracting simultaneously. There are regular reversals of the heartbeat in the blowfly, *Calliphora vicina*, with a sequence of backward and forward pulses (Wasserthal, 2015). Contraction may begin occasionally at both ends and meet near the middle. Reversals are unpredictable, and the reverse pulse rate was higher than the forward rate during flight that may be an adaptation to pump CO_2-enriched hemolymph into the thorax for release at excurrent metathoracic spiracles (Wasserthal, 2015; Wasserthal et al., 2018). Compared to the 15 days-old adults, aging of *A. gambiae* slows down the heart rate, while increases the rate of reverse pulse (Doran et al., 2017). Certain experimental treatments, such as amputations of parts of the heart, lateral pressure on the dorsal vessel, and ligation near the middle of the heart may alter

冈比亚按蚊取食血粉后的活性，发现其主要在分化中的原血细胞和已分化的粒细胞中表达（Ponderville 等，2020）。

6.2.3.4 心搏

心搏是心脏收缩波（舒张），一般起源于心脏后部末端，并向前波动。心脏收缩或心跳的速率因不同昆虫而有很大变化，并随生理状况、温度、品种、发育时期、神经活动和神经分泌变化而异。昆虫的心跳被认为是肌源性的，始发于肌肉本身，尽管许多其他无脊椎动物的心跳是神经性的。在深山锹幼虫中，心搏的次数可以低至 15 次 /min，而在其他一些昆虫中接近或高于 100 次 /min。在蚂蚁中，研发了一种可以不伤害蚂蚁而可以检测其心脏收缩的装置（Davis 等, 2021）。

美洲大蠊的心脏是最为复杂的神经支配系统之一，当其被切成无数小块后，每一块还能继续跳动，说明美洲大蠊背血管的不同部位都能发挥起搏器的作用。而起搏器通常位于心脏的后部，是支配性的，心脏收缩波（舒张）通常起源于心脏后部附近，并向前传递。收缩波在一些种类昆虫中移动非常慢，而在其他昆虫中的传播速率则非常快，以至于看起来整个血管都在同时收缩。绿头苍蝇（*C. vicina*）的心跳会发生有规律的逆转，伴随着一系列前、后搏动（Wasserthal，2015）。有时，收缩从两端开始，在接近中间的位置汇合。逆转是不可预测的，在飞行期间反向脉冲的速率比正向脉冲的速率要高很多，可能是将富含二氧化碳的血淋巴泵入胸腔并在后胸气门释放的一种适应（Wasserthal，2015；Wasserthal 等，2018）。与 15 日龄的成虫相比，老龄的冈比亚按蚊心率下降，同时反向脉冲发生的频率升高（Doran 等，2017）。某些实验处理，如部分心脏切除术、背血管侧压和心脏中部结扎可以改变心搏传导

conduction rates and cause reversal of contraction waves in some insects.

In insects, ion influences on heartbeat are variable and not well illustrated. Although potassium controls the resting potential of the American cockroach heart, whereas, it is not the controlling ion in other insects. Similarly, the heart of the American cockroach is insensitive to tetrodotoxin, a poison that completely blocks sodium channels in nerves, but does not stop the myogenic heartbeat. Calcium has been implicated in depolarization of the heart in some insects, but not others. Of note, saline containing high concentrations of magnesium stops the heart of *P. americana*, a typical response of animal hearts to high magnesium ion concentrations. The heart of *H. cecropia*, however, cannot beat without magnesium in the saline (Nation, 2022).

速率，导致一些昆虫心脏收缩的逆转。

在昆虫中，离子对心跳的影响是变化不定的，且尚未进行很好的研究。虽然 K^+ 控制美洲大蠊心脏的静息电位，但是其在其他昆虫中并不是控制离子。类似的，美洲大蠊的心脏对能够完全阻碍神经中 Na^+ 通道但不阻止肌源性心跳的河豚毒素也不敏感。Ca^{2+} 只在一些昆虫中对心脏的去极化有影响，对其他昆虫没影响。值得注意的是，含有高浓度镁离子的生理盐水会使美洲大蠊的心脏停止跳动，这是动物心脏对高浓度镁离子的典型反应。然而，鳞翅目昆虫惜古比天蚕的心脏在没有镁离子的盐水中则不能跳动（Nation，2022）。

6.2.4 Nerve supply to the heart

The heart receives innervation through lateral neurons from ventral ganglia in the abdomen or from fused thoracic and abdominal ganglia in some advanced insects, and from a chain of cardiac neurons that lie alongside the heart in some insects, notably Odonata, Orthoptera, Dictyoptera, and Hemiptera. Spontaneously active neurosecretory cells also are located in the cardiac chain, especially near junctions with the lateral segmental nerves. The cardiac neuron in the American cockroach are stretch-sensitive motor neurons. They fire at increasing rate as the myocardium stretches in diastole and cease firing at each systolic contraction (Fig. 6-5). Insects in the orders Diptera and Lepidoptera, and probably other orders, do not have cardiac neurons and a nerve chain paralleling the heart. The heart may still receive a nerve supply from the segmental ganglion.

6.2.4 心脏的神经控制

心脏接纳来自腹部的腹侧神经节或某些高等昆虫的胸腹融合神经节的侧神经元支配，在有些昆虫中则受到来自位于心脏旁边的心神经索的控制，如蜻蜓目、直翅目、网翅目和半翅目。自发的活性神经分泌细胞也位于心神经索内，特别是在靠近心神经索与节间侧神经交会处。美洲大蠊的心神经元是拉伸敏感型运动神经元。在心脏舒张期心肌伸展时，它们以递增的速率放电，在每次心脏收缩时停止传播（图 6-5）。双翅目和鳞翅目，或许还有其他目昆虫，没有心神经元及平行于心脏的心神经索，其心脏可能仍接受体节神经节的调控。

6.2.5 Cardioactive secretions

Various secretions and neurosecretory peptides act upon the heart to change the rate and amplitude of the heartbeat. There are more than 30 families of peptide neurohormones and many of them have some action on the heart and circulatory system, including acting as neurotransmitters and neuromodulators. Proctolin, a neuropeptide produced in motoneurons, interneurons, and neurosecretory cells located at various places

6.2.5 作用于心脏的分泌物

各种分泌物和神经分泌肽作用于心脏可以改变心搏的速率和振幅。目前，已发现有 30 多个家族的神经肽激素，其中包括许多神经传递物质和神经调节物质等对心脏和循环系统有作用的物质。直肠肽是一种由运动神经元、中间神经元以及位于不同昆虫各种不同位置的神

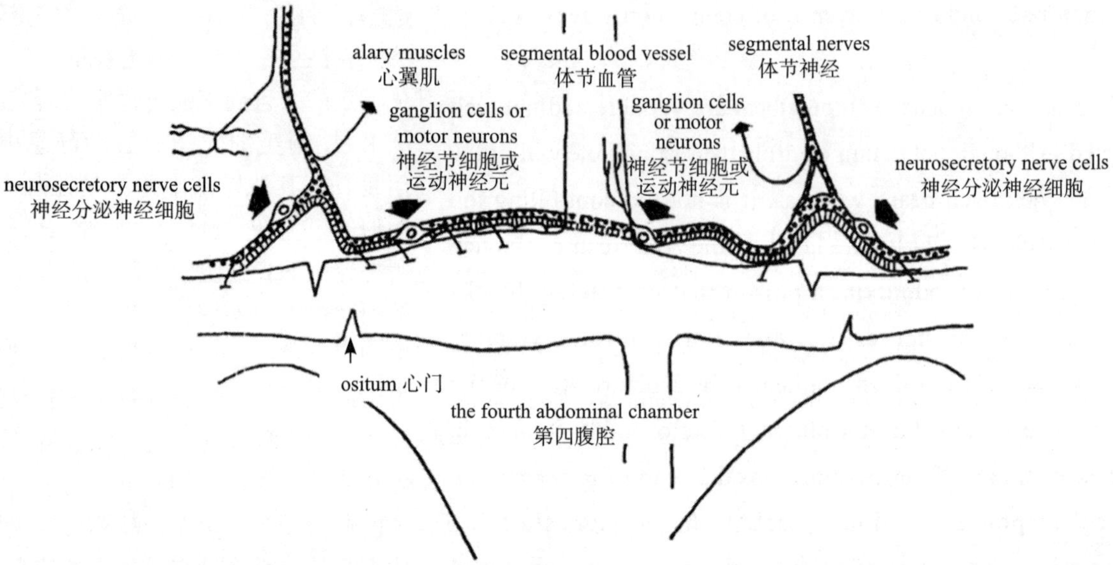

Fig. 6-5 A typical cardiac nervous system in the middle of abdomen of the cockroach *P. Americana*. One lateral cardiac cord is shown dorsal-side-down (modified from Miller, 1968).

图 6-5 典型的美洲大蠊沿腹部心内神经系统。一侧的神经节背部朝下方示意（仿 Miller, 1968）。

in different insects, typically in the brain, but sometimes in other ganglia, stimulates heart rate. Crustacean cardiactive peptide (CCAP), a cardioactive peptide composed of nine amino acids, was first isolated from a crab, and has also been isolated from *L. migratoria*. It stimulates the heart in Locusta and the stick insect *Carausius morosus*, but it also is known to have physiological actions unrelated to the heart (Marco et al., 2018). CCAP increases the rate of accessory hearts at the base of the antennae of *A. gambiae* (Suggs et al., 2016).

Several neurotransmitters modulate the heart contraction rate. 5-Hydroxytryptamine is a neurotransmitter in the nervous system, and it causes vigorous increase in heart beat rate at very low concentrations in isolated or semiisolated hearts of some insects. Even in those in which it increases the heart rate, it seems to have little effect when injected into living insects, but this could be because it is quickly deactivated. Octopamine are cardioacceleratory; glutamate is cardioacceleratory in some insects and has unclear effects in others. Moreover, octopamine may regulate heartbeat reversals, and glutamate has been implicated in the retrograde pacemaker of fruit flies but not mosquitoes.

经分泌细胞所产生的，通常位于脑内，但有时也位于其他神经节，对率起促进作用。甲壳动物纲心激肽（CCAP）是由 9 个氨基酸组成的心激肽，首次从蟹中分离，也在飞蝗中分离得到。它刺激蝗虫和竹节虫 C. *morosus* 的心脏，但也有其他与心脏无关的生理学作用（Marco 等，2018）。CCAP 可以增加位于冈比亚按蚊触角基部的辅助心脏的跳动速率（Suggs 等，2016）。

一些神经递质调节心脏的收缩速率。5- 羟基色胺是一种神经系统的神经递质，有些昆虫离体或半离体的心脏在其非常低的浓度下就会引起心搏速率的剧烈增加。即使上述那些 5- 羟基色胺能增加它们心跳速率的昆虫，当注射到活体时则几乎无效，这可能是因为失效迅速的缘故。章鱼胺具有加速心脏跳动的作用，谷氨酸盐在一些昆虫也是促进心搏的，但是在其他昆虫中作用尚不清楚。此外，章鱼胺可能调控心跳方向的逆转，谷氨酸盐则参与果蝇而非蚊子的起搏器的逆转。

6.2.6 Accessory pulsatile organ

Accessory pulsatile organs or accessory hearts manage the exchange of hemolymph between the hemocoel and the appendages. Accessory pulsatile organs have several shapes and variable morphology, but all are simple pulsatile, sac-like structures. They occur at various places in the body, but most commonly at the base of antennae and wings, at the thoracic base of legs, and within the leg, usually near the femurotibial joint, and in the dorsal region of the meso- and metathorax. They assist circulation of the hemolymph into and through the appendages. Based on location, they are usually referred to as leg, antennal, or wing hearts. They merely aspirate hemolymph into a sinus cavity through ostial openings and pump it out. A mesothoracic accessory heart of *B. mori* pumps hemolymph into the wings, and is not directly connected to the aorta, while in some other insects it is connected to the aorta. In primitive insects, hemolymph is supplied to antennae and cerci by arteries connected to the dorsal vessel. In higher insects, however, these arteries were decoupled and associated with autonomous pumps. To ensure hemolymph supply to legs, wings, and some other appendages, completely new accessory pulsatile organs were evolved.

Antennal hearts localize at the base of the antennae, and are common in many insects, but antennal ampullae in some insects do not have a direct connection with the aorta. Some wing circulatory organs are extensions of the dorsal vessel, while others are structurally independent and called wing hearts. Dorsal vessel-independent wing hearts can be paired organs, in which case each wing has its own pulsatile diaphragm, or an unpaired organ, in which case a single pulsatile diaphragm circulates hemolymph in both wings (Chintapalli et al., 2016). Leg hearts are common in some Odonata, Hemiptera, Homoptera, and higher Diptera. There may be a membranous septum between the ventral and dorsal regions of the tibia, and hemolymph is pumped into one channel and returns through the other, aided by muscle contractions in the leg (Hillyer et al., 2020; Nation, 2022).

6.2.6　辅搏器

辅助搏动器官（辅搏器）或辅助心脏掌管血腔与附肢之间血淋巴的交换。辅搏器有几种不同的形状和形态，但都是简单跳动的囊状结构。它们存在于虫体内数个位置，但多数普遍位于触角和翅的基部、在胸足的胸部、足内、腿节和胫节交界处以及背部的中胸和后胸区域。它们协助血液循环进入和通过附肢。根据其位置，它们常被称为足辅搏器、触角辅搏器或翅辅助心脏。它们仅仅是通过心门开口将血液吸入窦腔并泵出。家蚕的中胸辅搏器将血淋巴泵入翅膀，不与大动脉直接相连，但其他的一些昆虫是直接与大动脉连接的。原始昆虫通过动脉与背血管的连接向触角和尾须供应血液。然而，高等昆虫的这些动脉是分离的，并与自主性泵连接。为确保向足、翅和其他附肢供应血液，它们演变出全新的辅搏器。

触角心脏位于触角底部，在许多昆虫中很常见，但有些昆虫的触角壶状体与主动脉不直接相连。一些昆虫翅膀的循环器官是背血管延伸的，而另一些在结构上独立于背血管，称为翅心。独立于背血管的翅心可以是成对的器官，在这种情况下，每只翅膀都有自己的搏动隔膜，或者单个的翅心，这种情况下通过一个搏动隔膜推动两只翅膀中的血液循环（Chintapalli 等，2016）。足部心脏在蜻蜓目、半翅目、同翅目和更高等双翅目中较为常见。在胫骨腹侧和背侧区域之间可能有一膜质隔膜，血液淋巴在足部肌肉收缩的辅助下被泵入一个通道，并通过另一个通道返回（Hilly 等，2020; Nation, 2022）。

6.3 血细胞

6.3 Hemocyte

6.3.1 Classification of hemocytes

Hemocytes are free cells in circulatory system. They change their appearance and shape from time to time even in the same insect. Hemocytes dynamically change from circulating with the hemolymph (circulating hemocytes) to attaching to tissues (sessile hemocytes). Some sessile hemocytes attach to the heart called pericardial cells, and in certain adult insects, they aggregate in the extracardiac space that surrounds the ostia. The pericardial cells adhere to the locations of the body that experience the most hemolymph flow, thus increasing the probability of encountering and neutralizing pathogens, which exemplify the functional integration of the immune and circulatory systems.

The seven most common types of hemocytes found in insects are prohemocytes, plasmatocytes, granulocytes, spherulocytes, adipohemocytes, oenocytoids, and coagulocytes (Hillyer et al., 2020; Nation, 2022). Most insects that have prohemocytes, plasmatocytes, and granulocytes, but the presence of the other types is variable. Only prohemocytes, plasmatocytes, granulocytes, spherule cells or spherulocytes, and oenocytoids were described from the pink bollworm, *Pectinophora gossypiella* (Lepidoptera: Gelechiidae). Mature larvae of the grapevine moth, *Eupoecilia ambiguella* have granulocytes, oenocytoids, plasmatocytes, and spherulocytes in the hemolymph and the proportion of each type varies with diet and immune challenge, with granulocytes showing the greatest increase after immune challenge (Vogelweith et al., 2016). Eight classes of hemocytes including prohemocytes, plasmatocytes, granular hemocytes, coagulocytes, crystal cells, spherule cells, oenocytoids, and thrombocytoids were established by Lackie (1988).

Crystal cells, plasmatocytes, and lamellocytes have been described from larval of *D. melanogaster*. Plasmatocytes in *Drosophila* are not equivalent of plasmatocytes in Lepidoptera,

6.3.1 血细胞的分类

血细胞是循环系统中的游离细胞。即使在同一只昆虫身上，它们也会不时地改变外观和形状。血细胞从伴随血淋巴循环（循环血细胞）到附着于组织上（固着血细胞）是一个动态变化的过程。一些固着血细胞附着在心脏上被称为围心细胞；在某些成虫中，它们聚集在心门周围的心外空间。围心细胞附着在心脏位置上可以经历最多血淋巴的流过，从而增加遇到和中和病原体的概率，这也体现了免疫系统和循环系统功能的整合。

研究发现昆虫有7种最常见的血细胞类型，原血细胞、浆血细胞和粒血细胞、珠血细胞、脂血细胞、类绛色细胞和凝血细胞（Hillyer等，2020; Nation, 2022）。多数昆虫具有原血细胞、浆血细胞和粒血细胞，而其他类型的出现是可变的。在红棉铃虫（鳞翅目，麦蛾科）中，只有原血细胞、浆血细胞、粒血细胞、球形细胞或珠血细胞和类绛色细胞。葡萄蔓蛾的老幼虫血淋巴中有粒血细胞、类绛色细胞、浆血细胞和珠血细胞，其中粒血细胞在免疫应激后表现出最显著的增加（Vogelweith 等，2016）。Lackie（1988）建立了八种血细胞类型，分别为原血细胞、浆血细胞、粒状血细胞、凝血细胞、晶状体细胞、球形血细胞、类绛色血细胞和类血小板细胞。

果蝇幼虫有含晶细胞、浆血细胞和片状细胞。其浆细胞与鳞翅目浆血细胞不同，而片状细胞与鳞翅目浆血细胞极

but lamellocytes show the most similarity to lepidopteran plasmatocytes. Crystal cells in *Drosophila* have great similarity to lepidopteran oenocytoids. Cells that look like prohemocytes are rare. So, the classification of hemocytes is critically not unified.

其相似。果蝇的含晶细胞与鳞翅目类绛色细胞非常类似。像原血细胞的细胞是罕见的。因此，血细胞的分类严格来说是不统一的。

6.3.1.1 Prohemocytes

They are the smallest hemocytes and may have stem cells which some other hemocytes may develop. Prohemocytes are known to divide and they may differentiate into plasmatocytes, which in turn may give rise to granulocytes, and these may differentiate into sperulocytes. The prohemocytes may be one source of new hemocytes. Prohemocytes are typically round, 6 to 13 μm in diameter, with a relatively large nuclear (70% to 80% of volume). They contain ribosomes and mitochondria, but little endoplasmic reticulum and Golgi membranes. They are not mobile and do not participate in phagocytosis. The main function of hemocytes may be to divide and give rise to new hemocytes.

6.3.1.1 原血细胞

它们是最小的血细胞，可能是干细胞，并可以发育成一些其他的血细胞。已知原血细胞可分裂分化成浆血细胞，进而生成粒血细胞，这些细胞再分化成珠血细胞。原血细胞可能是新血细胞的来源之一。原血细胞通常为圆形，直径为 6 ～ 13 μm，核相对较大（占体积的 70% ～ 80%）。其含有核糖体和线粒体，但几乎没有内质网和高尔基体膜。原血细胞不运动，也不参与吞噬作用。其主要功能可能是分裂，并产生新的血细胞。

6.3.1.2 Plasmatocytes

Plasmatocytes are small-to-large, polymorphic cells up to 40 to 50 μm in size, granular or agranular, and round-to-spindle shaped in wet suspensions, although they lose this shape when dries on a slide. "Young" plasmatocytes can be confused with prohemocytes. They may be binucleate. They contain lysosomal enzyme and are usually the most numerous of circulating cells. They are phagocytic and participate in encapsulation, nodule formation and wound healing (Fig. 6-6; Appendix color graphic Fig.6-6).

6.3.1.2 浆血细胞

浆血细胞是由小到大的多形细胞，其大小为 40 ～ 50 μm，颗粒或非颗粒状，在溶液中呈圆形到椭圆形，但是在载玻片上干燥后则失去该形状。“年轻”的浆血细胞可以与原血细胞融合。浆血细胞可能具有双核，并含有溶酶体酶，一般是数量最多的循环细胞。浆血细胞具有吞噬作用，参与包被和成瘤的形成及伤口愈合（图 6-6；附彩图 6-6）。

6.3.1.3 Granulocytes

Granulocytes are variable in size, spherical or oval, and up to 45 μm in size. The nucleus is usually small and the cytoplasm is regular. On the basis of histochemical tests, the granules are thought to be glycoproteins and mucopolysacharides. Granulocytes may arise from plasmatocytes. The precise function of granulocytes is unproven, but they may serve storage and possibly secretory functions. They may be involved in cellular defensive function in various insects, and may be phagocytic in some insects but in others, neither of these functions is established.

6.3.1.3 粒血细胞

粒血细胞大小不一，呈球形或椭圆形，其大小可达 45 μm。在一般情况下，核小、胞质均匀。基于组织化学测试，粒血细胞被认为是糖蛋白和黏多糖。粒血细胞可能由浆血细胞产生，其确切功能未被证实，但它们可能具有贮存和分泌的功能。在不同的昆虫体内，粒血细胞可能参与细胞防御功能，有些昆虫中可能参与吞噬，但在其他昆虫中任何这些功能都未被证实。

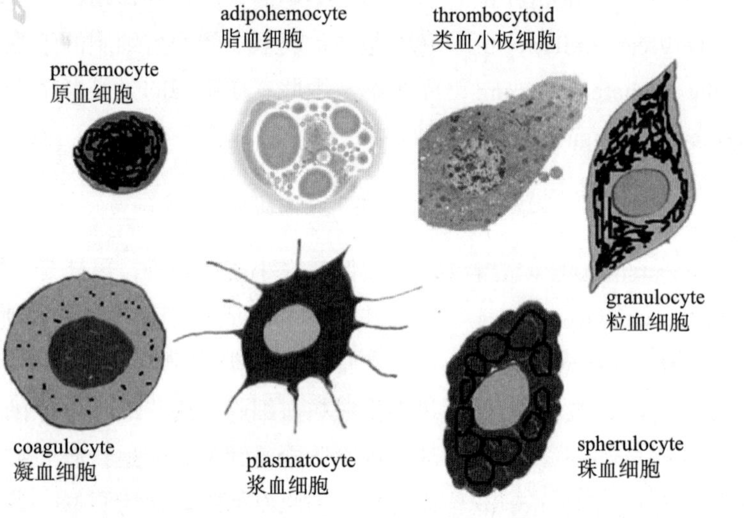

Fig. 6-6 Types of hemocytes found in many insects. Arrows indicate that prohemocytes transform into plasmatocytes, some plasmatocytes may give rise to granulocytes, and granulocytes may transform into sperulocytes (Nation, 2022).

图 6-6 多数昆虫中存在的典型血细胞类型。箭头显示原血细胞转变成浆血细胞，一些浆血细胞可能产生粒血细胞，而粒血细胞可能转变成珠血细胞（Nation, 2022）。

6.3.1.4 Spherulocytes

Spherulocytes are ovoid to round cells up to about 25 μm in length. They may contain few to many small spherical inclusions that stain for apical mucopolysaccharides. Their function is unknown, but they may participate in phagocytosis.

6.3.1.5 Adipohemocytes

Adipohemocytes may be small or large, spherual to oval, and contain lipid droplets. They might be plasmatocytes that are filled with lipids under certain physiological condition.

6.3.1.6 Oenocytoids

Oenocytoids are variable in size, often large, may be binucleate, and lyse easily, but do not cause hemolymph coagulation when they lyse. They are nonphagocytic. Some evidence indicates that they contain prophenoloxidase, an inactive form of phenoloxidase. Oenocytoids should not be confused with oenocytes, cells found among fat body cells and scattered among epidermal cells in many insects. Oenocytes are not blood cells.

6.3.1.7 Coagulocytes

Coagulocytes also have been called hyaline hemocytes and cystocytes. These cells rupture within seconds after injury or after taking a hemolymph sample from an insect, and initiate the clotting process. The cells may contain granules. In phase contrast they are nearly transparent, hence, the name hyaline hemocytes. The

6.3.1.4 珠血细胞

珠血细胞呈卵形至圆形，长直径达 25 μm。它们含有少量到很多的球形包含物，其顶端黏多糖可着色，但功能不明，可能参与吞噬作用。

6.3.1.5 脂血细胞

脂血细胞可小可大，呈球形或椭圆形，含有脂滴。它们可能是在某种生理条件下充满脂质的浆血细胞。

6.3.1.6 类绛色细胞

类绛色细胞大小不一，通常较大，可能有双核，细胞体容易被溶解，但在被溶解时不引起血淋巴的凝固。它们是非吞噬性的。有些实验证明其含有酚氧化酶的非活化形式，酚氧化酶酶原。类绛色细胞不能和绛色细胞混淆。类绛色细胞在脂肪体细胞间发现，分布在许多昆虫真皮细胞之间。绛色细胞不是血细胞。

6.3.1.7 凝血细胞

凝血细胞也被称为透明血细胞和囊血细胞。在受到损伤或从一个昆虫采集了血淋巴样品后，这些细胞在数秒内破裂并启动凝血过程。凝血细胞可能含有颗粒，在相差显微镜下它们近于透明，因

hemolymph of some insects, for example, hemolymph of larval honeybees, *A. mellifera*, does not coagulate.

6.3.1.8 Thrombocytoids

Thrombocytoids are peculiar filamentous cells. They are binucleated cells, very fragile and few in number. The "explosive corpuscle" present in insects that are involved in clotting and melanization.

In general, the nucleus of all hemocytes except prohemocytes, plasmatocytes and granulocytes deviated from the central, and nearly almost of the hemocytes are variable in size. Crystal cells and thromocytoids might considered as the products derived from coagulocytes (Nation et al., 2022).

6.3.2 Functions of hemocytes

Hemocytes play major protective roles to defense against invading microorganisms and the eggs or larvae of parasitoids by production of antibiotics, encapsulation, phenoloxidase (PO) production and release, and phagocytosis. Hemocytes participate in wound healing by aggregating at the wound site, where some cells phagocytize cellular debris or foreign organisms, such as bacteria, and hemocytes activate polyphenoloxidase (PPO). Particularly, coagulocytes and possibly granulocytes in some insects participate in the coagulation of plasma to help plug a wound. Entrapment of hemocytes in the coagulum helps plug the wound (Aprelev et al., 2019). In *A. gambiae*, phagocytic hemocytes function in complement recognition of Plasmodium invading ookinetes and reduce ookinete survival by the production of PO (Kwon et al., 2019).

Some hemocytes contain enzymes that aid in detoxication, including detoxication of some insecticides. Hemocytes participate in nodule formation and encapsulation of foreign objects. Invasion of bacteria may be attacks by nodule formation in which the bacteria are aggregated into nodules containing hemocytes and coagulum. Larger objects or invading parasites may be encapsulated. Plasmatocytes and granulocytes aggregate to form thin sheets of cells and plaster themselves around nodules, internal parasites, or other foreign objects. Sometimes encapsulated objects become melanized by action of the phenoloxidase cascade of enzymatic action on tyrosine or other phenolic compounds.

此命名为透明血细胞。有些昆虫，如蜜蜂幼虫 *A. mellifera* 的血淋巴是不凝固的。

6.3.1.8 类血小板细胞

类血小板细胞是异常纤细的细胞。它们是双核细胞，非常脆弱且数量稀少，是一种涉及昆虫血淋巴凝结和黑化的“易爆血球”。

总体来说，血细胞的核除了原血细胞、浆血细胞和粒血细胞，都偏离了中央，几乎所有的血细胞大小都是可变的。含晶细胞及类血小板细胞被认为是凝血细胞的衍生产物（Nation 等，2022）。

6.3.2 血细胞的功能

血细胞通过产生抗生素、包被、酚氧化酶（PO）的产生和释放以及吞噬等作用在抵御入侵微生物和寄生物的卵或幼虫的过程中发挥着重要的保护作用。血细胞通过聚集在创面部位参与创面的愈合，在此，细胞吞噬细胞碎片或外来生物，如细菌，血细胞激活多酚氧化酶（PPO）。特别是凝血细胞和某些昆虫的粒血细胞参与血浆的凝结，以帮助伤口愈口。凝结物中的血细胞有助于堵塞伤口（Aprelev 等，2019）。在冈比亚按蚊中，吞噬血细胞通过产生 PO 从而在抵御入侵的疟原虫动合子中发挥补体识别和减少动合子存活率的功能（Kwon 等，2019）。

有些血细胞含有的酶类有助于脱毒，包括某些杀虫剂的解毒。血细胞参与结节的形成和对外源物质的包被。侵入的细菌可能被聚集到含有血细胞和凝结物的结节内。大型物质或侵入的寄生虫可能被包被。浆血细胞和粒血细胞积聚而形成细胞薄层，将其自身紧贴在结节、内部寄生虫或其他外源物质的周围。有时包被的物质因为酚氧化酶级联的酶作用于酪氨酸或其他酚类物质而变

Encapsulated objects often become attached to various tissues or organs in the body. The ability to encapsulate parasitoids seems to be dependent on an evolutionary race between the parasitoid and its host. A parasitoid that is not well adapted to a particular host is likely to have a high percentage of its eggs encapsulated, while a parasitoid well adapted to its host may avoid significant encapsulation. Similar to nodule formation, the encapsulated object is surrounded by many layers of plasmatocytes that form sheets of cells around the object. Often, but not always, the whole mass becomes melanized, a process that also may be toxic to the offending organism and assist in killing it (Hillyer et al., 2020; Nation, 2022).

黑。包被的物质常常附着在体内各种组织和器官上。包被寄生虫的能力取决于寄生虫和宿主之间的进化竞争。拟寄生物如果没有被特殊的宿主很好适应的话，其卵被包被的概率很高，而适应了宿主的拟寄生物可逃脱有效包被。与结节的形成相似，包被的物质被多层浆血细胞所包围，在其周围形成细胞层。在大多数情况下，也并不总是整个细胞团变黑，这可能是对攻击型微生物有毒并有助于将其杀死的过程（Hillyer 等，2020; Nation, 2022）。

Bacterial invasion of some insects results in synthesis and secretion into the hemolymph of combinations of several antibiotic peptides and proteins. Fat body tissue is the usual site of synthesis of these proteins, but hemocytes and a variety of other tissues also contribute to the synthesis. Nearly almost all insects have their physical barriers that prevent pathogens from entering the body, the organs and tissues that regulate immune responses as well as the mechanistic and molecular bases of immunity (Hiller and Pass, 2020).

细菌入侵导致有些昆虫合成并分泌抗菌肽和抗菌蛋白到血淋巴中。脂肪体组织通常是合成这些蛋白质的主要场所，但血细胞和一系列其他组织也具有合成功能。几乎所有的昆虫都具备防止病原物进入其体躯的物理屏障和调节免疫反应的器官以及机制和分子的基础（Hiller 和 Pass, 2020）。

6.3.3 Hemocytopoietic tissues and organs

6.3.3 造血组织和器官

The number of hemocytes in the circulating insects is quite variable from species to species, and even within the same individual at different times depending on its physiological state. Sex, age, stage of development, and activity are known to influence the observed number of cells in some insects. Also, some hemocytes are sessile, attached to various tissue, at least for much of the time, but may be released into the circulation under certain circumstances. Different species contain widely differing numbers. For example, the American cockroach, *P. americana*, can have as many as 70,000 to 120,000 cells/μL hemolymph; the tobacco hornworm, *M. sexta*, has been reported to have 8,200 cells/μL hemolymph; wax moth larvae, *G. mellonella*, have 35,000 cells/μL hemolymph; silkworm *B. mori,* have 7,000 to 8,000 cells/μL hemolymph; and *R. prolixus,* only have 300 to 5,000 cells/μL hemolymph.

昆虫循环中的血细胞数量因物种而异，即使同一个体不同时间，也因其生理状态不同而存在差别。众所周知，性别、龄期、发育阶段和活动会影响一些昆虫可观察到的血细胞数量。有些血细胞也是固定的，至少很长时间内附着于各种组织，在某种环境下可被释放进入循环系统。不同物种含有的血细胞数量很不相同，如美洲大蠊每微升血淋巴中含有多达 7 万～ 12 万个；据报道烟草天蛾每微升为 8 200 个；大蜡螟幼虫每微升血淋巴中含有 35 000 个；家蚕每微升血淋巴中含有 7 000 ～ 8 000 个；吸血蝽每微升血淋巴中仅有 300 ～ 5 000 个血细胞。

During embryonic development, hemocytes develop from mesodermal tissue. In larvae and adults of some insects, circulating hemocytes are known to be divided and differentiate into other hemocytes. The rate of division of existing hemocytes in larvae of

在胚胎发育时期，血细胞由中胚层组织发育而来。有些昆虫的幼虫和成虫期，循环中的血细胞可被分裂和分化成其他血细胞。大蜡螟幼虫血细胞的分裂

the wax moth, *G. mellonella*, seem to be sufficient to account for the numbers of circulating hemocytes.

There is evidence for hemocytopoietic tissues in some insects, for example, in the cricket, *G. binaculatus*. Larvae of Lepidoptera have mass of loosely connected cells in a capsule located near the prothoracic spiracles that seem to give rise to hemocytes. Hemocytes are released from the capsules through gaps in the covering surface. The organs disintegrate in pupae and release large numbers of hemocytes into the body. Small clusters of cells near the wing imaginal disks in the commercial silkworm, *B. mori*, and the tobacco hornworm, *M. sexta*, give rise to hemocytes during larval life. Hemocytes also arise from division of existing hemocytes in the circulating hemolymph in the silkworm.

Similar structures without the capsule enclosure are reported to occur in some dipterous larvae, such as the housefly. The lymph gland in larvae of *D. melanogaster* gives rise to hemocytes, with about 95% of the cells becoming plasmatocytes (Blanco-Obregon et al., 2020). The Adult flies also have hematopoietic tissues. Some Orthoptera and some cockroaches have complex hemocytopoietic organs composed of hollow sacs at the anterior end of the heart. Cells inside the sac are phagocytic, and some divided into stem cells that can further differentiate into several different types of hemocytes. Small groups of cells between pericardial cells located near the dorsal diaphragm are believed to produce hemocytes in *Locusta* and *Melolontha*. More than 60% of the cultured prohemocytes from larvae of *B. mori* differentiated into plasmatocytes or granulocytes, and some granulocytes later differentiated into sperulocytes. Some prohemocytes divided into new hemocytes, but no oenocytoids cells were produced in culture. It suggested that oenocytoids may come from a different stem cell line.

Plasmatocytes in *Drosophila* larvae are phagocytic cells, but they can transdifferentiate into lamellocytes, whose primary function is encapsulation of foreign objects (Csordás et al., 2021). Probably insects are quite variable in how hemocytes are produced and whether they have hemocytopoietic tissues.

速率足以满足需要的循环血细胞的数量。

研究表明，在一些昆虫中存在造血组织，例如蟋蟀 *G. binaculatus*。鳞翅目幼虫前胸气门附近的囊包内有松散连接的细胞团，似乎能产生血细胞。血细胞是从囊包表面的缝隙中被释放的。蛹期器官解离时，释放出大量血细胞进入体腔。家蚕和烟草天蛾翅原基附近的小细胞串在幼虫期产生血细胞。在家蚕中，血细胞也可以由循环系统中现存的血细胞分裂而来。

据报道有些双翅目幼虫也有相似的结构，如家蝇，但没有囊包包裹。果蝇幼虫的淋巴腺也可以产生血细胞，其中 95% 的细胞是浆血细胞（Blanco-Obregon 等，2020）。苍蝇成虫同样也有造血器官。有些直翅目和一些蜚蠊有复杂的造血器官，由心脏前末端空泡组成。空泡内的细胞是吞噬性的，但有些分裂成干细胞，能进一步分化成几种不同类型的血细胞。飞蝗和金龟子背膈附近的围心细胞间的小群细胞，被认为可产生血细胞。培养的家蚕幼虫原血细胞，60% 以上分化成浆血细胞或粒血细胞，而有些粒血细胞可进一步分化成珠血细胞。有些原血细胞分裂成新的血细胞，但培养中没有产生类绛色细胞。这暗示类绛色细胞可能来自不同的干细胞系。

果蝇幼虫的浆细胞主要为吞噬细胞，但浆细胞可转化为片状细胞，其主要功能是包被异物（Csordás 等，2021）。昆虫可能在血细胞如何产生和是否存在造血组织的差别很大。

6.4　血淋巴的化学成分

6.4　Chemical Composition of the Hemolymph

Hemolymph contains many dissolved inorganic and organic substances, colloidally suspended proteins, and lipoproteins.

血淋巴包含许多溶解的无机和有机物质、胶质悬浮蛋白和脂蛋白。水分约占

It is about 90% water and 10% solids. Most components of hemolymph contribute to the osmotic pressure and specific gravity of hemolymph. Specific gravity is usually slightly greater than 1. Osmotic pressure in some species is not correlated with age, sex, developmental stage, diet, or systematic position among orders. In general, saline solutions that are equivalent to 0.9% to 1.6% sodium chloride will encompass the range of osmotic pressure that in insect hemolymph, but such a solution of NaCl will not be balanced ironically with respect to the composition of hemolymph.

90%，固态物占10%。血淋巴的大部分成分对其渗透压和比重都有影响。比重通常略大于1。某些种类昆虫的渗透压与年龄、性别、发育阶段、饮食或不同目之间的系统地位无关。一般来说，0.9%～1.6%的NaCl溶液涵盖了昆虫血淋巴的渗透压范围，但这种NaCl溶液相对于血淋巴的成分而言，它不能保持离子的平衡。

6.4.1 Inorganic ions

The ionic composition of hemolymph plasma is highly variable in different insects. Sodium, potassium, calcium, and magnesium are typical cations, and chloride, phosphate, amino acids, and sometimes bicarbonate are present as anions. Chloride and phosphate are the major anions in hemolymph of honeybee larvae, but these are not usually the major anions balancing the cations in hemolymph of most insects. Chloride accounts for about 7% of the anions in larvae of the horse botfly *G. asterophilus intestinalis*, 12% in larvae of silkworm, *B. mori,* and up to 39% of anions in larvae of the southern armyworm, *S. eridania.* Sodium and potassium concentrations, and Na^+/K^+ ratio in hemolymph are variable.

In the species from Odonata, Diptera, carnivorous Coleoptera, Dictyoptera, and Orthoptera, sodium: potassium ratio was relatively high in the plasma, while calcium:magnesium ratio is equal to about 1. It is considered to approximate the early evolutionary condition in insects. Lepidotera, phytophagous Coleoptera, Hemiptera, Homoptera, and Hymenoptera have a relatively low sodium:potassium ratio that is only a few multiples of 1 or even less than 1, some as low as 0.3 to 0.1, or lower. Some insects, at least, have the ability to regulate ion composition. Ions are able to bind with macromolecules in hemolymph in some insects. In *P. americana*, macromolecules bind more than 20% of the sodium and magnesium, about 16% of the calcium, and about 10% of the chloride in whole hemolymph, whereas, potassium is not bound. Larvae of *Chironomus riparius* are able to modulate ion concentrations in hemolymph in response to ion-poor conditions through the regulation of total net transport of Na^+, Cl^-, and H^+ by the anal papillae (Nguyen et al., 2010). In *Ostrinia furnacalis*, exposure to sublethal doses of azadirachtin (Az), dramatically increased the concentrations of Na^+, NH_4^+, K^+, Mg^{2+}, and Ca^{2+} in hemolymph of larvae (Shi et al., 2006). These reports show that the distribution and concentration of

6.4.1 无机离子

各种昆虫血浆的离子成分是高度变化的。钠、钾、钙、镁是典型的阳离子，而含有的氯化物、磷酸盐、氨基酸以及有时碳酸盐作为阴离子。蜜蜂幼虫血淋巴的主要阴离子是氯化物和磷酸盐，但这些并非是多数昆虫血淋巴主要平衡阳离子的阴离子。氯化物占肠胃蝇幼虫阴离子的7%，占家蚕幼虫阴离子的12%，而占南方黏虫阴离子的39%。血淋巴钠和钾的浓度以及钠钾比是可变的。

蜻蜓目、双翅目、肉食性鞘翅目、网翅目和直翅目的血浆中钠钾比相对较高，钙镁比大约等于1。这被认为是接近于昆虫进化早期的情况。鳞翅目、植食性鞘翅目、半翅目、同翅目和膜翅目的钠钾比相对较低，仅是1的数倍或低于1，有些低至0.3～0.1或更低。有些昆虫至少具备调节离子成分的能力。有些昆虫的离子可以与血淋巴中大分子物质的结合。在美洲大蠊中，大分子物质结合了整个血淋巴中20%以上的钠和镁以及16%左右的钙和大约10%的氯，然而不结合K^+。摇蚊（*C. riparius*）幼虫在离子缺乏的情况下由肛乳通过调节Na^{2+}、Cl^-和H^+的总转运净量调节体内血淋巴中的离子浓度（Nguyen等，2010）。在玉米螟中，暴露于亚致死剂量的印楝素后会导致血淋巴中Na^+、NH_4^+、K^+、Mg^{2+}和Ca^{2+}五种阳离子浓度急剧增加（Shi等，2006）。

ions in hemolymph are affected and regulated by the various physiological conditions of insects.

6.4.2 Free amino acids

Of note, insect hemolymph contains very large amounts of free amino acids, much more than that in the body fluids of other animals. The amino acids contribute to the osmotic pressure of hemolymph, and account for a substantial portion of the cations and anions of hemolymph. The hemolymph probably is a reservoir of amino acids for protein synthesis, much of which occurs in fat body cells.

Three amino acids, proline, *α*-alanine and glutamine, represented between 48 and 67% of the total amino acids in hemolymph of *Pyrrhocoris apterus*, in addition, proline and α-alanine most significantly contributed to the seasonal increase of amino acids in insects (Koštál et al., 2011). Proline and alanine also have high content and significantly increased after blood meal in female *Culex pipiens pallens*.

6.4.3 Proteins

Hemolymph contains a wide variety of different proteins. Many protein bands can be detected by electrophoresis and these may change with physiological state, age, sex, and other factors. Some insects synthesize large quantities of one or more storage proteins during late larval life. Calliphorin, has a molecular weight of 540 kDa and is composed of subunits of about 85,000. It accounts for up to 60% of the proteins in mature blowfly larvae. It is synthesized by the fat body, released into the hemolymph, and finally reabsorbed back into the fat body and stored for use during pupation and adult development. Approximate 40% of total proteins were composed by those with molecular weight between 55-100 kDa in hemolymph of larvae in *B. mori*, 90% of which are storage proteins and provide amino acid source for the biosynthesis of vitellogenins during pupation (Wang et al., 2020). Hemolymph proteins include many different enzymes, such as lysozyme, antibacterial and antifungi proteins, and transport proteins that carry hydropholic substances, such as juvenile hormone, cholesterol, diglycerides, hydrocarbons, and other lipoidal substances through the aqueous medium of hemolymph.

Phenoloxidase plays important role in sclerotization of the

这些研究表明，昆虫血淋巴中离子的浓度受到多种生理情况的影响与调控。

6.4.2 游离氨基酸

值得注意的是，昆虫血淋巴中游离氨基酸的水平颇高，比其他动物体液内的氨基酸含量都高。这些氨基酸有助于维持渗透压，占血淋巴中阳离子和阴离子的很大部分。血淋巴可能是用于合成蛋白质的氨基酸贮存库，其大部分合成于脂肪体细胞。

脯氨酸、*α*-丙氨酸和谷氨酸是始红蝽血淋巴中主要的三种氨基酸，占到血淋巴中总氨基酸的 48% ～ 67%。此外，脯氨酸和 α-丙氨酸还是其血淋巴中氨基酸总量季节性增长的主要来源（Koštál 等，2011）。脯氨酸和丙氨酸也是淡色库蚊含有的主要氨基酸种类，并在雌蚊取食血粉后显著增高。

6.4.3 蛋白质

血淋巴含有多种不同的蛋白质。电泳可以检测到许多蛋白质带，这些蛋白质带可随生理条件、龄期、性别和其他因子变化。有些昆虫在末龄幼虫期会合成大量的一种或多种储藏蛋白。丽蝇蛋白质的分子质量为 540 kDa，由大约 85 000 个亚基组成。它占成熟丽蝇幼虫蛋白质的 60%，由脂肪体合成并释放到血淋巴中，最后由脂肪体重吸收贮存，供化蛹和成虫发育利用。在家蚕幼虫的血淋巴中，约 40% 的总蛋白质分子质量为 55 ～ 100 kDa，其中 90% 左右为储藏蛋白，它们为卵黄蛋白原的合成提供了氨基酸来源（Wang 等，2020）。血淋巴蛋白质包含多种不同的酶，如溶菌酶、抗细菌和抗真菌蛋白，运载疏水性物质的转运蛋白，如保幼激素、甘油二酯、碳氢化合物和通过血淋巴的水相介质运载的其他类脂物质。

酚氧化酶在表皮骨化和保护昆虫免

cuticle and in protecting the insect from foreign invaders. It exists in the hemolymph and hemocytes as a proenzyme that can be converted to active PO. PO converts a variety of phenols to quinones. The enzyme and some of its products are important in tanning the cuticle. PO activity often causes melanization of hemolymph and eventually become dark brown or black at wound sites. Moreover, antimicrobial peptides (AMPs) such as cecropins, moricins, defensins, proline- and glycine-rich peptides were biosynthesized by immune organs, and secreted into hemolymph to defense the non-self agents (Wojda et al., 2020).

受外源物入侵中起着重要作用。它以酶原形式存在于血淋巴和血细胞中，能转化成活性的酚氧化酶。PO将多种酚类转化成醌类。酶及其一些产物对表皮的鞣化是重要的。PO活性常造成血淋巴黑化，最终使伤口部位变成棕褐色或黑色。抗菌肽（AMPs），如天蚕素、moricins、防御素以及富含脯氨酸和甘氨酸的多肽等被免疫器官合成后释放进入血淋巴，用以抵御异己因素（Wojda等，2020）。

6.4.4 Other organic constituents

Many organic compounds have been found in the hemolymph of insects. One of the main functions of circulating hemolymph is that of a major transport medium to move metabolic compounds synthesized in fat body cells or hormones synthesized in a variety of neurohemal organs to sites where they are needed. For example, triacylglycerols stored in fat body cells are released as diacyglycerols and these are transported by lipoproteins to thoracic flight muscles in Lepidoptera and other insects.

Insect hemolymph is notable for the large concentration of trehalose in solution, from 0.5 g to as much as 2.5 g/100mL in some cases. Trehalose is the principle hemolymph sugar in most insects and is rapidly converted to glucose for immediate metabolism in glycolysis and the Krebs cycle. EH (Ecdysteroids), JH (juvenile hormones), PTTH (prothoracicotropic hormone), PBAN (pheromone biosynthesis activating neuropeptide), AKH (adipokinetic hormone), HTH (hypertrehalosemic hormone), diuretic hormone, and a series of other neuropeptides and hormones are transported from the site where they are released into the hemolymph to their targets. The titer of these in the hemolymph varies continuously with the physiological state of insect.

Hemolymph is a major transporter of metabolic waste products for excretion or storage. Uric acid is transported by hemolymph from the main site of synthesis, the fat body, to the Malpighian tubules for excretion in many insects or for storage in various sites by some insects. Uric acid is poorly soluble in aqueous solutions, such as hemolymph, and its concentration in hemolymph is low. Possibly potassium and sodium urate account for much of the uric acid transported in hemolymph because the salts are much more soluble in an aqueous medium than the free acid. A possibility that has not been studied in detail is whether transport may be aided by

6.4.4 其他有机组分

在昆虫血淋巴中发现很多有机化合物。循环中的血淋巴主要功能之一是作为重要的运送介质，转运脂肪体细胞合成的代谢化合物或多种神经体液器官合成的激素到其需要的部位。比如，鳞翅目和其他昆虫贮存在脂肪体的甘油三酯以甘油二酯的形式释放，这些物质通过载脂蛋白运送到胸部飞行肌。

昆虫血淋巴中的海藻糖的浓度很高，在某些情况下，从0.5g/100mL到高达2.5g/100mL。海藻糖是多数昆虫的主要的血淋巴糖，在糖酵解途径和克雷布斯循环的直接代谢中被快速转换成葡萄糖。EH（蜕皮甾类）、JH（保幼激素）、PTTH（促前胸腺激素）、PBAN（信息素生物合成活化神经肽）、AKH（激脂激素）、HTH（高海藻糖激素）、利尿激素以及其他神经肽和激素从释放部位进入血淋巴，被转送到其靶点。这些激素在血淋巴中滴度随着昆虫生理状态的不同持续发生着变化。

血淋巴是排泄或贮存代谢废物的主要运送者。尿酸被血淋巴从其主要合成部位——脂肪体运送到马氏管，供许多昆虫排泄或在一些昆虫中贮存在各种不同的部位。尿酸在水溶液，如血淋巴中很难溶解，因而在血淋巴中浓度较低。血淋巴运送的尿酸钠和尿酸钾占尿酸的大部分，因为这些盐比游离酸在水溶液中溶解得更多。血淋巴中尿酸盐与蛋白质的结合后，辅助运输的可能性尚未进

binding of urates to proteins in the hemolymph.

行详细的研究可有助于运送。

6.4.5 Hemolymph pH and hemolymph buffers

The great majority of insects have hemolymph that is slightly acid, but a few have hemolymph that is as alkaline as pH 7.5 or slightly greater. Hemolymph pH typically falls into a pH range of 6 to 7.5. Measurement within the same species may vary by up to about 0.7 pH units. There is no strong correlation with sex, diet, stage of development, or taxonomic position. The greatest changes occur with metamorphosis; during the pupae stage, or sometimes just prior to pupation, there is a slight increase in acidity, but usually no more than about 0.3 pH units. Body temperature is involved in regulating hemolymph pH regulation. Hemolymph pH remains constant at temperatures up to 25℃ and then decreases about 0.017 pH units per degree Celsius above 25℃ in in two orthopterans, the two-striped grasshopper, *Melanoplus bivittatus*, and the locust, *Schistocerca nitens*.

Locusts and grasshoppers and possibly other insects regulate hemolymph pH in at least two different ways: ① by transfer of acid equivalent to the gut when acidosis is caused by a titratable acid; ② by increased ventilation of the tracheal system when acidosis is caused by an increase of CO_2 in the hemolymph, as in vigorous muscular activity. Nearing 90% of the buffering capacity in hemolymph of these locusts is due to bicarbonate and proteins, with a small additional contribution by inorganic and organic phosphates. Amino acids, although in high concentrations in insect hemolymph, have dissociation constants well removed from typical hemolymph pH.

(Tian Ling)

6.4.5 血淋巴 pH 和血淋巴缓冲

昆虫血淋巴绝大多数是微酸性的，但极少数血淋巴呈碱性 pH 为 7.5 或略高。通常，血淋巴 pH 范围为 6 ～ 7.5。同种昆虫内测定变化可达 0.7 个 pH 单位。与性别、食物、发育时期或分类地位没有很强的相关性。最大的变化是随变态发育发生；蛹期阶段或化蛹之前的某个时间，酸度轻微增加，但通常不超过 0.3 个 pH 单位。体温参与了血淋巴 pH 的调节。在两个直翅目昆虫双带黑蝗（*M. bivittatus*）和亮沙漠蝗（*S. nitens*）中，当温度在 25℃以下时，血淋巴 pH 保持不变；当身体温度在 25℃以上时，每增加 1℃，血淋巴的 pH 下降约 0.017 个 pH 单位。

蝗虫和蚱蜢也可能有其他昆虫，至少通过两种途径调节血淋巴 pH：①当可滴定酸引起酸毒症时，向中肠转移相当的酸；②当肌肉剧烈活动，血淋巴中 CO_2 的增加引起酸毒症时，气管系统的通风增加。这些蝗虫血淋巴将近 90% 的缓冲能力归结于碳酸氢盐和蛋白质，加上无机磷酸盐和有机磷酸盐的小部分额外贡献。尽管昆虫血淋巴氨基酸浓度很高，但其具有从典型血淋巴 pH 中很好去除的离体常数。

（田 铃）

Exercises

1. Describe the location, structure, and function of the dorsal vessel in insects.
2. Describe the flow of hemolymph into and out of the heart.
3. What are accessory pulsatile organs and how are they involved in circulation?
4. Describe the major types of hemocytes.
5. What are the main components in insect hemolymph?
6. How did cells in the insect body protect themselves from ions and other chemicals in the open circulatory system?

思考题

1. 描述昆虫背血管的位置、结构及功能。
2. 描述血淋巴进入和流出心脏的流向。
3. 什么是辅助起搏器？它们在循环中的功能是什么？
4. 描述血细胞的主要类型。
5. 昆虫血淋巴内的主要成分是什么？
6. 昆虫体内细胞是如何保护免受开管式循环系统中的离子和其他化学物质影响的？

第7章 昆虫免疫

Chapter 7 Insect Immunity

Key points

1. When a foreign intruder breaks through the insect's first physical line of defense, which is made up of the outer skin, a series of immune responses can be activated in the body. The innate immunity of insects consists of cellular immunity and humoral immunity.

2. Cellular immunity is a kind of hemocytes-mediated immune mechanism including phagocytosis, nodulation and encapsulation.

3. Humoral immunity clears the microbial infection by a series of effector molecules such as antimicrobial peptides, lysozyme, polyphenol oxidase and ROS, which were produced followed microbial recognition by pattern recognition receptor and activation of signaling pathway.

4. Antimicrobial peptides, known as broad-spectrum and potent activity, are important antibacterial effector molecules. Their productions are mainly controlled by Toll, Imd and JAK-Stat signaling pathways

知识点

1. 当外来物突破昆虫第一道由外表皮组成的物理防线后，体内产生一系列的免疫应答就可被激活。昆虫的先天免疫由细胞免疫和体液免疫两部分组成。

2. 细胞免疫是直接由血细胞介导的免疫机制，包括吞噬、集结和包囊等。

3. 体液免疫是通过宿主模式识别受体识别病原微生物，激活免疫信号通路产生抗菌肽、溶菌酶、多酚氧化酶和活性氧等一系列的效应分子，以清除外源微生物的感染。

4. 抗菌肽具有广谱高效的杀菌活性，是一类重要的抗菌效应分子。其表达主要受Toll、Imd和JAK-Stat等信号通路的调控。

Insects and microorganisms coexist within the biosphere in numerous ways. Regularly, insect larvae develop in decaying organic matter, and adults often serve as vectors for microorganisms causing plant and animal diseases. Therefore, insects have evolved sensitive mechanisms for recognition of pathogens and an array of strategies to defend themselves against attacks by bacteria, fungi, parasites, and viruses. The first defense of insects against microbial organisms and fungi is the sclerotized cuticle that covers the body, and cuticle that surrounds the trachea, parts of the internal reproductive tract, foregut, and hindgut. When organisms succeed in passing through the cuticular barrier, insects rapidly activate innate immune response to fight against invading organisms.

昆虫和微生物在生物圈中以多种方式共存。昆虫的幼虫通常在腐烂的有机物中发育，而成虫往往成为传播动植物疾病的媒介。因此，昆虫已经进化出识别病原体的灵敏机制和一系列防御细菌、真菌、寄生虫和病毒攻击的策略。昆虫抵抗微生物和真菌的第一道防线是覆盖身体的硬化表皮以及围绕气管、部分内生殖道、前肠和后肠的表皮。当微生物成功通过表皮屏障，昆虫迅速激活先天免疫对抗入侵的微生物。

The innate immune of insects includes a combination of cellular and humoral responses. Schematic overview of innate immunity in insects was shown (Fig.7-1). The cellular arm of the insect innate immune system includes mechanisms that are directly mediated by hemocytes (e.g., phagocytosis, nodulation and encapsulation). Melanization accompanying coagulation, clot formation and wound healing, nodulation and encapsulation processes leads to the formation of cytotoxic melanin precursors and reactive oxygen species. In addition, antimicrobial substance of production and secretion into the hemolymph is another important portion of humoral immunity. However, it is not scientific to separate simply cellular and humoral immune responses into two distinct categories. This is because many humoral factors affect hemocytes functions and hemocytes themselves are an important source of many humoral molecules. There is also a considerable overlap between cellular and humoral immune functions that span from recognition of foreign intruders to clot formation.

昆虫先天免疫包括细胞免疫和体液免疫，其组成部分如图 7-1 所示。细胞免疫是直接由血细胞介导的机制（如吞噬、集结和包囊）。伴随凝结、血凝块形成和伤口愈合、集结和包囊的黑化过程导致细胞毒性黑色素前体和活性氧的形成。此外，抗菌物质的合成与分泌是体液免疫的另一个重要组成部分。然而，简单地将细胞免疫反应和体液免疫反应区分为两种截然不同的类型，这是不科学的，因为许多体液因子影响血细胞的功能，并且血细胞本身也是许多体液分子的重要来源。从识别外来物到血凝块的形成，细胞免疫和体液免疫功能之间也有相当多的交叉重叠。

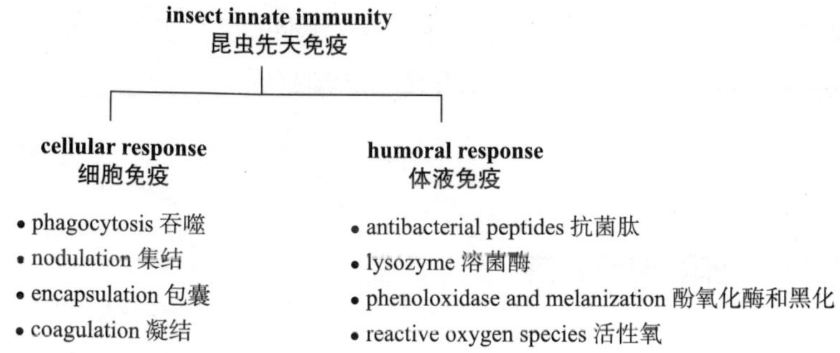

Fig. 7-1 Schematic overview of innate immunity in insects.
图 7-1 昆虫先天免疫组成概览。

7.1 抵抗入侵的物理屏障

7.1 Physical Barriers Against Invasion

The frontline of insect defense against pathogens is the cuticular coverage of epidermic body wall that acts as an effective physical barrier for many microorganisms and fungi, although some organisms have chitin and protein digesting enzymes that help them penetrate the cuticle and epidermis. Perhaps the thick cuticle is the main physical barrier, but in insects there are numerous thin areas of epidermis, particularly at abdominal intersegmental boundaries and in the lining of the tracheal system. Insects ingest

昆虫抵御病原体的第一道防线是覆盖于体壁真皮层的表皮，它是许多微生物和真菌的有效物理屏障，尽管有些生物有几丁质酶和蛋白质消化酶帮助它们穿透表皮和真皮。也许厚厚的表皮是主要的物理屏障，但在昆虫中，存在许多薄的真皮区，特别是腹部节间界限和气管系统的围线。昆虫在进食的同时摄入

many microorganisms and some pathogens along with their food. Oral infection is the main route of microorganism into insects. On the lumen surface of the foregut and hindgut epithelial cells, there is an attached cuticle lining that protects the epithelial cells, but the midgut is more vulnerable. The midgut may have an unattached protective layer, the peritrophic matrix is composed of chitin and protein that protects the delicate brush border on midgut cells from harsh food particles as well as ingested microorganisms.

The peritrophic matrix has holes or pores, through which small digested molecules are absorbed by midgut. These pores are about 9 nm in diameter in flies and probably are too small to allow the invasion of the parasites or microorganisms through these pores. In some mosquitoes, the pores are up to 200 nm in size, and may allow passage of some arboviruses rather than bacteria or other larger potential invaders. Some insects form the peritrophic matrix only after food is ingested, and orally ingested microorganism might contact the surface of midgut epithelial cells before the matrix is completely formed. The malaria parasite at mature ookinete stage produces chitinase to aid in penetrating the midgut peritrophic matrix.

许多微生物和一些病原菌，经口感染途径是微生物进入昆虫的主要入口。前肠和后肠的肠壁上皮细胞内腔表面有表皮围线，对上皮细胞有保护作用，而中肠则较为脆弱。中肠可能含有非接触性的保护层，围食膜由几丁质和蛋白质组成，可保护中肠细胞上纤细的刷状缘免受粗糙食物颗粒的损伤和食下微生物的感染。

围食膜上有孔洞，消化后的小分子可以通过这些孔洞被中肠吸收。果蝇围食膜上的孔洞直径大小约为 9 nm，这对入侵寄生物和微生物来说可能太小而不能通过。某些蚊子围食膜孔径大小可达 200 nm，虫媒病毒可以通过，细菌或其他较大的潜在外来物无法通过。有些昆虫仅在食物消化后才形成围食膜，经口消化的微生物在围食膜完全形成前可接触中肠上皮细胞的表面。疟疾寄生虫在动合子成熟期产生几丁质酶，有助于其穿透中肠围食膜。

7.2 细胞免疫

7.2 Cellular Immunity

The cellular immunity is hemocyte-mediated immune responses. Hemocytes include plasmatocytes, granulocytes, lamellocytes and oenocytoids/crystal cells, which possess a variety of immune functions and their responses occur once an intruder is recognized as a non-self in the insect body. The most obvious hemocytes-mediated immune responses are phagocytosis, nodulation, encapsulation and coagulation once upon microbial or parasitic infection. Encapsulation is mediated by lamellocytes, and nodulation, phagocytosis and coagulation by plasmatocytes. Crystal cells and oenocytoids are involved in melanization. Cytotoxic species released in epithelia include reactive oxygen species (ROS), reactive nitrogen species (RNS) and quinones. These processes depend upon recognition of the target followed by

细胞免疫是指血细胞介导的免疫反应。血细胞包括浆血细胞、粒血细胞、片状细胞和类绛色细胞 / 含晶细胞，它们具有多种免疫功能，一旦外来物在昆虫体内被识别为非己成分，它们就会做出反应。血细胞应答微生物或寄生虫感染的最明显的免疫反应是吞噬、集结、包囊和凝结，片状细胞执行包囊的功能、浆血细胞参与集结、吞噬和凝结、含晶细胞和类绛色细胞参与黑化反应。上皮细胞释放的细胞毒性物质包括活性氧（ROS）、活性氮（RNS）和醌类化合物。这些过程依赖

activation of downstream signaling and effector responses. Some foreign entities are recognized by humoral pattern recognition receptors (PRRs). Granulocytes appear to be involved in the initial stages of recognition of non-self. It seems possible that PGRPs, GNBP and βGRP are present in granules of the granulocytes. As a consequence of the interaction of the granulocytes with the invading organism, plasmatocytes are attracted to the site and both are commonly observed in capsules in lepidopterans, while in *Drosophila*, lamellocytes were observed most frequently in capsules. Hemocytes also secrete a variety of enzymes, which include PPO cascade components, lysozymes, nitric oxide (NO), etc. In addition to mediating bacterial killing, it is also possible that secreted molecules may be involved in coordinating the immune response. Upon infection, the fat body induces the production of clotting factors and serine proteases. The latter regulates the melanization response. Microbial recognition induces the expression of peptidoglycan recognition protein (PGRP), which regulates the activity of proteases that control the prophenoloxidase (PPO) cascade and melanization response (Fig.7-2; Appendix color graphic Fig.7-2).

于对靶标物的识别，随后是对下游信号和效应分子应答的激活。一些外来物被体液模式识别受体（PRRs）识别，粒血细胞似乎参与了识别非己的初始阶段，粒血细胞的颗粒中可能存在PGRPs、GNBP 和 βGRP。粒血细胞与外来生物相互作用的结果是浆细胞被吸引到这个位置。这两种细胞在鳞翅目昆虫的包囊中非常常见，而在果蝇中，片状细胞在包囊中最为常见。血细胞还分泌多种酶，包括 PPO 级联组分、溶菌酶和一氧化氮（NO）等。除了介导杀菌作用，分泌的分子也可能参与协调免疫反应。当昆虫受到感染时，脂肪体诱导产生凝血因子和调节黑化反应的丝氨酸蛋白酶。微生物的识别诱导了肽聚糖识别蛋白（PGRP）的表达，从而调控一些蛋白酶的活性来调节酚氧化酶原（PPO）级联和黑化反应（图 7-2；附彩图 7-2）。

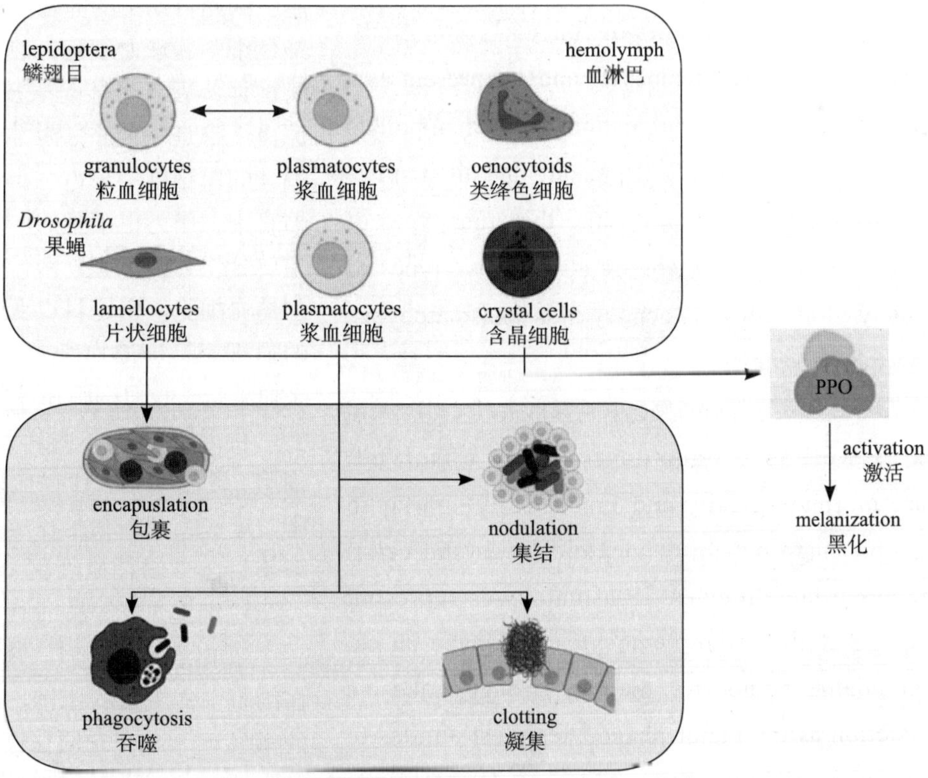

Fig. 7-2 Cellular immunity mediated by hemocytes and melanization (modified from Eleftherianos et al., 2021).
图 7-2 血细胞介导的细胞免疫及黑化反应（仿 Eleftherianos 等，2021 ）。

7.2.1 Hemocytes that involve in insect immunity

The classification of hemocytes is based on two major criteria: cell morphology and expression of marker genes. So far, seven types of hemocytes have been identified in various insects: prohemocytes, plasmatocytes, granulocytes, oenocytoids, spherulocytes, coagulocytes and adipohemocytes. However, most insects do not have all the seven hemocytes types and not all cell types involve in insect immunity.

In lepidopterans, such as *Anticarsia gemmatalis*, *B. mori*, *M. sexta*, as in several other models, five cell types are present: prohemocytes, plasmatocytes, granulocytes, spherulocytes and oenocytoids. In these insects, granulocytes are the most abundant hemocytes type in circulation as a major phagocyte. They are morphologically distinguished by the granules in their cytoplasm, their ability to strongly adhere and spread symmetrically on foreign surfaces. Plasmatocytes are usually larger than granulocytes, spread asymmetrically on foreign surfaces and are the main capsule-forming hemocytes. Non-adhesive hemocytes in larval Lepidoptera include oenocytoids that contain PO precursors and spherule cells that are potential sources of cuticular components. Prohemocytes are non-adhesive hemocytes and are present as a small proportion of hemocytes in circulation. In hematopoietic organs, prohemocytes, as progenitor cells, mature and differentiate primarily into plasmatocytes and partly into the other types of hemocytes. In Coleoptera also, there are at least five types of hemocytes, namely prohemocytes, oenocytoids, plasmatocytes, granulocytes and spherulocytes.

In *Drosophila* larvae, plasmatocytes, crystal cells and lamellocytes as well as prohemocytes were observed. These types are morphologically and functionally similar to granulocytes (= plasmato cytes), oenocytoids (=crystal cells) and plasmatocytes (=lamellocytes). Plasmatocytes represent a major percentage of all mature hemocytes and make up ca. 90%~95% of circulating hemocytes, they are strongly adhesive *in vitro*, and function as the major phagocytes most similar to the mammalian monocyte/ macrophage lineage. Crystal cells are non-adhesive cells that account for a small proportion of approximately 5% of the hemocytes population found in

7.2.1 涉及昆虫免疫的血细胞

对血细胞的分类主要基于两个标准：细胞形态和标记基因的表达，目前，在各种不同的昆虫中已经鉴定的血细胞类型有 7 种：原血细胞、浆血细胞、粒血细胞、类绛色细胞、珠血细胞、凝血细胞和脂血细胞。然而，大多数昆虫并没有全部七种血细胞类型，也并非所有类型都参与昆虫免疫。

鳞翅目昆虫，如刺蛾、家蚕、烟草天蛾以及其他几种模式昆虫存在五种细胞类型：原血细胞、浆血细胞、粒血细胞、珠血细胞和类绛色细胞。在这些昆虫中，粒血细胞是循环的血细胞中最为丰富的细胞类型，是主要的吞噬细胞。在形态上，粒血细胞区别于其他细胞类型的特征在于细胞质中的颗粒，它们在异物表面具有很强的黏附和对称扩散的能力。浆细胞通常比粒细胞大，在异物表面不对称分布，是主要的成囊血细胞。鳞翅目幼虫的非黏附血细胞包括类绛色细胞和珠血细胞，前者含有酚氧化酶前体，后者可能是表皮成分的来源。原血细胞是一种非黏附性的血细胞，在循环的血细胞中只占一小部分。在造血器官中，作为祖细胞的原血细胞成熟并主要分化为浆细胞，部分分化为其他类型的血细胞。鞘翅目也至少有五种类型的血细胞，即原血细胞、类绛色细胞、浆血细胞、粒血细胞和珠血细胞。

在果蝇幼虫体内可以观察到浆血细胞、含晶细胞、片状血细胞和原血细胞。这些类型的细胞在形态和功能上与鳞翅目昆虫的粒血细胞（= 浆血细胞）、类绛色细胞（= 含晶细胞）和浆血细胞（= 片状血细胞）相似。浆血细胞在成熟血细胞中占主要的比例，占循环血细胞的 90% ～ 95%，它们在体外具有很强的黏附性，与哺乳动物单核 / 巨噬细胞谱系最为相似，是主要的吞噬细胞。含晶细胞是一种非黏附细胞，只占循环血细

circulation. Crystal cells are recognized by their rounded morphology and the expression of PO cascade components and release of chemicals that encase microbes in a hardened gel. Lamellocytes are identified as large, flat, adhesive cells; their main function is encapsulation of parasitoids and other large foreign targets. Lamellocytes are virtually absent in naïve *Drosophila* larvae, but rapidly differentiate from prohemocytes following attack by parasitoid wasps and during metamorphosis. Prohemocytes mainly reside in the lymph gland where they originate from preprohemocytes; however, a small number is in circulation. The three hemocytes types are thought to be differentiated from prohemocytes. However, recent developments suggest that *Drosophila* hematopoiesis enfold a striking plasticity. For example, new crystal cells in third instar *Drosophila* larvae originate through a Notch-dependent process of plasmatocyte transdifferentiation.

胞总数的 5% 左右。含晶细胞可通过圆形的形态、表达 PO 级联反应组分和释放的化学物质将微生物包裹在硬化的胶质中使其被识别。片状血细胞是大的、扁平的、黏附的细胞，它们的主要功能是包裹寄生物和其他较大的外来异物。正常的果蝇幼虫中几乎不存在片状血细胞，但在受到寄生蜂攻击时和变态期，可迅速地从原血细胞分化而来。原血细胞主要定居于淋巴腺中，它们起源于前原血细胞；然而，有少量原血细胞存在于循环中。浆血细胞、含晶细胞和片状血细胞这三种类型的血细胞是由原血细胞分化而来的。然而，最近的研究表明，果蝇的造血具有很强的可塑性。例如，在 3 龄果蝇幼虫中，新的含晶细胞来源于一个 Notch 依赖的浆血细胞转分化过程。

In mosquitoes, comparative studies of hemocytes types based on antigenic, functional and morphological approaches revealed that both *A.aegypt*i and *A. gambiae* have in circulation the three hemocytes types of granulocytes, oenocytoids and prohemocytes. Granulocytes are strongly adhesive and phagocytic. They are also the most abundant circulating cell type in mosquito, comprising 80%–95% of the circulating hemocytes population. Oenocytoids and prohemocytes comprise a small percentage of the total hemocytes population (≤ 10%). Oenocytoids are non-adhesive and constitutively express PO; however, granulocytes express inducible PO following immune challenge. In contrast to *Drosophila* or lepidopterans, mosquitoes apparently do not produce any hemocytes type specialized for capsule formation. Hemocytes kill pathogens *via* phagocytosis, melanization and lysis.

在蚊子中，基于抗原、功能和形态学方法的血细胞类型比较研究表明，埃及伊蚊和冈比亚伊蚊都有三种血细胞类型，即粒血细胞、类绛色细胞和原血细胞。粒血细胞具有很强的黏附性和吞噬能力，是蚊子中最丰富的循环细胞类型，占循环血细胞总数的 80% ～ 95%。类绛色细胞和原血细胞只占全部血细胞的一小部分（ ≤ 10% ）。类绛色细胞是非黏附性的，组成性地表达 PO，而粒血细胞在免疫刺激后诱导 PO 的表达。不同于果蝇或鳞翅目昆虫，蚊子中不存在任何专门形成包囊的血细胞类型，血细胞通过吞噬、黑化和裂解等方式杀死病原体。

There are two kinds of hemocytes in the body cavity of insects, like that of all arthropods, is filled with a circulating hemolymph that contains both free-floating, which have an immune surveillance role in circulation, and sessile hemocytes generally attached to internal organs or tissues. The number of circulating hemocytes in insects varies according to the age, developmental stage, physiological conditions and infection status. Sessile hemocytes are released into the circulation when immunity is activated (Fig. 7-3; Appendix color graphic Fig.7-3). The density/number of circulating hemocytes reported in different

和所有的节肢动物一样，昆虫体腔中充满循环的血淋巴，其中的血细胞有两种类型：一类是自由漂浮在开放式血腔的血细胞，它们具有免疫盯梢的功能；另一类是固着血细胞，通常附着于某些器官或组织。循环血细胞的数量因年龄、发育阶段、生理条件和感染状态而异，在免疫激活的条件下，固着血细胞被激活进入循环的状态（图 7-3；附彩图 7-3）。不同种类昆虫的循环血细胞

insect species is variable. For instance, adult *D. melanogaster* contains between 1,000 and 2,000 circulating hemocytes, adult mosquitoes have a range of 500-4,000 circulating hemocytes.

密度/数目不同。黑腹果蝇成虫的循环血细胞含量为 1 000 ～ 2 000 个，蚊子成虫的循环血细胞含量为 500 ～ 4 000 个。

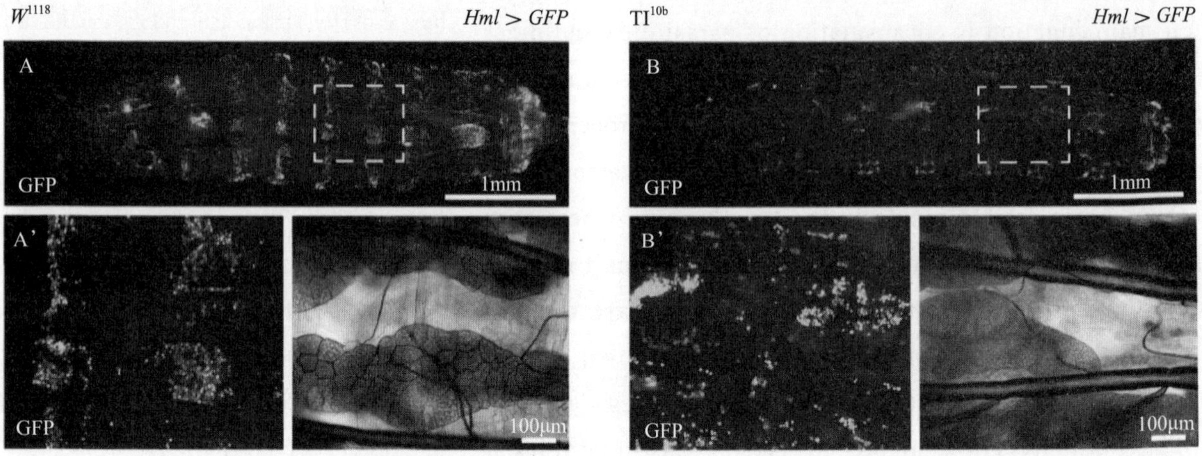

Fig. 7-3 Hemocytes distribution in wild type *Drosophila* larva and a gain-of-function Toll receptor mutant Toll^{10b}. *Hml* (Hemolectin) is a hemocytes-specific driver whose expression is visualized by GFP fluorescence after crossing to the UAS-Gal4. A, A' .Sessile banded pattern was seen in Toll wild-type; B，B' .In Toll^{10b} genetic background, banded pattern was lost B, B^1. The demarcated areas in A and B are shown enlarged in panels A' and B', bright-field images to the right (modified from Schmid et al., 2014).

图 7-3 野生型和免疫激活型 Tl^{10b} 果蝇幼虫中的血细胞分布。*Hml*（Hemolectin）. 血细胞特异性的启动子，与 UAS-Gal4 杂交后通过 GFP 荧光显示其表达。A, A' . Toll 野生型幼虫显示固着血细胞呈带状分布；B, B'.Toll^{10b} 遗传背景的果蝇幼虫中带状消失。A' 和 B' 是 A、B 中指示区域的放大图像，右边是白光明场图（仿 Schmid 等，2014）。

7.2.2 Phagocytosis

Phagocytosis is a widely conserved and ancient immune defense response performed by single-celled organisms and different metazoan cell types. Russian scientist Elia Metchnikoff first discovered the existence of internal phagocytosis in starfish larvae in 1882. He found special cells attacking small thorns placed into starfish larvae. Based on these findings, he later moved into immunology and championed the concept of cellular immunity. He was awarded the Nobel Prize in 1908. He shared the prize with Paul Ehrlich, a supporter of humoral immunity. Together they provided the bases for modern immunology.

Metchnikoff coined the term, phagocyte, from the Greek words, "phages", meaning "to eat", and "cite", meaning "cell". By phagocytosis, individual cells recognize, internalize and ingest invading microbial pathogens and apoptotic cells (≥0.5μm). Phagocytic hemocytes can function individually in the process of

7.2.2 吞噬

吞噬作用是一种保守而古老的吞噬免疫防御反应，广泛存在于单细胞生物和后生动物不同类型的细胞中。吞噬现象由俄国科学家 Elia Metchnikoff 于 1882 年首次在海星幼虫中被发现，他发现一种特殊的细胞会攻击置于海星幼虫体内的小刺。基于这些发现，他后来进入免疫学领域，并倡导细胞免疫的概念。1908 年，他被授予诺贝尔生理学或医学奖，与体液免疫的支持者 Paul Ehrlich 共同获得了该奖项，他们共同为现代免疫学奠定了工作基础。

Metchnikoff 创造的吞噬细胞（phagocyte）这个词来自希腊语，"phages" 意为 "吃"，"cite" 意为 "细胞"。吞噬是由单个细胞识别、内化和摄入入侵微

phagocytosis and the phagocytic ability declines with age. A small number of bacteria, fungal spores or protozoans are phagocytosed by the phagocytes. The phagocytic hemocytes in Lepidoptera are granulocytes and plasmatocytes, whereas the latter is the key component in *Drosophila*. Phagocytosis involves several receptors that have been well-documented in *Drosophila* and many of which have mammalian homologs. Recognition of the target object is either direct by binding of the hemocytes to its surface or mediated *via* opsonic ligands marking the object to be phagocytosed.

Phagocytosis by insect hemocytes can be envisioned as a specialized form of receptor-mediated endocytosis and may be assumed to be similar to those in other animals, differences may occur in some steps (Fig.7-4; Appendix color graphic Fig.7-4). This multiple step process involves: ① binding of the phagocyte to the target particle; ② particle internalization

生物病原体和凋亡细胞（≥ 0.5 μm）的过程。吞噬细胞在吞噬的过程中独立行使功能，随着年龄的增长细胞的吞噬能力下降。少数细菌、真菌孢子或原生动物可被吞噬细胞吞噬。鳞翅目昆虫的吞噬细胞是粒血细胞和浆血细胞，而果蝇中主要是浆血细胞。吞噬作用涉及若干已在果蝇中得到充分证明的受体，其中许多在哺乳动物中有同源物。吞噬靶标的识别可通过血细胞的直接结合或由调理配体介导靶向被吞噬的目标。

昆虫血细胞吞噬可被视为一种特殊形式的受体介导的内吞作用，与其他动物的内吞类似，只是在某些步骤可能有不同。吞噬是一个多步骤的过程（图 7-4；附彩图），包括：① 吞噬细

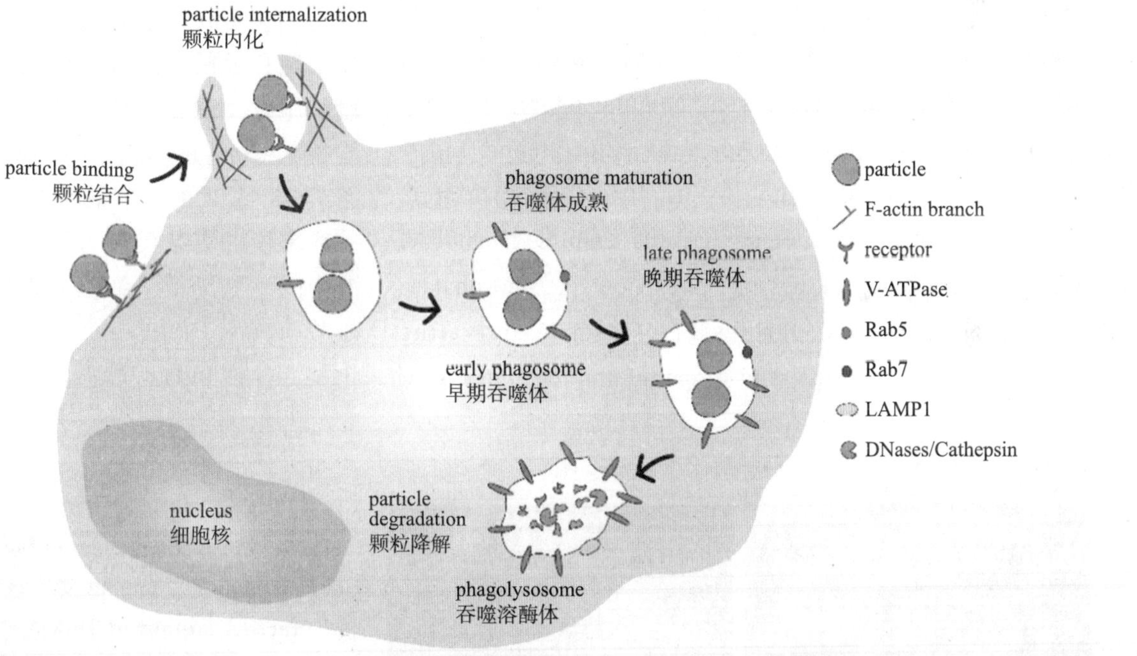

Fig. 7-4 The phagocytosis process in *Drosophila*. particle binding: *Drosophila* specialized phagocytic cell surface receptors recognize and bind target particles; particle internalization: F-actin branches off at the phagocytosis site, forming the phagocytic cup; Actin polymerizes around the granule until phagocyte processes fuse at the edge, producing a newly formed phagosome; maturation of phagosomes: the newly formed phagosomes then undergo a series of fission and fusion events with organelles; particle degradation: in this last step, the phagosome obtains the important components of the final particle destruction step, such as DNase and protease (modified from Melcarne et al., 2019).

图 7-4 果蝇的细胞吞噬过程。颗粒结合：果蝇专业吞噬细胞表面的特异受体识别并结合目标颗粒；颗粒内化：F- 肌动蛋白在吞噬位点形成分支，形成吞噬杯；肌动蛋白在颗粒周围聚合，直到吞噬细胞突起在边缘融合，产生一个新形成的吞噬体；吞噬体成熟：新形成的吞噬体随后与细胞器发生一系列裂变和融合事件；颗粒降解：在这最后一步中，吞噬体获得最终破坏颗粒的重要成分，如 DNase 和蛋白酶（仿 Melcarne 等，2019）。

and phagosome formation, through remodelling of the plasma membrane; ③ phagosome maturation; ④ particle destruction in the phagolysosome. Low pH environment in these compartments enhances killing of living intruders. There is some evidence that insect hemocytes generate reactive oxygen species (ROS) and reactive nitrogen species (RNS), which are released into the phagosome or extracellularly and are toxic to a variety of microbes.

胞与目标颗粒的结合；②颗粒的内化，通过质膜重构形成吞噬体；③吞噬体成熟；④颗粒在吞噬溶酶体中被破坏。吞噬溶酶体中的低 pH 环境增强了对外源入侵活体的杀灭作用。证据表明，昆虫血细胞产生活性氧（ROS）和活性氮（RNS）并释放到吞噬体中或细胞外，对多种微生物具有毒性。

7.2.3 Nodulation

Nodulation is the process of forming nodules, Which are multicellular hemocytes aggregates recruited to the site of infection that entrap large number of bacteria or fungal spores. Nodule formation can be induced by certain pathogen associated molecular patterns (PAMPs) of the invader. The process involves entrapping bacteria or foreign objects in the coagulum formed from granules and released by granulocytes, with subsequent melanization of necrotic hemocytes. Nodular aggregates may adhere to host tissues (Fig.7-5; Appendix color graphic Fig.7-5) and the larger nodules may eventually be encapsulated. Nodulation reactions clear infecting bacteria from the haemocoel in insects, generally. Cytokines like phosphoserine phosphatase (PSP), eicosanoids, NO, various PPO cascade-activating components, apolipophorins, and biogenic amines signal and/

7.2.3 集结

集结是形成结节的过程，结节是被聚集到感染部位的血细胞形成的多细胞聚集体，其中捕获了大量的细菌或真菌孢子。结节的形成可由外来异物的某些病原相关分子模式（PAMPs）诱导。这一过程涉及捕获细菌或异物，将它们包裹在由颗粒形成的，并由粒血细胞释放的血凝块中，随后坏死的血细胞发生黑化反应。结节聚集物可能黏附在宿主的组织上（图 7-5；附彩图 7-5），较大的结节最终可能被包裹。一般来说，集结反应可以清除昆虫血液中的感染细菌。细胞因子如磷酸丝氨酸磷酸酶（PSP）、类花生酸、一氧化氮、各种 PPO 级联激活组分、载脂

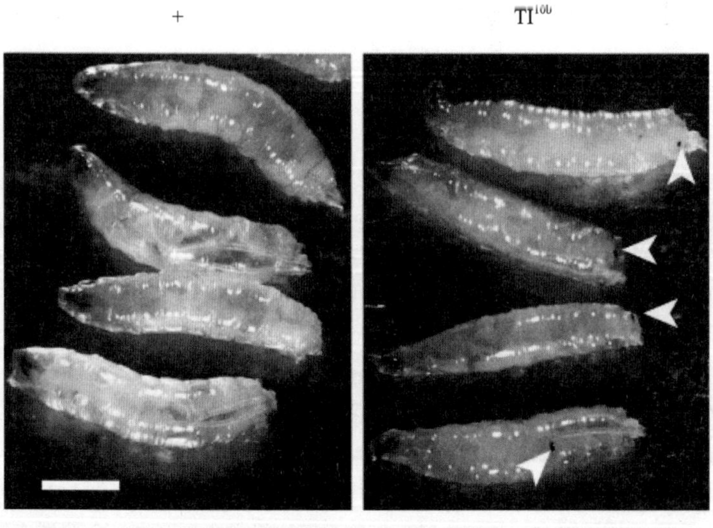

Fig. 7-5 Melanotic nodules in the larvae of *Drosophila* Tl^{10b} mutant. Tl^{10b} is a gain-of-function mutant of Toll. Melanotic nodules are seen black spots in the larvae of *Drosohila* indicated as the arrows. A. +, wild type; B. Tl^{10b} (Schmid et al., 2014).

图 7-5 果蝇 Tl^{10b} 突变体幼虫黑化的结节。Tl^{10b} 为免疫激活型突变，黑化的结节如图中箭头所示 A.+, 野生型；B. Tl^{10b} (Schmid 等，2014)。

or control the aggregation of plasmatocytes. The intensity of the response varies, depending on the pathogenicity of the invading organisms. A reeler domain (found in some extracellular matrix proteins)-containing protein, Noduler, regulates nodule formation in *Antheraea mylitta*. Certain entomopathogenic bacteria like *Xenorhabdus nematophila, X. bovienni, X. poinarii* and *Photorhabdus luminescens* induce host immunosuppression and inhibit nodule formation *via* inhibiting phospholipase A2 (PLA2) acting in eicosanoid biosynthesis.

蛋白和生物胺信号和 / 或控制浆细胞的聚集。反应的强度因入侵生物的致病性而异。在印度柞蚕中，一种含有卷轴结构域（存在于某些细胞外基质蛋白中）的蛋白质 Noduler 调控着结节的形成。某些昆虫病原细菌，如嗜线虫致病杆菌和发光杆菌通过抑制磷脂酶 A2 在类花生酸生物合成中的作用，诱导宿主免疫抑制和抑制结节的形成。

7.2.4 Encapsulation

When foreign invading objects, like metazoan parasites, protozoan, eggs or larvae and foreign invading organisms or abiotic particles are too large to be phagocytosed, they become encapsulated by multiple layers of haemocytes and/or a melanin coat. Two types of encapsulation are recognized in insects: cellular encapsulation, mainly described in Lepidoptera, and melanotic humoral encapsulation, more typical in some dipterans like *Drosophila*. Cellular encapsulation can occur without any sign of melanization. In contrast, melanotic encapsulation involves PO activity. In cellular encapsulation, granulocytes and plasmatocytes of Lepidoptera, and plasmatocytes and lamellocytes of *Drosophila*, are the main haemocyte types involved in encapsulation. In Lepidoptera, an inner layer of granulocytes and an outer one of plasmatocytes were reported to surround the encapsulated objects. Taking the nomenclature of *Drosophila* haemocyte types into account, the same sequence of plasmatocytes (equivalent to granulocytes of Lepidoptera) followed by lamellocytes (equivalent to plasmatocytes of Lepidoptera) may occur in capsule formation. Crystal cells of *Drosophila* and oenocytoids of other insects also play a role in melanization of the capsule.

7.2.4 包囊

当昆虫受到后生动物寄生虫、原生动物、虫卵或幼虫以及外来入侵生物或因非生物颗粒太大而无法被吞噬时，它们会被多层血细胞和 / 或一层黑色素外壳包裹。昆虫存在两种不同的包囊类型：一种是细胞型包囊，主要见于鳞翅目昆虫；另一种是黑色素为特征的体液包囊，主要见于一些双翅目昆虫，如果蝇。细胞型包囊可以在没有任何黑化迹象的情况下发生。相反，黑色素型包囊涉及酚氧化酶活性。在细胞型包囊中，鳞翅目的粒血细胞和浆血细胞以及果蝇的浆血细胞和片状血细胞是参与包囊的主要血细胞类型。鳞翅目围绕包囊异物的内层为粒血细胞，外层为浆血细胞。由于果蝇血细胞类型的命名不同，在形成包囊的过程中，浆血细胞（相当于鳞翅目的粒血细胞）在内层，片状血细胞（相当于鳞翅目的浆血细胞）在外层。果蝇的含晶细胞和其他昆虫的类绛色细胞在包囊的黑化中也起着重要作用。

7.3 黑化反应

7.3 Melanization Reaction

Insects make the fastest immune response, when insect is invaded by pathogens or their epidermal barriers are damaged. There will be melanin deposition at the injury site, where hemocytes aggregate. This kind of immune response is called melanization reaction. Melanization is a major humoral immune response in insects, while it often occurs accompanying cellular immune processes such as coagulation, phagocytosis and encapsulation.

当昆虫受到病原物的入侵或真皮屏障被破坏时，它们做出最快的免疫应答，在损伤部位产生了黑色素沉积和血细胞聚集出现黑化的现象，这就是黑化反应。黑化是昆虫的一种主要体液免疫反应，经常伴随着凝结、吞噬和包囊等细胞免疫的过程。

7.3.1 Simplified process of melanization reaction

Upon wounding or infection, phenoloxidase is activated from its zymogen form PPO to the active form PO, as a part of the innate immune response. The PPO-activation system may be elicited by the PAMPs such as PGN and β-1,3-glucan. This system is a proteolytic cascade comprising several serine proteases and their inhibitors. Microbial PGN and β-1, 3-glucan react with PRRs (PGRP or βGRP), which consequently induce activation of initiator serine proteases of the PPO-activation system; Next, the initiator protease triggers a protease cascade, activating terminal serine proteases termed as PPO-activating enzyme (PPAE), PPO-activating protein (PAP) or PPO-activating factor (PPAF) to cleave PPO and form active PO. This active form has a prominent role in melanin biosynthesis, and the formed melanin is deposited in specific sites of haemolymph clots of wounds, nodules and capsules. This process appears to be important in insect immune defences through its action in wound healing and pathogen sequestration in nodules and capsules and in their killing by resulting ROS and RNS.

7.3.1 黑化反应的基本过程

在受伤或感染时，作为先天免疫反应的一部分，酚氧化酶被激活，从其酶原形式 PPO 到活性形式 PO。PPO 活化系统由病原相关分子模式肽聚糖和 β-1,3- 葡聚糖触发，该系统是一个包括多个丝氨酸蛋白酶及其抑制剂的蛋白质水解级联反应。首先，微生物的肽聚糖和 β-1,3- 葡聚糖被模式识别受体（PGRP 或 βGRP）识别，进而诱导 PPO 激活系统中起始丝氨酸蛋白酶的激活；其次，起始蛋白酶触发蛋白酶级联反应，最终激活末端丝氨酸蛋白酶，它们被称为 PPO- 活化酶（PPAE）、PPO 激活蛋白（PAP）或 PPO- 活化因子（PPAF），切割 PPO 形成具有活性的 PO。这种活性形式在黑色素生物合成中具有重要作用，形成的黑色素沉积在伤口、结节和包囊等血淋巴凝块的特定部位。这一过程通过在伤口愈合结节以及包囊中的病原体隔离发挥作用，并通过产生 ROS 和 RNS 杀死病原体，在昆虫的免疫防御中发挥重要作用。

7.3.2 Proenzyme PPO synthesis, release and activation

7.3.2.1 PPO synthesis

Insect PPO synthesis are metallated zymogens belonging to the family of type-3 copper proteins, and have two copper ions and three histidine residues in each active site pocket. The PPOs are biosynthesized mainly by haemocytes. In lepidopterans, oenocytoids are identified as the cell-type producing PPO, whereas it is mainly produced in the crystal cells of *D. melanogaster*, suggesting that crystal cells and oenocytoids may be functionally homologous cell types. In mosquitoes, PPO is synthesized in both oenocytoids and granular hemocytes.

7.3.2.2 PPO release

The enzyme PO is in most cases released in its pro-form either by cell rupture after activation of the c-Jun N-terminal kinase (JNK) pathway as seen in *Drosophila* crystal cells or lepidopteran oenocytoids or possibly by other still not characterized mechanisms of continuous release. Oenocytoids from different species vary in stability. Some are spontaneously ruptured upon injury of the body or infection, leading to the high local concentrations of PPO. Oenocytoid lysis may occur *in vivo* in the absence of injury or infection, to maintain a steady state level of PPO in plasma. In lepidopterans, the bulk of PPO appears in plasma, but in insects with incomplete metamorphosis hemimetabolous insects such as locusts and cockroaches, PPO appears to be mostly retained in haemocytes until microbial exposure or injury stimulates its release.

7.3.2.3 PPO activation

The enzyme activating PPO in insects is known as prophenol-oxidase-activating enzyme (PPAE) or PPO activating Serine protease (PAP). PPAEs/PAPs were first purified from cuticle of *B. mori* larvae, then identified and cloned from *D. melanogaster*, *M. sexta*, *Holotrichia diomphalia* and *Plutella xylostella*. PAPs consist of one or two clip domains at the N-terminus and a serine protease catalytic domain at the C-terminus. The activation of PPO into PO is accomplished by PPAEs or PAPs, and Serine proteinase homologs (SPHs) are also required as basic cofactors to help

7.3.2 酶原 PPO 的合成、释放和激活

7.3.2.1 PPO 的合成

昆虫 PPO 的合成是金属化酶原，属于 3 型含铜蛋白质家族，每个活性位点有 2 个铜离子和 3 个组氨酸残基。PPOs 的生物合成主要是由血细胞执行的。鳞翅目昆虫中，类绛色细胞是合成 PPO 的细胞类型，而黑腹果蝇中主要在含晶细胞合成，表明含晶细胞和类绛色细胞属于在功能上同源的细胞类型。在蚊子体内，PPO 在类绛色细胞和粒血细胞中合成。

7.3.2.2 PPO 的释放

在多数情况下，PO 以前体形式释放，在果蝇含晶细胞或鳞翅目昆虫类绛色细胞中 c-Jun N- 末端激酶（JNK）通路激活后细胞破裂引起 PPO 释放，也可能转存在其他尚未发现的 PPO 持续释放机制。不同种类的类绛色细胞的稳定性不同，有些细胞在机体损伤或感染时自发破裂，损伤或感染部位产生 PPO 的血细胞裂解可导致局部出现高浓度的 PPO。在没有损伤或感染的情况下，体内也可能发生类绛色细胞溶解，以维持血淋巴中 PPO 水平稳定。此外，在鳞翅目昆虫中，PPO 主要存在血淋巴中，但在蝗虫和蟑螂等不完全变态昆虫中，PPO 似乎主要留存于血细胞中，受到微生物感染或创伤刺激时才促使 PPO 释放。

7.3.2.3 PPO 的激活

昆虫 PPO 的激活酶被称为酚氧化酶原激活酶（PPAE）或 PPO 激活丝氨酸蛋白酶（PAP），是酚氧化酶原激活过程中的一个关键酶和限速酶。PPAEs/PAPs 最早从家蚕幼虫表皮层中纯化，后来在果蝇、烟草天蛾、斜纹夜蛾、东北大黑鳃金龟和小菜蛾等昆虫中也相继被鉴定和克隆。PAPs 由 N- 端一个或两个夹结构域和 C- 端一个丝氨酸蛋白酶催化结构域组成。PPO 活化成 PO 由 PPAEs/PAPs 完成，在某些情况下也需要缺少酶活性

PAAEs/PAPs accurately cleave PPO into active PO. Cleavage of a conserved Arg^{51}-Phe^{52} bond about 50 residues from the N-terminus results in the activation of insect PPOs. There are two *PPO* genes (*PPO1* and *PPO2*) in silkworm, and PPAEs/PAPs cut *PPO1* and *PPO2* at Arg^{51}-Phe^{52} to become active PO1 and PO2.

The zymogen PPO can also be activated by interaction with amphipathic molecules. Both types of activation by specific proteolysis or interaction with amphipathic molecules must involve a conformational change in the protein, making the active site accessible to substrate, or rearranging the active site to a structure required for catalyzing oxidation reactions.

7.3.3 Melanin biosynthesis

Melanin biosynthesis is a multiple-step process, involving a number of intermediate metabolites and reactions (Fig.7-6). Phenylalanine is initially hydroxylated to tyrosine *via* phenylalanine hydroxylase (PAH), a step aided by tetrahydrobiopterin (BH4) synthesized by dihydropteridine reductase and guanosine triphosphate cyclohydrolase. Tyrosine can be transformed into dihydroxyphenylalanine (DOPA) either by PO or tyrosine hydroxylase. DOPA is converted to dopaquinone *via* PO or to dopamine *via* dopa decarboxylase (DDC). The reactive intermediate dopaquinone undergoes spontaneous intramolecular cyclization to form cyclodopa. Redox cycling of dopaquinone with cyclodopa produces DOPA and dopachrome. 5,6-dihydroxyindole (DHI) can be formed as the major product from both synthesis pathways by dopachrome conversion enzyme (DCE) which promotes the decarboxylation and rearrangement of dopachrome to DHI. Following some reaction steps involving the formation of indole-5,6-quinone *via* PO, dopamine and DHI are finally converted into melanin. Dopaquinone also reacts with thiol compounds to form cysteinyl and glutathionyl conjugates and benzothiazinylalanine compounds that produce pheomelanins (yellow to reddish-brown pigments). Both the brown to black eumelanin and the lighter yellow to reddish-brown pheomelanin pigments are redox-active polymers produced in insects and both are derived from the common precursor—dopaquinone.

的辅因子（丝氨酸蛋白酶同源物）帮助PPAEs/PAPs准确地把PPO切割成有活性的PO。大多数PPOs的激活都需要从PPO中保守的Arg^{51}-Phe^{52}位点进行剪切，去掉N-端大约50个氨基酸。家蚕有2个*PPO*基因（*PPO1*和*PPO2*），PPAEs/PAPs将*PPO1*和*PPO2*从Arg^{51}-Phe^{52}处剪切，成有活性的PO1和PO2。

酶原PPO也通过与两亲分子相互作用激活，这两种激活都必须涉及蛋白质构象的改变，活性位点接近底物，或将活性位点重新排列成催化氧化反应所需的结构。

7.3.3 黑色素的生物合成

黑色素的生物合成是一个多步骤的过程（图7-6），涉及许多中间代谢物和反应。苯丙氨酸通过苯丙氨酸羟化酶（PAH）羟基化为酪氨酸，该步骤由二氢蝶啶还原酶和鸟苷三磷酸环水解酶合成的四氢生物喋呤（BH4）辅助。酪氨酸可以通过PO或酪氨酸羟化酶转化为多巴。多巴通过PO转化为多巴醌或通过多巴脱羧酶（DDC）转化为多巴胺，反应中间体多巴醌进行自发的分子内环化形成环多巴。多巴醌与环多巴的氧化还原循环产生多巴和多巴色素，多巴色素转换酶（DCE）促进多巴色素的脱羧或多巴色素的重排均可以生成5,6-二羟基吲哚（DHI），进一步通过吲哚醌合成真黑色素。多巴醌还与硫醇化合物反应，形成半胱氨酸结合物、谷胱甘肽结合物以及苯并噻唑丙氨酸化合物，产生嗜黑色素（黄色至红褐色的色素）。棕色到黑色的真黑色素和淡黄色到红棕色的嗜黑色素都是昆虫产生的氧化还原活性聚合物，它们都来自共同的前体——多巴醌。

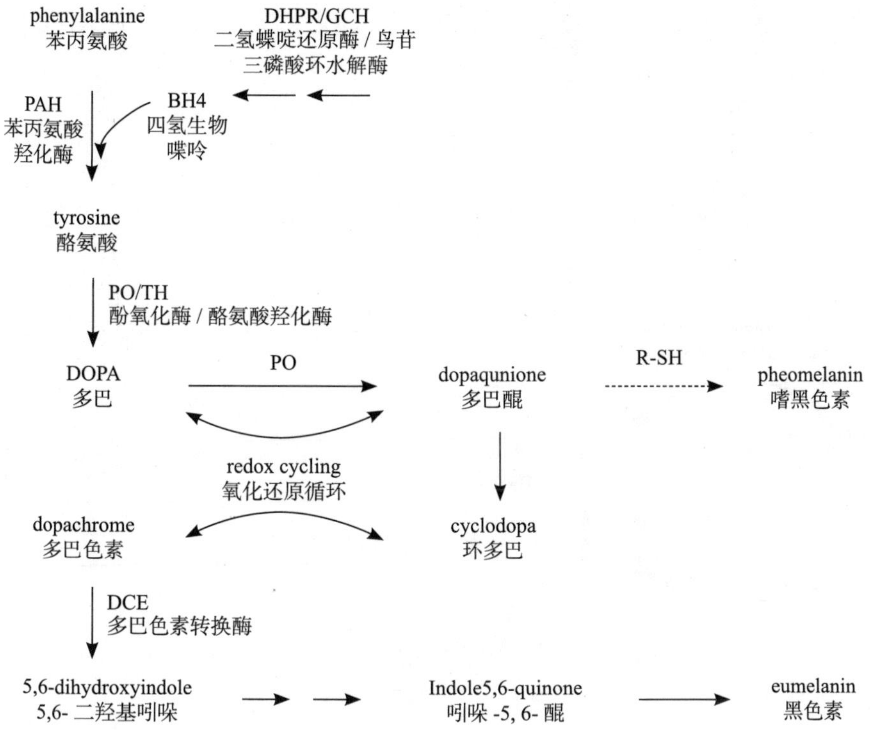

Fig. 7-6 Biosynthesis metabolism of insect melanin.

图 7-6 昆虫黑色素的生物合成代谢。

7.3.4 Regulatory mechanism of PPO activation

Melanization has a detrimental effect on the insect itself, so it is strictly spatial and temperal-controlled. Uncontrolled melanism produces excessive amounts of toxic intermediates and ultimately leads to the death of the host.

PPO is cleaved by a hemolymph serine protease containing a N-terminal CLIP domain, and then activated after the 50-aa peptide is removed. Regulation of PPO activation occurs at spatial and temporal levels. PPO activation first depends on the recognition of foreign microbial cell wall components or other triggering factors, thus ensuring that melanization occurs only at the specific sites where these factors are present. Recognition of triggers leads to activation of the protease cascade. Because these proteases are already inactive zymogens, PPO can be activated very quickly where melanization is needed. An additional benefit of activating PPO by protease cascade is signal amplification, in which serine protease inhibitor (serpin) is also involved in multiple steps of regulation. The general mechanism of PPO activation regulation is shown in Fig.7-7.

7.3.4　PPO 激活的调控机制

黑化反应对于昆虫本身具有损伤作用，因此在时间和空间上都受到严格的调控。不受控制的黑化反应会产生过量的毒性中间产物，最终导致宿主死亡。

PPO 由 N- 端 CLIP 结构域的血淋巴丝氨酸蛋白酶切割，切去 50 个 aa 的肽段后激活。PPO 的激活发生在空间和时间水平，首先激活依赖于对外源微生物细胞壁组分或其他触发因子的识别，由此保证了黑化只在这些因子存在的部位发生。黑化触发因子的识别导致了蛋白酶级联反应的激活。由于这些蛋白酶本来就以非活性的酶原形式存在，PPO 能够在需要黑化的地方很快地将它们激活。通过蛋白酶级联反应激活 PPO 的另外一个好处是信号的放大，其中丝氨酸蛋白酶抑制剂 serpin 也参与多个步骤的调控。PPO 激活调控的普通机制见图 7-7。

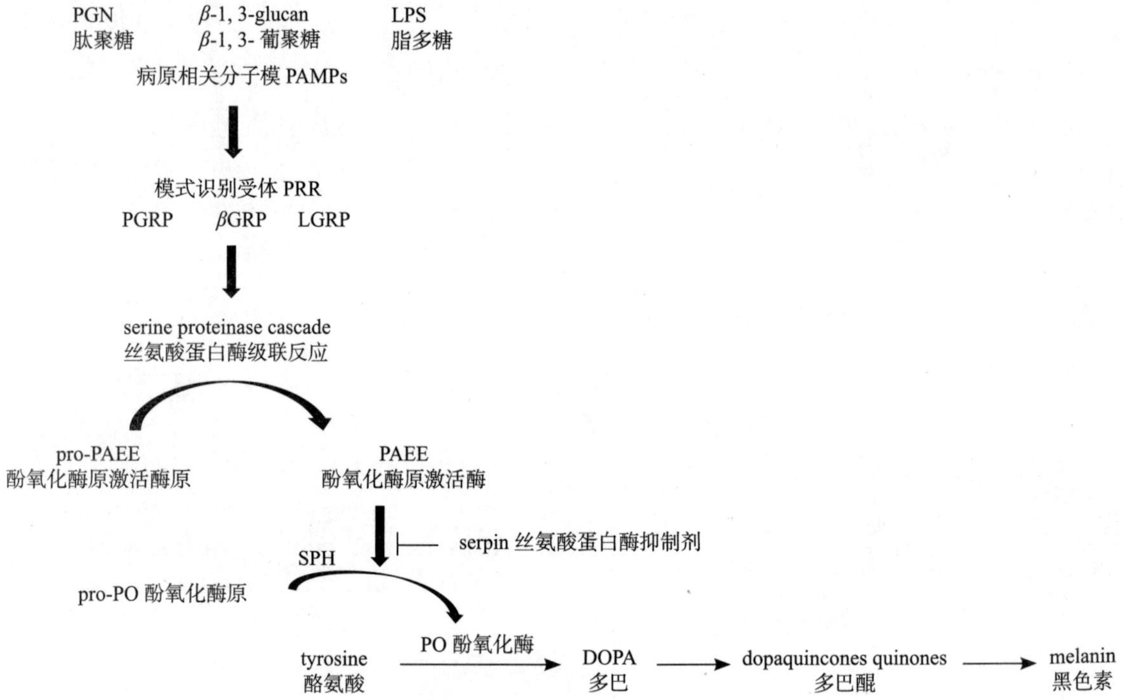

Fig. 7-7 Regulation model of PO activation in insects.

图 7-7 昆虫 PO 激活的调控模式。

7.4 抗菌肽

7.4 Antimicrobial Peptides

Insect immune response is characterized by the synthesis of anti-microbial peptides (AMP). In 1976, Hans G. Boman, professor of Microbiology at Stockholm University and his colleagues used blood from diapausing pupae of the moth *H. cecropia* as the main source for purification to reveal the biochemistry of the proteins and peptides underlying the potent antimicrobial defence in insects. In 1981, the primary structure of cecropin—the first insect antimicrobial peptide, was published on *Nature*. The field of insect immunity began to expand and a series of insect antimicrobial peptides and other immune proteins became characterized. 3,425 AMPs has been reported and 339 AMPs are from insects. AMPs have a wide range of inhibitory effects against bacteria, fungi, parasites and viruses. The emergence of antibiotic-resistant microorganisms and the increasing of concerns about the use of

昆虫免疫反应的特点是合成抗菌肽。1976 年，为了揭示昆虫体内潜在的有效抗菌防御的蛋白质和肽的生物化学特征，斯德哥尔摩大学微生物学教授 Hans Boman 和他的同事们以惜古比天蚕滞育蛹的血淋巴为材料进行抗菌物质的纯化。1981 年，从昆虫中纯化的第一个抗菌肽——天蚕素的一级结构发表在《自然》杂志上，之后昆虫免疫领域的研究开始拓展，一系列昆虫抗微生物肽和其他免疫蛋白质相继被鉴定。目前已经报道的抗菌肽数目达到 3 425 个，其中 339 个来自昆虫。抗菌肽对细菌、真菌、寄生虫和病毒有广泛的抑制作用。耐药微生物的出现和人们对抗生素使用的日益关注

antibiotics resulted in the development of AMPs, which have a good application prospect in medicine, food, animal husbandry, agriculture and aquaculture.

7.4.1 Features of insect AMPs

The antimicrobial peptides in different insects are greatly diverse. In general, insect antimicrobial peptides share common features as follows: ① AMPs has relatively small molecular weight. The amino acid number of AMPs is usually less than 60, thus most of AMPs are less than 10 kDa, with the exception of the attacin (20-23 kDa) and the Gloverin (~14 kDa). ② Most of AMPs are cationic peptide, which have positive net charge at physiological pH. Few AMPs are anionic in nature. ③ AMPs exhibit activities against bacteria and/or fungi, as well as some viruses and parasites.

7.4.2 Classification of insect AMPs

At present, insect antimicrobial peptide has been reported from various insect species of at least six orders, including Lepidoptera, Diptera, and Coleoptera, Hemipteran, Hymenoptera, Orthoptera etc. The majority of AMPs are from Lepidoptera and Diptera. Types and numbers of AMPs varied greatly between insects. *Drosophila* adults produce more than ten different molecules, including cecropin and defensin, while some insects produce relatively few molecules, only single peptide defensin in the dragonfly *Aeschna cyanea*. The distribution of various kinds of antimicrobial peptides was different. For example, defensin, an ancient peptide, has been identified in all insects (even nearly all living organisms). There are 349 defensins from various organisms in the APD31 database. While Drosomycin has only been found in *Drosophila*. Similarly, silkworm moricin and gloverin have been reported so far only in lepidoptera insects. To date, more than three hundred AMPs was identified from insects. These AMPs can be classified based on their structure, amino acid composition or the spectrum of activity.

促进了抗菌肽的研究和发展，未来抗菌肽在医药、食品、畜牧业、农业和水产养殖等领域具有良好的应用前景。

7.4.1 昆虫抗菌肽的特征

昆虫体内的抗菌肽种类多，差异大。但在一般情况下，昆虫抗菌肽具有共同特征：① 抗菌肽分子质量相对较小。除了樗蚕素（20 ～ 23 kDa）和葛佬素（～ 14 kDa），多数抗菌肽氨基酸残基数通常小于 60 个，分子质量小于 10 kDa。②大部分抗菌肽为阳离子肽，在生理 pH 条件下带正电荷，极少数抗菌肽为阴离子型多肽。③抗菌肽对细菌和 / 或真菌具有抑制活性，对某些病毒和寄生虫也具有活性。

7.4.2 昆虫抗菌肽的分类

目前，已在鳞翅目、双翅目、鞘翅目、半翅目、膜翅目和直翅目等至少 6 个目的昆虫中发现了抗菌肽的报道，其中以鳞翅目和双翅目的抗菌肽占多数。不同昆虫之间抗菌肽的种类和数量差异巨大，果蝇成虫产生天蚕素、防御素等十多种不同的分子，而有些昆虫产生的分子相对较少，蜻蜓中只有一种抗菌肽，即防御素。不同种类的抗菌肽分布范围不同，如防御素是一个古老的抗菌肽，几乎存在于所有的昆虫，甚至所有的生物中，在抗菌肽数据库 APD31 中有 349 个防御素；而果蝇抗真菌肽 Drosomycin 目前只发现存在于果蝇属。相似地，家蚕抗菌肽 moricin 和葛佬素暂时只在鳞翅目昆虫有报道。截至目前，已经从昆虫中鉴定的抗菌肽已经超过 300 个，这些抗菌肽可以根据其结构、氨基酸组成或抗菌活性谱进行分类。

7.4.2.1 Classification by the spectrum of activity

The activity and spectrum between AMPs are greatly varied. According to the main activity of AMPs, these categories can be summarized as antibacterial, antifungal, antiviral, antiparasitic, and anti-tumor peptides. The first small immune peptide from insect was named cecropin, meaning the antibiotics from *Cecropia*. Since cecropin was characterized for its activity on bacteria, the peptide was called antibacterial peptide. In addition to antibacterial peptides, insects produce cysteine-rich peptides with antifungal properties. Drosomycin was the first inducible antifungal peptide to be isolated from insects and it has a potent antifungal activity while it is ineffective against yeasts and bacteria. To date, seven types of AMPs have been identified in *Drosophila* and multiple homologous genes of one type are tandemly arrayed or clustered within the *Drosophila* genome. For example, Attacin multigene family has four members, and Drosomyicn has seven. These seven peptides showed diverse activity, so they can be grouped into three families based on their main biological targets, gram-positive bacteria (defensin), gram-negative bacteria (cecropin, drosocin, attacin, diptericin) or fungi (drosomycin, metchnikowin) (Tab. 7-1). In addition, cecropin has some activity on fungi and metchnikowin was active on some bacteria. The simultaneous presence of several antimicrobial peptides acting in synergy can provide insects with a more powerful defense against harmful invaders such as bacteria, fungi, virus and protozoa. In addition, andropin is a male-specific AMP which is uniquely expressed in the ejaculatory duct of adult males.

7.4.2.1 依据抗菌活性谱进行分类

不同抗菌肽之间的抗菌活性和抗菌谱差异很大，根据抗菌肽的主要活性可将其分为抗菌肽、抗真菌肽、抗病毒肽、抗寄生虫肽和抗肿瘤肽等。第一个昆虫小分子免疫多肽被命名为天蚕素，意思是来自天蚕的抗生素。由于天蚕素主要活性是抗细菌，因此该肽被称为抗菌肽。除了抗菌肽，昆虫体内还发现了具有抗真菌特性的富含半胱氨酸多肽。Drosomycin是从黑腹果蝇中分离得到的第一个诱导型昆虫抗真菌肽，具有较强的抗真菌活性，但对酵母和细菌没有活性。至今在果蝇中鉴定了7种抗菌肽，且同一种抗菌肽存在多个序列相似的拷贝，在基因组中串联或成簇分布，如樗蚕素家族有4个成员，果蝇抗真菌肽Drosomycin家族有7个成员。这7种抗菌肽具有不同的活性，根据它们的主要活性谱可以分为三组：抗革兰氏阳性菌（防御素）；抗革兰氏阴性菌（天蚕素、果蝇肽、樗蚕素、双翅肽）或抗真菌（果蝇抗真菌肽Drosomycin、碧蟀金属肽）（表7-1）。此外，天蚕素对真菌有一定的活性，碧蟀金属肽对某些细菌也有活性。同时存在的多种抗菌肽协同作用，可以为昆虫提供更强大的保护，如抵御细菌、真菌、病毒和原生动物等有害入侵物。此外，andropin是一种雄性特异性抗菌肽，在成年雄性果蝇的射精管中特异表达。

Tab. 7-1 Seven classes of antimicrobial peptides in *Drosophila* and their Characteristics

表7–1 果蝇中发现的7种抗菌肽及其特性

AMPs 抗菌肽	genes 基因数	main acitivity 主要活性	secondary structure 二级结构	modification 修饰
attacin 樗蚕素	4	gram-negative bacteria 革兰氏阴性菌	linear 线性	
cecropin 天蚕素	4	gram-negative bacteria 革兰氏阴性菌	α-helix α-螺旋	
diptericin 双翅肽	2	gram-negative bacteria 革兰氏阴性菌	linear 线性	
drosocin 果蝇肽	1	gram-negative bacteria 革兰氏阴性菌	linear 线性	glycosylation 糖基化

continued table 续表 7-1

AMPs 抗菌肽	genes 基因数	main acitivity 主要活性	secondary structure 二级结构	modification 修饰
defensin 防御素	1	gram-positive bacteria 革兰氏阳性菌	α-helix and β-sheet α- 螺旋和 β- 折叠	
drosomycin 果蝇抗真菌肽	7	fungi 真菌	α-helix and β-sheet α- 螺旋和 β- 折叠	
metchnikowin 碧蝽金属肽	1	fungi 真菌	linear 线性	

7.4.2.2 Classification by the structure and amino acid composition

AMPs are diverse in nature. Based on their secondary structure and unique sequences, they can be divided into four categories, including linear α-helical peptides, proline-rich peptides, cysteine-rich peptides, and glycine-rich proteins.

1. α -helical peptide

The largest and most widely studied AMPs are α-helical peptides. In the α-helix conformation, the distance between two adjacently attached amino acids is about 0.15 nm distance and from top view, the angle among them from the center is about 100°. Examples of α-helical peptides include cecropin and moricin, whose structure was shown in Fig. 7-8A and Fig.7-8B; Appendix color graphic Fig.7-8. In an aqueous solution, these peptides are unstructured but upon interaction with microbial membrane or in contact with trifluoroethanol, detergents/surfactants above critical micellar concentration such as sodium dodecyl sulfate (SDS), AMPs fold themselves into α-helical structure. Most of the α-helical peptides kill microorganisms by pore formation.

2.proline-rich peptides

The proline-rich peptides are linear, therefore, these peptides cannot fold into typical secondary structure. In an aqueous solution, these peptides are highly flexible. The peptides penetrate into the cell membrane of the organism and exert their antimicrobial activities through interaction with different intracellular proteins such as the heat shock proteins.

3.Cysteine-rich peptides

In this kind of peptides, at least 2 β-strands attach to each other

7.4.2.2　依据结构和氨基酸组成进行分类

抗菌肽在自然界中多种多样。根据抗菌肽的二级结构和氨基酸序列，可将其分为四类：线性 α- 螺旋多肽、富脯氨酸多肽、富半胱氨酸多肽和富甘氨酸蛋白质。

1.α- 螺旋肽

α- 螺旋肽是被研究得最多和最广泛的一种抗菌肽。在 α- 螺旋构象中，两个相邻的氨基酸之间的距离约为 0.15 nm，从顶部看，它们之间从中心的角度约为 100°。α- 螺旋多肽的例子包括天蚕素 cecropin 和家蚕抗菌肽 moricin，结构见图 7-8A 和图 7-8B；附彩图 7-8。在水溶液中，这些肽没有任何结构构型，但在与微生物膜相互作用时，或者与超过临界胶束浓度的三氟乙醇、洗涤剂 / 表面活性剂接触时，如十二烷基硫酸钠（SDS），抗菌肽会折叠成 α- 螺旋结构。大多数 α- 螺旋肽通过在膜上形成孔洞的方式杀死微生物。

2. 富含脯氨酸的多肽

富含脯氨酸的多肽是线性的，因此这些多肽不能折叠成典型的二级结构，在水溶液中高度灵活可变。这些多肽可以穿透细胞膜，并通过与不同的细胞内蛋白质，如热休克蛋白相互作用发挥其抗菌活性。

3. 富含半胱氨酸的多肽

在这类多肽中，至少有 2 条 β- 链

with the help of disulfide bonds. This structure involves an α-helix linked to a two or three-stranded β-sheet and includes the well-known cysteine stabilized *α*-helix and β-sheet (CS*α*β) motif. Typical examples of cysteine-rich peptides include drosomycin and defensin, whose structure were shown in Fig. 7-8C and Fig.7-8D. Drosomycin is stabilized by four disulfide bridges (Cys^1-Cys^8, Cys^2-Cys^4, Cys^3-Cys^6 and Cys^4-Cys^7), while Defensin contains six cysteine residues forming three disulfide bridges (Cys^1-Cys^5, Cys^2-Cys^4 and Cys^3-Cys^6). The arrangement of the cysteine residues in *β*-sheet is so crucial that replacement with other amino acids (such as aspartic acid and alanine) leads to inactivation of these peptides.

通过二硫键相互连接。这种结构包括一个与二链或三链β-片相连的α-螺旋，包括众所周知的半胱氨酸稳定的α-螺旋β-片（CS*α*β基序）。典型的富含半胱氨酸的多肽包括果蝇抗真菌肽和防御素，结构见图7-8C和图7-8D。Drosomycin有四个二硫链（Cys^1-Cys^8, Cys^2-Cys^4, Cys^3-Cys^6和Cys^4-Cys^7）稳定的多肽，防御素含有六个半胱氨酸残基，形成三个二硫桥（Cys^1-Cys^5, Cys^2-Cys^4和Cys^3-Cys^6）。*β*-折叠片中半胱氨酸残基的排列如此重要，以至于当以其他氨基酸（如天门冬氨酸和丙氨酸）进行替代时会导致这些肽的失活。

4.Glycine-rich proteins

These kind of Examples of glycine-rich proteins include attacin and gloverin. The former has been reported in many Lepidoptera insects such as *H.cecropia*, *M. sexta*, *B. mori*, *H. armigera* and dipteran such as *D. melanogaster*, *Glossina morsitans* and *Hermetia illucens*, while gloverin has only been found in Lepidoptera insects. The molecular weight of Attacin was 20 kDa, and it was mainly active against gram-negative bacteria. Gloverin, which is approximately 14 kDa in size, acts mainly on gram-negative bacteria and is active against certain gram-positive bacteria and fungi such as candida albicans. Circular dichroism showed that attacin and glvoerin showed mainly random coiled structures in aqueous solution, but changed into helical structures in hydrophobic solvent.

4. 富含甘氨酸的蛋白质

该类抗菌蛋白分子质量相对较大，包括樗蚕素和葛佬素，前者在鳞翅目的许多昆虫，如惜古比天蚕、烟草天蛾、家蚕和棉铃虫等以及双翅目果蝇、刺台蝇和黑水虻等均有报道，而葛佬素目前只发现存在于鳞翅目昆虫。樗蚕素的分子质量大小为20 kDa，主要对革兰氏阴性菌有活性。 葛佬素大小约为14 kDa，主要作用于革兰氏阴性菌，对某些革兰氏阳性菌和真菌，如白色念珠菌也有活性。圆二色谱显示，樗蚕素和葛佬素在水溶液中主要呈无规卷曲，而在疏水溶剂中转变为螺旋结构。

A cecropin B *Hyalophora cecropia* 惜古比天蚕
B moricin *Manduca sexta* 烟草天蛾
C defensin *Anopheles gambiae* 冈比亚按蚊
D drosomycin *Drosophila melanogaster* 黑腹果蝇

Fig. 7-8 Different structures of AMPs. A.Cecropin adopts two α-helical conformations (PDB : 2IGR); B.Moricin has a typical α-helical conformation (PDB : 2JR8); C.Defensin (PDB : 2NY8); L; DL.Drosomycin (PDB : 1MYN) adopts both α-helix and β-sheet conformation.

图7-8 不同抗菌肽的结构。A. 天蚕素呈现两个α-螺旋构象（PDB : 2IGR）；B.Moricin具有一个典型的α-螺旋构象（PDB : 2JR8）；C. 防御素（PDB : 2NY8）；D. 果蝇肽（PDB : 1MYN）采用α-螺旋和β-折叠混合的构像。

7.4.3 Signaling pathways controlling antimicrobial genes

Antimicrobial peptides can be used as efficient markers of the immune response, as their expression is significantly up-regulated following infection. AMPs are mainly synthesized in the fat body, an equivalent of the mammalian liver. Hemocytes is also involved in the antimicrobial peptide synthesis. The research on activation mechanism of insect antimicrobial peptide genes has made great progress in the recent decades, indicating AMPs gene regulation involves multiple signaling pathways and a large number of signaling molecules in the model organism *Drosophila*.

In response to microbial infection, insects activate expression of antimicrobial peptide genes which is mainly controlled by Toll, Imd and JAK-Stat signaling pathways. The process can be divided into three steps. The first step is recognition of microorganisms by insects, which is the first step to initiate immune response. Insect pattern recognition receptor (PRR) recognizes a specific molecule on the surface of microorganism, which is termed pathogen, associated molecular patterns (PAMP). PAMP are molecules with conserved motifs that are associated with pathogen infection that serve as ligands for host pattern recognition molecules. Generally, receptors of transmembrane proteins in immune signaling pathways cannot directly recognize PAMP. So mediator molecules known as PRR are needed. PRR mainly includes peptidoglycan recognition protein (PGRP) and Gram-negative bacteria binding protein (GNBP). PGRP can be divided into L-type (PGRP -LA , LB , LC , LD , LE) and S-type (PGRP-SA , SB , SC, SD) based on their structure. The second step is signal transduction, which activates the downstream signaling pathway through the transmembrane receptor Toll or PGRP. The third step is the immune response. Whether it is Toll or Imd or AK-STAT signaling pathways, NF-κB transcription factor is ultimately promoted to enter the nucleus and induce the expression of corresponding antimicrobial peptides in response to infection by different microorganisms (Fig.7-9; Appendix color graphic Fig.7-9).

7.4.3 调控抗菌肽基因表达的信号通路

抗菌肽基因的表达在感染后显著上调，因此抗菌肽常常被作为免疫反应的有效标记物。抗菌肽主要在昆虫的脂肪体中（相当于哺乳动物的肝脏）合成，血细胞也参与抗菌肽的合成。近几十年来，昆虫抗菌肽基因的激活机制研究取得了很大的进展，揭示抗菌肽基因的调控在模式生物果蝇中涉及多个信号通路和大量的信号分子。

昆虫应答微生物的感染，主要通过 Toll、Imd 和 JAK-Stat 信号通路激活抗菌肽的表达，这个过程大致可以分为三个步骤：第一步是昆虫对微生物的识别，通过模式识别受体（PRR）识别微生物表面的特异分子，即病原相关分子模式（PAMP）。PAMP 是与病原体感染相关的具有保守基序的分子，是宿主模式识别分子的配体。通常免疫信号通路中的跨膜蛋白受体不能直接识别这些分子，因此必须有中介分子参与，即模式识别分子（PRR）。PRR 主要包括肽聚糖识别蛋白（PGRP）和革兰氏阴性菌结合蛋白（GNBP）等。根据结构，PRGRP 可分为长型（PGRP -LA , LB , LC , LD , LE）和短型（PGRP-SA , SB , SC , SD）。第二步是信号转导，通过跨膜受体 Toll 或 PGRP 激活下游的信号通路。第三步是免疫应答，无论是 Toll 或是 Imd 和 JAK-Stat 信号转导通路，最终都是促使 NF-κB 转录因子进入细胞核，诱导相应抗菌肽的表达以应对不同微生物的感染（图 7-9 ；附彩图 7-9）。

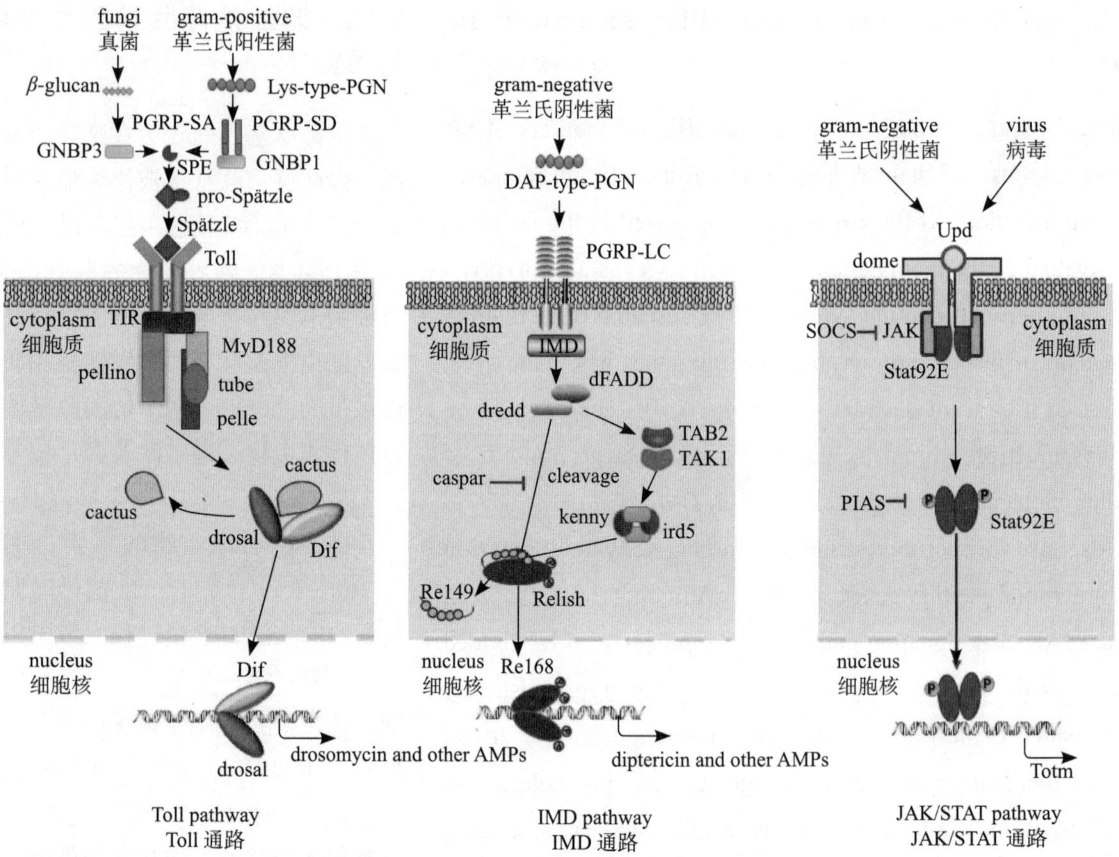

Fig. 7-9 Signaling pathways regulating expression of antimicrobial peptide genes in *Drosophila*.

Lys-type-PGN. Lys-tye-peptidoglycan; GNBP3. Gram-negative binding protein 3; GNBP1. Gram-negative binding protein 1; SPE. spätzle processing enzyme; TIR. Toll/IL-1R homologous region; dmyD88. Drosophila homology of differentiation primary response protein; Tube: adaptor protein containing death domain; Pelle. kinase protein; dif. NF-κB transcription factor; DAP-type-PGN. diaminopimelic acid-tye peptidoglycan; PGRP-LC. peptidoglycan recognition protein LC; dFADD. drosophila Fas-associated with death domain; Dredd. homology of cysteine protease; TAK1. growth factor β-activated kinase; TAB2. TAK1 binding protein; Upd. unpaired cytokines; Stat92E. signal-transducer and activator of transcription protein at 92E.

图 7-9 果蝇调控抗菌肽表达的土要信号传导通路。

Lys-type-PGN. 赖氨酸型肽聚糖；GNBP3. 革兰氏阴性菌结合蛋白 3； GNBP1. 革兰氏阴性菌结合蛋白；SPE. 一种加工 spätzle 的丝氨酸蛋白酶；PGRP-SA, SD. 肽聚糖识别蛋白 SA, SD; TIR. Toll/IL-1 受体同源域； dMyD88. 果蝇骨髓分化初级反应蛋白质同源物 Tube. 含有死亡域接头蛋白； Pelle. 蛋白激酶 cactus. IκB (NF-κB inhibitor protein 抑制蛋白); Dif. NF-κB 转录因子；DAP-type-PGN. diaminopimelic acid-type peptidoglycan; PGRP-LC. 肽聚糖识别蛋白 LC; IMD: immune deficiency protein; dFADD. 含死亡结构与的 Fas 相关蛋白质 dredd. 半胱氨酸蛋白酶同源物；TAK1. 生长因子 β 活化酶 1; TAB2. 生长因子 β 活化酶 1 结合蛋白；IKK complex. IKK 复合体 (ird 5, key); Relish. NF-κB 转录因子；Upd. 不对称因子、信号转导和转录激活因子 92E.

7.4.3.1 Toll signaling pathway

In 1980, German scientist Nusslein-Volhard et al. identified a gene determining the dorso ventral axis of the *Drosophila* embryo, which was named Toll. In 1988, Hashimoto et al. found that the *Toll* gene encodes a transmembrane protein and elucidated the structure of Toll protein. In 1991, Gay et al. found that Toll protein was structurally homologous to interleukin 1 receptor

7.4.3.1 Toll 信号通路

1980 年，德国科学家 Nusslein-Volhard 等在研究果蝇胚胎发育过程中发现有一个基因决定着果蝇的背腹极性发育，将其命名为 Toll。1988 年，Hashimoto 等发现 *Toll* 基因编码一种跨膜蛋白质，并阐明了 Toll 蛋白的结构。1991 年，Gay

(IL-1R), an innate immune function molecule in mammals, and their intracellular region was similar, suggesting that Toll might be related to immunity. Toll-like receptors in humans were first reported by Nomura et al. in 1994, at that time the immunological function of Toll was not elucidated yet. Two years later, in 1996, French scientists Jules A. Hoffmann and his colleague Lemaitre et al. established the immunological significance of Toll by discovering that it plays an important role in the immunity to fight against fungal infection in fruit flies. It promotes the discovery and the related research of mammalian toll-like receptor (TLR4) function involving in adaptive immunity.

Toll receptor is a type I transmembrane protein receptor, which has typical extracellular leucine-rich repeats (LRRs) and intracellular Toll/ interleukin-1 receptor homology domain (TIR). Späetzle (Spz) is a ligand for Toll, a cytokine-like ligand in the form of a non-active precursor protein (pro-Spz). When Drosophila is infected by microorganisms, a series of enzymatic reactions in hemolymph are triggered following the binding of pattern recognition receptors (PRR) and pathogen associated molecular patterns (PAMP). When infected with Gram-positive bacteria, *Drosophila* PGRP-SA, PGRP-SD (Peptidoglycan recognition protein-SA,-SD), and Gram-negative bacteria binding protein (GNBP-1) is involved in recognizing Lysine-type PGN (Peptidoglycan) (Lys-PGN)on positive bacteria, while *Drosophila* GNBP-3 is involved in recognizing or stimulating enzymatic reaction through proteases during yeast or fungal infection. When pro-Spz is cleaved by protease, mature Spz containing about 100 amino acids binds to the Toll receptor extracellular domain, forming Toll dimer, The intracellular TIR Domain binds to Myeloid differentiation primary response gene 88 (MyD88), which binds to the Myeloid differentiation primary response gene 88 (MyD88) through the Death Domain (DD) interacts with Tube, another molecule containing DD, which in turn recruits the kinase Pelle. The two DDs of the Tube bind to MyD88 and Pelle, respectively, forming the MyD88/Tube/Pelle heterotrimer. Pelle kinase phosphorylates the IκB inhibitor family protein cactus and promotes its degradation, which results in the release of dorsal or *Dif* from the dorsal or Dif complex. Nuclear localization signals are exposed and translocated into the nucleus. Dorsal/ Dif enters the nucleus and binds to the *κB* factor DNA element

等发现 Toll 蛋白在结构上与哺乳动物中一种天然免疫功能分子——白细胞介素 1 受体（IL-1R）具有同源性，胞内部分相似，提示 Toll 可能和免疫有关。1994 年，Nomura 等首先报道了人类中的 Toll 样受体，当时 Toll 的免疫学功能没有得到阐明。两年之后的 1996 年，法国科学家 Jules A. Hoffmann 和他的同事 Lemaitre 等发现 Toll 在果蝇对真菌感染的免疫中起着重要作用，从而确立了 Toll 的免疫学意义，推动了哺乳动物 Toll 样受体（TLR4）参与适应性免疫的功能发现及其相关的研究。

Toll 受体属于 I- 型跨膜蛋白受体，具有胞外典型的亮氨酸富集区和胞内 Toll/ 白介素 -1 受体的类似结构域（TIR）。Späetzle（Spz）是 Toll 的配体，以没有活性的前体蛋白（pro-Spz）形式存在。当果蝇感染微生物时，通过模式识别受体（PRR）与病原相关分子模式（PAMP）结合而激活一系列的酶联反应，激活下游的信号传导通路。当受到革兰氏阳性菌侵染时，果蝇肽聚糖识别蛋白（PGRP-SA, -SD）和革兰氏阴性菌结合蛋白（GNBP-1）参与识别阳性菌上的赖氨酸型肽聚糖（Lys-PGN），而酵母或真菌感染时，果蝇 GNBP-3 参与识别或通过蛋白酶而激发酶解反应。pro-Spz 被蛋白酶切割后释放出含有 100 个氨基酸左右的成熟 Spz，结合 Toll 受体胞外结构域，形成 Toll 二聚体，使胞内 TIR 结构域与接头蛋白 MyD88 结合，MyD88 通过死亡结构域（DD）与另一个含有 DD 的分子 Tube 相互作用，Tube 进而招募激酶 Pelle。Tube 的两个 DDs 分别与 MyD88 和 Pelle 结合，形成 MyD88/ Tube /Pelle 异源三聚体。而后 Pelle 激酶磷酸化修饰 IκB 抑制家族蛋白 cactus 并促进其降解，cactus 的降解导致 dorsal 或 *Dif* 从复合物中释放，露出核定位信号而转移到细胞核中。Dorsal/Dif 进入细胞核与抗菌肽基因启动子上的

on the antimicrobial peptide gene promoter, and the expression of antimicrobial peptide genes such as Defensin, Drosomycin and Metchnikowin is up-regulated (Fig. 7-7).

7.4.3.2 Imd signaling pathway

When fruit flies are infected by gram-negative bacteria or some DAP (meso-diaminopimelic acid)-PGN typed gram-positive bacteria (e.g. *Bacilli* spp.), PGRP-LC recognize DAP-PGN (Meso-diaminopimelic acid peptidoglycan) as the pattern recognition receptor. The binding of extracellular region of receptor and PGN promotes its intracellular region to interact with IMD protein, initiating the downstream cascade reaction. IMD is 30 kDa protein with Death domain (DD), which can recruit the adaptor molecule FADD (fas-associated death domain-containing protein). FADD binds to caspase-related protein dredd (death-related ced-3/Nedd2-like protein) and subsequently recruits TGFβ-activated kinase 1 (TAK1-TAB2). The TAK1-binding protein 2 complex activates the IKK kinase complex composed of ird5 (Kinase immune response deficient 5) and Kenny, which phosphorylates Relish. Relish, which is a transcription factor belonging to nuclear factor κB (NF-κB) family, containing an RHD domain (Rel homology domain) at the N-terminus and an NF-κB inhibitory domain (IκB) at the C-terminus. Under non-infectious conditions, the IκB domain inhibits RHD activity and relish locates in the cytoplasm. When Relish is phosphorylated and the Ankyrin repeat (IκB) in the Relish C-terminal NF-κB repressor domain is cleaved by the IMD-Fadd-Dredd complex. Inhibition of IκB on the N-terminal Rel homology domain was diminished, then the cleaved Relish translocate from the cytoplasm to the nucleus and bind to κB DNA cis-regulatory elements on the promoters of target genes. Subsequently diptericin, cecropin, defensin and attacin, then were up-regulated(Fig. 7-9B).

7.4.3.3 JAK–Stat signaling pathway

JAK-Stat signaling pathway is mainly involved in the immune response to gram-negative bacteria and viruses. This pathway is principally composed of transcription factor Stat92E, JAKase Hopscotch (Hop), receptor Domeless and three Upd ligands. Upd1 and Upd2 are predominantly responsible for growth and development, while Upd3 is mainly involved in the immune

κB 因子 DNA 元件结合，导致 Defensin、Drosomycin 和 Metchnikowin 等抗菌肽基因表达上调（图 7-7）。

7.4.3.2 Imd 信号通路

当果蝇受到革兰氏阴性菌或某些革兰氏阳性菌（如芽孢杆菌）侵染时，模式识别受体 PGRP-LC 识别 DAP-PGN（内消旋二氨基庚二酸型的肽聚糖），胞外区结合 DAP-PGN 后促使其胞内区与 IMD 蛋白质相互作用，启动下游的级联反应。IMD 为 30 kDa 具有死亡结构域（DD）的蛋白质，可以招募接头分子 FADD（Fas 相关死亡结构域蛋白质）。FADD 和 caspase 相关蛋白质 dredd（death-related ced-3/Nedd2-like protein）相结合，随后募集 TGFβ-activated kinase 1（TAK1-TAB2），TAK1- 结合蛋白 2 复合物，并激活由激酶 ird5（免疫应答缺陷 5）和 Kenny 组成的 IKK 激酶复合体，使核因子 κB 家族的转录因子 Relish 发生磷酸化。Relish 蛋白 的 N- 端含有 RHD 结构域，C- 端含有 NF-κB 抑制结构域（IκB）。在非感染条件下，IκB 结构域可以抑制 RHD 的活性使其存在于细胞质中。在感染条件下，Relish 转录因子被磷酸化，而后 Relish C- 端的 NF-κB 抑制结构域中 Ankyrin 重复序列（IκB）被 IMD-Fadd-Dredd 复合物切割，解除 IκB 对 Relish N- 端 RHD 结构域的抑制，从而使其从细胞质转移到细胞核，结合到靶基因启动子上的 κB DNA 顺式调控元件，进而上调双翅肽、天蚕素、防御素和樗蚕素等抗菌肽的表达（图 7-9B）。

7.4.3.3 JAK–Stat 信号通路

JAK-Stat 信号通路主要参与革兰氏阴性菌和病毒的免疫应答。该通路主要由转录因子 Stat92E、JAK 酶 Hopscotch（Hop）、受体 Domeless 和 3 个 Upd 配体组成，其中 Upd1 和 Upd2 主要负责生长发育，Upd3 主要参与对革兰氏阴性菌的免

response to gram-negative bacteria. Upd binds to Domeless dimer, then activates the intracellular JAK enzyme Hop, and promotes the phosphorylation of tyrosine residues of Stat92E to form homodimers, which are transported to the nucleus and regulate the expression of downstream effector genes such as Attacin *C* (Fig. 7-9C).

疫应答。Upd 与 Domeless 二聚体结合，激活胞内的 JAK 酶 Hop，使得 Stat92E 的酪氨酸残基被磷酸化，形成同源二聚体，并被转运至细胞核中，调节下游效应基因，如樗蚕素 C 的表达（图 7-9C）。

7.4.3.4 Crosstalk and feedback of signal pathways

7.4.3.4 信号通路的互作与负反馈

1.Multiple immune pathways in insects

In addition to Toll, Imd and JAK-STAT signaling pathways, which mainly involved in insect immunity as described above, there are JNK and other immune signaling pathways. C-Jun N-terminal kinase (JNK) is a family member of mitogen-activated protein kinases (MAPKs). When *Drosophila* is stimulated by immune stress, the signal is transmitted by the upstream serine protease cascade, which activates JNK and initiates the downstream immune pathway. Toll-7 receptor can also recognize the glycoprotein of vesicular stomatitis virus and activate Toll pathway. Other non-infectious conditions such as nutrition depletion, ecdysone (20E) and DNA damage signal can regulate the expression of antimicrobial peptides. It can be seen that insect immunity is a complex physiological process.

1. 昆虫同时存在的多个免疫通路

除了上述介绍的主要参与免疫的 Toll、Imd 和 JAK-STAT 信号通路之外，还存在 JNK 等信号通路也与先天免疫相关。C-Jun N- 末端激酶（JNK）是丝裂原活化蛋白激酶（MAPKs）的家族成员，果蝇受到免疫刺激，信号经由上游丝氨酸蛋白酶级联通路逐级传递，激活 JNK 蛋白，启动下游免疫通路。Toll-7 受体还能识别水疱性口炎病毒糖蛋白，激活 Toll 通路。其他非感染条件如营养缺乏、蜕皮激素信号（20E）和 DNA 损伤信号等均可以调控抗菌肽的表达。可见，昆虫的免疫是一个复杂的生理过程。

2.Interaction between multiple signaling pathways

Different signaling pathways are not isolated, but interconnected, and there is a complex interaction mechanism. For example, Imd pathway can activate JNK pathway through TAK1-TAB2 complex, affect wound repair and regulate antimicrobial peptide gene expression in Drosophila.

2. 多个信号通路之间的互作

不同的信号通路不是孤立的，而是相互联系，存在着复杂的互作机制。如 Imd 通路可以通过 TAK1-TAB2 复合物激活 JNK 通路，影响果蝇的伤口修复并调节抗菌肽基因的表达。

3.Feedback of immune signaling pathway

For the activation of any immune signaling pathway, there must be a certain negative feedback mechanism to avoid excessive activation of immunity and damage to the body. For example, in the Imd pathway, PGRP-LF competitively binds PGRP-LC and blocks its recognition of PGN. Pirk protein is a negative feedback regulator in Imd. PGRP-LB can hydrolyze PGN with aminase activity and downregulate Imd signal. PGRP-SC can cooperate with PGRP-LB and Pirk protein to clear PGN and negatively regulate Imd pathway.

3. 免疫信号通路的负反馈

任何一条免疫信号通路都具有一定的激活机制，也必定存在一定的负反馈机制，以避免免疫的过度激活和对机体的损伤。如在 Imd 通路中，PGRP-LF 竞争性地结合 PGRP- LC 并阻断其对 PGN 的识别；Pirk 蛋白是 Imd 中的负反馈调节因子，PGRP-LB 具有酰胺酶活性水解 PGN 而下调 Idm 信号，PGRP-SC 可协同 PGRP-LB 和 Pirk 蛋白清除 PGN 从而负调控 Imd 通路。

7.4.4 Action mechanism of AMPs

AMPs are diverse in structure and activity spectrum, hence they have various action mechanisms. According to their action mechanisms, AMPs are divided into two groups, membrane disruptive and non-membrane disruptive AMPs. Membrane disruptive AMPs interact with the cell membrane of the target organism leading to lesions on membrane or permeabilize phospholipid bilayer that leads to cell death. A classic view is that cationic AMPs such as Cecropin first attach to the bacterial surface with negative nets by electrostatic and hydrophobic interactions. Amphipathic configurations allow the binding of Cecropin molecules to the membrane of bacteria, leading to the formation of holes in the cell membrane, the release of intracellular cations, and finally the cell cleavage and death. Toroidal-pore, barrel-stave, carpet and aggregate models have been proposed to explain the mechanisms of action of membrane-targeting AMPs. Because of the differences in cell surface charge and cell membrane composition, AMPs have no apparent toxicity to mammalian cells (Fig.7-10; Appendix color graphic Fig.7-10).

Non-membrane disruptive AMPs do not target bacterial membranes, e.g., the glycine-rich AMPs attacins and gloverins

7.4.4 抗菌肽的作用机制

不同抗菌肽具有不同的结构和不同的抗菌谱，因此具有不同的作用机制。根据作用机制，抗菌肽可分为膜破坏型和非膜破坏型两类。膜破坏型抗菌肽通过与靶标细胞膜相互作用，使膜损伤或渗透磷脂双分子层，导致细胞死亡。一个经典的观点是阳离子抗菌肽，如天蚕素首先通过静电作用与疏水作用附着在带负电荷的细菌表面，两亲性的结构允许其分子结合到细菌的细胞膜中，导致细胞膜上孔洞的形成，胞内内容物外泄，最终细胞裂解死亡。环孔模型、桶壁模型和地毯模型已经被提出用于解释抗菌肽靶向细胞膜的作用机制。由于哺乳动物细胞表面的电荷和细胞膜组成不同，因此抗菌肽对哺乳动物没有明显的毒性（图 7-10；附彩图 7-10）。

非膜破坏型 AMPs 并不以细菌膜为靶点，例如，来自惜古比天蚕富含甘氨酸

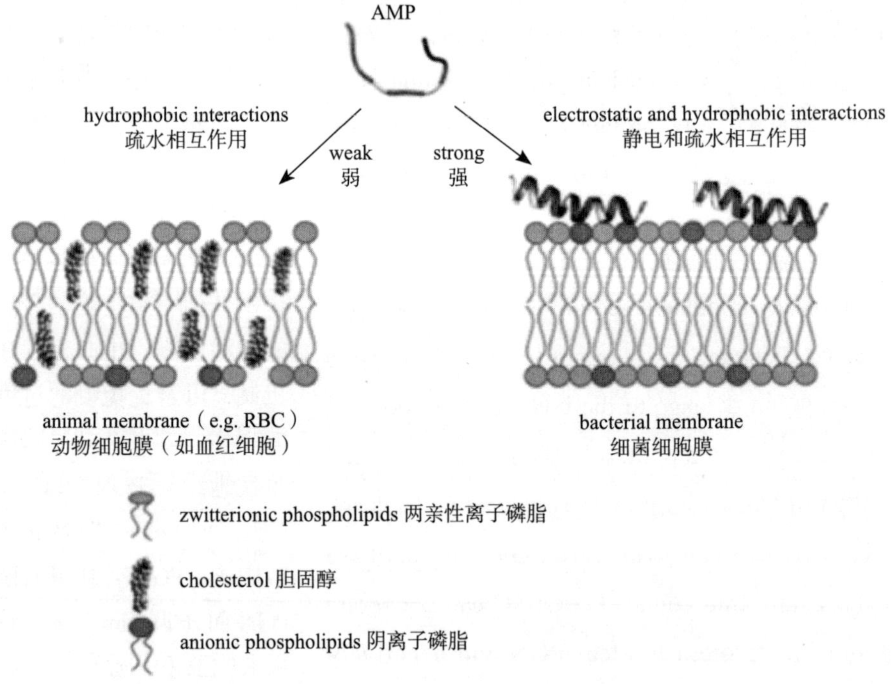

Fig. 7-10 Action mechanism of α-helical AMPs and its cytotoxic selectivity (Kumar et al., 2018）.
图 7-10 α- 螺旋抗菌肽的杀菌机制及其选择时的细胞毒性（Kumar 等，2018）。

from *Hyalophora cecropia* can inhibit synthesis of several proteins of the outer membrane of *Escherichia coli* after binding to the bacterial lipopolysaccharides (LPS). The proline-rich AMPs inhibit protein synthesis *via* binding to ribosomes.

的樗蚕素和葛佬素与细菌脂多糖（LPS）结合后，可以抑制大肠杆菌外膜几种蛋白质的合成。富含脯氨酸的抗菌肽通过与核糖体结合从面抑制蛋白质的合成。

7.4.5 Potential application of AMPs

7.4.5 抗菌肽的潜在应用

Antibiotic resistance is projected as one of the greatest threats to human health in the future. Antimicrobial peptides (AMPs) are considered as good candidates to overcome the alarming problem of acquired bacterial drug-resistance and of emergence of opportunistic pathogens since they present several advantages: ① they rapidly kill target microorganisms; ② they have broad activity spectra; ③ they have selective toxicity, meaning they are ineffective against mammalian cells at the concentration harmful for microorganisms; ④ they have a mode of action that should restrict the selection of resistant strains. In addition, expression of AMPs in transgenic vectors such as mosquitoes is a new approach for killing parasites or blocking parasite transmission.

Many AMPs such as Magainin, a 23-residue peptide extracted from the skin of *Xenopus laevis*, and Indolicidin, a 13-residue antimicrobial peptide isolated from the cytoplasmic granules of bovine neutrophils have reached clinical trials. To date, some AMPs have been approved by the US Food and Drug Administration (FDA). Gramicidin A, Gramicidin S, Colistin A, Polymyxin B, and Cubicin (daptomycin) are in clinical use. Lysozyme, nisin A, and pediocin PA-1 (AcH) are used as food preservatives.

Antimicrobial peptides have been engineered in plants to confer resistance to bacterial and fungal pathogens. For example, Cecropin is expressed in a variety of transgenic plants to confer resistance to bacterial and fungal pathogens. Transgenic rice with Cecropin B has enhanced resistance to bacterial fusarium wilt. After introducing Cecropin D gene into tomato and mulberry plants, the transgenic plants showed bacterial wilt resistance. The transgenic pepper of *Antheraea pernyi* Cecropin D has the antibacterial ability. The transgenic pepper is considered to be safe by isogeneity test. Proline-rich peptide Metchnikowin

抗生素耐药性将是未来对人类健康的最大威胁之一。抗菌肽（AMPs）被认为是抗生素的优良替代品，可以克服细菌耐药性和条件致病菌的出现，因为它们具有以下几个优势：① 能迅速杀死目标微生物；②具有广泛的活性谱；③具有选择性的毒性，这意味着在对微生物有害的浓度下对哺乳动物细胞没有作用；④不同于抗生素的作用机制，应该可以限制耐药菌株的产生。此外，抗菌肽在转基因媒介昆虫，如蚊子中表达是杀灭寄生虫或阻断寄生虫传播的新途径。

许多抗菌肽，如在非洲爪蟾皮肤中发现的 23 个残基抗菌肽爪蟾抗菌肽和从牛嗜中性粒细胞的细胞质颗粒中纯化 13 个残基的抗菌肽 Indolicidin 已经进行了临床实验，部分抗菌药物已获美国食品及药品监督管理局（FDA）批准。目前，已经在临床使用的是黏菌素 A、短杆菌肽 A、短杆菌肽 S、克利斯汀 A、多黏菌素 B 和克必信 (daptomycin)。另外，溶菌酶、nisin A 和 pediocin PA-1（AcH）被用作食品防腐剂。

抗菌肽也被应用于植物的转基因抗病育种，以赋予植物对细菌和真菌病原体的抗性。天蚕素在多种转基因植物中表达赋予细菌和真菌病原体的抗性，研究表明天蚕素 B 的转基因水稻对细菌引起的水稻叶枯萎病具有增强的抗性；将天蚕素 D 基因导入番茄和桑树植株后，转基因植株显示青枯病抗性；柞蚕抗菌肽 D 转基因辣椒具有抗菌能力，实质等同性检测认为转基因辣椒是安全的。富含脯氨

has been expressed in transgenic barley to enhance resistance. Transgenic expression of an insect defensin (*G. mellonella* Gallerimycin) and Cecropin (sarcotoxin-IA) in tobacco confers resistance to pathogenic fungi. Chimeric peptides by combining active regions of two AMPs have also been used in transgenic plants to enhance resistance or broaden spectrum of resistance to pathogens.

(Deng Xiaojuan)

酸肽 Metchnikowin 已在转基因大麦中表达以增强抗性。在烟草中转基因表达昆虫防御素（大蜡螟防御素，*G.mellonella* Gallerimycin）和天蚕素（sarcotoxin-IA）使烟草对病原真菌产生抗性。为增强抗菌活性或拓宽抗菌谱，结合两个抗菌肽活性区域的嵌合肽也被应用于植物的转基因工作。

（邓小娟）

Exercises

1. Describe the component of innate immunity in insects.
2. Describe the components and their functions of insect cellular immunity.
3. The activation process of insect humoral immunity by microbial infection.
4. What is the general characteristics of insect antimicrobial peptides ?
5. Describe the main signal transduction pathways regulating the expression of antimicrobial peptide genes.

思考题

1. 试述昆虫先天免疫的组成。
2. 试述昆虫细胞免疫的组成及各部分的功能。
3. 试述细菌感染激活昆虫体液免疫的基本过程。
4. 抗菌肽的基本特征有哪些?
5. 试述昆虫体内调控抗菌肽表达的主要信号转导通路。

第 8 章 神经解剖与神经肽

Chapter 8 Neuroanatomy and Neuropeptide

Key points

1.The composition of the brain and nervous system of insects has species diversity, but the anatomical morphology and characteristics are basically the same, such as brains and nerve cords, neurons, glial cells, nerve neurolemma, etc., forming sensory neurons, association neurons and motor neurons.

2.A more subtle understanding of the mushroom body, central body complexes and neural connections in the protocerebrum of insects is a highlight of insect physiology research in recent years.

3.The cognitive ability of insects, the use of the environment and the surrounding magnetic field are intelligent models of insects' long-distance foraging and directional long-distance migration. Insects can use the solar compass to navigate during their lives, and use polarized light.

4.More and more insect neuropeptides have been discovered through in-depth research, which are almost involved in all physiological and biochemical reactions and behaviors of insects.

知识点

1. 昆虫的脑和神经系统的组成具有物种多样性，但解剖学形态和特征大体相同，如脑和神经索、神经元、神经胶质细胞、神经鞘等，形成感觉神经元、联络神经元和运动神经元。

2. 对昆虫前脑中的蕈状体和中心体复合体及神经连接等更细微的认识，是近年昆虫生理学研究的一大亮点。

3. 昆虫的认知能力，对环境和周围磁场的利用是昆虫远距离觅食和定向长途迁徙的智能模式。昆虫在生活期间能使用太阳罗盘导航，并利用偏振光等。

4. 深入研究发现，越来越多的昆虫神经肽几乎都参与昆虫的所有生理生化反应和行为。

Insect behavior and movement are dominantly controlled under nerve system which directly or indirectly influence on endocrine system, and regulate insect growth and development. The environmental information was detected by varies sensory receptors of insects and highly integrated through brain and segmental nervous cord. It elaborately regulates physiology of insects. Insect nervous system is derived from ectodermal tissue in the embryo, and structurally classified as central nervous system (CNS), peripheral nervous system and sympathetic nervous system; and functionally it could be classified as sensory neurons, association neurons (interneurons) and motor neurons.

昆虫的神经系统支配昆虫的行为和运动，且直接或间接影响内分泌系统，对昆虫的生长发育具有调控作用。昆虫的各种感受器官感受的环境信息通过脑和各体节神经索的高度综合，精细地调节昆虫生理。昆虫神经系统由外胚层发育而来，按结构分为中枢神经系统（CNS）、外周神经系统和交感神经系统。神经细胞按功能分为感觉神经元、联络神经元（中间神经元）和运动神经元。

Bionics is one of the hot topics in entomological research. Zoologists have studied the nervous system of insects to study the principle of biological brain computing and its potential impact on machine learning and artificial intelligence. Specifically, they analyzed how insects learned to associate sensory information in their surroundings with food rewards, and how insects could recall this information later to solve complex tasks such as finding food. The results show that transforming sensory information into memory in the brain can stimulate future machine learning and artificial intelligence applications to solve complex tasks.

仿生学是昆虫学研究的热门之一。动物学家研究了昆虫的神经系统，以研究生物大脑计算的原理以及对机器学习和人工智能的潜在影响。他们特别分析了昆虫如何学会将周围环境中的感官信息与食物奖励相关联以及昆虫如何可以回忆起这些信息以解决诸如寻找食物之类的复杂任务。结果表明，将感觉信息转换为大脑中的记忆可以激发未来的机器学习和人工智能应用，以解决复杂的任务。

8.1 中枢神经系统

8.1 The Central Nervous System (CNS)

The central nervous system (CNS) of insects consists of the brain, the ventral ganglia, and the ventral nerve cord (Fig. 8-1). The brain is in the head in most insects, but it is located several body segments posterior to the anterior end of the body in dipterous larvae. Some insects experimentally rendered headless, or merely brainless, can live for long periods of time and may even walk, mate and lay eggs.Removal of the brain, entire head, or other parts of the nervous system, while keeping the insect alive, has been a useful approach in the study of hormones produced by

昆虫的中枢神经系统由脑、腹部神经节和腹部神经索组成（图 8-1）。昆虫的脑大多数位于头部，但有些双翅目昆虫的脑位于体节的后部到前端位置。实验性地使一些昆虫没有头部或只是没有脑同样能存活一段时间，甚至可以行走、交配和产卵。移除脑、整个头部或神经系统的其他部分而保持昆虫存活，对研究神

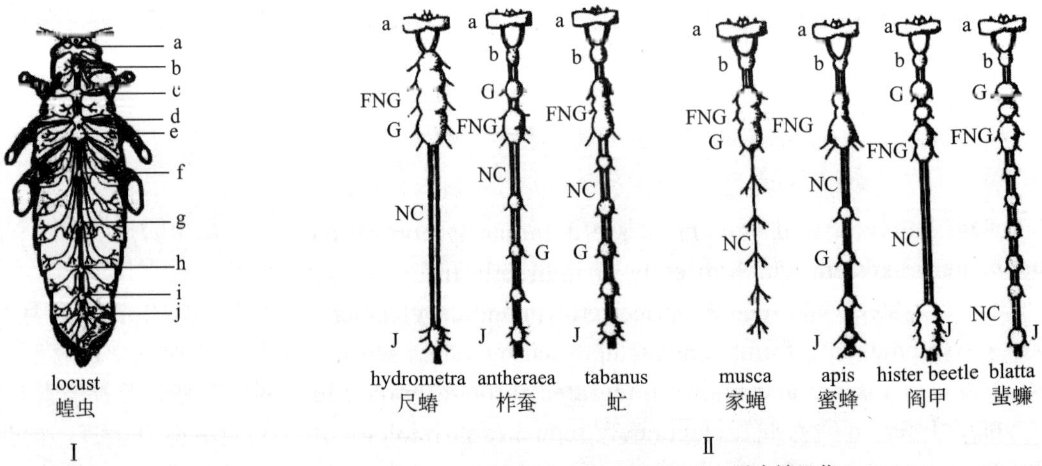

a. brain 脑
b. subesphageal ganglion 咽下神经节
c-e. pro-,meso- and meta-thoracic ganglion 前、中、后胸神经节
f-j. the 1st to 5th abdominal ganglion 第 1~5 腹神经节

FNG. fused nervous ganglion 融合神经节
J. TAG.terminal abdo-minal ganglion 腹末神经节
NC. nerve cord 神经索
G. ganglia 神经节

Fig. 8-1 Insect nervous system structure (modified from Wang Yinchang et al., 1991).

图 8-1 昆虫的神经系统结构（仿王荫长等，1991）。

the nervous system.

Each insect species is adapted for a specific habitat, which results in a rich diversity of morphology, body size, physiology, and behavior.

The insect brain consists of six neuromeres, which are arranged in two different ways. In hemimetabolous insects (such as locusts and cockroaches), and in beetles (holometabolous), the brain is separated into the cerebral ganglia (CRG) and the gnathal ganglia (GNG). The CRG, containing the protocerebrum (PR), deutocerebrum (DE), and tritocerebrum (TR), lie above the esophagus, whereas the mandibular, maxillary, and labial ganglia, comprising the GNG, are below the esophagus and are separated from the CRG through connectives. In many holometabolous insects, such as flies, bees, and moths, however, the CRG and GNG are fused into one ganglionic mass. Each neuromere contains neuropils defined as synapse-rich areas and a cell body rind surrounding the neuropils. The CRG consists of the cerebrum between two visual processing centers, the optic lobes (OL). The cerebrum contains several prominent neuropils: the antennal lobe (AL), the mushroom body (MB), the lateral horn (LH), the anterior optic tubercle (AOTU), and the antennal mechanosensory and motor center (AMMC) in each brain hemisphere, and the central complex (CX) connected to two lateral complexes in the center. Other areas, separated by fiber bundles and the prominent neuropils, are named based on relative position in the cerebrum, such as the superior (SNP), ventrolateral (VLNP), inferior-, ventromedial- and periesophageal neuropils.

A ganglion typically contains a mass of neuronal cell bodies of inter- and motoneurons at the periphery and a central region, the neuropil, where synapses occur. The cell bodies of sensory neurons of insects are usually near the site of sensory stimulus reception, and consequently, many sensory neurons cell bodies are located peripherally in the cuticle and in or on internal organs.

8.1.1 The brain

The brain consists of three fused ganglionic masses: the protocerebrum, the deutocerebrum, and the tritocerebrum which typically rest on top of the esophagus and passes posteriorly between the circumoesophageal connective to the subesophageal ganglion. Together, the three parts are sometimes called the supraesophageal ganglion.

经系统产生的激素是有用的途径。

每种昆虫都适应特定的栖息地，这导致了形态、体型、生理和行为的丰富多样性。

昆虫的大脑由 6 个神经原节组成，以两种不同的方式排列。在不完全变态昆虫（如蝗虫和蟑螂）以及在甲虫（完全变态）中，大脑分为脑神经节（CRG）和颚神经节（GNG）。包含前脑（PR）、中脑（DE）和后脑（TR）的 CRG 位于食道上方，而构成 GNG 的下颌、上颌和唇神经节位于食道下方，通过结缔组织与 CRG 分离。然而，在许多完全变态昆虫中，如苍蝇、蜜蜂和蛾，CRG 和 GNG 融合成一个神经节团。每个神经原节包含定义为富含突触区域的神经毡和围绕神经毡的细胞体。CRG 由两个视觉处理中心视叶（OL）之间的大脑组成。大脑包含几个突出的神经毡：触角叶（AL）、蕈状体（MB）、侧角（LH）、前视结节（AOTU）和每个大脑半球的触角机械感觉和运动中心（AMMC），以及连接到中心两个侧面复合体的中心复合体（CX）。其他区域由纤维束和突出的神经毡分隔，根据大脑中的相对位置命名，例如，上神经毡（SNP）、腹外侧神经毡（VLNP）、下神经毡、腹内侧神经毡和围咽神经毡。

典型的神经节包括外周的一群联络神经和运动神经的细胞体和中心区的神经毡，是突触发生的部位。昆虫感觉神经元的细胞体通常位于感受器刺激接收点附近，因此，许多感觉神经元细胞体位于表皮的外周和内部器官上或内部器官内。

8.1.1　脑

脑由三个神经节群组成：前脑、中脑和后脑，常位于食道上方，从咽下神经节及围咽神经连接间的后部通过。有时三部分一起统称为咽上神经节。

8.1.1.1 Protocerebrum

The protocerebrum is the site of major integrative centers that process incoming information from many sensory sources. It mainly consists of the mushroom body, central complex, and the optic lobes. The optic lobes contain several neuropil regions related to visual processing, which process information from the compound eyes. The protocerebrum also receives input from ocelli *via* the ocellar nerves. Small paired nerves, the nervous corporis cardiac I and II, link the protocerebral neurosecretory cells with the corpora cardiaca and corpora allata.

1.The mushroom bodies (MBs)

MBs are large, bilateral integrative centers in the protocerebrum (Fig.8-2). The MBs are divided into the calyx, the peduncle, and the root. The mushroom bodies (MBs) are paired and lobed neuropils in the lateral protocerebrum with mushroom- like shapes. They serve as multimodal integration centers for the processing of olfactory and other sensory information and have a well-established role in olfactory learning and memory. The locust MB consists of six major parts, a calyx, an accessory calyx, a pedunculus, a vertical lobe, the spur and a medial lobe. The calyx is a single cup-shaped neuropil in the superior protocerebrum and extends across the complete anterior–posterior dimension of the brain. Posteriorly, the rim of the calyx is interrupted, giving rise to the accessory calyx, which is subdivided into a bulbous neuropil, the accessory calyx bulb in the posterior part of the calyx, a central part between accessory calyx bulb and pedunculus termed accessory calyx core and a ring-shaped neuropil termed accessory calyx ring forming a collar around the pedunculus on its posterior, lateral and medial

8.1.1.1 前脑

前脑是昆虫对许多感觉器官收集到的信息进行综合的主要中心。主要由蘑菇状的蕈状体、中心体和视叶组成。视叶含几个与视觉有关的神经毡区，处理来自复眼的视觉信息。前脑也接收通过单眼神经传入的单眼信息。成对的小神经，心侧体神经 I 和心侧体神经 II 把脑神经分泌细胞、心侧体和咽侧体相连。

1. 蕈状体（MBs）

MBs 是前脑内对称的综合中心（图 8-2）。蕈体分蕈冠、蕈柄和蕈根三部分。蕈状体（MBs）是位于外侧前脑的成对且浅裂的神经毡，形状类似蘑菇。它们是嗅觉和其他感觉信息处理的多种模式的整合中心，在嗅觉学习和记忆中具有公认的作用。蝗虫 MB 由蕈冠、副蕈冠、蕈柄、垂叶、距和内侧叶六大部分组成。蕈冠是位于上前脑的一个杯状神经纤维网，延伸至整个大脑的前后维度。从后面看，蕈冠边缘被打断，产生副蕈冠，副蕈冠又分为球茎状神经毡，蕈冠后部的副蕈冠球，副蕈冠球和蕈柄之间的中心部分称为副蕈冠核和一个环形神经毡称为副蕈冠环，在其后部、外侧和内侧的蕈柄周围形成一个项圈，从

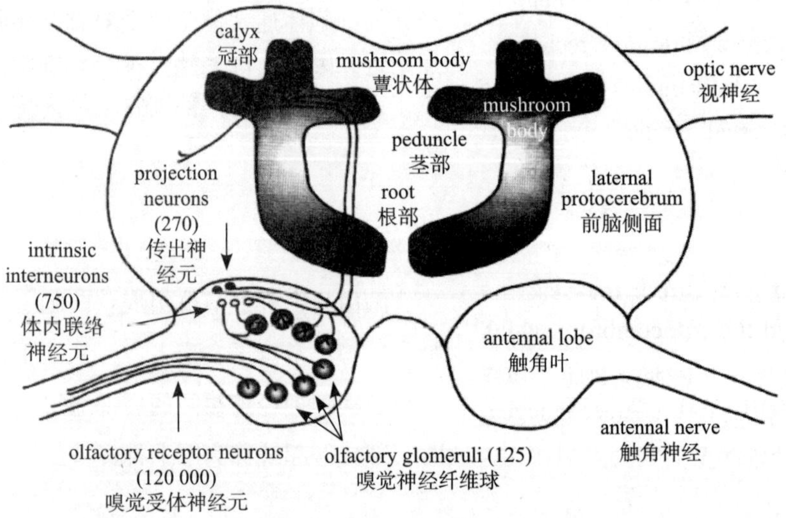

Fig. 8-2 A frontal section through the brain of a female cockroach illustrating the antennal lobes and mushroom bodies (modified from Lemon et al., 1999).

图 8-2 展示雌蜚蠊脑部触角叶和蕈状体的前切面图（仿 Lemon 等，1999）。

side as it emerges ventrally from the calyx. The pedunculus arises between calyx and accessory calyx core, extends in anterior ventral direction and connects the calyx neuropils *via* the mass of Kenyon cell fibers with the two lobes. Internally, the pedunculus consists of two large medial strands flanked by two smaller lateral strands of fibers. At the bifurcation site of the pedunculus into the vertical lobe and medial lobe, a lateral protrusion, the globular spur, arises. The vertical lobe is located anterio-lateral to the central complex and is subdivided into six tube-like structures while the medial lobe emerges ventrally from the pedunculus. It extends toward the midline of the brain dorsal to the antennal lobe and anterior from the lateral complex and is built by up to three strands. The size of the protocerebrum dedicated to the MBs varies in insects. The peduncle contains fibers of neurons going to and from the calyx, a synaptic region.

蕈冠向腹侧伸出。蕈柄位于蕈冠和副蕈冠核之间，向前腹侧延伸，通过大量的肯扬细胞纤维与两个叶连接蕈冠神经毡。在内部，蕈柄由两条较大的内侧纤维束和两条较小的外侧纤维束组成。在蕈柄进入垂叶和内侧叶的分叉部位出现一个外侧突起，即球状突起。垂叶位于中心复合体的前外侧，并细分为六个管状结构，而内侧叶从蕈柄向腹侧伸出。它向触角叶背侧和外侧复合体前方的大脑中线延伸，由多达三股组成。不同昆虫蕈状体大小不一，并且决定了前脑的大小。蕈柄包含的神经纤维进出突触区蕈冠。

MBs have an ensemble of synaptic connections which play important role in experience-dependent learning and several higher cognitive functions. These integrative centers are believed to be involved in olfactory learning through connections and input from the olfactory lobe integrative centers in the deutocerebrum, the region that receives olfactory input from the antennae. MB requires neurotransmission for an efficient flow of information across synapses with different flexibility to meet the demand of the dynamically changing environment of an insect. Neurotransmitter transporters coordinate appropriate changes for an efficient neurotransmission at the synapse. The MBs calyces of ants and other Hymenopterans receive direct inputs from the optic lobes, and involve in visual navigation to learned locations. The vertical lobes, which are innervated by the calyces, are also required for visual navigation. In ants, MBs are required for learned visual navigation, but not for innate visual behavior. Lesions of MB calyces impair ants' visual navigation to a remembered food location yet leave their innate responses to visual cues unaffected. However, fruit flies appear to depend upon the central complex to complete a place learning task and some other visual learning paradigms, although the MBs have been shown to have a role for some visual associations. This suggests that the locus of visual memories within the insect brain is likely task dependent or may differ among insects from different orders.

蕈状体由一整套突触连接，在经验依赖性学习和一些高级认知功能中起着重要作用。通过与中脑的连接和接收来自中脑嗅叶综合中心的信息，这些综合中心被认为与昆虫的嗅觉认知有关。蕈状体需要神经传递以实现信息在突触间的有效转导，并具有不同的灵活性，以满足昆虫应对动态变化环境的需要。神经递质转运体协调适当的变化，以便在突触上进行有效的神经传递。蚂蚁和其他膜翅目昆虫的蕈状体的蕈冠接收来自视叶的直接输入，参与了学习定位的视觉导航。由蕈冠支配的垂叶也是视觉导航所必需的。蕈状体是蚂蚁学习视觉导航所必需的，但不是先天视觉行为所必需的。蕈冠的损伤削弱了蚂蚁对记忆中食物位置的视觉导航，但不影响它们对视觉信号的先天反应。然而，果蝇似乎依赖中心复合体来完成位置学习的任务以及其他一些视觉学习模式。这表明昆虫大脑中的视觉记忆轨迹可能与任务有关，或者在不同目昆虫中可能有所不同。

In addition, the mushroom bodies were the first neuropils to be considered as critical regulators of insect sleep. The pars intercerebralis and pars lateralis followed, the neuropeptide

此外，蕈状体最先被认为是昆虫睡眠关键调节器的神经毡，随后，是大脑间部和外侧部的时钟神经元单独表达色

pigment-dispersing factor (PDF)- expressing clock neurons alone are wake-promoting components of the *Drosophila* sleep circuit. Finally, ellipsoid body, and fan-shaped body of the central body is part of the *Drosophila* sleep homeostat.

2.The Central body complex

Another major neuropil region in the central part of the protocerebrum is the central body complex, which is located between the bases of the stalks of the mushroom bodies. The central complex is a group of closely interconnected neuropils located across the brain midline. It consists of four neuropils in *Schistocerca*: the central body, subdivided into an upper division (CBU) and a lower division (CBL) (Fig.8-3A and Fig.8-3B; Appendix color graphic Fig.8-3.); however, it composed of an ellipsoid and fan-shaped body in flies (Fig.8-3D), the protocerebral bridge (PB) and the paired noduli. Internally, the PB, CBL and CBU are subdivided by columnar neurons into eight vertical slices per hemisphere. In addition, the central body subdivisions are composed of horizontal layers. Three layers have been distinguished in the CBU and six layers in the CBL. The most dorsal and posterior neuropil is the PB, shaped like a handlebar with ends bent posterior-ventrally. The central body is located more anterior-ventrally, while the nodules are ventrally attached to the central body. The nodules have been divided into four layers. The anterior lip is closely connected to the central complex. It is a slender horizontal neuropil extending across the brain midline, located ventrally to the CBU and anteriorly from the CBL. Lateral to the central body, a group of four closely connected neuropils forms the lateral complex. The anterior upper shell and the posterior lower shell of the lateral accessory lobe, are separated by several loosely bundled fascicles, the isthmus tracts. Partly embedded and surrounded by fibers of the isthmus tracts are two small microglomerular neuropils, the lateral and medial bulbs. The bulbs are connected to the CBL *via* tangential neurons of the isthmus tracts. In *Drosophila*, a small synapse-rich area at the superior lateral tip of the lateral accessory lobe has been recognized and termed gall. In *Schistocerca*, the gall cannot be distinguished from surrounding lateral accessory lobe neuropil by synapsin immunofluorescence but as in the fly can be identified as a specific area within the lateral accessory lobe supplied by axonal projections from columnar neurons of the CBL. On the basis of principal differences in arborization patterns, four categories

素分散因子（PDF）的神经肽，是果蝇睡眠回路的促醒成分。最后，果蝇中心体的扇形体和椭球体是维持其睡眠稳态的一部分。

2. 中心复合体

另一个前脑中心部分的主要神经毡区是中心复合体，位于蕈状体的两个蕈柄基部之间。中心复合体是一组紧密相连的神经毡，位于大脑中线。它由四个神经毡组成：中心体，分为上区（CBU）和下区（CBL）（图 8-3A 和图 8-3B；附彩图 8-3）；但在果蝇中为椭球体和扇形体（图 8-3D），前脑桥（PB）和成对结节。在内部，PB、CBL 和 CBU 由柱状神经元细分为每个半球八个垂直片。此外，中心体细分由水平层组成。CBU 中有三层，CBL 中有六层。最背部和后部的神经毡是 PB，形状像把手，末端向后弯曲。中心体位于更前面的腹侧，而结节腹侧附着于中心体。结节分为四层，前唇与中心复合体紧密相连。它是一个细长的水平神经毡，横跨大脑中线，位于 CBU 腹侧和 CBL 前方。在中心体的外侧，一组由四个紧密相连的神经毡组成外侧复合体。外侧副叶的前上壳体和后下壳体，由几个松散成束的小束即峡部束分隔。峡部束的纤维部分嵌入并包围着两个小的微珠状神经毡，即外侧球和内侧球。球茎通过峡部束的切向神经元连接到 CBL。在果蝇中，在外侧副叶的上外侧顶端有一个小的富含突触的区域，被认为是瘿。在沙漠蝗中，通过突触素免疫荧光法不能将瘿与周围的外侧副叶神经末梢区分开来，但在果蝇中，可以确定为外侧副叶内的一个特定区域，由 CBL 柱状神经元的轴突投射提供。根据树状结构模式的主要差异，四类神经元可以被区分：切向神经元、无长突神经元、脑桥神经元和柱状神经元。前脑桥的每一片

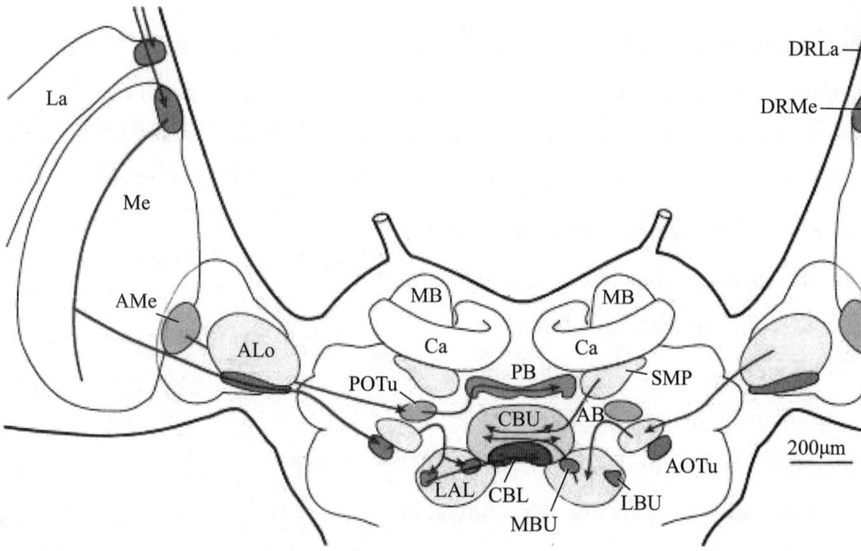

A.illustrating fiber connections of the central complex
A. 显示中心复合体的神经纤维连接

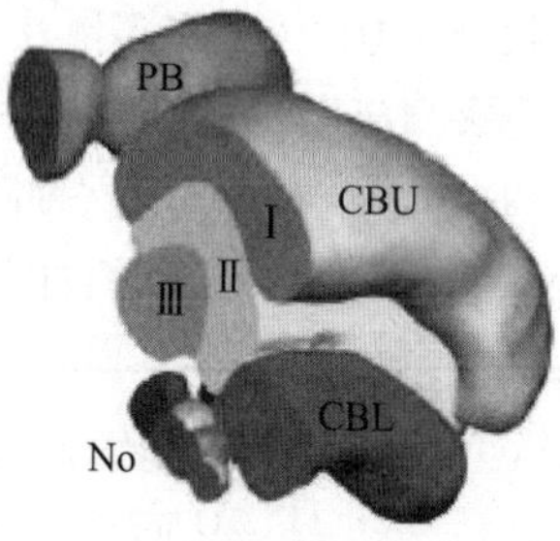

B.Oblique three-dimensional view of the locust central complex
B. 蝗虫中心复合体的倾斜三维视图

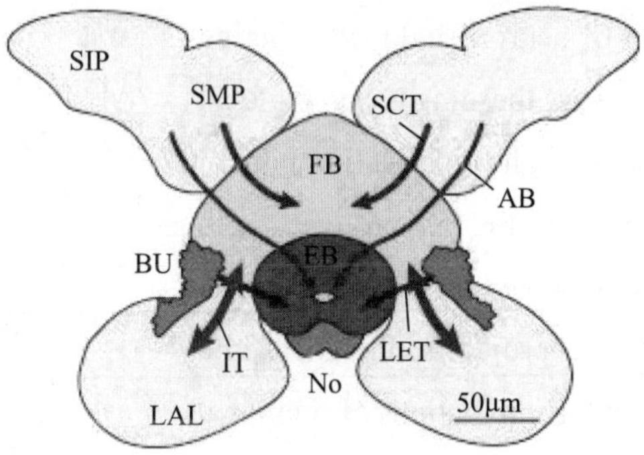

C.Fiber connections of the central body of the blow fly *Lucilia sericata*
C. 丝光绿蝇中心体的神经纤维连接

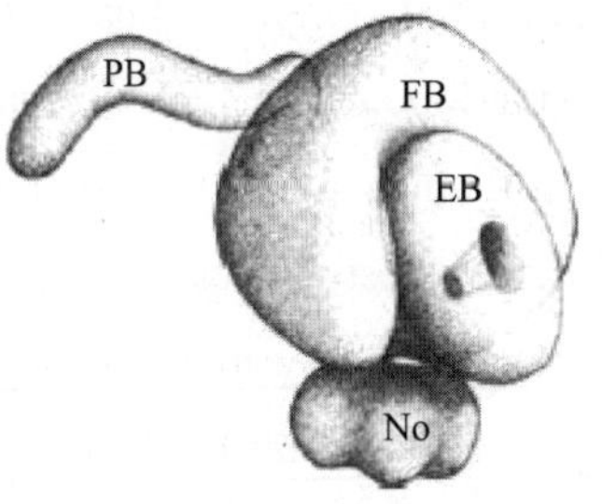

D.Oblique lateral view illustrating the subunits of the central complex of the fruit fly, *Drosophila melanogaster*
D. 侧视图显示果蝇中心复合体的亚单位

Fig. 8-3. Subunits and connections of the insect central complex. A.illustrating fiber connections of the central complex. The anterior polarization pathway connects the dorsal rim area of the lamina and medulla (DRLa, DRMe) *via* the ventral layer of the anterior lobe of the lobula (ALo), the lower unit of the anterior optic tubercle (AOTu), and the medial (MBU) and lateral bulb (LBU) to the lower division of the central body (CBL). A parallel pathway originating in the ALo connects *via* the upper unit of the AOTu and lateral accessory lobe (LAL) to the upper division of the central body (CBU). The superior medial protocerebrum is connected to the CBU *via* the anterior bundles (AB). Finally, projections from the accessory medulla (AMe) extend to the posterior optic tubercle (POTu) and likely target tangential neurons entering the protocerebral bridge (PB). B.Oblique three-dimensional view of the locust central complex. A sagittal cut reveals the positions of the PB, the three layers of the CBU (Ⅰ , Ⅱ , Ⅲ), the CBL, and the four · subunits of the noduli (NO). C.Fiber connections of the central body of the blow fly *Lucilia sericata*. The bulbs (BU) are connected through the lateral ellipsoid tracts (LET) to the ellipsoid body (EB). The isthmus tracts (IT) provide connections between the LALs and the fan-shaped body (FB). The superior intermediate protocerebrum (SIP) and superior medial protocerebrum (SMP) are co1mected to the FB *via* the AB and the superior central-body tract (SCT). D.Oblique latera l view illustrating the subunits of the central complex of the fruit fly, *Drosophila melanogaster* (modified from Von Hadeln et al., 2018).

图 8-3. 昆虫中心复合体的亚单位和连接。A. 显示中心复合体的神经纤维连接。前极化通路通过小叶前叶腹侧层（ALo）、前视突下单元（AOTu）、内侧球（MBU）和外侧球（LBU）连接椎板和髓质的背缘区域（DRLa，DRMe）和中心体下部（CBL）。起源于 ALo 的一条平行通路通过 AOTu 上部单位和外侧副叶（LAL）与中心体上部（CBU）相连。前脑上内侧通过前束（AB）与 CBU 相连。副髓质（AMe）的投射神经元延伸至后视突（POTu），并可能以进入前脑桥（PB）的切向神经元为靶点。B. 蝗虫中心复合体的倾斜三维视图。矢状切面显示了 PB、CBU 的三层（Ⅰ、Ⅱ、Ⅲ）、CBL 和结节（NO）的四个亚单位的位置。C. 丝光绿蝇中心体的神经纤维连接。球茎（BU）通过外侧椭球束（LET）与椭球体（EB）相连。峡部束（IT）在 LAL 和扇形体（FB）之间提供连接。上中间前脑（SIP）和上内侧前脑（SMP）通过 AB 和上中心复合体束（SCT）与 FB 相连。D. 侧视图显示果蝇中心复合体的亚单位（仿 Von Hadeln 等，2018 年）。

of neurons can be distinguished: tangential neurons, amacrine neurons, pontine neurons, and columnar neurons. Each slice of the protocerebral bridge is interconnected with a slice of the ellipsoid body *via* so called E-PG neurons. It also receives input from the optic lobes. There are also paired lateral neuropil regions in the protocerebrum. Internal commissures or fiber tracts connect parts of the protocerebrum and deutocerebrum with each other.

The central complex is an integration center for visual and motor information and therefore provides the neural machinery necessary for processing angular-dependent orientation behavior. The E-PG neurons not only encode landmarks in walking flies but also process visual cues while the insect is flying.

The most prominent pathway for sending compass information to the central complex comprises a set of neurons indirectly linking the optic lobe to the CBL. In many insects, these cells carry global compass information derived from the skylight polarization pattern and other celestial cues. The neural substrate for this proposed path integration memory are columnar neurons of the CBU, which connect the protocerebral bridge and the noduli to project to the CBU in a regular pattern. These cells encode specific combinations of rotational and translational movements and thereby encode future trajectories. They converge in a brain region called the lateral accessory lobe, which is known to contain premotor neurons in the silk moth *B. mori*. These moth neurons show so-called flip-flop activity that is characterized by sustained periods of alternating high and low firing rates. The transition between the two states is initiated by sensory information.The most commonly short pheromone pulses, can also be triggered by light flashes. As neck motor neurons show flipflop activity as well and head-turns often precede body-turns, the flip-flop circuit is considered a control network for turning movements during moth odor plume tracking. Mirroring the neural activity, this tracking behavior features typical zigzagging movements, during which the moth passes through the odor plume multiple times and reverses flight direction after each exposure to the plume.

3. The superior neuropils

The superior neuropils occupy dorsal regions of the brain. They lie ventrally from the calyces, occupy the dorsal space between the pedunculus and vertical lobe of the mushroom body and extend medially and laterally around the pedunculus. The superior neuropils

都通过被称为E-PG的神经元与椭球体的一片相连。它也能接收来自视叶的信息。前脑也有成对的侧神经毡区。内部接合处或纤维束将前脑和中脑相互连接起来。它在处理和整合与空间方向相关的感官信息方面发挥着重要作用，涉及天空罗盘方向、路径整合、物体的空间表示和识别以及位置学习。

中心复合体是视觉和运动信息的整合中心，因此为处理角度相关的定向行为提供了必要的神经机制。E-PG神经元不仅在行走中的果蝇进行地标编码，而且处理昆虫飞行时的视觉信号。

向中心复合体发送罗盘信息的最重要途径是一组神经元，这些神经元将视叶间接连接到CBL。在许多昆虫中，这些细胞携带来自天空偏振光模式和其他天体信号的地球罗盘信息。这种路径整合记忆的神经基础是CBU的柱状神经元，它连接前脑桥和结节，以规则的方式传送到CBU。这些细胞编码旋转和平移运动的特定组合，从而编码未来的轨迹。它们汇聚在一个称为侧副叶的大脑区域，已知家蚕的该区域包含前运动神经元。这些蛾类神经元表现出所谓的触发器活动。其特征是持续的高、低频交替放电。这两种状态之间的转换是由感觉信息启动的。最常见的短信息素脉冲也可以由闪光触发。由于颈部运动神经元表现出触发器活动，并且头部转动通常先于身体转动，因此触发器回路被视为蛾类气味缕流跟踪期间转向运动的控制网络。与神经活动相呼应，这种跟踪行为反映了神经活动，具有典型的曲折运动特征，在此期间，蛾多次穿过气味缕流，每次接触到气味缕流后都会反转飞行方向。

3. 上神经毡

上神经毡占据大脑的背侧区域。它们位于蕈冠的腹侧，占据蕈柄和蕈状体垂叶之间的背侧空间，并围绕蕈柄向内侧和侧向延伸。上神经毡由上

consist of the superior lateral protocerebrum, the superior intermediate protocerebrum, the superior medial protocerebrum and the anterior bridge.

4.The inferior neuropils

The inferior neuropils occupy the region ventral to the superior neuropils and partly surround the lobes and pedunculus of the mushroom body as well as the central body. The inferior neuropils consist of the crepine (CRE), the clamp, divided into the superior clamp and inferior clamp, the medial accessory lobe, the inferior bridge, the antler, the neck and the ocellar root.

5. The lateral horn

The lateral horn is a roughly spheroidal neuropil in the lateral cerebrum and is defined by massive projections from the medial antennal lobe tract.

6.Neuropils in the ventrolateral cerebrum

Neuropils in the ventrolateral cerebrum consist of the ventrolateral protocerebrum, the posterior lateral protocerebrum, the wedge and the anterior optic tubercle. The anterior optic tubercle is involved in the processing of visual information. It is divided into an ovoid upper unit that can be further subdivided into two distinct parts and a smaller, similarly shaped lower unit.

7.The ventromedial neuropils

The ventromedial neuropils are located ventral to the central complex and cover the whole posterior–medial cerebrum. They include the ventral complex consisting of vest and epaulet, the gorget, the posterior slope and the posterior optic tubercle. The vest is positioned anterior to the posterior slope, ventral of the great commissure, and ventro-lateral of the neck and is framed lateral by the wedge, epaulet and antennal mechanosensory and motor center.

8.The periesophageal neuropils

The periesophageal neuropils are located lateral of the esophagus. They include the antennal lobe (AL) with the median crescent, the AL mechanosensory and motor center (AMMC), the glomerular lobe and the tritocerebrum. The AMMC comprises several neuropils that are closely connected to each other: the dorsal (DAMMC), lateral (LAMMC) and medial (MAMMC), as well as the ventral area of flagellar afferents. The AL has a cup-shaped appearance with its opening pointing anterior-ventrally. The AL is the termination site of olfactory afferents of the antennal flagellum. Its frontal opening, like the mushroom body calyx, receives the cell body fibers from local and projection neurons, whose cell bodies are clustered in the soma rind anterior to the AL neuropil.

外侧前脑、上中间前脑、上内侧前脑和前桥组成。

4. 下神经毡

下神经毡占据上神经毡腹侧区域，部分环绕蕈状体的叶和蕈柄以及中心体。上神经毡由网穗（CRE）、夹钳组成，分为上夹钳和下夹钳、内侧副叶、次桥、鹿角、颈部和眼状根。

5. 外侧角

外侧角是外侧大脑中一个大致呈球状的神经毡，由内侧触角叶束的大量投射神经所定义。

6. 大脑腹外侧的神经毡

大脑腹外侧的神经毡由腹外侧前脑、后外侧前脑、楔状突起和前视结节组成。视前结节参与视觉信息的处理。它分为一个卵形上部单元，可进一步细分为两个不同的部分和一个较小、形状相似的下部单元。

7. 腹内侧神经毡

腹内侧神经毡位于中心复合体的腹侧，覆盖整个后内侧脑。它们包括由坎肩和肩袢、饰领、后斜坡和后视结节组成的腹侧复合体。坎肩位于后斜坡的前部、大接合处的腹侧和颈部的腹外侧，由楔形物、肩袢和触角机械感觉和运动中心侧向固定。

8. 围咽神经毡

围咽神经毡位于食管外侧。它们包括有中间新月形触角叶（AL）、触角机械感觉和运动中心（AMMC）、小球叶和后脑。AMMC 包括几个紧密相连的神经：背侧（DAMMC）、外侧（LAMMC）和内侧（MAMMC）以及鞭状传入的腹侧区。AL 有一个杯形外观，开口点在前腹部。AL 是触角鞭毛嗅觉传入的终点。它的额叶开口，如蕈状体冠部，接收来自局部和投射神经元的细胞体纤维，其神经元的细胞体聚集在体细胞外皮前端到 AL 神经毡。

9.The glomerular lobe

The glomerular lobe is a prominent spheroidal neuropil medioventrally from the AL. It is densely supplied by afferent input, likely from gustatory fibers, of the maxillary and labial palps and is specifically connected *via* the glomerular lobe-calycal tract to the accessory calyx of the mushroom body.

10.The lateral complexes

The lateral complexes (LXs) are bilaterally paired neuropils in the insect brain that mediate communication between the central complex (CX), a brain center controlling spatial orientation, various sensory processing areas, and thoracic motor centers that execute locomotion. The LX of the desert locust consists of the lateral accessory lobe (LAL), and the medial and lateral bulb. Analyzing the morphology of neurons connecting the CX and the LX revealed likely feedback loops in the sky compass network of the CX *via* connections in the gall of the LAL and a newly identified neuropil termed ovoid body. In addition, 16 different types of neurons connect the LAL with other areas in the brain. Eight types of neurons provide information flow between both LALs, five types are LAL input neurons, and three types are LAL output neurons. Among these are neurons providing input from sensory brain areas such as the lobula and antennal neuropils. Brain regions most often targeted by LAL neurons are the posterior slope, the wedge, and the crepine. Two descending neurons with dendrites in the LAL were identified. The neuronal network involved in processing of sensory information and generation of appropriate behavioral output for goal directed locomotion.

9. 小球叶

小球叶是一个突出的球状神经毡，位于腹内侧。它由上颌骨和唇须的传入神经纤维，可能来自味觉纤维，密集传入，并通过小球叶 - 蕈冠束与蕈状体的副蕈冠特异性连接。

10. 外侧复合体

外侧复合体（LXs）是昆虫大脑中的双侧成对神经毡，介导中心复合体（CX）之间的通信，中心复合体是控制空间方向的大脑中心、各种感觉处理区域和执行运动的胸部运动中心。沙漠蝗虫的 LX 由外侧副叶（LAL）以及内侧和外侧球茎组成。分析连接 CX 和 LX 的神经元的形态，发现 CX 的天空罗盘网络中可能存在反馈回路，通过 LAL 的瘿和一个新鉴定被称为卵圆形体的神经毡连接。此外，16 种不同类型的神经元连接 LAL 与大脑其他区域。八种类型的神经元在两个 LAL 之间提供信息流，五种类型是 LAL 输入神经元，三种类型是 LAL 输出神经元，包括从感觉脑区，如小叶和触角神经提供输入的神经元。LAL 神经元最常靶标脑区是后倾角、楔状体和网穗。在 LAL 中发现了两个具有树突的下行神经元。该神经网络涉及感官信息的处理和为目标定向运动产生适当的行为输出。

8.1.1.2 Deutocerebrum

The deutocerebrum receives sensory input from mechano- and chemosensory receptor neurons on the antennae, and sends motor signals to muscles of the antennae. There are separate neuropil regions in the deutocerebrum that process the information from the chemosensory and mechanosensory neurons; that is, the axons from the two types of receptors project (send axons) to separate areas within the deutocerebrum. Chemosensory input goes to the antennal lobe (AL) neuropil; while the antennae mechanosensory and motor center (AMMC) receives the mechanoreceptor input and sends motor information out. Each of these centers is represented on the left and right sides of the brain, with the AMMC located posterior and ventral to the AL.

8.1.1.2 中脑

中脑接受来自触角的机械和化学感受器神经元输入的信息，并向触角的肌肉发送运动信号。中脑有独立的神经毡可处理来自化学感受器和机械感受器的信息，即两类受体神经元的轴突分别传送（发送轴突）入中脑内的独立区域。化学感受器输入的信息至触角叶（AL）神经毡；而触觉机械感受器和运动中心（AMMC）接受机械感觉的信息输入并发送出运动信息。每个中心有位于触角叶后部和腹面的 AMMC，分别出现在脑的左边和右边。

1.The antennal mechanosensory and motor center (AMMC) neuropil

AMMC contains arborizations of both motor and sensory neurons. Motor centers controlling muscles and glands of the head are in the deutocerebrum (with some additional ones controlling these structures also in the tritocerebrum). The deutocerebrum sends motor neurons to the antennal muscles and muscles of the labrum.

As a sensory center, the AMMC primarily receives mechanoreceptor neuron terminals from Böhm's organ, Johnston's organ, Janet's organ, and other mechanosensory structures located on the two basal segments of the antenna (the scape and pedicel) of various insects. In addition to terminal arborizations in the AMMC, some arborizations from mechanosensory cells project to the protocerebrum, the subesophageal ganglion, and into thoracic ganglia, indicating widespread distribution of some mechanosensory information.

2.The Antennal Lobe (AL)

The AL is the first-order olfactory center in insects. Each AL receives sensory information from chemoreceptors on the flagellum of the antenna on the ipsilateral (same) side of the body. Within the chemosensory neuropil of the AL, there are two neuropil divisions in some insects: one concerned with food, host, and perhaps other general environmental odors; and a macroglomerular complex (MGC) that is sex specific in males of Dictyoptera (cockroaches) and Lepidoptera (moths) as the sensory neuropil for the sex pheromone receptor neurons. Axons from the antennal receptor cells pass through the antennal nerve and terminate in AL neuropil regions in structures called glomeruli. Glomeruli are somewhat cup-shaped masses of axonal terminals. Typically, a glomerulus is approximately 50 to 100 μm in diameter, and each is separated from other glomeruli by layers of glial cell membranes. The number of glomeruli is species specific, with approximately 10 in *A. aegypti* mosquitoes, 200 in *F. pratensis*, and 60 in *M.sexta.* The glomeruli house axonal terminals of antennal olfactory receptor cells, neurites of local and output (projection) neurons, and terminals of centrifugal neurons projecting to the antennal lobe from other sites in the brain. A given glomerulus, the converging sensory axons may come from various sites on the antenna where there are receptors sensitive to a specific odor.

Inside the glomeruli are the soma of one to several interneurons, whose terminals make synaptic connections with the incoming sensory axons. Three classes of interneurons are associated with the

1. 触觉机械感受器和运动中心（AMMC）神经毡

AMMC 包含了运动和感觉神经元的分枝。头部控制肌肉和腺体的运动中心位于中脑（少数例外，也位于后脑），中脑即向触角肌肉和上唇肌肉发出运动神经元。

作为感觉中心，AMMC 最先接收到机械感受器神经元末梢信息，来自不同昆虫、位于触角茎节和梗节的 Böhm's 器，Johnston's 器、Janet's 器和其他机械感受器。除了 AMMC 的末梢分枝外，有些机械感觉细胞的分枝投射到前脑、咽下神经节和胸神经节，这表明一些机械感觉信息分布广泛。

2. 触角叶（AL）

触角叶是昆虫第一级嗅觉中心，每个触角叶可接受虫体同侧触角纤毛上的化学感受器的信息。在触角叶的化学感受器神经毡内，某些昆虫有两个神经毡分区：一个区关系到食物、寄主和环境中的其他气味；另一个区是大神经纤维球复合体（MGC），在蜚蠊和鳞翅目（蛾）雌虫内是性专一性的，因为感觉神经毡是性信息素受体神经元。从触角感受器细胞来的轴突通过触角神经，终止于触角叶神经毡区，结构上称为神经纤维球。神经纤维球在某种程度上像杯形的轴突群末梢。通常直径为 50～100 μm，彼此之间被数层神经胶细胞膜隔开。神经纤维球数量是物种特异性的，埃及伊蚊有 10 个，草地蚁多达 200 个，烟草天蛾 60 个。神经纤维球容纳触角嗅觉感受器细胞轴突末梢、局部和输出神经元的树突，从脑的另一边投射到触角叶的离心神经元末梢。特定的神经纤维球汇聚了可能来自触角不同部位的感受器轴突，那里的受体对特殊气味敏感。

神经纤维球内是一种到几种联络神经元的细胞体，其末端与传入感觉轴突发生突触连接。三类联络神经与神经纤维球

glomeruli: ① local interneurons that interconnect areas of the AL but remain within the AL; ② projection neurons that have dendrites located in the AL with axons projecting into the mushroom bodies and lateral parts of the protocerebrum to which they relay information; ③ centrifugal neurons whose axons project into the AL but cell bodies and dendrites are located outside the AL.

有关：①与 AL 区域互连，但留在 AL 内的局部联络神经元；②位于触角叶内，有树突，其轴突转送信息，投射进入蕈状体和前脑侧面部分的投射神经元；③其轴突投射进入触角叶，但细胞体和树突位于 AL 外的离心神经元。

8.1.1.3 Tritocerebrum

The tritocerebrum sends motoneurons to muscles in the labrum and pharynx, and innervates the stomato-gastric nervous system, a system of several small ganglia, including the frontal ganglion, hypocerebral ganglion, and ingluvial ganglia having control of foregut muscles. In some insects, sensory axons from sensory receptors on the head terminate in the tritocerebrum; A commissural connective from each side of the tritocerebrum passes around the esophagus and provides cross-communication between the two halves, and lateral pharyngeal nerve cord connect each half of the tritocerebrum with the subesophageal ganglion (Fig.8-4).

The unpaired frontal ganglion, lying on top of the esophagus and anterior to the brain, is connected to the tritocerebrum by lateral connectives. The recurrent nerve and other small nerves arise from the frontal ganglion and carry motoneurons to muscles of the gut wall. Nerves from the frontal ganglion innervate the pharynx. Posteriorly, the median recurrent nerve runs along the surface of

8.1.1.3 后脑

后脑向唇和咽部肌肉发出运动神经元，使口道神经系统受神经支配。这个系统有几个小神经节，包括额神经节、脑下神经节和控制前肠肌肉的嗉囊神经节。有些昆虫头部感觉受体的感觉轴突终止于后脑；后脑每侧的神经结缔组织经过食道周围，为两瓣膜提供交叉连接，咽侧神经索将后脑的各半部与咽下神经节连接（图 8-4）。

不成对的额神经位于食道顶端和脑的前面，通过侧神经结缔组织与后脑相连。回神经和其他小神经从额神经节发出，将运动神经元送到肠壁肌肉。咽喉受到额神经节神经支配。之后，中间回

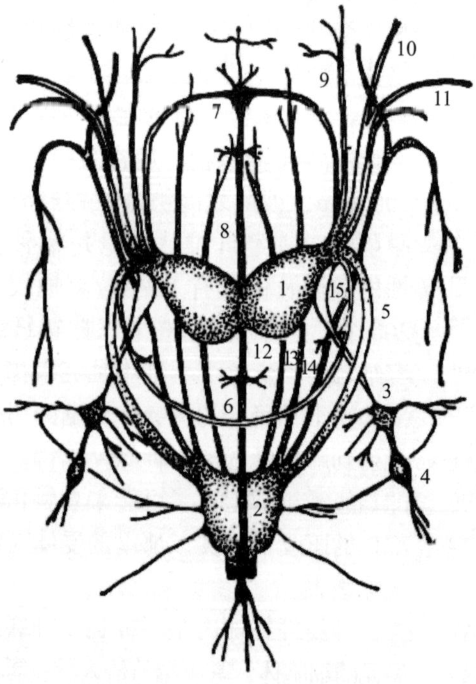

Fig. 8-4 A dorsal view of the brain and associated head ganglia and nerves of a silkworm larvae. 1. Brain; 2. Subesophageal ganglion; 3. Corpora cardiaca; 4. Corpora allata; 5. Lateral pharyngeal nerve cord; 6. Recurrent nerve; 7. Frontal nerve ganglion; 9. Labral nerve; 10. Antennal nerve; 11. Optic nerve; 12. Mandibular nerve; 13. Maxillary nerve; 14. Labial nerve; 15. Lateral pharyngeal sympathetic nerve（诸星静次郎, 1984）.

图 8-4 家蚕幼虫脑及其相关的脑神经节和神经的背面观。1. 脑；2. 咽下神经节；3. 心侧体；4. 咽侧体；5. 咽侧神经索；6. 围咽神经；7. 额神经节；8. 回神经；9. 上唇神经；10. 触角神经；11. 视神经；12. 上颚神经；13. 下颚神经；14. 下唇神经；15. 咽侧交感神经（诸星静次郎，1984）。

the esophagus, passes beneath the brain, and connects behind the brain with the hypocerebral ganglion, also lying on the surface of the esophagus. The small, unpaired hypocerebral ganglion innervates the corpora cardiaca. The small, paired ingluvial ganglia send nerves to the posterior foregut. The subesophageal ganglion is formed from the fusion of three pairs of ganglia. It has sensory and motor connections to sensory structures and muscles of the mouthparts, salivary glands, neck receptors in some insects, and neck muscles. Axons from neurons in the subesophageal ganglion project forward to the brain and posterior to the thoracic ganglia. The subesophageal ganglion has influence over motor patterns involved with walking, flying, and breathing.

神经沿咽表面游走，通过脑的正下方，与脑下神经节在脑的后方相连，其他位于咽喉表面上。不配对的脑下小神经节支配心侧体，成对的嗉囊小神经节向前肠后部发出神经。咽下神经节由三对神经节融合而成。有感觉和运动神经连接到感觉结构和口器、唾液腺肌肉、某些昆虫的颈受体和颈肌肉。咽下神经节神经元的轴突向前伸入脑部，向后进入胸神经节。咽下神经节对运动模式有影响，包括步行、飞翔和呼吸。

8.1.2 Insect cognition

Synthesize findings from neuroscience, psychology, and behavioral biology to show that some key features of cognition in the neuron-based brains of vertebrates are also present in the insect-based swarm of honey bees.

The sophistication of honeybee cognition was first suggested by von Frisch's pioneering studies of honeybee foraging. He demonstrated not only that scout have the ability to translate their experience of finding a nectar source into a sophisticated set of signals, the 'dance', the observers of this dance have the ability to translate it into a sequence of navigational maneuvers. These abilities are suggestive of the presence of explicit memory, a high-level cognitive function that involves memory of places, facts and events.

Two distinct types of memory-based guidance, include alignment image-matching and positional image-matching, called snapshot memories of the surrounding panorama. These encodings probably extract elementary features (such as oriented edges) from the retinal image and preserve their spatial relationships. Positional image-matching can provide guidance from novel locations and in novel directions. It will be effective so long as the panoramas at the novel and goal locations share sufficient elements in common. The third mechanism, path integration, uses cues derived from an insect's own movement. In each case, an insect compares its current sensory input with a memory of the desired sensory input to produce a difference signal. This difference signal is then transformed to generate a desired heading direction. Together, these guidance mechanisms allow insects such as ants and bees to forage

8.1.2 昆虫的认知

综合神经科学、心理学和行为生物学的发现，表明脊椎动物基于神经元的大脑中的一些关键认知特征也出现在基于昆虫的蜂群中。

von Frisch 对蜜蜂觅食的开创性研究首次提出了蜜蜂认知的复杂性。他不仅证明了侦察蜂有能力将它们发现花蜜源的经验转化为一组复杂的信号，即“舞蹈”，这种舞蹈的观察蜂有能力将其转化为一系列导航动作。这些能力表明外显记忆的存在，这是一种高级认知功能，涉及对地点、事实和事件的记忆。

两种不同类型的基于记忆的制导包括校正图像匹配和位置图像匹配，它们被称为周围全景的快照记忆。这些编码可能从视网膜图像中提取基本特征（如定向边缘），并保留其空间关系。位置图像匹配可以从新位置和新方向提供导航。只要新位置和目标位置的全景图共享足够的共同元素，它就会有效。第三种机制是路径整合。它使用来自昆虫自身运动的信息。在每种情况下，昆虫将其当前的感觉输入与所需感觉输入的记忆进行比较，以产生差异信号。然后转

far from their nests, and with experience to learn complicated visually-guided routes. They obtain directional information principally from the sun compass and distance information from monitoring the optic flow or the proprioceptive input generated by their movements. Both path integration and positional image-matching can guide travel across novel terrain, and can be used the first time an insect returns to its nest or to a newly discovered food site. In contrast to the two goal-directed forms of guidance, the memories used for alignment image-matching take longer to acquire, but can provide high levels of accuracy along an entire route. With the ability to remember visual scenes and to remember path integration information, insects have two components that could potentially be brought together into what can be described as a path integration-based cognitive map.

However, in Mediterranean desert ant *C. fortis*, ant's current path integration state can be read out by releasing it with food in an unfamiliar test area. On release, the ant performs a 'home vector': travelling in the direction and distance to where the path integration system indicates the nest should be. To test for resetting, ants were trained along routes so that specific locations were repeatedly experienced with a consistent path integration state. These experiments argue strongly that ants (and quite possibly insects more generally) do not have path integration-based cognitive maps.

One brain structure that may have adapted to this life style is the mushroom bodies, which are impressively large and elaborate in the social Hymenoptera. These prominent areas are involved in learning and memory in a wide range of insects. Microglomeruli density in the mushroom bodies was linked with an individual's ability to learn and retain information, suggesting that neuroanatomy may be associated with functional differences in insect cognition. Notably, some apparently complex insect behaviors may involve little information processing and minimal neural investment suggesting there may be multiple evolutionary pathways to complex cognition.

换该差异信号以生成所需的航向。总之，这些导航机制使蚂蚁和蜜蜂等昆虫能够在远离巢穴的地方觅食，并凭借经验学习复杂的视觉导航路线。它们主要从来自太阳罗盘获得方向信息，从监测视觉流或其运动产生的本体感觉输入中获得距离信息。路径整合和位置图像匹配都可以引导穿越新地形的旅行，并且可以在昆虫第一次返回其巢穴或新发现的食物地点时使用。与两种目标导向的制导形式相比，用于校正图像匹配的记忆需要更长的时间才能获取，但可以在整个路线上提供更高的精度。昆虫具有记忆视觉场景和路径整合信息的能力，两个组成部分可以潜在地结合在一起，形成一个基于路径整合的认知地图。

然而，地中海沙漠蚁的当前路径整合状态可以通过在不熟悉的测试区域释放食物来读取。释放时，蚂蚁执行“归巢航线”：沿着路径整合系统指示的巢穴位置方向和距离移动。为了测试重置，沿路径训练蚂蚁，以便在一致的路径整合状态下重复经历特定位置。这些实验有力地证明，蚂蚁（很可能还有更普遍的昆虫）没有基于路径整合的认知地图。

一个可能已经适应这种生活方式的大脑结构是蕈状体。在社会性膜翅目中，蕈状体大得令人难以想象，而且非常精细。这些突出的区域涉及广泛昆虫的学习和记忆。蕈状体内的神经纤维小球密度与个体学习和保留信息的能力有关，这表明神经解剖学可能与昆虫认知的功能差异有关。值得注意的是，一些明显复杂的昆虫行为可能涉及较少的信息处理和最小的神经投入，这表明复杂的认知可能有多种进化途径。

8.2 神经节

8.2 The Ganglia

8.2.1 Thoracic ganglion

Many insects have three thoracic ganglia, the prothoracic ganglion, the mesothoracic ganglion, and the metathoracic ganglion located, respectively, in each of the thoracic segments. In some insect groups, particularly Hemiptera and Diptera, the meso- and metathoracic ganglia have fused, often with fusion also of some of the abdominal ganglia into a large thoracic ganglion. Each thoracic ganglion sends motor axons to the leg muscles of its segment, and receives sensory axons from sensory receptors in the tarsi and leg joints. The meso- and metathoracic ganglia supply motor nerves to the wing muscles, respectively, of the mesothoracic segments and the metathoracic segment.

8.2.1　胸神经节

许多昆虫一般有 3 对胸神经节，前胸神经节、中胸神经节、后胸神经节分别位于昆虫的前胸、中胸和后胸体节。有些昆虫，尤其是半翅目和双翅目昆虫，中后胸神经节愈合，有的还进一步和腹部神经节愈合形成一个大的胸神经节。每个胸神经节发出的运动神经元轴突至本体节肌肉和腿肌，从跗骨和腿关节中的感觉受体接收感觉轴突。中胸和后胸神经节分别支配中胸和后胸体节翼肌的运动神经。

8.2.2 Abdominal ganglia

The number of abdominal ganglia is highly variable in different orders. One ganglion in each segment was the evolutionarily primitive condition, in all living insects, some fusion of abdominal ganglia has occurred. Some Apterygota still have eight abdominal ganglia, some Odonata larvae have seven, and Orthoptera have five or six. There may be fewer than five in some insects; and in some highly evolved dipterans and hemipterans, all abdominal ganglia have fused with thoracic ganglia into the one large metathoracic ganglion. In locusts, *S. gregaria*, for example, the first three abdominal ganglia are fused in the adult with the large metathoracic ganglion. The large 6th abdominal ganglion in the American cockroach is the terminal abdominal ganglion (TAG) and is a fusion of neuromeres (neuromeres are ganglionic masses associated with a particular segment) from the posterior segments. Numerous nerves radiate out from the TAG to the posterior structures in the body, and it is well supplied with tracheae. Cercal nerves from the cerci carry large numbers of sensory axons into

8.2.2　腹神经节

不同昆虫的腹神经节数量高度变异。在昆虫进化的原始条件下，每个体节均有一个神经节，现在几乎所有成虫的腹神经节都发生了不同程度的融合。无翅亚纲昆虫仍有 8 个腹神经节，有些蜻蜓目昆虫有 7 个，直翅目昆虫 5 ～ 6 个，而有些昆虫可能少于 5 个；有些高度进化的半翅目和双翅目昆虫，所有的腹神经节和胸神经节融合成一个大的后胸神经节。例如，蝗虫的前三个腹神经节在成虫期融合成一个大的后胸神经节。美洲大蠊的第六腹部神经节是腹末神经节（TAG），是后体节神经原节（神经原节是特定体节相连的神经节聚集）的融合，从 TAG 向虫体后部结构放射状伸出大量神经，并有发达的气管分布。尾须上的尾须神经传送大量的感觉轴突进入

the TAG and synapse with giant axons, functioning as a fast escape mechanism when the cockroach is threatened. The giant axons pass forward through abdominal ganglia without synapsing to eventually synapse with interneurons connecting to leg motoneurons in the thoracic ganglia. Thus, information can be sent from the cerci to the legs very rapidly.

TAG，与巨型轴突通过突触连接；当蜚蠊受到威胁时，起到快速逃离的作用机制。巨型轴突向前通过腹神经节，与联络神经元的终端突触没有突触连接，而与胸部的腿运动神经元连接。因此信息能从尾须非常快速地传递到腿部。

Thoracic and abdominal ganglia tend to be divided into three anatomical and functional divisions: a dorsal motor neuropil, a middle integrative neuropil, and a ventral sensory neuropil. Sensory information tends to come into the ventral portion of a ganglion from lateral nerves and from nerve tracts in the ventral nerve cord. Motor output more often occurs from the dorsal portion of a ganglion after associative interneurons allow the two regions to communicate. Internally, the middle layer (or associative zone) appears to contain the most complex array of variously shaped interneurons. Intersegmental ventral nerve cord connectives also tend to maintain the pattern of dorsal motor nerves and ventral sensory nerves.

胸腹部神经节在解剖学和功能上可划分为 3 类：背部运动神经毡、中部综合神经毡和腹部感觉神经毡。感觉信息常常从侧神经和腹神经的神经束传入神经节的腹部区域。运动神经元的输出信息更加频繁地出现在神经节的背部区域，经过相关的中间神经元使两部分相通。在内部结构上，中层（或关联区）显示出包含有排列最复杂的各种形状的中间神经元。节间腹神经索结缔组织也常保持背部运动神经和腹部感觉神经的模式。

8.2.3 Lateral nerves

8.2.3 侧神经

Ganglia give rise to variable numbers of nerve tracts in different groups of insects. Most of the nerve tracts that arise from the brain or ventral ganglia contain mixed nerves; that is, the tract carries both sensory and motor neurons. An exception is the ocellar nerve connecting the ocelli with the protocerebrum. It is purely sensory, carrying interneurons with sensory information inward to the protocerebrum. The antennal nerves are mixed, carrying sensory fibers from olfactory and mechanoreceptors inward, and motoneurons leading to tentorial ptilinal muscles and tentorial frontal muscles. The labro-frontal nerves and maxillary-labellar nerves are mixed, bringing sensory information from receptors on the mouthparts and carrying motoneurons to the muscles of the mouthparts.

不同种类昆虫的神经节发出的神经束数量不同。大多数由脑或腹部神经节发出的神经束包含混合型的神经，即神经束有感觉神经元，也有运动神经元。一个例外就是连接单眼和前脑的单眼神经，它是纯感觉神经，协同联络神经将感觉信息传入前脑。触角神经是混合型的，通过感觉纤维来传送从嗅觉和机械感受器传入的信息，运动神经元通向幕骨额胞肌和幕骨前肌。上颚额神经和下颚唇瓣神经是混合型的，传入从口器感受器而来的感觉信息，向口器肌肉传递出运动神经元。

Lateral nerves from the thoracic ganglia innervate the wing musculature and return sensory information from receptors associated with wing orientation. Motor neurons from the thoracic ganglia innervate the musculature of the legs and tarsi, and thoracic ganglia receive mechanosensory information from the legs and probably mechano- and chemosensory information from the tarsi

胸神经节发出的侧神经控制翅部肌肉，并反馈与翅飞行方向相关受体传来的感觉信息；胸部神经节的运动神经元刺激腿节和跗节的肌肉，大多数昆虫胸神经节接收来自腿的机械感觉信息，可能还有来自跗节的机械和化学感觉信

in most insects. Motor neurons and sensory neurons do not directly synapse with each other, but each makes synaptic connections with interneurons (probably with many interneurons). Interneurons make synaptic connections with many other interneurons, and thus every part of the nervous system is potentially in communication. Incoming sensory information is received in various neuropils, interpreted, and motor commands are formulated to send to muscles or glands.

息。运动神经元和感觉神经元相互之间并没有突触连接，而是各自和中间神经元连接（可能和多个中间神经元）。中间神经元与其他多个中间神经元突触连接，因此神经系统的每个部分都可能相通。传入的感觉信息被各种神经毡接收和整合，然后向肌肉和腺体发出运动指令。

8.2.4 Giant axons in the insect central nervous system

In contrast to the very small diameter of most neurons, some insects have "giant" axons of varying size running through the abdominal ganglia and into the thoracic ganglia. The ganglionic junctional synapses are electrical rather than chemical, and much faster conducting than chemical synapses. The large diameter of giant axons also promotes more rapid rate of impulse conduction. In the cockroach, the somata of the giant axons are located in the last abdominal ganglion, and the giants make synaptic contacts within the neuropil of the ganglion with sensory neurons from the paired cerci, the short appendages from the last segment of the insect. The giant axons run without synapsing through the abdominal ganglia and synapse in the thoracic ganglia with motoneurons going to the legs. These axons provide a rapid pathway for sensory information from the cerci to activate the legs in an escape behavior. A puff of air on the cerci causes a cockroach to make rapid escape movements. Some of the smaller giants also pass through the thorax and synapse in the brain. The giant axons in the CNS of dragonfly larvae (*Anax* spp.) run from the last abdominal ganglion to the thoracic ganglia, but these giants have synapses in each abdominal ganglion with motoneurons to abdominal muscles as well as with motoneurons in the thoracic ganglia to the legs. The escape reaction in dragonfly larvae involves raising the legs, releasing the hold on the substrate, and simultaneous contracting abdominal muscles to force water out of the large rectum in a jet propulsion mechanism that propels the insect forward.

8.2.4　中枢神经系统内的大轴突

相对于直径细小的神经元而言，有些昆虫从腹部神经节到胸部神经节有大小不一的“大轴突”。神经节的突触连接是电突触而不是化学突触，传导比化学突触快得多。大直径的大轴突也启动更快速的脉冲传导。蜚蠊的大轴突细胞体位于腹末神经节内，大轴突在神经节的神经毡内与一对尾须感觉神经元突触连接，尾须是昆虫最后一个体节的短附属器官。大轴突沿腹神经节延伸而无突触，与胸部进入腿部的运动神经元突触连接。这些轴突在逃避行为中为尾须感觉信息到激活腿部提供了一条快捷途径。尾须的空气流动引起蜚蠊快速的逃逸运动。有些稍小的大轴突也通过胸部在脑内发生突触连接。蜻蜓幼虫（麻斑晏蜓）中枢神经的大轴突沿腹末神经节向胸神经节延伸，但这些大轴突在每个腹部神经节均有腹部肌肉运动神经元突触相连接，这与胸神经节到腿部的运动神经元连接一样。蜻蜓幼虫的逃跑反应牵涉举腿，松开被抓牢的基质，同时收缩腹部肌肉迫使水从大回肠喷射的推进，推动虫体前进。

8.3 神经毡

8.3 Neuropil

The main mass of any ganglion is a central region of synaptic connections between arborizations of sensory, motor and interneurons called the neuropil. The largest and most complex neuropil occurs in the brain in such associative centers as the mushroom bodies, central body, and optic lobes, but all ganglia contain neuropil regions. The neuropil is concentrated in the center of a ganglion and is surrounded with a shell of cell bodies, the somata of motoneurons, interneurons, and glial cells. The neuropil continues to grow in size with new synaptic connections in the milkweed bug *Oncopeltus fasciatus*, in *D.melanogaster*, and possibly in most or all insects, even after the nervous system as a whole stops growth. This suggests the importance of new information processing and integration that an insect uses to acquire food, mate, lay eggs, and survive in its environment.

任何神经节的主要汇聚都是感觉神经元、运动神经元和联络神经元树突之间彼此通过突触连接的中心区域，称为神经毡。最大和最复杂的神经毡位于脑内汇聚中心，如蕈状体、中心体、复合体和视叶，但所有神经节都包含神经毡区。神经毡集中在神经节中心，被细胞体骨架，运动神经元、联络神经元细胞体和胶细胞包裹。美洲脊胸长蝽、果蝇，甚至可能大多数昆虫的神经毡的大小都随着新的突触连接而继续增长，这一过程持续到整个神经系统停止生长后。这表明信息处理和整合新信息在昆虫的取食、交配、产卵和应对环境方面起着重要作用。

8.4 血 - 脑的屏障

8.4 Hemolymph-brain Barrier

At the hemolymph interface, the brain, ventral connectives, ventral ganglia, and large nerves are protected from direct contact with the hemolymph by a selectively permeable barrier, the hemolymph-CNS barrier, or blood-brain barrier. The barrier consists of an outer acellular layer called the neurolemma (also neural lemma, neural lamella) and an inner cellular layer, the perineurium, consisting of the perineural cells. The entire acellular and cellular barrier is thin. Some insects have a continuous fatty sheath of variable thickness composed of fat body cells exterior to the neurolemma that covers the entire brain and ventral nerve cord. When present, the fatty envelope also acts as a barrier to ions and

在血淋巴的界面，脑、腹神经结缔组织、腹神经节和大神经都受到一层高度选择透过性的屏障保护，以避免与血淋巴直接接触，即血淋巴 - 中枢神经屏障或血淋巴 - 脑屏障。该屏障由一层被称为神经膜（也称为神经鞘、神经片层）的外部无细胞层和一层由围神经细胞组成的内部细胞层，神经束膜组成。整个非细胞层和细胞层屏障很薄。有些昆虫在整个脑和腹神经索外覆盖着一层连续的、由脂肪体细胞组成的厚薄不均

osmolites in the hemolymph

The acellular neurolemma of some insects is composed of several layers identifiable in the electron microscope, and contains fibrils of collagen-like material that are embedded in a matrix of glycoaminoglycans. The neurolemma is relatively permeable to large molecules, such as insulin and methylene blue, and in itself does not appear to be much of a barrier to hemolymph substances. This acellular neurolemma is probably secreted by the outermost layer of cells, sheath cells, located just below the surface of the neurolemma.

The cellular layers beneath the neurolemma, the perineural cells, constitute the main hemolymph-CNS barrier. The perineural layer is several cells in thickness. The perineural cells have extensive couplings between the cells *via* tight and gap junctions, which probably account for much of the impermeability of the cell layer; large molecules and ions would have great difficulty in passing between the cells because of the between-cell barriers.

The open circulatory system with high hemolymph concentrations of trehalose and amino acids, some of which may function as neurotransmitters, and the high potassium to sodium ratio in some phytophagous insects may have greatly influenced the evolution of the blood-brain barrier. The blood-brain barrier protects not only the brain but all ganglia and major nerves from direct contact with the hemolymph.

的脂肪鞘。需要时，脂肪包被血液中的离子和渗透物质也起着屏蔽作用。

有些昆虫的非细胞层神经膜在电镜下由可辨识的几层组成，包括胶原蛋白类物质的原纤维，包埋在甘油氨基聚糖的基质中。神经膜对大分子具有相对的渗透性，如胰岛素和亚甲蓝，其本身并不太像血液物质的屏障，这一非细胞的神经膜也许由最外层的细胞，鞘细胞所分泌，位于神经膜的正下方的表面。

神经膜下方的细胞层，即神经束膜细胞，组成主要的血淋巴 -CNS 屏障。神经束膜层约为几个细胞的厚度。神经束膜细胞之间有大量的紧密连接和间隙连接，这也许可以认为是细胞层不通透的缘由；因为细胞间的屏障，大分子和离子将很难通过细胞间隙。

开放式循环系统含有高浓度的海藻糖和氨基酸，其中有些可能起着神经递质的作用，而植食性昆虫的高 K^+/Na^+ 值对血淋巴-脑屏障的进化影响很大。血淋巴-脑屏障不仅保护脑，而且对所有神经节和主要的神经都有保护作用，避免与血淋巴的直接接触。

8.5 神 经 元

8.5 Neurons

Neurons have one or more dendrites, a cell body called the soma (perikaryon) that contains the nucleus, and an axon (Fig.8-5). Sometimes, the axon has collateral branches. Extensive arborization of axonal and dendritic processes is typical. Within the CNS, the arborizations enable synaptic contact with many other neurons, thus making possible communication throughout the nervous system. A dendrite is defined as any process conducting electrical excitation from the site of stimulation toward the soma, and the axon is the process conducting excitation away from

神经元有一个或多个树突，一个含有细胞核被称为胞体（核周质）的细胞体和一个轴突（图 8-5）。有时，轴突有侧支。轴突和树突的突起有大量树状末梢是其典型结构。在 CNS 内，树状末梢能和其他许多神经元通过突触相连，这样才能使联络遍及神经系统。树突被定义为从刺激部位传导电兴奋到胞体的任何过程，轴突是传导兴奋离开胞体的过程。一个轴突可以向 CNS 传导

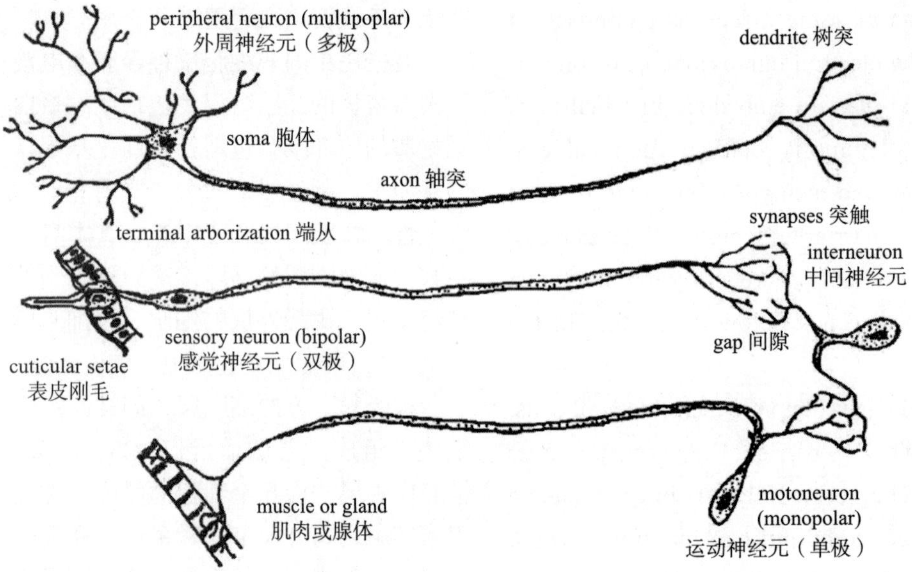

Fig. 8-5 A schematic drawing of sensory, interneuron and motor neuron to illustrate axon, dendrite, soma, arborization and effector (muscle or gland) (Nation, 2022; Feng lichun, Shen Weide, 2015).

图 8-5. 感觉神经元、中间（联络）神经元和运动神经元示意图，说明轴突、树突、细胞体、端丛和效应器（肌肉或腺体）（Nation，2022；冯丽春、沈卫德，2015）。

the soma. An axon may conduct excitation toward the CNS (sensory neuron) or away from the CNS (motoneuron) toward an effector, such as a muscle or gland. Some of the arborizations from certain neurons in insects are known to receive input in the neuropil (i.e., act like dendrites), while others deliver output (act like axons), and morphology cannot be used to discriminate between these.

There are a number of morphological types of cells in the nervous system, they can be grouped functionally into sensory neurons, motoneurons, interneurons, neurosecretory cells, and glial cells.

兴奋（感觉神经元）或从 CNS 传出兴奋（运动神经元）到效应器，比如肌肉和腺体。昆虫某个神经元的有些树状末梢已知可接收神经毡输入的信息（即像树突一样起作用），而另一些即发送信息（像轴突一样起作用），在形态学上无法区分这些。

神经系统有些不同形态特征的细胞。根据其功能，它们被分为：感觉神经元、运动神经元、联络神经元、神经分泌细胞和胶细胞。

8.5.1 Afferent or sensory neurons

Sensory neurons carry nerve impulses toward the CNS. The somata of insect sensory neurons, with few exceptions, are located very near the site of stimulus detection, that is, at the cuticular surface for external receptors and as part of various internal organs. The dendritic processes are very short, and the axonal processes tend to be relatively long. Insect sensory neurons to be bipolar. The short dendritic processes of insect receptor cells probably do not generate or conduct spikes, but develop graded electrical activity that spreads from the site of the stimulus. When the graded electrical activity is strong, it is likely to spread to the region where spikes are generated. The number of sensory neurons may change with molts and metamorphosis. For example, mechanoreceptors (innervated hairs) on the abdominal cerci of *A. domesticus*

8.5.1 传入或感觉神经元

感觉神经元向 CNS 传导神经脉冲。昆虫感觉神经元胞体，除少数例外，都非常靠近刺激检测部位，即位于表皮表面的外部受体或成为内部器官的一部分。树突的突起非常短，而轴突的突起常常相对较长。昆虫的感觉神经元为双极。昆虫受体细胞的短树突的突起也许不会产生或传导峰电位，但会从刺激部位扩展时产生分级电活动，在分级电活动很强时，就跟产生峰电位的区域扩展相似。感觉神经元的数量在蜕皮和变态期会发生变化。如家蟋蟀的腹部尾须上的机械感受器（感觉毛）在生长和蜕皮

increased from 50 to 750 during growth and molting.

期间从 50 条增加到 750 条。

8.5.2 Efferent or motor neurons

Motoneurons have their somata located in the peripheral region of a ganglion and are usually monopolar in insects. The somata of motoneurons are usually very large, up to 100 μm in diameter. Motoneurons are usually paired on each side of a ganglion. Dorsal unpaired medial (DUM) motoneurons are not paired, but have neurites in each half of a ganglion and an axon on each side of the ganglion that emerges through lateral nerve of the ganglion to innervate an effector gland or muscle.

A few motoneurons have branches on both sides of a ganglion, and the axon may even emerge from the contralateral side. Usually, the single process that arises from the peripheral soma of a motoneuron passes into the neuropil of its ganglion, branches off a network of arborizations neurites that make many synaptic connections with neurites of other interneurons, and the axon exits the ganglion on the ipselateral side through a large nerve and continues to its target site of gland or muscle. Some axons exit into the ventral connectives between ganglia and terminate in another part of the central nervous system, or eventually pass out through the lateral nerve of a different ganglion than that of its origin.

8.5.2 传出或运动神经元

运动神经元有位于神经节的外周的胞体，通常为单极。运动神经元胞体常常很大，直径达 100 μm。常成对位于神经节的外周两侧。背内侧不配对的运动神经元不成对，但其神经节的每一半都有神经突，神经节每边都有轴突，通过神经节的侧神经支配效应器腺体或肌肉。

少数运动神经元在神经节两侧有分支，轴突甚至可从对侧出现。通常，单个过程从运动神经元外周胞体发出进入其体节神经毡，分支成神经突树状末梢网络，与许多其他中间神经元树突发生突触连接，同侧神经节出去的轴突通过一大神经，继续通向其腺体或肌肉的靶点位置。一些轴突进入神经节之间的腹侧连接并终止于中枢神经系统的另一部分或最终通过与源头神经节不同的神经节外侧神经传出。

8.6 中间神经元

8.6 Interneurons

Interneurons, also called association neurons, may be located entirely within a ganglion or they may send intersegmental (axonal) processes through the ventral nerve cord to make synaptic connections in other ganglia. Interneurons make extensive synaptic connections with other interneurons, with incoming afferent (sensory) neurons, and with motoneurons. With their vast array of neurites and interconnections, interneurons are extremely important in coordinating communication between sensory and motor systems and within the central nervous system. The somata of interneurons lie in the peripheral region of a ganglion.

中间神经元也称为联络神经元，可能完全位于神经节内，或通过腹部神经索发出节间（轴突）突起与另一个神经节突触连接。中间神经元与其他中间神经元、传入（感觉）神经元和运动神经元有大量的突触连接。有着大量的轴突和相互连接，中间神经元在协调感觉和运动系统之间以及中枢神经系统内的传递极其重要。中间神经元胞体位于神经节外周区域。

Local interneurons are those that make connections within a ganglion and do not exit from it. Interneurons that exit a ganglion to make connections in other ganglia are called intersegmental interneurons. Local interneurons are further classified into spiking and nonspiking interneurons, depending on whether the neuron transmits a spike or only graded electrical activity. Spiking local interneurons are most common in the optic neuropils, mushroom bodies of the protocerebrum, and in the antennal lobes of the deutocerebrum, but they can be found in thoracic ganglia and a few in other parts of the nervous system. In general, local interneurons are paired in the two halves of a ganglion, but the dorsal unpaired median (DUM) local interneurons, as the name indicates, are not paired. The somata of these DUM neurons are located in the periphery of a ganglion near the midline, and extensive neurites from each DUM spread into both halves of the ganglion. DUM neurons may be local, intersegmental, or efferent DUM neurons.

局部中间神经元指在神经节内连接而不离开神经节的神经元。离开神经节与另一个神经节连接的称为节间联络神经元。局部中间神经元根据其传送峰电位或只传送分级电活动进一步分为峰电位和非峰电位联络神经元。峰电位局部中间神经元在视神经毡、前脑蕈状体和中脑的触角叶内最普遍，但在胸神经节和少数其他神经系统内也有发现。总体来说，局部中间神经元在神经节两边成对存在，但 DUM 局部联络神经元，如名称所示是不配对的。这些 DUM 中间神经元胞体位于神经节中线附近的外周内，每个 DUM 中大量的神经突延伸至神经节的两边。DUM 神经元有局部的、节间和传出的 DUM 神经元。

Anatomically, insect exhibit, for the most part, anterior-posterior and dorsal-ventral asymmetries, whereas their left and right body sides are mirrored (for exceptions, for example, mandibles; genitalia). The macroscale organization of their nervous system is similar: interconnected segmental ganglia, symmetrical on the lateral axis and asymmetrical along the longitudinal one. Despite this general left-right symmetrical architecture, the function of the nervous system is often asymmetric. For example, sensory perception requires distinct activities of bilateral sensory centers in order to maintain information about asymmetric environmental cues, and turning behavior requires distinct motor outputs to each side of the body. These kinds of asymmetrical activities arise from symmetrical neuronal networks, capable of differentially activating bilateral parallel parts of the nervous system. Therefore, these asymmetries are transient and can interchange in a matter of milliseconds.

从解剖学上看，昆虫在很大程度上表现出前后不对称和背腹不对称，而身体左右两侧是镜像的（例外情况，例如，下颌骨、生殖器）。它们的神经系统的宏观结构相似：相互连接的节间神经节，在横轴上对称，在纵轴上不对称。尽管有这种左右对称的结构，但神经系统的功能通常是不对称的。例如，感官感知需要双侧感觉中心的不同活动，以保持关于不对称环境信号的信息，而转向行为需要向身体两侧输出不同的运动。这些不对称活动来自对称的神经元网络，能够差异激活神经系统的双侧平行部分。因此，这些不对称是暂时的，可以在几毫秒内交换。

There may be several different neurotransmitters involved with spiking interneurons. Some show inhibitory action and evidence of release of, γ-aminobutyric acid (GABA), a common neurotransmitter at inhibitory synaptic endings. DUM-neurons are not only a potential source of inhibition in the prothoracic network, but also the target of inhibitory influences. Other spiking interneurons have excitatory action at the postsynaptic connections, and they must release a stimulatory neurotransmitter. Efferent DUM interneurons appear to be octopaminergic (releasing octopamine) and probably exert neurohormonal control over muscle or gland action.

峰电位中间神经元有几种不同的递质。γ-氨基丁酸（GABA）在抑制性的突触终端是一种普遍存在的递质。DUM 神经元不仅是前胸神经网络中潜在的抑制源，而且是抑制影响的靶点。其他峰电位中间神经元在突触后膜连接处具有兴奋作用，必须释放刺激性的递质。传入 DUM 中间神经元似乎是多巴胺类的（释放多巴胺），也许会对肌肉和腺体活动施加神经激素控制。

Nonspiking interneurons have only a vast array of neurites projecting into the neuropil of a ganglion. They are contained within a single ganglion. They release their neurotransmitter at the synaptic junctions without transmitting spikes. They also occur in the neural network controlling walking behavior in cockroaches and other insects, and in the TAG where they process signals from certain mechanoreceptors on the cerci.

非峰电位中间神经元只有大量的神经突投射至神经节的神经毡内。它们被包裹在单独的神经节内，在神经突触连接处释放递质而不传送峰电位。在蜚蠊和其他昆虫体内，它们也出现在神经网络中控制其行走，出现在 TAG 中的它们处理来自尾须的某些机械感受器信息。

8.7 神经胶细胞

8.7 Neuroglial Cells

Neuroglial cells include: ① wrapping glia of the peripheral nervous system; ② surface glia, which make the brain hemolymph barrier, is further subclassified into perineural glia (small cells lying on the ganglionic surface) and subperineural or basal glia (which are large sheet-like cells connected with septate junctions that form the barrier); ③ cortex glia that contacts neuronal cells somata in the CNS, with each glial cell establishing contacts with many neurons; ④ neuropil glia, which are located in the neuropil and cover axons and synapses. The neuropil glia is further subdivided into ensheathing or fibrous and astrocyte-like glia, which forms a perisynaptic glial cover; ⑤ tract glial cells cover axonal tracts connecting different neuropils.

神经胶细胞包括：①包裹外周神经系统的胶细胞；②构成脑，血淋巴屏障的表面胶细胞，进一步细分为外周神经胶细胞（位于神经节表面的小细胞）和亚外周神经或基底胶细胞（与形成屏障的间隔连接相连的大片状细胞）；③与中枢神经系统中的神经元细胞体接触的皮质细胞，每个细胞与许多神经元建立接触；④神经毡胶细胞，位于神经毡内，覆盖轴突和突触。神经毡胶细胞进一步细分为鞘状或纤维状和星形胶质细胞样胶细胞，形成突触周胶质覆盖层；⑤束神经胶细胞覆盖连接不同神经毡的轴突束。

Perineural glia and primary surface glia of the ventral nerve cord are derived from embryonic neuroglioblasts. Both types of primary glial cells increase in size, but not in number, during the larval stage.

腹神经索的神经周围胶质和初级表面胶质来自胚胎神经胶质母细胞。在幼虫期，这两种初级胶细胞的大小增加而数量并未增加。

Glial cells are always present in nervous systems and are in intimate contact with neurons. Glial cells ensheathing the axonal and neurite processes, as well as the somata, of neurons. Glial cells serve neurons and the nervous system in a number of ways; they provide structural support, nutritive and metabolic functions, and protection from outside chemical and ionic influence. They also may provide regenerative guidance for regrowth of severed or damaged neuronal processes, repair and maintenance of neurons,

胶细胞通常出现在神经系统内，与神经元密切相连。胶细胞将神经元轴突和树突纳入鞘中，也包裹着神经元的胞体。胶细胞用许多方式对神经元和神经系统起作用；它们提供结构支持，营养和代谢功能，保护神经系统使之免受外界化学物质和离子的影响。它们也为切断或损伤的神经突起提供再生引导，修

and may possibly function in signaling and information processing. Multiple layers of glial cells help provide the blood-brain barrier that protects the CNS from direct contact with chemical components in the hemolymph.

Typically, in insects, many axons may be ensheathed by the same glial cell. For example, in *A.domesticus*, from 10 to 35 contiguous, small-diameter axons may share a glial cell in the dorsolateral bundles of the metathoracic nerve, as well as in the cercal nerves of adult crickets. In the house cricket, an axon must have a diameter greater than 1μm to have its own glial sheath. Glial cells serve protective functions by isolating neurons or their processes from each other, and probably have a role in nutritive support of neurons. They may remove and degrade excess neurotransmitter, or transmitter precursors, and possibly other toxins, and store nutritive macromolecules.

复和维护神经元，且可能具有信号传导和信息处理功能。胶细胞的多层有助于提供保护 CNS 不与血液中的化学成分直接接触的血 - 脑屏障。

通常，昆虫的许多轴突都包裹在相同的胶细胞内，如家蟋蟀的后胸背侧神经束，与其成虫的尾须神经一样，被包裹的近邻小直径轴突数量有 10 ～ 35 根。家蟋蟀轴突直径必须大于 1 μm，其外周才有自己的胶细胞鞘。胶细胞通过将神经元或其突起彼此分开包裹的方式进行保护，也许对神经元有营养支持的作用。它们也能移除和降解过量的神经递质或递质前体以及其他可能的毒素，并贮存营养大分子。

8.8 神经分泌细胞和中枢神经系统的分泌产物

8.8 Neurosecretory Cells and Neurosecretion Products from the CNS

Neuropeptides are secretory peptides that are initially translated as large biologically inactive prepropolypeptides or preprohormones that undergo post-translational processing (cleavage) in the endoplasmic reticulum-Golgi network and electron-dense secretory granules. Neuropeptides play crucial roles throughout an insect's life. They are involved in physiological and developmental processes and in the modulation of many essential behaviors, including metabolism, homeostasis, development, reproduction and behavior. Neuropeptides that influence muscle contraction, contraction of foregut, midgut and/or hindgut influences feeding and/or digestion, contraction of skeletal and flight muscle influences locomotion, contraction of heart muscle influences hemolymph circulation, contraction of ovarian muscle affects reproduction and regulate behavioral cascades such as ecdysis. There are several peptide families that exert their influence on muscle contraction in insects. These include the FMRFamide-related peptides (FaRPs), pyrokinins, tachykinins,

神经肽是一种分泌肽，最初翻译为大的生物非活性前多肽原或前激素原，在内质网—高尔基体网络和电子致密分泌颗粒中进行翻译后转运加工（裂解）。神经肽在昆虫的一生中起着至关重要的作用。它们参与生理和发育过程以及许多基本行为的调节，包括代谢、体内平衡、发育、生殖和行为。神经肽影响肌肉收缩，前肠、中肠和 / 或后肠的收缩影响进食和 / 或消化，骨骼肌和飞行肌的收缩影响运动，心肌的收缩影响血淋巴循环，卵巢肌的收缩影响生殖并调节行为级联反应，如蜕皮。有几个肽家族对昆虫的肌肉收缩产生影响。主要包括酰胺化 FMRF 相关肽（FaRPs）、焦激肽、速激肽、直肠肽、心脏加速肽（CAPA）、黑化诱导素、orcokinins 和咽侧体抑制素。在昆虫中，FaRPs 家族包括 N- 末端

proctolin, cardioacceleratory (CAPA) peptides, corazonin, orcokinin and allatostatins. In insect FaRPs family includes N-terminally extended FMRFamides (thus far identified only in Dipterans) and FL/IRFamides, the myosuppressins, sulfakinins and neuropeptide Fs (NPFs). Myosuppressin action on cockroach hindgut contraction, mealworm neuromuscular junction and locust oviduct contraction. Several neuropeptides being stimulators of muscle contraction, such as proctolin, crustacean cardioactive peptide and corazonin were found. Proctolin functions as a myostimulator with insect specific actions on foregut, midgut, hindgut; skeletal muscle, heart and muscles of the oviduct.

In silkworm larva, subpharyngeal ganglion (SOG) was connected to the brain by a pair of long circumoesophageal connectives, while in the adult, SOG was completely fused with the tritocerebral lobes of series of neurosecretory cell axons. Histomorphologically, four groups of neuronal cells were identified in the brain areas of adult *A. mylitta*, viz. medial, lateral, posterior, and ventral neurosecretory cells. Six groups of neurosecretory cells in larval SOG, *viz*. mandibular, maxillary, labial, paired antero-lateral, paired mid-lateral and paired postero-lateral neurosecretory cells were identified.

延伸的酰胺化 FMRF（迄今为止仅在双翅目昆虫中鉴定）和酰胺化 FL/IRF，心肌抑制素、硫激肽和神经肽 Fs（NPFs）。心肌抑制素对蟑螂后肠收缩、粉虱神经肌肉接头和蝗虫输卵管收缩有作用。几种刺激肌肉收缩的神经肽，如直肠肽、甲壳类动物心激肽和黑化诱导素被发现。昆虫直肠肽是一种肌肉刺激剂，对前肠、中肠、后肠、骨骼肌、心脏和输卵管肌肉具有特殊的作用。

在蚕的幼虫中，咽下神经节（SOG）通过一对长的围咽结缔组织与大脑相连，而在成虫中，SOG 与一系列神经分泌细胞轴突的后脑叶完全融合。在组织形态学上，柞蚕成虫的脑区鉴定出四组神经元细胞，即内侧神经分泌细胞、外侧神经分泌细胞、后部神经分泌细胞和腹侧神经分泌细胞。在幼虫 SOG 中鉴定出六组神经分泌细胞，即下颌神经分泌细胞、上颌神经分泌细胞、唇神经分泌细胞、成对前外侧神经分泌细胞、成对中外侧神经分泌细胞和成对后外侧神经分泌细胞。

8.8.1 Neurosecretory Cells (NSC)

Neurosecretory cells usually have a very large soma, are usually monopolar, and occur in all ventral ganglia and the brain of insects. The somata are located peripherally in a ganglion. They are usually characterized by their large size and staining properties. Axonal processes from neurosecretory neurons often project to the periphery of the body, and staining suggests that they carry the neurosecretory products to functional sites. Except for a few neurotransmitter and modulating chemicals, most neurosecretory hormones are peptides or small proteins.

Neurosecretion, the secretion and release of products that may function as hormones and as neuromodulators, is one of the major functions of the nervous system. Neurosecretion is ideally suited for control of physiological and biochemical processes in which sustained stimulation is needed, such as secretion of PTTH over several days in some insects in order to stimulate the prothoracic glands to begin to produce the molting hormone.

8.8.1 神经分泌细胞（NSC）

神经分泌细胞通常由一个很大的单极细胞体，出现在昆虫的脑和所有腹部神经节中。胞体位于神经节外周。它们通常是特征化的大胞体大及其染色的特点。神经分泌细胞来的轴突的突起常投射到胞体外周，染色表明神经分泌产物向功能部位的输送。除了少数神经递质和调质外，大多数神经激素都是肽类或小分子蛋白质。

神经分泌是神经系统的主要功能之一。其分泌和释放的产物在功能上起激素和神经调质的作用。在需要持续的激发时，神经分泌完美地控制昆虫生理和生化过程。比如，有些昆虫分泌促前胸腺激素结束后几天会促使前胸腺开始产生蜕皮激素。

Physiologically active molecules secreted by the nervous system of animals have been peptides or small proteins (except neurotransmitters, such as acetylcholine, γ-aminobutyric acid, and some biogenic amines), and they are usually called neuropeptides. Usually, the neuropeptides found in insects have been small, composed of 10 to 15 or fewer amino acids. The identified peptides have been grouped into about 20 families based on structure similarity. Some of the neuropeptides have described functions as hormones (such as PTTH, AKH, eclosion hormone, and diuretic hormone). They are highly likely that some neuropeptides function as neurotransmitters and as neuromodulators that could modify the input or output from neural connections. There is isolated evidence in insects of the co-localization of neuropeptides and neurotransmitters in the same nerve terminals, and neuropeptides are known in some cases to be released simultaneously with neurotransmitters. If some neuropeptides do indeed work in this way, a single neuron in a network may be able to regulate many variations on a basic behavior by modulation with neuropeptides. As an example, the cardio acceleratory peptides (CAPs) in *M. sext* modulate four different behaviors at different periods in the life of the tobacco hornworm, with two modifying feeding, ingestion, and nutrition, and two others regulating heart activity in relation to wing inflation and flight. Neuromodulators might alter response characteristics of neurons, including such activities as feed-back, feed-forward, motor output, and muscle or gland response to nervous activity. Neuropeptides may have roles in embryonic development and as cytokines in non-self-recognition and response.

动物神经系统分泌的生理活性分子都是肽或小分子蛋白质（除神经递质外，如乙酰胆碱、γ-氨基丁酸和一些生物胺），它们通常被称为神经肽。昆虫内被发现的神经肽一般，由10～15个或甚至更少的氨基酸组成。根据结构的相似性，已鉴定为肽被分成大约20个族。有些神经肽已被描述为激素功能（如PTTH、AKH、利尿激素）。它们非常像能调整神经连接处输入和输出的神经递质和神经调质，有时神经肽被认为是与神经递质同时释放的。如果某些神经肽的确以这种方式起作用，神经网络中的某个神经元就可能通过调整肽来调节昆虫基本行为的许多变化。比如，烟草天蛾心脏激速肽（CAPs），在烟草天蛾的4个发育阶段分别调节4个不同的行为，其中两个调节取食、消化和营养，另两个则调节翅的煽动及与飞行相关的心脏活动。神经调质可改变神经元的反应特性，包括反馈、前馈、运动输出和肌肉或腺体对神经活动的反应。神经肽在胚胎发育中也有作用，并在异己识别和反应中起着细胞因子类的作用。

8.8.2 Adipokinetic hormone (AKH)

AKH, first isolated from the CC of *L. migratoria* is a decapeptide with the amino acid (coded) sequence pQLNFTPNWGTamide. The first lepidopteran AKH peptide was identified from the tobacco hornworm, *M. sexta*, and the nonapeptides with the same sequence (pELTFTSSWG-amide) were later found in several other lepidopterans, including *B. mori* and *H. zea*. Some are octapeptides instead of decapeptides, usually they begin with pyroglutamate and have an amide function at the carboxy-terminal end.

Based on peptide sequence similarity hypertrehalosaemic hormone is part of a large family of peptides that collectively are referred to as the adipokinetic hormones (AKHs). AKHs are

8.8.2 激脂激素（AKH）

AKH最早从飞蝗心侧体中分离出来，是一种氨基酸（编码）序列酰胺化为QLNFTPNWGT的十肽。第一个鳞翅目AKH从烟草天蛾中被鉴定出来，具有相同序列的九肽（酰胺化ELTFTSSWG），后来在其他几种鳞翅目昆虫中发现，包括家蚕和美洲棉铃虫。有些是八肽而不是十肽，通常从焦谷氨酸开始，在羧基末端具有酰胺功能。

基于肽序列的相似性，高血糖激素是肽大家族的一部分，被统称为激脂激素（AKHs）。AKHs在心侧体腺叶的神

produced in the neurosecretory cells of the glandular lobe of the CC that functions as a neurohaemal organ, storing and releasing material from the brain and products of its intrinsic secretory activity. Neuropeptides originating from neurosecretory cells of the brain send axons to the glandular lobe that may regulate the release of AKHs. In locusts, AKH release appears to be stimulated by neuropeptides proctolin, crustacean cardioactive peptide and the desert locust tackykinin. AKHs primary function is to regulate metabolism of energy stores. In response to flight, or other energy expensive activities, AKHs mobilize energy reserves from storage organs such as the fat body to the hemolymph. AKHs have additional roles that may be complementary to energy metabolism through direct interaction with the nervous system. DUM neurons release the biogenic amine octopamine which act on muscles of the ventilatory system and on the heart.

经分泌细胞中产生，作为神经血器官发挥作用，贮存和释放大脑中的物质及其内在分泌活动的产物。来自脑神经分泌细胞的神经肽将轴突发送到腺叶，可调节 AKHs 的释放。在蝗虫中，AKH 的释放似乎受到神经肽直肠肽、甲壳类动物心激肽（CCAP）和沙漠蝗虫速激肽的刺激。AKHs 主要功能是调节贮存能量的代谢。为了应对飞行或其他耗能巨大的活动，AKHs 将能量储备从脂肪体等贮存器官动员至血淋巴。AKHs 具有额外的作用，可能通过与神经系统的直接相互作用补充能量代谢。DUM 神经元释放生物胺章鱼胺，作用于气管系统的肌肉和心脏。

AKH has functional activity on mobilization of lipids or carbohydrates, acceleration of heart beat, myoactivity, and inhibition of fatty acid and protein synthesis. AKH-like immunoreactive (AKH-LI) cells are present in the CC, brain, and subesophageal ganglion of many insects. The brain of the blowfly (*Calliphora*) contains about 50 AKH-LI neurons scattered in proto-, deuto-, and trito-cerebrum, and several hundred in the medulla of each optic lobe. One very prominent AKH-LI neuron on each side of the protocerebrum of *Calliphora* has arborizations that cover most of the superior protocerebrum on the ipsilateral side.

AKH 有动员脂类和碳水化合物、加速心跳、肌肉活动以及抑制脂肪酸和蛋白质合成的功能活性。AKH 样免疫反应性（AKH-LI）细胞出现在很多昆虫的心侧体、脑和咽下神经节。丽蝇的脑有 50 个 AKH-LI 神经元广泛分布在前、中和后脑，每个视叶髓质即有几百个。丽蝇前脑每侧的一个非常突出的 AKH-LI 神经元都有树状末梢，覆盖同侧上前脑的大部分。

8.8.3 Proctolin

Proctolin was the first insect neuropeptide to be sequenced. It contains the amino acid sequence arginine-tyrosine-leucine-proline-threonine. Proctolin has action on skeletal, heart, and visceral muscle. In crustaceans, it controls central pattern-generating nerve networks that regulate feeding and ventilatory behaviors. There are about 40 cell bodies that are proctolin-immunoreactive in the brains of cockroaches, a similar number in Colorado potato beetle, from 80 to 90 in blowfly brain, and about 100 in each lobulus (a specific neuropil region) of the optic lobe in blowflies. Proctolin in the insect brain may act as a neurohormone and as a modulator of responses within central synaptic neuropil. In *M.* sexta, NSC that are proctolin immunoreactive send branches and arborizations to the CA, and in Colorado beetle such terminals are found in the CC.

8.8.3 直肠肽

直肠肽是最早被测序的昆虫神经激素。它包括的氨基酸序列为精氨酸 - 酪氨酸 - 亮氨酸 - 脯氨酸 - 苏氨酸。直肠肽作用于昆虫骨骼肌、心肌和内脏肌。在蜚蠊脑内有 40 个细胞体对直肠肽有免疫反应，科罗拉多马铃薯甲虫的数量与之相同，丽蝇脑有 80 ～ 90 个细胞体，而丽蝇视叶的小叶（专门的神经毡区）中大约有 100 个。昆虫脑内的直肠肽在中心突触神经毡内可能作为一种神经激素和反应的调质。在烟草天蛾中，NSC 是具有直肠肽免疫反应的，其发出的分枝和树状末梢到咽侧体，而在科罗拉多马铃薯甲虫中类似的终端则在心侧体。

8.8.4 FMRFamide-related peptides (FaRPs)

Neuropeptides with the general structure of FMRFamides are widely represented throughout the Metazoa, and are characterized by the amino acid sequence phenylalanine-methionine-arginine-phenylalanine amide (FMRFamide) at the carboxy-terminal end. Because of their wide distribution and important actions in many groups, the FaRPs are the best studied of all the neuropeptides. FaRPs have been isolated and sequenced from *L. cinerea*, *M. sexta*, *S. gregaria*, *Aedes sp.*, and *C. erythrocephata*. There is striking diversity in the distribution and number of FMRFamides within insects, as there is in other groups. In *Calliphora*, 13 different sequences are found, of which CalliFRMFamide 5 is the most abundant and is located in the ventral nerve cord. *D. melanogaster* has 13 known FMRFamides, while a closely related species, *D. virilis*, has only 10; five are shared by the two species. One of those in *D. melanogaster*, a heptapeptide, is the same as (Calli FMRFamide 11 in *Calliphora*. The distribution of neurons with sequences that show a positive FMRFamide immunoreaction is very widespread in the brain of a number of insects, including the Colorado potato beetle, *Drosophila*, a blowfly, *M. sexta*, and the honeybee. About 240 cell bodies that react positive are located in the proto-, deuto-, and tritocerebrum and suboesophageal ganglion of *Drosophila*. Few functions of FMRFamide peptides have been determined in insects, but CalliFMRFamide 1, CalliFMRFamide 2, and CalliFMRFamide 3 induce salivation from blowfly salivary glands at nanomolar concentrations. FaRRs may have multiple physiological effects in different tissues.

8.8.4 酰胺化 FMRF 相关肽（FaRPs）

具酰胺化 FMRF 普遍结构的神经肽广泛分布在整个动物界（也称后生动物），其特征是在羧基末端有氨基酸序列苯丙氨酸 - 蛋氨酸 - 精氨酸 - 苯丙氨酸（酰胺化 FMRF）。由于其分布的广泛性及其在许多群体中的重要作用，使得 FaRPs 是所有神经肽中研究最透彻的。蜚蠊、鳞翅目昆虫烟草天蛾、直翅目沙漠蝗、双翅目依蚊和丽蝇中的 FaRPs 已被分离并进行了序列测定。像其他群体一样，昆虫酰胺化 FMRF 的数量和分布具有显著的多样性。丽蝇有 13 种不同的序列，其中丽蝇酰胺化 CalliFMRF5 丰度最高，位于腹部神经索。黑腹果蝇已知具有 13 种酰胺化 FMRF, 而其近邻种黑果蝇只有 10 种；两个品系共享 5 种。黑腹果蝇的其中一种七肽与丽蝇酰胺化 CalliFMRF11 相同。具有酰胺化 FMRF 序列免疫反应阳性的神经元在许多昆虫的大脑中分布非常广泛。果蝇的前脑、中脑、后脑和食道下神经节大约有 240 个胞体显示为阳性。少数丽蝇的功能已阐明。如丽蝇酰胺化 CalliFMRF1, CalliFMRF2 和 CalliFMRF3 在纳摩尔浓度能诱导丽蝇唾液腺分泌唾液。酰胺化 FMRF 肽在不同的组织中可能具有多种生理学效应。

8.8.5 Tachykinins

Tachykinins are a large family of peptides in lower vertebrates and mammals, and now found in some insects. There are four locusta tachykinins (LomTK Ⅰ, LomTK Ⅱ, LomTK Ⅲ, and LomTK Ⅳ) isolated from the brain and CC of *L. migratoria*. There is about 50% homology of the sequence of LomTK I with a vertebrate tachykinin, physalaemin. Another group of insect kinins are the leucokinins, first discovered in the cockroach *Leucophe maderae*, but now known also from a cricket *Acheta domesticus* and a locust. The leucokinins show myotropic action on visceral muscle and influence ion transport in Malpighian tubules. Eight

8.8.5 速激肽

速激肽是在低等脊椎动物和哺乳动物中发现的一大肽家族，现在许多昆虫中也被发现。从飞蝗脑和心侧体分离到四种蝗激肽（LomTK Ⅰ、LomTK Ⅱ、LomTK Ⅲ和 LomTK Ⅳ）。其中 LomTK Ⅰ与脊椎动物速激肽，泡蛙肽有 50% 的序列同源性。另一个昆虫激肽群是蜚蠊激肽，最早在马德拉蜚蠊中发现，但现在已知家蟋蟀和蝗虫也有。蜚蠊激肽影响心肌和马氏管内离子的转运。已知

known leucokinins are octapeptides (LK Ⅰ to Ⅷ); there are five known achetakinins and one locusta-kinin. Neurons that show immunoreactivity for these peptides are present in the brains of the insects indicated by the names of the peptides. It is thought that both tachykinins and kinins act as neuromodulators, and the kinins may also have important roles as neurohormones. In *D. melanogaster*, two different tachykinin receptors have been isolated by recombinant DNA procedures. One tachykinin receptor protein has been identified and cloned from *Drosophila*, and it was expressed when put into mouse NIH-3t3 cells, causing the cells to increase synthesis of inositol trisphosphate (IP_3) in response to locusta-tachykinin II.

8 种蜚蠊激肽是八肽激素（LKⅠ至Ⅷ）；其中 5 种是蟋蟀激肽，1 种是蝗虫速激肽。对这些肽类显示免疫反应的神经元位于昆虫脑内，用肽的名称表明。速激肽和激肽都被认为起着调质的作用，激肽也可能作为神经激素发挥重要作用。在果蝇中，通过 DNA 重组技术分离得到两种不同的速激肽受体。其中一个速激肽受体蛋白已在果蝇中被鉴定并克隆，该基因被导入鼠细胞系 NIH-3t3 中表达，导致细胞对蝗激肽Ⅱ反应增加三磷酸肌醇的合成。

8.8.6 Leucokinins (LKs)

Leucokinins (LKs) constitute a neuropeptide family first discovered in a cockroach and later identified in numerous insects and several other invertebrates. In insects, the number and complexity of LK-expressing neurons vary, from the simple pattern in the *Drosophila* larva where the entire CNS has 20 neurons of 3 main types, to cockroaches with about 250 neurons of many different types. Common to all studied insects is the presence or 1-3 pairs of LK-expressing neurosecretory cells in each abdominal neuromere of the ventral nerve cord, that, at least in some insects, regulate secretion in Malpighian tubules. The diverse functional roles of LK include regulation of ion and water homeostasis, feeding, sleep-metabolism interactions, state-dependent memory formation, as well as modulation of gustatory sensitivity.

Eight LKs (sequence-related paracopies) were identified in *L. maderae,* which share the C-terminus pentapeptide FXSWGamide. LKs are encoded by a typical peptide precursor and act on a G-protein-coupled receptor (GPCR). LKs stimulate contractions in visceral muscles of the cockroach, and that they act in vitro on Malpighian tubules of several insect species to induce secretion. Moreover, in *Drosophila*, LK has potent diuretic activity. Immunocytochemical localization of LKs in the CNS of insects such as cockroaches and locusts suggest further modulatory roles in interneuronal systems in the antennal lobe, central complex, biological clock circuits, visual system, as well as hormonal roles by means of neurosecretory cells in the brain and abdominal

8.8.6 蜚蠊激肽（LKs）

蜚蠊激肽（LKs）是一个神经肽家族，最初在蜚蠊中被发现，后来在许多昆虫和其他几种无脊椎动物中也相继发现。在昆虫中，LK 表达神经元的数量和复杂性各不相同，从果蝇幼虫中的简单模式，整个中枢神经系统有 3 种主要类型的 20 个神经元，到蜚蠊中的许多不同类型的约 250 个神经元。所有被研究昆虫的共同点是在腹神经索的每个腹神经原节中出现或 1～3 对表达 LK 的神经分泌细胞，至少在某些昆虫中调节马氏管的分泌。LK 的功能多样性包括调节离子和水稳态、取食、睡眠-代谢相互作用、状态依赖性记忆形成以及调节味觉敏感性。

在马德拉蜚蠊中鉴定出 8 个 LKs（序列相关副拷贝），它们共有 C- 端酰胺化五肽 FXSWG。LKs 由典型的肽前体编码，并作用于 G 蛋白耦联受体（GPCR）。LKs 刺激蟑螂内脏肌的收缩，并在体外作用于几种昆虫的马氏管以诱导分泌。此外，在果蝇中 LK 具有强大的利尿活性。昆虫如蟑螂和蝗虫等中枢神经系统 LKs 的免疫细胞化学定位表明，触角叶、中心复合体、生物钟回路和视觉系统的中间神经系统中的 LK 存在进一步的调节作用以及通过大脑和腹神经节中的神经分泌细胞发挥激素作

ganglia. In these insects, many of the LK neurons have bilateral arborizations supplying neuropils in both brain hemispheres. The LKs are C-terminally amidated and vary in length between 6 and 21 amino acids in common model insects. However, there seem to be some LKs that are even longer. In insects, the LKs are characterized by an FXSWGamide or FXPWGamide C-terminus.

The cockroach LK neurons are more diverse than those in *Drosophila* and seem to underlie distributed functions in different brain/ganglion regions. Such functions may include neuromodulation in the olfactory system, visual system, central complex, mushroom bodies, circadian clock, tritocerebral neuropil, circuits of the thoracic ganglia, and the frontal ganglion (regulation of feeding). The two pairs of protocerebral descending LK neurons, which span the entire ventral nerve cord, may provide a pathway for linking protocerebral and olfactory systems to regulate ganglionic activity. In addition, there are three types of neurosecretory cells producing LKs, namely, lateral and median neurosecretory cells, and abdominal ganglion LK neurons, which probably release LKs into the circulation to target peripheral organs such as Malpighian tubules, heart, and visceral muscle.

8.8.7 Allatotropin (AT) and allatostatin (AST)

Allatotropins (ATs) and allatostatins (ASTs) are neuropeptides isolated from nervous and some non-neural tissues that either stimulate or inhibit, respectively, the corpora allata. Several allatotropins have been discovered based on bioassays showing stimulation of corpora allata, the only one of known structure is Mas-AT from *M. sexta*. Mas-AT has the structure Gly-Phe-Lys-Asn-Val-Glu-Met-Met-Thr-Ala-Arg-Gly-Phe-NH_2. It also has been isolated from the lepidopterans *Spodoptera frugiperda* (fall armyworm) and *Lacanobia olerace* (tomato moth). Physiological stimulation of juvenile hormone synthesis by ATs has generally been demonstrated in adult females, but Mas-AT stimulates, and Mas-AST inhibits, the larval corpora allata of the tomato moth.

There are three identified allatostatins: Mas-AST from *M. sexta* (pGlu-Val-Arg-Phe-Arg-Gln-Cys-Tyr- Phe -Asn-Pro-Ile-Ser-Cys-Phe-OH), Dip-AST from the cockroach *D.punctata*; and an Grb-AST from the cricket *G. bimaculatus* that is similar

用。在这些昆虫中，许多LKs神经元具有双侧树状末梢供给两个脑半球神经毡。在常见的模式昆虫中，LKs是C末端酰胺化的，长度在6～21个氨基酸之间变化。然而，有一些昆虫的LKs似乎更长。在昆虫中，LKs以C-末端酰胺化FXSWG或酰胺化FXPWG。

蟑螂LK神经元比果蝇中的神经元更为多样，似乎是不同脑/神经节区域功能分布的原因。这些功能可能包括嗅觉系统、视觉系统、中心复合体、蕈状体、昼夜节律时钟、后脑神经毡、胸神经节回路和额叶神经节的神经调节（取食调节）。横跨整个腹侧神经索的两对前脑下行LK神经元可能提供连接前脑和嗅觉系统的路径以调节神经节的活动。此外，有三种类型的神经分泌细胞产生LKs，即外侧和中间神经分泌细胞和腹神经节LK神经元，它们可能将LKs释放到循环系统中，以靶向周围器官如马氏管、心脏和内脏肌。

8.8.7 促咽侧体素（AT）和咽侧体抑制素（AST）

促咽侧体素（ATs）和咽侧体抑制素（ASTs）是神经系统和非神经组织分离到的神经肽，对咽侧体分别具有促进和抑制作用。已知几种AT通过生物学分析发现对咽侧体有促进作用，但只有一种来自烟草天蛾Mas-AT已知结构。Mas-AT的结构是Gly-Phe-Lys- Asn-Val-Glu-Met-Met-Thr-Ala- Arg-Gly-Phe-NH_2。这种肽在鳞翅目昆虫秋黏虫（草地黏虫）和草安夜蛾（马铃薯夜蛾）中也被分离得到。ATs对保幼激素合成的生理促进作用一般证实是在雌性成虫中，但Mas-AT刺激和Mas-AST抑制马铃薯草安夜蛾幼虫的咽侧体。

有三种已鉴定的咽侧体抑制素：Mas-AST来源于烟草天蛾，序列是Glu-Val-Arg-Phe-Arg- Gln-Cys-Tyr-Phe-Asn-Pro-Ile-Ser-Cys-Phe-OH；Dip-AST来源于

to the cockroach family of ASTs and also a different AST. Mas-AST and Dip-AST have physiological action on larvae and adults, and in the case of Dip-AST, in the embryo. Action of the cricket AST has been demonstrated only in adults. The *Diploptera* ASTs are a family of 13 allatostatins, each of which has physiological action in inhibiting the corpora allata, but with strikingly different effectiveness. Allatotropins and allatostatins are widely distributed in various tissues of insects and in other invertebrates, and they have physiological action on tissues other than the corpora allata. It appears highly likely that they have fundamental actions unrelated to juvenile hormone synthesis.

蜚蠊；Grb-AST 来源于蟋蟀，序列与蜚蠊 AST 族类似，也有一个不同的 AST。Mas-AST 和 Dip-AST 对幼虫和成虫都有生理作用，Dip-AST 的也作用于胚胎，而蟋蟀 AST 证实只作用于成虫。折翅蠊 ASTs 是一个族的 13 种，都能抑制咽侧体的生理功能，但效果迥然不同。ATs 和 ASTs 在不同昆虫组织和其他无脊椎动物中分布广泛，它们对组织的生理学功能比咽侧体强。高度表明其有些基本功能与保幼激素合成无关。

8.8.8 Crustacean cardioactive peptide (CCAP)

A cardioactive peptide with the sequence PFCNAFTGCamide has been isolated from *L. migratoria* and a crab. Similar or possibly identical peptides appear to occur in *T. molitor* and in *M. sexta*. As the name implies, one action may be to stimulate the heart, but its true function(s) in insects is not known. Corazonin is another cardioactive peptide that has been isolated from the cockroach *Periplaneta*. Corazonin antiserum D reacts with lateral neurosecretory cells in the protocerebrum and in two descending neurons in the blowfly, suggesting the same or a very similar molecule.

8.8.8 甲壳纲心激肽（CCAP）

从飞蝗和蟹中分离到的肽，序列为酰胺化 PFCNAFTGC。在黄粉甲和烟草天蛾内出现相似或可能相同的肽。顾名思义，其可能的作用之一是激发心脏，但在昆虫体内的功能并不知道。黑化诱导素是另一个从蜚蠊分离得到的心激肽。黑化诱导素抗血清 D 与前脑侧神经分泌细胞和丽蝇两个下行神经元反应，表明它是一个相同或非常类似的分了。

8.8.9 Corazonin (Crz)

Crz is a highly conserved neuropeptide composed of 11 amino acids with a different seventh amino acid residue within the mature peptide. It is widely spread in Insecta and plays various roles in physiological processes such as heartbeat acceleration, pigment migration, metabolism homeostasis, developmental and circadian rhythmic regulation, ecdysis delay and spinning reduction in different insect species. In *Drosophila*, the endocrine corazonergic pathway is known to alter stress responses and metabolism. In addition, Crz signaling is important for the development of moth larvae, and induces morphological changes in locusts, which are associated with behavioral transitions. Crz is an important regulator of the sexual and behavioral differentiation in the ponerine ant *H. saltator*. Higher Crz levels were found in workers compared to queens (or pseudoqueens). Moreover, worker-specific hunting

8.8.9 黑化诱导素（Crz）

Crz 是一种高度保守的神经肽，由 11 个氨基酸组成，在成熟肽中具有不同的第 7 个氨基酸残基。它广泛分布于昆虫纲中，在生理过程中发挥着各种作用，如加速心跳、色素迁移、代谢稳态、发育和昼夜节律调节、蜕皮延迟和减少吐丝。在果蝇中，内分泌黑化诱导素能途径可以改变应激反应和代谢。此外，Crz 信号对蛾幼虫的发育很重要，并诱导蝗虫的形态变化，这与行为转换有关。黑化诱导素是猛蚁的性和行为分化的重要调节者。与蚁后区（或假蚁后区）相比，工蚁的黑化诱导素水平更高。此外，工蚁特定的追踪行为可以通过实验增加黑化诱导素水平诱导。

黑化诱导素神经元活性的抑制增加

behavior could be induced by experimental increase of Crz levels.

Inhibition of Crz neuronal activity increased pupal size, whereas it hardly affected pupariation timing. This phenotype resulted from enhanced growth rate and a delay in ecdysteroids elevation during the mid-third instar larval. Crz neurons contact both PG cells and PTTH neurons. The Crz-PTTH neuronal axis modulates ecdysteroids biosynthesis in response to octopamine, uncovering a regulatory neuroendocrine system in the developmental transition from growth to maturation. Crz neurons negatively control systemic growth by regulating basal ecdysteroids biosynthesis *via* PTTH neurons. However, Crz neurons affect PTTH neurons only during the mid-L3 stage. In addition, Crz release is under neural regulation rather than transcriptomic regulation.

了蛹的大小，但几乎不影响化蛹的时间。这种表型是由于幼虫3龄中期生长速率加快和蜕皮甾类升高延迟所致。黑化诱导素神经元同时与PG细胞和PTTH神经元相连。黑化诱导素-PTTH神经轴调节蜕皮甾类的生物合成以响应多巴胺，揭示从生长到成熟的发育过渡中的调节性神经内分泌系统。黑化诱导素神经元通过PTTH神经元调节基本的蜕皮甾类生物合成，从而负调控系统生长。然而，黑化诱导素神经元仅在幼虫3龄中期影响PTTH神经元。此外，黑化诱导素的释放受神经调节而不是转录组调控。

8.8.10 Neuropeptide F (NPF) and short NPFs

Insect NPFs play crucial roles in regulation of growth and development, physiology, behavior and co-evolution with environment. In insects, most of NPFs are consisted of 36-40 amino acids and the sequence of amino acids at C-terminal end is relatively conservative with RPRFamide, which is important to maintain functional activity of NPFs, but short NPFs with 7-16 amino acid residues. NPFs in most insects are widely distributed in brain neurons of larvae and adults, subesophageal ganglia, as well as foregut, midgut and hindgut. NPFs are a group of multifunctional peptides and broadly but differentially present in different tissues and cells, suggesting that an intimate relationship between location and function. For example, the expression of *npf* in midgut is related to feeding behavior, and its expression in brain (clock neurons, mushroom body) is related to daily circadian rhythm and sleep. Its expression in subesophageal ganglia and mouthparts is closely related to sensibility of gustatory tissues. In *B. mori*, the *npf 1* and *npf 2* expressed in the frontal ganglion and the midgut regulate feeding and digestion, and the *npfr* expressed in midgut, brain, fat body, Malpighian tubules, testicles and ovaries plays an important role in feeding.

Insect sNPFs act on G-protein coupled receptors (GPCR) and have been shown to stimulate ovarian development, to be involved in adult diapause regulation, induce host seeking behavior in female mosquitoes, and regulate feeding and growth.

8.8.10 神经肽F（NPF）和短NPFs

昆虫NPF在调节生长发育、生理、行为以及与环境的协同进化中起着至关重要的作用。在昆虫中，大多数NPFs由36～40个氨基酸组成，C末端的氨基酸序列相对保守，这对于维持NPFs的功能活性很重要，但短NPFs只有7～16个氨基酸。大多数昆虫的NPFs广泛分布于幼虫和成虫的脑神经元、食道下神经节以及前肠、中肠和后肠。NPFs是一组多功能肽，广泛但不同地存在于不同的组织和细胞中，这表明位置和功能之间存在密切关系。例如，中肠*npf*的表达与进食行为有关，而其在大脑（时钟神经元、蕈状体）的表达与每日昼夜节律和睡眠有关。其在食道下神经节和口器中的表达与味觉组织的敏感性密切相关。在家蚕中，额叶神经节和中肠中表达的*npf*1和*npf*2调节摄食和消化，中肠、脑、脂肪体、马氏管、睾丸和卵巢表达的*npfr*在取食中起重要作用。

昆虫sNPFs作用于G蛋白耦联受体（GPCR），并已被证明可刺激卵巢发育，参与成年滞育调节，诱导雌性蚊子的宿主寻求行为，并调节取食和生长。

8.8.11 Elevenin

The insect body color is modulated by endocrine control. Corazonin, elevenin, and pigment dispersing factor (PDF) are thought to be involved in body-color control in insects and crustaceans.

Elevenin is a newly discovered novel neuropeptide. Knockdown of either elevenin or G-coupled receptor NlA42 transcript expression by RNA interference caused severe cuticle melanization in the brown planthopper (BPH). Injection of a synthetic elevenin peptide not only rescued the body color phenotype in dselevenin-pretreated individuals but also suppressed melanization of black insects grown in natural conditions. Real-time quantitative PCR results revealed that elevenin expression levels were highest in the brain and salivary gland. Immuno- histochemistry analysis confirmed that a precursor peptide of elevenin was generated in the salivary gland, suggesting that the salivary gland might be an important neurosecretory tissue in addition to the brain in BPH. Furthermore, double-strand RNA-mediated silencing of elevenin and NlA42 resulted in down-regulation of arylalkylamine-*N*-acetyltransferase and up-regulation of tyrosine hydroxylase, whereas elevenin peptide injection resulted in up-regulation of *N*-beta-alanyldopamine synthase and aspartate 1-decarboxylase, indicating a complex regulation network for cuticle pigmentation. In addition, functional characterization demonstrated that NlA42 is a cognate receptor for elevenin, and couples to Gq and Gs proteins, triggering both PLC/Ca^{2+}/PKC and AC/cAMP/PKA signaling pathways in response to elevenin treatment. These findings suggest that the elevenin signaling functions control BPH body color through the tyrosine-mediated cuticle melanism pathway.

Elevenin has been reported as another peptide that contributes to body-color change in insects. The precursor of elevenin is composed of 99-170 amino acid residues with highly conserved disulfide bridges in the mature peptide. In the brown planthopper, *N.lugens*, elevenin is expressed mainly in the brain, salivary gland, and abdominal integument. In *N. lugens*, injection of elevenin induces the inhibitory regulation of cuticle melanization. Knockdown manipulations of elevenin and its receptor NlA42 cause the formation of a dark cuticle, which is involved in the melanin biosynthetic enzymes, *N*-*β*-Alanyldopamine (NBAD) synthase and aspartate 1-decarboxylase at the transcriptional level. Except for body color, elevenin does not affect any morphological

8.8.11 Elevenin

昆虫的体色受到内分泌控制调节。黑化诱导素、elevenin 和色素分散因子（PDF）被认为与昆虫和甲壳类动物的体色控制有关。

Elevenin 是一种新发现的神经肽。RNA 干扰下调 elevenin 或 G 耦联受体 NlA42 转录表达，导致褐飞虱表皮严重黑色素化。注射一种合成的 elevenin 肽不仅挽救了经 dselevenin 预处理的个体的体色表型，而且抑制了在自然条件下生长的黑色昆虫的黑色素化。实时定量 PCR 结果显示，脑和唾液腺中的 elevenin 表达水平最高。免疫组化分析证实，唾液腺中产生了一种 elevenin 前体肽，表明唾液腺可能是除大脑外的重要神经分泌组织。此外，双链 RNA 介导的 elevenin 和 NlA42 沉默导致芳烷基胺 -*N*- 乙酰转移酶的下调和酪氨酸羟化酶的上调，而注射 elevenin 肽导致 *N*-*β*- 丙氨酰多巴胺合酶和天冬氨酸 1-脱羧酶的上调，证明表皮层色素沉着的复杂调控网络。此外，功能表征表明，NlA42 是 elevenin 的同源受体，与 Gq 和 Gs 蛋白耦联，触发 PLC/Ca^{2+}/PKC 和 AC/ cAMP/PKA 信号通路以响应 elevenin 处理。这些发现表明，elevenin 信号功能通过酪氨酸介导的表皮层黑色素化途径控制褐飞虱的体色。

Elevenin 是另一种参与昆虫体色变化的肽。Elevenin 的前体由 99 ～ 170 个氨基酸残基组成，在成熟肽中具有高度保守的二硫键。在褐飞虱中，elevenin 主要表达于大脑、唾液腺和腹壁。在褐飞虱中，注射 elevenin 可诱导对表皮层黑色素化的抑制调节。Elevenin 及其受体 NlA42 的敲除操作导致形成深色表皮层，该表皮层在转录水平上参与黑色素生物合成酶、*N*-*β*- 丙氨酰多巴胺（NBAD）合酶和天冬氨酸 1- 脱羧酶。除体色外，elevenin 不影响

changes such as wing size, which can be observed according to different rearing conditions in *N. lugens*.

任何形态变化，如翅膀大小，这可以根据褐飞虱的不同饲养条件观察到。

8.8.12 Orcokinins (OK)

Orcokinins and homologues as myoactive peptides acting on gut tissues was firstly found, which can be classified as three types of neuropeptides now. The *OK* genes of insects exhibit two types of transcripts encoding *OK-A* and *OK-B*, respectively. In other groups of Arthropoda, the *OK* genes encode both *OK-A* and *OK-B* or orcomyotropin in the same precursor. The latter two are related but shorter peptides (FDAFTTGF- NH_2 and FDAFTTGFGHS).

In *B. mori*, OKs are found in midgut endocrine cells and different neurons of the brain and ventral nerve cord, whereas immunohistochemical analysis using an antibody against an OK-A peptide localize this peptide family only in the nervous system. In *B. mori,* OK is co-expressed with *Bombyx* FMRFa in one pair of the neurosecretory cells of the prothoracic ganglion, and OK peptides are delivered to the surface of the prothoracic gland (PG) through the neurons projecting to the PGs. *OK-A* mRNA was detected in brain, whereas *OK-B* mRNA was detected both in brain and midgut in *B. germanica. In vivo* silencing of OK precursors suggests the involvement of *OK* gene products in the regulation of vitellogenin expression in the fat body, an action that appears to be independent of juvenile hormone.

The effects of OKs are remarkably diverse and include (*in vitro*) stimulation of prothoracicotropic activity, regulation of pigmentation in the lepidopteran *Bombyx mori*, and stimulation of oogenesis in *B. germanica.* OKs have also been implicated in the control of innate behaviors, such as ecdysis in *R.prolixus*, circadian activity in the cockroach *L.maderae*, and escape responses in *T.castaneum.* RNAi-mediated knockdown of the *OK* gene caused a disinhibition of male courtship behavior, including the occurrence of male to male courtship, which is rarely seen in wildtype flies. In addition, *OK* gene silencing caused a reduction in egg production. OK is emerging as an important neuropeptide family in the regulation of the physiology of insects from different orders.

8.8.12 亲肌神经肽（OK）

Orcokinins及其同系物最先作为一种作用于肠组织的肌肉活跃肽被发现，现可分为三种神经肽。昆虫的*OK*基因表现出两种类型的转录物，分别编码*OK-A*和*OK-B*。在其他节肢动物中，*OK*基因在同一前体中编码*OK-A*和*OK-B*或orckyotropin。后者是两种相关但较短的肽（FDAFTTGF- NH_2 和FDAFTTGFGHS）。

在家蚕中，在中肠内分泌细胞以及大脑和腹侧神经索的不同神经元中发现了OKs，而使用抗OK-A肽的抗体进行的免疫组织化学分析仅将该肽家族定位在神经系统中。在家蚕中，OK与家蚕FMRFa在前胸神经节的一对神经分泌细胞中共同表达，OK肽通过投射到前胸腺的神经元传递到前胸腺（PG）表面。在德州小蠊的大脑中检测到*OK-A* mRNA，而在大脑和中肠中均检测到*OK-B* mRNA。OK前体的体内沉默表明*OK*基因产物参与调节脂肪体内卵黄蛋白原的表达，这一作用似乎独立于保幼激素。

OKs的作用具有显著的多样性，包括（体外）刺激促前胸活性、调节鳞翅目家蚕的色素沉着以及刺激德州小蠊的卵子发生。OKs还参与了先天行为的控制，例如吸血蝽的蜕皮、马德拉蜚蠊的昼夜节律活动和赤拟谷盗的逃逸反应。RNAi介导的*OK*基因敲除导致雄性求偶行为的解除抑制，包括发生雄性对雄性求偶，这在野生型果蝇中很少见。此外，*OK*基因沉默导致产卵量减少。OK是一个重要的神经肽家族，在调节不同目昆虫的生理过程中发挥着重要作用。

8.8.13 CCHamides

CCHamides are newly identified insect neuropeptides, which are

8.8.13 酰胺化CCH

酰胺化CCH是一种新发现的昆虫神

widely occurring in most insects. The two peptides of CCHamide-1 and -2 have similar structures and, at least in *Drosophila*, each has its own specific receptor. The localization of the CCHamides in both neurons and gut endocrine cells. Characterization of signaling indicated that Bombyx CCHamide-1 and -2 are specific endogenous ligands for BNGR-A15 and BNGR-A14, respectively. Upon activation by CCHamide -2, BNGR-A14 elicited significant increases in CRE (crepine)-driven luciferase activity, intracellular Ca^{2+} mobilization and ERK1/2 phosphorylation in a G_q inhibitor-sensitive manner, while BNGR-A15 was activated by CCHamide-1, thus leading to intracellular accumulation of cAMP, Ca^{2+} mobilization, and ERK1/2 phosphorylation in a Gs and G_q inhibitor sensitive manner. CCHamides are considered to require intrachain disulfide bonds to activate their respective receptor in the physiological concentration range. CCHamide-1 is more likely to serve as a brain-gut peptide to regulate feeding behavior and growth through BNGR-A15, whereas the CCHamide-2 signaling system might play an important role in the control of multiple physiological processes.

经肽，广泛存在于大多数昆虫中。两种酰胺化 CCH 肽 CCH-1 和 CCH-2 具有相似的结构，至少在果蝇中，每种肽都有自己的特定受体。酰胺化 CCH 定位在神经元和肠道内分泌细胞中。信号表征表明，家蚕酰胺化 CCH-1 和酰胺化 CCH-2 分别是 BNGR-A15 和 BNGR-A14 的特异性内源性配体。在被酰胺化 CCH-2 激活后，BNGR-A14 以 G_q 抑制剂敏感的方式引起 CRE（网穗）驱动的荧光素酶活性、细胞内 Ca^{2+} 活化和 ERK1/2 磷酸化的显著增加，而 BNGR-A15 被酰胺化 CCH-1 激活，从而以 Gs 和 G_q 抑制剂敏感的方式导致细胞内 cAMP 的积累、Ca^{2+} 活化和 ERK1/2 磷酸化。酰胺化 CCH 被认为需要链内二硫键在生理浓度范围内激活其各自的受体。酰胺化 CCH-1 更可能作为一种脑 - 肠肽，通过 BNGR-A15 调节摄食行为和生长，而酰胺化 CCH-2 信号系统可能在控制多种生理过程中发挥重要作用。

8.8.14 Myosuppressin

A myosuppressin (MS) precursor with a length of 143-174 AA is present in all Polyneopteran orders. All of these precursors contain a highly conserved MS motif with a C-terminal amidation site.

MS gene in *R. prolixus* encodes one spacer peptide and one conserved peptide with the sequence QDIDHVFMRFamide. It has unique features among the MSs described so far, since it contains an Ile residue in the third position instead of the conserved Val, and the C-terminal FMRFamide, as opposed to the usual FLRF-amide. *Rhopr* MSs are brain-gut neuropeptides, detected in CNS and posterior midgut. *In vitro* assays demonstrated that *Rhopr* MS inhibited the heart rate and the amplitude and frequency of spontaneous contractions from the anterior midgut and hindgut.

8.8.14 心肌抑制素

一种长度为 143 ～ 174 个氨基酸的心肌抑制素（MS）前体存在于所有的多新翅类昆虫中。所有这些前体都包含一个高度保守的带有 C- 末端酰胺化位点的 MS 基序。

吸血蝽的 *MS* 基因编码一个间隔肽和一个保守肽，序列为酰胺化 QDIDHVFMRF。它在迄今为止所描述的心肌抑制素中具有独有的特征，因为它在第三位置包含 Ile 残基替代了保守的 Val 以及 C- 末端酰胺化 FMRF，而不是通常的酰胺化 FLRF。吸血蝽 MSs 是脑 - 肠神经肽，在中枢神经系统和中肠后部检测到。体外实验表明，吸血蝽 MS 抑制心率、中肠和后肠自发收缩的幅度和频率。

8.8.15 Capa neuropeptides

Capa neuropeptides are secreted from specific secretory neurons

8.8.15 Capa 神经肽

Capa 神经肽由中枢或外周神经系

in the central or peripheral nervous system and released into the hemolymph regulating insect excretion, and acting on the Malpighian tubules to elicit either diuretic or anti-diuretic effects on fluid secretion. Each neuropeptide has a complex mode of action, modulating both cell signaling and ion transport pathways to elicit the required secretion response. Capa peptides were first identified as cardioacceleratory peptide (CAP) peptides CAP1 and 2 from the ventral nerve cord of *M. sexta*. Then it was demonstrated that CAP1 and 2 increased heart rates in adult moths during wing inflation and then again during flight. CAP2 was also shown to modulate heart rate in periplaneta americana and *L. migratoria* and to act *via* inositol 1,4,5 trisphosphate in regulation of heart beat. However, CAP2 was also shown to affect hindgut contraction in wandering *M. sexta* caterpillars, and modulated gut physiology in *L. migratoria. D. melanogaster Capa* encodes Drome-Capa-1 Drome-Capa-2 Drome-Capa-3 or Drome-PK-1. Capa-1 and Capa-2 are similar to *M. sexta* CAP2b, sharing the PRV-NH_2 (amide) C-terminal motif; while Drome-PK-1 is a member of the pyrokinin/pheromone biosynthesis-activating neuropeptide/diapause hormone family. In a word, Capa peptides stimulate diuresis in adult Dipteran tubules.

Moreover, the impact of Capa on fluid homeostasis and therefore insect desiccation tolerance, at least in Diptera, suggests that understanding capa signaling will be of real value in understanding desiccation responses in flies which by extension, may provide new routes for insect control.

统中的特定分泌神经元分泌，并释放到血淋巴中调节昆虫排泄。作用于马氏管，对液体分泌产生利尿或抗利尿作用。每种神经肽都有一个复杂的作用模式，调节细胞信号和离子传输途径，以引起所需的分泌反应。Capa 肽最初被鉴定为心脏加速肽（CAP）CAP1 和 CAP2，来源于烟草天蛾腹侧神经索。然后证明 CAP1 和 CAP2 在成虫的翅膀膨胀期间增加其心率，并在飞行期间再次增加。CAP2 还可调节美洲大蠊和蝗虫的心率，并通过肌醇 1,4,5- 三磷酸调节心跳。然而，研究表明 CAP2 也会影响游走期烟草天蛾幼虫的后肠收缩，并调节沙漠飞蝗的肠道生理功能。果蝇 *Capa* 编码 Drome-Capa-1、Drome-Capa-2、Drome-Capa-3 或 Drome-PK-1。Capa-1 和 Capa -2 与烟草天蛾 CAP2b 相似，共有 PRV-NH_2（酰胺）C- 末端基序；Drome-PK-1 是焦激肽 / 信息素生物合成激活神经肽 / 滞育激素家族的成员。总之，Capa 肽可刺激成年双翅目昆虫马氏管的多尿。

此外，至少在双翅目昆虫中，Capa 对体液稳态和昆虫脱水耐受性的影响表明，了解 Capa 信号将对理解果蝇的脱水反应具有实际意义，进而可能为昆虫控制提供新途径。

8.8.16 Prothoracicotropic hormone (PTTH)

PTTH, a brain derived neuropeptide hormone, is the primary promoter of ecdysteroidogenic activity in the prothoracic gland (PG). PTTH neurons evaluate and integrate various environmental and developmental inputs to determine when to transition to the next developmental stage. A range of environmental and internal factors, such as photoperiod, temperature, and the nutritional state of each animal affect PTTH release, which concurrently affects larval development or behavioral characteristics. Additionally, PTTH neurons in the brain may receive many diverse regulatory inputs from different upstream neurons. There may also be other complex regulatory mechanisms at the level of PTTH neurons, which form an upstream control center that regulates the

8.8.16 促前胸腺激素（PTTH）

PTTH 是一种脑源性神经肽激素，是前胸腺（PG）蜕皮甾类生成活性的主要促进剂。PTTH 神经元评估并整合各种环境和发育的输入信息，以确定何时过渡到下一个发育阶段。一系列环境和内部因素，如光周期、温度和每只动物的营养状态，都会影响 PTTH 的释放，同时影响幼虫的发育或行为特征。此外，大脑中的 PTTH 神经元可能接收来自不同上游神经元的多种不同的调节输入信息。在 PTTH 神经元水平上也可能存在其他复杂的调节机制，形成上游控制中

physiological activity of the PG.

Two pairs of PTTH-releasing neurons are located in each hemisphere of the central brain in *Drosophila* larvae. They project and secrete PTTH neuropeptide into the PG. PTTH activates a canonical mitogen-activated protein kinase pathway *via* its receptor, *torso*, to promote ecdysone production for regulating larval maturation.

Several upstream neural pathways of PTTH neurons have been discovered, demonstrating that a synthetic regulatory mechanism exists at the level of PTTH neurons. Both allatostatin A (AstA)-producing neurons and corazonin (Crz) neurons form synaptic connections with PTTH neurons in the larval brain and increase PTTH release by activating PTTH neurons to promote larva-to-pupa transition. AstA neurons control PTTH secretion through the AstA receptor 1 (AstAR1), which is expressed on PTTH neurons, to start metamorphosis. Silencing *Ast*A expression in the larval brain-delays the onset of maturation and extends the growth period. Hence, AstA/AstAR1 signaling contributes to the balance of development and maturation. Crz neurons are another kind of upstream neuron, and function only during the mid-third instar larval (L3) stage. The inhibition of Crz neuronal activity increases pupal size by enhancing the growth rate and delaying ecdysteroids elevation during the mid-L3 stage, but has almost no effect on pupariation timing.

Furthermore, pigment-dispersing factor (PDF) producing neurons are considered to be upstream neurons that influence larval phototaxis and pupal eclosion in *Drosophila*. The axonal projection regions of PDF neurons remain fairly close to PTTH neurons in both larval and adult brains. Furthermore, rhythmic changes in PTTH transcript levels indicate that pacemaker neurons could temporally control PTTH production. Therefore, PTTH neurons receive circadian information from PDF neurons *via* the inhibitory short neuropeptide F precursor (sNPF) and then transmit time information to the PG *via* PTTH. The synchronization of peripheral clock in the PG with central brain clock could be essential for larval normal development. All of the neurochemical molecules that directly act on PTTH neurons belong to neuropeptides, including AstA, Crz, and sNPF.

In addition, various neurotransmitters affect both steroid

心，以及调节前胸腺的生理活性。

果蝇的两对 PTTH 释放神经元位于幼虫大脑中心的每个半球。它们将 PTTH 神经肽投射并分泌到前胸腺。PTTH 通过其受体 *torso* 激活经典的丝裂原活化蛋白激酶途径，促进蜕皮激素的产生，以调节幼虫成熟。

已经发现了 PTTH 神经元的几种上游神经通路，表明在 PTTH 神经元水平上存在一种综合调节机制。产生咽侧体抑制素 A（AstA）的神经元和黑化诱导素（Crz）的神经元都与幼虫大脑中的 PTTH 神经元形成突触连接，并通过激活 PTTH 神经元增加 PTTH 的释放，促进幼虫到蛹的过渡。AstA 神经元通过在 PTTH 神经元表达的 AstA 受体 1（AstAR1）控制 PTTH 的分泌，从而开始变态。在幼虫大脑中沉默 *AstA* 的表达可以延缓成熟的开始并延长生长期。因此，AstA/AstAR1 信号有助于发育和成熟的平衡。Crz 神经元是另一种上游神经元，仅在 3 龄中期的幼虫（L3）阶段起作用。在 L3 中期，抑制 Crz 神经元活性可通过提高生长速度和延缓蜕皮甾类的升高来增加蛹的大小，但对化蛹的时间几乎没有影响。

此外，产生色素分散因子（PDF）的神经元被认为是影响果蝇幼虫趋光性和蛹羽化的上游神经元。在幼虫和成虫的大脑中，PDF 神经元的轴突投射区域与 PTTH 神经元相当接近。此外，PTTH 转录水平的节律性变化表明，起搏神经元可以暂时控制 PTTH 的产生。因此，PTTH 神经元通过抑制性短神经肽 F 前体（sNPF）从 PDF 神经元接收昼夜节律信息，然后通过 PTTH 将时间信息传输到 PG。PG 中的外周生物钟与中枢生物钟的同步可能对幼虫的正常发育至关重要。所有直接作用于 PTTH 神经元的神经化学分子都属于神经肽，包括 AstA、Crz 和 sNPF。

另外，各种神经递质影响其他昆

hormone release and the time of metamorphosis in other insects, such as lepidopteran insects. For example, cholinergic transmission might directly regulate PTTH release from neurosecretory cells in *B. mori*. PTTH neurons produced spontaneous action potentials at a low firing frequency *in vivo*. In *Drosophila* larvae, the neurotransmitters acetylcholine (ACh) affects larval development by modulating the activity of PTTH neurons. Downregulating the expression of different subunits of nicotinic ACh receptors in PTTH neurons, pupal volume was significantly increased, whereas pupariation timing was relatively unchanged. PTTH neurons were excited by ACh application *ex vivo* in a dose dependent manner *via* ionotropic nicotinic ACh receptors. Relatively low doses of octopamine caused increased Ca^{2+} levels in PTTH neurons, whereas higher doses led to decreased Ca^{2+} levels. PTTH neurons produced spontaneous activity *in vivo*, which provides the possibility of the bidirectional regulation, coming from neurons upstream of PTTH cells in *Drosophila* larvae.

虫（如鳞翅目昆虫）的类固醇激素释放和变态时间。例如，胆碱能的传递可能直接调节家蚕神经分泌细胞的 PTTH 释放。PTTH 神经元在体内以较低的放电频率产生自发动作电位。在果蝇幼虫中，神经递质乙酰胆碱（ACh）通过调节 PTTH 神经元的活性来影响幼虫的发育。下调 PTTH 神经元中烟碱型乙酰胆碱受体不同亚单位的表达，蛹体积明显增大，而化蛹时间相对不变。ACh 以剂量依赖性方式通过嗜离子烟碱型乙酰胆碱受体在体外激活 PTTH 神经元。相对低剂量的章鱼胺导致 PTTH 神经元的 Ca^{2+} 水平升高，而高剂量导致 Ca^{2+} 水平降低。PTTH 神经元在体内产生的自发活动，这为果蝇幼虫 PTTH 细胞上游的神经元的双向调节提供了可能。

8.8.17 Pyrokinin/DH/PBAN

Pyrokinins (PKs) are pleiotropic neuropeptides with significant roles in invertebrate physiology. Peptides of the PK/DH/PBAN family are myotropic and promote pheromone biosynthesis in different insect orders. In addition, they regulate other physiological process in Lepidoptera, such as melanization, induction of embryonic diapause and termination of pupal diapause, puparium formation and ecdysteroidogenesis in prothoracic glands.

PKs constitute a group of peptides characterized by a C-terminal FxPRLamide motif and exerting muscle contracting (myotropic) activity. The peptides are classified into three subfamilies, each having diverse biological roles in insects: ① pyrokinin (PK) includes the pheromone biosynthesis activating neuropeptide (PBAN) and the diapause hormone (DH); ② the capability (CAPA) peptides (peptides produced from the *capa* gene); ③ the ecdysis-triggering hormone (ETH). Particular PKs are the PBAN where precursor sequences have only been identified in lepidopterans. The precursor of the silk moth (*B. mori*) contains in addition to PBAN, the DH, a PK, too. Whereas in the silk moth diapause hormone and PBAN are coded for by the same gene. In the fruit fly, *D.melanogaster*, two genes encode the PK peptides: the *hugin* and capability (*capa*) genes. The *hugin* gene encodes PK/PBAN while the *capa* gene encodes a

8.8.17 焦激肽 /DH/PBAN

焦激肽（PKs）是一种多效性神经肽，在无脊椎动物生理学中具有重要作用。PK/DH/PBAN 家族的肽是亲肌的，可促进不同目昆虫的信息素生物合成。此外，它们还调节鳞翅目昆虫的其他生理过程，如黑色素化、胚胎滞育的诱导和蛹滞育的终止、蛹的形成和前胸腺中的蜕皮甾类的生成。

焦激肽由具有 C- 末端酰胺化 FxPRL 基序特征的一组肽组成，并发挥肌肉收缩（亲肌）活性。这些肽分为三个亚类，每个亚类在昆虫中具有不同的生物学作用：①焦激肽（PK）包括信息素生物合成激活神经肽（PBAN）和滞育激素（DH）；②性能肽（CAPA）（由 *Capa* 基因产生的肽）；③蜕皮触发激素（ETH）。特别的焦激肽是 PBAN，其前体序列仅在鳞翅目昆虫中发现。蚕娥除了含有 PBAN 的前体外，还含有滞育激素 DH，也是一种焦激肽。而在蚕蛾中，滞育激素和 PBAN 是由同一基因编码。在果蝇中，有两个基因编码焦激肽：*hugin* 和 *Capa* 基因。*hugin* 基

PK peptide and two cardioacceleratory peptides (periviscerokinin, PVK or CAP2b). *Hugin* is exclusively expressed in perikarya of the subesophageal ganglion. These neurons project into the central brain, to the ring gland (corpus cardiacum, corpus allatum), and to the pharyngeal muscle. The hugin neurons in fly larvae receive signals from several sensory organs. The PBAN neuropeptide proprotein of the moth *Agrotis ipsilon* includes a diapause hormone, where the leucine-amide is replaced by isoleucine-amide, PBAN, and two additional pyrokinins; in the oriental tobacco budworm (*H.assulta*) DH and PBAN and two further pyrokinins are present in the precursor. In two *H. halys* PRXamide genes, a pyrokinin gene encode three PK2 peptides (PK2-1, PK2-2, and PK2-3), but no PK1 (DH) peptide, and the *Capa* gene encode two periviscerokinin peptides (Capa-PVK1 and Capa-PVK2) and a Capa-DH (PK1-like) peptide.

Originally PKs were found due to their myotropic activity. They further influence contractions of the locust oviduct, diapause of the eggs of the silk moth (DH), acceleration of pupation in larvae of flesh flies, or melanization and reddening in larvae of moths. PBAN stimulates *via* the PBAN receptor, a heptahelical GPCR; calcium influx and cAMP increase in cells of the pheromone gland. Therefore, the acetyl CoA carboxylase activity is enhanced in some species, and in other ones the reduction of fatty acids to aldehydes or alcohols is stimulated, which in turn induced pheromone biosynthesis and release.

因编码 PK/PBAN，而 *Capa* 基因编码一个焦激肽和两个促心肌肽（围脏激肽，PVK 或 CAP2b）。*Hugin*（果蝇焦激肽）仅在食管下神经节的细胞核内表达。这些神经元投射到大脑中心、环腺（心侧体、咽侧体）和咽部肌。果蝇幼虫的 hugin 蛋白质神经元接收来自几个感觉器官的信号。球菜夜蛾的 PBAN 神经肽蛋白原包括滞育激素，其中酰胺化亮氨酸被酰胺化异亮氨酸、PBAN 和另外两种焦激肽取代；在烟青虫中，DH 和 PBAN 以及另外两种焦激肽存在于前体中。在两个茶翅蝽 PRX 酰胺化基因中，一个焦激肽基因编码三个 PK2 肽（PK2-1、PK2-2 和 PK2-3），但不编码 PK1（DH）肽以及编码两个围脏肌肽（Capa-PVK1 和 Capa-PVK2）和一个 Capa-DH（PK1 类）肽的 *Capa* 基因。

最初发现的焦激肽是由于其亲肌活性。它们进一步影响着蝗虫输卵管的收缩、蚕蛾卵的滞育、肉蝇幼虫化蛹的加速或蛾类幼虫的黑化和变红。PBAN 通过 PBAN 受体（一种七螺旋受体 GPCR）刺激；信息素腺细胞中钙离子的内流和 cAMP 的增加。钙内流和 cAMP 增加。因此，某些物种的乙酰辅酶 A 羧化酶活性增强，而在其他物种中刺激脂肪酸还原为醛或醇，进而诱导信息素的生物合成和释放。

8.8.18 SIFamide peptide

AYRKPPFNGSIFamide, isolated from the grey flesh fly *Neobellieria bullata*, was the first member of this family to be sequenced and described. Since then, SIFamides have been identified in a variety of holometabolous insects, including dipteran, hymenopteran, coleopteran, and lepidoteran insects. SIFamides are extremely conserved, as the isoforms do not only share the eponymous C-terminal Ser-Ile-Phe-amide sequence but in fact differ only in the N-terminal amino acid residue, which can be Ala, Thr, or Gly.

In the larva these neurons each project a major axon to the ventral neuromeres, but during metamorphosis they develop

8.8.18 酰胺化 SIF 肽

酰胺化 AYRKPPFNGSIF 是从灰肉蝇中分离出来的，是该家族中第一个被测序和描述的成员。自那时起，已在多种完全变态昆虫中鉴定出酰胺化 SIF，包括双翅目、膜翅目、鞘翅目和鳞翅目昆虫。酰胺化 SIF 是极为保守的，因为异构体不仅共有同名的 C-末端酰胺化丝氨酸-异亮氨酸-苯丙氨酸序列，而且实际上仅在 N-末端氨基酸残基上不同，可以是 Ala、Thr 或 Gly。

在幼虫中，这些神经元各自向腹侧神经原节投射一个主轴突，但在变态过

extensive arborizations within the entire central nervous system. As in the fleshfly, *R. maderae, Rhodnius* and *Drosophila* only four large cells in the pars intermedia of the brain were found to produce the peptide. In cockroaches and *Schistocerca* there are other additional SIFamide immunoreactive neurons, only in the cockroach R.maderae and *Rhodnius* have SIFamide immunoreactive projections been found in the corpora cardiaca, suggesting an endocrine function for SIFamide in those species.

SIFamide plays an important and perhaps preponderant role in reproduction and sexual behavior. SIFamide and its receptor GPCR have now been shown to be also important for the correct execution of other behaviors as well, such as sleep in *Drosophila* and aggression in a decapod crustacean. While in the tick *I. scapularis* SIFamide innervates the salivary gland and in *Rhodnius* it is released during feeding and increases meal size as well as the rate of heart beat.

In some insect species the SIFamide gene has a paralog. IMFamide is a SIFamide paralog that was isolated from the silkworm *B.mori*, but seems to be generally present in Lepidoptera. Another independent SIFamide gene duplication occurred in the termite *Z.nevadensis* and the locust *L.migratoria* where the paralog peptide is SMYamide. Whereas the primary sequence of IMFamide shows significant differences from SIFamide, SIFamide and SMYamide peptides appear more similar.

(Li Wenchu)

程中，它们在整个中枢神经系统内形成广泛的树突状结构。与肉蝇的一样，马德拉蜚蠊、吸血蝽和果蝇的大脑也只有中间部四个大细胞能产生这种肽。在蟑螂和沙漠蝗中，还有其他额外的酰胺化 SIF 免疫反应神经元，只有在马德拉蜚蠊和吸血蝽中，才在心侧体中发现酰胺化 SIF 免疫反应投射神经元，这表明这些物种具有内分泌酰胺化 SIF 的功能。

酰胺化 SIF 在生殖和性行为中发挥着重要且可能占主导地位的作用。酰胺化 SIF 及其受体 GPCR 现已被证明对正确执行其他行为也很重要，例如，果蝇的睡眠和十足甲壳动物的攻击性。在肩突硬蜱中，酰胺化 SIF 促进唾液腺活动；在吸血蝽中，它在进食过程中被释放，增加了食物量和心跳频率。

在某些昆虫物种中，酰胺化 SIF 基因具有旁系同源物。酰胺化 IMF 是一种从家蚕中分离的酰胺化 SIF 旁系同源物，但似乎普遍存在于鳞翅目昆虫中。另一个独立的酰胺化 SIF 基因拷贝出现在白蚁和蝗虫中，其中的旁系同源物肽为酰胺化 SMY。虽然酰胺化 IMF 的一级序列与酰胺化 SIF 有显著差异，但酰胺化 SIF 和酰胺化 SMY 肽似乎更相似。

（李文楚）

Exercises

1. Write from memory about the composition of insect nerve system.
2. What are the anatomical features and the functions of the mushroom body and central body of insect brain? Especially that of *D.melangaster*, *B.mori* and *S.gregaria* as model animal.
3. What's NSC? Make sure yourself understood neuropeptides and their functions.
4. How some insects use their cognition in hunting and returning back nests?

思考题

1. 默写昆虫神经系统的组成。
2. 昆虫脑的蕈状体和中心体有哪些解剖学特征和功能？特别是模式动物果蝇和沙漠蝗。
3. NSC 是什么？掌握神经肽及其功能。
4. 一些昆虫是如何利用它们的认知来狩猎和返回巢穴的？

第 9 章 昆虫神经生理生化

Chapter 9 Insect Nerve Physiology and Biochemistry

Key points

1.The function of neurons in insects is completed through neuronal network of neuritis and synapses. Many behavior and motion of insects gradually formed a complex model, neuronal regulated and controlled program in the evolution processes.

2.Nervous electric activities are the basic physiology of insect behavior, involving in ion channels on the cell membrane and the change of potential.

3.Signals are conducted along axons from sensory neurons to associate neurons, and to motorneurons that control muscles and gland activities through which the information are integrated in the brain or central nervous system.

知识点

1. 昆虫神经系统的功能都是通过神经突和突触形成的神经网络来完成的。昆虫的许多行为和动作在长期的演化过程中逐渐形成复杂的模式，即神经调控程序。

2. 神经电活动是昆虫行为的生理学基础。神经电活动会牵涉细胞膜上的离子通道及电位的变化。

3. 信号沿着轴突从感觉神经元传导到联络神经元，再从联络神经元传导到运动神经元，通过大脑或神经中枢对信息的整合，产生控制肌肉和腺体的行为。

9.1 神经细胞对刺激的反应

9.1 Nerve Cell Responses to Stimuli

A number of different types of electrical responses (potentials) from a neuron are possible in response to a stimulus. When an experimental electrical stimulus is delivered to a nerve cell membrane, even if it is too weak to excite the cell into action, it causes a passive alteration in the membrane potential called electrotonus (Fig.9-1) because cells are conductors of electricity. The electrotonic effect passively spreads along the length of the cell as far as the natural tissue resistance and capacitance will allow. If the tissue is excitable, and the electrical stimulation is

在对刺激的反应中，神经元可能出现一些不同类型的电反应（电位反应）。由于细胞是电导体，当一个实验性的电刺激被传递到神经细胞膜上时，即使该刺激很弱，不能使细胞兴奋起来，它也会引起膜电位的被动改变，这种变化被称为电紧张（图 9 1）。在自然组织的电阻和电容允许下，电效应沿着细胞的长度被动地扩散。如果组织是兴奋性的，

strong enough, the neuron responds with a graded membrane response, often called a local potential. An even stronger stimulus can cause the neuron to respond by changing the graded response into an all-or-none spike, the action potential (Fig. 9-2). A graded response always precedes a spike. In some neurons, any part of the neuron may conduct a spike; but typically in insect neurons, spikes are conducted by axonal processes.

且电刺激够强，那么神经元就会做出分级的膜反应，这种反应通常被称为局部电位。更强的刺激能引起神经元将分级反应改变为全有或全无峰电位，即动作电位（图 9-2）。分级反应常常发生在峰电位之前。有些神经元的任何部分都能产生峰电位，但昆虫神经元的峰电位通常由轴突的突起传导。

Electrically excitable cells have a characteristic membrane threshold that must be exceeded in order to generate a response. In general, a stimulus must be strong enough to cause a change of about 10-15 mV in the axon membrane potential to exceed the threshold and thereby generate a spike.

电兴奋细胞具有特征性的膜阈值，必须超过阈值才能产生反应。总体上，轴突膜电位的刺激强度要达到 10～15 mV，引起的变化才能跨越阈值从而产生峰电位。

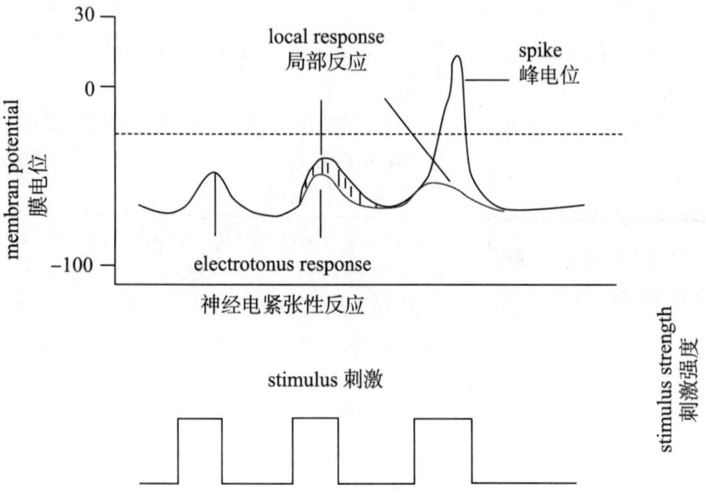

Fig.9-1 A conceptual diagram to illustrate the difference between passive electrotonic response, graded potential response, and spike response of a nerve cell membrane subjected to stimuli by increasingly large square-wave current pulses (modified from Nation, 2022).
图 9-1 说明神经细胞膜受到增大的矩形波电流脉冲刺激时的被动质子反应、分级电位反应和尖峰反应之间区别的概念图（仿 Nation, 2022）。

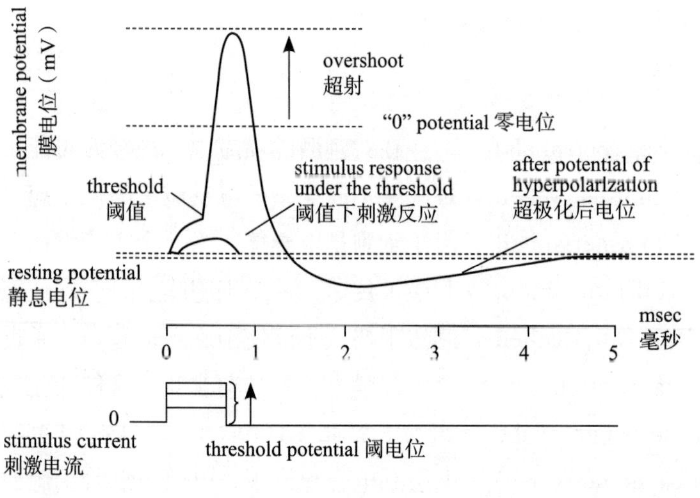

Fig.9-2 Diagram to illustrate the action potential. The overshoot is indicated by the height of the spike above the "0" potential. Stimulus response under the threshold and after-potential of hyperpolarization are also indicated in the figure.
图 9-2 动作电位图解。超射是以峰电位高于"0"电位的高度表示的，还显示了阈值下的刺激反应和超极化的后电位。

9.1.1 Graded Responses

Graded responses are very important responses that nerve cells generate, and some neurons (e.g., some interneurons in the ventral nerve cord) only make graded responses. Graded responses are often named after the site of their occurrence, such as synaptic potentials, receptor potentials, pacemaker potentials, or local potentials. Graded potentials are always localized and usually do not spread very far from their origin; they are propagated decrementally, becoming weaker the farther away they travel from their origin, and they soon are extinguished by resistance of the tissue to current flow. Their intensity is proportional to the strength of the stimulus. They rise and fall slowly in comparison with the 1 to 3ms time duration of a spike. Therefore, graded potentials are sometimes called slow potentials.

9.1.1 分级反应

分级反应是神经细胞产生的非常重要反应，有些神经元（如腹神经索的一些联络神经元）只产生分级反应。分级反应常常以产生部位命名，如突触电位、受体电位和起搏器电位或局部电位。分级电位常会局部化，通常从起点开始并不会传播到很远的距离；其传播是递减的，离起点越远变得越弱，很快就因为组织对电流的电阻效应而消失。它们的强度与刺激强度成正比。与峰电位发生在 1～3 ms 相比，它们上升和下降都很缓慢。因此，分级电位有时也被称为慢电位。

9.1.2 Spike potentials

All-or-none spike or action potentials rise and fall very rapidly, self-generating along the axon, and are propagated without decrement. The size of the spike is not propor tional to the stimulus strength, provided the stimulus exceeds the spike threshold. Within the same neuron over a short period of time, spikes may be about the same size. But different neurons develop spikes of different size.

9.1.2 峰电位

全有或全无的峰电位或动作电位上升和下降非常快，沿轴突自主产生，且传播时没有衰减。只要刺激超过了阈值，峰电位的大小就不与刺激强度成正比。在同一神经元且在很短的时间内，峰电位的大小可能大致相同。但不同的神经元会出现不同大小的峰电位。

9.2 神经元对刺激反应的生理生化基础

9.2 The Physiological and Biochemical Basis for Neuronal Responses to Stimuli

9.2.1 Membrane ion channels: bioelectric potential

Stimuli cause changes in membrane permeability of excitable cells. The major ions involved in the transmembrane potential of nerve cells are K^+, Na^+, Cl^-, Ca^{2+} and large negatively charged organic ions (primarily proteins). The membrane has different permeabilities to each of these ions, and the permeabilities are very different in a resting neuron than in one undergoing stimulation. There is virtually no permeability to large, negatively charged protein ions at any time in a healthy neuron, and permeability to the other ions is

9.2.1 膜离子通道：生物电位

刺激引起兴奋性细胞膜的通透性发生改变。涉及神经细胞跨膜电位的主要离子是 K^+、Na^+、Cl^-、Ca^{2+} 和带负电的大分子有机离子（主要是蛋白质）。膜的通透性对这些离子都不相同，静息神经元和正受到刺激的神经元差异很大。健康的神经元对大分子和带负电的蛋白

variable and depends on the physiological state and voltage across the membrane. The permeabilities to K^+ and Na^+ in a resting neuron can change extremely rapidly upon stimulation. Chloride ions tend to follow the dictates of the distributions of the positive ions. In nerve cells calcium is primarily involved with the release of neurotransmitter at the presynaptic terminals. The distribution of sodium, potassium, chloride, and negatively charged proteins that influence the membrane potential can be represented as follows:

质几乎没有渗透性，对其他离子的通透性是可变的，依赖于其生理状态和跨膜电压。静息神经元对 K^+ 和 Na^+ 的通透性可根据刺激发生极快的变化。Cl^- 有助于遵循阳离子的分布规律。在神经细胞内，钙主要牵涉突触终端神经递质的释放。影响膜电位的钠、钾、氯和带负电大分子蛋白质的分布表示如下：

$$K^+_{\text{inside 内侧}} + Na^+_{\text{inside 内侧}} = \text{Protein}^-_{\text{inside 内侧}} + Cl^-_{\text{inside 内侧}}$$

$$K^+_{\text{outside 外侧}} + Na^+_{\text{outside 外侧}} = \text{Protein}^-_{\text{outside 外侧}} + Cl^-_{\text{outside 外侧}}$$

Despite the potential difference across the cell membrane, electrical neutrality is maintained on each side. The total negatively charged ions on a given side equals the total positively charged ions on the same side. The "outside" in the second equation above is the small space, the mesaxon between the nerve cell membrane and the membrane of the protective glial cell. The concentration of protein is normally very low in the mesaxon space, so most of the negatively charged ions in the mesaxon space are chloride ions. Thus, $C1^-$ concentration is normally greater outside the neuron than inside because the inside negative charge is shared by $C1^-$ and large, nondiffusible negative ions, such as proteins$^-$. During depolarization and repolarization of a nerve cell, ions move through microscopic pores in the membrane. Transmembrane proteins (Fig. 9-3) control the movement of the ions by forming narrow, hydrophilic channels through the cell membrane.

虽然有跨膜电位差，但细胞的某一侧是电中性的。给定的某侧带负电离子的总和等于其同侧带正电离子的总和。上述第二个等式的“外侧”是个小间隙，是神经细胞膜和保护性胶细胞膜之间的轴突系膜。轴突系膜间隙正常的蛋白质浓度非常低，因此轴突系膜间隙中大多数带负电的离子是 Cl^-。所以神经元 Cl^- 浓度通常外侧比内侧大，因为内侧的负电荷由 Cl^- 和大分子和难溶的带负电离子共有，如带负电的蛋白质。在神经细胞的去极化和复极化过程中，离子可穿过膜上的微孔。跨膜蛋白（图 9-3）通过形成狭窄而亲水的通道穿过细胞膜，控制这些离子的运动。

These pores, which are capable of opening and closing rapidly, are frequently described as gates, channels, gated channels, and ion channels. Some ion channels are very selective for a particular ion, while others are relatively nonselective. Potassium channels and sodium channels in nerve cell membranes are generally very selective for the named ion. Numerous ion channels occur in biological tissues, but the most important ones for nerve cell function and muscle cells are channels for Na^+, K^+, Cl^-, and Ca^{2+}.

这些能够快速闭合的孔道，常被称为门、通道、门控性通道和离子通道。一些离子通道对某种特定离子具有高度选择性，另一些相对无选择性。神经细胞膜的 K^+ 和 Na^+ 通道一般对指定的离子具有高度选择性的。生物组织存在很多离子通道，但对神经细胞和肌肉细胞功能最重要的是 Na^+、K^+、Cl^- 和 Ca^{2+} 通道。

Channels are ligand gated (Fig. 9-4) if a neuro transmitter or other molecule controls the opening of the gates, or voltage gated if the membrane voltage controls the opening of the channels.

如果神经递质或其他分子控制着通道的开放，那么通道就是配体门控（图 9-4）；如果膜电压控制着通道的开放，通道就是电压门控。

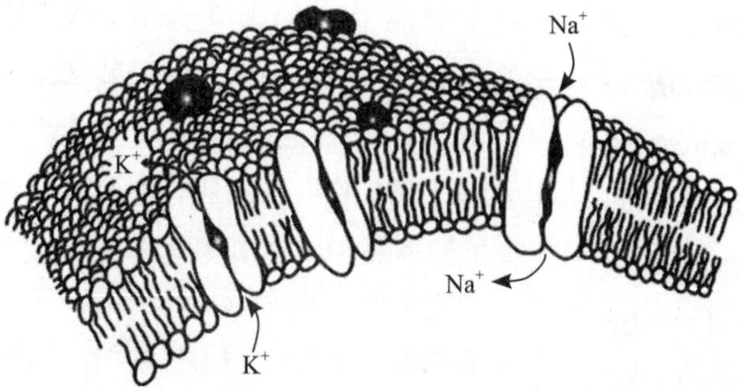

Fig.9-3 A conceptual illustration of transmembrane proteins that function as ion channels. Some channels are ion selective, while others may allow passage of several different ions (modified from Nation, 2022).

图 9-3 跨膜蛋白起离子通道作用的概念图解。有些通道是离子选择性的，而另一些允许几种离子通过（仿 Nation, 2022）。

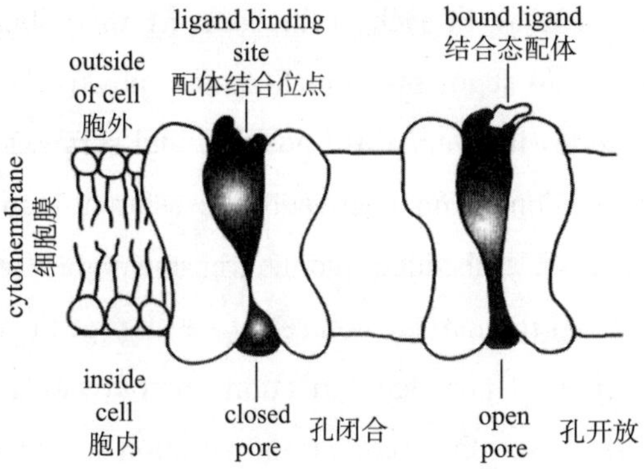

Fig.9-4 Illustration of an ion channel controlled by the binding of a ligand. The hypothetical channel in this diagram is composed of four transmembrane proteins. The ligand is shown being bound to the stippled transmembrane protein and opening the channel (modified from Nation, 2022).

图 9-4 配体结合控制的离子通道图解。图中假设的通道由四个跨膜蛋白组成。配体与点绘的跨膜蛋白结合，打开通道（仿 Nation, 2022）。

9.2.1.1 Voltage gated channel

The voltage gated channels are controlled by mesaxon potential, and usually include Na^+, K^+, Ca^{2+} and Cl^- channel. The transmembranc protein is composed of three parts: ① selectivity filter. It is located at the outside of the cell bilayer membrane and is selective to the ions. ② Voltage sensor. It is located between the cell bilayer membrane which is very sensitive to the variation of membrane voltage. It is the switch of gate controlling. The voltage sensor is formed by four antiparallel transmembrane α-helices, S1–S4 packed in a counterclockwise fashion seen from the cxtracellular side, and the S4 helix of a subunit is within proximity of S5 from the adjacent subunit nearest in the clockwise direction. ③ Gate. It is consist of an intercellular globe at the opening of the inner membrane channel. The globe response rapidly when membrane voltage change so as to open or close the channel which results in variation of ion concentration change inside and outside of membrane and active axon action potential.

1.The typical model of sodium channels

The principal subunit of the voltage-gated sodium channel is a large 260 kDa polypeptide and has four homologous domains (DI to DIV) which together form a functional channel, and contain six α-helical transmembrane segments (S1-S6) and 4 different β-sheet

9.2.1.1 电压门控通道

电压门控通道由轴突系膜电位控制。常见的有 Na^+、K^+、Ca^{2+} 和 Cl^- 通道。电压门控通道的跨膜蛋白由 3 部分组成：①选择性滤器。位于细胞双层膜的外侧，对离子具有选择性。②电压感受元件。位于双层膜中间，对膜电位变化非常敏感，是控制闸门的开关。电压感受元件由 4 个反平行的跨膜 α-螺旋形成，S1 ～ S4 从胞外看是逆时针方向组装，S4 螺旋亚基靠近邻近亚基 S5，几乎是顺时针方向。③闸门。由膜内通道口上的一个胞内球构成，膜电位变化时，能迅速做出反应从而开启或关闭通道，造成膜内外离子浓度的改变，从而激发轴突的动作电位。

1. 钠通道结构的典型模式

电压门控钠通道的主要亚基是一个大约 260 kDa 的多肽，有 4 个同源的结构域（DI ～ DIV），包含 6 个跨膜的 α-螺旋片段（S1 ～ S6），和 4 个不同的 β

(30-40kDa). In each of the domains, the voltage sensor is located in the S4 segments, which contain positively charged aminoacid residues in every third position and serve as voltage sensors to initiate ion channel activation. A reentrant loop between helices S5 and S6 is embedded into the transmembrane region of the channel to form the narrow, ion-selective filter at the extracellular end of the pore. Upon depolarization, the outward movement of the S4 helices and their concurrent rotation initiates a conformational change that opens the sodium-channel pore (Fig.9-5; Appendix color grophic Fig.9-5). This is sliding helix or helical screw model of voltage sensor. The wider intracellular end of the pore is formed by the four S6 segments. Small extracellular loops connect the transmembrane segments, with the largest ones connecting the S5 or S6 segments to the membrane reentrant loop. The intracellular short loops among DIII to DIV are hydrophobic and play a key role in inactive gate. β subunit has dual functions of regulation of ion channel gate and reaction among cells. It is composed of a large molecular domain with extracellular glycosylation, a separated transmembrane segment and a small intracellular domain.

The first sodium ion channel, *para* locus was cloned from *Drosophila* (Loughney et al., 1989). Subsequently, the para genes were cloned from *H. virescens, A. aegypti, B. germanica, B. mori* and *M. domestica*. The *para* sodium ion channel in the silkworm, *B. mori* is 4,882 base pairs. The cDNA fragment is virtually

层（30 ～ 40 kDa）。每个结构域中的电压感受器都位于 S4 片段，每个域的第三位置都含有带正电的氨基酸残基，对离子通道的活性起着电压感受器的作用。S5 ～ S6 的再入式环肽链被掩埋在通道的跨膜区域，在孔道末端的胞外形成一个狭窄的、离子选择性滤器；在去极化的时候，S4 螺旋向外运动，同时完成的旋转启动了构象的变化，使 Cl^- 通道开启。这就是电压感受器的滑动螺旋或螺旋扭转模型（图 9-5；附彩图 9-5）。孔道较宽的分子内末端由 4 个 S6 片段组成。胞外小环与跨膜片段连接，最大的环连接到 S5 或 S6 片段到膜的再入式环肽链。DIII ～ DIV 的分子内短环是疏水性的，充当失活门的角色。β- 亚基有双重功能，调节离子通道门和帮助细胞间反应。它们由一个胞外糖基化的大分子结构域、一个单独的跨膜片断和一个胞内小分子结构域组成。

昆虫的 Na^+ 通道最早被克隆得的果蝇 *para* 基因（Loughney 等，1989），其后相继在棉铃虫、埃及伊蚊、德国小蠊、家蚕和家蝇等昆虫中克隆得到。家蚕的 *para* Na^+ 通道为 4 882 bp，α 亚

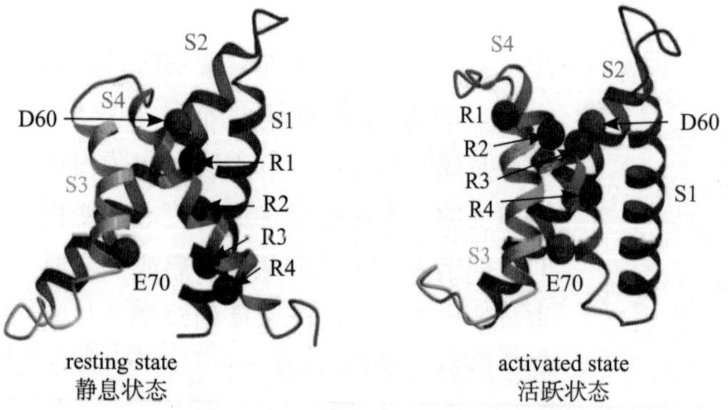

Fig.9-5 Voltage sensing models in both resting and active states. The charged arginine in fragments S1 ～ S4 is R1 ～ R4 in blue, while D60 and E70 are marked in red color. In the resting state, the first arginine gating charge (R1) of the S4 segment is predicted to interact with D60 in the S2 segment. On activation of the voltage sensor, the model predicts that the S4 segment moves outward along a spiral pathway, and the third arginine gating charge (R3) moves into close proximity to D60 in the S2 segment in its activated state. D60 is in an analogous position to negatively charged aspartate or glutamate residues or hydrophilic asparagine residues (modified from DeCaen et al., 2008).

图 9-5 静息和活动状态下的电压感受模型。在 S1 ～ S4 片段中带电精氨酸为 R1 ～ R4，蓝色，D60 和 E70 标记为红色。在静息状态下，预测 S4 片段的第一精氨酸门控电荷（R1）与 S2 片段中的 D60 相互作用。在激活电压感受元件时，该模型预测 S4 片段沿着螺旋路径向外移动，并且第三精氨酸门控电荷（R3）在 S2 片段的激活状态下移动到非常接近 D60。D60 与带负电荷的天冬氨酸或谷氨酸残基或亲水性天冬酰胺残基的位置相似（仿 DeCaen 等，2008）。

identical to the *para* sodium channel α subunit gene amplified from other insects. Amino acid homology of the cloned fragment with *para* sodium channel α subunit genes from the other insects was more than 77% (He et al., 2011). Among 36 exons, One optional exon is unique to *Bm*Nav, while the others are conserved in other insect sodium channel genes. Interestingly, the expression of the mutually exclusive exons is developmentally regulated (Shao et al., 2009).

基的 cDNA 片段与其他昆虫的几乎一致。克隆到的片段氨基酸同源性与其他昆虫同源性高于 77%（He 等，2011）。在 36 个外显子中，有一个随机外显子 *Bm*Nav 是家蚕独有的，而其他外显子在其他昆虫钠通道基因中是保守的。有趣的是，多个单独外显子的表达受到发育调控（Shao 等，2009）。

2.Potassium ion channel

A conceptual model for potassium ion channel function is known as the ball and chain model (Fig.9-6). The model consists of several transmembrane protein domains, and an intracellular ball and chain sequence of amino acids. A molecular conformation allows the ball portion to swing in to block the channel, or swing out to open it. Channel proteins may be composed of a variable number of transmembrane segments and protein subunits.

Drosophila has been important in the study of potassium ion channels. For example, the *shaker* gene codes for the proteins that make the potassium channel in *D. melanogaster*, and it was the first potassium channel gene to be cloned from any biological tissue and studied in detail. The *shaker* potassium channel polypeptide includes seven hydrophobic transmembrane segments that loop back and forth across the membrane. Six segments of the molecule exist as α-helices that span the membrane, while one segment, a β-hairpin loop within the membrane, forms the pore and is capable of conformational changes that allow or occlude passage of K^+. The ball portion of the model is thought to consist of amino acid residues 1 to 20, and the chain, residues 23 to 40. The functional gate consists of a tetramer of four polypeptide subunits. K^+ channels serve a variety of functions in excitable cells. A neuron often expresses multiple types of K^+ channels that accomplish various tasks in cellular signaling (Li et al., 2006).

2.K^+ 通道

已知 K^+ 通道功能的概念模型为球链模型（图 9-6）。这个模型由几个跨膜蛋白域和分子内的球链氨基酸序列组成。分子构象允许球部分向内摆动以阻塞通道，或向外摆动打开通道。通道蛋白质由不同数量的跨膜部分和蛋白质亚基组成。

果蝇对于研究 K^+ 通道方面一直很重要。例如，*shaker* 基因编码构成黑腹果蝇 K^+ 通道的蛋白质，它是第一个从生物组织中克隆出来并被详细研究的 K^+ 通道基因。*Shaker* K^+ 通道多肽包括 7 个疏水的跨膜片段，以回环来回跨膜。分子的 6 个片段以跨越膜的 α- 螺旋存在，其中一个片段 β- 发夹环位于膜内形成孔道，能通过构象的变化允许或阻隔 K^+ 的通行。球体部分由第 1 ～ 20 个氨基酸残基组成，而链部分由第 23 ～ 40 个氨基酸残基构成。功能性闸门则由 4 个多肽亚基的四聚物形成。K^+ 通道在兴奋性细胞内提供多种功能。一个神经元常表达多个类型的 K^+ 通道，以完成胞内多种任务（Li 等，2006）。

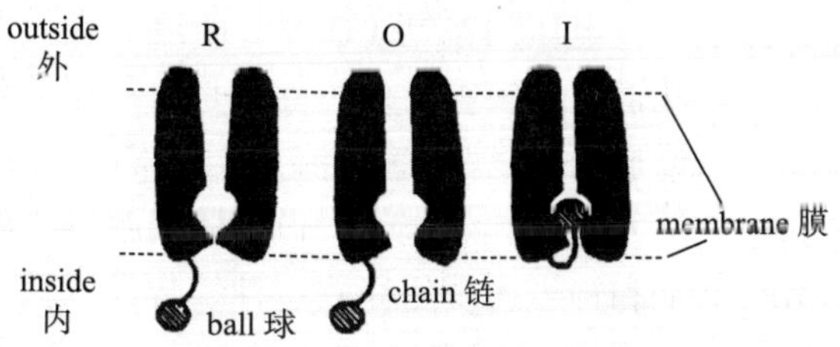

Fig.9-6 The ball and chain model illustrated as a mechanism for opening or closing the shaker potassium channel in *D.melanogaster*. R.resting state of the channel; O.open state; I.inactivated state (modified from Miller, 1991; Nation, 2022).

图 9-6 果蝇钾通道 *shaker* 的球链模型开闭机制。R. 通道的静息状态；O. 打开状态；I. 非活化状态（仿 Miller，1991；Nation，2022）。

A number of other potassium channel genes have been cloned from *D.melanogaster*, including three genes, *Shaw*, *Shab*, and *Shal*, that are similar to *Shaker*, all of which code for channel proteins that form voltage-gated channels. Ganetzky et al. (1993) cloned two additional *Drosophila* potassium channel genes, *slowpoke* (*slo*) and *ether-à-go-go* (*eag*), two genes that code for calcium-activated potassium ion channels in muscle cells. The *Drosophila eag* gene has been shown to regulate neuronal excitability, olfaction, associative learning and larval locomotion. The gene for *slo* is expressed in neurons of the CNS and peripheral system, in muscle cells, in some cells in the midgut, and in tracheal cells of *D. melanogaster*. It is likely that the gene controls ion movements in cells with very different functionalities.

Drosophila gene probes have been used to isolate potassium channel genes from other insects and from vertebrates. For example, a homology to *slo* was isolated from a mosquito, *Aedes* sp., with >90% similarity to the *Drosophila* gene in coding for amino acid sequence. Both *slo* (Ganetzky et al., 1993) and *eag* (Warmke et al., 1993) homologues have been cloned from mouse and human tissues. The proteins coded by the mouse and human genes have 71% and 48% identity, respectively, with *Drosophila eag* protein. The *slo* proteins from mice and humans have about 70% identity to the *Drosophila* proteins coded by *slo*. Certain drugs, insecticides, and naturally occurring poisons, however, can block various ion gates opening or closing them.

3.Calcium channels

Calcium channels are involved in functions including membrane excitability, synaptic transmission, regulated secretion, and cell differentiation. They conduct currents with heterogeneous conductances, kinetics, and pharmacological sensitivities. Calcium channels are heterooligomeric assemblies of α1, α2, δ, β, and γ subunits, the α1 subunit forms the calcium conducting pore of the channel (Zheng et al., 1995). A number of studies indicate that γ subunits modulate calcium current in skeletal muscle and neurons. The calcium channel γ subunit family consists of eight genes that are predicted to produce proteins with four transmembrane segments and intracellular N- and C- terminals.

Voltage-gated calcium (CaV) channels form membrane pores that open in response to membrane depolarization to allow the influx of extracellular calcium ions. They mediate a wide range of critical intracellular processes, including muscle contraction, hormone and neurotransmitter release, neurotrans mission, and regulation of enzymatic activities and patterns of gene

从果蝇中已克隆得到大量的 K^+ 通道基因，包括 3 个与 *shaker* 相似的 *shaw*，*shab* 和 *shal* 基因都能够编码形成电压门控性通道蛋白。Ganetzky 等（1993）克隆了另外两个果蝇 K^+ 通道基因，即 *slowpoke*（*slo*）和 *ether-à-go-go*（*eag*），这两个基因是编码肌肉细胞中 Ca^{2+} 激活的 K^+ 通道。果蝇 *eag* 基因已被证明可调节神经兴奋性、嗅觉、学习相关和幼虫运动。*slo* 基因在果蝇 CNS 神经元和外周神经系统、肌细胞、中肠的某些细胞和气管细胞都有表达。该基因很可能控制着细胞内具有显著不同的功能性离子运动。

果蝇基因探针被用于分离其他昆虫和脊椎动物的钾通道基因。如分离的伊蚊 *slo* 基因编码的氨基酸序列与果蝇的同源性大于 90%。在鼠和人类组织中克隆到 *slo*（Ganetzky 等, 1993）和 *eag*（Warmke 等, 1993）基因，其编码的蛋白质与果蝇 *eag* 蛋白质同源性分别达到 71% 和 48%。鼠和人类的 *slo* 蛋白质与果蝇 *slo* 编码的蛋白质同源性约为 70%。然而，某些药物、杀虫剂和天然产生毒素能阻断不同离子门，打开或关闭它们。

3.Ca^{2+} 通道

Ca^{2+} 通道参与的功能包括膜的兴奋性、突触传递、分泌调节和细胞分化。其传导的电流电导系数、动力学和药理敏感性多种多样。Ca^{2+} 通道是 α1, α2, δ, β 和 γ 亚基装配的异源寡聚体，α1 亚基形成通道的钙传导孔道（Zheng 等, 1995）。大量研究表明，γ 亚基在平滑肌和神经元内调节钙电流。Ca^{2+} 通道 γ 亚基族由 8 个基因组成，预计产生具有四个跨膜段和细胞内 N- 和 C- 终端的蛋白质。

电压门控钙（CaV）通道形成膜上孔道，对膜去极化的反应是打开通道允许胞外 Ca^{2+} 通道的流入。它们介导广泛的关键性内过程，包括肌肉收缩、激素和神经递质的释放、神经传递以及酶活性和基因表达模式的调节。CaV 通

expression. CaV channels are divided into two broad superfamilies based on their voltage- dependence of activation: low-voltage-activated (LVA) CaV channels are activated by small membrane depolarizations and show rapid voltage-dependent inactivation; whereas high-voltage-activated (HVA) CaV inactivate more slowly. The quaternary structure of HVA CaV channels is more complex than that of voltage-gated sodium (NaV) and potassium (KV) channels and typically comprises 4-5 subunits: ① a pore-forming α1 subunit of 170-250 kDa; ② an extracellular α2 subunit; ③ a transmembrane δ subunit that is covalently linked to α2 *via* a disulfide bond to form an 170 kDa α2-δ complex; ④ an intracellular 50-78 kDa β subunit; ⑤ in some cases, a transmembrane γ subunit of 25-36 kDa.

In contrast, LVA CaV channels appear to consist primarily of just the pore-forming α1 subunit, with little evidence of regulation by additional subunits. In HVA Cav channels, the α2-δ, β, and γ subunits modulate properties of the α1 subunit such as activation/inactivation kinetics and the voltage-dependence of activation. The β subunit additionally helps traffic α1 to the plasma membrane. Channels are only activated by larger depolarizations. The *D. melanogaster* genome encodes only three α1 subunits, one β subunit, three α2-δ complexes, and possibly one γ subunit. The Cav channel structure of *Drosophila* only classified into *L*, non-*L* and *T* type. The pore-forming α1 subunit of CaV channels and the α subunit of NaV channels together constitute a superfamily of structurally related voltage-gated channels. The CaV α1 subunit comprises four homologous but non-identical domains (I-IV) connected by cytoplasmic linkers. Each domain contains six putative α-helical transmembrane segments (S1-S6) plus a membrane-penetrating (re-entrant) loop between S5 and S6 that forms the ion selectivity filter. The four domains are arranged in a circular fashion such that the S5 and S6 helices and the re-entrant loops from each domain form the channel pore, with the reentrant loop containing several conserved glutamate residues that are essential for Ca^{2+} selectivity (Catterall, 2000). The cytoplasmic C-terminal loop of CaV1 and CaV2 (but not CaV3) channels contains a consensus sequence motif (designated the "IQ motif") that provides a docking site for Ca^{2+}/ calmodulin (Fig. 9-7; Appendix color graphic Fig.9-7). The β subunit is a variant of the membrane-associated guanylate kinase family of proteins. It comprises an N-terminal SH3 domain tightly associated with an C-terminal guanylate kinase-like domain.

道根据其活化的电压依赖性分为两个广义的超级族：低电压激活（LVA）的 Ca^{2+} 通道通过微弱的膜去极化活化，并表现出快速电压依赖性失活；而高电压激活（HVA）的 Ca^{2+} 通道失活更缓慢。HVA CaV 的四级结构比电压门控钠通道（NaV）和 K^+ 通道（KV）更为复杂，一般由 4 ～ 5 个亚基组成：①形成孔道的 170 ～ 250 kDa α1 亚基；②一个胞外 α2 亚基；③跨膜 δ 亚基通过双硫键与 α2 共价连接，形成 170 kDa α2-δ 复合体；④一个胞内 50 ～ 78 kDa β 亚基；⑤有时含一个跨膜的 25 ～ 36 kDa γ 亚基。

相比之下，LVA CaV 通道似乎主要由形成孔的 α1 亚基组成，几乎没有证据表明有其他亚基的调节。在 HVA Cav 通道中，α2- δ，β 和 γ 亚基调节 α1 亚基的特性，如激活 / 失活的动力学和激活的电压依赖性。β 亚基另外还协助将 α1 输送到仅在较大去极化时才被激活的质量通道。果蝇基因组编码只有 3 个 α1 亚基、1 个 β 亚基、3 个 α2-δ 复合体，可能还有 1 个 γ 亚基。果蝇 CaV 通道仅分为 *L* 型，非 *L* 型和 *T* 型。CaV 通道形成孔道的 α 亚基和 NaV 通道的 α 亚基一起组成结构上与电压门控通道相关的超家族。CaV α1 亚基由 4 个同源但不相同的结构域（I ～ IV）组成，通过胞质接头连接。每个域含有 6 个公认的 α- 螺旋跨膜片段（S1 ～ S6），加上 S5 和 S6 间横穿膜的环（再入式），形成的离子选择器。四个结构域以环形方式排列使 S5 和 S6 螺旋和每个结构域的再入环形成通道孔，再入环含有几个对 Ca^{2+} 通道选择性至关重要的谷氨酸残基（Catterall, 2000）。CaV1 和 CaV2（但不是 CaV3）通道的胞质 C 末端环含有相同序列的元件（称为“IQ 元件”），给 Ca^{2+}/ 钙调蛋白提供锚定位点（图 9-7；附彩图 9-7）。β 亚基是与膜相关的鸟苷酸激酶蛋白族。它由 N- 端 SH3 域牢固地连接到 C- 端的鸟苷酸激酶样结构。

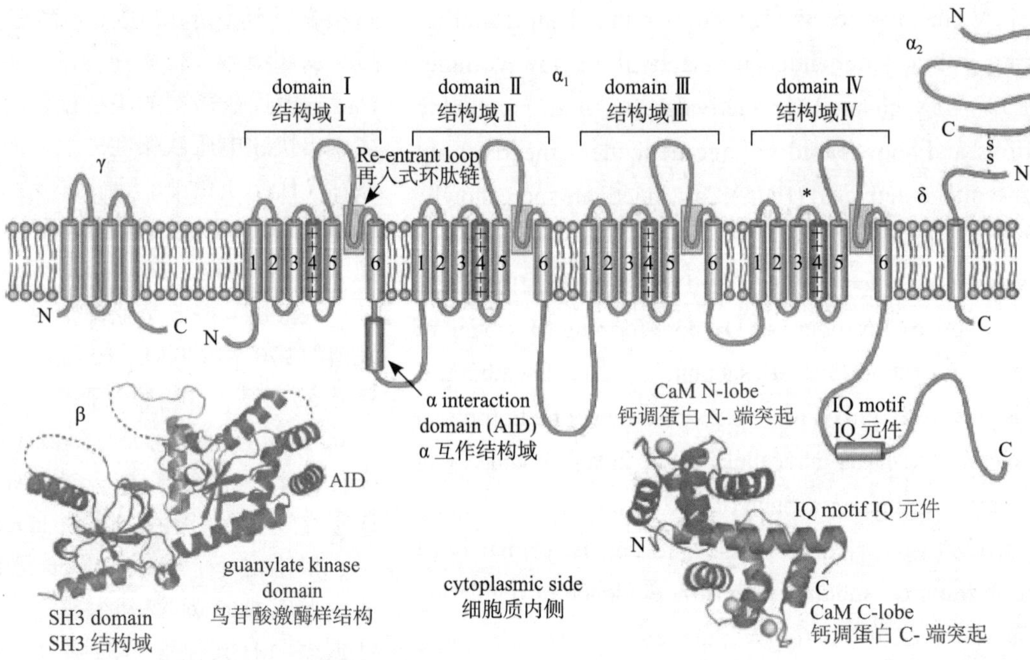

Fig. Cartoon of the molecular structure and membrane topology of CaV channel subunits. HVA CaV channels usually comprise a single copy of each subunit (α1, α2, δ, β, and γ), whereas LVA CaV channels consist of only the pore-forming α1 subunit. CaM. Calmodulin; the "IQ motif". conserved sequence; AID. α-interaction domain (modified from King, 2007).

图 9-7 电压门控钙离子（CaV）通道亚基的分子结构草图和膜的拓扑图。HVA CaV 通道常由单拷贝亚基（α1, α2, δ, β, 和 γ）组成，而 LVA CaV 通道仅由形成通道的 α1 亚基组成。Domain I~IV. 结构域 I~IV；CaM. 钙调蛋白；IQ 元件. 保守序列；AID. α- 互动结构域（仿 King, 2007）。

9.2.1.2 Ligand gated channel

Ligand-gated ion channels superfamily of receptors contains an intrinsic ion channel, which undergoes a conformational change following the binding of an extracellular ligand to its site on the channel. This conformational change allows for the opening of the ion pore. The channel may open and close several times while the ligand is attached. This gating will cease when the ligand disassociates from the binding site. The ligand-gated ion channels group includes the nicotinic acetylcholine receptor (nAchR), the *γ*-aminobutyric acid receptors, the glycine receptor (GlyR), ionotropic glutamate receptors (iGluR), and the P2X receptors, which are ATP-gated membrane ion channels with multifarious roles, including afferent sensation, autocrine feedback loops, and inflammation (Roberts et al., 2006).

9.2.1.2 配体门控通道

配体门控离子通道超家族的受体包含一个内在的离子通道，在细胞外配体与其在通道上的位点结合后发生构象变化。这种构象变化使离子孔道打开。在配体结合于其上时，通道可打开和关闭几次。当配体从结合位点解离时，门控将停止。配体门控离子通道群包括烟碱型乙酰胆碱受体（nAchR）、*γ*- 氨基丁酸受体、甘氨酸受体（GlyR）、促离子型谷氨酸受体（iGluR）以及 P2X 受体，这是一种 ATP 门控的膜离子通道，具有多方面的功能，包括感觉传入、自分泌反馈回路和炎症（Robert 等，2006）。

9.2.2 The resting potential

The resting potential is the potential difference between the inside and the outside of the cell at rest; the inside of the cell is

9.2.2 静息电位

静息电位是指神经细胞在静息状态下的跨膜电位差；此时，细胞膜对 Na^+

negative relative to the outside. At rest, the cell membrane is nearly impermeable to sodium ions and to the large charged protein molecules, but relatively permeable to potassium and chloride ions. The space outside a nerve cell is the mesaxon space between the neuron membrane and the protective glial cell membrane, and potassium ions are typically at low concentration here, but sodium ion concentration is usually high. Thus, both an electrical gradient and a concentration gradient act on any movement of sodium and potassium ions.

The negative charge on the inside of a resting neuron will attract positively charged ions, such as K^+ and Na^+. The concentration gradient acting on K^+ will tend to force it outward until the attraction of the negative charge (the electrical gradient) on the inside just balances the concentration force. Any change in the membrane potential immediately results in a redistribution of K^+ ions until a new equilibrium is achieved. Although both a concentration gradient and an electrical gradient act on Na^+ to promote inward movement when a channel is open, the membrane is nearly impermeable to Na^+ entry at rest, but only very slowly leaks in with time.

和带电荷的蛋白质大分子几乎没有通透性，Na^+ 只能缓慢渗透；而对 K^+ 和 Cl^- 即具有较强的通透性。神经细胞外层空隙是神经元膜与保护性胶细胞膜之间的轴突系膜间隙，一般此处 K^+ 浓度相对较低，而 Na^+ 浓度相对较高。因此，电梯度和浓度梯度都作用于 Na^+ 和 K^+ 的任何运动。

静息神经元内部的负电荷将吸引带正电的离子，如 K^+ 和 Na^+。在内部负电荷（电梯度）的吸引力正好平衡其浓度驱动力之前，作用于 K^+ 的浓度梯度趋向于驱动其外移。膜电位的任何变化可立即引起 K^+ 离子的重新分布直至达到新的平衡。虽然浓度梯度和电梯度均作用于 Na^+，在通道打开时促使其内流，但膜在静息时对 Na^+ 几乎是不通透的，但随时间的推移 Na^+ 有非常缓慢的渗入。

9.3 神经元的电活动

9.3 Electric Activities of Neurons

9.3.1 Action potential (Sodium activation)

Upon stimulation that causes depolarization, a nerve cell undergoes remarkable permeability change to both Na^+ and K^+. The most dramatic change is that the excited membrane becomes explosive permeable to Na^+, and a small number of sodium ions, acted upon by both electrical and concentration gradients, rush into the membrane causing depolarization of the membrane. The sudden increase to Na^+ influx is called sodium activation. The rush of Na^+ for 2 to 3 ms into the neuron when the sodium channels are completely open dominates membrane physiology and its electrical properties. The sodium ions carry an inward current, and this is usually recorded on an oscilloscope or similar device as a spike or action potential, and described as the all-or-none response of a neuron.

The rate at which the membrane potential changes is very fast.

9.3.1 动作电位（钠活化）

在引起去极化的刺激下，神经细胞对 Na^+ 和 K^+ 的通透性发生了显著变化。最引人注目的变化是兴奋性细胞膜对 Na^+ 爆发性的通透性，在电位差和离子浓度梯度共同作用下，少数 Na^+ 快速涌入膜内，引起膜的去极化。这种 Na^+ 电流的瞬间增加称之为 Na^+ 的活化。当钠通道完全开放时，2～3 ms 的时间 Na^+ 即可涌入神经元，主导了膜的生理和电特性。Na^+ 携带的内流常在示波器或类似设备上记录为峰电位或动作电位，并被描述为神经元的全有或全无反应。

膜电位的变化速率非常快。在动作

The rapid inward movement of sodium ions during an action potential allows the membrane potential to overshoot zero, and the inside becomes positive to the outside over that portion of the membrane surface affected by the action potential. The reversal is called the overshoot potential. In a cockroach giant axon, overshoot potentials of about +35 mV have been recorded. The total magnitude of the action potential is stated as the sum of the absolute values of the resting potential and the overshoot potential. If a neuron has a resting potential of -70 mV and an overshoot potential of +35 mV, then the action potential is 105 mV.

电位产生过程中，Na^+ 的快速流入使膜电位超射零电位。在动作电位影响下，与膜表面的膜外部位相对的膜内变为正，这种电位反转称为超射电位。在蜚蠊的大轴突内，测得的超射电位是 +35 mV。动作电位的总幅度表示为静止电位绝对值和超射电位绝对值之和。如果某个神经元的静息电位是 -70 mV，超射电位是 +35 mV，那么动作电位就是 105 mV。

During the spike, the membrane is absolutely refractory to further stimulation. A neuron is partially or relatively refractory for a further few to many milliseconds. Only a very strong stimulus will elicit a new response during the relatively refractory period. Thus, although the absolute refractory period and the relative refractory period are short, they set an upper limit, typically about 100 impulses/s, on how many separate nerve impulses a neuron can transmit in 1 second.

在峰电位期内，膜对进一步刺激处于绝对不应期。神经元在接下来的几毫秒到若干毫秒内部分或相对不应。只有特别强的刺激才能在相对不应期内才引起新的反应。因此，尽管绝对和相对不应期都很短，但它限定了单个神经元在 1 s 内传导神经脉冲的上限，大约是每秒 100 个脉冲。

Sodium concentration in the mesaxonal space is very important to the action potential, and low sodium concentrations result in small action potentials. Although the concentration of sodium ions is relatively high in the mesaxon space, only a small volume of solution containing the ions occurs in this restricted space. The clefts in the mesaxon channel around a cockroach giant axon contain only enough free sodium ions for about 20 to 30 action potentials if no corrective pump action occurs. Thus, without some way to restore the ionic composition of this microchannel around an axon, it would soon fail to fire. In reality, axons are capable of firing continuously for many minutes because at the inner membrane surface of the glial cell coupled Na^+-K^+ pumps pump Na^+ from the glial cell into the mesaxon channel, and pump K^+ from the mesaxon channel into the glial cell. Accumulation of K^+ in the mesaxon would also be detrimental to continued function. The axon membrane also has a coupled Na^+-K^+ pump that works to pump Na^+ out of the axon and bring K^+ back into the axon.

轴突系膜空间内的钠浓度对动作电位非常重要，低浓度 Na^+ 引起小动作电位。尽管轴突系膜的 Na^+ 浓度相对较高，但由于空间有限，只有一小部分含 Na^+ 的溶液出现。没有离子泵的补偿作用，蜚蠊大轴突的轴突系膜间隙的 Na^+ 仅够产生 20～30 次动作电位。因此，微小通道内若没有其他途径重储离子组分，轴突很快就会失活。事实上，轴突可连续不断的触发很多分钟，因为胶细胞的内膜表面耦联有 Na^+-K^+ 泵，可将 Na^+ 从胶细胞泵入轴突系膜通道，而将 K^+ 从轴突系膜泵入胶细胞。轴突系膜内 K^+ 的积聚也是对连续活跃十分不利的。轴突系膜上也耦联有 Na^+-K^+ 泵，用于将 Na^+ 泵出轴突，而将 K^+ 带回进轴突。

9.3.2 Sodium inactivation and repolarization

9.3.2 钠失活和复极化

The excited membrane state is normally a very transitory event, and the sodium channels have a time dependent closing mechanism, called sodium inactivation. The time required for complete sodium inactivation to occur is variable, but can be a few milliseconds or several hundred milliseconds. During sodium inactivation, the sodium channels close.

膜的兴奋状态是个非常短暂的事件，钠通道具有时间依赖性关闭机制，称为钠失活。钠完全失活的发生所需时间是可变的，但可以是几毫秒或几百毫秒。钠失活时，钠通道关闭。

Membrane permeability to potassium changes immediately after the spike develops, and potassium starts to move outward across the axonal membrane, but it moves slowly at first, and the membrane is dominated by the inwardly directed sodium ion movements for a few milliseconds. The outwardly directed potassium current is counter to the inwardly directed current flow carried by sodium. Only when the sodium channels have partially closed, thus restricting the inward flow of Na^+, does the outward flow of potassium begin to bring the membrane potential back toward the resting value.

As K^+ continues to move out and the sodium channels close, the membrane potential begins to return to its resting condition in which the inside is negative to the outside. The negative pole of this small biological battery now attracts the positively charged potassium ions and slows their outward movement. Net outward flux ceases when the membrane potential has become sufficiently negative enough to attract potassium and counterbalance the concentration gradient that drives it outward. The neuron has recovered its resting value; it is repolarized and ready to respond to a new stimulus.

对钾的膜通透性在峰电位产生后迅速改变，钾开始跨越轴膜往外运动，但开始时移动缓慢，膜在几毫秒内受到内流方向的 Na^+ 运动控制。外向的钾电流与钠携带的内向电流大致相当。只有在钠通道已经部分关闭，借此限制了 Na^+ 内流，钾的外流才开始促进膜电位回流至静息值。

当 K^+ 继续向外运动，钠通道关闭时，膜电位开始恢复到其内负外正的静息状态。这个小生物电池的负极现在吸引带正电的 K^+，使其向外的运动减缓。当膜电位负电位足以吸引 K^+，驱动其外流的浓度梯度达到平衡时，外向的净电流停止。神经元恢复到了其静息值；神经元恢复到其静息值，并准备对新刺激产生应答。

9.3.3 Conduction of the action potential: the local circuit theory

After a spike arises, it is self-propagating and travels rapidly along the length of the axon. Local electrical disturbances (Fig.9-8;Appendix color graphic Fig.9-8) in the nerve set up a pattern of local currents or local circuits around the excited region of membrane. The propagated impulse does not go backward over the same length of axon it has just passed because that part of the membrane is still too refractory to be depolarized again by the local currents.

According to the local circuit theory, the currents flowing into the

9.3.3 动作电位的传导：局部电路理论

峰电位上升后，它沿着轴长自动而迅速传导。神经内的局部电扰流（图 9-8；附彩图 9-8）建立起局部电流模式或在膜兴奋区域周围建立起局部电路。传导的脉冲在其刚经过的同节段轴突不向后传，因为那部分膜仍处于不应期，不能再被局部电流极化。

根据局部电路理论，流进轴膜的

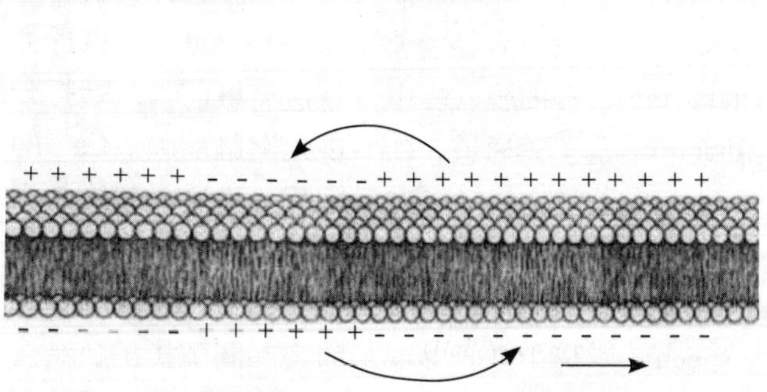

Fig.9-8 Local currents and all illustration of how they function to propagate a spike along the axon. Small currents flowing ahead of the active region open voltage-gated Na^+ channels in the axon. The currents also flow back into the region over which the impulse has just passed, but that region is still partially refractory, and the local currents are not strong enough to open Na^+ channels (modified from Nation, 2022).

图 9-8 局部电流及其如何沿轴突产生峰电位的功能完全图解。流出活跃区域之前的小电流打开轴突电压门控 Na^+ 通道，电流也回流进脉冲刚刚通过的区域，但那个区域仍部分处于不应期，局部电流强度不足以打开 Na^+ 通道（仿 Nation，2022）。

axon membrane just ahead of the active region open voltage-gated Na^+ channels leading to depolarization. The repolarization then allows local currents to flow into the next part of the membrane and cause a spike to develop in that new location, and these actions are repeated along the neuron as the depolarized region moves along the axon. The membrane immediately behind the active region also receives the local circuits, but this part of the membrane is in a state of incomplete recovery (in the relative refractory period) and the local circuits are not strong enough to cause a new spike; thus, nerve transmission is normally unidirectional.

电流正好在打开的电压门控 Na^+ 通道活跃区域的正前方，引起极化。之后的复极化使局部电流进入膜的下半部分，在那个新的位置引起峰电位产生，这些活动在极化区域沿轴突移动时顺着神经元重复。位于活跃区域后方的膜很快也接受到局部电路，但这部分膜处于不完全应期状态（处于相对不应期内），局部电路的强度不足以引起新的峰电位，因此，神经传导通常是单向的。

9.4 突触生理和生化

9.4 Physiology and Biochemistry of the Synapse

9.4.1 Physiology and biochemistry of the synapse

A synapse is any site where one nerve cell influences another neuron. In insects, synapses typically occur between the axon of one cell and a neurite of another neuron within the neuropil of a ganglion. A single neuron can have synaptic contacts with many other neurons. The axon-to-muscle contact is also sometimes described as a synapse. Some synaptic contacts, in addition to the axon to dendrite (neurite), are known in other animals, but they are either not known in insects or are very uncommon; for example, axon-to-soma synapses are very common in vertebrates, but none have been observed in insects.

The transmitter chemical is contained within membrane-bound synaptic vesicles, usually from 200 to about 400 Å in diameter, near the presynaptic membrane. The arrival of spikes at the presynaptic terminals opens calcium channels and Ca^{2+} enters the presynaptic membrane. The entry of Ca^{2+} promotes the fusion of transmitter vesicle membranes with the presynaptic membrane at the synaptic cleft, thus releasing the transmitter chemical into the synaptic cleft. The released chemical diffuses across the synaptic cleft, a distance of some 100 to 200Å, and binds to highly specific receptor proteins on the postsynaptic membrane surface. The binding of the transmitter to receptor opens ion channels in the postsynaptic membrane, and a postsynaptic membrane potential (a graded potential) develops (Fig. 9-9).

9.4.1 突触生理和生化

突触是一个神经细胞影响另一个神经元的任何部位。在昆虫中，突触通常发生在一个细胞的轴突和神经节的神经毡内另一个神经元的神经突之间。一个单神经元能和多个其他神经元有突触连接。轴突至肌肉间的连接有时也描述为突触。有些突触连接，除了轴突到树突（神经突）在其他动物中广为人知，其在昆虫中要么不为所知，要么非常罕见。例如，轴突到胞体的突触在脊椎动物中很普遍，但在昆虫中还没有任何发现。

化学递质被包含在膜结合突触囊包内，直径通常为 200 ～ 400 Å，靠近突触前膜。峰电位到达突触前膜终端打开了钙通道，Ca^{2+} 进入突触前膜。Ca^{2+} 的进入启动了突触间隙处递质囊包膜与突触前膜的融合，因此，向突触间隙释放出化学递质。释放的化学递质跨越距离 100 ～ 200 Å 的突触间隙扩散，结合到突触后膜表面上的高度特异性受体蛋白。递质和受体的结合打开了突触后膜的离子通道，产生突触后膜电位（分级电位）（图 9-9）。

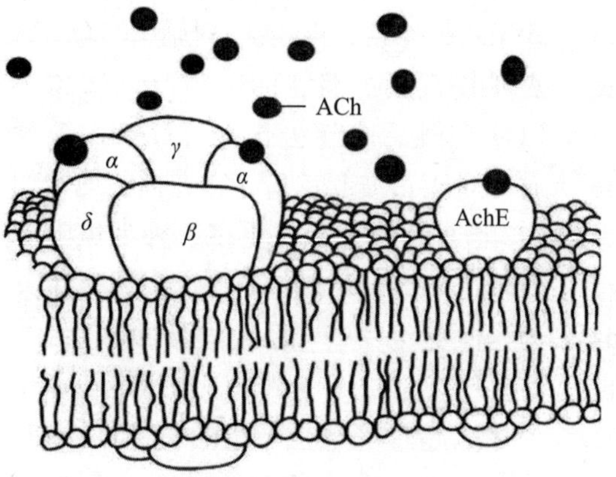

Fig.9-9 A conceptual model for the postsynaptic membrane in which acetylcholine (Ach rounds) may encounter its receptor (the subunits of the sodium channel, left) or acetylcholine esterase molecules (Achase, right). When Ach binds to each of the α subunits, the channel formed by the proteins opens and sodium ions enter the postsynaptic neuron, resulting in a postsynaptic (graded) potential. The Ach receptor quickly releases a bound Ach molecule and it may bind to another receptor and repeat the action. Ach encountering AchE is hydrolyzed into acetic acid and choline (modified from Nation, 2022).

图 9-9 突触后膜乙酰胆碱（Ach 圆形）可能碰到其受体（钠通道的 α 亚基，左）或乙酰胆碱酯酶分子（AchE，右）。当 Ach 与亚基每个 α 结合，蛋白质形成的通道被打开，钠进入到突触后膜神经元，产生突触后膜电位（分级电位）。Ach 受体快速释放出结合的 Ach，又和另一个受体结合并重复作用。Ach 遇到 AchE 分解成为乙酸和胆碱（仿 Nation, 2022）。

A particular synapse is stimulatory or inhibitory, depending on the neurotransmitter released. Stimulatory synapses give rise to excitatory postsynaptic potentials (EPSPs), while inhibitory synapses generate inhibitory postsynaptic potentials (IPSPs). EPSPs and IPSPs are graded potentials. Their magnitude, rate of rise, duration, and spread will depend on the amount of transmitter chemical released, which in turn might depend on the number of spikes arriving per second at the presynaptic terminals. Increasing release of neurotransmitter causes larger postsynaptic potentials. If the postsynaptic potential is a stimulatory potential and is strong enough to spread over the postsynaptic membrane surface and reach the pike generation region of the postsynaptic neuron, then spikes may be generat ed. If the transmitter chemical is an inhibitory chemical, spike generation will be suppressed or prevented in the postsynaptic neuron. This could mean reduction or cessation of spike generation in a postsynaptic neuron that spontaneously is active. There is evidence that aetylcholine (Ach) is a synaptic transmitter at stimulatory synapses in the central nervous system (CNS), and γ-aminobutyric acid (GABA) is an inhibitory transmitter in the CNS of insects. These two transmitter chemicals are not likely to be the only neurotransmitters in the insect CNS, but others remain to be positively identified. The transmitter at nervemuscle junctions (neuromuscular synapses) in insects is *L*-glutamic acid and possibly *L*-aspartic acid. There is fragmentary evidence from insects for a number of additional putative transmitter chemicals, including 5-hydroxy-tryptamine, catecholamines, octopamine, and some peptides.

某个特定的突触是激发性的还是抑制性的，依赖于释放的神经递质。激发性的突触产生兴奋性突触后电位（EPSPs），而抑制性突触产生抑制性的突触后电位（IPSPs）。EPSPs 和 IPSPs 是分级电位。其幅度、上升速率、持续时间和传播依赖于释放的化学递质，反过来又依赖于突触前膜终端每秒到达的峰电位数量。神经递质的释放增加导致突触后电位增大。如果突触后电位是激发性的，强度足以传播到整个突触后膜表面并达到突触后神经元峰电位产生区域，那么就会产生峰电位。如果化学递质是抑制性化学物质，突触后神经元峰电位的产生将会受到抑制或阻碍。这可能意味着减少或停止自发活动的突触后神经元的峰电位产生。有证据证明乙酰胆碱（Ach）是中枢神经系统（CNS）内激发性突触的突触递质，而 γ- 氨基丁酸（GABA）是昆虫中枢神经系统的抑制性递质。这两种化学递质不可能是昆虫 CNS 中唯一的神经递质，但其他的递质仍有待确定。昆虫神经 – 肌肉连接处（神经肌肉突触）的递质是 *L*-谷氨酸，也可能是 *L*-天冬氨酸。少量的证据表明，昆虫中还有一些假定的化学递质，包括 5- 羟色胺、儿茶酚胺、辛胺和一些肽。

Inhibition is a vital process in nervous systems and acts like a brake on the system. Synapses in the 6th abdominal ganglion of *P. americana*, for example, exhibit a high degree of sensitivity to applied GABA, and 1.05×10^{-3} mol/L is sufficient to hyperpolarize postsynaptic membranes. When GABA is released in response to a wave of excitation, it selectively increases permeability to chloride ions, allowing more of these ions with a negative charge to enter the postsynaptic membrane. This causes the inside negative potential to become slightly more negative. A neuron that has been hyperpolarized requires a greater than normal stimulus to change the membrane potential enough to exceed the characteristic membrane threshold for spike generation.GABA is synthesized in insects from glutamate by action of the enzyme *L*-glutamic acid decarboxylase, an enzyme widely distributed in high titers in insect nervous tissue. Following secretion at the nerve end ings, GABA is probably rendered inactive by uptake at the synaptic terminal and/or by glial cells. It also may be oxidized to succinic acid.

抑制是神经系统的一个重要过程，就好像系统的刹车一般。例如，美洲大蠊第六腹神经节的突触，对施用的GABA表现出高度敏感性，1.05×10^{-3} mol/L 就足以使突触后膜超极化。当 GABA 被释放，对兴奋波起反应，它选择性地增加 Cl^- 的通透性，允许更多这种带负电的离子进入突触后膜。这引起内负电位变为稍微更负一些。已经超极化的神经元需要比正常更大的刺激，将膜电位改变到足以超过峰电位产生的特征性阀值。昆虫 GABA 是谷氨酸通过 *L*- 谷氨酸脱羧酶的作用合成的，这种酶广泛存在于昆虫神经组织内，并具有高滴度。随着神经终端的分泌，GABA 通过突触终端和 / 或胶细胞的吸收使之失去活性，它也可能被氧化成琥珀酸。

9.4.2 Acetylcholin-mediated synapses

Numerous studies have shown the presence of the necessary components of a cholinergic system in the CNS of insects; that is, acetylcholine (Ach); choline acetyltransferase that synthesizes Ach, acetylcholineesterase (AchE) (Fig.9-10) that breaks down Ach after its secretion into the synapse; and receptors in the CNS for Ach. Cholinergic receptors are located only within the CNS in insects, and are not found at neuromuscular junctions as in vertebrates. Applications of Ach in isolated insect preparations have not always produced spike activity at physiologically relevant Ach concentrations, possibly because the CNS is protected from Ach applied in a bathing saline by the hemolymph-CNS barrier and fatty sheath surrounding the brain, ventral ganglia, and connectives. A very active acetylcholineesterase also acts quickly to destroy applied Ach. However, the high sensitivity of insect neurons to Ach is revealed by application of the Ach with a microsyringe, or iontophoretically, to neurons inside a ganglion(Callec,1985). Acetycholine administered iontophoretically into thc 6th abdominal ganglion of *P. americana* was stimulatory at a dilution of 1.31×10^{-13} mol/L Ach. Pretreatment of nervous tissue with inhibitors of AchE further increases sensitivity of esheathed ganglia to Ach.

9.4.2 乙酰胆碱介导的突触

大量研究表明，昆虫的中枢神经系统内胆碱能系统必须存在的成分，即乙酰胆碱（Ach）；合成 Ach 的胆碱乙酰转移酶，分泌进突触后降解 Ach 的乙酰胆碱酯酶（AchE）（图 9-10）及中枢神经系统内 Ach 的受体。胆碱能受体仅位于昆虫 CNS 内，在脊椎动物的神经肌肉连接处没有被发现。纯化的昆虫样品应用 Ach 并不能一直在相应的生理浓度产生峰电位活性，可能是因为通过血液 –CNS 屏障、脑、腹神经节和结缔组织周围的脂肪鞘保护，使 CNS 免受浸浴 NaCl 溶液中施用的 Ach 的影响。十分活跃的乙酰胆碱酯酶也起着快速破坏施用的 Ach 的作用。然而，昆虫神经元对 Ach 的高度敏感性可通过对神经节内的神经元显微注射 Ach 或离子渗入法显示出来（Callec，1985）。乙酰胆碱在 1.31×10^{-13} mol/L 的稀释度下对美洲大蠊的第 6 腹腔神经节有刺激作用。

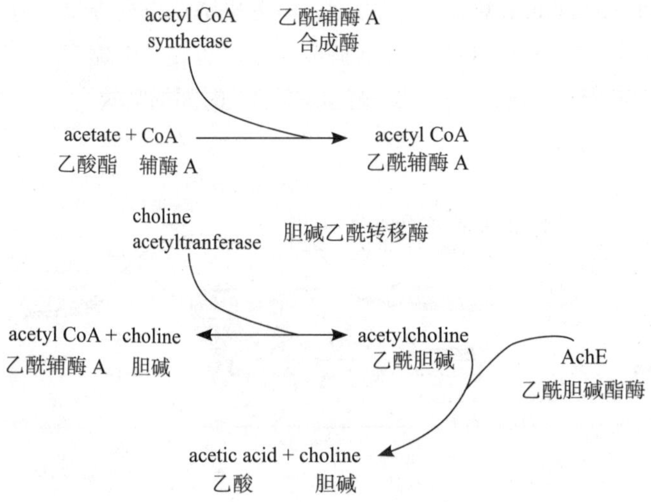

Fig.9-10 Acetylcholine synthesis and degradation pathway of insects.
图 9-10 昆虫乙酰胆碱的合成与降解途径。

9.4.2.1 Action of acetylcholine at the synapse

A volley of spikes arriving at the presynaptic terminal increases permeability of the presynaptic membrane to calcium, which diffuses into the terminal, and through a second messenger sets up a cascade of actions that facilitate attachment of synaptic vesicles to the synaptic membrane. The synaptic vesicles fuse with the membrane and release quanta or packets of Ach into the synaptic cleft. Electron micrographs have shown that the presynaptic membrane expands slightly with the incorporation of vesicular membranes. When released into the synaptic cleft, the Ach molecules diffuse randomly, with some contacting and attaching to acetyl choline receptors and others encountering acetylcholineesterase. Ach is rapidly released from the receptor and may randomly collide with another receptor to repeat the process. When Ach is bound to its receptor, Na^+ channels are opened. When large numbers of Na^+ channels are opened in the postsynaptic membrane, the inward movement of Na^+ depolarizes the postsynaptic membrane with production of an EPSP that is conducted decrementally away from the site of origin. If the stimulation is strong enough, the excitation may spread to the region of the axon that generates spikes (Fig.9-11).

Ach molecules seem to have about the same probability of encountering the enzyme acetylcholineesterase, which is also bound to the postsynaptic membrane, as they do to encounter a receptor molecule. An Ach molecule encountering acetylcholinesterase is hydrolyzed to acetic acid and choline, neither of which has any physiological action at the synapse. Both break-down products diffuse out of the synapse and/or are taken up by the presynaptic neuron, and may be used to synthesize new Ach through action of the enzyme choline acetyltransferase.

9.4.2.1　乙酰胆碱在突触处的作用

峰电位到达突触前膜终端增加了突触前膜对 Ca^{2+} 的通透性，Ca^{2+} 融合进入终端后，通过第二信使产生级联反应，促进突触囊泡附着在突触膜上。突触囊泡与融合膜，并释放 Ach 量子或小包进入突触间隙。电子显微照片显示，突触前膜随着囊泡膜的加入而略微膨胀。当释放到突触裂隙中时，Ach 分子随机扩散，有些接触并附着在乙酰胆碱受体上，有些则遇到乙酰胆碱酯酶。Ach 从受体快速游离出来，可随机地与其他受体相撞以重复这一进程。当 Ach 结合到受体上时，Na^+ 通道被打开。突触前膜上大量 Na^+ 通道被打开之后，Na^+ 的内流极化了突触前膜，产生 EPSP，从其起点不间断地传导。如果刺激够强，兴奋沿轴突区域就会扩散并产生峰电位（图 9-11）。

乙酰胆碱分子遇到乙酰胆碱酯酶的可能性似乎与遇到受体分子的可能性相同，后者也结合在突触后膜上。Ach 分子遇到乙酰胆碱酯酶水解的乙酸和胆碱在突触处都没有任何生理功能。两个裂解产物自突触处扩散和 / 或被突触前神经元吸收，可能通过胆碱乙酰转移酶用于合成新的 Ach。乙酰胆碱酯酶具有保

Acetylcholin esterase is protective in function, and poisoning it with molecules such as organophosphate insecticides results in prolonged stimulation at synaptic sites throughout the CNS of insects.

护功能，用有机磷杀虫剂等分子对其进行毒害会导致昆虫整个中枢神经系统的突触部位受到长时间的刺激。

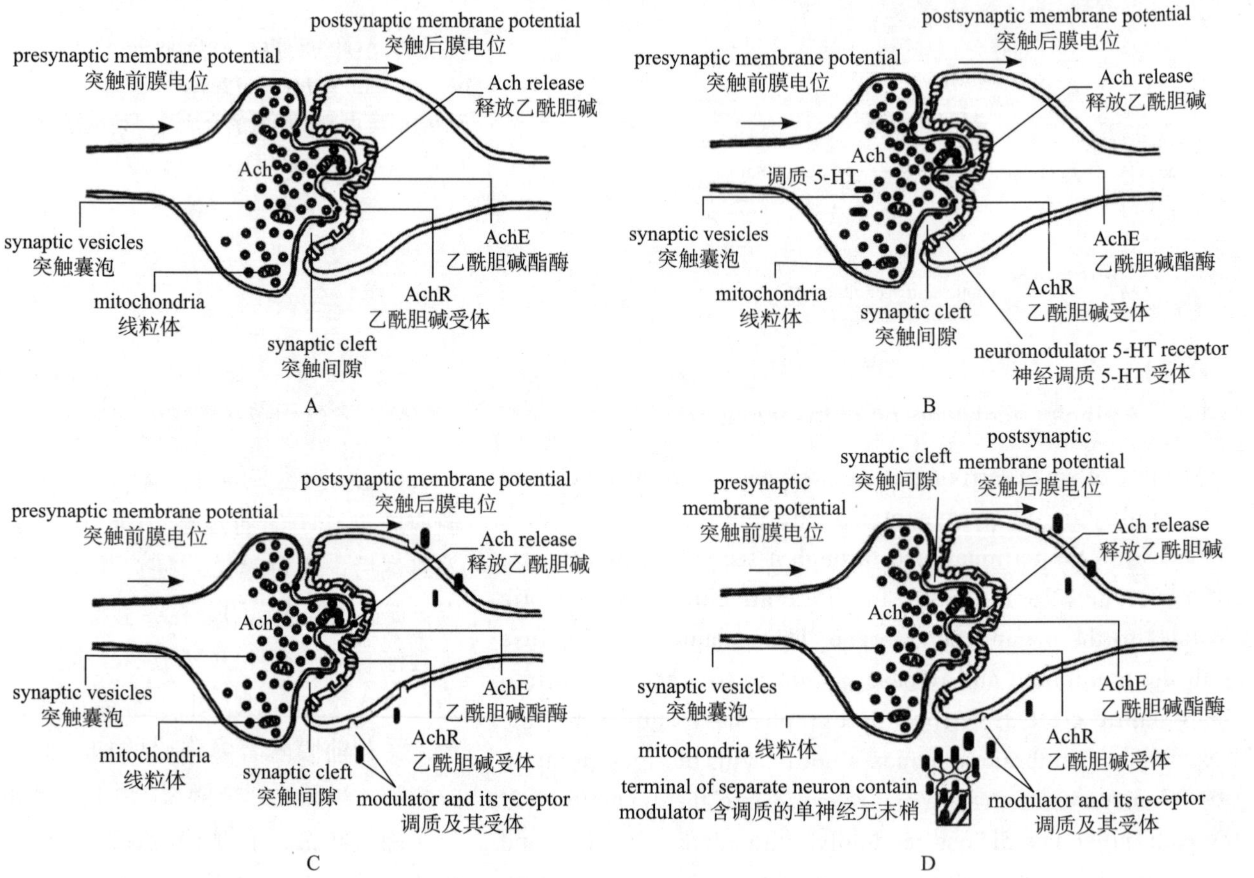

Fig.9-11 A schematic diagram for a synapse in an insect. Synaptic vesicles store the synaptic transmitter near the presynaptic membrane. Arrival of a nerve impulse opens voltage-gated calcium channels; calcium enters the presynaptic membrane and promotes the fusion of synaptic vesicles with the presynaptic membrane at the synaptic cleft region. In a stimulatory neuron in which the transmitter chemical is acetylcholine (Ach), the released Ach diffuses across the synaptic cleft and has about equal probability of encountering an acetylcholine receptor (AchR) or the enzyme acetylcholine esterase (AchE). A. normal secretion of Ach; B. parasynaptic secretion of modulator; C. paracrine secretion of modulator; D. paracrine secretion with terminal of separate neuron containing modulator (modified from Nation, 2022; Chapman, 1998).

图9-11 昆虫突触。靠近突触前膜处的突触囊泡贮存有突触递质。神经脉冲到达时打开电压门控钙通道，钙进入到突触前膜，引发突触间隙处突触囊泡和突触前膜的融合。兴奋性的神经元化学递质是乙酰胆碱（Ach），释放的 Ach 跨越突触间隙扩散，遇到乙酰胆碱受体（AchR）或乙酰胆碱酯酶（AchE）可能性相等。A. 乙酰胆碱正常分泌；B. 调质的副突触分泌；C. 调质的旁路分泌；D. 有单神经元末梢的调质旁路分泌（仿 Nation, 2022; Chapman, 1998）。

Poisoned insects typically show uncontrolled leg tremors, buzzing of the wings without control for flight, and eventual death. Probably many other physiological and biochemical processes are disrupted, such as release of neurohormones and depletion of energy reserves in uncontrolled muscular actions, and all of these contribute to the death of a poisoned insect.

中毒的昆虫通常表现出不受控制的腿部颤抖、翅膀嗡嗡作响无法控制飞行并最终死亡。可能还有许多其他的生理和生化过程被破坏，如神经激素的释放和失控肌肉行为中的能量储备消耗，所有这些导致了中毒昆虫的死亡。

9.4.2.2 Nicotinic and muscarinic cholinergic receptors in insects

In insects, as in vertebrates, more than one cholinergic receptor type exists, and that in insects have been characterized as nicotinic, muscarinic, and mixed receptors (Callec, 1985). At very low concentrations, nicotine and muscarine mimic the action of Ach; at higher concentrations, they block the receptor. In insect central nervous tissue has more nicotinic receptors that muscarinic ones, and Ach is not the synaptic mediator at insect neuromuscular junctions, so insect muscle tissue has neither type of Ach receptor. Nicotinic cholinergic receptors in insects, as in vertebrates, are sensitive to the inhibitory action of a very potent toxin—α-bungarotoxin. The data indicate that nicotinic Ach receptors are present in the CNS of *D.Melanogaster*, *P. americana*, *M. domestica*, and *M. sexta*.

Ach receptors that specifically bind muscarine, a very potent poison obtained from an *Amanita* sp. of mushroom, are classified as muscarinic-type receptors. Muscarinic receptors have been demonstrated in the head of *D. melanogaster* and in the last abdominal ganglion of the cricket, *A. domesticus*. Insects appear to have a much higher concentration of nicotinic receptor types than muscarinic receptor types.

9.4.2.2　昆虫烟碱型和蕈毒碱型类胆碱功能受体

与脊椎动物一样，昆虫不止存在一种胆碱能受体类型，已鉴定昆虫中的有烟碱、蕈毒碱和混合受体（Callec,1985）。在非常低浓度下，烟碱和蕈毒碱的作用酷似 Ach；在高浓度下，它们却阻碍受体。昆虫中枢神经组织烟碱受体比蕈毒碱的多，Ach 不是昆虫神经肌肉连接处的突触介质，因此，昆虫肌肉组织没有任何类型的 Ach 受体。昆虫烟碱胆碱能受体与脊椎动物一样，对一种非常强效的毒素——α- 银环蛇毒素的抑制作用敏感。资料表明，烟碱 Ach 受体存在于果蝇、美洲大蠊、家蝇和烟草天蛾的中枢神经系统中。

特异性结合蕈毒碱的 Ach 受体被归类为蕈毒碱受体，蕈毒碱是一种从鹅膏菌属伞菌中获得的非常强效的毒药。证实果蝇头部和蟋蟀最后一个腹部神经节有蕈毒碱受体。昆虫的烟碱受体类型似乎比毒蕈碱受体类型浓度高得多。

9.4.3 Electric transmission across synapses

Transmission of impulses across some synapses is electrical. In insects, some or possibly all the synapses within giant fiber systems of the ventral nerve cord are electrical. The spike crosses the electrical synapse without involvement of a chemical transmitter. Electrical synapses allow faster transmission of spikes (a message) than a network containing several synapses. The cercal nerve-giant axon complex provides the circuitry for a startle and escape reaction in cockroaches. An escape reaction starts with the reception of stimuli at the mechanoreceptors on the cerci. Strong stimuli result in large receptor potentials and a series of spikes that travel over the cercal nerve to the 6th abdominal ganglion. In the neuropil of the 6th ganglion, Ach is released at synapses with one or more of the giant axons. There is a delay in giant axon response of 0.68 ms to the released Ach, followed by a slow rise time of EPSPs over about 2 to 3 ms. The EPSPs are only about 2 to 5 mV in amplitude, but they give rise to spikes in the giant axon that do not

9.4.3　跨突触电传导

某些突触之间的脉冲传导是电传导。昆虫某些或可能全部腹部神经索的大型神经纤维系统的突触都是电传导。峰电位跨越电突触而没有化学递质的参与。电突触让峰电位（一个信息）的传导要比含有几个突触的神经网络快得多。尾须神经 - 巨轴突复合体供给蜚蠊惊跳和逃逸反应的电路系统。逃逸反应开始于尾须上的机械感受器对刺激的感觉。强刺激产生大受体电位，一系列峰电位在尾须神经上向腹部第六神经节传导。在第六神经节的神经毡内，Ach 在一个或多个巨轴突的突触处被释放出来。释放 Ach 后，巨轴突有 0.68 ms 的延迟反应，接着是兴奋性突触后电位 EPSPs 的缓慢上升期，超过 2 ～ 3 ms。EPSPs 幅度仅仅 2 ～ 5 mV，但

have to cross additional chemically mediated synapses until they synapse in thoracic ganglia with mononeurons to the leg muscles. Much of the time delay in escape can be attributed to the slowness of chemically mediated synapses in the 6th abdominal ganglion and in the thorax. In general, synaptic transmission, and especially a circuit with multiple synapses, appreciably slows the speed of communication within the nervous system. The giant fibers actually represent multiple neurons that have anastomosed together, and the points of fusion are electrical synapses that a spike crosses without a chemical mediator. Probably, selection for speed of transmission was a major evolutionary force acting on the development of electrical synapses in the giant fiber system as part of control for an escape mechanism.

(Sang Wen Qiu Baoli)

其在巨轴突内上升到峰电位，在胸部神经节内与单神经元发生突触传到腿肌之前没有必要跨越额外的化学介导的突触。逃逸过程中延迟的时间归因于第六腹部神经节和胸部化学介导突触的迟钝。总之，突触传送，特别是多突触电路使神经系统内的传递明显减缓。巨神经纤维确实象征着连续的网状多神经元和峰电位无化学介质的电突触融合点。也许，传输速率的选择是一种主要的进化力量，它作用于巨神经纤维系统中电突触的发展，是控制逃逸机制的一部分。

（桑 文 邱宝利）

Exercises

1. What are the three main components of the transmembrane proteins of voltage-gated channels and what are their approximate roles?

2. What are the main ions involved in the transmembrane potential of nerve cells？

3. How exactly are nerve signals conducted in insects?

思考题

1. 电压门控通道的跨膜蛋白主要由哪三部分组成？它们大致的作用是什么？

2. 牵涉神经细胞跨膜电位的主要离子有哪些？

3. 昆虫的神经信号具体是如何传导的？

第 10 章 昆虫激素

Chapter 10 Insect Hormones

Key points

1. Insect hormones mainly include neurohormones, ecdysone, and juvenile hormone synthesized by neurosecretory cells, prothoracic gland, and corpus allatum, respectively.

2. Insect growth and development are synergistically regulated by ecdysone and juvenile hormone. Ecdysone facilities the larval molting and metamorphosis. Juvenile hormone antagonizes the effect of ecdysone to maintain the larval character.

3. The 20-hydroxyecdysone-EcR/USP complex induces the expression of primary responsive transcription factor genes, such as *BR-C*, *E75*, *E74* and *E93*, and then the primary transcription factors regulate the expression of secondary responsive transcription factor genes, such as *βFTZ-F1*, *HR3*, *HR4*, *HR39* and *E78*. The primary and secondary responsive transcription factors further regulate the expression of target genes directly related to molting and metamorphosis.

4. Juvenile hormone binds to its nuclear receptor complex Met/Taiman (SRC or FISC), and then activates the expression of transcription factor *Krüppel Homolog 1* (*Kr-h1*). *Kr-h1* further inhibits the transcript of genes related to the ecdysteroidogenesis and downstream signaling of ecdysone to antagonize metamorphosis.

知识点

1. 昆虫激素主要包括由脑神经细胞分泌的神经激素、前胸腺分泌的蜕皮激素及咽侧体合成的保幼激素等。

2. 昆虫的生长发育主要受蜕皮激素与保幼激素的协同调控。蜕皮激素主导幼虫蜕皮与变态，保幼激素则在幼虫蜕皮期拮抗蜕皮激素作用而维持幼虫性状。

3. 蜕皮激素与其核受体复合物EcR/USP结合后，先诱导初级应答转录因子基因，如*BR-C*、*E75*、*E74* 和 *E93* 等的表达，初级转录因子又调节次级应答基因，如*βFTZ-F1*、*HR3*、*HR4*、*HR39* 和 *E78* 等的表达，初级与次级应答因子再调控与蜕皮和变态直接相关的靶基因的表达。

4. 保幼激素与其核受体复合物Met/Taiman（SRC 或 FISC）结合后，激活转录因子 *Krüppel Homolog* 1（*Kr-h1*）的表达，*Kr-h1* 进一步抑制蜕皮甾类生成及其下游信号通路相关基因的转录，进而抑制变态。

Hormones are the trace chemical messengers secreted by insect tubeless glands or cells and affect a wide variety of physiological processes including embryogenesis, postembryonic development, behavior, water balance, metabolism, caste determination polymorphism, mating, reproduction and diapause, etc. Insect hormones contain endohormones produced by endocrine system and pheromones produced by specific glands in female or male

激素是由昆虫体内无管腺体或细胞所分泌的微量化学物质，影响昆虫的众多生理过程，例如，胚胎发育、胚后发育、行为、水分平衡、代谢、等级决定多态性、交配、繁殖和滞育等。昆虫激素包括由内分泌系统分泌的内激素和雌性或雄性成虫一些特有的腺体分泌的昆

adults. Hormones work along with the nervous system to provide the necessary communication between all the cells in insects.

Understanding insect hormones and their actions is not only helpful to elucidate the molecular regulatory mechanism of growth, development and metamorphosis, but also of great significance in application. For example, hormones, their analogues and antagonists can be used to artificially regulate insect growth and development.

虫信息素等。昆虫激素协同神经系统为昆虫所有细胞提供必要的交流。

了解昆虫激素及其作用不仅有助于阐释昆虫生长、发育及变态的分子调控机理，而且在实际应用上也有重要意义。例如，可利用激素、激素类似物及其拮抗剂人为调节昆虫的生长与发育。

10.1 昆虫内分泌系统

10.1 Insect Endocrine System

The classic endocrine glands in insects include neurosecretory cells (NSCs), prothoracic gland (PG), corpora cardiac (CC), corpora allata (CA), and every kind of accessory glands, etc (Fig.10-1A). The hormones secreted by the endocrine glands transport by the hemolymphe and act on receptor-bearing target tissues or cells elsewhere in the body.

昆虫体内主要的内分泌腺体包括神经分泌细胞（NSCs）、前胸腺（PG）、心侧体（CC）、咽侧体（CA）及各种附腺等（图 10-1A）。内分泌腺分泌的激素通过血淋巴运输，作用于身体表达其受体的靶组织或靶细胞。

10.1.1 The neurosecretory cells

The neurosecretory cells (NSCs) are clusters of relatively large cells with secretory function and locate at the middle and lateral region of insect protocerebrum, suboesophageal ganglion, and other thoracic and ventral ganglia (Fig.10-1B). They are mainly composed of three parts: ① cell body, which is the main synthetic site of neurosecretory particles; ② axons, which transport neurosecretory particles and transmit nerve impulses; ③ expanded terminals of nerve, which store and release of the neurosecretory particles. In general, NSCs have the dual functions of transmitting impulses and secreting hormones.

10.1.1 神经分泌细胞

神经分泌细胞（NSCs）是一类体积较大且具有分泌功能的细胞。其位于前脑中区与侧区、咽下神经节及其他胸腹部神经节中（图 10-1B）。神经分泌细胞主要由 3 部分组成：①细胞体是神经分泌颗粒的合成部位；②轴突可转运神经分泌颗粒，并传递神经脉冲；③膨大的神经末梢，贮存并释放神经分泌颗粒。总体而言，神经分泌细胞具有传递脉冲和分泌激素的双重功能。

10.1.2 The corpora cardiac

The corpora cardiac (CC) is located at the bilateral sides of esophagus and dorsal vessel at the back of the brain (Fig.10-1B). It is neurohemal organ of brain NSCs and is differentiated from the ectoderm to the inside. It is a paired or single structured globoid. It consists of three parts: ① the intrinsic NSCs that bave their cell bodies and axons located entirely within the CC; ② the axon terminals that are from the NSCs of the brain and full of secretory particles; ③ glial cells

10.1.2 心侧体

心侧体（CC）位于脑后方的食道与背血管的两侧，是脑神经分泌细胞的神经血器官，由外胚层向内分化而成（图 10-1B）。心侧体一般是成对的或融合成一个的球状体，主要由 3 部分组成：①心侧体本身的神经分泌细胞；②起源于脑神经分泌细胞的轴突末梢，内部充满神经分泌颗粒；③包围神经分

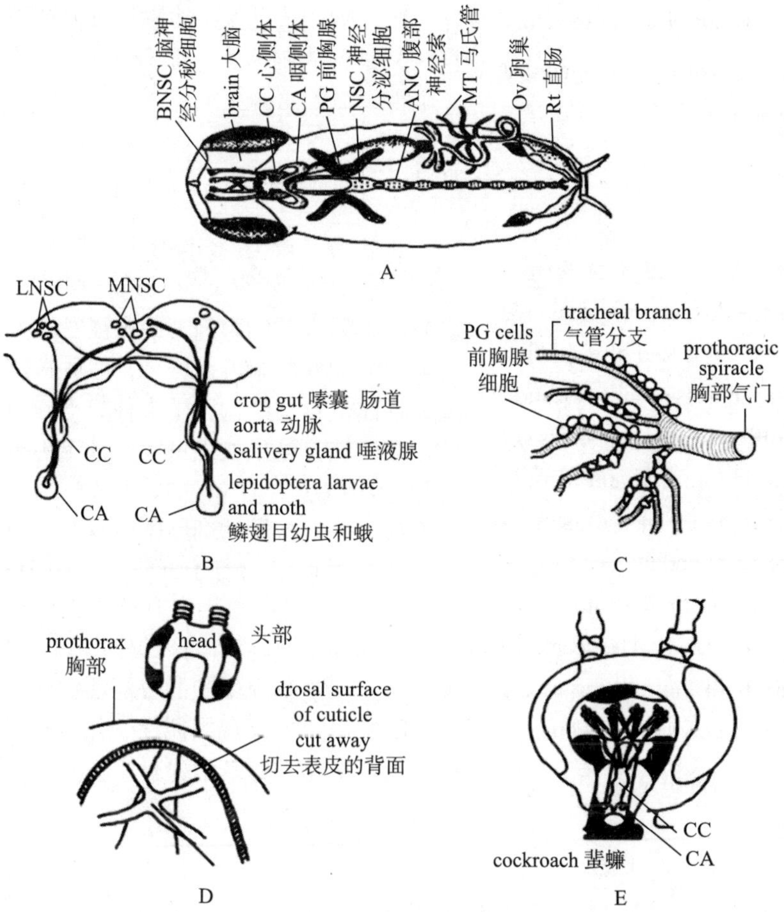

Fig. 10-1 Neuroendocrine structure in different insect species.
A. insect endocrine system. B and C. general positioning of median and lateral NSC and their axons to CC and CA typically in Lepidopteran larvae, and adult PG cells clustered around the larger tracheal branches near the prothoracic spiracle. D. the typical "X" shaped, and nearly transparent PG in an American cockroach nymph. E. view of the head of cockroach to show the fused CC and lateral CA lying on top of the aorta, with NSCs in the protocerebrum with axons leading to CC and CA (modified from Nation, 2022).

图 10-1 不同种类昆虫内分泌系统的结构。A. 昆虫内分泌系统。B 和 C. 鳞翅目幼虫和蛾的中央与侧群神经分泌细胞及其通往心侧体和咽侧体的轴突及聚集在靠近前胸气门较大气管分支周围的成虫前胸腺细胞。D. 美洲大蠊若虫典型的"X"状且近乎透明的前胸腺。E. 展示了蜚蠊位于动脉正上方的融合的心侧体和侧面的咽侧体，前脑的神经分泌细胞由轴突通向心侧体和咽侧体（仿 Nation，2022）。

that encircle NSCs to form pericardium and provide nutrients for NSCs. The main function of CC is to store and release the neurohormones, such as prothoracicotropic hormone, diuretic hormone, and antidiuretic hormone, etc. Additionaly, the NSCs of CC also secrete certain neurohormones such as adipokinetic hormone and myotropins, etc.

10.1.3 The cardiaca allata

The corpora allata (CA) is the ectodermal epithelial gland derived from the differentiation of maxillary and the first mandibular segments during embryogenesis. In Lepidoptera, the paired CA is the small globular organ located in the posterior region of brain and below the CC, and connects the brain *via* the CC nerve cord to form brain-CC-CA system (Fig.10-1B). In Embiotera, Hemiptera, Dermaptcra, and Pscoptera, the left and right CA are fused ventral to the aorta as a central gland. In the higer Diptera, CA, CC and PG are fused into a ring gland, and it is located dorsally to the aorta. CA consists of surrounding membrane and secretory cells. CA contains the axons and secretory

泌细胞的胶质细胞，形成一个围膜，为神经分泌细胞提供营养。心侧体的主要功能是贮存和释放神经激素，如促前胸腺激素、利尿激素和抗利尿激素等。同时，其本身的神经分泌细胞也分泌某些神经激素，如激脂激素和亲肌肽等。

10.1.3 咽侧体

咽侧体（CA）是由胚胎发育时上颚节和第一下颚节间的外胚层芽体发育形成的。在鳞翅目中，咽侧体是位于脑后方心侧体下方的一对球状小器官，通过心侧体神经与脑相连，形成脑 - 心侧体 - 咽侧体系统（图 10-1B）。在纺足目、半翅目、革翅目及啮虫目中，左右的咽侧体在主动脉腹侧融合成一个中央腺。在高等双翅目中，咽侧体、心侧体与前胸腺在主动脉背侧合并为一个环腺。咽侧体主要由围膜和分泌细胞组成。咽侧体有来自神经分泌细胞的轴突及其分泌

substances from the neurosecretory cells, so it is not only the secretory organ of juvenile hormone (JH), but also another neurohemal organ that releases the brain hormones (such as PTTH) as well.

物，所以咽侧体既是保幼激素的分泌器官，又是释放脑激素（如 PTTH）的另一个神经血器官。

10.1.4 The prothoracic gland

The prothoracic gland (PG) develops from the ectodermal invagination of the second mandibular segments during embryogenesis and it is a typical hormone secreting gland. PG is named after its first discovery near the thoracic spiracles in *B. mori*, but its specific location and structure are later found to vary with different insect species. In the lower insects such as locusta, it is located in the head and densely organized. In Lepidopteran larvae, it is composed of a cluster of loose cell string located near the thoracic spiracles and attached to the trachea (Fig.10-1C). In Diptera, it has been incorporated into a ring gland. PG produces ecdysone under the action of PTTH. The secretory activity of the PG cells changes periodically with the molting of the larvae during the larval stage and the PG cells generally degenerate and decompose during the adult stage.

10.1.4 前胸腺

前胸腺（PG）是由第二下颚节的外胚层内陷发育而来的，是典型的激素分泌腺体。前胸腺因其最先被发现位于家蚕前胸气门附近而得名，但后来发现其具体位置与结构因昆虫种类不同而异。在较低等昆虫如蝗虫中，其位于头内且组织致密；在鳞翅目昆虫幼虫体内，其由一群松散的细胞串组成并位于前胸气门附近且附着于气管上（图 10-1C）。在双翅目昆虫中，其合并在环腺中。前胸腺主要在 PTTH 的作用下分泌蜕皮激素。前胸腺细胞的分泌活性在幼虫期随幼虫蜕皮呈现周期性变化且一般在成虫期退化并分解消失。

10.2 促前胸腺激素、蜕皮激素和保幼激素的互作

10.2 The Interaction of PTTH, Ecdysone, and Juvenile Hormone

Insects' growth and development, such as molting and metamorphosis, is under the control of PTTH from the brain neurosecretory cells (NSCs), ecdysone from the prothoracic gland (PG), and JH from the corpora allata (CA).

By secreting PTTH, the brain stimulates the PG to synthesize and secrete ecdysone. Ecdysone combines with the nuclear receptor that consists of a heterodimer to form the ecdysone receptor complex and then binds to DNA to activate or inactivate the transcription of certain genes. Transcripts of these genes turn on an active cascade of genes, ultimately resulting in cell division of epithelial cells, secretion of molting fluid, secretion of a new cuticle, and numerous structural changes in morphology and physiology of internal organs, such as the nervous system, midgut, and reproductive organs. The timing and quantity of JH secretion in target cells influences the type of molting, such as larva-larva,

昆虫的生长与发育，例如蜕皮与变态，主要受到脑神经分泌细胞（NSCs）分泌的 PTTH、前胸腺（PG）分泌的蜕皮激素和咽侧体（CA）分泌的 JH 调控。

通过分泌 PTTH，脑刺激前胸腺合成并分泌蜕皮激素。蜕皮激素与细胞核受体二聚体形成蜕皮激素受体复合体，复合物再与 DNA 结合以激活或抑制一些基因的转录。这些基因的转录打开了基因活跃的级联，最终导致上皮细胞的分裂、蜕皮液的分泌、新角质层的分泌及根据发展阶段导致内部器官（如神经系统、中肠和生殖器官）的形态结构和生理变化。在靶细胞中，JH 分泌的时间和数量通过调节蜕皮激素的作用来影

larva-pupa or pupa-adult, by modifying the action of ecdysone. JH also determines whether major changes occur in the internal organs, such as usually little or no changes occur during the larval molts, whereas dramatic changes occur during the molts of metamorphosis.

In hemimetabolous insects, such as *Nauphoeta cinerea*, JH is secreted earlier than ecdysone in every instar and its titers gradually decrease at the end of the instars, allowing secretion of another cuticle. JH titers are low or undetectable in the final nymphal instar, which is important to allow the nymphs molt into adults (Fig.10-2A). The premolt pulse of ecdysone secretion near the end of the last nymphal instar is crucial for inducing cell division of wing pads, starting the development of flight muscle, and initiating changes of female accessory glands before the molting starts. Subsequently, a major peak of ecdysone in the continued absence of JH causes cuticle secretion of adults, accompanied by changes in internal structure and the appearance of physiology characteristic of adults.

In holometabolous insects, such as *Manduca sexta*, JH titers also tend to be high, relative to ecdysone, in the early larval instars and falls only moderately just prior to each molting. The interaction of JH and ecdysone causes the epidermal cells to secret another larval cuticle. However, JH titers are reduced to a low level before the molting in the final larval instar, partly in response to a decline of synthesis. JH titers sharply drop below the detectable level after entering into the pupal stage (Fig.10-2B).

There are six distinct defined pulses of ecdysone secretion during the development of *D.melanogaster* (Fig.10-2C), each of which precedes by a pulse of PTTH secretion. JH is secreted in conjunction with each peak of ecdysone except for the last one when the pupa molts to the adult. JH is produced in the early part of last larval instar, but its titers decrease to a low or undetectable level in the late larval instar. As part of many physiological and morphological changes induced by the 4th secretory pulse of ecdysone, body-shortening and pupariation are the most obvious. About 12 h after pupariation, a 5th pulse of ecdysone induces the secretion of a pupal cuticle. Finally, a 6th broad pulse of ecdysone is secreted in the presence of low titers of JH in the pupal stage, initiating the adult development.

响蜕皮的类型，即幼虫—幼虫、幼虫—蛹或蛹—成虫。JH 也决定了内部器官是否发生重要变化，例如，在幼虫间蜕皮时，其变化通常很小或没有；但在变态蜕皮时，其变化剧烈。

在不完全变态昆虫如蜚蠊幼龄时，JH 的分泌早于蜕皮激素，其滴度在龄末逐渐下降，允许分泌另一个若虫的角质层。JH 滴度在末龄若虫中非常低或测定不出，这对于允许若虫蜕变为成虫非常重要（图 10-2A）。在接近若虫末龄时，蜕皮激素分泌的预蜕皮脉冲对于诱导翼板的细胞分裂、启动飞行肌的发育及蜕皮真正开始前雌虫附腺的起始变化至关重要。随后，在持续缺乏 JH 的情况下，蜕皮激素主分泌峰导致成虫角质层分泌，并伴随着内部结构变化和成虫生理特征的出现。

在完全变态昆虫中，如烟草天蛾，每一幼龄初期的 JH 滴度相对于蜕皮激素而言往往较高，在每次蜕皮前略有下降。JH 与蜕皮激素的互作致使真皮细胞分泌下一龄幼虫的角质层。然而，在末龄幼虫期，JH 滴度在蜕皮前下降到较低水平，部分原因是合成的下降。进入蛹期后，JH 水平急剧降低到可检测水平之下（图 10-2B）。

在果蝇的发育过程中，存在 6 个清晰的蜕皮激素分泌脉冲，每个都在 PTTH 分泌脉冲之前。JH 的分泌处于每个峰值的结合部，只有蛹蜕变为成虫时期例外。JH 在末龄幼虫的早期阶段产生，但在龄末期下降到非常低，甚至无法检测的水平（图 10-2C）。蜕皮激素的第 4 个分泌脉冲引起众多生理和形态学上的变化，最明显的表现为虫体缩短和化蛹。化蛹后约 12 h，蜕皮激素的第 5 个脉冲引起蛹几丁质的分泌。最后，在蛹期 JH 滴度低的情况下，第 6 个宽幅的蜕皮激素分泌脉冲启动成虫的发育。

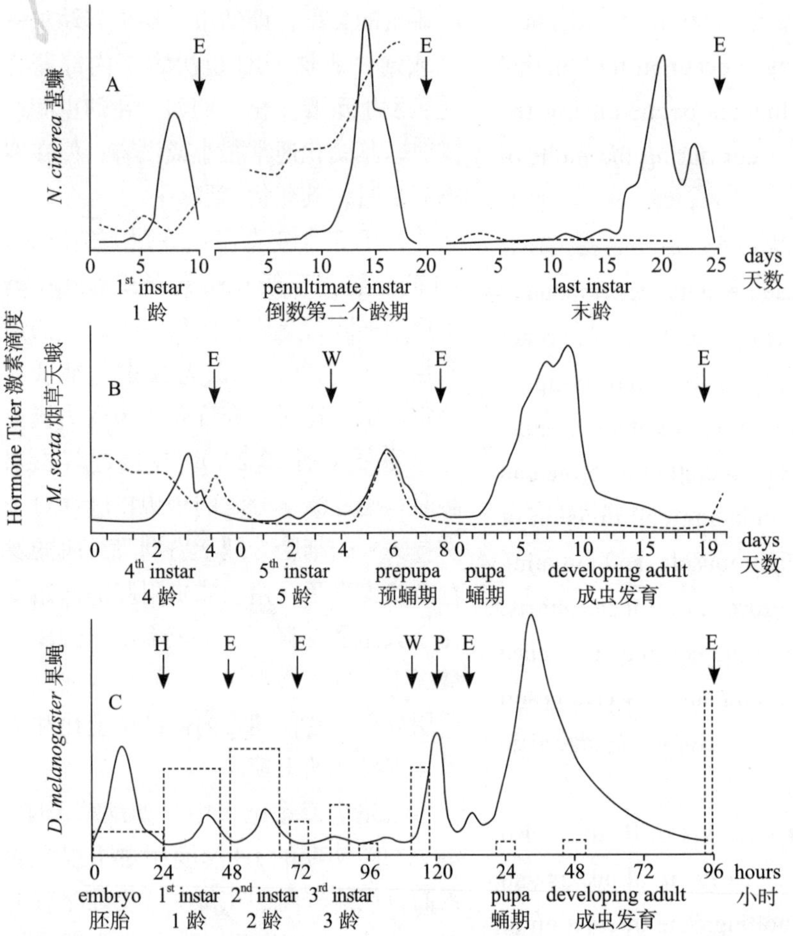

Fig.10-2 Ecdysone and JH titers during the development of model insects. The cockroach, *N. cinerea*, belongs to holometabolous, while *M. sexta* and *D. malanogaster* belong to hemimetabolous. In each case, the solid line indicates ecdysone titer and the broken line indicates JH titer. Key: E.ecdysis; W.wandering larva; P.pupation; H.hatching (modified from Riddiford, 1994).

图 10-2 模式昆虫发育过程中蜕皮激素和保幼激素滴度变化。蜚蠊为不完全变态，而烟草天蛾和黑腹果蝇为完全变态。每个图中的实线表示蜕皮激素滴度，虚线表示保幼激素滴度。E. 蜕皮；W. 游走期；P. 化蛹；H. 孵化（仿 Riddiford, 1994）。

10.3 脑神经分泌细胞与促前胸腺激素

10.3 Brain Neurosecretory Cells and Prothoracicotropic Hormone

10.3.1 Source of hormone and chemistry

Prothoracicotropic hormone (PTTH) is the first insect hormone to be discovered but the last major hormone to be structurally identified. PTTH is produced in brain neurosecretory cells (NSCs). In some insects, only a few large brain NSCs are involved in producing PTTH. Only two pairs of NSCs in each hemisphere of the protocerebrum in *B.mori* are pinpointed as producers using immnunofluo-rescent antibodies against PTTH. PTTH is not released directly from the brain, but passes down monopolar axons as electron-dense granules to a neurohemal organ, the general name for a structure from which neurosecretory hormones are released into the circulating hemolymph. The

10.3.1 激素的来源及其化学

促前胸腺激素（PTTH）是最早被发现的昆虫激素，但其结构是最后被解析的。PTTH 由脑神经分泌细胞（NSCs）产生。在一些昆虫中，仅少量大型的 NSCs 产生 PTTH。PTTH 的免疫荧光抗体检测结果显示，家蚕前脑的每个半球仅有两对 NSCs 被精确定位为 PTTH 分泌细胞。PTTH 不是直接由脑释放的，而是通过单极的神经轴以电子密度颗粒传送到神经血器官。神经血器官是一个通用名称，是指神经分泌激素被释

neurohemal organs that release PTTH in most insects are corpora cardiaca (CC). Howerve, in *M. sexta*, and perhaps in other or even all Lepidoptera, PTTH is released from the paired corpora allata (CA). The CC and CA are paired bilaterally in most insects, but they are fused in some insects. Initially identified PTTHs as multiple forms from *B. mori* include two different molecular sizes, a relatively small polypeptide of about 4.4 kDa and a large form of about 30 kDa. They often are designated as 4K PTTH and 30K PTTH, respectively.

The 4K PTTH are further separated by electrophoresis into three species, designated as PTTH-I, PTTH-II and PTTH-III. The 4K PTTH is a peptide belonging to the insulin superfamily based on its amino acids and it is expressed in four pairs of medial brain NSCs whose axons terminate in the CA. It can stimulate diapause pupae of *Papillo machaon* and brainless pupae of *Samia cynthia ricini* to develop into adults, but can't induce the molting of the 4th instar larvae lapped at the neck and the adult development of debrained diapause pupae. Thereafter, 4K PTTH is renamed bombyxin. The 30K PTTH has the true PTTH activity and it can stimulate the PG to produce ecdysone. The 30K PTTH exists as a homodimer with two monomer unit glycosylated and held together by three intramolecular disulfide bonds and a single intermolecular cysteine-cysteine bond (Fig.10-3).

PTTH of different insects is diverse and species-specific. For example, the molecular weight of PTTH in *Drosophila* is larger than that in moths and it has more glycosyl. The sequences of the PTTHs in 7 Lepidoteran species, *M. sexta*, *Agrius convolvuli*, *B. mori*, *S. cynthia ricini*, *A.pernyi*, *H.cecropia*, *and H.virescens*, show 53%-58% identity, but they possess the characteristics of PTTH: 7 cysteine residues and an *N*-glycosylation site.

放进循环血淋巴的结构。大多数昆虫释放 PTTH 的神经血器官是心侧体（CC）。在烟草天蛾，也许还有其他或全部鳞翅目昆虫中，PTTH 是由成对的咽侧体（CA）释放的。CC 和 CA 是成对且双极的，但在有些昆虫中是融合的。最早分离鉴定的家蚕 PTTH 包含一个分子质量较小的 4.4 kDa 多肽和一个约 30 kDa 的大分子，分别被定义为 4K PTTH 和 30K PTTH。

4K PTTH 可进一步分为 3 种电泳形态：PTTH-I、PTTH-II 和 PTTH-III。按氨基酸组成划分，4K PTTH 是胰岛素超家族的一种肽类，其表达位于脑中区轴突，终止于 CA 的 4 对 NSCs 中。4K PTTH 可刺激金凤蝶滞育蛹和蓖麻蚕无脑蛹发育为成虫，但无法诱导颈部结扎的 4 龄家蚕幼虫的蜕皮和无脑休眠蛹发育为成虫。因此，4K PTTH 之后被重新命名为家蚕素。30K PTTH 真正具有 PTTH 的活性，可刺激前胸腺产生蜕皮激素。30K PTTH 是一个同源二聚体，其单体被糖基化并通过 3 个分子内二硫键与 1 个分子间半胱氨酸 - 半胱氨酸键连接在一起（图 10-3）。

不同昆虫的 PTTH 具有多样性与物种特异性。例如，果蝇的 PTTH 比蛾类的分子质量大且带有更多的糖基；7 种鳞翅目昆虫，烟草天蛾、白薯天蛾、家蚕、蓖麻蚕、柞蚕、惜古比天蚕蛾及烟芽夜蛾的 PTTH 相似性仅为 53%～58%，但具有 7 个保守的半胱氨酸残基和 1 个氨基端糖基化位点。

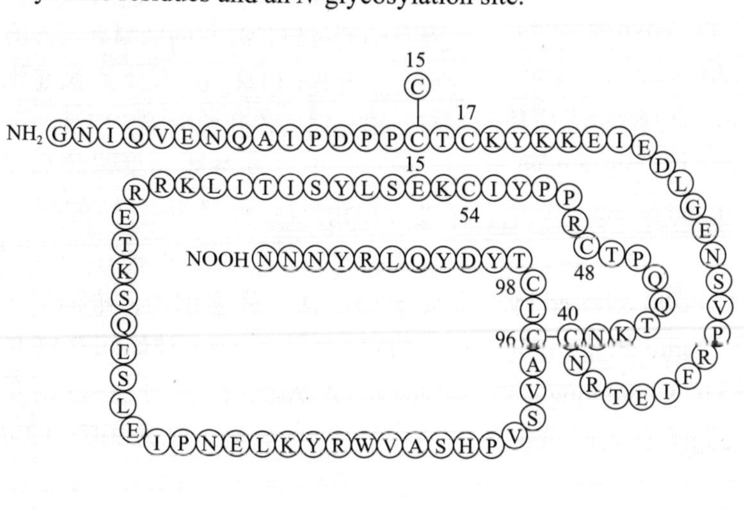

Fig.10-3 The amino acid structure of PTTH. Only one of the two identical chains in the homodimer is shown (modified from Nagata, 2005).

图 10-3 PTTH 的氨基酸结构。显示的是同源二聚体中的一条链（仿 Nagata, 2005）。

10.3.2 Secretion of PTTH

Release of the neuropeptide PTTH takes place at particular developmental stages and is under control of a series of physiological factors, such as insects' nutrition, and certain environmental cues including photoperiod, temperature, time of day and nervous stimuli.

Firstly, the insects must reach a critical size or critical weight. For example, about one-third of the 4th instar larvae of *M. sexta* reared at 25℃ under a 12 h light : 12 h dark photoperiod don't secrete PTTH until they reach the critical size after 36 h of the 4th instar. These larvae that begin to secrete PTTH in the first hour after the beginning of the artificial darkness are designated as Gate I larvae. The remaining two-thirds of the larvae that have not grown to the critical size of Gate I must wait another 24 h to secrete PTTH even if they reach the critical size before the next artificial darkness, being described as Gate II larvae.

Secondly, the filled gut activates stretch receptors that send a message to the brain. For example, in *Rhodnius*, the abdominal distension resulting from a large blood meal triggers stretch receptors that then send a message to the brain to secrete PTTH. Abdominal stretching also is the activator of the brain in another rediviidae bug, *D. maximus*, and the receptors are located in the abdominal nerves that became stretched up to one and a half times the normal length during feeding.

Thirdly, PTTH secretion is regulated by photoperiod in a wide range of insects, occurring during a circadian window. In *Drosophila*, circadian influence is believed to be mediated by input from the short neuropeptide F (sNPF) produced by clock neurons, which synapse with the PG innervating NSCs.

Fourthly, cold temperature exposure, and other environmental stimuli can activate the secretion of PTTH in the brain. The brain of diapause pupae in *H. cecropia* is "made ready" to secrete PTTH by a prolonged exposure to cold temperature and the pupae begin to secret PTTH only after returning to the room temperature for a few weeks.

Besides, *Drosophila* insulin-like peptides 8 (DILP8) released by growing discs acts on presynaptic *via* growth-coordinating Lgr3-expressing (GCL) neurons of the larval brain thereby inhibiting the release of PTTH. Allatostatin A (AstA) released by AstA neurons

10.3.2 PTTH 的释放

神经肽 PTTH 的释放发生在特定的发育阶段并受到一系列生理因素（如昆虫的营养状况）和特定环境因素（如光周期、温度、生物钟和神经刺激）的控制。

第一，昆虫身体必须达到体型临界大小或临界体重。例如，在 25℃和 12 h 光照：12 h 黑暗的饲养条件下，1/3 的烟草天蛾 4 龄幼虫在 36 h 后达到临界大小才开始分泌 PTTH。这些在人工暗期 1 h 开始分泌 PTTH 的幼虫被称为门控 I 型幼虫；剩余的 2/3 幼虫达不到门控 I 型的临界大小，即使在下一个人工暗期前达到临界值，也须再等 24 h 才开始分泌 PTTH，这些幼虫被称为门控 II 型幼虫。

第二，充填的肠道或腹部活化了张力感受器进而将信息传递给脑。例如，吸血蝽取食大量血粉引起的腹胀会激活牵引受体进而刺激脑分泌 PTTH。另一种猎蝽科昆虫腹部牵引感受器也能活化脑，饲喂时，其腹部神经伸长为正常长度的 1.5 倍。

第三，大多数昆虫的 PTTH 分泌受光周期的调节，主要发生在昼夜节律窗内。在果蝇中，昼夜节律对 PTTH 分泌的影响被认为是由生物钟神经元产生的神经肽 sNPF，进一步突触连接 PG 支配的神经分泌细胞所介导的。

第四，低温处理和其他环境刺激均可激活昆虫脑分泌 PTTH。惜古比天蚕蛾滞育蛹通过较长时间的低温处理才能"准备"分泌 PTTH，且天蚕蛹仅在恢复至室温几周后才开始分泌 PTTH。

此外，生长盘释放的果蝇胰岛素样肽 8（DILP8）通过幼虫脑的生长协调 Lgr3（GCL）表达的神经元作用于突触前的神经分泌细胞，进而抑制 PTTH 的释放。由 AstA 神经元释放的 Allatostatin A (AstA) 作用于突触前的 PTTH 神经分泌细胞，促进 PTTH 的释

acts on presynaptic PTTH neuosecretory cells to promote the release of PTTH. The expression of amino acid transporter in glia is required for proper PTTH expression. Retinoids released by damaged discs inhibit *Ptth* expression, but ecdysone promotes *Ptth* expression in a feedback loop contributing to the metamorphosis-triggering surge of PTTH and Ecdysone. JH also affects PTTH signaling to delay the secretion of ecdysone.

放。氨基酸转运体在神经胶质的表达是PTTH正确表达所必需的。受损的成虫盘释放的类维生素A会抑制*Ptth*的表达。蜕皮激素则可反馈促进*Ptth*的表达进而有助于触发变态期PTTH和蜕皮激素的迅速增加。JH也能影响PTTH的信息传递，从而延缓蜕皮激素的分泌。

10.3.3 PTTH mode of action

After its secretion into the hemolymph, PTTH is transported through the hemolymph and binds a receptor Torso located at the cell membrane surface of PG (Fig. 10-4) to initiate a cascade of reactions, resulting in the synthesis and secretion of ecdysone. The PTTH-Torso complex opens the *L*-type or *T*-type Ca^{2+} ion channel to calmosulin, causing an increase of intracellular Ca^{2+}

10.3.3 PTTH的作用模式

PTTH分泌后经血淋巴运输至PGs，并与PG细胞表面的受体酪氨酸激酶Torso结合（图10-4），开始启动级联反应，以促进蜕皮激素的合成与分泌。PTTH-Torse复合物先开启*L*-型或*T*-型Ca^{2+}通道，胞内Ca^{2+}浓度增加，活化了钙调蛋

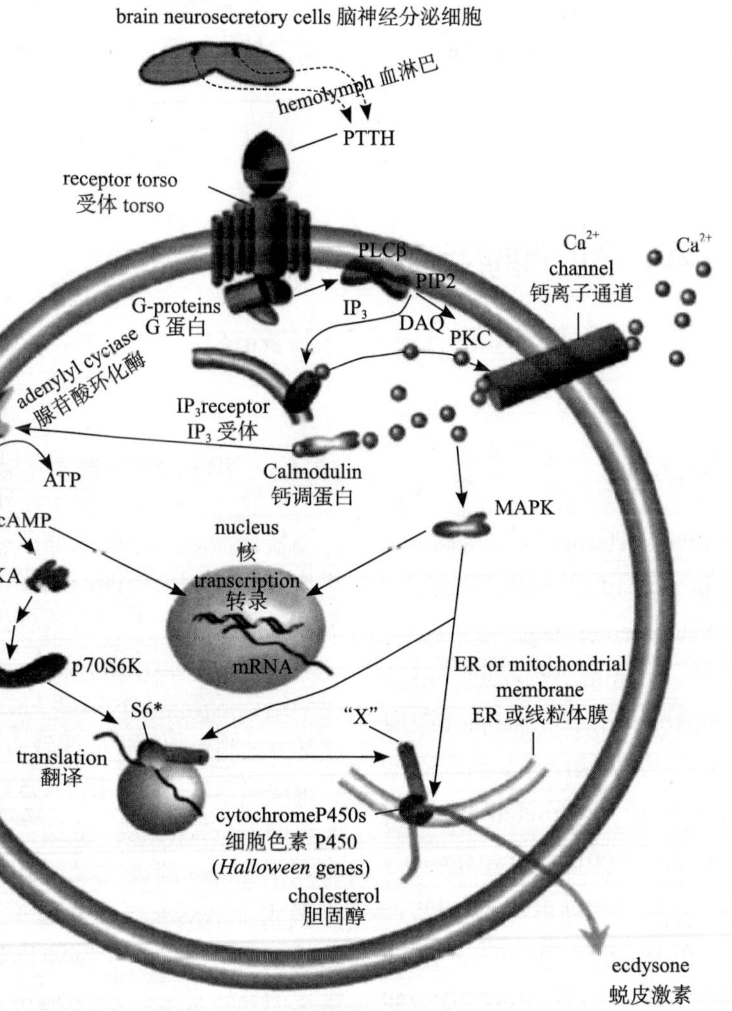

Fig. 10-4 The PTTH signal transduction cascade in prothoracic gland cells. ER. endoplasmic reticulum; IP_3. inositol triphosphate; PLCβ. phosphorlipase Cβ; PIP2. phosphatidyl-inositol-4,5-bisphosphate; DAG. diacylglycerol; PKA. protein kinase A; PKC. proteinkinase C; ATP. adenosine triphosphate; MAPK. mitogen-activated protein kinase (modified from Gilbert, 2005).

图 10-4 前胸腺细胞中PTTH的信号转导级联图。ER. 粗面内质网；IP_3. 肌醇三磷酸；PLCβ. 磷酸化酯酶Cβ；PIP2. 磷脂酰肌醇-4, 5-二磷酸；DAG. 二酯酰甘油；cAMP. 环磷酸腺苷；PKA。蛋白激酶A；PKC. 蛋白激酶C；ATP. 腺苷三磷酸；MAPK. 分裂素活化的蛋白激酶（仿Gilbert, 2005年）。

concentration and activation of calmosulin. Then the Ca^{2+}/calmodulin complex forms, resulting in a rapid increase of cyclic adenosine monophosphate (cAMP) in the cells with the help of a Ca^{2+}-calmodulin dependent adenylate cyclase (CaM-AdCyc). Futhermore, cAMP activates several protein kinases such as p70S6K and initiates a series of protein phosphorylation, such as ribosomal protein S6. The phosphorylated protein regulates the protein translation of specific mRNAs. The final signal transduction step in PTTH-regulated ecdysone production is the rapid translation of several proteins required for gland growth and periods of high ecdysone synthesis associated with an elevated protein tureover, such as a 70 kDa heat shock protein and β-tubulin. In addition, PTTH stimulates the transcription and post-translational phosphorylation of genes involved in the ecdysone synthesis.

It is generally believed that PTTH stimulates the PG cells of the moths to synthesize the ecdysone by increasing the concentration of intracellular cAMP, whereas the synthesis of ecdysone in Diptera does not depend on the cAMP as the second messenger, but on the Ca^{2+}.

白，并与之形成 Ca^{2+}/ 钙调蛋白复合体。接着，Ca^{2+}- 钙调蛋白敏感性腺苷酸环化酶（CaM-AdCyc）促进了胞内环腺苷酸（cAMP）的快速合成，导致 cAMP 依赖性蛋白激酶，如 70 kDa S6 激酶的活化和随后一系列蛋白质的磷酸化，如核糖体蛋白 S6，这些磷酸化的蛋白质进一步参与特异性 mRNA 的翻译。PTTH 调控蜕皮激素产生信号转导的最后一步是腺体的生长并促进蜕皮激素大量合成相关蛋白质转化所需的几种蛋白质的快速翻译，如热激蛋白 70 和 β- 微管蛋白。此外，PTTH 还可激活蜕皮激素合成相关基因的转录及其翻译后的磷酸化。

一般认为，蛾类前胸腺细胞受 PTTH 刺激时，通过胞内 cAMP 浓度的升高来提高蜕皮激素的合成，但在双翅目昆虫体内，蜕皮激素合成不依赖于 cAMP 作为第二信使，而是依赖于 Ca^{2+}。

10.4 前胸腺与蜕皮激素

10.4 Prothoracic Gland and Ecdysone

10.4.1 Identification of insect ecdysone

Ecdysone is the general name for related steroid hormones that are primarily involved in the molting process of arthropods but also have wideranding effects on every developmental stage.

Ecdysone is the first hormone to be chemically structured in insects. In 1934, Fraenkel G. demonstrated the presence of the molting hormone active substances in *Calliphora vicina Robineall* by ligature experiment and bioassay. In 1954, Butenandt A.F. J and Karlson P. purified 25 mg of the ecdysone from approximately 500 kg *B. mori* pupae and named it ecdysone, also known as α-ecdysone. Subsequently, 20-hydroxyecdysone, also known as β-ecdysone, was purified. Additionlly, 3-dehydroecdysone and 2-dehydroecdysone were identified in *M. sexta*. Then the

10.4.1 昆虫蜕皮激素的鉴定

蜕皮激素是一种主要参与节肢动物蜕皮过程的相关类固醇激素的总称，在节肢动物的每个发育阶段均发挥作用。

蜕皮激素是昆虫体内第一个被鉴定出化学结构的激素。1934 年 Fraenkel G. 利用结扎方法与生物测定证明红头丽蝇体内存在蜕皮激素活性的物质。1954 年 Butenandt A. F. J 和 Karlson P. 从 500 kg 家蚕蛹中分离得到了 25 mg 的蜕皮类固醇，命名为蜕皮激素（ecdysone），也称 α- 蜕皮激素。后续又纯化出 20- 羟基蜕皮酮，也称 β- 蜕皮激素。此外，还从烟草天蛾中鉴定出 3- 脱氢蜕皮酮与 2- 脱氢蜕皮酮；在双

higher titer of makisterone was found to exist in the Dipteran, Hymenopteran and Isopteran. The 2-deoxyecdysone and 26-hydroxyecdysone were found in the eggs and embryos of insect. The above sterols all have the function of ecdysone (Hoffman, 1986).

翅目、膜翅目和异翅目昆虫中还发现了高滴度的罗汉松甾酮 A；在昆虫卵与胚胎内还发现有 2- 脱氧蜕皮酮与 26- 羟基蜕皮酮。这些甾醇类物质均具有蜕皮激素的功能（Hoffman，1986）。

10.4.2 The prothoracic gland and ecdysone

The main organ for the synthesis of ecdysone is the PG in most insects. The PG in Lepidoptera consists of loose clusters of cells widely scattered in the prothorax, but may even reach into the head. In *H. cecropia* larvae, the PG consists of loose clusters of large cells (about 47 μm×22 μm in 4th instars) with a close association with the tracheal branches near the prothoracic spiracle (Fig. 10-1C). The cells receive neurons from the prothoracic ganglion and subesophageal ganglion. There are 220 cells in the PG of *M. sexta*. The PG in cockroaches presents an "X" shape (Fig. 10-1D), while in *B. mori*, it likes a reversed letter "Y". When activated by PTTH, the PG secrete ecdysone or a closely related ecdysteroid. The PG in the Hemipteran, *R. prolixus*, has an endogenous photosensitive circadian oscillator that regulates ecdysone synthesis after the release of PTTH caused by blood feeding, which acts as an entraining agent. The PG of Blattaria doesn't have an endogenous circadian oscillator, but the rhythmicity of ecdysone secretion during the photophase is controlled by secretion of PTTH during the scotophase. The brain in the larvae of *C. vicina*, contains extractable prothoracicotropic and prothoracicofstatic compounds, and the ecdysone synthesis seems to be regulated by the interaction and reverse interaction of several factors. The PG in most insects, degenerates during or soon after metamorphosis. Therefore, the ecdysone is synthesized by the follicular epithelial cells of ovarie and male accessory gonad in the pupae and adults.

10.4.2 前胸腺与蜕皮激素

大多数昆虫的前胸腺是合成蜕皮激素的主要器官。鳞翅目昆虫的前胸腺由松散的细胞束组成，广泛分布在前胸，甚至到达脑部区域。惜古比天蚕的前胸腺由松散的大型细胞（4 龄期约 47 μm × 22 μm）沿着前胸附近的气管主要分支分布（图 10-1C）。这些细胞受到前胸和咽下神经节的控制。烟草天蛾的前胸腺约有 220 个细胞。蜚蠊的前胸腺呈"X"形（图 10-1D），而家蚕的呈倒字母"Y"形。前胸腺被 PTTH 激活并分泌蜕皮激素或密切相关的甾类化合物。半翅目吸血蝽的前胸腺内有内源性光敏感节律振荡器，在血粉摄食导致 PTTH 释放后调节外皮甾类化合物合成。蜚蠊前胸腺细胞内没有内源性的节律振荡器，其在光期蜕皮激素的分泌节律由暗期 PTTH 分泌控制。红头丽蝇幼虫脑内含有可抽提的促前胸腺和抑前胸腺混合物，其蜕皮激素的合成似乎受到几种混合因子的相互作用和反向作用调控。多数昆虫的前胸腺在变态过程中或变态之后很快离解。因此，昆虫蛹期和成虫的蜕皮酮主要是由卵巢滤泡上皮细胞和雄性附性腺合成的。

10.4.3 Synthesis of ecdysone and its regulation

The PG sequesters cholesterol from the circulating hemolymph and converts it into ecdysone or a closely related ecdysteroid (Fig. 10-5). The cholesterol that insects can't synthesize should be obtained from the diet. The first step in the synthesis of ecdysone is the endocytosis of cholesterol by PG cells and their transport by the transporter protein (NPC1) to the endoplasmic reticulum, where the cholesterol is oxidized to 7-dehydrocholesterol by the neverland (Nvd). Then, 7-dehydrocholesterol is transported to mitochondria

10.4.3 蜕皮酮的生物合成及其调控

在前胸腺螯合循环血淋巴中，甾醇转换成蜕皮激素或蜕皮甾类（图 10-5）。昆虫自身不能合成胆甾醇，必须从食物中获取。蜕皮酮合成反应的第一步是前胸腺细胞内吞胆甾醇并经转运蛋白（NPC1）转运至内质网，内质网上的胆甾醇被 neverland（Nvd）氧化为 7- 脱氢胆固醇。7- 脱氢胆固醇被转运至

and catalyzed by the short chain dehydrogenase (Shroud) and the cytochrome P450 hydroxylase (Spookier/Spook and Cyp6t3) to produce 5*β*-ketodiol. The process of this step has been not yet fully understood and it is the so called "Black Box". Then the cytochrome P450 hydroxylase Phantom (Cyp306a1), Disembodied (Cyp302a1), and Shadow (Cyp315a1) further converts 5*β*-ketodiol into 2, 22-ketotriol, 2-deoxyecdyone, and low bioactive *α*-ecdyone, respectively.

PTTH promotes the PG to synthesize the ecdysone, wherease prothoracicostatic peptide (PTSP) damages the Ca^{2+} channel protein of PG cell membrane and inhibits the effect of PTTH to regulate the synthesis of ecdysone. Insulin-related peptide binds their receptor in PG and activates the PI3K-Akt signaling to upregulate the ecdysone synthesis. Nitric oxide (NO) regulates the expression of *Phantom* and *Disembodied* in PG to suppress the synthesis of ecdysone. The high titer of 20-hydroxyecdysone inhibits the ecdysone synthesis through the feedback regulation. JH is found to inhibit the growth of PG cells, thereby inhibiting the ecdysone synthesis.

线粒体，由短链脱氢酶（Shroud）和细胞色素 P450 羟化酶（Spookier/Spook 和 Cyp6t3）催化生成 5*β*- 酮二醇，该过程仍未完全清楚，因此，被称为“Black Box”。细胞色素 P450 羟化酶 Phantom (Cyp306a1)、Disembodied (Cyp30 2a1) 及 Shadow (Cyp315a1) 进一步分别逐步催化 5*β*- 酮二醇生成 2, 22- 酮三醇、2- 脱氧蜕皮酮及无生物活性的 *α*- 蜕皮酮。

PTTH 可促进前胸腺合成蜕皮激素，抑前胸腺肽（PTSP）可损坏前胸腺细胞膜的 Ca^{2+} 通道蛋白进而通过抑制 PTTH 作用来调控蜕皮激素合成。昆虫胰岛素类似物与前胸腺中的胰岛素受体结合，激活 PI3K-Akt 信号上调蜕皮激素合成。一氧化氮（NO）可调节前胸腺中 *Phantom* 和 *Disembodied* 的表达进而抑制蜕皮激素的合成。高浓度的 20- 羟基蜕皮酮通过反馈抑制蜕皮激素合成。保幼激素抑制前胸腺细胞生长进而抑制蜕皮激素合成。

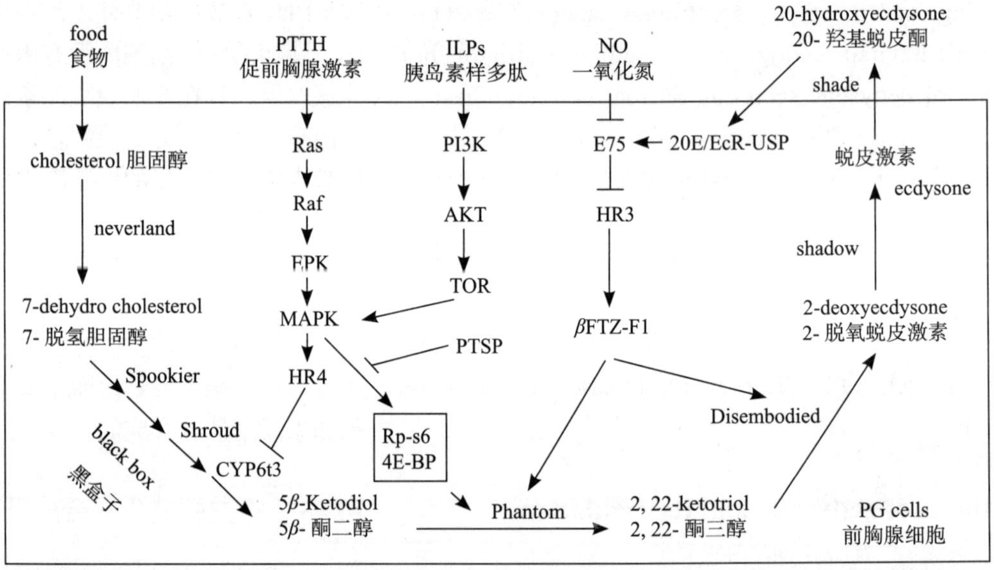

Fig. 10-5 Biosynthetic pathway of ecdysone and its regulation in insects. As the conversion steps between 7-dehydrocholesterol and 5*β*-ketodiol have not been chemically characterized, they are collectively called the "Black Box." Notably, Spookier and Spook are paralogous to each other. Although genetic and physiological analyses have suggested that Noppera-bo plays a role in cholesterol intake or trafficking in PG, endogenous substrates of Noppera-bo have not been identified. Therefore, the role of Noppera-bo is indicated by the dashed arrow in this illustration.

图 10-5 昆虫蜕皮激素的生物合成途径及其调控。由于 7- 脱氢胆固醇和 2, 22, 25- 酮二醇之间的转化步骤尚未被化学鉴定，所以它们被统称为“黑盒子”。值得注意的是，Spookier 和 Spook 的作用是平行的。虽然遗传与生理分析表明 Noppera-bo 在前胸腺细胞吸收胆固醇中发挥作用，但 Noppera-bo 的内源性底物尚未确定，因此用虚线箭头表示 Noppera-bo 的作用。

10.4.4 Conversion of ecdysone into 20-hydroxy-ecdysterone

The *α*-ecdyone seems to have low hormonal activity itself. Thus the *α*-ecdyone is imported into the vesicles and exocytotically released into the hemolymph from PG and enters the target cells with the fluid circulation, where it is catalyzed by cytochrome P450 hydroxylase Shade (Cyp314a1) and 20-hydoxymono-oxygenase in the cytoplasm to produce 20-hydroxy-ecdysterone that has more hormonal bioactivity (Fig. 10-6). The 20-hydoxymono-oxygenase is rich in the malpighian tubules, gut and fat body. However, it is not present in PG. 20-hydroxyecdysone also is known as *β*-ecdysone, ecdysterone, and crustecdytersone. Although "textbook" nuclear-hormone ligands enter cells *via* simple diffusion, ecdysone diffusion into receptive cells must be facilitated by the transporter protein ecdysone importer (EcI) in *Drosophila*, which is also required for the passage of ecdysone across the blood-brain barrier into the brain.

10.4.4 蜕皮激素转换成 20- 羟基蜕皮酮

α- 蜕皮激素本身的激素活性很低。被输入囊泡的 *α*- 蜕皮素从前胸腺释放至血淋巴并随着体液循环进入靶细胞，在细胞质内被细胞色素 P450 羟化酶 Shade (Cyp314a1) 和 20- 羟单氧化酶催化生成具有生物活性的 20- 羟基蜕皮酮（图 10-6）。马氏管、中肠和脂肪体都富含 20- 羟单氧化酶，但这种酶在前胸腺中并不存在。20- 羟基蜕皮酮也称为 *β*- 蜕皮激素、蜕皮甾酮和甲壳蜕皮酮。虽然"教科书"式的核 - 激素配体可通过简单扩散少量进入细胞，但在果蝇中发现，蜕皮激素需要通过蜕皮激素转运蛋白（EcI）才能进入受体细胞，同样这也是蜕皮激素通过血脑屏障进入脑所必需的。

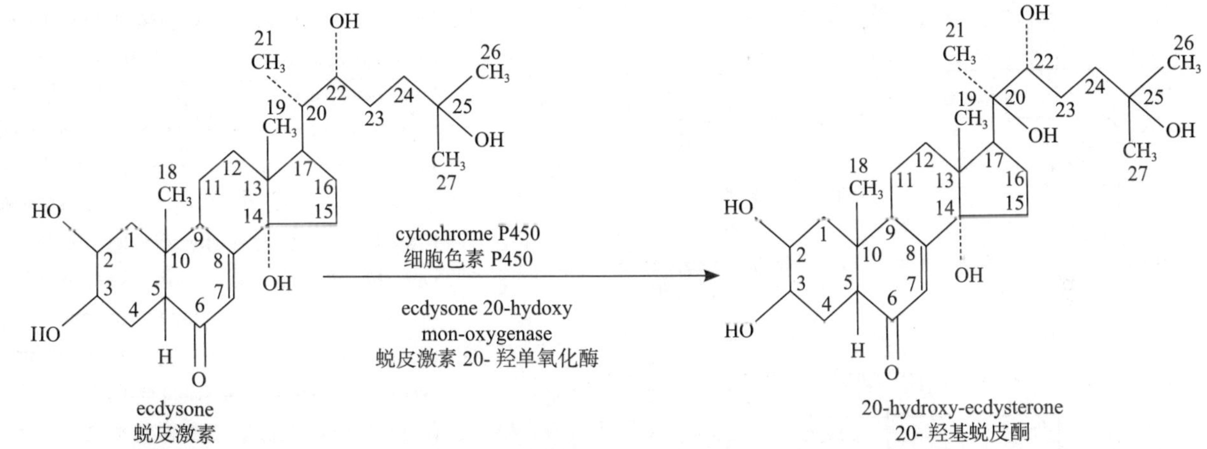

Fig. 10-6 Structure of ecdysone and 20-hydroxyecdysterone. Active form 20-hydroxyecdysone is converted from the principal prohormones ecdysone which is produced by the PG. The reaction is characteristic of many tissues, but does not occur in PG cells.

图 10-6 蜕皮激素和 20- 羟基皮酮。活跃形式 20- 羟基蜕皮酮由前胸腺产生的主要激素前体物蜕皮激素转化而来。这个反应是许多组织的特征，但前胸腺细胞内不发生该反应。

10.4.5 Degradation of ecdysone

There are mechanisms to destroy and remove ecdysone rapidly after the hormone has accomplished its intended function. The halflife period of most ecdysone is very short, a few hours at most. Excretion is one route to remove ecdysone from the body, and ecdysone has been detected in the fecal wastes as conjugates with glucose, glucuronic acid, and sulfates. These conjugates

10.4.5 蜕皮激素的降解

在蜕皮激素完成预期的任务后，体内机制很快将其破坏和清除。大多数蜕皮激素的半衰期很短，最长也只有数小时。从体内排泄是移除蜕皮激素的途径之一，因此，在粪便中可检测到蜕皮激素与葡萄糖、葡萄糖酸及硫酸盐结合

render the molecule some of physiologically inactive. Oxidation or modification of the functional groups on the molecule also reduce or destroy its biological activity. An active enzyme in the larval midgut of *M. sexta* converts ecdysone to inactive compounds by 3-epimerization, i.e., the conversion of the 3β-hydroxy-group to a 3-oxo-group or to a 3α-hydroxy-group. Conversion of 20E to 20, 26-hydroxyecdy-sterone is also a degradation pathway. The 20E that don't bind to the receptor is converted rapidly to 20, 26-hydroxy-ecdysterone. The 20, 26-hydroxy-ecdysterone is then metabolized to 20-hydroxy-ecdysterone-26-aldehyde, with the aldehyde group subsequently oxidized to 20-hydroxy-ecdysterone-26-oic acid (Fig. 10-7). These converted products have only little or no molting hormone activity. The high level of molting hormone secretion pulses during each molt preparation of immature insects may induce the enzymes necessary for hormone removal, which is a kind of self-destruct mechanism.

物。这些结合物使分子在某些生理过程中上无活性。分子上某些功能基团的氧化或修饰也还原或破坏了其生物活性。烟草天蛾幼虫中肠存在一种活跃的酶，其可将蜕皮激素通过 3- 差向异构化转化成无活性化合物，即将 3β- 羟基 - 基团转换成 3- 氧络 - 基团或 3α- 羟基基团。将 20E 转化成 20, 26- 羟蜕皮酮是另一条降解途径。未结合受体的 20E 酮快速转换成 20, 26- 羟蜕皮酮，之后进一步代谢成 20- 羟蜕皮酮 -26- 醛，醛基氧化转化成 20- 羟蜕皮酮 -26- 酸（图 10-7）。这些转化产物的活性很低或没有。未成熟昆虫每次蜕皮准备时，高水平的蜕皮激素分泌脉冲可诱导清除激素所需的酶，这是一种自毁机制。

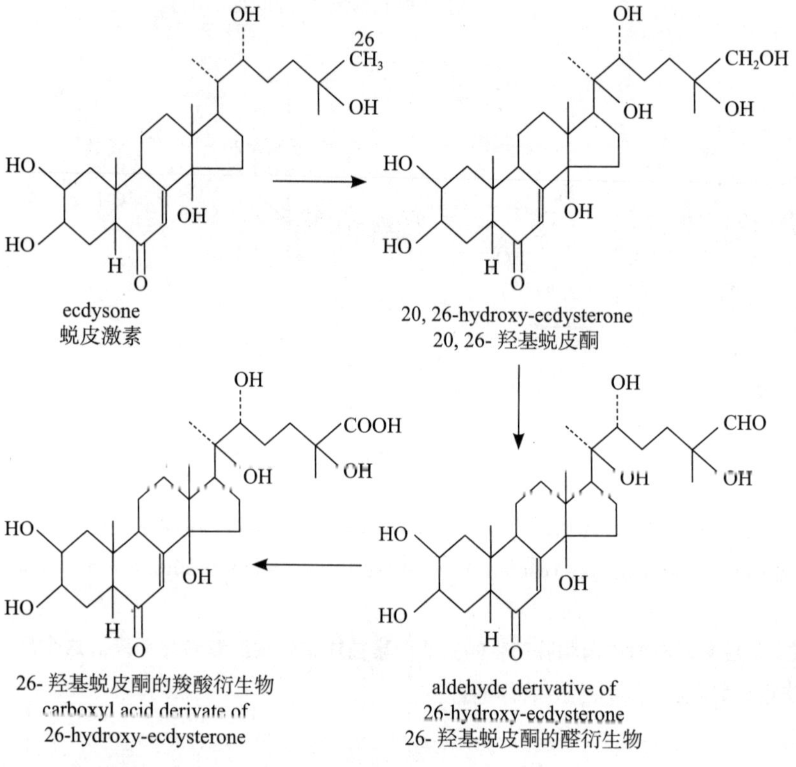

Fig. 10-7 One of the ways that the molting hormone is inactivated in *Spodoptera littoralis*. Ecdysone is usually converted to the more active 20-hydroxy-ecdysterone. 20-hydroxyecdysone that does not bind the receptors is converted rapidly to 20, 26- hydroxy-ecdysterone with low hormonal activity. 20, 26-hydroxy-ecdysterone is then metabolized to the 26-aldehyde derive, with the aldehyde group subsequently oxidized to a carboxyl group. The latter two compounds do not have hormonal activity and are rapidly excreted (modified from Chen et al., 1994).

图 10-7 棉贪夜蛾蜕皮激素失活途径之一。蜕皮激素常转换成活跃的 20- 羟基蜕皮酮。没有结合受体的 20- 羟基蜕皮酮快速转换成活性很低的 20, 26- 羟基蜕皮酮，之后进一步代谢成 26- 羟基蜕皮酮的醛衍生物，而醛基氧化成羧基，后两种化合物没有激素活性，很快被排泄（仿 Chen 等，1994）。

10.4.6 Function of ecdysone

Ecdysone mainly regulates insects' growth and development, molting, metamorphosis and reproduction. In addition, molting and related physiological processes during embryonic development are also regulated by ecdysone.

During larval growth and development, the secretion activity of

10.4.6 蜕皮激素的功能

蜕皮激素主要调控昆虫生长与发育、蜕皮变态及生殖。此外，在胚胎发育过程中的蜕皮及其相关的生理过程也受到蜕皮激素的调控。

在幼虫生长发育过程中，蜕皮激素

ecdysone changes periodically with instar. In both holometabolous and hemimetabolous insects, the titer of ecdysone increases rapidly when the larvae molt. Usually, the titer of ecdysone reaches the peak at the time of the dissolution and separation of the epidermis, and then the new epidermis forms. Its titer gradually decreases to the lowest value during the time of molting. Most lepidopterous larvae have two ecdysone peaks in their final instar. The first peak is smaller, and the titer ratio of ecdyone to 20E is 1 : 1, which mainly causes insects to stop feeding, enter the wandering stage and search for a suitable place for pupation. The second peak is larger, and the titer ratio of ecdyone to 20E is 1 : 5, which mainly causes the pupariation. Generally, the molting frequency of larvae is fixed, but starvation, low temperature and hormone treatment can increase the molting frequency. Therefore, ecdysone analogs or ecdysone antagonist can be used to interfere the insect growth and molting, so as to control pest.

Ecdysone also regulates the regeneration of tissue cells and their new functions, the proliferation and differentiation of cells and the programmed death of some cells.

的分泌活性随着龄期呈现周期性变化。在完全变态与不完全变态发育昆虫中，每次幼虫蜕皮时，蜕皮激素的滴度迅速升高，通常蜕皮激素的滴度在真皮层溶解分离时达到峰值，此后新真皮形成，此时滴度逐渐下降，至蜕皮时下降到最低值。大多数鳞翅目幼虫末龄期存在两个蜕皮激素峰：第一个峰较小，蜕皮激素与 20E 的滴度比为 1 : 1，主要使昆虫停止进食进入游走期并寻找合适的化蛹场所；第二个峰较大，蜕皮激素与 20E 的滴度比为 1 : 5，主要导致幼虫化蛹。一般来说，幼虫的蜕皮次数是固定的，但饥饿、低温及激素处理可增加蜕皮次数。因此，可利用蜕皮激素类似物或拮抗物干扰昆虫的生长与蜕皮，从而进行害虫防治。

蜕皮激素还对组织细胞的重建与新功能的产生、对细胞的增殖与分化及某些细胞的程序化死亡均有调控作用。

10.5 蜕皮激素基因水平上的作用模式

10.5 Mode of Action of Ecdysone at the Gene Level

10.5.1 Chromosomal puffs

Ecdysone, as a typical steroid hormones, can easily diffuse into cells. They directly affect gene expression, causing the activation or inactivation of certain genes and resulting in the synthesis of inhibition of enzymes and other regulatory peptides. The evidence that ecdysone influences genes' transcription comes from studies of polytene chromosomes in Diptera *D. melanogaster* and *Chironomidae tentans*. The polytene chromosomes are formed by replication rather than division and the polytene chromosomes in a single cell may consist of up to 213 chromatids. Alternating bands of condensed and decondensed chromatin occur along the length of the polytene chromosomes (Fig.10-8A; Appendix color graphic Fig.10-8), and they show characteristic puff patterns (Fig.10-8B) with light microscopy. The puffs that are sometimes visible represent sites of active genes' transcription, and puffing patterns

10.5.1 染色体膨突

蜕皮激素作为一种典型的类固醇激素，易扩散至细胞。它们直接影响基因的表达，包括激活或失活某些基因，从而合成抑制酶和其他调节肽。蜕皮激素影响基因转录的证据来自对双翅目摇蚊和果蝇唾液腺多线染色体的研究。它们的多线染色体是通过复制而非分裂形成的，一个单细胞的多线染色体含有 213 个染色单体。光学显微镜下可清晰看到浓密的和稀松的染色质带交替出现在多线染色体上（图 10-8A；附彩图 10-8），并表现出特征化的膨突（图 10-8B）。膨突的出现表示这一位点正进行

are correlated with the developmental stage of insect and ecdysone titers. There is a characteristic sequence of puffs in the salivary gland polytene chromosomes of last larval instar *Drosophila*, which is induced by an ecdysone pulse in the late instar larvae. When the active substance of ecdysone (20E) was injected into last instar larvae of *C. tentans*, it induces chromosomal puffing within 2 hours. Within a few minutes after addition of 20E to a culture medium containing isolated 3rd instar *Drosophila* salivary glands, six "early" puffs develop at sites. The early puffs grow larger over a period of about 1 hour, persist for about 4 hours, then begin to regress and disappear in about 6 hours. A few hours after the early puffs regress, up to 100 puffs can be observed in other chromosomal regions and these are called "late" puffs. Based on the above phenomenon, Ashburner (1974) proposed a model that hormone-receptor complex consists of 20E and an ecdysone receptor. The binding of the hormone-receptor complex at the early puff gene sites stimulates their transcription but inhibits the formation of the late genes. The early gene products subsequently inhibit the early genes but stimulate the transcription of late genes. Therefore, the genes associated with the early puffs are the regulators of late gene expression, and the late genes consequently play a direct role in salivary gland morphogenesis. Many of these genes have been identified as transcription factors. This model

活跃的基因转录，且膨突模式与昆虫的发育阶段和激素滴度相关。末龄果蝇的唾液腺的多线染色体上有一个特殊的膨突序列，这是由幼虫后期的激素脉冲诱导的。将蜕皮激素活性物质 20- 羟基蜕皮酮（20E）注射到摇蚊的末龄幼虫中，2 h 内可诱生唾液腺多线染色体膨突。在培养的 3 龄果蝇唾液腺的培养基中加入 20E 后数分钟内，出现有 6 个“早期”膨突；早期膨突在 1 h 后长大并维持 4 h，然后开始消退，约 6 h 内消失；早期膨突消退后数小时，在其他染色体区域可观察到 100 个膨突，这些膨突被称为“后期”膨突。Ashburner（1974）据此提出模型，蜕皮激素与其受体复合物结合形成复合体，复合体与早期基因结合并激活它们的转录，同时抑制晚期基因的转录；早期基因产物随后抑制早期基因的转录，但激活了晚期基因的转录。因此，与早期膨突相关的基因是晚期表达基因的调控因子，而晚期基因在唾液腺形态发生中直接发挥作用。这一

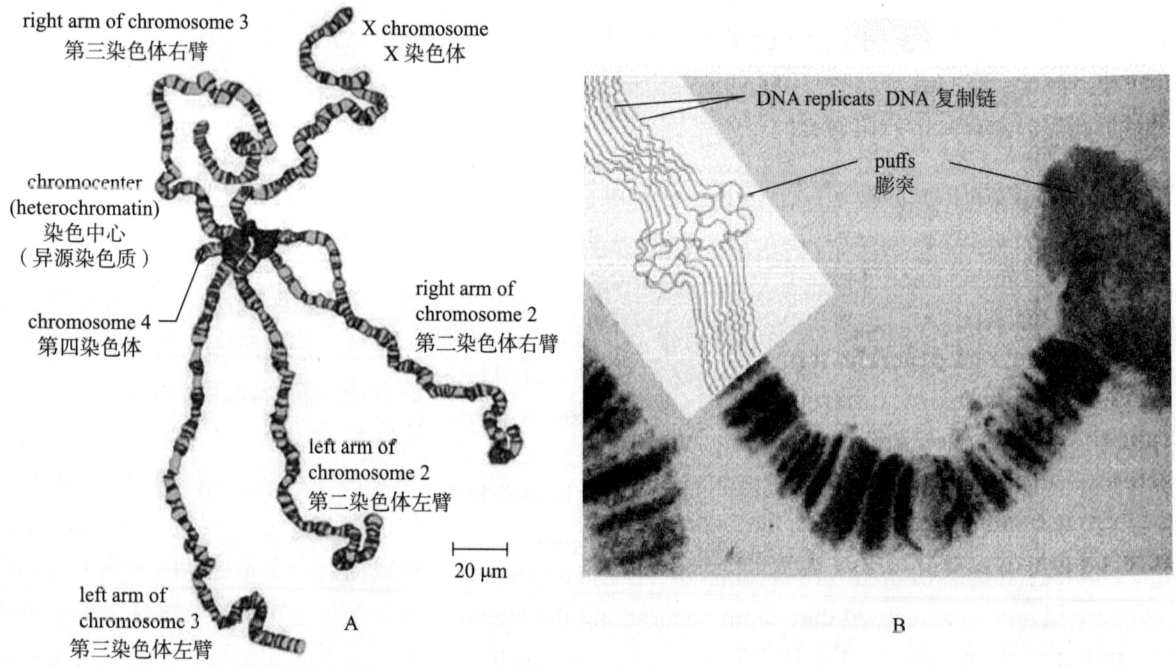

Fig. 10-8 Polytene chromosomes of *D. melanogaster* (modified from other authors). A.The entire set of polytene chromosomes; B. The light micrograph of puffs in polytene chromosome.

图 10-8 果蝇多线染色体（仿其他作者）。A. 果蝇多线染色体全部详情；B. 多线染色体膨突部分的光学显微镜图像。

demonstrates the cascade of gene activity that is involved in salivary gland morphogenesis. Besides, inhibition of early genes and induction of late genes require exposure of the salivary gland chromosomes to hormone for a critical period of time. After only short exposure, early puff regression and late puff development are aborted. These suggest that the early genes transcriptional products have to be made in sufficient quantity to compete with the 20E receptor complex for binding sites on the DNA.

模型印证了在唾液腺形态发生过程中基因活性的级联放大效应。此外，早期基因的抑制和后期基因的激活需要在临界期将唾液腺染色体浸泡在激素中，若浸泡时间很短，早期膨突的衰退和后期膨突的产生会中断，这表明早期基因转录产物必须有足够的量，才能保证 20E 受体复合体与 DNA 位点的结合。

10.5.2 Ecdysone receptor and its signaling

Ecdysone receptor (EcR) is first cloned in *D. melanogaster* and has high affinity both with DNA and 20E. EcR is abundant in the cells, but 20E can't function when it is alone. Subsequently, it is found that EcR must bind the ultraspiracle protein (USP) to form a heterodimer, and then 20E binds EcR to form a 20E-ECR/USP complex, being a functional receptor of ecdysone.

EcR has been cloned in Lepidoptera, Homoptera, Orthoptera, Blattaria and Diptera. It has the structural features common to members of the nuclear receptor gene family, and it consists of A/B-domain (transactivation domain), C-domain (DNA-binding domain, DBD), D-domain (hinge region), E-domain (ligand binding domain, (LBD) and F-domain from the N-terminal end (Fig. 10-9). The DNA-binding domain contains two zinc fingers and hinge domain is necessary for the binding between EcR and USP. USP has the same structural characteristics as EcR except for the absence of F domain belonging to nuclear receptor superfamily. The ligand binding domain of EcR is highly helical, and EcR binds 20E only after the formation of heterodimer between EcR and USP. 20E-ECR/USP further binds the EcR response element (EcRE: 5'-AGGTCA-3' or 5'-AGAACA-3') in the upstream of the target

10.5.2 蜕皮激素受体及其信号转导

蜕皮类固醇激素受体是最早在果蝇中被克隆出来的，其对 DNA 和 20E 均有很高的亲和力。EcR 在细胞中大量存在。但当其单独存在时，20E 不能发挥作用。随后发现，EcR 必须与超气门蛋白（USP）结合形成异质二聚体时，20E 才能结合到 EcR 上，形成 20E-EcR/USP 复合物，进而发挥蜕皮激素受体功能。

EcR 已在鳞翅目、同翅目、直翅目、蜚蠊目和双翅目等诸多昆虫得到克隆。该受体有着核受体基因家族的成员具有的共同结构特征，从 N- 端起，包含 A/B 域（转录激活域）、C 域（DNA 结合域）、D 域（铰链域）、E 域（配体结合域）和 F 域 5 个部分组成。其中 DNA 结合域包含了 2 个锌指（图 10-9）。铰链区是 EcR 和 USP 形成二聚体必须的。USP 除缺少 F 域，其他结构特征与 EcR 相同，二者均属核受体超家族。EcR 的配体结合域高度螺旋化，需与 USP 形成异源二聚体产生变构效应后才可结合 20E。20E-EcR/USP 进一步结合靶基因上游的 EcR 响应元件（EcRE：5'-AGGTCA-3' 或 5'-AGAACA-3'）调控

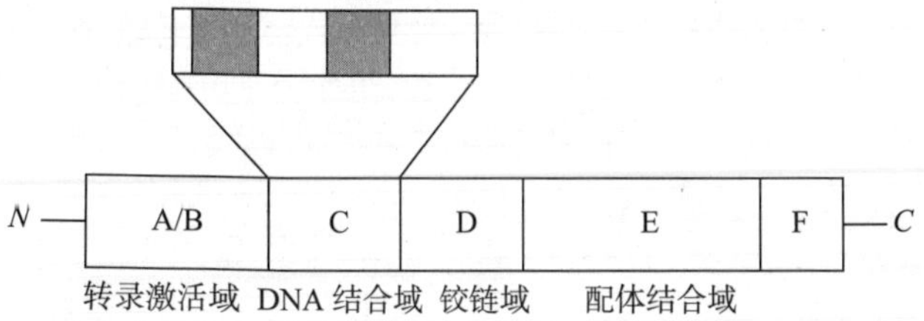

Fig.10-9 The domain of ecdysone receptor EcR.
图 10-9 蜕皮激素受体 EcR 的结构域。

genes to regulate their transcriptions.

Ecdysone transported to target tissues through hemolymph binds the nuclear receptor complex EcR/USP to activate the expression of primary responsive transcription factors, and then the complex activate the expression of secondary responsive transcription factors. The primary and secondary responsive transcription factors further regulate the expression of target genes, completing the signal transduction process (Fig. 10-10). The early response transcription factors include BR-C, E74A, E75B and E93, while the secondary response transcription factors mainly include HR3, HR4, HR39, E78 and βFTZ-F1.

它们的转录。

蜕皮激素通过血淋巴运输至靶组织，进入细胞内与核受体复合物EcR/USP结合后先激活初级应答转录因子的表达，接着与初级应答转录因子一起再激活次级应答转录因子的表达，初级与次级应答转录因子进一步调控靶基因的表达，从而完成了激素信号转导过程（图10-10）。早期应答转录因子包括BR-C、E74A、E75B及E93，次级应答转录因子主要包括HR3、HR4、HR39、E78及βFtz-F1等。

10.5.3 Differential tissue and cell response to ecdysone

The ecdysone action on genes involves the activation of certain genes, repression of certain genes, and indirect action through transcriptional products. For example, some larval tissues (such as salivary glands, muscles and intestines) are degraded, the imaginal discs directly develops into the adult structure, and the larval fatbody and central nervous system are transformed into adult structures regulated by the ecdysone during the metamorphosis. How do different tissues respond differently to the same hormonal

10.5.3 不同组织和细胞对蜕皮激素的响应

蜕皮激素对基因的作用牵涉到一些基因的活化、一些基因的消退及转录产物的间接作用。比如，在变态过程中，一些幼虫组织（如唾液腺、肌肉和肠）在蜕皮激素的作用下将会退化，成虫盘在蜕皮激素的作用下将直接发育为成虫结构；幼虫脂肪体和中枢神经系统等在蜕皮激素的作用下会转化为成体改造

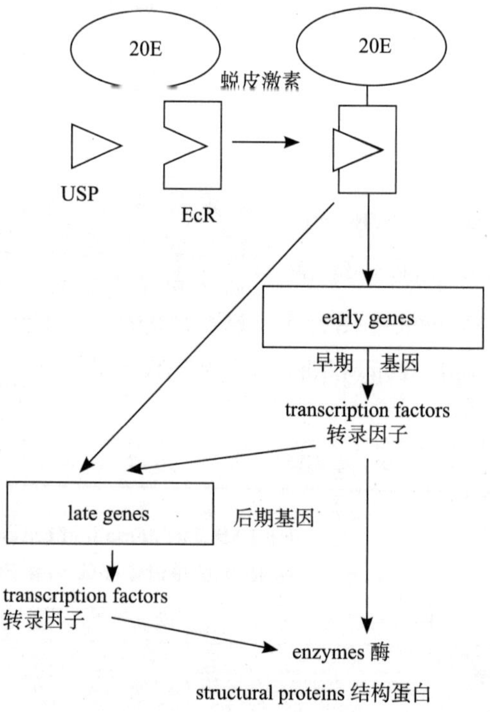

Fig. 10-10 Signal transduction pathway of ecdysone.

图10-10 蜕皮激素的信号转导途径。

signaling? Current research suggests the answer may be: ① the presence of different isoforms of the ecdysone receptor; ② different number and/or the binding of receptors in different tissues; ③ the presence and interaction of specific tissue factors, and hormone induced transcription factors acting in conjunction with the hormone-receptor complex.

EcR encodes three protein subtypes, EcRA, EcRB1 and EcRB2. EcRA and EcRB are produced by selective transcriptional initiation, while EcRB1 and EcRB2 are produced by alterative splicing. The amino acids difference of the isomers is within the A/B region at the N-terminal, while the amino acids of the E domain bound the ligand, and the C domain bound the DNA element are identical. The concentration of different isoforms is different depending on the tissues and cells. For example, EcRA is highly expressed in the imaginal discs and differentiated neurons. The expression of EcRB1 is high in the degraded larval tissues, whereas EcRB2 plays a role in the formation and development of maplighian tube in *Drosophila*. Differences in gene expression of different cells may be due to the binding of 20E to the different receptors.

Another possible mechanism for different response is the dimerization of EcR and U1 traspiracle (USP). In the absence of 20E, the EcR/USP receptor complex could bind the promoter region of ecdysone responsive genes, but the genes' transcription could not be activated due to the presence of co-repressor. In the presence of 20E, EcR/USP dissociates from the repressor, and then interacts with coactivators to initiate the transcription of target genes. In addition, different receptor heterodimer combinations also play the different functions. There are two subtypes of USP, USP1 and USP2. The 20E-EcRB1/USP2 complex activates the transcription of *HR*3 in *M. sexta*, whereas the 20E-EcRA/USP1 complex inhibits its transcription.

Different tissue and cell responses may be also related to the presence of tissue-specific factors and transcription products that modify hormone-receptor action. For example, two forms of the transcription factor designated E74 in *D. melanogaster* express sequentially during pupariation, possibly with different 20E titers during pupariation.

In a word, DNA-binding sites for 20E-receptor complex may not function as response elements (i.e., not all DNA-binding sites

物。不同组织是如何对同一激素信号产生不同的反应机制呢？目前的研究显示其答案可能是：①存在有不同的蜕皮激素受体分子；②受体的二聚作用和 / 或异源二聚作用；③特异性组织因子的存在及其相互作用和激素诱导的转录因子作用于激素 - 受体复合体的结合部位。

EcR 编码 3 个蛋白质亚型，EcRA、EcRB1 和 EcRB2。EcRA 和 EcRB 是由选择性启动转录造成的，EcRB1 和 EcRB2 是由选择性剪接造成的。异构体的氨基酸差异在 N- 端的 A/B 区域内，而与配体结合的 E 区和与 DNA 结合的 C 区的氨基酸序列都是相同的。不同的同源异构体在不同的组织细胞中的含量不同，例如，EcRA 主要在成虫盘和分化中的神经元中有较高的表达；EcRB1 主要在变态中退化或消失的幼虫组织中高水平表达；EcRB2 在果蝇马氏管形成与发育过程中发挥作用。不同细胞中基因表达差异可能是 20E 与不同受体结合的结果。

存在不同反应的另一种可能的机制是 EcR 与超气门蛋白 USP 的二聚作用。20E 通过与 EcR 和 USP 组成的异源二聚体结合后调控下游基因的表达。在无 20E 时，虽然 EcR/USP 受体复合体能结合到蜕皮激素反应基因的启动子区域，但共阻遏蛋白的存在导致基因转录不能被激活。在 20E 存在时，EcR/USP 与阻遏因子分离并与共激活因子互作后激活靶基因的转录。此外，不同的二聚体组合也可发挥不同的功能。USP 存在 2 个亚型：USP1 和 USP2。20E-EcRB1/USP2 复合体对烟草天蛾 *HR*3 的转录起激活作用，而 20E-EcRA/USP1 复合体对其则起抑制作用。

组织和细胞的反应不同也可能与修饰激素 - 受体作用的组织特异性因子和转录产物有关。比如，果蝇中命名为 E74 的转录因子以两种形式在化蛹过程中依次出现，可能是在组织化蛹期间随着不同的 20E 滴度而表达。

综上所述，并非所有 20E- 受体复合体的 DNA 结合位点在功能上都是应

enable the 20E-receptor complex to act as a promoter, repressor, or modifier of genes). Additional transcription factors, possibly specific to particular cells, may be required to turn a binding site into a response element.

Therefore, when details of these and perhaps other control mechanisms are elucidated, it may be clear why some the late 3rd instar tissues exposed to the same strong pulse of ecdysone undergo histolysis (e.g., midgut, salivary glands), while others (imaginal discs) grow and form adult structures.

答因子（即并非所有 DNA 结合位点都能使 20E- 受体复合体发挥启动、抑制或修饰基因的作用）。由结合位点转换成响应因子可能需要特定细胞有特异性的其他转录因子。

因此，当这些细节和其他可能的控制机制明了后，为什么有些三龄末幼虫组织在同等强度的蜕皮激素释放脉冲时产生组织溶离（如中肠、唾液腺），而成虫盘等却分化形成成虫结构的问题答案就逐渐清晰了起来。

10.5.4 Regulatory Proteins: Broad and E93

Ecdysone induces the molting, but the direction of the molt is determined by JH. In the presence of JH, larvae molt and become larvae; in the absence of JH, ecdysone causes the switching in gene expression necessary for metamorphosis and the larvae molt first into pupae and then into adults. The ecdysone signaling transcription factor Broad and E93 is the key determinant of pupal and adult metamorphosis, respectively (Fig.10-11; Appendix color graphic Fig.10-11).

Broad (Br) is a transcription factor consisting of the core region of BTB (Broad-Tram-Track-Bric-a-brac) domain involved in protein interaction, and the zinc finger DNA-binding domain. The difference of zinc finger DNA-binding domain causes the production of Br isoforms. Broad-Complex (Br-C) is predominantly involved in the signaling of 20E and JH to regulate a variety of developmental processes, including embryogenesis, larval molting, metamorphosis, and reproduction. The PKA phosphorylation

10.5.4 调节蛋白：Broad 和 E93

蜕皮激素引起蜕皮，但蜕皮的方向由 JH 决定。当 JH 存在时，幼虫蜕皮成为幼虫；当缺乏 JH 时，蜕皮激素打开变态发育基因表达必需的开关，幼虫首先蜕皮变成蛹，再蜕皮变为成虫。蜕皮激素信号通路的转录因子 Broad 和 E93 分别是蛹变态发育和成虫变态发育的关键决定因子（图 10-11；附彩图 10-11）。

Broad（Br）是含有参与蛋白质-蛋白质相互作用的 BTB（Broad -Tram-Track-Bric-a-brac）结构域的核心区和锌指 DNA 结合域组成的转录因子。锌指 DNA 结合域的不同形成了 Br 的同源异构体。Broad 复合体（Br-C）主要通过参与 20E 和 JH 的信号传导，昆虫生长发育所必需的关键激素 20E 和 JH 的信号转导进而调节各种发育过程，包括胚胎发生、幼虫蜕皮、变态和生殖。PKA

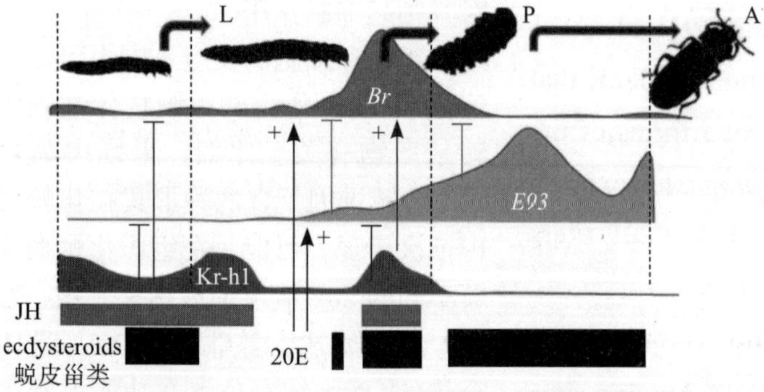

Fig. 10-11 Expression change of *Broad* (*Br*) and *E93* to the normal progression from larva (L) to pupa (P) to adult (A) (modified from Truman et al., 2020).
图 10-11 *Broad* (*Br*) 和 *E93* 在幼虫 - 蛹 - 成虫发育过程的表达变化（仿 Truman 等，2020）。

event suppresses the transcriptional activation of silkworm Br-C target genes, and this inhibition is caused by the repression of the binding of Br-C and its DNA targets. Therefore, the PKA-mediated phosphorylation inhibits silkworm Br-C activity by interfering with its binding to DNA, and 20E signaling relieves PKA-mediated phosphorylation of Br-C, thereby maintaining its transcriptional activity. The expression of *Br* in *D. melanogaster* starts from the middle of last instar larvae that is the critical period of *D. melanogaster*, initiating larval-pupal metamorphosis, and the lack of *Br* causes the larvae death due to their failure of pupariation. The *Br* in *Manduca* begins to express in the pupal commitment, and the initiation and increase of its expression are accompanied by the emergency of pupal cells indicating that *Br* plays a key role in the pupariation. The expression of *Br* is induced by 20E during pupariation, but JH inhibits its expression during the larval stage, thus antagonizing 20E and inhibitng the pupariation. Therefore, *Br* is considered to be the first molecular marker of pupal metamorphosis initiation (Fig. 10-11).

In the hemimetabolous insect, milkweed bug, *O. fasciatus*, *Br* is shown to be necessary for the heteromorphic changes during development. *OfBr* is expressed in two phases. The first phase of *ofBr* expression occurs during germ band invagination and segmentation when *OfBr* is expressed ubiquitously in the embryonic tissues. The second phase of *OfBr* expression appears during the pronymphal phase of embryogenesis. It persists through nymphal differentiation and declines just before hatching. RNAi studies in *T. castaneum* shows that *TcBr* is required for midgut remodeling during larval-pupal metamorphosis. *Br* also plays an important role in the regulation of wing primordial growth in hemimetabolan.

E93 is an early gene in the ecdysone signaling cascade that is specifically expressed in late prepupae of *D. melanogaster*. *E93* encodes a helix-turn-helix transcription factor, and E93 protein has the RHF domains that are significantly similar to pipsqueak motifs which is found to be a key player in the degeneration process of the salivary glands during *D. melanogaster* metamorphosis. However, the action of *E93* in metamorphosis is not restricted to the regulation of degeneration processes, it also plays a role in morphogenesis. *E93* is widely expressed in the cells of *D. melanogaster* pupae and it causes *Distal-less* (*Dll*)

磷酸化事件抑制家蚕 *Br*-C 靶基因的转录激活，这种抑制是由 Br-C 与其 DNA 靶结合的抑制引起的。因此，PKA 介导的磷酸化通过干扰家蚕 Br-C 与 DNA 的结合来抑制其活性，而 20E 信号传导可减轻 PKA 介导的 Br-C 磷酸化，从而维持其转录活性。果蝇的 *Br* 从幼虫末龄的中期开始表达，而此时正是果蝇启动蜕皮变态的关键时期且 *Br* 的缺失会使幼虫不能进行变态而致死，烟草天蛾的 *Br* 在化蛹启动期开始表达且表达的起始和上升伴随着化蛹细胞的出现而增加，这表明 *Br* 在化蛹变态中起关键作用。*Br* 在化蛹变态时的表达受到 20E 的诱导，JH 则会抑制 *Br* 在幼虫期的表达进而拮抗 20E 作用并抑制蛹的变态发育。因此，*Br* 被认为是蛹变态启动的第一个分子标记（图 10-11）。

在不完全变态昆虫乳草虫，*Br* 是发育过程中产生形态变化所必需的。*OfBr* 在两个时期表达。第一个 *OfBr* 的表达期在胚带内陷和体节形成期，此时的 *OfBr* 在胚胎组织内普遍表达；第二个 *OfBr* 表达期在胚胎形成的预若虫期，并在若虫分化阶段持续，在刚要孵化前下降。赤拟谷盗 RNA 干扰的研究表明，*TcBr* 是幼虫 - 蛹变态过程中肠重塑所必需的。*Br* 还在不完全变态昆虫胚后的翅发育过程中起重要调控作用。

E93 是蜕皮激素信号级联的一个早期基因，其在黑腹果蝇预蛹后期特异性表达。该基因编码 helix-turn-helix 结构的转录因子，E93 蛋白质具有与 pipsqueak 基序极其相似的 RHF 结构域，而该基序被发现在果蝇变态时期唾液腺的退化过程中起关键作用。然而，*E93* 在变态中的作用并不仅仅局限于对器官退化的调控，它还在形态发生中发挥作用。*E93* 在黑腹果蝇蛹的成体细胞中广泛表达并能使蛹腿苞片细胞中的 *Distal-*

gene within bract cells of the pupal leg to become responsive to epidermal growth factor (EGF) receptor signaling. *E93*-depleted *D. melanogaster* larvae are able to pupate but die at the end of the pupal stage. In the beetle *T. castaneum*, decreased expression of *E93* prevents the pupal-adult transition, resulting in the formation of a supernumerary second pupa. Similar results are obtained in the cockroach *B. germanica*, where *E93* depletion in nymphs prevents the nymphal–adult transition, giving rise to reiterated supernumerary nymphal instars. Therefore, *E93* is considered to be a initiator of adult metamorphosis.

less（*Dll*）基因对真皮生长因子 (EGF) 受体信号通路产生响应。*E93* 缺失的黑腹果蝇幼虫能化蛹，但会在蛹末期致死。赤拟谷盗蛹期 *E93* 表达的降低阻止了蛹 - 成虫的转变，导致蜕皮后形成了第二个蛹。在德国小蠊若虫体内降低 *E93* 的表达阻止了若虫-成虫的过渡，使其产生了反复的多生若虫龄期。因此，*E93* 被认为成虫变态发育的启动因子。

10.6 咽侧体和保幼激素

10.6 The Corpora Allata and Juvenile Hormones

10.6.1 Major types of JHs and their synthesis

Juvenile hormones (JHs) are sesquiterpenoid compounds known to be produced by the cells of the corpora allatas (CA) just posterior to the brain. Several JH variants, including JH0, JH Ⅰ, JH Ⅱ, JH Ⅲ, 4-methyl-JH I, JH Ⅲ bisepoxide (JHB3), JH Ⅲ skipped bisepoxide (JHSB3) and Methyl farnesoate (MF), have been found in insects (Fig. 10-12).

Usually, mixtures of JHs are found in insects with a homologous function in different insects and they differ in the number and location of carbon and epoxy groups. JH Ⅲ is the principal component of JHs and exists in most insects in Lepidoptera,

10.6.1 保幼激素的主要类型及其合成

保幼激素（JHs）主要是由位于脑后的咽侧体（CA）细胞分泌的萜烯类化合物。昆虫体内已发现的保幼激素包括 JH0、JH Ⅰ、JH Ⅱ、JH Ⅲ、4-methyl-JHⅠ、双环氧化 JHⅢ（JHB3）、非环氧化 JH Ⅲ（JHSB3）和法尼酯甲酯（MF）（图 10-12）。

通常，在昆虫中的 JHs 具有同源功能，它们的区别在于碳和环氧基团的数量与位置。JH Ⅲ是保幼激素的主要成分，存在于鳞翅目、双翅目和膜翅目中；JH0 被发现存于鳞翅目中（如烟草

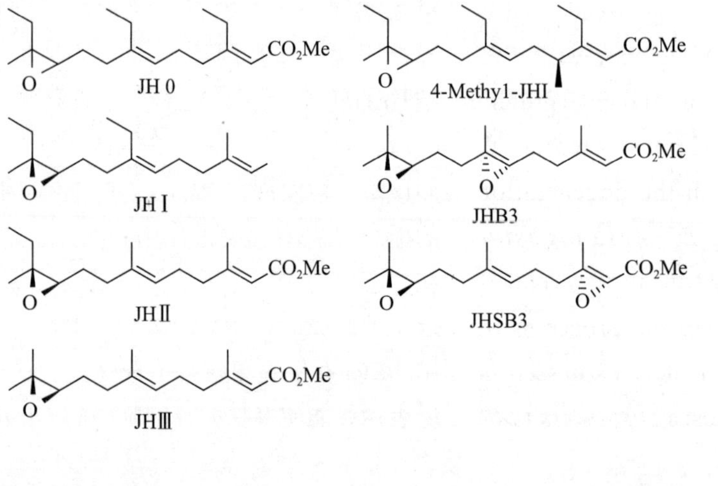

Fig. 10-12 Naturally occurring molecules with JH activity. JH bisepoxide is synthesized by the ring gland of some Diptera.

图 10-12 自然界存在的有保幼激素活性的分子。双环氧化保幼激素由一些双翅目昆虫环腺合成。

Diptera, and Hymenoptera. JH0 is found in Lepidoptera (i.e., *M. sexta* and *H. cecropia*) and mainly regulates embryonic development. 4-methyl-JHⅠ is identified in developing embryos of *M. sexta*. JH I exists in Orthoptera (e.g., *Locusta migratoria*, *Gryllus bimaculatus*) and Lepidoptera (e.g., *M. sexta*, *H. cecropia*, *Spodoptera frugiperda*), and regulates larval development. Occurrence of JHⅡ is reported in the 5th instar larvae of, *S. frugiperda*. JHB3 and JH Ⅲ are found in higher Diptera (e.g., *D. melanogaster*, *P. regina*, *C. vomitoria*, *L. cuprina*, and in four species of mosquitoes), and the biosynthesis rate of JHSB3 is higher than JHⅢ. In Hemiptera, JHSB3 is synthesized to replace the regulation of JHⅢ. MF as an immediate non-epoxidation biosynthesis the precursor of JH Ⅲ regulates reproduction in Crustacea, while it also exists in insect hemolymph. Besides, JH acids, the non-methylated derivatives of JH, also constitute a relevant part of JH compounds circulating in hemolymph, which may have hormonal functions that cannot be performed by JH itself.

JH is a class of sesquiterpenoids synthesized by mevalonate pathway. Taking JH Ⅲ, which is found in most insects, as an example, its synthesis mainly consists of five steps: Acetyl coenzyme A→ mehydroxyvalerate → Isopentenyl pyrophosphate (IPP) → Farnesyl pyrophosphate (FPP) → Farnesoic Acid (FA) →JH Ⅲ (Fig. 10-13). In general, acetyl-coA is catalyzed by acetoacetyl-CoA thiolase (AACT), 3-hydroxy-3-methylglutaryl CoA synthease (HMGS), HMG-CoA reductase (HMGR), mevalonate kinase (MVK), phosphomevalonate kinase (PMK), diphosphomevalonate decarboxylase (MVD), IPP isomerase (IPI) and FPP synthetase (FPPS), respectively, to form FPP (Mevalonate pathway). Then FPP is converted into FA by the catalytic action of the Farnesyl diphosphate synthase (FDPS), Farnesol oxidase (FO), and Farnesol dehydrogenase (FD), respectively (Farnesoic pathway); Finally, FA methyltransferase (FAMET) transfers methyl to FA to produce MF, which is epoxized by MF epoxidase to form JHⅢ in the presence of oxygen. However, in Lepidoptera, FA is first epoxized by FA epoxidase, forming JHA. JH acid methyl transferase (JHAMT)

天蛾和惜古比天蚕蛾）且主要调控胚胎发育；4-methyl-JHⅠ被发现存在于烟草天蛾胚胎中；JHⅠ被发现存在于直翅目（如东亚飞蝗与双刺灰蟋蟀）及鳞翅目（如烟草天蛾、惜古比天蚕蛾与草地贪夜蛾）中且主要调控幼虫的生长发育。JHⅡ被发现存在于草地贪夜蛾的5龄幼虫中。JHB3和JHⅢ共存于较高等的双翅目（如黑腹果蝇、伏蝇、反吐丽蝇、绵羊蝇铜绿蝇和4种蚊虫）等体内，且JHB3的合成率高于JHⅢ。半翅目通过合成JHSB3来代替JHⅢ的调控作用。作为JHⅢ非环氧化生物合成前体的法尼酯甲酯不仅可调控甲壳类的生殖，也存在于昆虫血淋巴中。此外，JH的非甲基化衍生物JH酸也构成了在血淋巴中循环的JH化合物的相关部分，其可能具有JH本身不能发挥的激素功能。

JH是通过甲羟戊酸途径合成的一类倍半萜化合物。以昆虫中普遍存在的JH Ⅲ为例，其合成途径主要为5大步骤：乙酰辅酶A→甲羟戊酸→异戊烯醇焦磷酸（IPP）→法尼焦磷酸（FPP）→法尼酸（FA）→JH Ⅲ（图6-13）。总体来说，乙酰辅酶A依次经过乙酰辅酶A酰基转移酶（AACT）、3-羟甲基戊二酰辅酶A合成酶（HMGS）、3-羟甲基戊二酰辅酶A还原酶（HMGR）、甲羟戊酸激酶（MVK）、磷酸甲羟戊酸激酶（PMK）、焦磷酸甲羟戊酸脱羧酶（MVD）、异戊烯醇焦磷酸异构酶（IPI）及法尼焦磷酸合成酶（FPPS）依次催化形成法尼焦磷酸（Mevalonate途径）；接着，法尼焦磷酸分别经法尼焦磷酸合成酶（FDPS）、法尼脂氧化酶（FO）及法尼醛脱氢酶（FD）催化生成法尼酸（Farnesoic途径）；最后，法尼酸甲基转移酶转移甲基到法尼酸生成MF，接着MF在有氧情况下由甲基法尼酯环氧酶环氧化生成JHⅢ。不同的是，在鳞翅目昆虫中，法尼酸可先由法尼酸环氧酶环氧化生成JHA，JHA甲基转

then transfers methyl to JHA to produce JH (JH pathway). In addition, cytochrome oxidase CYP15C1 converts the methyl FA into JHA while CYP15A1 catalyzes the formation of JH from MF.

移酶再转移甲基到 JHA 生成 JH（JH 途径）。此外，还发现细胞色素氧化酶 CYP15C1 可催化甲基法尼酸生成 JHA，CYP15A1 可催化 MF 形成 JH。

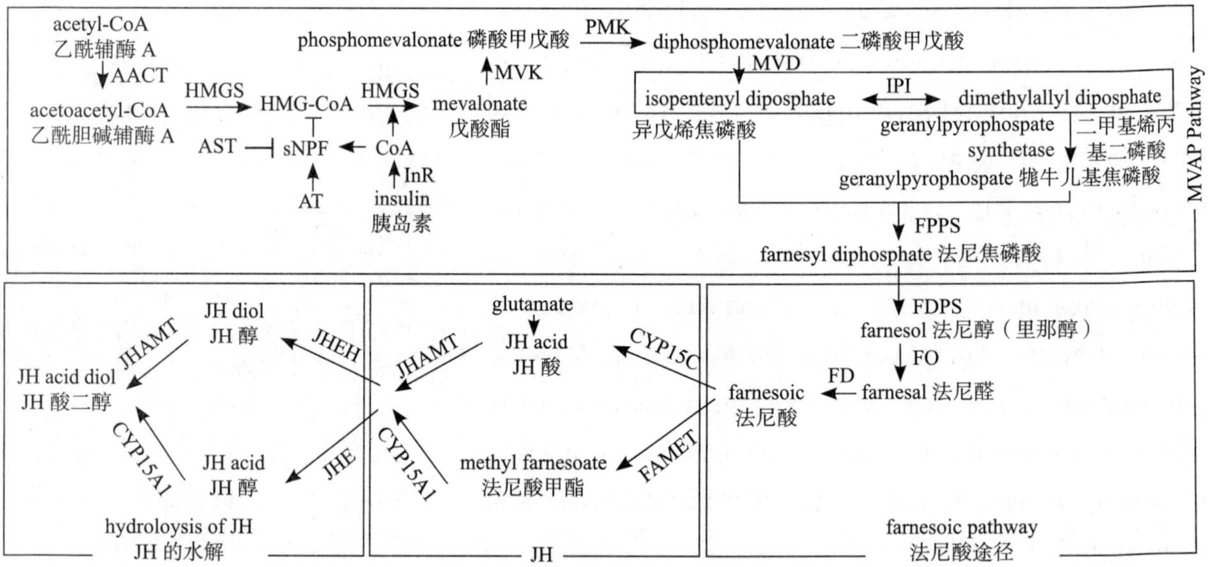

Fig. 10-13 Regulation of juvenile hormone biosynthesis.
图 10-13 保幼激素生物合成调控途径。

All kinds of JH play important roles in different physiological altivities of insects. Pathways of biological degradation involve the inactivation or structural alteration of JHs (Fig.10-13). JHA is produced at certain times by the CA of some insects, and its inactivation is caused by the loss of methyl ester at C-1 position. The inactive JH diol is produced by the loss of the epoxide at the C-10, 11 position. The three unsaturated positions at C-1, C-10 and C-11 modulate the biological activity of JH. Saturation of the 2-ene double bond drastically reduces activity, while the saturation of the 6-ene double bond increases the activity. Loss of the methyl branches along the chain reduces the activity. An isoprenoid chain length of at least 10 —CH_2— units is required to maintain the JH biological activity, and 14 to 16 —CH_2— units is optimal. Given the importance of JHs to many physiological processes in insects and their inactivity in vertebrates, fish, and many aquatic crustaceans, JH is considered as a new insecticide. However, since the native isoprenoid JH is unstable and unsuitable for application, several JH analogens have been developed and

各类 JH 在昆虫的不同生理活动中均发挥重要功能。生物降解途径涉及 JHs 的失活或结构的改变（图 10-13）。JHA 是由某些昆虫咽侧体在特定时期产生的，而其失活是由 C-1 位甲酯丢失造成的。无活性的 JH 二醇是由 C-10 和 C-11 位环氧化合物缺失产生的。C-1、C-10 和 C-11 这 3 个不饱和位点调节着 JH 的生物活性。2- 烯双键的饱和会降低其活性，而 6- 烯双键的饱和则会提高其活性，链上甲基分支的缺失也会降低其生物活性。至少 10 个—CH_2—单元的类异戊二烯链长度才能维持 JH 的生物活性，包含 14 ～ 16 个—CH_2—单元的 JH 是最稳定的。鉴于 JH 在昆虫许多生理过程的重要性，但其在脊椎动物、鱼类和许多水生甲壳类动物中无活性，因此其被认为可成为一种新型的杀虫剂。由于天然类异戊二烯 JH 是不稳定的，所以人们已开发了几种 JH 类似

proposed as insecticides or insect growth regulators. For example, the agonist Methoprene was registered for application in 1975, followed by the development of Hydroprene and Fenoxycarb. JH agonists (analogues) have been applied to control agents mosquito, fly, flea and fire ant, and they have also been applied to increase the silk production of silkworms.

物作为杀虫剂或昆虫生长发育调节剂，例如，激动剂甲基异戊二烯于 1975 年被注册使用，随后又开发了氢丁二烯和苯氧威。JH 激动剂（类似物）已被用于防治蚊子、苍蝇、跳蚤及火蚁，也被用于提高家蚕的产丝量。

10.6.2 Control of JH biosynthesis

The titer of JH in the hemolymph and tissues is a function of biosynthesis by the CA balanced against degradation in the tissues.

Since JH is not stored in the CA, its release depends on synthesis, and this synthesis is rigidly controlled by several avenues. Environment stimuli (e.g., photoperiods) and endogenous factors (e.g., mating and nutritional state) are transmitted to the brain, thereby regulating CA activity through the neurosecretory hormone. The regulatory effect may be either activation or inhibition.

The brain clearly has some control over the secretion of JH by the CA. A number of allatotropins (ATs) and allatostatins (ASTs) peptides isolated from the nervous system of several different species and several orders of insects, have effect on JH biosynthesis. ATs stimulate JH synthesis, whereas ASTs inhibit its synthesis. Besides, an allatoinhibin identified from the brain in *Manduca* inhibits the CA nonreversibly, unlike the reversible action of allatostatins. Short neuropeptide F (sNPF) inhibits the MVA pathway in the JH synthesis, while AT positively regulates sNPF, but AST negatively regulates sNPF.

The nutrition and reproductive states of insects also affect the synthesis of JH. The unrestricted diet in *Calliphora vomotoria* shows high production of JH in the first gonadotrophic cycle. In the laboratory, colonies that a restricted diet (i.e., sugar, water, meat) is given, the JH production is less than 50%. The diet only affects biosynthesis of JH in female, *Neobellieria bullata*, but not the males. Starvation causes low degradation of JH in their haemolymph of *B. dorsalis* and *C. vomotoria*. The synthesis rate of JH maintains at a high level for 3 days after the starvation during the early 5th larval instar, while it decreases rapidly in the last larval instar in normal condition.

Insulin signaling facilitates JH synthesis. JH synthesis in

10.6.2 JH 生物合成的调节

组织和血淋巴中的 JH 滴度是 CA 的生物合成与组织内 JH 降解的相对平衡。

由于 CA 不能贮存 JH，JH 的释放依赖于合成且其合成受到几个途径的严格调控。环境刺激（如光周期）和内源因素（如交配和营养状态）均传递到脑，脑进一步通过神经分泌激素调节 CA 活性。这种调控作用可能是激活，也可能是抑制。

脑对咽侧体 JH 的分泌有着明显的控制作用。从多个物种和目的昆虫神经系统分离得到的促咽侧体神经肽（ATs）和抑咽侧体神经肽（ASTs）均对 JH 的生物学合成有影响。ATs 刺激 JH 的合成，而 ASTs 则抑制 JH 的合成。此外，还从烟草天蛾脑中鉴定出一种抑制 CA 活性的神经肽且该抑制作用是不可逆的。短神经肽 F（sNPF）会抑制 JH 合成中的 MVA 途径，同时 AT 正向调控 sNPF、而 AST 负向调控 sNPF。

昆虫的营养和生殖状态也影响 JH 的合成。在不限制饮食的情况下，反吐丽蝇成虫在第一个促性腺生长周期显示出较高的 JH 合成速率。在实验条件下，饮食限制（如糖、水和肉）会导致合成的 JH 减少一半。在蛐蜓中，饮食仅影响雌性的 JH 合成，对雄性无影响。饥饿会导致桔小实蝇和反吐丽蝇血淋巴中的 JH 降解。烟草天蛾 5 龄早期幼虫断食后，JH 的合成速率维持在较高的水平并维持 3 天之久，而正常情况下，咽侧体合成 JH 的速率在最后一龄是迅速下降的。

胰岛素信号促进 JH 的合成。黑腹

adult *D. melanogaster* decreases after insulin signaling pathway is blocked. Insulin receptor (InR) is found to be involved in the regulation of *CoA* expression, which further regulates JH synthesis *via* sNPF and HMGR.

Additional agents have been shown to stimulate or inhibit CA to procluce JH in some insects, including neuro- transmitters (e.g., dopamine and octopamine), second messenger systems (cAMP, inositol trisphos-phate, diacylglycerol, and calcium signaling), location, kind, number of potential receptors, and ecdysone action. In addition, glutamate regulates JH synthesis by regulating JHAMT.

10.6.3 Regulation of JH products' activity

Once secreted by the CA, JH is transported to target tissues through the hemolymph. Because of its lipophilic nature, JH must first bind other molecules to move through hemolymph. Pherhaps more important, binding can also protect the hormone from degradation by nonspecific tissue-bound esterases. There are a variety of such molecules that bind JH for transportation and protection, namely the juvenile hormone binding proteins (JHBPs). Most of the circulating JH is bound these JHBPs, leaving little JH in free circulation. They not only move the lipophilic JH throughout the hemolymph, but also may be storage sites for hemolymph JH.

Three types of JHBPs have been indentified in insects. The high affinity, and low molecular weight JHBPs are synthesized by the fatbody, and consist of polypeptides containing one JH binding site. Upon JH binding, the protein undergoes a conformational change and protects the hormone from ester hydrolysis. Another group of high affinity and high molecular weight binding proteins are also known as lipophorins that shuttle many different lipids between tissues within the insects. Lipophorin is composed of two apoproteins of 230-250 kDa and 70-80 kDa, and it has multiple JH binding sites. The third type of JHBP with lower concentrations than the others is generally known as hexamerin, composing of six subunits and multiple JH binding sites. Lipophorin is a primary JH carry in heminetabolous insects. In two cockroach species, lipophorin and vitellogenin are the two principal JHBPs. In *M. sexta*, a JHBP of 32 kDa has been purified and JH binds its hydrophobic pocket to protect from the

果蝇的胰岛素信号通路被阻断后，成虫 JH 的合成下降。胰岛素受体（InR）被发现参与调控还原性辅酶 A 基因的表达，而原性辅酶 A 通过 sNPF 和 HMGR 进一步调控 JH 的合成。

有些昆虫还有额外的作用剂刺激或抑制咽侧体合成 JH，包括神经递质（如多巴胺和章鱼胺）、第二信使系统（cAMP、三磷酸肌醇、二酰甘油和钙信号）、电位受体的位置、种类和数量及蜕皮激素的作用。此外，谷氨酸盐通过调控 JHAMT 来调控 JH 合成。

10.6.3 JH 产物的活性调节

JH 一旦从 CA 分泌出来就会通过血淋巴运输至目标组织。由于其亲脂性，JH 必须首先和其他脂分子结合才能通过血淋巴运转，这样还可保护 JH 免受非特异性组织结合酯酶的降解。结合 JH 并进行运输和保护的分子称为保幼激素结合蛋白（JHBPs）。大部分流通的 JH 与这些 JHBPs 结合在一起，自由流通的 JH 则很少。结合蛋白不仅在血淋巴中运输亲脂性 JH，而且可能还是血淋巴 JH 的贮存场所。

昆虫已经鉴定出 3 种 JH 结合蛋白。由脂肪体合成的高亲和力、低分子质量的结合蛋白是由包含　个 JH 结合位点的多肽组成的。其在结合 JH 后，蛋白质构象发生变化可保护 JH 免受酯的水解。另一类高亲和力、高分子质量的结合蛋白也被称为载脂蛋白，它可在昆虫的组织中运送许多不同的脂质。载脂蛋白是由 230 ～ 250 kDa 和 70 ～ 80 kDa 的两个载脂蛋白组成，具有多个 JH 结合位点。第 3 类浓度较低的结合蛋白通常被称为六聚氰胺，由 6 个亚基和多个 JH 结合位点组成。在不完全变态昆虫，载脂蛋白是主要的 JH 携带者。在两种蜚蠊中，载脂蛋白和卵黄蛋白是两种主要的 JHBP。在烟草天蛾中分离出一种 32 kDa 的 JH 结合蛋白质，JH 结合到该蛋白质

esterase attack.

The onset of metamorphosis following the last larval instar requires the clearance of JH in the hemolymph to cause epidermal cells to switch their commitment to the developmental pathway that leads to the production of pupal and adult cuticle in the presence of 20E. Although JH titer is reduced when the CA ceases its synthesis, the reduction of circulating JH titer is accomplished by degrading the JH that already been produced. There are two major routes by which degradation of JH occurs. Specific JH esterases (JHE) recognizes the JH-JHBP complexs and hydrolyzes the methylester of JH to produce the JH acids. When interacting with the JH-JHBP, JHE may lead to a conformational change in the JHBP, causing JH to be released from its binding pocket. High levels of JHE are correlated with low JH titer, clearing any residual JH activity in the hemolymph once the CA stops. Non-specific esterases are also present that are able to hydrolyze both JH and other general substrates, but these are probably less important in the regulation of JH levels than specific esterases. JH epoxide hydrolase (JHEH) hydrates the epoxide to produce the JH diol, but its ability to degrade JH is reduced when JH binds the JHBP. Thus, it only has effect on unbound JH or JH acid (Fig. 10-13). Only one, or both actions, occurs in some insects. The metabolic activity not only eliminate all or most of the hormonal activity of the molecules, but the molecules become more water soluble, and can be excreted by the Malpighian tubules.

A main factor contributing to degradation of JH in *M. sexta* is the large amount of specific JHE synthesis in the later part of each larval molt and in the final larval instar. Although the JHBPs protect JH from esterase attack, the high titer of esterase just prior to a molt destroys any free JH. Hydrolysis of undissociated JH continues to pull the equilibrium of bound JH toward dissociation and degradation, allowing the molt to the next instar. Thus, fluctuations in the level of JHBP in the hemolymph at different developmental stages is a factor that affects JH titer. Overall, although JH titer is influenced by the rate of synthesis, rate of breakdown, sequestering in some target tissues, presence of JHBP and excretion, the two main processes appear to be the rate of synthesis and rate of breakdown. The metabolic pathway of JH and its interaction with ecdysone are shown as Fig. 10-14.

的袋状疏水区可使其免受酯酶的降体。

最后一个幼虫龄期后的变态起始需要清除血淋巴中的 JH，以便真皮细胞在 20E 存在的情况下转换到导致蛹和成虫表皮产生的发育途径。虽然当 CA 停止合成时，JH 滴度会降低，但循环中 JH 滴度的降低是通过降解已经产生的 JH 来完成的。JH 的降解有两种主要途径。特异性 JH 酯酶（JHE）识别 JH-JHBP 复合物并水解 JH 的甲酯生成 JH 酸。当 JHE 与 JH-JHBP 相互作用时，JHE 还可能引起 JHBP 构象发生变化，导致 JH 从其束缚的口袋中被释放。高水平的 JHE 与低的 JH 滴度有关，一旦 CA 停止合成 JH，JHE 会清除血淋巴中任何残留的 JH 活性。非特异性酯酶也可以水解 JH 和其他普通底物，但在调控 JH 水平方面可能不如特异性酯酶重要；JH 环氧水解酶（JHEH）使环氧化物水合生成 JH 二醇，但当 JH 与 JHBP 结合时，水解酶降解 JH 的能力降低，因而仅对未结合的 JH 或 JH 酸有一定的降解作用（图 10-13）。有些昆虫中仅仅发生一种或两种作用。代谢活动不仅清除所有或大多数分子的激素活性，分子也变得更易溶于水并可通过马氏管排出体外。

烟草天蛾 JH 降解的主要因素是每个龄期的后阶段和末龄幼虫特异性 JHE 的大量合成。虽然 JHBP 保护了 JH 免受酯酶攻击，但蜕皮前酯酶的高滴度破坏了所有游离的 JH。非游离 JH 的水解继续将结合态 JH 推向游离和降解的平衡方向，使之蜕皮成为下一龄幼虫。因此，不同发育时期血淋巴中 JHBP 水平的波动实际上是影响 JH 滴度的一个因素。总体上，尽管 JH 滴度受到合成的速率、分解的速率、有些组织中的螯合剂、存在的 JHBP 及排泄的影响，但最主要还是与合成和分解的速率有关。JH 的代谢及其与蜕皮激素的互作关系如图 10-14 所示。

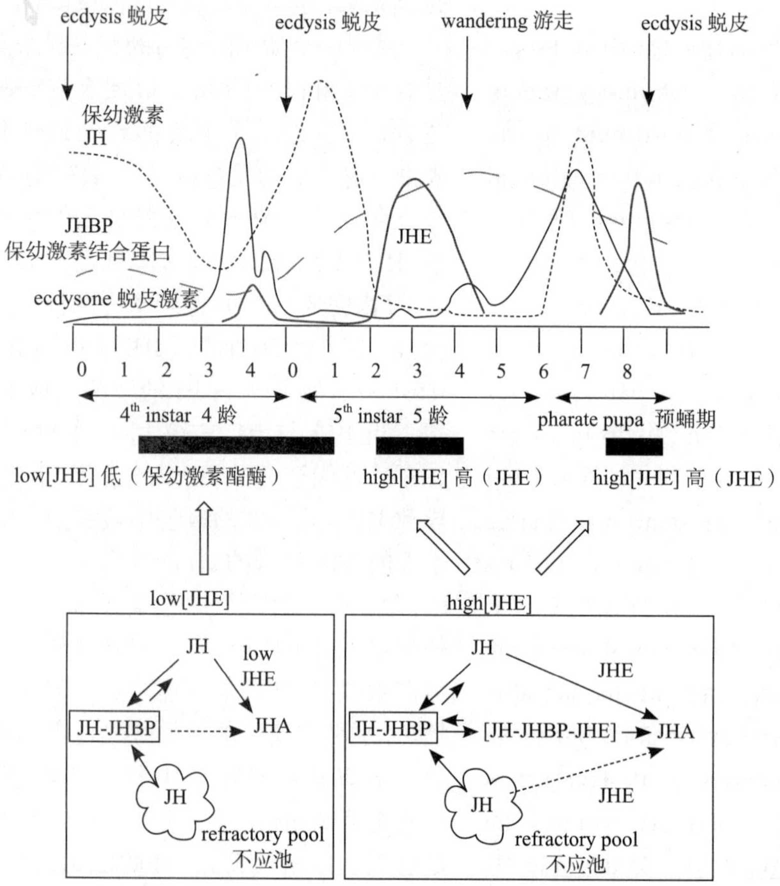

Fig. 10-14 Metabolic pathway of JH in 5th instar of *M.sexta* larvae and its interaction with ecdysone regulates insect development. At low and high JHE concentration, the rate of hydrolysis of JH is determined by the rate of dissociation JH from JHBP and increasing of JHE concentration, respectively. JHBP protects from nonspecific hydrolytic processes, while still allow highly specific JHE to hydrolysis JH by recognizing JH-JHBP complex. JHBP efficiently extract JH from lipophilic depots (refractory pool), resulting in more rapid degradation of JH by JHE (modified from Touhara et al., 1995).

图 10-14 JH 代谢途径及其与蜕皮激素互作推动昆虫发育。当低浓度和高浓度 JHE 时，JH 的水解速率分别由 JH 从结合态解离速率和 JHE 浓度的增加决定。JHBP 保护 JH 避免非特异性水解反应，仍然允许高度特异性 JHE 通过识别 JH-JHBP 复合体水解 JH。JHBP 从亲脂性储藏处（不应池）高效获取 JH，经 JHE 导致 JH 快速降解（仿 Touhara 等，1995）。

10.6.4 Cellular receptors for JH and JH response element

Recent studies have shown that *Methoprene-tolerant* (*Met*), which encodes a transcription factor of the basic helix-loop-helix Per-Arnt-Sim (bHLH-PAS) family, is the nuclear receptor of JH (JHR), and the dimer formed by Met and Ncoa (Ncoa/SRC/P160, FISC or Taiman) belonging to the bHLH-PAS transcription factor family, may be the functional receptor of JH. Met is first reported in the study on resistance of *Drosophila* to methoprene, a JH analogue. *Met* mutants are 100 times more tolerant to methoprene and JH III than the normal individuals, and inhibited the formation of pseudotumor induced by methoprene in larvae and the oocytes development are inhibited in the mutants. *In vitro* translated *Drosophila* Met shows a physiological binding activity with JH III, and JH also induces the transcription of UAS linked reporter gene by the fusion protein of Met and GAL4-DNA binding domain. However, the embryonic development of *Drosophila* mutants with *Met* deficiency are basically normal,

10.6.4 JH 的细胞受体与 JH 反应元件

近年研究表明，编码基础螺旋-环肽链-螺旋的 Per-Arnt-Sim（bHLH-PAS）转录因子家族的 *Methoprene-tolerant*（*Met*）是 JH 的细胞核受体（JHR），而 Met 与同属 bHLH-PAS 转录因子家族的 Ncoa (NCoA/SRC/ p160、FISC 或 Taiman) 形成的二聚体则可能是保幼激素的功能受体。Met 最先报道于果蝇对 JH 类似物烯虫酯的抗性研究。*Met* 突变体对烯虫酯和 JH III 的耐受力是正常个体的 100 倍，并能抑制烯虫酯诱导的假瘤在幼虫中的生成和雌虫卵母细胞的发育。离体转录翻译的果蝇 Met 与 JH III 有生理水平上的结合活性，JH 也能诱导 Met 与 GAL4-DNA 结合域的融合蛋白对 UAS 连接报告基因的转录。然而，*Met* 缺失的果蝇突变体除了成虫产卵延后与生殖力下降外，胚后发育却

except for the delay of oviposition and decline of fecundity in adults, which do not match the phenotype of Met as JH receptor.

In *Drosophila*, *gce* (*germ-cell expressed*), a homologue of *Met*, is found, and the amino acids in the bHLH and PAS domains of Gce show more than 60% identity with Met. Both Met and Met or Met and Gce can form homodimers or heterodimers in the absence of JH. In addition, there is partialy functional overlap between Met and Gce. Gce significantly increases the sensitivity of *Met* mutants in response to methoprene, and only Met and Gce double mutants die during the larval-pupal transition, resembling what is seen in the JH deficient animal. *In vitro* translated Gce also has a physiological affinity with JH III. In *T. castaneum*, reduction of *Met* expression mediated by RNAi results in the precocious metamorphosis. Importantly, both *D. melanogaster* Met/Gce and *T. castaneum* Met binds the JH at the physiological concentrations *in vitro*, strongly suggesting that Met/Gce is a bona fide JH intracellular receptor.

Subsequent genetic evidence in *Drosophila* suggests that the binding of Met/Gce to JH is absolutely required *in vivo* for the receptor to sustain development of *D. melanogaster*. In hemimetabolan, such as the linden bug, *Pyrrhocoris apterus* and the German cockroach *B. germanica*, Met has been shown to be also a factor required for JH signal transduction. So far, Met (or Gce) as the nuclear receptor of JH has been widely recognized by the academic community.

Several types of JH-response elements (JHREs) have been identified in JH-regulated insect genes. *Krüppel homolog 1* (*Kr-h1*) encoded a transcription factor with eight C_2H_2 zinc fingers, has been identified as a JH primary response gene activated by the JH receptor Met/Gce in various insect species. JHRE in the *Kr-h1* promoters of a number of insects have been identified as or a C-box sequence (CACGCG, an E-box-like motif), being essential for JH-induced gene expression through Met. Furthermore, a study in *A.aegypti* reveals that a consensus 9 bp Met binding motif, CACG(C/T) G(A/G)(T/A)G, is conserved in the promoter regions of many JH primary-response genes.

In addition to JH intracellular signaling pathways, studies in a variety of insects have found that JH putative membrane receptors rapidly exert non-genomic effects. For example, in *A. aegypti*, JH binds an unidentified Receptor Tyrosine Kinase (RTK) to activate

基本正常，这与 Met 作为 JH 受体应有的表型相互矛盾。

进一步发现，果蝇有一个 *Met* 的旁系同源基因 *gce* (*germ-cell expressed*)，它们编码的蛋白质在 bHLH 和 PAS 结构域内 60% 以上的氨基酸具有一致性。Met 与 Met 或 Met 与 Gce 在 JH 缺失的情况下都能形成二聚体。此外，Met 和 Gce 具有部分功能重叠性。Gce 能显著提高 *Met* 突变体对烯虫酯的敏感性且有 Met 和 Gce 缺失的双突变体才和 JH 缺失的突变体一样在预蛹期死亡。离体转录翻译的 Gce 与 JH III 也有生理水平上的亲和性。在赤拟谷盗中，RNAi 降低 *Met* 的表达导致变态提早发生。更重要的是，果蝇 Met/Gce 和赤拟谷盗 Met 在体外生理浓度下都能与 JH 结合，表明 Met/Gce 是 JH 真正的细胞内受体。

随后果蝇遗传学上的证据表明，Met/Gce与JH的结合是维持果蝇生长发育所必需的。在半变态昆虫（如始红蝽和德州小蠊）中，Met也被证明是JH信号转导所必需的。至此，Met（或 Gce）是JH的细胞核受体已被学术界广泛认可。

目前，已在受到 JH 调控的基因中鉴定出了几种 JH 的反应元件（JHREs）。已在多种昆虫中证实，*Krüppel homolog 1* (*Kr-h1*) 编码的含有 8 个 C_2H_2 型锌指结构的转录因子是 JH 通过受体 Met/Gce 介导的初级响应基因。多数昆虫 *Kr-h1* 启动子上存在与 Met 功能复合体结合的元件：经典的 E-box（CACGTG）或 C-box（CACGCG E-box 类似基序）。这些元件是 JH 通过 *Met* 诱导基因表达必需的。此外，还在埃及伊蚊的 JH 初级响应基因的启动子上鉴定到一个保守的包含 9 个碱基的 Met 结合位点（CACG(C/T) G(A/G)(T/A)G）。

除 JH 胞内信号通路外，在多种昆虫中的研究发现，JH 存在假定的膜受体迅速发挥非基因组效应，如在埃及伊蚊中，JH 与未被鉴定的受体酪氨酸激酶（RTK）结合后，激活磷脂酶

the phospholipase C (PLC) pathway, leading to phosphorylation and activation of calcium/calmodulin -dependent protein kinase Ⅱ (CaMK Ⅱ) and protein kinase C (PKC). Then CaMK Ⅱ and PKC are involved in the phosphorylation of Met and Taiman, thereby regulating the intracellular JH signaling. Meanwhile, JH membrane signals activate phosphatidylinositol-3-kinase (PI3K) and protein kinase B to regulate the alternative splicing of *Taiman*. In most insects, JH membrane signal facilitates patency of ovarian follicular cells to help the oocytes absorb vitellogenin. JH membrane signal also promotes the protein synthesis in male accessory glands of *Drosophila*.

C（PLC）通路，导致钙调蛋白/依赖性蛋白激酶Ⅱ (CaMK Ⅱ) 和蛋白激酶 C (PKC) 的磷酸化与激活。CaMK Ⅱ和 PKC 进一步参与 Met 和 Taiman 的磷酸化，从而调节胞内 JH 信号。同时，JH 膜信号激活磷脂酰肌醇-3-激酶（PI3K）和蛋白激酶 B，调节 *Taiman* 的选择性剪接。在大多数昆虫中，JH 膜信号通过促进卵巢滤泡细胞的开放行为使卵母细胞吸收卵黄原蛋白。JH 膜信号在果蝇雄性附性腺中还可促进蛋白质的合成。

10.6.5 Action mode of JH

JH has been identified to regulate various physiological functions in insects, such as metamorphosis, reproduction, color morphs and caste differentiation.

The role of JH in "*status quo*" action and the prevention from metamorphosis during the post-embryonic development is determined. In hemimetabolan, JH and its primarily effective factor Kr-h1 inhibit the transcription of the adult specifier *E93* during the nymphal instar to keep the insects in the nymphal state. The high expression of E93 and the low expression of *Kr-h1* caused by the low titer of JHs during the last nymphal instar facilitate the transition from the nymphs to adults. The larval instar of holometabolan is similar to the early nymphal instar of hemimetabolan. Because of the inhibited transcription of the pupal specifier *Br* by the existence of JH and Kr-h1, the larvae maintain the larval state after molting, while the decrease of JH titer and *Kr-h1* expression results in the enhance of *Br* transcription, completing the transition from the larvae to pupae. The pupal stage of holometabolan is similar to the nymphal stage of hemimetabolan. At this time, the high expression of E93 inhibits the transcription of *Br* and *Kr-h1*, promoting adult emergence (Fig. 10-15).

JH also plays an important role in reproduction, including vitellogenesis and oogenesis, mating behavior, ovulation and pheromone production. In many female insects, JH has been found to act as a gonadotropin that activates vitellogenesis by regulating the synthesis and absorption of Vg. In *L. migratoria*, JH signaling promotes the endoreplication of adult fat body cells to prepare the female for vitellogenesis and is indispensable for basal oocyte development. In *T. castaneum*, JH indirectly

10.6.5 JH 的作用模式

JH 调控昆虫的众多生理活动，包括变态、生殖、变色及等级分化等。

JH 在昆虫胚后发育中起“维持原状”或阻止变态发育的作用。对于不完全变态昆虫，在若虫前期，由于 JH 及其关键效应分子 Kr-h1 的存在抑制了成虫特异分子 *E93* 的转录，使得昆虫维持若虫的状态；在末龄若虫阶段，高表达的 E93 和低滴度的 JH 导致的 *Kr-h1* 低表达使若虫羽化为成虫。对于完全变态昆虫而言，其幼虫阶段类似于不完全变态昆虫的若虫前期。由于 JH 和 Kr-h1 的存在抑制了蛹期特异分子 *Br* 的转录，幼虫蜕皮之后仍然可以维持幼虫状态；在末龄幼虫阶段，随着 JH 滴度和 *Kr-h1* 表达的下降，并导致 *Br* 的转录增强进而完成幼虫至化蛹的变态；完全变态昆虫的蛹期类似于不完全变态昆虫的末龄若虫期。此时，高表达的 E93 抑制 *Br* 和 *Kr-h1*，导致了成虫的羽化（图 10-15）。

JH 在生殖过程中也发挥着重要作用，包括卵黄发生与卵子生成、交配、排卵及信息素合成等。在许多雌性昆虫中，JH 被发现作为一种促性腺激素，通过调控 Vg 的合成与吸收来激活卵黄发生。在东亚飞蝗中，JH 信号促进成虫脂肪体细胞的多倍化，进而为卵子形成所必需的卵黄形成做准备。在赤拟谷盗

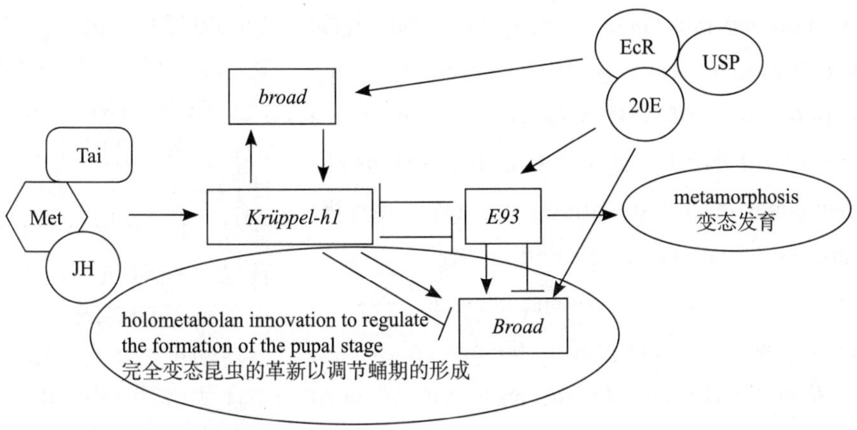

Fig. 10-15 The MEKR-E93 pathway in hemimetabolan and holometabolan. In the nymph-nymph transitions of JH (through Met-Tai) induces the expression of *Kr-h1*, which, in turn, represses the expression of *E93*. In contrast, the fall of JH production in the final juvenile stage, interrupts the expression of *Kr-h1* and allows a strong induction of *E93* by the ecdysone signaling, which triggers metamorphosis. The main difference between the hemimetabolan and holometabolan mode is the regulation (and function) of *Broad*: in the hemimetabolan mode, it is mainly involved in promoting the growth of wing primordia, whereas in the holometabolous mode, *Broad* triggers the formation of the pupa (modified from Belles, 2019).

图 10-15 不完全变态和完全变态发育昆虫中的 MEKR-E93 通路。在不完全变态的若虫 - 若虫转换过程中，JH（通过 Met-Tai）诱导 *Kr-h1* 的表达，进而抑制 *E93* 的表达。相反，JH 滴度在幼虫最后阶段的下降，中断了 *Kr-h1* 的表达，并允许蜕皮激素信号强烈诱导 *E93* 表达，从而触发变态反应。半变态模式和完全变态模式的主要区别在于 *Broad* 的调节（和功能）：在不完全变态昆虫中，*Broad* 主要参与促进翅的生长与发育，而在完全变态发育昆虫中，*Broad* 触发蛹的形成（仿 Belles, 2019）。

stimulates Vg production through insulin signaling, while insulin-insulin growth factor signaling promotes JH synthesis and then activates yolk genesis in cockroaches. In *Drosophila*, Kr-h1, a primary response factor of JH signaling, regulates ovarian development through coordination with 20E signaling response genes *E93*, *E74* and *E75A*. JH signaling in ovarian muscle cells promotes ovarian muscle contraction to facilitate ovulation. JH is more closely associated with reproduction in lower termites, while JH plays an important role in sex determination in higher termites. In social insects, JH has been known to be involved in the polyethism (i.e., age polyethism and castle polyethism).

JH interacts with 20E to regulate metamorphosis. Some transcription factors (e.g., Met, Kr-h1, USP, Taiman/SRC /FISC, E75A, etc.) are involved in the interaction between JH and 20E. As a receptor for JH, Met can bind the 20E receptor complex EcR/USP and participate in 20E signaling. The heterodimer formed by Tai/SRC/FISC and Met is the functional receptor of JH as a coactivator, while Tai/SRC/FISC can also bind the EcR/USP, the receptor complex of 20E. USP binds the EcR or Met in the

中，JH 通过胰岛素信号间接刺激 Vg 产生，而在蜚蠊中，胰岛素 - 胰岛素生长因子信号促进 JH 的合成进而激活卵黄发生。在果蝇中，JH 信号通路的关键因子 Kr-h1 通过与 20E 信号应答基因 *E93*、*E74* 和 *E75A* 的协调作用调节卵巢的发育。卵巢肌肉细胞的 JH 信号还通过促进肌肉收缩来调控排卵。在低等白蚁中，JH 与生殖联系更紧密，而在高等白蚁中，JH 在性别决定中发挥重要作用。在社会性昆虫中，JH 参与调控了多聚性（即年龄多聚性与领地多聚性）。

JH 与 20E 发生交互作用调控变态发育。一些转录因子（如 Met、Kr-h1、USP、Taiman/SRC/FISC、E75A 等）参与 JH 与 20E 的互作。作为 JH 的受体，Met 可与 20E 的受体复合物 EcR/USP 结合并参与 20E 的信号传递。Tai /SRC/FISC 作为共激活因子与 Met 结合形成的异源二聚体是 JH 的功能受体，同时

phosphorylated and non-phosphorylated states to participate in the 20E or JH signal pathway, respectively. JH membrane signaling regulates the phosphorylation of USP. Besides, JH promotes the transcript of 20E-induced *E75A*, and *E75A* further inhibits the expression of other genes in 20E signaling pathway, so that the larvae can maintain the larval state after molting. Kr-h1 directly inhibits the transcript of 20E primary response genes *Br*, *E93* and *E75* in the fatbody and wing discs to prevent the metamorphosis. During the larval stage in *Drosophila*, the high expression of Kr-h1 induced by JH in the PGs inhibits the expression of cell cycle genes to suppress the size of the PG. Additionally, Kr-h1 directly inhibits the expression of ecdysone synthetase related genes, resulting in the inhibition of ecdysone synthesis and arrest of metamorphosis. JH biosynthesis is prevented by 20E in the CA and the other portion of the ring gland that produces JH to ensure the occurrence of metamorphosis (Fig. 10-16).

Tai/SRC/FISC 也可与 20E 的受体复合物 EcR/USP 结合。USP 可分别以磷酸化状态和非磷酸化状态与 EcR 和 Met 结合，参与 20E 和 JH 的信号传递。保幼激素膜信号可促进 USP 的磷酸化。此外，JH 可促进 20E 诱导的 *E75A* 的转录，*E75A* 进一步抑制 20E 信号通路中其他基因的表达，使得幼虫蜕皮后维持幼虫的状态。Kr-h1 可直接抑制脂肪体和翅原基中 20E 初级响应基因 *Br*、*E93* 及 *E75* 的转录从而阻止变态。在果蝇幼虫期时，JH 诱导的 Kr-h1 一方面抑制细胞周期基因表达，阻遏前胸腺细胞生长；另一方面直接抑制前胸腺中蜕皮激素合成酶相关基因的表达，从而抑制蜕皮激素的合成，阻碍变态发育进程。20E 可抑制咽侧体中 JH 的生物合成，而环腺其他部分产生的 JH 则确保了变态的发生（图 10-16）。

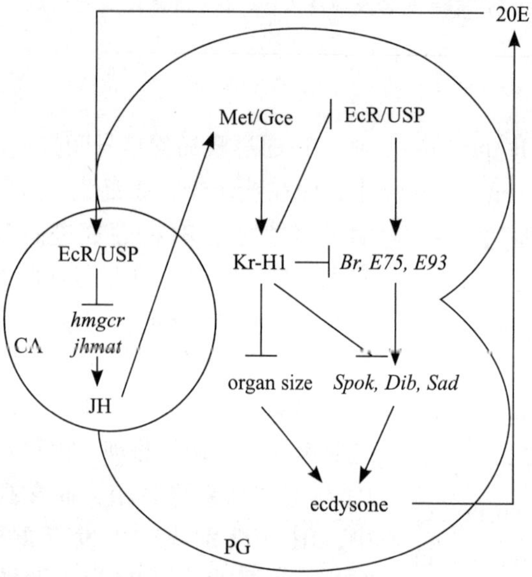

Fig. 10-16 Antagonistic action of JH and 20E within the ring gland determine the developmental transitions in *Drosophila* (Liu et al., 2018; Zhang et al., 2018).

图 10-16 果蝇环腺中 JH 与 20E 的拮抗作用决定了发育的转变（Liu 等，2018; Zhang 等，2018）。

10.6.6 Insect growth regulator

An insect growth regulator (IGR) is a chemical substance that inhibits the life cycle of an insects. IGRs are typically used to control populations of pests, such as cockroaches and fleas. Hormonal IGRs typically work by mimicking or inhibiting JH, one of the two major hormones involved in insect molting. IGRs can also inhibit the action of ecdysone, which triggers the insect to

10.6.6 昆虫生长调节剂

昆虫生长调节剂（IGR）是抑制昆虫生活周期的化学物质。IGRs 普遍用于控制害虫种群，如蟑螂和跳蚤。激素型 IGRs 常常通过模仿或抑制牵涉昆虫蜕皮的两个主要激素之一的 JH 而发挥作用。IGRs 也能抑制促使昆虫蜕皮

molt. If JH is present at the time of molting, the insect molts into a larger larval instar; if it is absent, it molts into a pupa or adult. IGRs that mimic JH can produce premature molting at immature stages, disrupting larval development. They can also act on eggs, causing sterilization, disrupting behavior or disrupting diapause. IGRs that inhibit JH production can cause insects to prematurely molt into a nonfunctional adult. IGRs that inhibit ecdysone can cause pupal mortality by interrupting the transformation of larval tissues into adult tissues during the pupal stage. As such, IGRs don't need necessarily show toxicity to the targets, instead, they cause various abnormalities that lead to the relatively rapid death of the insects. Accordingly, compared with the classic toxic insecticide, IGRs provide a physiological approach to pest control, which would be safe for the use in indoor environment.

激素发挥作用。如果蜕皮时有JH出现，昆虫蜕皮成为较大龄期的幼虫；如果缺乏JH，蜕皮则进入蛹期或成虫期。IGRs可模仿JH，使未成熟幼虫产生早熟，进而打断幼虫发育。它们也能作用于卵，引起绝育、中断行为或打破滞育。抑制JH产物的IGRs还会造成昆虫早熟而蜕皮，使其成为无功能性的成虫。抑制蜕皮激素功能的IGRs，因打断幼虫组织转化成为成虫组织而导致蛹期死亡。因此，IGRs不一定对靶标产生毒性，取而代之的是引起各种畸变体，进而导致昆虫相对快速地死亡。据此，与传统的昆虫剧毒杀虫剂比较，IGRs提供了害虫控制的生理学途径，在室内环境中可安全使用。

However, another insect growth disrupters (IGDs) do not regulate endocrine processes, but disrupt normal endocrine activities. Moreover, some chemicals such as chitin synthesis inhibitors (CSIs) are not directly involved in endocrine processes. These chemicals are a type of insecticides that disrupt the normal activity of the endocrine system, resulting in influences on growth, development, metamorphosis and reproduction of the target pests, and have slower mode of action as compared to the synthetic chemical insecticides. So far, basically three types of IGRs are commercially available, including juvenile hormone analogues, ecdysone agonists, and chitin synthesis inhibitors.

然而，另一种昆虫生长干扰物（IGDs）并不调节内分泌过程，而是扰乱正常的内分泌活动。此外，一些化学物质，如甲壳素合成抑制剂（CSIs），也并不直接参与内分泌过程。这些化学物质是一种破坏内分泌系统正常活动的杀虫剂，对目标害虫的生长、发育、变态和繁殖产生影响。与合成的化学杀虫剂相比，其作用方式较慢。目前，基本上有三种类型的IGR是商品化的，即保幼激素类似物、蜕皮激素激动剂和几丁质合成抑制剂。

The general assumption about JHAs is that they mimic the action of naturally occurring JH and produce the following functions: ① interference of normal metamorphosis of last instar larva, resulting in larva-pupal intermediates; ② JHAs block embryonic development at blastokinesis stage and induce ovicidal effects; ③ induction of sterility in adults; ④ termination of reproductive diapauses.

关于JHA的一般假设是它们模拟自然发生的JHAs并产生以下功能：①干扰末龄幼虫正常变态，导致幼虫蛹的中间产物；② JHAs在胚胎分裂期阻断胚胎发育并诱导杀卵作用；③成虫不育诱导；④生殖滞育的终止。

The anti-JH agents are compounds that have property of inhibiting the biosynthesis of JH in insects, eventually leading to halting of biological processes under the control of JH. The sublethal affects include inhibition of growth and development, deranged morphogenesis, precocious metamorphosis, lower rates of adult emergence and reduced survival of adults. These compounds also possess antigonadotropic activity, affecting oocyte maturation, oviposition and reproductive capacity in insects. Most of these compounds induce precocious

抗JH药剂是具有抑制昆虫JH生物合成特性的化合物，最终导致JH控制下的生物进程停止。亚致死影响包括生长发育抑制、形态发生紊乱、早熟变态、成虫羽化率降低和成虫存活率的降低。这些化合物还具有抗促性腺激素活性，影响卵母细胞的成熟、产卵和昆虫的生殖能力。这些化合物大多可诱

metamorphosis. Precocenes, flumevalonic acid (FMeV), terpenoid Imidazoles, and derivative of fungi and bacteria anti-juvenile hormone compounds, benzoate and methyl dodecanoate compounds, bisthiolcarbamate and sulphoxides are commercially available.

Ecdysone antagonists include chromafenozide, halofenozide, tebufenozide and methoxyfenozide etc.

(Deng Huimin)

导性早熟变态。早熟激素、氟甲戊酸（FMeV）、萜类咪唑、真菌和细菌衍生的抗保幼激素化合物、苯甲酸酯和十二酸甲酯化合物、双硫代氨基甲酸酯和亚砜等已在市面上出售。

蜕皮激素拮抗剂包括嗜铬酰肼、氟虫酰肼、虫酰肼和甲氧虫酰肼等。

（邓惠敏）

Exercises

1. What are insect hormones?
2. What does insect endocrine organ have and what is the hormone that it secretes?
3. What are the pupal and adult specific transcription factors in the ecdysone signaling pathway?
4. How to use hormones to control pests?
5. What physiological processes do insect hormones regulate in addition to growth and development?

思考题

1. 什么是昆虫激素？
2. 昆虫内分泌器官主要有哪些及其分泌的激素是什么？
3. 蜕皮激素信号通路蛹期与成虫特异性转录因子分别是什么？
4. 如何利用激素防治害虫？
5. 除了调控生长发育外，昆虫激素还调控哪些生理进程？

第 11 章 昆虫感觉系统

Chapter 11 Insect Sensory System

Key points

1.Insect sensor receptors are usually classified according to their functions, but many researchers also classify them according to the morphology of the electron microscope. Among them, there are the most abundant types of mechanoreceptors, which are related to insects' perception of environmental information, navigation, living habits, behavior, etc.

2.Insect sensory mechanisms are all related to insect nervous system and neurons. For example, chemical receptors only have taste receptors or tactile receptors and olfactory receptors, which transmit information through odor receptors (ORs), ion receptors (IRs) and taste receptors (GRs). Although ORs are seven transmembrane protein domains, in contrast to GPCR, they have cytoplasmic N- terminal and extracellular C terminal, and form complexes with odor receptor co-receptor (Orco), playing the role of ion channel proteins.

3.Insect temperature and humidity receptors, geomagnetic receptors, and gravitational receptors are related to insect solar compass navigation, foraging, and migration behaviors.

知识点

1. 昆虫感受器通常是根据其功能进行分类，但也有较多研究人员根据电镜形态分类。其中机械感受器类型最为丰富，与昆虫感受环境信息、导航、生活习性和行为等有关。

2. 昆虫感受机制都与昆虫的神经系统、神经元相关联，如化学感受器仅有味觉或触觉感受器和嗅觉感受器，通过气味受体（ORs）、离子受体（IRs）和味觉受体（GRs）传导信息。其中ORs虽然是七个跨膜蛋白结构域，但与GPCR相反，ORs具有细胞质N-末端和细胞外C-末端，与气味受体辅受体（Orco）形成复合体，起着离子通道蛋白的作用。

3. 昆虫温湿度感受器、地磁感受器、地球引力感受器与昆虫太阳罗盘导航、觅食和迁徙行为相关。

Like other animals, insects have five main sensory organs—vision, hearing, smell, touch and taste. The antennae of insects are important sensory organs of insects, which have the functions of smell, touch and hearing. This skill makes them the best insect catchers in the insect world.

与其他动物一样，昆虫具有五大主要的感觉器官——视觉、听觉、嗅觉、触觉和味觉。昆虫的触角是昆虫重要的感觉器官，有嗅觉、触觉和听觉的功能。这种本领使得它们成为昆虫界的捕虫高手。

Insects have a surprisingly diverse array of sensory receptors that feed them information about their internal and external environment. The sensory neurons of insects have their cell bodies, with only rare exceptions, located very near the stimulus site.

昆虫的感觉受体为它们提供内、外部环境信息，具有丰富的多样性。其感觉神经元除了极少数外都有细胞体，都位于

Many receptors detect changes occurring at the cuticular surfaces and the cell bodies are located peripherally just beneath the cuticle. Most sensory neurons are bipolar, with a few multipolar ones, and the dendritic terminals are usually very short compared to the relatively long axon leading to the CNS. The axons from many sensory neurons pass into the brain prior to synapsing, and are specified as primary or type Ⅰ sensory neurons. Secondary or type Ⅱ sensory neurons synapse prior to entering the brain.

刺激位点附近。许多受体能检测到表皮表面的变化，而细胞体则位于外周表皮的正下方。大多数感觉神经元都是双极的，少数是多极的，与导向 CNS 的长轴突相比，其树突末端通常很短。许多感觉神经元的轴突在发生突触连接之前先进入大脑，称为原发性或Ⅰ类感觉神经元。感觉神经元在进入大脑之前发生突触连接，称二级或Ⅱ类感觉神经元。

A common characteristic of all types of sensory neurons is that they transducer the stimulus energy, such as light, heat, chemical, or mechanical energy, into a slow, or graded, electrical potential. The receptor process can be divided into three steps: ① absorption of the stimulus energy; ② transduction into the receptor potential; ③ repetitive impulse discharge from the axon portion of the receptor neuron.

所有的感觉神经元的共同特征是把刺激能量（如光、热、化学能或机械能量）向一个慢电位或分级电位转导。接收过程可分为 3 步：①吸收刺激能量；②向受体电位传导；③受体神经元轴突部分的反复脉冲释放。

Adaptation (Fig. 11-1), to a steady stimulus is a characteristic feature of many receptor neurons. During adaptation, the receptor potential falls from its initial response level to some lower level or perhaps even to a silent rate. Receptors that adapt rapidly to continuing steady stimuli are phasic receptors, while those that adapt slowly are tonic receptors.

对稳定刺激的适应（图 11-1）是许多感受器神经元的特征。在适应期内，感受器电位从起初反应水平下降至稍低水平或甚至到静息速率。快速适应持续刺激的感受器是相位感受器，而那些缓慢适应的感受器是紧张性感受器。

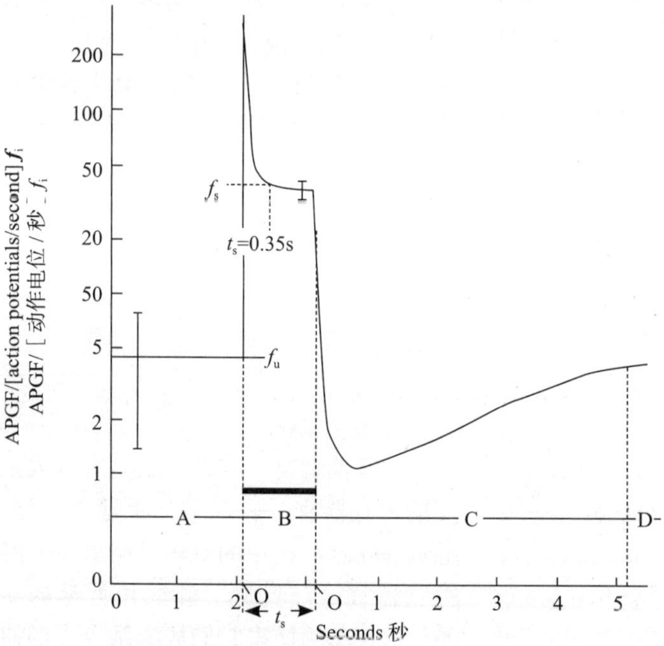

Fig.11-1 The response of a chemoreceptor on the labellum of the blowfly *Phormia regina* to a solution of 0.4 M KCl. The stimulus duration is indicated by the dark bar at B; Periods A and D show the unstimulated spontaneous activity of the receptor. The receptor neuron initially makes a strong response to the stimulus by increasing spike output to a rate greater than 200 spikes/s., but adaption occurs rapidly, and, after about 0.35 s., the output rate falls to a tonic output of about 40 impulses/s. The post stimulation response or "off" response is shown in C. t_s. Time elapsed since the commencement of stimulation; f_s. APGF at t_s= 0.35 s representing "steady state" of response; f_u. APGF when receptor is not stimulated (modified from Ress,1968).

图 11-1 丽蝇的化学感受器对 0.4 M KCl 的反应。B. 刺激持续时段；A，D. 非刺激时段。受体神经元起初对刺激产生很强的反应，峰电位输出速率大于 200 峰值 /s，但很快就出现适应。在大约 0.35 s 后，输出速率下降成为约 40 脉冲 /s 的相位性输出。刺激后的反应或反应"关闭"期如图中的 C 所示。t_s. 刺激起始延续的时间；f_s.APGF 在 t_s=0.35 s 时的值，代表反应"稳定状态"；f_u. 受体未受刺激时的 APGF(仿 Ress,1968)。

11.1 感受器的分类及其功能

11.1 Classification of Sensory Receptors and General Function

11.1.1 Functional classification of sensory receptors

Receptors can be classified functionally with respect to the type of energy they transduce. For example, they might be classified as light and/or visual detectors, mechanoreceptors (including tactile, vibration, and sound detectors), chemoreceptors (including contact receptors: gustatory or taste receptor and olfactory receptors), humidity receptors, temperature receptors, magnetic receptors, geographical receptors.

11.1.1 感受器的功能分类

根据其转换的能量类型，感受器可从功能上分类。例如，它们可分为光和/或视觉感受器、机械感受器（包括触觉、振动和声音探测器）、化学感受器（包括接触感受器：味觉感受器和嗅觉感受器）、湿度感受器、磁场感受器和地理感受器。

11.1.2 Anatomical classification of insect receptors

More researchers prefer to classify according to the morphological features of receptors obtained by scanning electron microscopy, such as *S. mammilliformia, S. cylindrical* and *S. aurcillica* etc. It is mainly divided into the following categories:

11.1.2 感受器的解剖学分类

更多的研究人员比较喜欢根据感受器电镜扫描得到的形态特征进行分类，如乳形感受器，圆柱形感受器和耳形感受器等。其主要分为以下几类。

11.1.2.1 *Sensilla trichoidea*

Sensilla trichoidea are sensory hairs or setae and their elaborations, including hair plates. These sensilla are widely distributed among insects and other arthropods. Many of the hairs on insects are innervated by sensory neurons. The hair might not have any openings (tactile hairs, hair plates), may be perforated by only one pore (gustatory contact chemoreceptors), or by many pores (olfactory receptors).

11.1.2.1 毛形感受器

毛形感受器是感觉毛或刚毛及其精细结构，包括毛板。这些感受器广泛分布在昆虫和节肢动物中。昆虫的许多体毛受感觉神经元支配。感觉毛可能没有任何开口（触毛、毛板），可能只有一个孔道开口（味觉触化感受器）或多孔（嗅觉感受器）。

11.1.2.2 *Sensilla chaetica*

Sensilla chaetica are sensory spines or bristles that are stouter than *S.trichoidea* and often are located singly rather than in groups. *S.chaetica* consists of an innervated hair in a flexible socket. The hair may have a single pore (gustatory or some mechanoreceptor) or pores may be absent (most mechanoreceptors).

11.1.2.2 刺形感受器

刺形感受器是刺状或鬃毛状感受器，比毛形感受器更粗，常单个分布而不是成团的。刺形感受器由神经支配的毛组成，有韧性窝。毛可能单孔（味觉或一些机械感受器）或无孔（多数是机械感受器）。

11.1.2.3 *Sensilla squamiformia*

Sensilla squamiformia are flattened hairs or sensory scales. They

11.1.2.3 鳞形感受器

鳞形感受器是扁平的毛或鳞片感受

are common on the wings of Lepidoptera, although not every wing scale is innervated as a sensory organ.

器。在鳞翅目的翅上很普遍，但并不是每个翅基板都像感官一样受神经支配。

11.1.2.4 *Sensilla basiconica*

Sensilla basiconica are sensory pegs, cones, or stumpy hairs. They may be either thick-walled or thin-walled, and lack pores (thermo- and hygroreceptors), or have a single core (contact chemoreceptors and gustatory receptors), or have many pores (olfactory receptors). *S.basiconica* may contain from a few to many neurons.

11.1.2.4 锥形感受器

锥形感受器是钩形、圆锥形或粗短毛感受器。厚壁或薄壁，无孔（温湿度感受器），有一单孔（触化感受器和味觉感受器）或多孔（嗅觉感受器）。锥形感受器含有少数到多个神经元。

11.1.2.5 *Sensilla coeloconica*

Sensilla coeloconica are cones or pegs set in small depressions or pits in the cuticle. Both thick-walled and thin-walled sensilla occur. Some have multiple pores and serve as olfactory receptors, and some have no pores and appear to serve as thermo-and hygro-receptors.

11.1.2.5 腔锥形感受器

腔锥形感受器是小凹陷内的圆锥形或钩形感受器或表皮内陷，厚、薄壁都有。一些有多孔，起着嗅觉感受器作用，有些无孔，可能起着温湿度感受器的作用。

11.1.2.6 *Sensilla styloconica*

Sensilla styloconica consist of elevated cones that may be located in pits or at the cuticle surface. *S.styloconica* typically serves a gustatory function and have a single pore at the tip.

11.1.2.6 栓锥形感受器

栓锥形感受器由隆起的圆锥组成，可能位于凹陷内或体表面。典型的栓锥形感受器用于味觉功能，顶部有单孔。

11.1.2.7 *Sensilla ampullacea*

Sensilla ampullacea are sensory tubes. They may be cone-shaped and rest on long tubes in sunken pits. They occur on the antennae of bees and other Hymenoptera.

11.1.2.7 坛形感受器

坛形感受器是管状感受器。可能是圆锥形，支撑在凹槽的长管上。它们出现在蜜蜂和其他膜翅目的触角上。

11.1.2.8 *Sensilla placodea*

Sensilla placodea are multiporous plate structures with an olfactory function. *S.placodea* have many neurons with dendritic terminals ending on thin plates covering a fluid-filled canal. These sensilla occur on antennae of several insect orders.

11.1.2.8 板形感受器

板形感受器为多孔的板形结构，有嗅觉功能。板形感受器有许多神经元，薄板上的树突状终末端覆盖着充液的管道。这些感器出现在几个昆虫目的触角上。

11.1.2.9 *Sensilla campaniformia*

Sensilla campaniformia include various structures, Hick's organs, cupola organs, dome organs, umbrella organs, bell organs and sensory pores. *S.campaniformia* is innervated by a single neuron that terminates in a small dome. Sometimes there is a scolopale at the center of the dome where nerve contact occurs. These sensilla may be recessed in cuticular. *Sensilla campaniform* are often arranged in small groups along the long axis of a limb. It seems likely they are important in maintaining the stance of an insect, in geotaxis, and in coordinating walking or running.

11.1.2.9 钟形感受器

钟形感受器包括各种结构，如希氏器、塔形器、圆顶器、伞形器、铃形器和微孔感受器。钟形感受器由终止于小圆顶内的单神经元支配，偶尔有感橛位于圆顶中心与神经接触。这些感受器可能凹入表皮。钟形感受器常沿翅腿长轴小群分布。其重要性可能在于保持昆虫姿态和趋地性，与行走或奔跑有关。

11.1.3 Simplified classification

A simplified classification scheme was recommend based primarily on the presence of pores, and number of pores in external sensilla, with general functional significance when known. The sensory receptors were recommended three major groupings as receptors with multiple pores, a single pore and without pores.

11.1.3.1 Receptors with multiple pores

Receptors with multiple pores in the external cuticular structure tend to be olfactory receptors that detect airborne chemicals. The external cuticular structure may take the form of a hair (*Sensilla trichodea*), plate (*S.placodea*), peg (*S. basiconica*), or peg-in-a-pit (*S.coeloconica*). Olfactory receptors are probably present on the antennae of all insects, and occur elsewhere on the body of some insects. Functionally, the openings or pores through the cuticular covering allow airborne molecules to enter the sensillum. The external structure tends to be thin-walled and the cuticular socket inflexible.

11.1.3.2 Receptors with single pores

Receptors that have a single pore near or at the tip of the structure usually have a gustatory function (i.e., taste receptor) and detect chemical substances in solution. Gustatory receptors also are called contact chemoreceptors. The external appearance may be that of a hair (*S.chaetica or S.trichodea*), peg (*S.basiconica*), dome (*S.styloconica*). The external cuticular structure is thick-walled, and sockets may be flexible or inflexible. Gustatory receptors are numerous on the mouthparts, as well as on the tarsi and ovipositor of some insects.

11.1.3.3 Receptors with no pores

The lack of any pore in the external cuticular part of a sensillum is typical of mechanoreceptors that detect vibrations in the air (insect "ears" and vibration receptors), water, or substrate on which an insect's rests. Mechanoreceptors may consist of a dome (*S.campaniformia*), or hair (*S.chaetica*) set in a flexible socket. Humidity and temperature receptors, which may have the form of a peg (*S.basiconica*), or peg-in-a-pit (*S.coeloconica*) in an inflexible socket, also usually lack a pore.

Although the structural unit is the sensillum, but the functional units are the neurons within the sensillum. Multiple neurons in the same sensillum may serve multiple physiological roles. For

11.1.3 简单分类法

简化的分类方法主要根据有孔或无孔，外部感受器孔的数量和已知功能。感受器可分为 3 类：多孔感受器、单孔感受器和无孔感受器。

11.1.3.1 多孔感受器

多孔感受器体表外的结构往往是嗅觉感受器受体，侦测空气中传播的化学物质。其体表结构为毛形（毛形感受器）、板状（板形感受器）、钩形（锥形感受器）或坑钩状（腔锥形感受器）。嗅觉感受器既在所有昆虫的触角上，也可以在其他地方。在功能上，表皮覆盖层上的开口或孔允许空气中传播的化学分子进入感受器。外表结构上为薄壁，表皮孔穴无韧性窝。

11.1.3.2 单孔感受器

感受器在表皮结构附近或顶部有 1 个单孔，通常具有味觉功能（即味觉感受器），可探测溶液中的化学物质。味觉感受器也称为触化感受器。其外观可能是毛形（刺形或毛形感受器）、板状（板形感受器）和圆顶（栓锥形感受器）。外表皮结构是厚壁形，可能有韧性窝或非韧性窝。有些昆虫味觉感受器在口器上的数量和跗节及产卵器上的一样多。

11.1.3.3 无孔感受器

感受器的体表外部分无孔是典型的机械感受器，用于侦测空气（昆虫的"耳"，振动感受器）、水或支撑昆虫的载体的振动。机械感受器可由圆顶形（钟形感受器）或毛形（毛形感受器）在一个韧性窝内组成。湿度感受器和温度感受器通常也无孔，形状上有钩形（锥形感受器）或坑钩（腔锥形感受器），有韧性窝。

尽管其结构单位是感受器，但感受器内的功能单位是神经元。同一个感受器内的多神经元起到多种多样的生理作用。比如感受器内的某个神经元功能上

example, one neuron may function as a chemoreceptor while another neuron in the sensillum functions as a mechanoreceptor.

是化学感受器，而另一个神经元功能上是机械感受器。

11.2 机械感受器

11.2 Mechanoreceptor

Mechanoreceptors are involved in detection of airborne sounds, substrate vibrations, appendage movement and orientation (proprioceptors), flight speed, gravity, and heat detection. It including tactile sensilla vibration sensilla, acoustical sensilla, and stretch receptor.

机械感受器牵涉到空气中传播的声音侦测、基质振动、附肢移动和方向（本体感受器）、飞行速率、重力和热探测。机械感受器包括了触觉感受器、振动感受器、听觉感受器以及牵引张力感受器。

Mechanoreception is the sense that allows insects to detect movements of objects in the environment and sounds or provides proprioceptive cues (sensory input about the position or orientation of the body and its appendages). A bipolar sensory neuron is closely opposed to the base of the sensory cuticle, and its dendrite contains tightly packed microtubules constituting a distal structure called the tubular body. Cuticular mechanoreceptors can be subdivided into three major groups, hairs, campaniform sensilla, or chordotonal organs. Hairs can be in the form of long, thin, or shorter hairs (*trichoid sensilla*) or pegs and are functionally divided into filiform hairs and bristles. The bristle seems to be the most common type of mechanoreceptor, usually responding to direct touch, while filiform hairs or trichobothria are sensilla deflected by faint air currents and low-frequency sound, or vibrations of the surrounding environment. Chordotonal receptors are generally found further beneath the integument and can be connected to the cuticle by attachment structures. They serve several functions, including hearing, joint movement detection and spatial perception mediated by insect antenna and cerci mechanosensory system.

机械感觉是一种允许昆虫检测环境中物体的运动和声音，或提供本体感觉信息（关于虫体及其附件的位置或方向的感觉输入）。双极感觉神经元与感觉表皮层的基底紧密对向相连，其树突包含紧密排列的微管，构成了一种称为管状体的远端结构。表皮机械感受器可分为三大类：毛板、钟形感受器或弦音感受器。毛板可以是长、细或短的毛（毛状感觉器）或钉状，在功能上分为丝状毛和刚毛。刚毛似乎是最常见的机械感受器类型，通常对直接触摸做出反应，而丝状毛或感觉毛则是被微弱气流和低频声音或周围环境的振动偏斜的感受器。弦音感受器通常位于体壁下方深层，可通过附着结构连接到表皮层。它们具有多种功能，包括听觉和关节运动检测以及通过昆虫触角和触须机械感受系统进行的空间感知。

11.2.1 Proprioceptors

11.2.1 本体感受器

Receptors providing information about internal body conditions are called proprioceptors. Proprioceptors are present in the connective tissue of the body, among muscles, and along the surface of the alimentary canal of all insects. The sensory neurons are frequently multipolar. Some types of proprioceptors function

提供虫体内部状况信息的感受器被称为本体感受器。本体感受器出现在虫体组织连接处，肌肉之间以及所有昆虫的肠道表面。多数感觉神经元是双极的，有些类型的本体感受器在功能上是机械感

as mechanoreceptors or stretch receptors to indicate gut filling, muscle tension, and general body movements, while others act as chemoreceptors relaying information about the chemical composition of the body. Proprioceptors generally adapt slowly to a constant stimulus, which clearly is adaptive if they must indicate body orientation, equilibrium, limb positioning, or fullness of the gut.

Drosophila has six major types of somatosensory neurons. Gustatory sensilla neurons detect food stimuli through the pores of external taste hairs. External sensilla neurons detect movement of the bristles caused by contact, wind, etc. Stretch receptor neurons contact neighboring leg joints to detect stretch and tension between them. The chordotonal organs have thin elongated structures that span across different parts of the exoskeleton to detect tension or vibration. *Campaniform sensilla* feature dome-shaped thin cuticles to detect deformation of the exoskeleton. Finally, multi-dendritic neurons extend complex dendrites spanning under the epidermis to detect diverse stimuli such as touch, pain, heat, and coldness.

受器或牵引张力感受器，指示填饱的肠道、肌肉的张力和整个身体的运动，而其他的即行使化学感受器的功能，传播虫体的化学组成信息。总体上，本体感受器必须标示虫体的方向、平衡、航行定位或肠道的饱填，它们对持续的刺激产生缓慢的适应，很明显具有适应性。

果蝇有六种主要的体感神经元。味觉感受器神经元通过外部味觉毛的毛孔检测食物刺激。外部感觉神经元检测由接触和风等引起的刚毛运动。牵引张力感受器神经元接触相邻的腿部关节，以检测它们之间的拉伸和张力。弦音感受器具有细长的结构，跨越外骨骼的不同部分以检测张力或振动。钟形感受器具有圆顶状的薄表皮，用于检测外骨骼的变形。最后，多树突神经元延伸表皮下的复杂树突，以检测不同的刺激，如触摸、疼痛、热和冷。

11.2.2 Tactile hair

The simplest mechanoreceptor, a sensory hair, contains a minimum of three cells, all derived from a common epidermal moth cell that divides to give rise to the trichogen, the tormogen, and the bipolar sensory neuron (Fig.11-2). Tactile hairs have two pores, a terminal and a basal one, in which the latter is in direct contact with the nerve cell. Just below the bristle, there is a cavity filled with lymph, which reinforces the signal captured by the cuticular structure from the movement of the bristle. As in other parts of the nervous system, the sheath cells insulate the neuron, may provide it with nutrients, and may help control concentrations of ions necessary for nerve function.

Cercal receptors detect a range of vibration in the air and substrate, and can act as sound receptors and vibration receptor. In cock roaches, and possibly in other insects, cercal receptors function in an escape mechanism in which the tactile hairs respond to sudden vibrations or loud sounds by sending spikes through the cercal nerve to connect with giant axons at synapse in the sixth abdominal ganglion. The giant axons pass uninterrupted to the thoracic ganglion where synaptic connections are made with motoneurons to the leg muscles. The system results in rapid transmission of stimuli that give rise to escape maneuvers. The stimulation of the bristles can be involved in detection

11.2.2 感觉毛

最简单的机械感受器，即感觉毛，包含从真皮的一个母细胞衍生而来的3个细胞，分裂毛原细胞、膜原细胞和双极感觉神经元（图 11-2）。感觉毛有两个孔，一个末端孔和一个基底孔，后者与神经细胞直接接触。刚毛下方有一个充满淋巴的空腔，它加强了表皮结构从刚毛运动中捕获的信号。与神经系统的其他部分一样，鞘细胞隔离神经元，可能供给营养，协助控制神经功能所必需的离子浓度。

尾须感受器能检测一定范围的空气和基质震动，行使声音感受器和震动感受器的功能。在蜚蠊体内，也可能在其他昆虫中，尾须感受器在逃跑机制中起作用；感觉毛对突然震动或响亮声音的反应是通过尾神经将峰电位传送到第六腹神经节的巨型轴突的突触实现的。巨型轴突不间断地传送到胸神经节，与腿肌运动神经元进行突触连接。系统导致刺激的快速传送而引起逃跑行为。蝗虫的刚毛被刺激可以检测强气流，蟋蟀和

of strong air currents in locusts and can detect wind, courtship song, and the wingbeats of a predator in crickets and cockroaches.

蟑螂也可以检测风、求偶鸣叫和捕食者的翼拍。

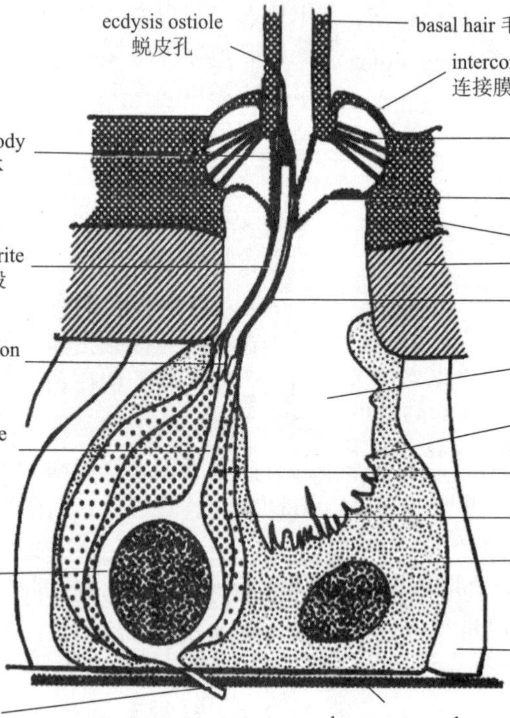

Fig. 11-2 The structure of tactile hair (modified from Chapman, 1998).
图 11-2 毛形感受器的结构（仿 Chapman, 1998）。

11.2.3 Hair plates

Hair plates are common at leg joints and at points of limb articulations with the body. They respond of touch, bending, and to joint flexion with output of nerve spikes. They adapt slowly, a characteristic of static receptors that indicate body orientation. These tactile structures enable an insect to know the position of its limbs with respect to the body, and probably enabling more precise movement of these large, heavily muscled parts of the leg. Hair plates also are common on sclerites at the back of the head and/or neck and on the anterior parts of the prothorax in mantids, locusts, and bees, and act as proprioceptors enabling insect to know its head orientation with respect to the body. They may also be important in some cases, at least, in flight ability because destroying the hairs on the cervical plates of locusts influences their equilibrium in flight. Sensory neurons, like other parts of the nervous system, have high demand for oxygen; the mechanoreceptors of honeybee cervical hair plates were shown to be very sensitive to oxygen deficiency, with spikes ceasing after 2 min and the receptor potential after 10 min.

11.2.3 毛板

毛板大多着生在腿接合和肢关节处。它们对触觉、弯曲和关节弯曲做出反应，并输出神经峰电位。毛板缓慢地适应指示身体方向的静态感受器的特征。这些触觉结构使昆虫知道其与虫体关联的肢体的位置，并且可以使它们在运动中更准确地移动这些大而肌肉发达的腿部。毛板在蟑螂、蝗虫和蜜蜂的头和 / 或颈部背面以及前胸的前部，起着本体感受器作用，使昆虫感知头部相对躯体的方向。在有些情况，至少在飞行中也是很重要的，因为破坏蝗虫颈背板上的毛会影响它们在飞行中的平衡。感觉神经元与神经系统的其他组成一样，它的需氧量很高。蜜蜂颈部毛板上的机械感受器表现出对负氧的高度敏感，缺氧后 2 min 峰电位停顿，10 min 后受体电位停止。

11.2.4 Chordotonal sensilla

A chordotonal sensillum is anatomically more complex than a tactile hair. It occurs at most exoskeletal joints, limb joints, and body segment joints. There are many morphological variations, but basically the sensory neuron is enclosed within parts of two or three other cells, including a characteristic scolopale cell and cap cell (Fig.11-3). There may also be additional sheath cells. Instead of the dendritic terminals being attached directly to the cuticle, they terminate within the cap cell, and the cap cell is attached to some internal structure or to the cuticle. Any stress or pull on the cap cell is then transmitted to the neuronal endings as a stimulus. Often, the cap cell extends well down over the scolopale sheath cell. A single sensory unit with the scolopale and cap structure is called a scolopidium or, alternatively, a chordotonal sensillum. Complex chordotonal organs, such as a tympanum or Johnston's organ, contain many scolopidia, and both external and proprioceptors contain scolopidia as the morphological unit. Examples of sensory organs containing scolopidia are: ① subgenual organs; ② tympanal organs; ③ Johnston's organ; ④ simple structures.

11.2.4 弦音感受器

弦音感受器在解剖学上比感觉毛更为复杂。多数位于外骨骼、肢体及体节接合处。形态上变化很大，但本质上是封闭在 2 ～ 3 个其他细胞中的感觉神经元，包括特有的感橛细胞和冠细胞（图 11-3），另外也有鞘细胞。树突末端不是直接与表皮相连，而是终止于冠细胞，冠细胞与一些内部结构或表皮相连。任何施于冠细胞的压力或拉力都可以作为刺激源传导至神经末梢。通常，冠细胞向下延伸覆盖住感橛器的鞘细胞。一个单独的感觉神经单位和感橛器及冠细胞形成的结构称为具橛细胞或弦音器。弦音器复合体，如鼓膜器或 Johnston's 器，含有许多感橛器，外部和本体感受器都含有感橛器当作形态学单元。包含感橛器的例子：①膝下器；②鼓膜器；③ Johnston's 器；④简单结构。

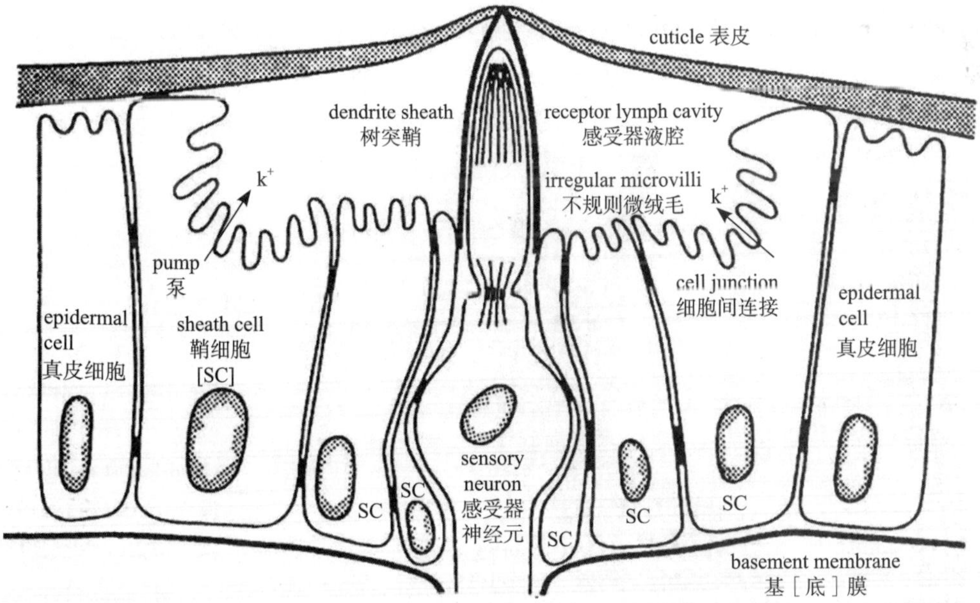

Fig. 11-3 Illustration of Chordotonal sensillum or mechanoreceptor. Any stress at the cuticular surface where the scolopale cap cell attached is transmitted to the sensory neuron beneath (modified from French, 1988).

图 11-3 弦音感受器或机械感受器的说明。表皮下的感橛细胞和冠细胞接收到来自表皮的压力信息时立即传送到其下的感觉神经元（仿 French, 1988 改绘）。

11.2.4.1 Subgenual organs

The subgenual organ is a complex chordotonal organ composed of multiple scolopidia. The term "subgenual" means below the knee, from Latin for *genu*, and this complex chordotonal organ usually is located near the joint between the femur and tibia (Fig.11-4). The organ contains as few as three scolopidia in some earwigs (*orficula* spp.), but contains more in most insects. It acts as a proprioceptor and detects vibrations of the substrate, and it has become specialized as a tympanal organ in some insects. The subgenual organ is especially well developed in crickets (Gryllidae) and katydids (Tettigoniidae) and is associated with the tympanal organ, with both organs located on the tibia. The two organs have separate innervation, however, and probably have separate functions.

In stick insects (Phasmatodea), the subgenual organ complex contains the subgenual organ and the distal organ located distally to the subgenual organ. The subgenual organ is a highly sensitive detector for substrate vibrations and also to airborne sound. The distal organ has a characteristic linear organization of sensilla, and is likely also to respond to substrate vibrations. By the position in the hemolymph channel, the distal organ possibly also detects vibrations transferred along the leg.

In arthropods, the detection of vibrational signals and stimuli is essential in several behaviours, including mate recognition and

11.2.4.1 膝下器

膝下器是一类由多个具橛感受器组成的复合弦音器。"膝下"表示膝盖下方，来自拉丁语"*genu*"，通常位于腿节和胫节接合处（图 11-4）。在革翅目昆虫蠼螋中，只有很少的 3 个具橛细胞，但大多数昆虫具橛细胞很多。膝下器具有本体感受器的功能，侦测基质的振动，但在有些昆虫中特化成鼓膜器。在蟋蟀和豆娘中，膝下器特别发达，并与鼓膜器相连，都位于胫节中。两个器官的神经分布是分开的，因此，其功能也不同。

在竹节虫（竹节虫目）中，膝下器复合体包含膝下器和位于膝下器远端的末梢器官。膝下器是对基底振动和空气传播声音高度敏感的探测器。末梢器官具有典型的线性感受器组织，也可能对基质振动做出反应。根据血淋巴通道中的位置，末梢器官还可检测沿腿部传递的振动。

在节肢动物中，振动信号和刺激的检测在几种行为中至关重要，包括配偶

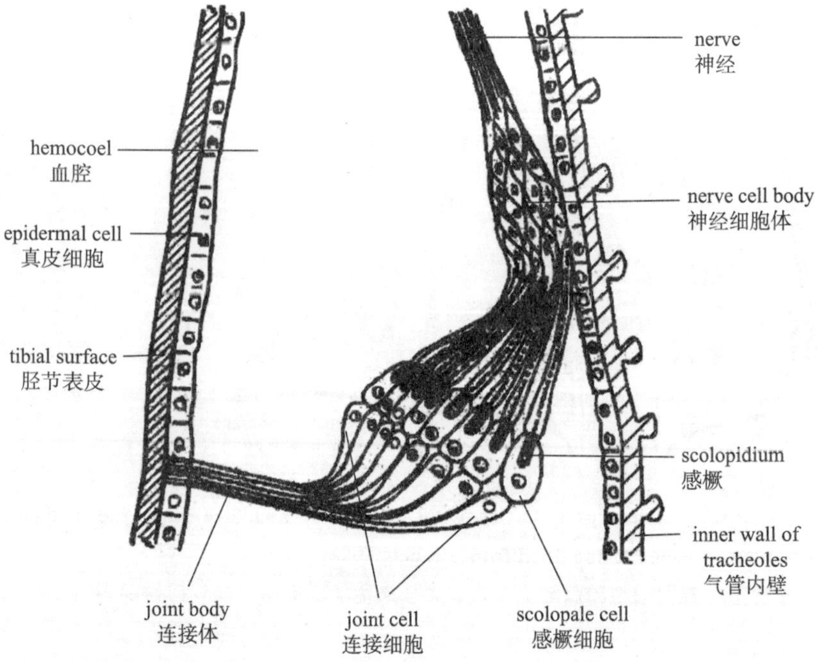

Fig. 11-4 Structure of subgenual organs from a lepidopteran (modified from Horridge, 1965).
图 11-4 鳞翅目昆虫的膝下器结构（仿 Horridge,1965）。

pair formation, prey detection, and predator evasion. Vibration stimuli are transferred over the animals' appendages and the body to vibrosensory organs. Ultimately, the stimuli are transferred to act on the dendrites of the mechanosensitive sensilla. These biomechanical processes have important roles in filtering and preprocessing of stimuli, which are not carried out by neuronal components of sensory organs. Also, the macromechanical transfer is posture-dependent and enables behavioural control of vibration detection. Diverse sensory organs respond to vibrations, including cuticular sensilla (*slit sensilla, campaniform sensilla*) and internal chordotonal organs. These organs provide various adaptations, as they occur at diverse body positions with different mechanical couplings as input pathways. Macromechanics likely facilitated evolution of vibrosensory organs at specific body locations.

识别和配对、猎物检测和逃避捕食者。振动刺激通过动物的附肢和身体转移到振动感觉器官。最终，刺激被转移到作用于机械敏感感受器的树突上。这些生物力学过程在刺激的过滤和预处理中具有重要作用，这些刺激不是由感觉器官的神经元成分执行的。此外，宏观机械传递依赖于姿态，并能够对振动检测进行行为控制。不同的感觉器对振动做出反应，包括表皮感器（狭缝感器和钟形感器）和内部弦音器。这些器官提供各种适应性，因为它们随不同的机械耦合作为输入路径而出现在身体的不同位置。宏观力学可能促进了特定身体部位可控振源器官的进化。

The transmission of vibrations includes successive steps: from the source of vibration generation through the environment and over the body of an animal receiver to the site of specific sensory organs. All vibrational behaviours require mechanosensory systems to detect and process the vibrational signals transmitted in the environment.

振动的传递包括连续的步骤：从振动产生的源头通过环境和动物接收器的身体到特定感觉器官的部位。所有振动行为都需要机械传感系统来检测和处理环境中传输的振动信号。

11.2.4.2 Tympanal organs

11.2.4.2 鼓膜器

Tympanal organs are chordotonal organs or "insect ears" that are specialized for high-frequency sound detection as opposed to low-frequency vibration detection. Acoustical communication in insects is important in several behaviors, including mate location and selection in singing insects and detection of possible predators. Often, multiple species may be singing at the same times and there may be other acoustical signals in the environment, so conspecifics need to be able to detect the song of its species. Cicadas, for example, are singing insects that depend on particular acoustical frequencies for mate location. Song structure, including pulse carrier frequency, pulse durations and pauses, and number of pulses per chirp, varies across species. In many species, the courtship song consists of brief pulses repeated at regular intervals (pulse song) interspersed with near- constant -frequency song (sine song).

鼓膜器是弦音器或昆虫的“耳朵”。与低频振动的侦测不同，它们是特化的高频率声音的侦测器。昆虫的听觉通信在几种行为中是重要的，包括在振鸣昆虫中择偶和选择交配场所以及探测可能的天敌。通常多个种在同一时间振鸣，环境也可能有其他听觉信号，因此，同种昆虫必须能侦测出其物种的鸣声。例如，蝉是依靠特别听觉频率寻找求偶场所的振鸣昆虫。虫鸣结构包括脉冲载波频率、脉冲持续时间和暂停以及每个虫鸣的脉冲数，这都因物种而异。在许多物种中，求爱鸣声由规则间隔重复的短暂脉冲（脉冲虫鸣）组成，其间穿插着接近恒定频率的虫鸣（正弦鸣声）。

Another function of sound production by some insects may be warning signals (aposematic signals) that help deter some predators. For example, sound production in *A. pernyi* usually proceeded defensive regulation, and the regurgitation. Arctiid moth, *C. tenera,* preferentially responds to bat attack calls as distinct from bat searching calls. In addition to tympanal organs, some insects

有些昆虫另一个发声的作用可能是警戒信号（防护信号），有助于阻止一些天敌。比如柞蚕发出的声音通常进行防御调节和反胃。灯蛾对蝙蝠的攻击信号最先反应，明显能区别蝙蝠的搜寻信

also hear some sounds with other organs, including the Johnston's organ, subgenual organs, scattered simple chorotodonal sensilla, and simple hair sensilla.

With respect to the sound-receiving structures, insect ears can be classified into antennal and tympanal ears. The sound-receiving cuticular region in some tympanal ears is not conspicuously thinned into a tympanal membrane. Anatomically, such ears might be classified as atympanal. Although the cuticle might nonetheless vibrate in response to sound and function as an eardrum.

Pressure-sensitive tympanal and atympanal ears operate at frequencies of up to some 300 kHz and detect communication sounds and bat echolocation signals over large distances. Particle-velocity-sensitive antennal ears. In contrast, particle-velocity-sensitive antennal ears are advantageous for intimate communication at close distances and usually operate at frequencies below 1 kHz. With the exception of particle-velocity sensitive hairs, all insect ears derive from chordotonal sensory organs and can be categorized into particle-velocity-sensitive antennal and pressure-sensitive tympanal or atympanal ears.

Tympanal organs are located at various places on the body of insects. Examples of locations are near the sternum of the first abdominal segment of Acrididae and Cicadidae, on the tibia of Tettigoniidae and Gryllidae, on the thorax of Notonectidae, and on the thorax or abdomen of some Lepidoptera. Insect in eight orders produce sounds, are sensitive to sounds, and utilize sounds in courtship, mating, prey location, and predator avoidance.

Energy radiates from a sound source as both high frequency airborne sound waves and lower frequency substrate vibrational waves. The two types of energy radiate as different waveforms and are detected generally by different types of receptors. Air pressure receptors occur in flies and mosquitoes in which the antennae oscillate and transmit impinging air pressure waves to Johnston's organ. Many other insects have a tympanum, a thin membrane stretched over an air chamber. The tympanum is deflected by sound pressure waves. In the desert locusts, *S.gregaria,* the tympanum converts acoustic energy into mechanical energy and directs specific vibrational frequencies to different neurons. Thus, The tympanum functions to detect sound and is involved in frequencies analysis as well. Although most of the energy from an insect that is calling from a perch on a leaf or stem radiates as airborne sound, some is nearly always transmitted to the substrate as a low-frequency vibration.

号。除鼓膜器外，有些昆虫也能用其他器官听到某些声音，包括 Johnston's 器、膝下器、零星的简单弦音器和简单的毛形感器。

就声音接收结构而言，昆虫的耳朵可分为触角耳和鼓室耳。一些鼓室耳，接收声音的表皮区域没有明显变薄为鼓膜。从解剖学上讲，这种耳朵可能被归类为无鼓室耳。尽管表皮可能会响应声音而振动，并起到鼓膜的作用。

压力敏感型鼓室耳和无鼓室耳的工作频率高达 300 kHz，可远距离检测通信声音和蝙蝠回声定位信号。相反，对粒子速率敏感的触角耳有利于近距离的亲密交流，通常工作频率低于 1 kHz。除了对粒子速率敏感的毛发外，所有昆虫的耳朵都来自弦音器，可分为对粒子速率灵敏的触角耳朵和对压力敏感的鼓室耳朵或无鼓室耳朵。

鼓膜器在昆虫的不同部位分布广泛。典型的位置是蝗科（蚱蜢）和蝉科（蝉）的第一腹节靠近胸片处；螽斯科（螽斯）和蟋蟀科（蝼蛄）的胫节；仰蝽总科（水生半翅目）的胸节和鳞翅目的胸节或腹节。八个目的昆虫产生声音，都对声音敏感。在求爱期、交配、捕食场所和逃避肉食动物时利用声音。

从声源发出的能量以高频的空气传播声波和低频的基质振动波辐射。两个类型能量传播的波形不同，通常被不同的感受器侦测到。果蝇和蚊虫有气压感受器，其触角将与气压波的碰撞振荡并传递给 Johnston's 器。许多其他昆虫有鼓室，由一层薄膜延伸覆盖气室。鼓室被气压波反弹。沙漠蝗的鼓室将声波能量转换成机械能，并引导特殊的振动频率到不同的神经元。因此，鼓室有监测声音的功能，也涉及频谱分析。虽然昆虫发出的多数能量是从叶片上或茎的栖息处以空气传播声波辐射，但有些经常以低频振动传播到基质。

Tympanal ears typically have a minimum of three components: ① a thin cuticular tympanum on the cuticular surface; ② an air sac or other tracheal structure behind the tympanum; ③ sensory neurons organized in scolopidia attached to the tympanal membrane or attached near it, so that they vibrate in response to the vibrations of the tympanum. Airborne sound waves cause the tympanum to vibrate and sensory neurons enclosed in the scolopale cell detect the vibrations and respond, first, by graded electrical potentials, followed by a burst of spikes in the axon. Some insects have a tympanum that can respond to sound waves striking it from the inside of the air chamber as well as from outside; such tympanal organs are pressure-difference receivers, and they are especially sensitive to directionality of the sound.

Crickets have a tympanic hearing organ on each front leg, which acts as a pressure gradient receiver since it receives sound directly on its external surface and indirectly on its internal surface through openings in the thorax and an auditory trachea. Grasshoppers have a tympanic hearing organ on each side of the abdomen. Afferents project to the metathoracic ganglion, which contains approximately 15 types of local neurons (LNs) with diverse morphologies and physiologies. These LNs then synapse onto approximately 20 ANs, which in turn project to the brain.

昆虫鼓膜耳一般至少有 3 个组成部分：①表皮层表面有一很薄的表皮质鼓膜；②鼓膜后有一个气囊或气管结构；③与鼓膜或鼓膜附近相连的有具橛感受器的神经元，因此它们对鼓膜室的振动做出反应而振动。空气中传播的声波引起鼓膜振动，具橛细胞包围的感觉神经元侦测到振动，先以分级电位响应，随后在轴突上暴发一个峰电位。有些昆虫的鼓膜对气室内部声波的冲击能和外部冲击的一样产生反应；这种鼓膜器是压力差接收器，它们对声音的方向性特别敏感。

蟋蟀的每只前腿上都有一个鼓膜听觉器官，该器官充当压力梯度接收器，因为蟋蟀直接在其外表面接收声音，并通过胸部和听觉气管的开口间接在其内表面接收声音。蚱蜢的腹部两侧各有一个鼓膜听觉器官。传入信息投射到后胸神经节，后胸神经节包含大约 15 种局部神经元（LNs），具有不同的形态和生理学功能。然后，这些局部神经元与大约 20 个触角神经元形成突触，反过来投射到大脑。

11.2.4.3 Johnston’s organ

Johnston’s organ is a large, complex chordotonal organ that may consist of several groups or a single grouping of scolopidia located between the second (the pedicel) and third joints of each antenna of most adult insects (Fig.11-5), although some Apterygota (Collembola and Diplura) do not have a Johnston’s organ. A simplified form of the organ occurs in some larvae. Johnston’s organ responds to several kinds of stimuli in different insects, including acting as a proprioceptor to indicate movement of the antennae, monitoring wing beat frequency in relation to flight speed in some Diptera, indicating gravity, detecting ripples at the water surface in gyrinid beetles, and receiving sound in mosquitoes and, perhaps, other insects.

Whereas crickets and grasshoppers have tympanal ears that are sensitive to sound pressure, flies detect sound using a feathery extension of the antenna called the arista, which is sensitive to particle displacements induced by the near-field component of sound or by wind. Antennal movements activate Johnston’s organ

11.2.4.3 Johnston’s 器

Johnston’s 器是一大型复合弦音器，由多组或单组位于多数昆虫成虫的触角第二（梗节）和第三节接合处的具橛感器组成（图 11-5），但无翅亚纲（弹尾目和双尾目）并没有 Johnston’s 器。有些幼虫出现简化型 Johnston’s 器。Johnston’s 器在不同昆虫中对几类刺激有反应，包括作为本体感受器指示触角运动，有些双翅目监测与飞行速率相关的翅膀拍击频率，指示重力，在鼓甲科甲虫监测水面的波纹以及在蚊子中接收声音，也许还有其他昆虫。

蟋蟀和蚱蜢的鼓室耳朵对声压很敏感，而果蝇则使用一种称为芒刺的羽状天线来探测声音，该天线对声音的近场分量或风引起的粒子位移很敏感。触角运动激活触角内的 Johnston’s

neurons (JONs) within the antenna. Three of five JON subsets respond to the range of antennal deflections representative of song carrier frequencies and project to distinct areas of the antennal mechanosensory and motor area (AMMC) in the brain. From the AMMC, known auditory pathways travel to the wedge, then to the ventrolateral protocerebrum (VLP), and on to the lateral protocerebral complex (LPC).

器神经元。五个 Johnston's 亚群中的三个对代表虫鸣载波频率的触角偏转范围做出响应，并投射到大脑中触角机械感觉和运动区（AMMC）的不同区域。从 AMMC 已知的听觉通路进入楔状体，然后到腹外侧前脑（VLP），再进入外侧前脑复合体（LPC）。

Insect tympanic hearing organs decompose sounds into their constituent frequencies using receptor cells whose physical location within the organ determines the receptor's most sensitive frequency. The correspondence of frequency sensitivity to physical location is called tonotopy and is generally maintained throughout ascending projections. Cross-frequency inhibition may further sharpen central tuning.

昆虫鼓膜听觉器官使用受体细胞将声音分解为其组成频率，受体细胞在器官内的物理位置决定了受体最敏感的频率。频率敏感性与物理位置的对应关系被称为频率拓扑，通常在整个上行投射中保持。交叉频率抑制可能会进一步加快中心调制。

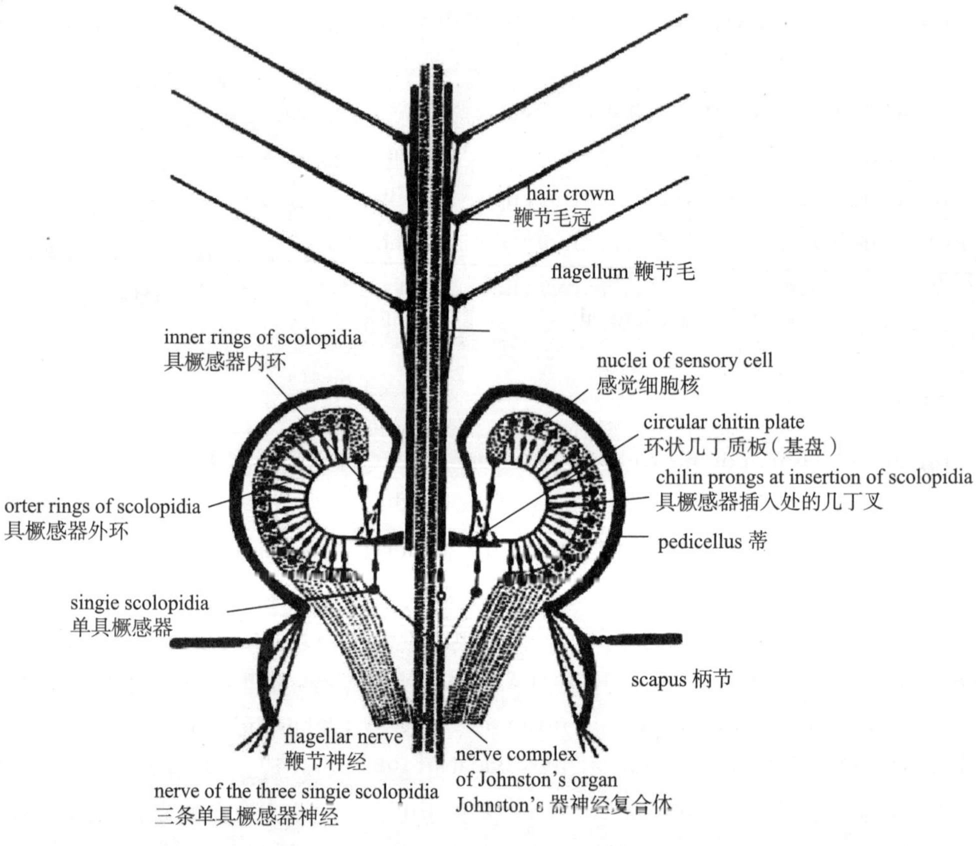

Fig.11-5 Johnston's organ on the antennae of male mosquito (modified from Autrum, 1963).

图 11-5 雄蚊触角上的 Johnston's 器（仿 Autrum,1963）。

11.2.4.4 Simple chordotonal organs

Simple chordotonal structures that have no specific name occur widely over the body of most orders of insects, including adults and larvae. The structures usually consist of only a few scolopidia. There are about 90 such small chordotonal organs

11.2.4.4 简单弦音器

简单弦音结构没有特定命名，广泛存在于大多数目昆虫体上，包括成虫和幼虫。这种结构通常只由少数具厥感受器组成。在果蝇幼虫体节上大约有 90

arranged along the length of *Drosophila* larvae. Relatively simple chordotonal organs occur in the legs (on the femur, and sometimes, on the tibia), on the wings at the base of the radial and subcostal veins, and within the lumen of the radial vein of many, but not all, insects. Some of the simple chordotonal sensilla respond to certain airborne frequencies, i.e., they are sound receptors, but they are not very sensitive and have a narrow response range. Similar simple chordotonal organs may have been the precursors of tympanal ears in the early evolution of insects.

个这种小弦音器。相对而言，简单弦音器存在于许多昆虫的腿（腿节上，有时在胫节上）、翅膀径向和亚前缘翅脉基部和径脉腔内。有些简单弦音器对某个声波频率有反应，即它们是声音感受器，但不是很灵敏且反应范围很窄。相类似的简单弦音器可能是昆虫鼓膜听觉的早期进化的初期形式。

11.2.5 *Sensilla campaniformia*

Sensillium campaniformia is composed of cuticular, hypodermal elements, and a primary sensory neuron. *Sensillia campaniformia* have a malleable cuticle, called dome, which is composed of three layers: a thin outer layer, a porous middle layer, and a more fibrous internal layer. The cuticle of these organs forms a socket and a dome to which the fan-shaped sensory dendrite attaches. Their morphology and orientation make them selectively sensitive to cuticular deformation along particular lines of mechanical stress. *S.campaniformia* is proprioceptive mechanoreceptors associated with the exoskeleton. The tip of the sensory process is firmly attached to a thin cap of exocuticle; mechanical depression of this cap, which probably occurs during walking movements, effectively stimulates the sensillum. *Campaniform sensilla* plays an important role in coordinating joint movement both within a leg and across legs, and that natural tripod gaits are shaped mainly by sensory feedback to motor circuits.

11.2.5 钟形感受器

钟形感器由表皮和真皮部分以及初级感觉神经元组成。钟形感器有一个可塑的表皮层，称为圆顶，由三层组成：薄的外层、多孔的中间层和纤维状较多的内层。这些感器的表皮形成一个窝和一个圆顶，有扇形的感觉神经树突附着在上面。其形态和方向使之对特殊垂直方向的机械压力造成的表皮变形选择性敏感。钟形感器是与骨骼相连的本体机械感受器。感觉突的顶端与外表皮的薄冠牢固地相连。在步行运动中经常发生的对这个冠部的机械压缩有效地刺激了感觉器。钟形感受器在协调腿内和腿间关节运动方面起着重要作用，三脚架步态特征主要是通过对运动回路的感觉反馈形成的。

11.2.6 Infrared receptors

Many species of insects have specialized infrared receptors and are attracted to forest fires, where they lay their eggs in fire-damage trees and bound-over debris. Most of the insects are beetles. *M. acuminata*, a beetle in the family Buprestidae, has been most intensively studied and is known to be attracted in large numbers to burning forests, where they mate and begin laying eggs in the burned wood. *M.acuminata* has paired pits adjacent to the mesothoracic coxa on the pleuroventral thorax (Fig.11-6). The pits are slightly variable in size.

Within each pit are 50 to 100 slightly oblong hollow domes and associated multiple pores wax glands. Each dome is innervated by one bipolar neuron and associated sheath cells (Fig.11-7). The

11.2.6 红外感受器

多种昆虫有特化的红外感受器被森林火灾吸引并将卵产于被大火烧毁的树上和过火的树木残骸上。多数甲虫是研究较为透彻的吉丁虫族甲虫，如松黑木吉丁虫。大火会吸引大量的吉丁虫交配并产卵于燃烧过的树木上。松黑木吉丁虫有一成对的凹陷与侧腹胸上的前胸基节邻接（图 11-6）。凹陷的大小有轻微的变化。

每个凹陷内有 50～100 个细长的长方形圆顶并与多孔的蜡腺相连。每个圆顶由一个双极神经元和相连的鞘细胞

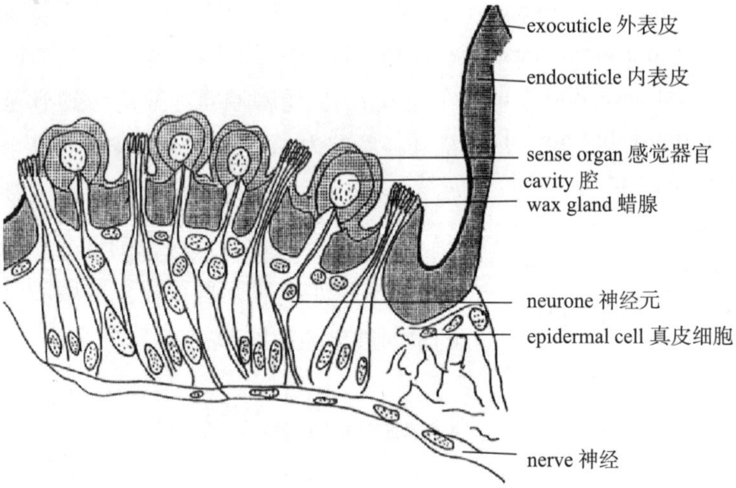

Fig.11-6 Infrared reception of Buprestidae (modified from Evans,1966).
图 11-6 吉丁虫的红外感受器（仿 Evans, 1966）。

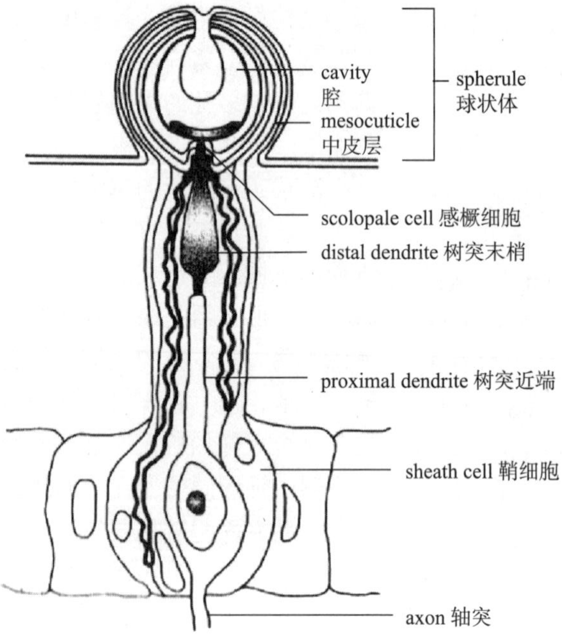

Fig. 11-7 The structure of one of the infrared receptors from *M. scuminata.* (modified from Vondran et al., 1995).
图 11-7 吉丁虫单个红外感受器的结构（仿 Vondran 等，1995）。

cell body lies just beneath the cuticle at the base of the pit, and its axon passes without synapsing into the metathoracic ganglion. The dendrite of the sensory neuron contains neurotubules and has a ciliary constriction near its midpoint. The distal tip of the dendrite is attached at the base of the spherical dome. The dome surface is thin and unsclerotized, possibly allowing the hollow sphere beneath the domed cuticle to change its volume due to absorption of infrared radiation, thus, mechanically stretch the dendritic tip of the neuron.

The beetles respond behaviorally to infrared radiation at 3 μm wavelength and infrared wave-length ranging from 2.5 to 4 μm are emitted by intense forest fires. Like *Melanophila*, the Australian "fire-beetle" *M.atrata* is attracted to forest fires, and the larvae can only develop if the eggs are laid in freshly burnt wood. The infrared detector is richly endowed with nerve cells and also mitochondria.

进行神经控制（图 11-7）。细胞体位于凹陷基部表皮的正下方，其轴突不经突触进入前胸神经节。感觉神经元的树突包含神经微管和中间点附近的毛状颈缩。树突的顶端与球状圆顶的基部相连。圆顶表面较薄，非骨质化，可能允许圆顶表皮下的空球体在吸收红外线对改变容积，因此机械地拉紧神经元的树突顶。

这种甲虫对红外波长为 3 μm 的红外辐射有行为反应，而猛烈的森林大火发出的红外波长辐射是 2.5 ～ 4 μm。像松黑木吉丁虫一样，澳洲"火堆甲虫"也被森林大火所吸引，卵只有产在新燃烧过的树木上，它的幼虫才能发育。其

The detector itself is an area of cuticle, with a characteristic honeycomb structure, quite unlike that found in *Melanophila*. The beetle to evolve infrared detection, Australian"little ash beetle" *A. nigricans*, is only distantly related to the two buprestid beetles, and unsurprisingly it has arrived at yet another solution to the detection of infrared radiation. After a short period of running around having alighted on the hot ash or burnt bark quickly burrows and then lays its eggs. The infrared organs are located in front of the first pair of legs and they consist of a sensory plate, richly endowed with nervous tissue and associated mitochondria. The plate is connected to the rest of the body by a narrow stalk, and is located above a cavity. It thus according to insect bolometer to detect inferred.

红外探测器含有丰富的神经细胞和线粒体。探测器本身是个表皮区域，具有典型的蜂窝式结构，与松黑木吉丁虫完全不同。进化了红外侦查器的甲虫是澳洲“灰烬小甲虫”，是唯一与吉丁科甲虫关系较远的，那么它通过其他办法检测红外辐射也就毫不奇怪了。它们在落在热灰烬或燃烧的树皮上进行短时间的奔跑后会迅速挖掘，然后产卵。红外感受器位于第一对腿的前面，由板形感器组成，富含神经组织及相关线粒体。板形感器通过一条窄柄与躯体其余部分连接，位于体腔上方。因此，它根据昆虫辐射检测器检测红外线。

11.2.7 Stretch receptors

Stretch receptor is proprioceptor which is composed of only one free multipolar dendrite. In Ephemera (Ephemeroptera), Dictyopterygella and Isoperla (Plecoptera), Periplaneta and Blaberus (Dictyoptera), Forficula (Dermaptera), and Dytiscus (Coleoptera) there is a pair of stretch receptors on each side in the dorsal region of most abdominal segments.

11.2.7 牵引张力感受器

牵引张力感受器是一种本体感受器，仅由一个具游离树突的多极神经元末梢组成。蜉蝣（蜉蝣目）、网翅目、石蝇（襀翅目）、大蠊属、巨型蟑螂（网翅目）、蠼螋科（革翅目）和龙虱属（鞘翅目）等，它们大多数腹部体节的背部区域两侧都有成对的牵引张力感受器。

11.2.7.1 Subepidermal plexus or innervate stretch receptors

Schrader et al (2007) suggested that insects possess two types of sensory neurons: ciliated type I sensory neurons that innervate external sensory organs and chordotonal organs, and type II sensory neurons that form a subepidermal plexus or innervate stretch receptors. Among stretch receptors, a dorsal longitudinal stretch receptor is highly conserved in insects, being found in all insect orders investigated. The stretch receptor strand (Fig. 11-8) of cockroaches is derived from the basement membrane of peripheral nerves. The tergal nerve runs in close juxtaposition to the neuron and the glial cell and it is not separated from them by extracellular material, but enveloped in the same sheath of extracellular matrix. Naked dendritic terminals that insert into different types of tissue are a common feature of insect sensory structures innervated by type II sensory neurons.

It also distributed in the foregut and hindgut to feel repletion degree of foregut in grasshopper and cicada, and provide information to adjust the amounts of ingestion. The stretch receptors in the hindgut could be used to detect the amounts of

11.2.7.1 皮下神经丛或神经性张力感受器

Schrader 等（2007）认为昆虫神经元有两类：有纤毛的 I 型感觉神经元支配外部感觉器和弦音器；II 型感觉神经元形成皮下神经丛或神经性牵引张力感受器。在牵引张力感受器中，背纵牵引张力感受器在昆虫中高度保守，在研究的所有昆虫目都有。蜚蠊的纤维束牵引张力感受器（图 11-8）从外周神经的底膜衍生而来。背神经与神经元和胶细胞紧密并列延伸，胞外物质不是使之分开，而是将其密封在相同的胞外基质鞘中。裸露的树突末梢插入不同种类的组织，由 II 型感受器神经元控制感受器结构是昆虫的共同特征。

牵引器在前、后肠也有大量分布，蝗虫嗉囊和蝉的前肠的牵引张力感受器，可用于检测饱胀程度，以调节取食量；后肠上的牵引张力感受器用于感受

food residue that might adjust excretion. In butterfly, the stretch receptors are located on copulatory pouch, its dendritic terminals ended on the pericardial membrane. Remating is inhibited when the insects feel dilation. In many insects, the stretch receptors are located on the basal part of gonapophysis that inspect the ovulation time of ovarian ducts so as to control release of germ cells in the spermatheca together with chordotonal organ.

肠腔内的食物残渣量，可能有调节排粪的作用；粉蝶交配囊上的牵引张力感受器的树突末梢终止于围膜，其感受到膨胀时，便抑制再交配；很多昆虫的牵引张力感受器位于生殖突基部，与弦音器一起共同检测输卵管的排卵时间，可以此控制受精囊内精子排放。

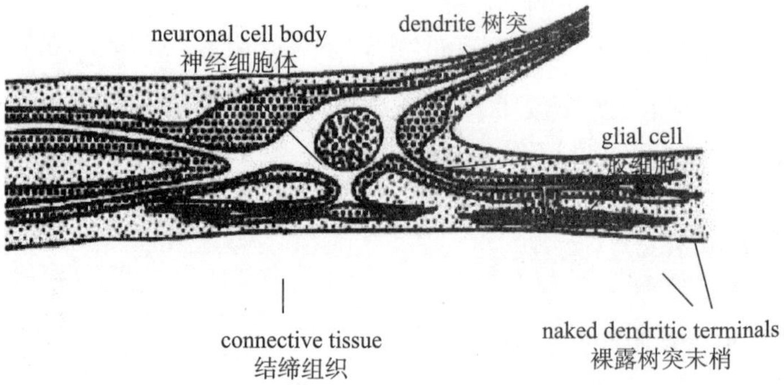

Fig.11-8 Fiber bundle stretch receptors in cockroach (modified from Osborne, 1970).
图 11-8 蜚蠊的纤维束牵引张力感受器（仿 Osborne,1970）

11.2.7.2 Connective-tissue stretch receptors

In *Limnophilus* and *Phryganea* (Trichoptera) there is a longitudinal receptor only. It consists of a muscle-fiber to which is attached a tube of connective tissue containing the dendrites of the sensory neuron. The caddis receptor resembles closely the lepidopteran receptor. No abdominal stretch receptors could be found in *Rhodnius* or Nepa (Hemiptera and Heteroptera), and in *Schistocerca* and *Locusta* (Orthoptera) no vertical receptors could be found. Finlayson et al (1958) observed the morphology of dorsal abdominal stretch receptors in representatives of the Orthoptera, Odonata, Hymenoptera and Lepidoptera, and described them on the basis of dissections, serial sections and whole mounts. The receptors consist of sensory neurons in association with connective tissue strands or muscles (Fig. 11-9). The organs are found to be slung between the intersegmental folds (Lepidoptera), between one intersegmental fold and the dorsal body wall (Orthoptera, Odonata, Hymenoptera), or between the dorsal body wall and a nerve branch in the abdomen (Odonata). The mode of anchorage of the organs and other morphological features suggests a proprioceptive function. At the junction between the fiber tract and the muscle cell are found longitudinally arranged bundles of dense connective tissue fibrils, the fiber bundles.

The stretch receptor encodes the amplitude of dis placement in the number of spikes and the instantaneous frequency of the burst in response to controlled mechan ical displacement of the subalar membrane. In response to rapid steps in mechanical wing elevation,

11.2.7.2 结缔组织牵引张力感受器

沼石蚕和石蚕（毛翅目）仅有纵向受体。它由肌纤维组成，系于含有感觉神经元树突的结缔组织管上。石蚕受体的装配与鳞翅目受体接近。长红猎蝽和 Nepa（半翅目和异翅亚目）没有腹牵引张力感受器，而沙漠蝗和蝗虫（直翅目）没有纵向感受器。Finlayson 等（1958）基于解剖、切片和全虫标本观察，描述了直翅目、蜻蜓目、膜翅目和鳞翅目背腹牵引张力感受器的形态。感受器由感受神经元及与之相连的结缔组织纤维束或肌肉（图 11-9）组成。感受器悬挂于节间折叠之间（鳞翅目）、一个节间折叠和背部体壁之间（直翅目、蜻蜓目、膜翅目）或背部体壁和腹部神经分支之间（蜻蜓目）。感受器锚定的方式和其他形态特征表明了其本体感受器功能。在纤维束和肌细胞之间发现有纵向排列及致密的结缔组织纤维束连接。

牵引张力感受器编码峰电位总量位移的幅度和对基膜机械位移反应暴发的瞬时频率。Gettrup（1962）对翅膀上举的快速反应研究表明，蝗虫的牵引受体在约 25 ms 的适应期内进行一个相位暴

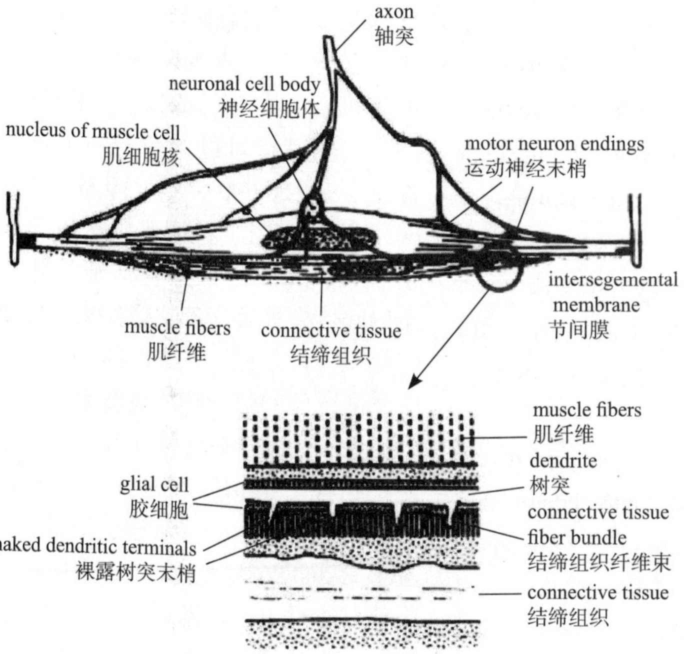

Fig. 11-9 Muscle stretch receptor of moth (modified from Osborn, 1970).
图 11-9 蛾类的肌肉牵引张力感受器（仿 Osborn，1970）。

Gettrup (1962) showed that the locust stretch receptor fires a phasic burst of spikes that adapts within approximately 25 ms to a tonic firing rate proportional to the degree of wing elevation. In time, the stretch receptor adapts to a tonic firing rate as the subalar membrane relaxes. Stretch receptor feedback probably tunes the flight motor pattern such that individual wing strokes are temporally precise and are of appropriate amplitude. An intriguing hypothesis about the ultimate function of the stretch receptor signal emerges from the fact that insects do not simply beat their wings up and down as they fly. Each translation, during which the wing is held at a nearly constant angle of attack, is followed by pronation or supination, during which the wing is rotated about a spanwise axis. The timing of these rotations within the stroke cycle is a major determinant of unsteady aerodynamic force production in several species of fly and possibly in all flying insects.

发的峰电位到紧张性的传导，其速率与翅膀提升的程度成比例。当肌膜松弛时，牵引受体及时适应了紧张性传导速率。牵引张力感受器反馈的信息可能调整了飞行运动的形式，如翅膀拍击精确的时间和合适的幅度。一个有趣的假说是关于牵引张力感受器信号的最终功能的一个有趣的假说来自这样一个事实，即其昆虫在飞行时并不是简单地上下拍击翅膀。在每次平移期间，翅膀维持几乎恒定的冲（迎）角，紧接着旋前或旋后，在这段时间内，翅膀沿翼展轴旋转。在翅膀拍击周期内，这些旋转时机是几种蝇，也可能是所有飞行昆虫产生的非稳定空气动力的主要决定因素。

11.3 化学感受器

11.3 Chemoreceptor

11.3.1 Specialists and generalists among chemoreceptors

Among both taste receptors and olfactory receptors (OR), there are some receptors specialized for detection of very specific

11.3.1 专化型和普通型感受器

在味觉感受器和嗅觉感受器（OR）中，有些受体专一性地检测非常特定的

chemical substances and others capable of responding to a wide variety of chemicals. These are commonly called "specialists" and "generalists," respectively. For example, each antenna of a male *T. prolyphenus* moth bears more than 60,000 sensilla containing about 150,000 sensory neurons. About 60% to 70% of these neurons are specialized for detection of the female-produced sex pheromone, about 20% respond to other odors, and the remainder serves a variety of sensory functions. Even the specialists, however, are usually not absolutely specific. A few other chemicals in high concentration may also stimulate them.

Chemoreceptors are highly well developed in insects, and play a critical role in their survivorship. It could be divided into olfactory and gustatory receptors according to the way of pheromone compounds molecules contact with insects. In many insects olfactory sensilla are present on two pairs of olfactory organs on the head, the antennae and the maxillary palps. Gustatory sensilla are present on several parts of the body including the external and internal mouthparts, the legs, the wing margins and the ovipositor in females.

Honeybee(OBP) ASP2 specifically interacts with imidacloprid, a neonicotinoid insecticide, which affects olfaction and cognition. Upregulation of OBP protein C01 expression in *Tribolium castaneum* resulted in reduced sensitivity to dichlorvos and carbofuran. An increase in affinity between OBPs and exogenous insecticides is one of the major mechanisms in the development of insecticide resistance in insects. The *OBP9* gene of *Spodoptera litura* was significantly upregulated after treatment with chlorpyrifos, and mortality increased significantly when this gene was deleted. These results indicate that the toxicity of insecticides may be reduced by their interaction with OBPs. Thus, this mechanism may play a role in insect resistance to insecticides. *OBP28* can be used as a new target gene for the surveillance of deltamethrin resistance in *Cx. ouinquefasciatus* cause of a correlation between the expression of *OBP28* and insecticide resistance in mosquitoes.

化学物质，而其他的能对多种化学物质做出反应。这些通常分别被称为专化型化学感受器和普通型化学感受器。例如，美国蛾雄蛾的触角有 60 000 个感器，含有 150 000 个感觉神经元，其中 60% ～ 70% 用于检测雌蛾产生的雌性信息素，20% 用于感受其他气味，其余的起着多种多样的感受作用。即使是专化型，常常也不是绝对的专一。其他一些高浓度的化学物质也能刺激它们。

昆虫的化学感受器高度发达，对其生存起着关键作用。根据信息化合物分子接触昆虫的途径，化学感受器分为嗅觉感受器和味觉感受器。许多昆虫的嗅觉感器成对出现在昆虫头部、触角和上颚须上。虫体几个部分有味觉感受器，包括口器外部和内部、腿、翅边和雌虫的产卵器。

蜜蜂气味结合蛋白（OBP）ASP2 与一种新烟碱类杀虫剂吡虫啉特异性相互作用，影响嗅觉和认知。拟谷盗中的 OBP 蛋白 C01 表达上调会导致对敌敌畏和呋喃丹敏感性的降低。OBPs 与外源杀虫剂之间亲和力的增加是昆虫产生抗药性的主要机制之一。斜纹夜蛾的 *OBP9* 基因在毒死蜱处理后显著上调，当该基因缺失时，死亡率显著增加。这些结果表明，杀虫剂的毒性可能因其与 OBPs 的相互作用而降低。因此，这种机制可能在昆虫对杀虫剂的抗性中发挥作用。*OBP28* 可作为监测致倦库蚊对溴氰菊酯抗性的新靶标基因，因为 *OBP28* 表达与蚊子杀虫剂抗性之间具有相关性。

11.3.2 Olfactory sensilla

Olfactory is a very important to most insects. Males (and sometimes females) depend on detecting and orienting to sex pheromones released by the opposite sex. Adult females use olfaction as well as other cues to locate suitable oviposition sites. Larvae of many insects feed on the host where the mother has laid eggs, but more mobile immatures, such as grasshoppers and

11.3.2 嗅觉感受器

嗅觉感受器在多数昆虫中非常重要。雄性（有时是雌性）昆虫依此侦测和确定异性释放的性信息素的方向。雌性成虫除了利用其他信号外，也能利用嗅觉确定适宜的产卵场所。多数昆虫幼虫取食母代产卵的寄主，但更多昆虫在

cockroaches, move frequently in search of food and olfaction, as well as taste, which is important to them.

Like other sensory neurons, chemoreceptor neurons are bipolar, with the cell bodies located peripherally near the stimulus site. Characteristically, the dendrite of an olfactory neuron comprises a relatively large inner segment connected by a narrow ciliary segment to a smaller outer process extending to the tip of the sensillum. In some sensillia, notably *S.basiconica*, there are many branches extending to the tip of the fine terminal process. Occasionally, the ciliary segment is missing. The large inner segment contains many neurotubules and mitochondria, suggesting a high rate of metabolic activity and utilization of oxygen. There may be a need for high metabolic pump activity in the very fine dendritic endings if there is only a small reserve of the ions necessary for nerve function. The function of the ciliary region, when it is present, is not clear. It contains nine pairs of neurotubules attached to the dendritic membrane. Although the two central neurotubules typical of ciliary structures are sometimes present, they are more often absent.

The ciliary organelles might function as organizers for regeneration of dendritic endings following molting. The cuticular walls of olfactory sensilla contain pores varying from 10 to 100 nm in diameter. The neurotubules extend into the outer dendritic segment. Microtubules lead from the inner wall of the pore inward and frequently seem to make direct contact with the dendritic endings. The walls of these microtubules are about 3 nm thick. Chemical molecules enter through the pores, are captured by odorant binding proteins, and are transported across the sensillum liquor to the dendritic endings. Olfactory receptors typically give a phasic-tonic response (Fig.11-10).

The axons of olfactory sensory neurons (OSN) usually are small, measuring 0.1 to 0. 2 nm in diameter, and they usually pass directly into the deutocerebrum without synapsing, though there are examples of axons from antennal receptors that coalesce or emerge by synapsing near the base of the antenna. The functions of olfactory involved in courtship, foraging and looking for suitable site for eggs laying. Sometimes olfactory sensory function as thermo- and hygroreceptors.

Aside from sensory neuron(s), whose axons may be enveloped by one or more glial cells, each sensillum is populated by a thecogen, tormogen, and trichogen cell. These cells are collectively referred to as accessory, auxiliary, or supporting cells. For instance, tormogen cells, in particular, were reported to express specific

未成熟时迁徙，如蝗虫和蜚蠊，频繁移动搜寻食物，嗅觉和味觉对它们来说都是重要的。

与其他感觉神经元一样，化学感受器神经元都是双极的，细胞体位于刺激位点附近的外周，神经元被鞘原细胞、膜原细胞和毛原细胞包围。其特征是嗅觉神经元的树突由相对较大的内部节片通过一较窄的纤毛区与感受器顶部较小的外部延伸区相连。在有些感受器中，最明显的是锥形感受器，其树突有很多的纤细末梢，但有时纤毛区会消失。树突较大的基段含有很多神经微管和线粒体，表明其高效率的代谢活动和氧气的利用。如果神经功能所需的离子仅有少量存储，也许在树突非常纤细的端段需要高效的代谢泵活性。当有纤毛区出现时，其功能却不明。它包含 9 对神经微管与树突膜相连，尽管纤毛结构中独特的两条中央微管有时出现，但通常是不存在的。

纤毛细胞器的功能可能是蜕皮后形成再生树突端段的组织者。嗅觉感器的表皮壁上含有直径为 10 ～ 100 nm 的微孔，神经微管延伸进入树突端段，从微孔的内壁导引使之向内，通常看起来与树突末梢直接接触。这些微管壁厚 3 nm，化学分子进入孔道，被气味结合蛋白捕获，跨过感液腔被转运到树突末梢。典型的嗅觉感受器产生相位-紧张性反应（图 11-10）。

嗅觉感受器神经元（OSN）轴突直径小，为 0.1 ～ 0.2 nm。它们通常直接进入中脑而无突触，也有例外，触角感受器的轴突通过触角基部附近进行突触汇聚或显露。嗅觉的功能涉及求偶、觅食以及寻找适宜的产卵场所，有时嗅觉感受器起着温湿感受器的功能。

除了感觉神经元（其轴突可能被一个或多个胶质细胞包裹）外，每个感器由一个鞘原细胞、膜原细胞和毛原细胞填充。这些细胞统称为附属细胞、辅助细胞或支持细胞。例如，据报道，膜原

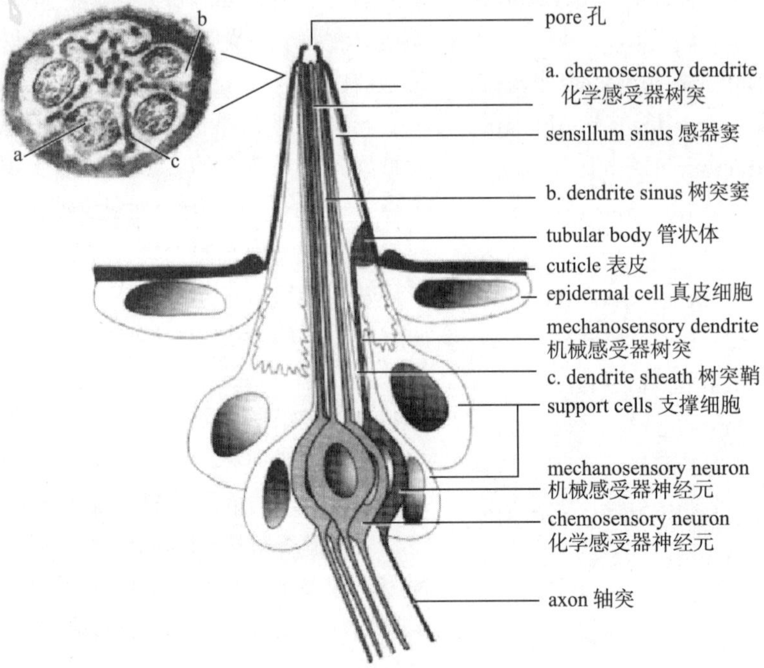

Fig.11-10 Cross-section through a typical contact-chemoreceptor sensillum or gustatory receptor (modified from Merritt, 2006).

图 11-10 典型的触化感受器或味觉感受器横切面图（仿 Merritt，2006）。

cytochrome P450 enzymes at their apical poles and display features of active transport across the apical portion, which is covered in microvilli. In the silk moth, two tormogen cells arise during development, where the inner tormogen cell degenerates within 2 days post-apolysis.

Atypically to the general rule of one support cell type per sensillum, *Drosophila coeloconic sensilla* are populated with two tormogen cells. A mature tormogen cell is distinct from the trichogen cell in that it extends a characteristically long, stalk-like protrusion that terminates below the level of the olfactory sensory neurons (OSN) soma, at least in *Drosophila*. Interestingly, all three support cell types occur broadly across extrasensory and *chordotonal sensilla*. For example, each scolopidium in the Johnston's organ is comprised of a triad of non-neuronal cells: a ligament cell, a scolopale cell, and a cap cell, which are equivalent to the triad found in olfactory tissues.

Mounting evidence points to the important role of many support cells in regulating sensory neuronal activity, transmission, and structural integrity. These sensory neurons individually express distinct receptors from a wide range of receptor families such as the odorant receptor (OR), gustatory receptor (GR), ionotropic receptor (IR), pickpocket (Ppk), and transient receptor potential (TRP) protein families. Support cells play a passive role in lymph maintenance by releasing various ''helper'' proteins into the sensillum lymph to aid in stimulus recognition. Among

细胞尤其在其顶端表达特定的细胞色素 P450 酶，并在覆盖微绒毛的顶端部分显示主动运输的特征。在蚕蛾的发育过程中，出现两个膜原细胞，其中内部膜原细胞在解离后 2 天内退化。

与一个感器一种支持细胞类型的一般规则不同，果蝇腔锥感器由两个膜原细胞填充。成熟的膜原细胞不同于毛原细胞，因为它延伸出一个特征性的长柄状突起，其终止于嗅觉感受器神经元（OSN）胞体水平以下，至少在果蝇中是这样。有趣的是，这三种支持细胞类型广泛存在于超感和弦音感器中。例如，Johnston's 器中的每个头节由三个非神经元细胞组成：韧带细胞、感橛细胞和冠细胞，其相当于嗅觉组织中的三个细胞。

越来越多的证据表明，许多支持细胞在调节感觉神经元活动、传递和结构完整性方面发挥着重要作用。这些感觉神经元分别表达来自广泛受体家族的不同受体，例如，嗅觉受体（OR）、味觉受体（GR）、促离子型受体（IR）、“pickpoket”蛋白（Ppk）和瞬时受体电位（TRP）蛋白质家族。支持细胞通过将各种“辅助”蛋白质释放到感器淋巴中以帮助刺激识别，在淋巴维持中

these are the odor-degrading enzymes (ODEs), a set of catabolic enzymes which play a role in odor degradation and clearance, and chemosensory proteins (CSPs) which act as extracellular retention of enzyme chaperones in their ability to bind and transport chemicals in aqueous environments. Support cells have also been shown to be essential for the correct biogenesis of functional sensilla. Support cells have likewise been shown to express sensory neuron membrane proteins (SNMPs) alongside neurons in various insect species and may play important roles in odor detection as membrane spanning molecules involved in the capture, transport, clearance, or presentation of odorants to ORs.

Release and uptake of K^+ are entangled with several dynamic processes, namely the generation of action potentials and dendritic potentials, as well as the further concentration and release of K^+ from the same or neighboring cells. Tormogen cells seem to exhibit a much quicker Ca^{2+} response than the K^+ sequestering occurring in thecogen cells, perhaps due to aforementioned changes in shape or due to quicker mobilization of intracellular Ca^{2+} stores, which may hint at the active role of tormogen cells in maintaining properties of the sensillum lymph space through mechanisms such as export of accessory proteins. At least in some moth species, the thecogen cell and neuron have been purported to contribute jointly to the electrical properties of the sensillum. The glia-like thecogen cells are reminiscent of glial cells of the tripartite synapse model, where glia are thought to modulate synaptic communication through K^+ ion buffering, dispersion, and redistribution.

发挥被动作用。其中包括气味降解酶（ODEs），一组在气味降解和清除中发挥作用的分解代谢酶以及作为细胞外保持酶伴侣的化学感受蛋白（CSPs），它们在水相环境中结合和运输化学物质。支持细胞对于功能性感器正确的生物发生也至关重要。支持细胞同样被证明与各种昆虫的神经元一起表达感觉神经元膜蛋白（SNMPs），并且可能在气味检测中发挥重要作用，作为跨膜分子，参与气味物质的捕获、运输、清除或把气味分子传给嗅觉受体。

K^+ 的释放和吸收与几个动态过程有关，即动作电位和树突状电位的产生以及 K^+ 从相同或相邻细胞的进一步富集和释放。与鞘原细胞中发生的 K^+ 络合相比，膜原细胞似乎表现出更快的 Ca^{2+} 反应，这可能表明于上述形状的变化或由于细胞内 Ca^{2+} 库的更快动员，这可能暗示了膜原细胞通过辅助蛋白的输出等机制在维持感器淋巴空间性质方面发挥的积极作用。至少在一些蛾类物种中，鞘原细胞和神经元被认为共同影响感器的电特性。胶质样鞘原细胞类似于三方突触模型中的胶细胞，在该模型中，胶质细胞被认为能通过 K^+ 缓冲、分散和再分配来调节突触通信。

11.3.3 Contact chemoreceptors-gustatory receptors (GR)

Gustatory receptors or taste receptors and they respond to stimulus molecules in solution. The labellar hairs of blowflies contain several neurons in each sensillum. The different neurons are sensitive to sugars, water, anions such as Cl^-, and, in addition, one neuron usually act as a mechanoreceptor.

Gustatory sensilla are typically uniporous sensilla responding to taste and tactile stimuli. One neuron is mechanosensory and ends at the base of the sensillum while the remaining sensory cells reach the terminal pore, and function as contact chemoreceptors that the dendrite sheath stops at the base of the sensory hair. In *D.melangaster*, gustatory neurons are found in sensory sensilla that are associated with 3 different anatomical structures: taste bristles, taste pegs, and taste pores. Taste bristles are the most common

11.3.3 触化感受器 – 味觉感受器（GR）

味觉感受器对溶液中的刺激物分子起反应。丽蝇唇瓣上每个感器都含有几个神经元。不同的神经元对糖、水分和阴离子，如 Cl^- 敏感，此外，常有一个神经元起着机械感受器的作用。

味觉感受器通常是典型的单孔感受器，对味觉和触觉刺激做出反应。一个神经元是机械感觉神经元，终止于感器的底部，而其余的感觉细胞到达端孔并作为接触化学受体发挥作用，其树突鞘终止于感觉毛的底部。果蝇味觉神经元存在于感觉感器中，与 3 种不同的解剖结

taste organ, found on legs, wings, and the proboscis. Each bristle is innervated by a sensillum composed of 2-4 gustatory neurons, a single mechanosensory cell, and 4 support cells. Taste pegs are found between the pseudotracheae of the labellum. Taste pores are the anatomical structures associated with the internal mouthpart organs that line the pharynx. They comprise the labral sensory organ, the ventral cibarial sensory organ, and the dorsal cibarial sensory organ, each containing 3-10 gustatory neurons.

Gustatory receptors are located on mouth part, circus, antennae, ovipositor, tarsus, as well as the surface of labrum and glossa, and also include internal gustatory receptor. The structure of the circuits that process gustatory compounds is providing insight into integration of different gustatory compounds. Although sugar detection drives acceptance behavior and bitter compounds drive avoidance, the response to bitter-sweet mixes depends on the relative quantities of bitter and sugar in the mix and the hunger state of the animal. Some bitter compounds suppressed the activity of sugar-sensing sensory neurons, while another mechanism by which bitter compounds inhibit sugar sensory cells is through feedback inhibition. Sugar and bitter sensory neurons activate GABAergic interneurons that in turn inhibit sugar sensory axons expressing the GABAB receptor. Presynaptic inhibition in the presence of bitter compounds thus dampens sugar responses.

The taste center is a fused neuropil including three main neuromeres, the mandibular, maxillary, and labial neuromeres. This neuropil contains synapses between sensory input neurons, central interneurons, and motor output neurons. These elements form the initial neural circuits processing the relevant taste information ensuring vital functions such as food detection, intake, and toxin avoidance. The gustatory receptor neurons (GRNs) in larvae are housed inside gustatory sensilla of distinct mouthparts and they project to the gnathal ganglion (GNG) *via* periphery nerves. In both larval and adult *Drosophila*, GRNs from different sensory organs target different regions of the GNG. For instance, in adults, GRNs in the external sensilla of the labellum project *via* the labial nerve to the labial gustatory center, while GRNs in the internal gustatory sensilla enter *via* the pharyngeal nerve toward the pharyngeal sensory center of the GNG. Another pattern in the fly GNG is formed by projections conveying aversive and attractive stimuli, respectively, which terminate in different parts of the GNG. The

构相关：味觉刚毛、味橛和味觉孔。味觉刚毛是最常见的味觉器官，存在于腿、翅膀和喙上。每根刚毛由 2 ～ 4 个味觉神经元、一个机械感觉细胞和 4 个支持细胞组成的感器支配。在唇瓣的假气管之间发现味橛。味觉孔是与沿咽喉排列的内部口器器官相关的解剖结构，包括唇侧感觉器、腹侧感觉器和背侧感觉器，每个器官包含 3 ～ 10 个味觉神经元。

昆虫的味觉感受器位于口器、尾须、触角、产卵器和跗节、上唇表面和舌上，也包括内部味觉感受器。处理味觉化合物的回路结构提供了对不同味觉化合物的综合直觉。虽然糖的检测驱动接受行为，苦化合物驱动回避，但对苦 - 甜混合物的反应取决于混合物中苦味和甜味的相对量以及动物的饥饿状态。一些苦味化合物抑制糖感应感觉神经元活性，而苦味化合物通过反馈抑制来抑制糖感觉细胞。糖和苦感觉神经元激活 GABA 能中间神经元，后者反过来抑制表达 GABAB 受体的糖感觉轴突。因此，存在苦味化合物时的突触前抑制减弱了糖反应。

味觉中心是一个融合的神经毡，包括三个主要的神经毡，即下颌、上颌和唇神经毡。该神经毡包含感觉输入神经元、中枢中间神经元和运动输出神经元之间的突触。这些元素形成了处理相关味觉信息的初始神经回路，确保重要功能，如食物检测、摄入和毒素避免。幼虫的味觉感受神经元（GRNs）位于不同口器的味觉感器内，通过周围神经投射到颚神经节（GNG）。在幼虫和成虫果蝇中，来自不同感觉器官的 GRNs 靶向 GNG 的不同区域。例如，在成虫中，唇外感器中的 GRNs 通过唇神经投射到唇味觉中心，而内部味觉感器中的 GRN 通过咽神经进入 GNG 的咽喉感觉中心。果蝇 GNG 中的另一个模式是通过分别传递厌恶和吸引刺激的投射形成的，这些刺激终止于GNG的不同部分。

interneurons in the GNG include local interneurons and projection neurons sending axons to the higher brain centers. The GNG is also a sensorimotor center and contains many motor neurons that are involved in feeding motor programs. The motor neurons have large somata in the GNG and axons in the GNG nerves. Their dendrites, which are spatially organized, are involved in different feeding subprograms.

GNG 中的中间神经元包括局部中间神经元和投射神经元，将轴突传递到更高的大脑中心。GNG 也是一个感觉运动中心，包含许多参与摄食运动程序的运动神经元。运动神经元在 GNG 中有较大的细胞体，在 GNG 神经中有轴突。它们的树突是空间组织的，参与不同的摄食子程序。

The types of taste receptors in lepidopterous larvae are more complex, with large numbers and varied functions. In Caterpillar, they are located on the maxillae, epipharynx, possibly in hypopharynx and deeper portion of the buccal cavity. Each maxillae has two lobes arising from the basal segment. The medial lobe is the galae and the lateral three-segmented lobe is the maxillary palpus. Each galae bears two elongated blunt protuberances which each support a uniporous peg, commonly referred to as the medial and lateral sensilla stylconica. Each sensillum contain 4 neurons, they are salt sensitive, sugar-sensitive, inositol-sensitive or sugar alcohol-sensitive, respectively, and one special neuron responding to plant compounds (repellent neuron). The third segment of the palpus bears a group of eight terminal *sensilla basiconica*, each innervated by 3-4 chemoreceptor neurons. Three of them have a grainy or pitted appearance.

鳞翅目昆虫幼虫的味觉感受器的种类复杂且数量很多，功能也不相同。它们位于毛虫的下颚、内唇、可能在喉咽部以及口腔的深处。每个下颚有 2 个产生于底膜的瓣膜。中叶为外瓣，侧面 3 个分开的侧叶是下颚须。每瓣具有 2 个延伸的钝形隆起，每个支撑一个单孔钩，通常称为中部和侧面栓锥感器。每个感器含 4 个神经元，分别为盐敏感神经元、糖敏感神经元、肌醇或糖醇敏感神经元和 1 个对特殊植物化合物响应的神经元（忌避神经元）。触须第三节具有 8 个一群的锥形感器末梢，每个被 3～4 个化学感受器神经元支配，其中 3 个呈粒状或凹陷状。

11.4 化学感受器感受机制及相关蛋白质

11.4 Chemosensory Mechanisms and the Related Proteins

Two-layered networks process odorants detected by the antennae: initial processing occurs at the antennal lobe (AL), and the information is conveyed further on to higher brain centres, including the calyx of the mushroom body (MB) and the lateral protocerebrum (LPC), for second-order processing. These circuits are connected by second-order neurons called the projection neurons (PNs), which run through multiple pathways. Each pathway exhibits unique features, such as the travelling path, targeting area, commissural innervation, and mode of odor processing. Population labelling of LPC output neurons (third-order olfactory neurons) that receive the AL output in the brain of

两层网络处理触角检测到的气味：初始处理发生在触角叶（AL），信息进一步传递到更高的大脑中心，包括蕈状体的蕈冠（MB）和外侧前脑（LPC），进行二阶处理。这些回路由称为投射神经元（PNs）的二阶神经元连接，并通过多条路径运行。每个通路都具备独有的特征，如行走路径、靶向区域、连合神经支配和气味处理模式。在家蚕大脑中，群体标记的外侧前脑输出神经元（三阶嗅觉神经元）接收触角输出。外侧前脑进一步分为下外侧前脑的 δ

a silkmoth, *B. mori*. The LPC is further divided into the delta area in the inferior lateral protocerebrum (ΔILPC) and the lateral horn (LH). PNs processing major pheromone components send their axons to the ΔILPC, which is in turn connected with a higher order center, the superior medial protocerebrum. A major portion of other PNs that process plant odor information sends their axon to the LH.

区（ΔILPC）和外侧角（LH）。处理主要信息素成分的 PNs 将其轴突发送到 ΔILPC，而 ΔILPC 又与一个高阶中心，即上内侧前脑相连。处理植物气味信息的其他 PNs 的主要部分将轴突发送到 LH。

11.4.1 Chemosensory mechanisms

Insect chemosensory systems detect a wide range of volatile and soluble chemicals and are very important for finding and assessing the quality of food sources, in addition to identifying mates and oviposition sites.

The basis of difference in sensitivity to different compounds from species to species and from receptor to receptor reflects variations in the molecular structure of the receptor substrate. In contrast, a comparison of the ligand specificities of individual odor receptors demonstrated that many odor receptors respond to common ligands, and thus one odor typically activates multiple receptors. OSNs are localized in the forehead appendages, antennae, and maxillary palps.

For recognition of olfactory signals, insects use several families of receptor proteins, including odorant receptors (ORs), ionotropic receptors (IRs) and gustatory receptors (GRs). Insect ORs form a greatly expanded phylogenetic lineage derived from insect GRs, and that they are not related to GPCRs.

In insects, the first synaptic processing of the olfactory input from the antennae occurs in the antennal lobe. Key components of the olfactory network in the antennal lobe are two main types of neurons: the local interneurons and the projection (output) neurons. Both neuron types have different physiological tasks during olfactory processing, which accordingly require specialized functional phenotypes. Based on morphological features, three main types of olfactory sensilla can be distinguished in cockroach: *perforated basiconic*, *trichoid*, and *grooved basiconic* sensilla, which correspond to the *sensilla basiconica*, *trichodea*, and *coeloconica* in the fruit fly and several moths.

Generally, olfactory receptors in the dendritic membrane of OSNs are responsible for the detection and discrimination of odors. In insects, the receptors for airborne volatiles belong to three classes of chemoreceptor proteins. In most species, the ability to detect a vast array of chemically diverse odorants (including

11.4.1 化学感受机制

昆虫化学感受系统可探测较大范围的挥发性和可溶性化学物质，除识别配偶和产卵场所外，对寻找和评估食物源的质量很重要。

种和种间、感受器和感受器之间对不同化合物敏感性的根本差异反映了受体底物分子结构的变化。相反，比较单个气味受体的特异性证实许多气味受体对共同的受体有反应，因此某个气味分子通常激活多个受体。OSNs 被定位于前额附器、触角和上颌触须。

为了识别嗅觉信号，昆虫使用了几个受体蛋白质家族，包括嗅觉受体（ORs）、促离子型受体（IRs）和味觉受体（GRs）。昆虫 ORs 形成了一个由昆虫 GRs 衍生而来的极大扩展的系统发育谱系，且它们与 GPCRs 无关。

在昆虫中，触角的嗅觉输入的第一个突触处理发生在触角叶。触角叶嗅觉网络的关键组成部分是两种主要类型的神经元：局部中间神经元和投射（输出）神经元。这两种神经元在嗅觉处理过程中具有不同的生理任务，因此需要特殊的功能表型。根据形态特征，蜚蠊嗅觉感受器可以分三种主要类型：孔锥形感受器、毛形和沟槽锥形感器，与果蝇和几种蛾的锥形、毛状和腔锥感器相对应。

通常嗅觉神经树突膜上的嗅觉受体负责检测和辨别气味。在昆虫中，空气中挥发物的受体属于三类化学受体蛋白质。在大多数物种中，检测大量化学多样性气味物质（包括信息素）的能力是

pheromones) is based on a class of seven-transmembrane domain proteins, the odorant receptors (ORs). The second class of olfactory receptors are the so-called ionotropic receptors (IRs) that are related to ionotropic glutamate receptors (iGluRs) and seem to be predominantly tuned to certain short-chain organic acids, amines and aldehydes. The third class of olfactory receptors comprises another small group of seven transmembrane domain proteins that are assigned to the GR family of insects based on sequence homology. These "olfactory" GR types mediate the detection of carbon dioxide in various insects and of certain pheromones in *D.melanogaster*.

Pheromone molecules enter an olfactory sensillum by pores in the cuticle. At the air/sensillum lymph interface, OBPs are take up and solubilize the hydrophobic pheromones in the aqueous sensillum lymph. The OBP/pheromone complex is diffuse towards the receptive membrane of the OR-expressing OSN where SNMP1s in the membrane may mediate the unloading of the OBP and pave the way to the OR. In this process, the SNMP1 ectodomain might trigger the release of the pheromone from the OBP, either *via* its ionization state and creation of a local pH that induces conformational changes in the OBP, or through a direct protein-protein interaction.

Subsequently, the tunnel-like ectodomain of SNMP1 could take up the pheromone and pass it on to a heteromeric OR/Orco complex that is either directly associated with SNMP1 or lies in close vicinity to SNMP1 in the membrane. In the first case (an association of SNMP1 and OR/Orco), the pheromone could be directly transferred via the SNMP1 ectodomain tunnel to the pheromone-detecting OR ligand–binding site, supposedly lying in the transmembrane domains of the OR. Considering the importance of SNMP1 for rapid activation and termination of pheromone-evoked responses, the SNMP1 ectodomain might pipe ligands to and from the binding site of an OR. In the second case (no association of SNMP1 and OR/Orco), the tunnel might release the hydrophobic pheromones into the lipid bilayer from where they reach the OR ligand–binding site by lateral diffusion in the membrane. A different function for non-neuronal SNMP2 types, possibly in sensillum lymph clearance processes, is suggested based on their expression in support cells and a localization in the apical membrane facing the sensillum lymph.

The ORs and GRs range in size from 350 to 500 amino acids

基于七种跨膜结构域的蛋白质，即嗅觉受体（ORs）。第二类嗅觉受体是所谓的促离子型受体（IRs），与促离子型谷氨酸受体（iGluRs）相关，似乎主要调节到某些短链有机酸、胺和醛。第三类嗅觉受体包括另一小群 7 个跨膜结构域蛋白质，这些蛋白质根据序列同源性被分配到昆虫的 GR 家族。这些"嗅觉"GR 类型介导各种昆虫中二氧化碳和果蝇中某些信息素的检测。

信息素分子通过表皮中的孔进入嗅觉感器。在空气 / 感器淋巴界面，OBPs 被吸收并溶解于水溶性感器淋巴的疏水性信息素中。OBP/ 信息素复合物向表达 OSN 的 OR 的接收膜扩散，其中膜中的 SNMP1 可能介导 OBP 的消除并为 OR 铺平道路。在这个过程中，SNMP1 胞外域可能通过其电离状态和诱导 OBP 构象变化的局部 pH 产生，或通过直接的蛋白质 – 蛋白质相互作用，触发 OBP 释放信息素。

随后，SNMP1 的通道样胞外域可以吸收信息素并将其传递到与 SNMP1 直接相关或在膜中靠近 SNMP1 处的异聚体 OR/Orco 复合物中。在第一种情况下（SNMP1 和 OR/Orco 的结合），信息素可以通过 SNMP1 胞外域通道直接转移到信息素检测或配体结合位点，据推测该位点位于 OR 的跨膜结构域。考虑到 SNMP1 对于信息素诱发反应的快速激活和终止的重要性，SNMP1 胞外域可能将配体输送至 OR 的结合位点。在第二种情况下（SNMP1 和 OR/Orco 没有关联），通道可能会将疏水性信息素释放到脂质双层中，通过膜中的横向扩散到达 OR 配体结合位点。基于非神经元 SNMP2 类型在支持细胞中的表达以及在面对感器淋巴的顶膜中的定位，表明它们在感器淋巴清除过程中具有不同功能。

ORs 和 GRs 的大小范围为 350 ～

and contain seven hydrophobic transmembrane domains (Fig.11-11;Appendix color graphic Fig.11-11). These two families were affiliated in the insect chemoreceptor superfamily and, for most insects, constitute the majority of their chemoreceptors. They have the opposite membrane topology of GPCRs that mediate chemoperception in many other animals, including nematodes and vertebrates, with their N-termini internal to the cell and their C-termini external. They function as ligand gated ion channels; for ORs, they consist of a highly conserved protein called the

500个氨基酸不等，包含7个疏水性跨膜结构域（图11-11；附彩图11-11）。这两个家族隶属于昆虫化学受体超家族，对大多数昆虫来说，它们构成了大多数化学受体。它们具有与GPCRs相反的膜拓扑结构，介导许多其他动物的化学感受，包括线虫和脊椎动物，其N-端位于细胞内部，C-端位于细胞外部。它们作为配体门控离子通道发挥作用；

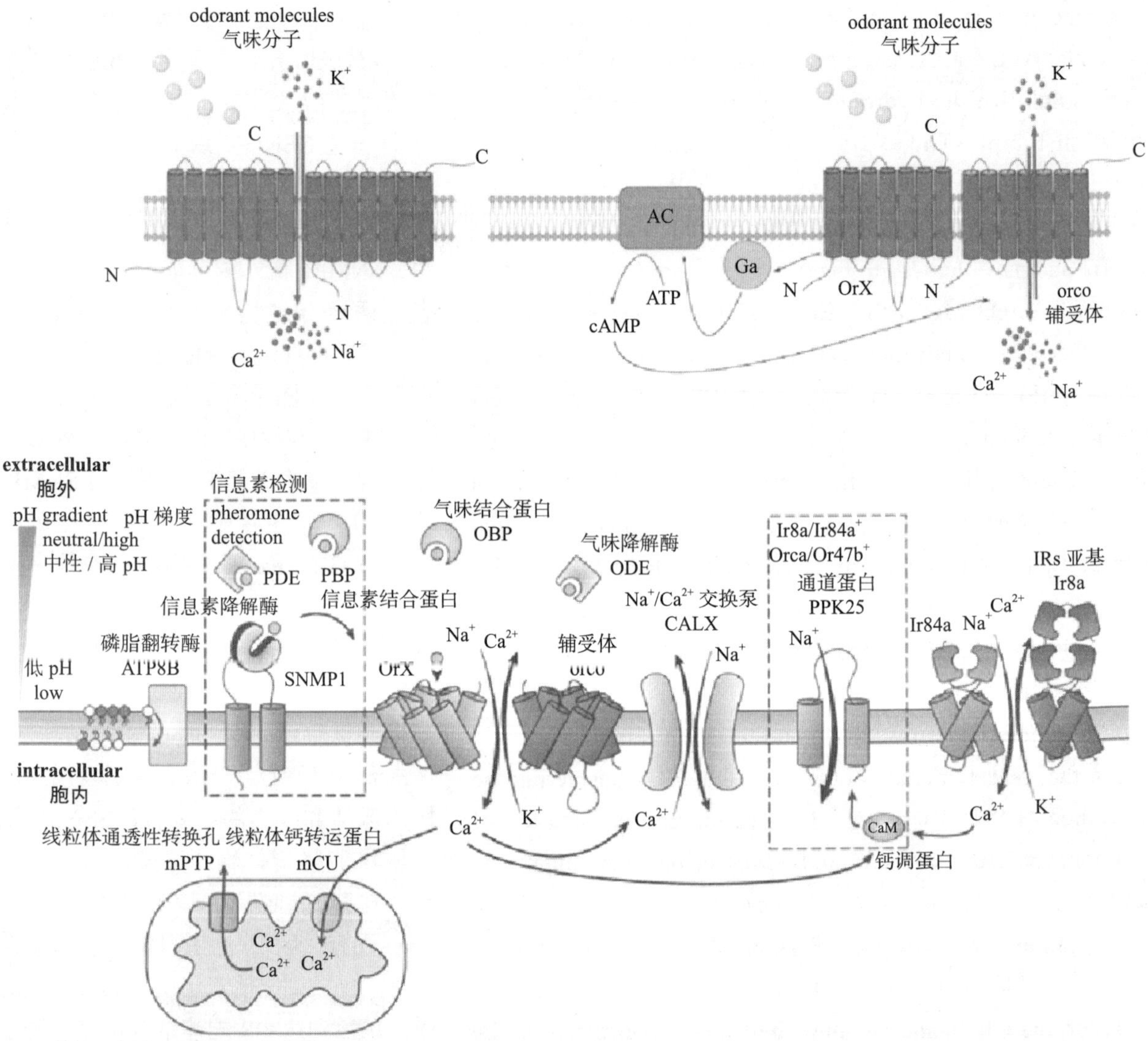

Fig.11-11 Activity modulation of olfactory receptors at the interface between the extracellular sensillum lymph and the cytoplasm (modified from Wicher and Miazzi, 2021). SNMP1. sensory neuron membrane protein 1; DEG/ENaC. degenerin/epithelial sodium.

图11-11 细胞外感器淋巴和细胞质界面嗅觉受体的活性调节（仿Wicher和Miazzi, 2021）。SNMP1. 感觉神经元膜蛋白1；DEG/ENaC. 退化蛋白 / 上皮钠。

odorant receptor coreceptor (Orco) and one of the many ligand-specific ORs.

ORs consist of seven associated α-helices, however, they differ from GPCR in terms of helice orientation with respect to the plasma membrane. Thus, insect ORs have a cytoplasmic N-terminus and an extracellular C-terminus. Although the OrX/Y-Orco heterodimer (OrX in the case of food odor recognition, also sensitive to odors of oviposition places, predators, toxic substances, etc., and OrY in the case of pheromone recognition) serves as an elementary functional unit of the insect olfactory system, the receptors in the membrane of olfactory neurons most likely function within large supramolecular ensembles. Orco forms tetramers having a "pinwheel" shape when viewed from above, perpendicular to the membrane plane. The pore of the non-specific cation ion channel is formed by transmembrane helical structures from each subunit that interact *via* their conserved C-termini (domains 4-7), which resembles acid-sensing ion channels and ATP-gated P2X channels. In Orco–OR complexes, the specific tuning OR alters cation permeability, and involved in current rectification, suggesting that both Orco and OR subunits form the central ion-conducting pore of the channel. Orco can also assemble as a homotetramer and is modulated by phosphorylation.

Structurally, GRs are related to ORs, making it quite likely that they also act as ligand-gated ion channels, although some *in vivo* studies have suggested the involvement of G protein subunits in taste transduction.

By using RT-PCR and immunohistochemistry, BmGr6 was detected in the midgut, central nervous system, and oral sensory organs, suggesting it acts as not only a GR but also a chemical sensor regulating gut movement, physiological conditions, and feeding. In addition, BmGr6 and BmGr9 were colocalized with cells in the brain, suggesting BmGr6 may be involved in the regulation of feeding behaviors in *B. mori* larvae. Membrane topology studies on BmorGr8 predicted an inverted topology relative to GPCR. BmGr8 functions independently in Sf 9 cells and responds in a dose-dependent manner to the polyalcohols *myo*-inositol and *epi*-inositol, suggesting it involves in essential nutrients regulation for some Lepidoptera.

By immunohistochemistry, BmGr9 was also detected the

对于 ORs，它们由一种称为气味受体辅助受体（Orco）的高度保守蛋白质和许多配体的特异性受体 ORs 之一组成。

ORs 由 7 个相关的 α-螺旋组成，然而，它们在相对于质膜的螺旋取向方面不同于 GPCR。因此，昆虫 ORs 具有细胞质 N-末端和细胞外 C-末端。尽管 OrX/Y-Orco 异二聚体（食物气味识别的 OrX，对产卵地、捕食者和有毒物质等气味也很敏感；OrY 识别信息素的）是昆虫嗅觉系统的基本功能单位，但嗅觉神经元膜上的受体很可能在大型超分子集合中发挥作用。Orco 形成四聚体，从上方垂直于膜平面观察，可见其具有"针轮"形状。非特异性阳离子通道的孔由每个亚单位的跨膜螺旋结构形成，这些结构通过其保守的 C-末端（结构域 4-7）相互作用，类似于酸感应离子通道和 ATP 门控 P2X 通道。在 Orco-OR 复合物中，特定的调节或改变阳离子渗透性，并参与电流整流，表明 Orco 和 OR 亚单位形成了通道的中心离子传导孔。Orco 也可以作为同型四聚体组装，并通过磷酸化进行调节。

在结构上，味觉感受器受体与嗅觉感受器受体相关，使之看起来非常像作用于配体门控离子通道，但有些活体研究表明味觉转导牵涉异源三聚体 G 蛋白亚基。

通过 RT-PCR 和免疫组织化学检测，BmGr6 在中肠、中枢神经系统和口腔感觉器官中均有表达，表明它不仅是 GR，而且是调节肠道运动、生理条件和进食的化学传感器。此外，BmorGr6 和 BmGr9 与大脑中的细胞共定位，这表明 BmGr6 可能参与了桑蚕幼虫摄食行为的调节。对 BmGr8 的膜拓扑研究预测了相对于 GPCR 的反向拓扑。BmGr8 在 Sf 9 细胞中单独发挥功能，并以剂量依赖的方式对内消旋肌醇和表肌醇等多元醇做出反应，证明它涉及一些鳞翅目昆虫必需营养物质的调节。

免疫组织化学还检测到 BmGr9 在

expression in cells of oral sensory organs including maxillary galea, maxillary palps, labrum, and labium as well as in putative neurosecretory cells of the central nervous system and brain, suggesting it is involved in promotion of feeding behaviors. Another GR43a ortholog from *B.mori*, *BmGr10* was detected expression in *B.mori* mouthparts such as maxillary galea, maxillary palp, and labrum. BmGR10 showed response to *myo*-inositol, suggesting it may play an important role in the *myo*-inositol recognition.

口腔感觉器官的细胞中的表达，包括上颌盖骨、上颌触须、上唇、上唇以及以及中枢神经系统和大脑中公认的神经分泌细胞，表明它参与促进进食行为。另一个来自家蚕的 GR43a 直系同源基因 *BmGr10* 在家蚕口器中，如上颌盖骨、上颌触须和唇部均有表达。BmGr10 对肌醇有反应，表明它可能在肌醇识别中起重要作用。

Bitter is a taste modality generally associated with toxic substances evoking aversive behaviors in insects. Type Ⅰ exhibits structural features, which are conserved across Lepidoptera species while Type Ⅱ and Type Ⅲ are intronless GRs that are less commonly found in the available Lepidopteran genomes. Insect GRs were predicted to have seven transmembrane domains, an intracellular N-terminus, and an extracellular C-terminus. However, two different topologies were detected from two Type Ⅲ bitter GRs, HarmGR35 and HarmGR50 in *H. armigera*. HarmGR35 has two intracellular termini while HarmGR50 has two extracellular termini. By using calcium imaging in mammalian cells, two Type Ⅰ bitter receptors (BmGr16 and BmGr18) and one Type Ⅱ bitter receptor (BmGr53) from *B. mori* were confirmed widely responded to structurally different feeding deterrents such as coumarin and caffeine, suggesting their roles in the host plant recognition. One *B. mori* Type Ⅰ bitter GR, *BmGr66*, was genetically mutated by using CRIRPR/Cas9 system *in vivo*. The mutated silkworm larvae lost their specificity for mulberry and started to feed on fruits and grain seeds, indicating a single bitter GR is a major factor affecting insect feeding behaviors.

苦味是一种味觉方式，通常与有毒物质有关，可引起昆虫的厌食行为。类型Ⅰ在鳞翅目物种中表现出保守的结构特征，而类型Ⅱ和类型Ⅲ是无内含子的 GRs，在现有鳞翅目基因组中不太常见。据预测，昆虫 GRs 具有七个跨膜结构域、一个细胞内 N- 端和一个细胞外 C- 端。然而，从Ⅲ型苦味 GRs 中检测到两种类型的不同的拓扑结构，棉铃虫 HarmGR35 和 HarmGR50。HarmGR35 有两个细胞内末端，而 HarmGR50 有两个胞外末端。通过在哺乳动物细胞中使用钙成像，证实了来自家蚕的两种Ⅰ型苦味受体（BmGr16 和 BmGr18）和一种Ⅱ型苦味感受器（BmGr53）对结构不同的进食阻碍素，如香豆素和咖啡因有广泛反应，表明它们在宿主植物识别中的作用。通过体内使用 CRIRPR/Cas9 系统，对一个家蚕Ⅰ型苦味受体，*BmGr66* 进行了基因突变。突变的家蚕幼虫失去了对桑叶的特异性，开始以水果和谷物种子为食，这表明单一苦味 GR 是影响昆虫取食行为的主要因素。

11.4.2 Odorant binding proteins (OBPs) and chemosensory proteins (CSPs)

11.4.2 气味结合蛋白（OBPs）和化学感受蛋白（CSPs）

Insect OBPs are small (13-17 kDa) water-soluble proteins, generally consisting of chains of 130-150 amino acids. Primary sequences of this type of proteins are highly divergent and amino acid identity between OBPs from the same species might be lower than 10%. The correct assignment of a protein into the OBP-family is mainly based on a pattern of six cysteines whose

昆虫 OBPs 是一种小的（13 ～ 17 kDa）水溶性蛋白质，通常由 130 ～ 150 个氨基酸组成。这类蛋白质的一级序列差异很大，同一物种的 OBP 之间的氨基酸同源性可能低于 10%。将蛋白质正确分配到 OBP 家族主要基于六个半胱氨酸的模式，其相对位置在所有昆虫目中

relative positions are well-conserved across all insect orders. OBPs are usually comprised of six α-helical domains folded into very compact and stable globular structures. The six cysteines form three interlocked disulfide bridges in an arrangement that provides great stability to the protein's 3-D structure and also seems to be a conserved feature among OBPs. This cysteine pattern has become a "signature" for insect classical OBPs. The first OBP to have its structure determined both alone and bound to its ligand bombykol, was BmPBP1 from *B. mori*. In this complex of BmPBP1-bombykol crystal grown at pH 8.2, BmPBP1 consists of a compact structure formed by a roughly conical arrangement of six α-helices knitted together by three disulfide bridges. Bombykol is bound in a completely enclosed hydrophobic cavity, allowing its transport through hydrophilic environments.

The C-terminal region of OBP acts as a "lid", which is held in place by a triad of well conserved hydrogen bonds that seems to work as a pH-sensitive hinge. It is likely that when the OBP-odorant complex reaches the proximity of dendritic membrane, where the pH is reduced, some of these interactions are disrupted due to carboxylate protonation, leading the C-terminal to move away from the binding cavity, thus "opening the lid" and promoting ligand release. OBP functions as： ① carriers of odorants; ② buffering agents for maximal olfactory sensitivity; ③ activators of olfactory receptors; ④ in the case of sex pheromones, OBP binding may contribute to specificity by selecting the odorants for delivery to the olfactory neurons. Whereas the pheromone-interacting OBPs are necessary for normal responses to pheromones.

Insect ORs adopt an inverse membrane topology when compared to classical GPCRs, with the N-terminal region located intracellularly. ORs are usually part of a heteromeric complex comprising at least two subunits: a ligand-specific OR and a highly conserved co-receptor (Orco). Orco is characteristic for each OR-expressing(OSN) and highly conserved across insect species and orders. Functionally, Orco is required for membrane targeting of canonical ORs. moreover, it appears to heteromerized with other ORs, forming odorant-gated ion channels. Molecules specifically bind the OR, so the Orco is not directly involved in odorant recognition. Heteromeric Orco complexes form ion channels induced by odorant binding to the OR, which leads to signal transduction.

IRs are related to ionotropic glutamate receptors, even though

都很保守。OBPs 通常由六个 α 螺旋结构域组成，折叠成非常紧凑和稳定的球状结构。六个半胱氨酸形成三个连锁的二硫键，其排列方式为蛋白质的三维结构提供了极大的稳定性，并且似乎也是 OBPs 中的一个保守特征。这种半胱氨酸模式已成为昆虫经典 OBPs 的“特征”。第一个单独确定其结构并与其配体桑蚕醇结合的 OBP 是来自家蚕的 BmPBP1。在 pH 8.2 下形成的 BmPBP1- 桑蚕醇复合体中，BmPBP1 由一个紧凑的结构组成，大致由三个二硫键连接在一起的六个 α-螺旋呈锥形排列。桑蚕醇结合在一个完全封闭的疏水腔中，允许其在亲水环境中运输。

OBP 的 C- 末端区域充当“盖”，由三个高度保守的氢键固定在适当位置，这些氢键似乎起着 pH 敏感铰链的作用。当 OBP– 气味分子复合物接近树突状膜时，pH 降低，其中一些相互作用可能因羧酸质子化而中断，导致 C- 端从结合腔中移出，从而“打开盖子”并促进配体释放。OBP 的功能：①气味剂的载体；②最大嗅觉敏感性缓冲剂；③嗅觉受体激活剂；④在性信息素的情况下，OBP 结合可能通过选择气味传递到嗅觉神经元而有助于特异性。而与信息素相互作用的 OBP 对于信息素的正常反应是必需的。

与经典的 GPCRs 相比，昆虫 ORs 采用反向膜拓扑结构，N- 端区域位于细胞内。ORs 通常是由至少两个亚单位组成的异聚体复合物的一部分：配体特异的 OR 和高度保守的辅受体（Orco）。Orco 是每个表达气味受体神经元（ORN）的气味受体的特征，在昆虫的种和目中高度保守。在功能上，Orco 是典型 ORs 膜靶向所必需的。此外，它似乎与其他 ORs 异构化，形成气味门控离子通道。分子特异性结合 OR，因此 Orco 不直接参与气味的识别。异聚 Orco 复合物通过气味剂与 OR 结合形成离子通道，从而导致信号转导。

both families are divergent given the fact that important residues for glutamate binding are not present in insect IRs. These receptors are expressed along with a co-receptor in sensory neurons that do not express ORs or Orco and, in contrast to ORs, there are different co-receptors for IRs. Although functional properties of IRs have not yet been studied in heterologous expression systems, their localization, and structural features suggest they are chemosensory receptors that might function as ligand induced ion channels.

The number of ORs actually expressed in a given insect species may vary considerably between sexes, distinct developmental stages or different olfactory tissues, for example, from the 66 OR-encoding genes of *B. mori*, 24 are expressed in larvae and 35 in adult antennae; for 25 OR genes, no transcripts could be verified by RT-PCR in adults or larval stages.

Insect odor binding proteins (OBPs) were first discovered in 1981 in the antenna of the moth *A. polyphemus* and have since been identified in a wide variety of insect species. Some OBPs are expressed in heromone-responsive sensilla, bind to pheromone molecules, and are therefore called pheromone binding proteins (PBPs). Diverse biochemical roles for OBPs and PBPs have been proposed, including the solubilization of odors in the sensillum lymph, the transport of odors through the lymph to the odor receptors, the removal of deleterious compounds from the lymph, and the deactivation of odors following receptor activation.

OBPs and CSPs are small soluble proteins. OBPs are divided into three groups: pheromone binding proteins (PBPs), general odorant binding proteins (GOBPs) and antennal binding proteins (ABPs), and each of them may have different binding abilities. PBPs and GOBPs are specific of Lepidoptera. Chemosensory Proteins may have the same function of solubilization and transport of chemosensory molecules. GOBP2 and GOBP1 are both expressed in male and female antennae, and that GOBP2 is in addition expressed in male legs. GOBP2 is expressed from fifth instar larva to the beginning of the pupal stage, whereas GOBP1 expression is restricted to the adult, suggesting a different role for these two classes of proteins.

In many moth species, some of the trichoid neurons on the male antenna are, as demonstrated electro-physiologically, sensitive to species-specific sex pheromones, such as bombykol in the case of *B. mori*. A family of at least 30 candidate odor receptors

IRs与离子型谷氨酸受体有关，尽管鉴于昆虫IRs中不存在谷氨酸结合的重要残基，因此这两个家族是不同的。这些受体与不表达ORs或Orco的感觉神经元中的辅受体一起表达，与ORs相反，IRs有不同的辅受体。虽然IRs的功能特性尚未在异源表达系统中被研究，但其定位和结构特征表明，IRs是化学感觉受体，可能作为配体诱导的离子通道发挥作用。

在给定昆虫物种中实际表达的ORs数量可能因性别、不同发育阶段或不同嗅觉组织而有很大差异，例如，从家蚕的66个ORs编码基因来看，24个在幼虫中表达，35个在成虫触角中表达；RT-PCR无法在成虫或幼虫阶段验证转录的有25个OR基因。

昆虫气味结合蛋白（OBPs）最初在1981年发现于多音天蚕蛾的触角上，至今已确认广泛存在于多种昆虫。有些OBPs表现为激素反应性感器，与激素分子结合，因此称之为信息素结合蛋白（PBPs）。OBPs和PBPs生化作用具有多样性，包括气味分子在感器液中的溶解性、气味分子通过感器液被转运到味觉受体、从感器液内去除有害化合物，以及受体活化后的气味分子失活。

OBPs和CSPs都是小分子可溶性蛋白质。OBPs分为三组：信息素结合蛋白（PBPs）、普通气味结合蛋白（GOBPs）和触角结合蛋白（ABPs），其中每个分子都有不同的结合能力。PBPs和GOBPs为鳞翅目特有的。化学感受蛋白在溶解性和转运化学感受分子上可能具有相同的功能。GOBP2和GOBP1在雌雄触角上均有表达，而GOBP2还在雄蛾腿上表达。OBP2从5龄幼虫直到蛹期开始一直表达，而GOBP1局限于成虫期，表明这两种蛋白质作用不同。

电生理学证实，许多种蛾类的某些雌蛾触角上的毛状神经元对某种特异性性信息素敏感，例如，家蚕醇。最近

was recently identified in the silk moth, *B. mori*. Pheromone-sensitive sensilla on the silk moth antenna contain one ORN that is responsive to the sex pheromone bombykol and one ORN that is responsive to bombykol, a secondary pheromone component that is released by the female pheromone gland along with bombykol. The *B. mori* odor receptor BmOR1 was identified as a male antenna-specific odor receptor and is expressed in the region of the antenna containing pheromone-sensitive *trichoid sensilla*. Heterologous expression of BmOR1in *X. laevis* oocytes resulted in a small but specific odor response to bombykol, and ectopic expression of BmOR1 in the antenna of the female moth conferred electrophysiological responses to bombykol, thus identifying BmOR1 as a bombykol receptor. A second male specific receptor, BmOR3, was subsequently identified as a bombykal receptor. In *S.litura*, the GOBP1 specifically express in the tissue of moth antenna, and their transcripts were more widely distributed in the cells bordering of the antennal surfaces where the chemosensory hairs are located.

在家蚕中鉴定出至少 30 种匹配的气味受体家族。蚕蛾触角上的信息素敏感型感器含一个对家蚕性信息素蚕蛾性诱醇响应的嗅觉感器神经元和一个对蚕蛾醛有反应的嗅觉感器神经元。蚕蛾醛是由雌蛾信息素腺体伴随蚕蛾信诱醇分泌的第二个信息素化合物。家蚕气味受体 BmOR1 被鉴定为雄蛾触角特异性气味受体，在对信息素敏感的毛形感器触角区表达。BmOR1 在非洲爪蟾卵巢细胞中的异种表达产生了一种分子质量小，但对家蚕性诱醇特异的气味分子；BmOR1 在雌蛾触角上的异位表达，对家蚕性诱醇产生了电生理学反应，因此确定了 BmOR1 是蚕蛾性诱受体。随后，第二种雄蛾特异性受体，BmOR3 被鉴定为蚕蛾醛受体。斜纹夜蛾的气味结合蛋白 GOBP1 特异性的在蛾的触角组织内表达，其转录更为广泛地分布在有化学感觉毛的触角表面的细胞内。

In the olfaction system, insects detect semiochemicals through interactions with various olfactory proteins, including chemosensory proteins (CSPs), odorant binding proteins (OBPs), odorant receptors (ORs), sensory neuron membrane proteins (SNMPs) and odorant degrading enzymes (ODEs). OBPs and CSPs can transport volatiles and lipophilic semiochemicals to ORs. CSPs are a family of acidic, small molecular weight (1-15 kDa), soluble olfactory proteins located in the chemosensory lymph of insect olfactory sensilla, and can recognize, bind, and transport hydrophobic chemicals from the environment to ORs, and probably play important roles in insect chemoreception. CSPs contain four conserved cysteine residues that form two disulphide bonds, displaying high levels of amino acid identity across insects. *CSP* genes from different insect orders are extensively expressed not only in chemosensory organs, including antennae, tarsus, labial palps and maxillary palps, but also in nonchemosensory organs, including legs, wings, thoraxes, abdomens, ovaries, pheromone glands and compound eyes. CSPs are involved in embryonic development, embryo maturation, oviposition and egg hatching, ecdysis in the larval limb, limb regeneration, brood pheromone transportation, and odor recognition. Therefore, CSPs perform many various tasks from behavior to several physiological and

在嗅觉系统中，昆虫通过与各种嗅觉蛋白的相互作用检测信息化学物质，包括化学感受蛋白（CSPs）、气味结合蛋白（OBPs）、气味物质受体（ORs）、感觉神经元膜蛋白（SNMPs）和气味物质降解酶（ODEs）。OBPs 和 CSPs 可以将挥发物和亲脂性信息化学物质传输到 ORs。CSPs 是一个酸性的、小分子质量（1 ～ 15 kDa）的可溶性嗅觉蛋白质家族，位于昆虫嗅觉感器的化学感受淋巴中，能够识别、结合并将疏水性化学物质从环境运输到 ORs，可能在昆虫的化学感受中发挥重要作用。CSPs 含有四个保守的半胱氨酸残基，形成两个二硫键，在昆虫中显示出高度的氨基酸一致性。来自不同目昆虫的 *CSP* 基因不仅广泛表达于化学感觉器官，包括触角、跗骨、唇须和上颌须，而且广泛表达于非化学感受器官，包括腿、翅、胸、腹、卵巢、信息素腺和复眼。CSPs 参与胚胎发育、胚胎成熟、产卵和卵孵化、幼虫肢体蜕皮、肢体再生、孵卵信息素运输和气味识别。因此，CSPs 执

biological processes (Zhou et al., 2020).

Silkworm CSPs may be involved in development. The majority of silkworm *CSP* genes are expressed broadly in tissues including the antennae, head, thorax, legs, wings, epithelium, testes, ovaries, pheromone glands, wing disks, and compound eyes. The number of *CSPs* in the honey bee genome is similar to that found in the sequenced dipteran species, but is much lower than the number of *CSPs* in the moth or in the beetle (around 20 each). These differences seem to be the result of lineage specific expansions. *CSPs* in a number of arthropods reveal a conserved gene family found in both Mandibulates and Chelicerates. Expressional profiling in diverse tissues and throughout development reveals broader than expected patterns of expression with none of the *CSPs* restricted to the antennae and one found only in the queen ovaries and in embryos. CSPs are multifunctional context-dependent proteins involved in diverse cellular processes ranging from embryonic development to chemosensory signal transduction. Some CSPs may function in cuicle synthesis, consistent with their evolutionary origins in the arthropods. Recently, expressed sequence tags (ESTs) of the maxillary galea of the silkworm were analyzed to identify proteins involved in food selection systems. From the 1,251 redundant genes of the *ESTs*, 7 odorant-binding protein-like genes (*BmOBPL*), 6 *takeout*-like genes (*BmTOL*), and 6 chemosensory protein genes (*BmCSPs*) were identified.

Quantitative RT-PCR analysis indicated that *BmOBPL1*, *BmOBPL2*, *BmOBPL3*, *BmOBPL5*, *BmTOL1*, *BmTOL3*, and *BmorCSP15* were predominantly expressed in the larval oral appendages, such as the maxilla, labrum, labium and antenna. Immuno-cytochemical analysis indicated that the proteins of *BmOBPL1*, *BmOBPL3*, and *BmTOL1* were localized in the gustatory chemosensilla on the maxillary galea and olfactory sensilla in the antenna. The proteins encoded by *BmOBPL1* and *BmOBPL3* were detected in the gustatory chemosensilla of the epipharynx. However, BmOBPL1 and BmTOL1 were also detected in tactile hairs and in the epidermis of several chemosensory organs. The BmOBPL2 protein was localized inside and in the epidermis around the chemosensilla, tactile hairs, and wide surface of the epipharynx.

行许多不同的任务，从行为到几个生理和生物过程（Zhou 等，2020）。

家蚕 CSPs 牵涉发育，其多数 *CSP* 基因在组织中广为表达，包括触角、头、胸部、腿部、翅膀、上皮、睾丸、卵巢、信息素腺、翅原基和复眼。蜜蜂基因组 *CSPs* 的数量与已测序的双翅目昆虫相同，但比蛾和甲虫的数量（每个约 20）少得多。这些差异似乎是谱系特异性扩展的结果。大量节肢动物的 *CSPs* 显示颚亚门和螯肢动物亚门都有保守基因家族。多种组织和整个发育过程的表达谱显示比预想的表达模式更广，任何 *CSPs* 都不仅局限于触角，仅发现一个在蚁后和胚胎期表达的 *CSPs*。CSPs 是环境依赖型多功能蛋白质，牵涉从胚胎发育到化学信号转导各种细胞进程。有些 CSPs 可能起表皮合成作用，与节肢动物的进化起源是一致的。最近，家蚕下颚盔状突起的表达序列标签（ESTs）用于鉴定涉及择食系统的蛋白质。在 1 251 个之多的 *ESTs* 基因中，鉴定出 7 个家蚕气味结合蛋白样基因（*BmOBPL*）、6 个 *takeout* 类基因（*BmTOL*），和 6 个化学感受蛋白基因（*BmCSPs*）。

定量实时 PCR 分析证明，*BmOBPL1*, *BmOBPL2*, *BmOBPL3*, *BmOBPL5*, *BmTOL1*, *BmTOL3* 和 *BmorCSP15* 主要在幼虫口器附属物上表达，如下颚、上唇、下唇和触角。免疫组织化学分析表明，*BmOBPL1*、*BmOBPL3* 和 *BmTOL1* 定位于下颚盔状突起上的味觉化感器和触角内的嗅觉感器。*BmOBPL1* 和 *BmOBPL3* 编码的蛋白质在内唇的味觉化感器内被检测到。但是在触须和几种化学感受器感觉毛和真皮内能检测到 BmOBPL1 和 BmTOL1。BmOBPL2 蛋白位于化感器、感觉毛、内唇宽表面的内部及其周围的真皮内。

11.5 温湿度感受器

11.5 Thermo-and Hygroreceptor

Receptors to function as thermo- and hygroreceptors often occur in the same sensillum on the antennae. The most common triad arrangement is that one neuron is sensitive to cold air, one to moist air, and one to dry air. The firing rate of cold receptor increases during rapid falling temperature and decreases during warming, the moist air receptor fires more frequently when the humi dity rises, and the dry air receptor fires in response to falling humidity. Triads have been found on the antennae of the American cockroach, *P.americana,* the migratory locusts, *L. migratoria*, the European walking stick, *T. infestans, C. morosus*, the honeybee, *A. mellifera* and a noctuid moth, *M. brassicae.* A warm receptor that fires in response to rising temperature has been found in the same sensillum with a cold receptor on the antennae of the mosquito, *A.aegypti.*

Typically, the cuticular portion of these sensilla has no pore and is set in an inflexible socket; *Sensilla trichodea, S. basiconica, S.coeloconica*, and *S.styloconica*, as well as other morphological structures, are known to house receptors believed to be thermo- and hygroreceptors. A few examples of either a thermoreceptor or a hygroreceptor associated with an olfactory receptor in a sensillum with multiple pores have been described. In the most common triad arrangement (Fig.11-12), typically the dendritic of the third neurons fill the lumen of the peg, while the dendrite of the third neuron has short branches, often forming lamellae, and ending beneath the peg. The former cells are called type Ⅰ cells and the latter is a type Ⅱ cell. In some cases, a type Ⅲ cell has been found in which the outer dendritic segment is slender, like the cilium portion, and ends much before the outer cuticular structure. A few sensilla have been found with four or five sensory neurons. The arrangement and some data suggest that type I cells are actually mechanoreceptors that respond to cuticular distortion due to changes of water content in the cuticular portion of the sensillum. The flexible socket, short cuticular projection, pit or collar arrangement that is common, and location beneath more massive mechanoreceptors seem likely to

触角上的同一个感受器受体通常起着温湿度受体的作用。最普遍的三元组合是一个对冷空气敏感、一个对潮湿空气、一个对干燥空气敏感。冷感受器在温度急剧下降时传导速率快，升温时传导慢；潮湿空气感受器常在湿度上升时启动；干燥空气感受器在湿度下降时启动响应。这类组合在美洲大蠊，飞蝗、骚扰锥、欧洲竹节虫、蜜蜂和夜蛾的触角上都有。在埃及伊蚊触角上，体温感受器在温度上升时启动响应，发现其与冷感受器在同一感器内。

这类感受器典型的表皮结构是无孔，着生于无韧性窝内的，在形态结构上可以是毛形感器、锥形感器、腔锥形感器和栓锥形感器与其他形态结构一样，其都已知可以是温湿度感受器。有的温度感受器或湿度感受器则与另一个嗅觉感受器共存于一个多孔的感器内。典型的三元体结构（图 11-12）是两个神经元的树突部分伸入栓腔，而第三个神经元的树突则有短而多的分支，通常形成薄片，中止于栓的下方。前一种称为Ⅰ型细胞，后一种称为Ⅱ型细胞。在有些情况下，发现还有Ⅲ型细胞，外部神经突部分纤细，像纤毛部分，大都在外表皮结构前中止。少数感器也发现四种或五种神经元。这种排列和部分数据表明Ⅰ型细胞的确是机械感受器，它们对由感器表皮的水分含量变化引起的表皮变形产生反应。韧性窝、短表皮质突起、凹陷或环形排列是常见的，位于更大规模的机械感受器下方，好像是将其

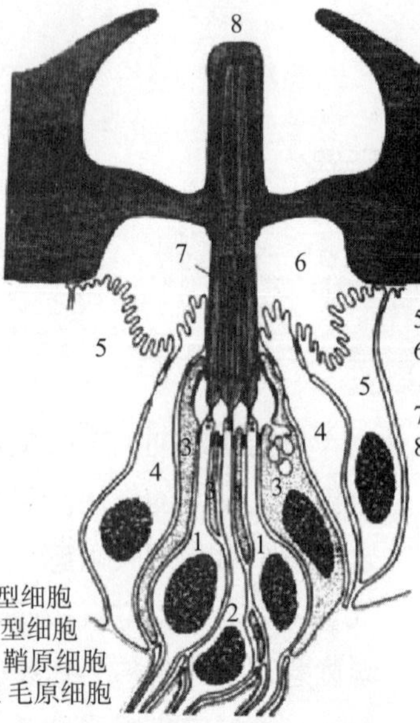

Fig. 11-12 The structure of thermo- and hygroreceptors in cockroach (modified from Altner and Loftus, 1985).
图 11-12 蜚蠊温湿度感受器的结构（仿 Altner 和 Loftus, 1985)

protect them from ordinary mechanical disturbance.

The arrangement of pairs of receptors, such as a warm and cold receptor, or a moist air receptor with a dry air receptor in the same sensillum may improve the ability of the system to discriminate changes in environmental conditions. Each will respond in firing rate to a change in temperature or humidity, but the change will be an opposite direction.

Thermal preference is not a singular response, but involves multiple systems relevant in different contexts. The transient receptor potential channel TRPA1 acts internally to control the slowly developing preference response of flies exposed to a shallow thermal gradient. The rapid response of flies exposed to a steep warmth gradient does not requireTRPA1; rather, the gustatory receptor GR28B(D) drives this behaviour through peripheral thermosensors. At the molecular level, GR28B(D) misexpression confers thermosensitivity upon diverse cell types, suggesting that it is a warmth sensor.

In *Drosophila* two sets of warmth sensing neurons control thermal preference: the anterior cell (AC) neurons, located inside the head, and the hot cell (HC) neurons, located peripherally in the arista. At the cellular level, primary importance has been attributed to either internal or peripheral warmth sensors. At the molecular level, the internal AC neurons sense warmth *via* TrpA1, which encodes a warmth activated transient receptor potential (TRP) channel, whereas the peripheral HC neurons seem to be TrpA1-independent.

保护起来免于机械失调。

温湿度感受器神经元常常成对存在，如温暖和寒冷感受器，湿度感受器和干燥感受器位于同一个感受器内，有助于提高对环境条件变化的辨别能力。每个感受器对温度或湿度的反应速率都非常快，但变化将是反向的。

热嗜好不是一个单一的响应，而是涉及在不同环境中相关的多个系统。瞬时受体电位通道 TRPA1 在内部起作用，控制暴露于弱温度梯度的果蝇缓慢发展的偏好反应。暴露在过高温度梯度下的果蝇的快速反应不需要 TRPA1 ；相反，味觉受体 GR28B（D）通过外围温度感受器驱动这种行为。在分子水平上，GR28B（D）的错误表达使不同类型的细胞对温度敏感，表明了它是一种温度感受器。

在果蝇中，有两组感温神经元控制热嗜好：位于头部内的前细胞（AC）神经元和位于触角芒外周的热细胞（HC）神经元。在细胞水平上，最重要的是内部或外围温度感受器。在分子水平上，内部前细胞神经元通过 TrpA1 感知温度，TrpA1 编码一个热激活的瞬时受体电位（TRP）通道，而外周热细胞神经元似乎是不依赖 TrpA1 的。

Marin et al (2020) identified two new target glomeruli in the antennal lobe, in addition to the five known ones, and the ventroposterior projection neurons (VP PNs) that relay thermo- and hygrosensory information to higher brain centers, including the mushroom body and lateral horn, seats of learned and innate behavior. A few mushroom body-intrinsic neurons solely receive thermosensory input from the lACA, while most receive additional olfactory and thermo-and hygroreleptor PN inputs. Furthermore, several classes of lACA-associated neurons form a local network with outputs to other brain neuropils, suggesting that the lACA serves as a hub for thermo-and hygroreleptor circuitry.

Marin 等（2020）在触角叶中发现了两个新的靶标神经纤维小球，另外还有五个已知的神经纤维小球和腹后投射神经元（VP PNs）将热和湿度感觉信息传递给更高的脑中心，包括蕈状体和侧角，这是学习和先天行为的场所。少数蕈状体固有神经元仅接收来自 lACA 的热感受器的输入，而大多数接收额外的嗅觉以及温和 / 或湿度感受投射神经元的输入。此外，几类 lACA 相关神经元形成了一个局部网络，输出到其他脑神经毡，这表明 lACA 是温湿度感受器循环回路的中枢。

11.6 地磁感受器

11.6 Magnetoreceptor

11. 6.1 Magnetoreception

Magnetoreception of honeybees has been proposed on the basis of numerous behavioral evidences. Free-flying honey bees are able to detect static intensity fluctuations as weak as 26 nT against the earth strength magnetic field. Free-flying honeybees can be trained in discrimination experiments to respond to local magnetic anomalies. Tactile mechanosensory input from the antennae, presumably processed by neurons of the antennal joint hair sensilla and the neck hair plates, enables bees to estimate both the direction relative to gravity and the length of the waggle phase.

11.6.1 地磁感受作用

蜜蜂的地磁感受作用是基于大量行为学证据提出来的。自由飞行的蜜蜂能检测到静态地球磁场强度弱于 26 nT 的强度变化。经过驯化的自由飞行蜂可在辨别实验中对局部磁场异常做出反应。触角的机械感觉输入可能被触角连接处的毛形感器和颈部毛板的处理，使蜜蜂能够估算与地球引力相关的方向和摆动的长度。

11.6.2 Honey bee dance

There are two major theories on how honey bee foragers communicate with other workers about a new food source: the honey bee dance and the odor plume. The honey bee dance is a way for bees to communicate with one another. Karl von Frisch, who earned the 1973 Nobel Prize in Physiology or Medicine, observed that a honey bee that discovers a new food source will tell other honey bees about its location through the honey bee dance. When a worker bee finds food close to the bee hive, it performs its simplest dance, the Round

11.6.2 蜜蜂舞蹈

找到食物的工蜂如何与其他工蜂交流新食物源的信息有两个理论：蜜蜂舞蹈和气味跟踪。蜜蜂舞蹈是蜜蜂彼此之间联络的途径。1973 年诺贝尔生理学或医学奖获得者 Kar von Frisch 观察到如果某个蜜蜂发现新的食物源，就会通过蜜蜂舞蹈告诉其他蜜蜂食物源的位置。当工蜂发现食物源离巢近时，表演

Dance (Fig.11-13; Appendix color graphic Fig.11-13). This dance doesn't provide much information, it is more of an arousal signal. The forager bee runs in a small circle, leaving a single cell inside it. Every one or two circles it suddenly reverses orientation and this goes on from seconds to minutes.

When the goal is further away, the bees need more sophisticated means of communication. If food is scarce, bees have been known to feed up to about 15 km from the beehive. In relation to the small size of this animal these distances are outstanding. Although a bee flying to a known source of food uses as references conspicuous landmarks in addition to the sun compass, it can only communicate information about the latter to fellow bees. The Tail-Wagging dance tells the other bees very accurately at what distance and in which direction the food is, so they can look for it by themselves. Some European honeybees start to perform it when the source of food is more than 100 meters away. Other bee species will do them for closer sources, up to just a few meters away in the case of some Indian bees. For intermediate distances there is a gradual transition between the round dance and the tail-wagging dance.

In a typical tail-wagging dance the honeybee (*A. melliphera*) runs straight ahead for a short distance, returns in a semicircle to the starting point, again runs through the straight stretch, describes a semicircle in the opposite direction and so on in regular alternation. The straight part of the run is given particular emphasis by a vigorous wagging of the body (Fig.11-14; Appendix color graphic Fig.11-14). In addition, during the tail-wagging portion of the dance it emits a buzzing sound. Interestingly, the dance followers can make the dancer pause and give them a taste of the nectar by using a squeaking sound. With increasing distance, the number of circuits (8 s) per unit time decreases and the length and duration of the individual circuits increases. For example, for a goal at 100 meters it makes 10 short circuits in 15 seconds but at 3 km only 3 long circuits in the same time (Fig.11-15; Appendix color graphic Fig.11-15). The duration of the wagging part has the best correlation with distance. The distance is calculated based on the expenditure of energy on the flight towards the source.

简单的圆圈舞（图 11-13；附彩图 11-13）。这种舞蹈不提供很多信息，更多的是一种挑逗信号。工蜂跑一个小圆圈，留一个蜂巢小室在内。每转 1 ～ 2 圈，它就会突然改变方向；这个过程持续数秒到数分钟。

当目标更远时，蜜蜂需要更加复杂的沟通方法。如果食物匮乏，蜜蜂知道如何到离巢 15 km 外取食。相对于这种小动物的体型来说，这个距离不可思议。虽然蜜蜂飞往已知的食物源除了利用明显的地标，还使用日光罗盘作参照物，但它只把后者的相关信息传送给跟随起舞的蜜蜂。尾部摇摆舞蹈非常准确地告诉其他蜜蜂食物的距离和方向，从而使它们能自己找到食物。有些欧洲蜜蜂在食物源远于 100 m 之外时开始表演。其他蜂群在食物源稍近时表演，有些印度蜂在食物源仅几米外时表演。中间距离在圆圈舞和尾部摇摆舞之间逐次变换。

蜜蜂（意大利蜜蜂）常见的尾部摇摆舞是向前直行一段小距离，半圆形返回到起点，再爬行一直线距离，往相反方向半圈，如此循环，定时交替。直行的部分通过躯体特别注重有力的摇摆（图 11-14；附彩图 11-14）。此外，舞蹈的尾部摇摆部分还发出嗡嗡声，有趣的是跟随起舞者能让跳舞的蜜蜂暂停，用啸叫声给它们花蜜的味道。随着距离的增加，单位时间内循环（8 s）数量减少，单个循环持续的时间增加。例如，100 m 处的目标，蜜蜂会在 15 s 内短循环 10 次，但 3 km 的目标相同时间内只有 3 个长循环（图 11-15；附彩图 11-15）。摆动持续时间与距离正相关。距离的计算基于飞往食物源消耗的能量。

Fig. 11-13 "Nearby" round dance (modified from Yahya, 2011).

图 11-13 "附近"的圆圈舞（仿 Yahya, 2011）。

Fig.11-14 Tail-wagging long distances dance (modified from Yahya, 2011).

图 11-14 远距离尾部摇摆舞（仿 Yahya, 2011）。

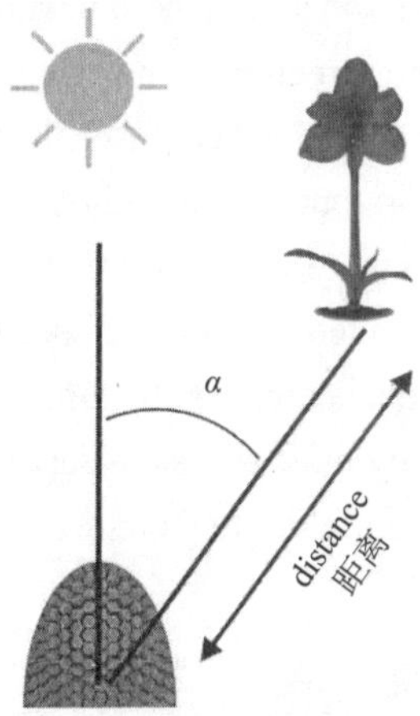

Fig. 11-15 The Tail-Wagging dance tells the other bee information of food distance and direction (modified from Brown, 2011).

图 11-15 工蜂传递给其他蜜蜂关于食物源距离和方向的信息（仿 Brown, 2011）。

11.7 地球引力感受器

11.7 Gravireceptor

The dancer always showed the other bees the direction to the food source in relation to the position of the sun. She did this by performing the tail-wagging portion of her dance on the honeycomb at an angle (α) away from vertical (the honeycomb typically hangs vertically) representing the angle away from the sun at which the recruit bee would need to fly to find the food source. For example, if the target food source or flower resided in the direction of the sun, the dancer's tail-wagging would point straight upward on the vertical honeycomb surface; otherwise, if the food source resided back of the sun, the dancer's tail-wagging would point straight downward. If the flower resided an angle of 40° to the left of the sun's position, the dancer would shift the straight "waggle" of the dance 40° to the left of the vertical. By always choosing to represent the angle away from the sun as an angle away from the vertical, the dancing bee could direct other bees to the scrumptious goal that she had just returned from even in the darkness of the hive. She could even account for the continuous daily motion of

跳舞的蜜蜂展示给其他蜜蜂食物源的方向与太阳的位置有关。这是它通过在蜂巢上与垂直方向的夹角（α）表演摇摆舞做到的（蜂巢一般是垂直悬挂的），那代表征召蜂要寻找食物源必须飞向与太阳位置相关的角度。例如，如果食物源或鲜花目标位于太阳的方向，摇摆舞者沿着蜂巢表面垂直向上；否则，如果蜜源背着阳光，摇摆舞者头朝下。如果花源向左偏离太阳 40°，舞者将向垂直方向偏左 40°"摇摆"。常常通过象征偏离太阳的角度作为垂直方向的角度，跳舞蜂即使蜂巢在黑暗中也能指示其他蜂前往其刚刚返回的美味的目的地。它甚至能通过其"内部钟"精确地计算出太

the sun across the sky with her precisely timed "inner clock", adjusting the direction of her waggle dance accordingly. Indeed, experiments intended to "confuse" the honeybees, in which their hives were placed horizontally instead of vertically, and in which the sky was screened so that the bees were not aware of their position in relation to the sun, resulted in very confused bees regardless of scent stimuli. However, a honeycomb can be turned continuously, and the honeybees, as long as they have access to some portion of sunlight or blue sky, will "adjust themselves to their new direction, like the needle of a compass."

(Li Wenchu)

阳跨越天空的持续日常活动，有根据地调整舞蹈的方向。事实上，有意地将蜂巢水平摆放替代垂直方向来“困惑”蜂，并遮盖天空使蜂不知道与太阳的相对位置的实验，导致困惑的蜂无视香味的刺激。但是继续调转蜂巢，只要它们能部分利用阳光或蓝天，将“像罗盘指针一样调整到新的方向。”

（李文楚）

Exercises

1. Outline the insect sensory receptor classifications according to their functions in detail.
2. What are the structural features and functions of Johnston's? Try to list any five types of mechanoreceptors.
3. How do insect odor receptors interact with their co-receptors to form a sensitive odor sensing mechanism?
4. What are the components of chemoreceptors? What is the molecular mechanism of odor sensing molecules?

思考题

1. 详细概述昆虫感受器的功能分类。
2. Johnston's 器有哪些结构特征和功能？试列举任意五种机械感受器。
3. 昆虫气味感受器如何与其辅助受体共同作用形成灵敏的气味感受机制？
4. 化学感受器由哪些部分组成？感受气味分子的分子机制是什么？

第 12 章 光和昆虫视觉感受器

Chapter 12 Light and Insect Visual Receptors

Key points

1.Insects often have several types of light-sensitive organs, including compound eye and ocelli. Ocelli are divided into dorsal ocellus and stemmata.

2.Color vision, form discrimination, and detection of planc-polarized light are well evolved in some insects and play important roles in their behavior.

3.Some hypotheses explaining phototropism have been proposed. The most important is the light stress hypothesis, which addresses light stress occurring in insects exposed to light. Overall, the positive and negative phototropism of insects is a passive behavioral response to physiological stress caused by light stress.

4.Bioluminescence is critical in insect reproduction, predation and prey, and hermaphroditic communication. A large amount of organs and nerve axon innervated the light-emission cell layers and provide oxygen. Numerous mitochondria and fluorescence particles are distributed within the cells. Luciferin and luciferase fullfilled within the particle.

5.The circadian clock can be entrained to light by the visual system and by non-visual, dedicated circadian mechanisms.

知识点

1. 昆虫通常有几类光敏感器官，包括复眼和单眼，单眼又分为背单眼和侧单眼。

2. 许多昆虫更进化出色觉、形状识别和偏振光分辨能力，这在它们的行为中起着重要作用。

3. 提出了几种向光性的假说。最重要的是光胁迫假说，它参与昆虫在光照下发生的光应激反应。总体来说，昆虫的正向和负向趋光性是它们在光照胁迫下生理应激后的被动行为反应。

4. 生物发光在昆虫的生殖、捕食和被捕食、两性通信等方面至关重要。大量器官和神经轴突使发光细胞层受神经支配并提供氧，这些颗粒内部充满荧光素和荧光素酶。

5. 生物钟可通过视觉系统和非视觉系统驯化以适应光的昼夜节律机制。

Insects often have several types of light-sensitive organs. The compound eyes allow insects to detect color and to form visual images. Compound eyes are especially sensitive to motion. Ocelli found on both immature and adult insects of somc species are located between the front compound eyes on the head and are called the dorsal ocellus. They can only perceive changes in light intensity and the direction of the light source, and reacl to certain wavelengths of light, but cannot perceive the shape of an object for imaging. The eyes of holometabolous insect larvae are

昆虫通常有几类光敏感器官。复眼使昆虫能够检测颜色并形成视觉图像。复眼是特别灵敏的运动检测器。某些昆虫的幼体和成虫都有单眼，位于头部前复眼之间的称为背单眼。它们只能感受光强度的变化和光源的方向，对某些波长的光产生反应，但不能感受物体的形状进而成像。完全变态昆虫幼虫的单眼，称为侧单眼。不

called stemmata. The immatures of hemimetabolous insects have compound eyes similar to those of adults. The overall structure of compound eyes, ocelli, and stemmata follow the same basic cellular plan, with a light-sensitive area called the rhabdomere. Some insects, particularly dipterous larvae, have cuticular light-sensitive cells located in various parts of the body without a rhabdomere region. Color vision, form discrimination, and detection of plane-polarized light are well-developed in some insects and play important roles in their behavior.

完全变态昆虫的幼虫具有与成虫复眼相似的复眼。复眼、单眼和侧单眼总体结构上都具有相同的基本细胞设计，有一个光敏感区域，感杆。有些昆虫，特别是双翅目幼虫，体表不同部位还具有非感杆结构的表皮光敏感细胞。某些昆虫的色觉、形状识别和偏振光检测高度发达，并在它们的行为中起着重要作用。

12.1 单眼和复眼的结构

12.1 Ocelli and Compound Eye Structure

12.1.1 Ocelli

Ocelli share some anatomical features with compound eyes, including a corneal lens, corneagenous cells that secrete the lens, retinular cells, and an ocellar nerve that leads into the protocere brum (Fig. 12-1). The retinular cells typically contain a rhabdomere region with microvilli. The corneal cuticle is probably transparent to various wavelengths of light, but few measurements have been

12.1.1 单眼

单眼具有与复眼相同的一些解剖特征，包括角膜晶状体、分泌晶体的角膜细胞、视网膜细胞以及与前脑相连的单眼神经（图 12-1）。视网膜细胞通常包含有微绒毛的感杆区。晶状体表皮可能对不同波长的光具有透过性，但相关的

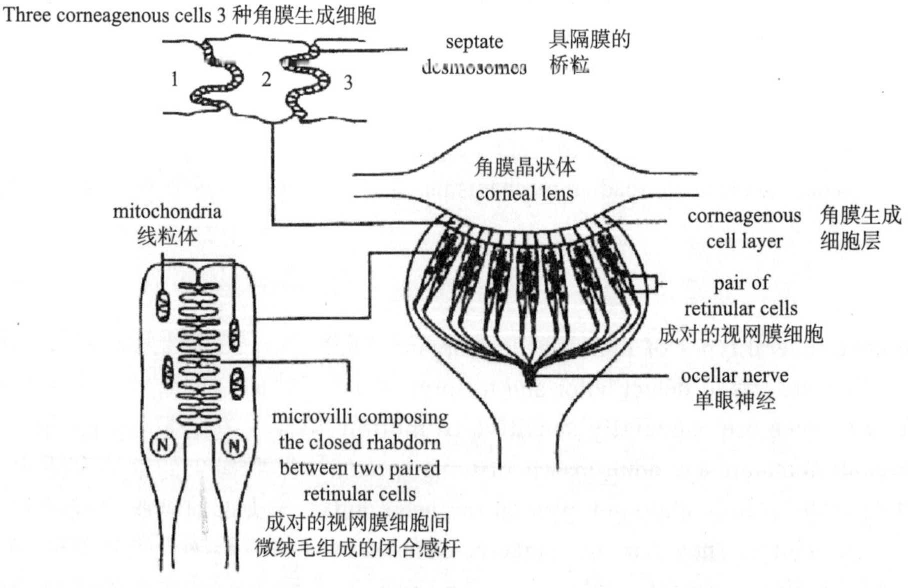

Fig.12-1 Diagram of the structure of an ocellus from a worker honeybee (modified from Toh and Kuwabara, 1974).
图 12-1 工蜂的单眼结构（仿 Toh 和 Kuwabara，1974）。

made. The cornea of the ocelli of *B. craniifer* was transparent to light from about 350 to 700 nm. The cornea of an ocellus covers a large visual field and appears to form an image, but the image is focused beneath the layer of retinular cells and is not transmitted to the brain. The retinular axons from ocelli are spontaneously active in the dark. Rather than depolarizing the ocellar retinular cells, light leads to a more stable or increased membrane potential, hence decreasing or even ceasing spike activity sent to the brain. The ocelli can signal light on/off information, intensity of illumination, and possibly in some insects may indicate the quality (wavelength) of light.

实验证据较少，如蜚蠊单眼晶状体能透过 350 ～ 700 nm 波长的光。单眼的晶状体覆盖了大部分视野，似乎能成像，但所成像聚焦于视网膜细胞层的下方，并不传递到大脑。因此，单眼能很好地聚光，并可对大范围的光强变化做出反应。在黑暗中，单眼的视网膜轴突能自发激活。光线则会导致膜电位更加稳定或提高，而不是使单眼的视网膜细胞去极化，因此，传送到脑的峰电位活性会降低或中断。单眼能传送光线有无或光源强度的信号，有些昆虫甚至可能传送光质（波长）信号。

12.1.2 Stemmata

The eyes holometabolous insect larvae are usually called stemmata. The number of stemmata varies from one in some insects to as many as six on each side of the head in Lepidoptera. Their general organization varies considerably among insect groups, but the most well-developed stemmata have an overlying transparent cuticle, a crystalline lens, and a few retinular cells with rhabdomere regions. Some stemmata do not have the crystalline lens. In some insect larvae, the stemmata have two separate rhabdoms: a distal one nearest the overlying cornea and a proximal one below. Each of these rhabdoms is formed by microvillar extensions from only a few cells, and although the dioptric apparatus forms an image that falls on the rhabdomere surfaces, it seems likely that resolution of any image is poor.

Caterpillars frequently move their head from side to side, which may be a behavior that aids them in obtaining a wider field of view with small, multiple stemmata. Some sawfly larvae and tortricid caterpillars have been reported to sense plane-polanzed light, but it is difficult to see how this could be adaptive to these larval insects.

12.1.2　侧单眼

完全变态昆虫幼虫的眼常称为侧单眼。昆虫侧单眼的数量差异很大，有些昆虫头部每侧 1 个鳞翅目昆虫的侧单眼多至 6 个。侧单眼的基础结构在昆虫不同种群中差异较大，但多数发育完整的侧单眼都覆盖透明的表皮、含有晶状体和少量带有感杆区域的视网膜细胞。有些侧单眼没有晶状体。有些昆虫的幼虫有两个分开的感杆束：顶部的一个靠近角膜，而近邻的在其下方。每个感杆束都是由少数几个细胞的微绒毛延伸形成的。尽管其折光结构形成的图像聚焦于感杆表面，但成像的分辨率却不高。

鳞翅目的幼虫经常左右摆动头部，这种行为可能有助于它们利用侧单眼获得更宽的视野。有些叶蜂和卷叶蛾的幼虫能感觉到平面偏振光，但尚不清楚这些昆虫幼虫是如何适应的。

12.1.3 Compound eyes

The compound eyes are composed of multiple functional units called ommatidia. Each ommatidium is composed of many cells and of functional parts that include dioptric structures, photosensitive cells containing photosensitive pigments, and shielding cells which usually also contain a variety of pigments (Fig.12-2). The small eyes of Thysanura contain only a few ommatidia (12 in

12.1.3　复眼

复眼由许多独立功能单位组成，称为小眼。每个小眼由很多细胞和功能部分组成，包括屈光结构、含有光敏色素的光敏细胞以及通常也包含有多种色素的保护细胞（图 12-2）。缨尾目昆虫的复眼很小，仅包括极少数单眼（衣鱼属

Lepisma spp.), while the very large eyes of dragonflies contain as many as 10,000 ommatidia. Adult Zygentoma, Siphonaptera, and Strepsiptera do not have compound eyes, but instead have simple eyes similar to ocelli.

昆虫 12 个），而蜻蜓的复眼很大，包括 10 000 个单眼。衣鱼目、蚤目和捻翅目成虫没有复眼，取而代之的是与单眼相似的简单眼。

There is considerable diversity in the structural detail of insect ommatidia among different insect groups. Two major variations in structure are represented in Figure 12-2, the photopic eye of a dipteran; and in Figure 12-3, the scotopic eye of a moth. Photopic eyes occur in diurnal insects, which are active during the day. The rhabdom extends from the cone to the basement membrane at the proximal end of the eye. There is little or no movement of the pigment in the shielding cells around the retinular cells. Light can strike the photosensitive pigment in the rhabdom only by entering axially i.e., directly from above after passing through the cornea and cone of the ommatidium. The pigment-containing cells surrounding the retinular cells effectively shield the rhabdom from any stray light from adjacent ommatidia. An image is focused on the rhabdom or rhabdomeres of retinular cells just below the cone. Scotopic eyes occur in nocturnal and crepuscular insects. The rhabdom is shorter than in photopic eyes and usually extends only about one-third the distance from the basement membrane to the

昆虫小眼的精细结构在不同的种群间存在明显的差异。结构上主要分为两类，如图 12-2 的双翅目明视型复眼和图 12-3 的蛾类暗视型复眼。明视型复眼出现在白天活动的日行性昆虫，感杆束从晶锥延伸到眼最近端的底膜。围绕视网膜细胞的保护细胞中的色素很少或不移动。光线仅在轴向进入，即通过单眼的角膜和晶锥时，才能照射到感杆束中的光敏感色素。包含色素的细胞包围住视网膜细胞，有效地保护感杆束免受近邻单眼散光的照射。图像聚焦在晶锥正下方的视网膜的感杆束或感杆上。仅夜行性和晨昏活动的昆虫拥有暗视型复眼。感杆束比光适应的要短，通常只占底膜到晶锥距离的 1/3。剩下的距离

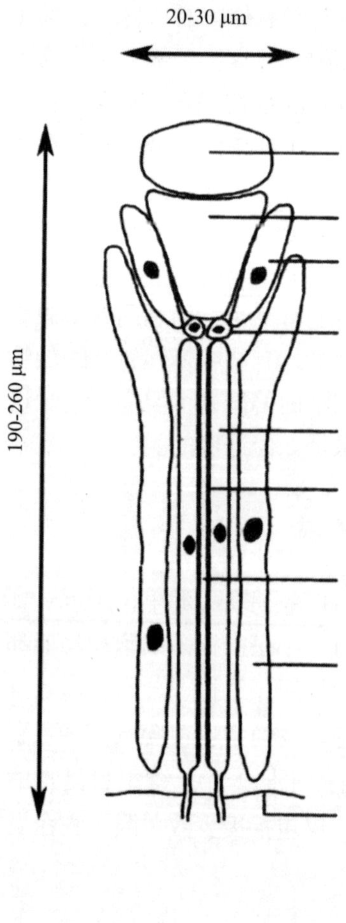

Fig.12-2 Diagrammatic representation of the structure of an ommatidium in the photopic compound eye of the tephritid *A. suspensa*. (modified from Agee *et al.*, 1977).
图 12-2 加勒比实蝇光适应复眼中的小眼结构图解（仿 Agee 等，1977）。

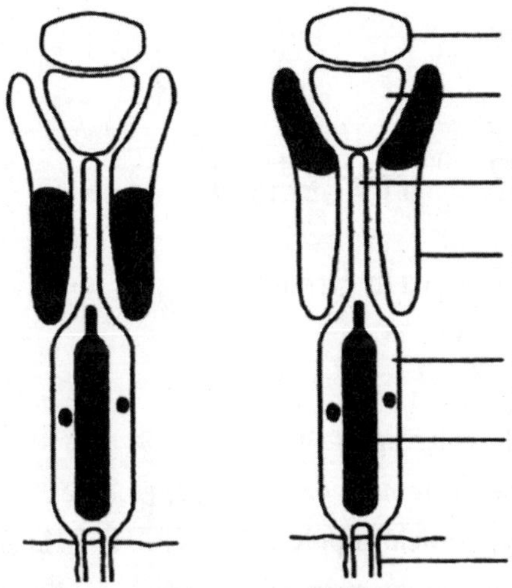

Fig.12-3 Diagrammatic illustration of the structure of an ommatidium in the scotopic eye of a nocturnal moth to show shielding pigment distribution in a light-adapted ommatidium and its distribution in a dark-adapted ommatidium. In the light-adapted eye, the dispersed pigment protect crystalline tract of rhabdomeres against light escaping from neighboring ommatidia, while in the dark-adapted state, migration of pigment to the periphery of the ommatidium allows the potential for light to enter from adjacent ommatidia. The latter condition may make for less sharp visual images, but probably allows better visual responses in dim light, which is likely an adaptation for night-flying moths (modified from Nation, 2022).

图 12-3 夜行性蛾类光适应眼的一个小眼的结构图解，展示保护性色素在光适应小眼和暗适应小眼内的分布。光适应眼色素分散，保护感杆晶束免受隔邻小眼遗漏光干扰；在暗适应状态，色素向小眼外周移动，允许潜在的光源从邻近小眼进入。在微弱的光线下，后者产生清晰的成像也许是较好的视觉反应，这可能是夜行性蛾类的一种适应性进化（仿 Nation, 2022）。

cone. The remaining distance to the cone may be filled with thin strands of thc rctinular cells forming the crystalline tract (Fig.12-3). The crystalline tract is surrounded by pigment cells that allow the pigment to contract into the distal part of the cells in dim light. This allows light from adjacent ommatidia to pass through and strike the rhabdom below. Pigment disperses throughout the shielding cells upon exposure to bright light or in response to higher environmental temperatures. This anatomical type of eye came is called a superposition eye. Migration of the pigment in bright light so as to shield the rhabdomeres may be a protective mechanism to prevent bright light from bleaching the visual pigment. Another morphological variation in eye structure, although not correlated with photopic or scotopic eyes, is that rhabdomeres may be fused into a rhabdom in some insects and remain unfused (called open) in others.

可能由视网膜细胞形成的晶束填充（图 12-3）。晶束被色素细胞包围，在暗淡的光线下，色素会向细胞顶部紧缩。这样，邻近单眼的光线就可以通过并照射到下面的感杆束。暴露在明亮的光线或较高温度的环境下，色素会沿保护细胞扩散。这种复眼在解剖学上被称为重叠眼。明亮的光线条件下，色素的移动也可能是防止光敏色素褪色的一种保护性机制。还有另一种眼结构的形态变化，即有些昆虫感杆会融合成感杆束而有些昆虫的感杆却不融合（称之为开放式），但这种眼结构的变化与光适应眼或暗适应眼无关。

12.2 折光结构

12.2 Dioptric Structure

The dioptric structures refract or bend the light entering the eye allowing it to be focused. A variety of dioptric structures exist in different insect eyes, including cornea, cone, corneal nipples, Crystalline tract, and layers of different density in one or more of these structures.

折光结构对进入眼睛的光线进行折射并使之聚焦。在不同的昆虫眼中，有多种多样的折光结构，如角膜、晶锥、角膜乳突、晶束以及这些结构中的一个或多个具有不同密度的细胞层。

12.2.1 Cornea

All compound eyes and ocelli have a corneal covering that transmits and refracts the light passing through. The cornea is composed of transparent cuticle secreted by corneagenous cells (highly specialized and modified epidermal cells). For effective vision, the cornea must transmit a significant portion of incident light. The cornea of *M. sexta* transmits about 90% of light between 400-650 nm wavelengths. Wavelengths shorter than 350 nm are strongly absorbed within the dioptric apparatus of *M. sexta*, with wavelengths shorter than 300 nm being almost completely blocked.

12.2.1 角膜

所有的复眼和单眼都有角膜覆盖，它可以透过和反射光线。角膜是由角膜生成细胞（高度特化和修饰化的真皮细胞）分泌的透明表皮。为获得有效的视觉，角膜必须使大部分照射光线能够透过。400～650 nm 波长的光有 90% 都能透过烟草天蛾的角膜，而波长低于 350 nm 的光线则会被烟草天蛾的角膜强烈吸收，低于 300 nm 光线不能透过角膜。

12.2.2 Cone

The cone is another part of the dioptric apparatus in most insects. The light transmission characteristics of the crystalline cone in *M. sexta* are similar to the properties of the cornea. The cone in insects is also very variable in structure and is formed in several ways. In general, there are four cells, sometimes called Semper's cells, that form the cone.

12.2.2 晶状体

晶状体是多数昆虫的另一个重要屈光器。烟草天蛾晶束的光线透过性和角膜类似。昆虫的晶状体结构变化很大，并通过多种方式形成。在通常情况下，有四种细胞，有时称为森帕氏细胞，森帕氏细胞可以形成晶状体。

12.2.3 Acone eyes

In insects without a cone, the cone cells themselves are transparent. They are not much modified in shape nor do they secrete crystalline products as in insects with cones in the eye. Acone eyes are considered the most primitive type, occuring in apterygote insects, and also in some Hemiptera, Coleoptera, and Diptera.

12.2.3 无晶眼

无晶眼昆虫的晶锥细胞本身具有透光性。其形状没有经过太多修饰，也不像眼里有锥体的昆虫那样分泌结晶产物。无晶眼被认为是最原始类型，出现在无翅亚纲以及有些半翅目、鞘翅目和双翅目昆虫中。

12.3.4 Eucone eyes

Eucone eyes are common and occur in most orders of insects. In these, the cone cells contain a clear, hard intracellular secretion that fills most of the cell volume. The remaining small shell of living cytoplasm in the cone cells is pushed towards the cell margins.

12.2.4　真晶眼

真晶眼普遍出现在多数昆虫中。真晶眼晶锥细胞含有一种清澈而坚硬的细胞内分泌物，填充了大部分细胞腔。在晶锥细胞中，余下的少量细胞质被挤压到细胞边缘。

12.3.5 Pseudocone eyes

They are common in many Diptera, including the Cyclorrhapha, and in some Odonata. The pseudocone is a cavity below the cornea filled with a gelatinous or liquid secretion. The cytoplasm of the four cone cells is squeezed into a thin layer beneath the pseudocone.

12.3.5　伪晶眼

伪晶眼在多数双翅目，包括环裂亚目以及一些蜻蜓目中较为普遍。伪晶体是角膜下充满胶质或液态分泌物的腔，四种晶锥细胞的胞质在伪晶体下方被挤压成薄薄的一层。

12.3.6 Exocone eyes

Exocone eyes occur in some Coleoptera, most notably in fireflies and related beetles. Exocone eyes are not considered homologous to any of the preceding three types. The cone is formed by an inward projection of the cornea, and thus it is cuticular in structure. The four Semper's cells form a short crystalline tract below the exocone, but they secrete no true crystalline core.

The compound eyes of some insects contain additional dioptric components much smaller in dimension than the cone and cornea. These components include a crystalline tract, corneal nipples, corneal layering, and periodic layering of tracheolar structures, commonly referred to as a tapetum.

12.3.6　外晶眼

外晶眼出现在一些鞘翅目昆虫中，主要是萤火虫和同种类的甲虫。外晶眼与前述三种类型无同源性。外晶眼的晶状体由角膜内陷而成，因此这是一种表皮类的结构。四个森帕氏细胞在外晶锥下形成短晶束，但不分泌真的晶核。

有些昆虫的复眼还包含一些额外的折光组分，尺寸比晶锥和角膜更小。这些组分包括晶束、角膜乳突、角膜层和周期性的气管反射层，常称为膜状层。

12.3.7 Crystalline tracts

Crystalline tracts are found in some butterflies, moths, and in fireflies. The diameter of a tract is one important parameter that determines whether the tract can transmit an image to the rhabdom below. The diameter varies from 2 μm in butterflies to 4μm in *M. sexta*, to 10μm in some fireflies. At least in the moths and butterflies, the diameters measured are considered to be too small to allow an image to be transmitted. Other factors are also important to image transmission and no studies of this nature have been reported from fireflies. Nevertheless, the tracts of moths and butterflies can transmit light to the rhabdom by functioning as a light guide if the refractive index of the tract is greater than that of the surrounding medium so that the light is kept inside the tract and not diffused to the surrounding tissue. The presence of shielding,

12.3.7　晶束

在一些蝴蝶、蛾和萤火虫中发现了晶束。其直径是决定晶束能否将图像传输到下面感杆的重要的参数。晶束直径在蝴蝶中为 2 μm，在烟草天蛾中为 4 μm，而在一些萤火虫中为 10 μm。在蝴蝶和蛾类中，所测得的晶束直径太小，以至于不足以透过影像。其他因素对图像的传送也很重要，但没有报道过萤火虫的相关研究。然而，如果晶束的折射率高于周围介质，那么蛾类和蝴蝶的晶束就能通过光导将光线传送到感杆束，以便将光线保留在晶束内而不扩散到周围组织。在这方面，填满色素的保

pigment-filled cells is important in this respect. Finally, to be effective, the light will have to be brought to focus on the rhabdom. On the basis of both theoretical and experimental data, Miller et al. (1968) concluded that the tracts act as light guides in dark-adapted eyes of *M. sexta, Elpinor* spp. (a sphingid moth), *Cecropia* and *Polyphemus* (silk moths), and a skipper, *Hobomok* spp.

护性细胞也非常重要。最后，光线必须有效地折射到感杆束聚焦，才能获得清晰的影像。根据理论和实验数据，Miller 等（1968）断定晶束在烟草天蛾（一种鞘类蛾）、天蛾、天蚕（一种蚕蛾）和美洲大蛾的暗适应眼内起着导光作用。

12.3.8 Corneal nipples

They are present at the corneal surface of the eyes of many nocturnal Lepidoptera. The nipples probably function as antireflection devices, which in turn should increase the transmission of light into the eye, possibly by as much as 5%. The nipples might also benefit the insect by reducing the reflection of light to predators and parasites, and by reducing or preventing internal reflection of light from the tapetum below the eye.

12.3.8 角膜乳突

角膜乳突多出现在夜行性的鳞翅目昆虫的角膜表面。乳突的功能可能是一种抗反射装置，能增加多达 5% 的入眼光线的传输。乳突也可能通过减少光在复眼表面的反射，从而伪装自身，使其不易被天敌或猎物发现。

12.3.9 The tapetum

It is a thick mat of tracheoles at or near the base of the eye. The shiny surface of the tracheoles reflects light back into the upper parts of the eye and creates eye shine in moths and some other insects. The glow disappears in moths with light-adapted eyes because the shielding pigments move down in the pigment cells enough to absorb most of the light. The reflection of light back into the distal parts of the eye is likely to create blurred images, but insects that have a large tapetum are those that are active at night when the light is dim anyway and not best suited for the formation of sharp images.

12.3.9 气管反射层

气管反射层是眼基部或附近的厚气管垫，气管明亮的表面把光线反射回去，进入眼的上部，在飞蛾和其他一些昆虫中形成眼睛的光泽。在具有光适应眼的飞蛾中，这种光亮的消失是因为屏蔽色素在色素细胞中向下移动，足以吸收大部分的光。光线反射回眼顶部很可能会产生模糊的图像，但有气管反射层的昆虫往往在夜间活动，此时光线暗淡，本就不适合形成清晰的图像。

12.3.10 Corneal Layering

The eyes of some dipterans, especially the Tabanidae, show patterns of bright and dark stripes, frequently in colors, in the cornea of the compound eyes. The pattern and colors are caused by the reflection of light from dense (refractive index 1.74) and thin layers (refractive index 1.40) in the corneal cuticle. Deerflies have as many as 20 layers in the cornea. Miller et al. (1968) suggested that the layers might serve as color contrast filters, enabling fast-flying flies to locate more easily a host against the background.

12.3.10 角膜层

有些双翅目，特别是虻科的眼在复眼内呈明暗相间的条纹图案，通常是彩色的。这种图案和色彩是由角膜内厚表皮层（反光系数 1.74）和薄表皮层（反光系数 1.40）引起的。鹿虻的角膜多到 20 层。Miller 等（1968）认为这些角膜层可能起着色彩反差滤镜的作用，使快速飞行的昆虫更容易锁定猎物。

12.3　视网膜细胞

12.3　Retinular Cells

12.3.1 The structures of retinular cells

The receptor neurons in the eyes are called retinular cells. They are primary receptors because they send their axons directly into the optic lobe of the brain without synapsing. Eight is the most common number of retinular cells in each ommatidium, arranged in a circle. The number is variable in different insects, however, and ten or eleven have been found in some Lepidoptera, nine in honeybees, four in the dorsal part of the eye of adult dragonflies, and six in the ventral part, and eight in dragonfly nymphs. In the most common situation, the eighth cell, or in honeybees the ninth cell, is short and can be found only in the more proximal region of an ommatidium.

The retinular cells are long slender cells extending over the greater part of the length of an ommatidium. They are 60-100 μm long in *D. melanogaster*, about 100 μm long in *P. americana*, and from 150 to >200 μm long in *A. suspensa*.

12.3.1　视网膜细胞的结构

眼内的感受器神经元称为视网膜细胞。因为其轴突直接进入脑的视叶中，中途无突触，因此属于初级感受器。在每个单眼内，最普遍的是 8 个视网膜细胞，排列成圆圈。但不同的昆虫的视网膜细胞的数量是不同的，如有些鳞翅目昆虫有 10 ～ 11 个；蜜蜂有 9 个；蜻蜓成虫眼的背部有 4 个、腹面有 6 个，蜻蜓若虫为 8 个。在大多数情况下，蜜蜂的第 8 个细胞或第 9 个细胞很短，而且在最接近单眼的区域才存在。

视网膜细胞是细长的细胞，延伸覆盖大部分单眼。在果蝇中的尺寸是 60 ～ 100 μm，在美洲大蠊中的尺寸是 100 μm，在加勒比实蝇中的尺寸是 150 ～ 200 μm 及以上。

12.3.2 Electrical activity of retinular cells

Resting potentials ranging from 25-70 mV have been recorded across the membranes of retinular cells. The inside of the cell is negative relative to the outside during resting conditions in the dark. In contrast to the polarizing effect of light on vertebrate eye receptors, light acts as a depolarizing stimulus for retinular cells in the compound eye.

Illumination of the eye causes a slow or graded potential that increases with the intensity of the stimulus. The slow potential is the receptor potential and shows the typical characteristics of a graded potential. At higher intensities of illumination, there appears a transient "on" component that has a sharp threshold and a refractory period. Moreover, the "on" component can be abolished by high external K^+ ion concentration. Thus, the "on" component has the characteristics of a spike (action potential). Both the

12.3.2　视网膜细胞的电生理活动

视网膜细胞的跨膜静息电位范围是 25 ～ 70 mV。在黑暗的静止条件下，细胞内部相对于外部的膜电位是负的。与脊椎动物眼睛受体的光极化效果相反，在复眼中，光对视网膜细胞起着去极化的刺激作用。

对眼的照明会引起慢电位或分级电位，随刺激的强度增加而增加。慢电位是受体电位，表现为典型的分级电位特征。高强度的照明会导致出现一个可短暂“打开”的部件，具有灵敏的临界点以及折返期。这个“打开”部件消除外部高浓度的 K^+。因此，“打开”部件具有峰电位的特征（动作电位）。瞬时反

transient response and the receptor potential are linear functions of the logarithm of light intensity. The transient "on" potential is a compound action potential of many individual axon responses. Increased illumination brings more and more axons into play and the "on" transient becomes larger.

应和受体电位都与光照强度的对数呈线性关系。瞬时"打开"电位是许多单个轴突反应的综合动作电位。照明强度的增加使越来越多的轴突参与反应，从而使"打开"瞬时电位变大。

12.4 感　　杆

12.4 Rhabdomere

12.4.1 Rhabdomeres

Each retinular cell contains a specialized region called the rhabdomere, which is the site of the light-sensitive pigment. Perpendicular to the long axis of the retinular cell, the cell membrane extends into thousands of microvilli, typically about 40-120 nm in diameter along all or part of the length of the cell. These microvilli make up the rh abdomere. In *D. melanogaster*, there are about 60,000 microvilli per cell. There are about 80,000 tubules in each rhabdomere of *P. americana*. The microvilli are the site of some 10^7 molecules of rhodopsin per cell. A rhabdomere extends over most of the length of a retinular cell, and is usually oriented toward the circle formed by the group of retinular cells in an ommatidium. The diameter of the rhabdomere is about 1.2 μm in *D. melanogaster* and about 2 μm in *P. americana*. Rhabdomeres fuse at the center of the circle to form a closed or fused rhabdom in honeybees, Lepidoptera, and many other insects. Alternatively, the rhabdomeres may remain free and face a central hollow cylinder to form open rhabdoms in some Diptera and Hemiptera. Electron microscope studies reveal that extracellular spaces exist within even the tightest, closed rhabdoms. In compound eyes, the microvilli typically are oriented perpendicular to the long axis of the retinular cell; but in simpler eyes of some arthropods, the direction of microvilli is in line with the long axis of the retinular cell.

12.4.1 感杆

每个视网膜细胞有个特殊的区域称为感杆，是光敏色素所在的场所。与视网膜细胞长轴垂直，细胞膜延伸成数以千计的微绒毛，通常直径为 40～120 nm，沿细胞的长度或部分长度分布。这些微绒毛组成感杆。果蝇每个细胞约有 60 000 根微绒毛。美洲大蠊每个感杆约有 80 000 条小管，每个细胞的微绒毛上约有 1 亿个视紫红质分子。感杆延伸覆盖视网膜细胞的大部分长度，常常呈圆环形走向，在小眼内形成一群视网膜细胞。果蝇的感杆直径约 1.2 μm，美洲大蠊约 2 μm。在蜜蜂、鳞翅目和其他许多昆虫中，感杆融合于圆环中央，即形成闭合型感杆束或融合感杆束。然而，在一些双翅目和半翅目昆虫中，感杆仍然是游离的，面向中央圆筒形空腔，即形成开放型感杆束。电镜学研究揭示，即使在最紧密、闭合的感杆束中，细胞外空隙仍存在。复眼中的微绒毛典型的方向是与视网膜细胞长轴垂直，但有些节肢动物的简单眼，微绒毛的方向与细胞长轴平行。

12.4.2 Neural connections in the optic lobe

The optic lobe of the brain contains three large neuropils (Fig. 12-4). The most distal one, the lamina ganglionaris, is

12.4.2 视叶内的神经联络

脑的视叶包含 3 个大的髓区（图 12-4）。最顶部的一个，神经节视叶是

the site of the first synaptic junctions of retinular cell axons. The laminais composed of groups of large monopolar neurons, association neurons, and glial cells. In dipterans such as *Musca* and *Calliphora*, which have groups of neurons form neural cartridges or neuroommatidia in the lamina ganglionaris. There are 3,000 cartridges in each lobe corresponding to the 3,000 ommatidia in the eye, which send axons into the cartridges for synaptic connections. Within each cartridge of *Musca* and *Calliphora*, there are four monopolar neurons, six branched α- and β-neurons (centrifugal neurons coming from the medulla to the lamina), and six axons coming from retinular cells in six different ommatidia. The axons from the retinular cells decussate (cross) before entering the lamina. The retinular neurons synapse with monopolar neurons. Some retinular axons pass through the lamina without synapsing.

视网膜细胞轴突发生突触连接的第一个位点（或称视网膜的神经纤维）。视神经片由大的单极神经元群、联络神经元和胶细胞组成。在像苍蝇和丽蝇这样的双翅昆虫中，神经元群在视神经纤维形成神经暗盒或神经单眼。每个视叶有 3 000 个暗盒对应于复眼中的 3 000 个单眼，将轴突伸入暗盒中形成突触连接。苍蝇和丽蝇的每个暗盒有 4 个单极神经元，6 个分支的 α- 神经元和 β- 神经元（来自薄层髓质的远心神经元）以及来自 6 个不同的单眼中的视网膜细胞的 6 个轴突。来自视网膜细胞的轴突在进入薄层前相互交叉。视网膜神经与单极神经元突触连接，有些直接通过薄层无突触连接。

Fibers leave the lamina, decussate, and enter the second optic ganglion, the medulla. The network of fibers between the lamina and medulla is called the external chiasma. Fibers leave the medulla, again decussate, and enter the lobuli, the third ganglionic mass. The lobuli consist of several, sometimes separated, masses of cell bodies and neuropil. In *Musca*, there are two masses known as the lobula and lobulus. Regular repeating units have not been found in the lobuli, and little is known about synaptic connections or input-output interactions. The large amount of crossing of fibers in the optic lobe of insects would appear to provide a great deal of backup security if some parts of either the external eye or brain receiving the input suffered damage.

神经纤维离开薄层，相互交叉，然后进入第二个视神经节，即髓质。薄层和髓质之间的网状纤维称为外部交叉。神经纤维离开髓质，再相互交叉，进入第三个神经节细胞群，即小叶。小叶是由几个（有时相互分离的）细胞体和神经髓组成的细胞群。苍蝇有两个细胞群分别称为左小叶和右小叶。小叶内没有发现有规律的重复单元，对突触连接或输入 - 输出的相互作用知之甚少。昆虫视叶中的大量神经纤维的交叉似乎提供了很大的备用安全性，以防外部眼或大脑的某些部分在接收输入时受到损伤。

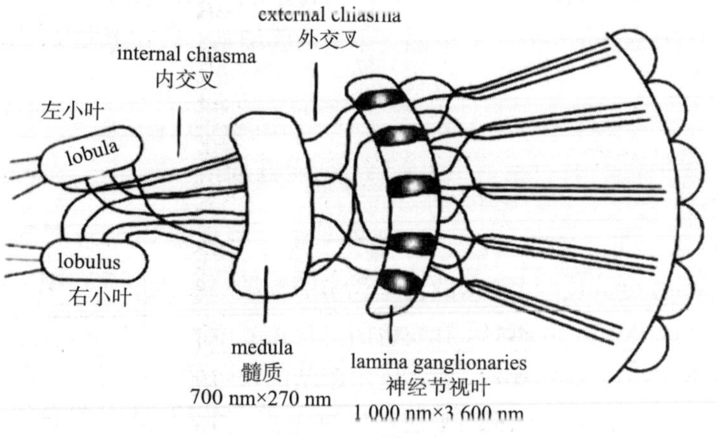

Fig.12-4 Basic structure of the optic lobe, illustrating the four major optic neuropils and neural cartridges in the most distal neuropil, the lamina ganglionaris (modified from Nation, 2022).

图 12-4 视叶的基本结构，显示多数神经髓顶端的 4 个主要视神经髓和神经芯、神经节视叶（仿 Nation, 2022）。

12.5 昆虫的视觉化学

12.5 Chemistry of Insect Vision

12.5.1 Dermal light sense

There are small pockets of photosensitive cells in the cuticle of larvae of Diptera, which have no obvious eyes on the surface of the body. These cup-shaped pockets contain a few cells whose axons run into the short, stubby central nervous system. Tests of avoidance responses made with monochromatic light indicate that 540 nm light is most effective in stimulating the photosensitive cells. The larvae try to avoid the light, and attempt to crawl into the food or under other debris or cover. Many insect larvae can respond to light when the eyes are blackened or otherwise occluded. Very little information is available on the possible receptors involved.

12.5.1 真皮光觉

双翅目幼虫表皮有小型的光敏细胞，在体表面没有明显的眼。这些杯形小区域含有少量细胞，其轴突进入短而粗的中枢神经系统。用单色光进行的回避反应测试表明，540 nm 的光对光敏细胞的刺激最有效。幼虫设法避开光线，尝试爬进食物或其他碎屑或覆盖物下。许多昆虫幼虫在黑暗时或以其他方式遮蔽眼睛时能对光做出反应。涉及的可能的受体可获得信息非常少。

12.5.2 Photoelectricity transduction

The first event leading to a visual image is a photochemi cal reaction. Light quanta are absorbed by the pigment rhodopsin in rhabdomere microvilli. This leads to a chemical change in rhodopsin, which in turn leads to the opening of Na^+ and Ca^{2+} channels, depolarization, and electrical activity in the axon of the retinular cell. Thus, light energy is transduced into chemical energy by the pigment, and then chemical energy is transduced into electrical energy by the neuron. Retinular cells are sensitive to one photon of light.

12.5.2 光电转导

视觉影像形成的第一个环节是产生光化学反应。光量子被感杆微绒毛的色素视紫红质吸收，视紫红质随即发生化学变化，导致 Na^+ 和 Ca^{2+} 通道的开放，去极化，视网膜细胞轴突会产生电活动。因此，通过光敏色素，光能被转换成化学能，化学能再通过神经元转换成电能。小视网膜细胞对一个光量子也是敏感的。

12.5.3 Rhodopsin

Rhodopsin is the name for the visual pigment in light- sensitive cells of all animals. It is composed of a protein, opsin, and a chromophore, retinal, or a closely related molecule. In insect eyes the chromophore can be retinal or 3-hydroxy retinal. Species-specificity and wavelength of light to which rhodopsin is maximally sensitive are dependent on both opsin and chromophore structure. Rhodopsin is usually characterized on the basis of its

12.5.3 视紫红质

视紫红质是所有动物光敏细胞中的视觉色素的称谓。由蛋白质、视蛋白和发色团、视黄醛或密切相关的分子组成。昆虫眼睛的发色团可以是视黄醛或密切相关的分子。昆虫眼内的发色团可以是视黄醛或 3- 羟基视黄醛。视紫红质的种属特异性和对光波长特殊的敏感性

maximum spectral sensitivity, for example, as a UV-sensitive rhodopsin, or a blue-green sensitive rhodopsin, etc.

It is one of a large family of G protein-coupled receptors involved with ligand binding and message transduction at the cell membrane. G proteins are intracellular proteins that are part of the second messenger signaling process. A characteristic of G protein-coupled receptors is that they are transmembrane proteins with seven α-helical transmembrane domains, and they bind intracellular G proteins when activated by a signal. Rhodopsin binds its ligand, *cis*-11-retinal, as a protonated Schiff base, with the ligand forming a covalent bond between the aldehyde group of retinal and the ε-amino group of a lysine residue (Lys-296) in transmembrane-seven of the protein. The proton of the Schiff base is counterbalanced by an induced negative charge on a glutamate residue in transmembrane three. The chromophore is bound to the portion of opsin that lies within the plasma membrane of the retinular cell. Amino acid substitutions in opsin, especially in transmembrane- three and transmembrane-six domains, shift the distribution of electrons about the retinal chromophore and within the opsin portion of rhodopsin, and alter the wavelength of maximum light absorption. Both extracellular and intracellular portions of the rhodopsin molecule are important to function. Disulfide bonds in some of the loops of the extracellular portion of the molecule are important in holding the molecule in its characteristic shape and orientation. Portions of the intracellular part of rhodopsin are the sites for G protein recognition and binding. When activated by light, rhodopsin couples mainly with the G protein designated G_t, although additional G proteins are known with which activated rhodopsin may couple.

主要依赖于视蛋白和发色团结构。视紫红质常常以其最大光谱敏感性来定性，如 UV- 视紫红质或蓝绿光视紫红质等。

视紫红质是 G 蛋白耦联受体大家族中的一员，牵涉受体结合和膜上的信号转导。G 蛋白是部分第二信使信号传导过程中的分子内蛋白质。G 蛋白耦联受体的特征之一是有 7 个 α- 螺旋跨膜区域，被信号激活后与分子内 G 蛋白结合。视紫红质与其受体，质子化的 Schiff 碱，顺式 -11- 视黄醛结合，在视黄醛的醛基和跨膜蛋白 -7 的赖氨酸残基 (Lys-296) 的 ε- 氨基基团之间形成共价键。Schiff 碱的质子被跨膜蛋白 -3 上的谷氨酸残基带有的感应负电荷平衡。发色团和位于视网膜细胞质膜内的视蛋白部分结合。替换视蛋白的氨基酸，特别是跨膜蛋白 -3 和跨膜蛋白 -6 区域，改变了视黄醛发色团和视紫红质的视蛋白部分的电子分布，也就改变了最大吸收光的波长。视紫红质的胞内和胞外部分在功能上都非常重要。在分子的胞外部分，有些环上的双硫键对保持分子特有的形状和方向非常重要，而胞内部分是 G 蛋白识别和结合的位点。当视紫红质被光激活时，虽然与其他 G 蛋白耦联，但主要与命名为 G_t 的 G 蛋白耦联。

12.5.4 The visual cascade

When a photon of light is absorbed by rhodopsin, 11- *cis*-retinal is isomerized to 11-*trans*-retinal.This initiates a conformational change in the protein portion of rhodopsin, and it becomes activated as metarhodopsin, a catalyst that activates a cascade of reactions. G protein, also called transducin, binds to a loop of the metarhodopsin at the inside surface of the plasma membrane. Transducin is composed of three subunits: T_α, T_β, and T_γ. Guanosine diphosphate (GDP) is bound to the T_α subunit as a normal part of its structure. The G protein metarhodopsin complex is in an activated state, enabling GTP (Guanosine triphosphate) to replace GDP at the T_α subunit without any additional energy requirement. Transducin, now in a higher

12.5.4　视觉级联反应

当光量子被视紫红质吸收时，11- 顺式 - 视黄醛变构成 11- 反式 - 视黄醛。这就引发了一种视紫红质的蛋白质部分发生构象变化，它被激活为后视紫红质，一种催化剂激活一连串的反应。G 蛋白，也称为转导素，由 T_α、T_β、T_γ 三个亚单位构成。鸟苷二磷酸（GDP）与 T_α 结合成为其结构的一部分。G 蛋白 - 变视紫红质复合体处于活跃状态，促使鸟苷三磷酸在没有任何能量需求的情况下取代 T_α 中的 GDP 位置。此时，转

energy state, breaks away from metarhodopsin, and the α-subunit with bound GTP separates from the *β*-*γ* subunits. The T_α-GTP protein is the functional G protein, and it activates phospholipase C which cleaves PIP2 (phosphatidyl inositol bisphosphate), a component in the membranes of the microvilli, into DAG (diacylglycerol) and IP_3 (inositol trisphosphate):

导素处于高能状态，从变视紫红质中解离，α 亚基绑定 GTP 从 *β*-*γ* 亚基中分离出来。T_α - GTP 蛋白是功能性 G 蛋白，它激活磷脂酯 C，分解微绒毛膜内的化合物 PIP2（磷脂酰肌醇二磷酸）成 DAG（甘油二酯）和 IP_3（肌醇三磷酸）：

$$PIP_2 \xrightarrow{\text{active phospholipase C}} DAG + IP_3$$

肌醇二磷酸　活化磷脂酶 C　甘油二酯 + 肌醇三磷酸

IP_3 promotes the release of calcium bound to smooth endoplasmic reticulum near the base of the microvilli in retinular cells. Free ionic calcium (Ca^{2+}), opens sodium channels that also admit Ca^{2+} in the base of the microvilli. The inward movement of a few sodium ions generates a receptor (graded) potential, and entry of more Ca^{2+} acts as an amplifying mechanism, resulting in more sodium entry.

If the receptor potential is strong enough, it may give rise to spikes in the retinular axons. Liberated DAG activates protein kinase C, but its role in the visual process is not clear. It may be involved in the adaptation process in the eye. The *β*-*γ* subunits may also have some role in signal transduction.

This signaling pathway is extremely fast, and can be turned on and off, and repeat many times per second. The cascade in *D. melanogaster* is the fastest G protein cascade that has been measured thus far, and it takes only several tens of milliseconds to proceed from light activation of rhodopsin to the generation of a receptor potential. The enzymatic character of the visual cascade at several steps results in amplification of the signal by as much as 100-1,000 times. Each metarhodopsin can catalyze GTP binding to as many as several hundred T_α subunits before it is deactivated, and similar amplification occurs in production of IP_3 and in release of Ca^{2+}. Thus, one photon light signal is turned into a barrage of neuronal impulses going into the lamina ganglionaris of the optic lobe.

IP_3 促使视网膜细胞微绒毛基部附近的滑面内质网上结合态钙的释放，游离的离子钙（Ca^{2+}）打开 Na^+ 通道，在微绒毛基部允许 Ca^{2+} 进入。少量 Na^+ 内流产生了感受器（分级）电位，更多 Ca^{2+} 的进入起着一种放大机制的作用，导致更多的 Na^+ 进入。

如果受体电位够强，在视网膜轴突上就会产生峰电位。释放的 DAG 活化了蛋白质激酶 C，但它在视觉反应中的作用尚不明确，可能参与眼的适应过程。*β*-*γ* 亚基在信号转导中也有某些作用。

信号转导途径极快，能够每秒钟启动和关闭并重复若干次。果蝇中的级联是目前已测定的 G 蛋白级联反应是最快的，目前测定的从光激活视紫红质到产生受体电位整个过程只需几十毫秒。视觉级联几步反应酶学特征导致信号被放大 1 00 ～ 1 000 倍。每个变视紫红质在失活前可催化 GTP 和几百个 T_α 亚单位结合，相似的放大也发生在 IP_3 的产生和 Ca^{2+} 的释放。这样，1 个光量子信号转换成密集的神经脉冲进入到视叶的神经层。

12.5.5 Regulation of the visual cascade

How is the visual cascade turned off? One way is that insect metarhodopsin can absorb another photon of light and the metarhodopsin/11-*trans*-retinal complex is reconverted to rhodopsin with 11-*cis*-retinal, a process called photoisomerization.

12.5. 5　视觉级联的调节

视觉级联是如何关闭的？一个途径是昆虫的变视紫红质能吸收另一个光量子，变视紫红质 /11- 反式 - 视黄醛复合体再转变回视紫红质和 11- 顺式 - 视黄醛，

This biochemical change terminates the visual response and allows the retinular cell to recover for the next event. In contrast, the metarhodopsin of vertebrate eyes loses its retinal chromophore upon absorption of a photon of light, and a new rhodopsin molecule must be synthesized.

A second way involves the binding of an inhibitory protein, called arrestin (ARR), to the metarhodopsin (MET). The metarhodopsin-arrestin complex (MET-ARR) will no longer bind to G proteins, so the amplification cascade is prevented or stopped upon binding of arrestin. Rhodopsin kinase phosphorylates the MET-ARR complex to give a MET-PO_4-ARR complex. Absorption of a photon of light by the MET- ARR-PO_4 complex causes release of ARR, and a phosphatase removes the phosphate to regenerate metarhodopsin (Fig. 12-5; Appendix color graphic Fig.12-5). There are also less well-known regulatory processes within the G protein cascade, and in activity of phospholipase C.

Although constant resynthesis of 11-*cis*-retinal and rhodopsin after each visual stimulus is not necessary in insect compound eyes because of photoisomerization of metarhodopsin to rhodopsin, visual pigment molecules do wear out and new ones must be synthesized by insects from time to time. Some insects also shed the rhabdomere portions of the retinular cells from time to time and

这个过程称为光异构化。这个生化变化终止了视觉反应，允许视网膜细胞为下一个事件恢复。相反，脊椎动物眼睛的视紫红质吸收一个光量子失去了视黄醛发色团，新的视紫红质必须合成。

第二个途径牵涉一种视紫红质抑制蛋白（ARR）与变视紫红质（MET）的结合作用。MET- ARR 复合体形成后，MET 不能再与 G 蛋白结合，因此，级联放大作用被抑制或停止。视紫红质激酶磷酸化 MET-ARR 复合体形成 MET-PO_4- ARR 复合体。这一复合体吸收一个光量子导致 ARR 的释放，磷酸酶进一步移除磷酸盐，重新产生视紫红质（图 12-5；附彩图 12-5）。G 蛋白级联和磷脂酶 C 活性的调节过程还知之甚少。

即使在每个视觉刺激后，由于变视紫红质光异构化变成视紫红质，昆虫的复眼也不需要持续不断地合成 11- 顺式 - 视黄醛和视紫红质，视色素分子的确会耗尽，因此昆虫必须不时地合成新的。有些昆虫视网膜细胞的感杆部分时不时地会脱落，因此也必须合成新的微绒毛和视紫红质。为合成新的视紫红质，从

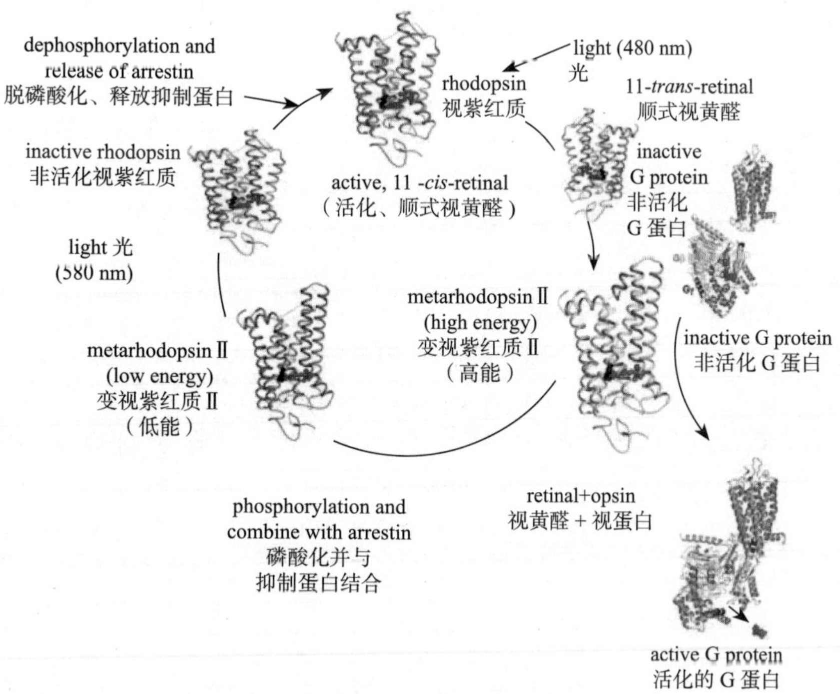

Fig.12-5 Regulation of visual cascade.

图 12-5 视觉级联的调节。

thus have to synthesize new microvilli and rhodopsin. To synthesize new rhodopsin, 11-*trans*-retinol obtained from β-carotene must be enzymatically converted to 11-*trans*-retinal. A photoisomerase enzyme in eye tissues of honeybees can be activated by absorbing a photon of light, and it then catalyzes the isomerization of 11-*trans*-retinal to 11-*cis*-retinal. In the dark part of the eye, 11-*cis*-retinal is enzymatically combined with opsin to produce rhodopsin.

β-胡萝卜素获得的 11- 反式 - 视黄醇必须酶促转化成 11- 反式 - 视黄醛。蜜蜂的眼组织中有一种光异构化酶，当其吸收光量子后被活化，然后可催化 11- 反式 - 视黄醛变构形成 11- 顺式 - 视黄醛。在眼组织的黑色部分，11- 顺式 - 视黄醛酶促地与视蛋白结合产生视紫红质。

12.5.6 Color vision

Color vision is refered to the ability to discriminate between two wavelengths of light. Many, and perhaps most, insects show color sensitivity, which means that they have two or more receptors in the eye that show absorption of various wavelengths of light. Sensitivity to, for example, blue and green wavelengths does not mean that the brain receiving this information orders behavior based on the information received. The receptor potentials from single visual cells of honeybee eyes exposed to monochromatic light between 318-650 nm to obtain spectral sensitivity curves suggesting four receptor pigments in honeybee eyes, each showing a different absorption maximum. The relative sensitivity of honey bee eyes to different wavelengths is UV > blue > violet > green > yellow. The presence of four pigments in the eyes of *M. sexta* adults, based on difference spectra obtained from bleaching experiments. The four maxima for light absorption are at approximately 350 nm (UV), 450 nm (blue-green), 490 nm (green), and 530 nm (yellow). After the eyes were bleached (exposed to strong light) at 21 ℃, pH 7.4 to 8.5, each of the last three pigments absorbed light at 370 nm, indicative of a mixture of bound and free 11-*cis*-retinal (i.e., the rhodopsin had been split into the protein portion and the aldehyde). A small amount of a pigment absorbing maximally at 325-330 nm, and possibly indicative of retinol (vitamin A), was also detected. The UV-absorbing visual pigment (wavelength maximum at 350 nm) bleached to an unknown product with an absorption maximum at 290-300 nm. Many flowers have UV reflecting patterns that may be important signals to insects, many of which are sensitive to UV wavelengths. Some butterflies may be among the few insects that have red-sensitive rhodopsins in their eyes.

12.5.6 彩色视觉

彩色视觉是昆虫区别两种不同波长光的能力。大量，也许多数昆虫表现出色彩敏感性，意味着它们眼里有两个或多个受体对不同波长的光有不同的吸收特性。例如，对蓝色和绿色波长的敏感性并不意味着大脑接收到这些信息后会根据所接收到的信息来命令行为。蜜蜂眼睛的单个视觉细胞暴露于 318 ～ 650 nm 的单色光受体电位，获得光谱敏感性曲线，表明蜜蜂眼睛中有四种受体色素，每个都显示不同的最大吸收峰。蜜蜂眼对不同波长光的相对敏感性为：UV> 蓝光 > 紫光 > 绿光 > 黄光。基于脱色实验得到的差异光谱表明，烟草天蛾成虫眼中出现 4 种色素。4 个最大光吸收峰大约在 350 nm（UV）、450 nm（蓝绿光）、490 nm（绿光）和 530 nm（黄光）。在 21 ℃、pH 为 7.4 ～ 8.5 的条件下对昆虫的眼睛进行脱色（暴露在强光下）后，后三种色素中每个都能吸收 370 nm 光，表现出 11- 顺式 - 视黄醛结合和游离的混合（即视紫红质已裂解成蛋白质部分和醛）。还检测到少量在 325 ～ 330 nm 处有最大吸收峰的色素，可能表明该色素是视黄醇（维生素 A）。吸收 UV 的视色素（最大吸收波长 350 nm）脱色成一种最大吸收峰在 290 ～ 300 nm 的未知产物。很多花都有 UV 反射方式，对昆虫而言可能是重要信号，许多昆虫对 UV 波长都敏感。一些蝴蝶可能是少数其眼中有红光敏感视紫红质的昆虫。

12.5.7 Detecting of plane-polarized light by insects

Many invertebrates are able to detect plane-polarized light in

12.5.7 平面偏振光的检测

很多无脊椎动物都能检测到空中的

the sky. Von Frisch demonstrated that honeybees used plane-polarized light for flight navigation. Houseflies, *P. pennsylvanicus* fireflies, Japanese beetles, and several species of ants orient to plane-polarized light under conditions that prevent them from using background reflections as orientation cues. An ant from the North African desert (*C. bicolor*) utilized plane-polarized light to travel a direct path to its nest in the ground after having wandered, with many turns, up to 100 meters from its nest in search of food. The cells sensitive to the plane of polarization are the UV-sensitive cells. Holding a UV-absorbing shield over ants in the field, thereby causing them to wander aimlessly, the ants unable to locate their underground nest. *F. polyctena*, the red wood ant of northern Europe, uses polarized light as a compass. The precise way in which these insects determine the plane of polarization and how they measure time lapse is not known. Tentative explanations involve the twisting of retinular cells (Fig.12-6) and the orientation of rhodopsin molecules in the rhabdomere microvilli with respect to the plane of polarization.

平面偏振光。Von Frisch 曾证实蜜蜂利用平面偏振光进行飞行导航。家蝇、萤火虫、日本金龟甲和几种蚂蚁，在没有背景物作参照时，都能利用平面偏振光定向。北非沙漠蚁在离巢 100 m 外的地方寻找食物，很多次转向迷途时，能利用偏振光找到路径回到巢中。对平面偏振光敏感的细胞是 UV- 敏感。通过 UV- 吸收罩将田里的蚂蚁盖住，使之无目的地爬行，它们就无法确定地面上巢穴的位置。北欧红木蚁也能利用偏振光作罗盘。不知道的是，昆虫如何精确确定平面偏振光以及它们如何测定时间的流逝。初步的解释包括视网膜细胞的扭转（图 12-6）以及感杆束微绒毛上视紫红质分子随平面偏振光的定向。

12.5.8 Hypotheses on the phototactic behavior of insects

Phototropism is one of the important behavioral characteristics of insects. The sensitivities and behavioral responses of insects to light vary among species due to their varied life histories and biology. The behavioral response of an insect to light is related to its adaptation to light, circadian periodicity, photoperiodism, and light toxicity. Why many insects exhibit particular phototactic behavior remains unclear. Over the decades, researchers have

12.5.8 昆虫趋光行为的假说

向光性是昆虫重要的行为特征之一。由于昆虫的生活史和生物学的不同，它们对光的敏感度和行为反应在不同的物种中不同。昆虫对光的行为反应与对光的适应性、昼夜节律周期、光周期和光毒性有关。许多昆虫为何表现出特殊的趋光性行为仍然不清楚。几十年来，

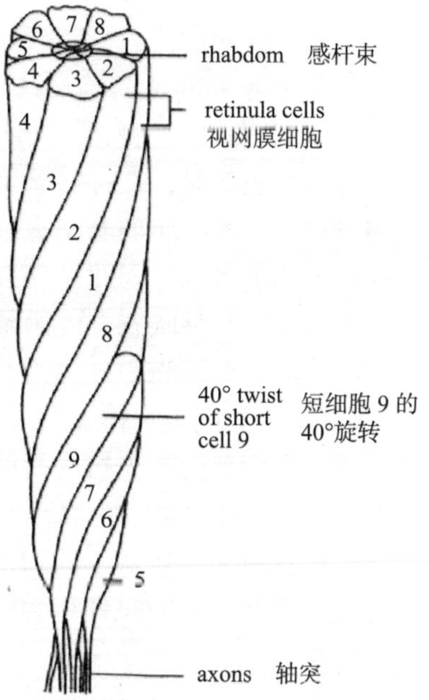

Fig.12-6 Ommatidial structure in the eye of the honeybee. About 5,500 ommatidia occur in each compound eye. Each ommatidia composed of eight elongated retinular cells, and the ninth is short and confined to the base of the eye. Half of the ommatidia are twisted clockwise and half are twisted counter clockwise; the twisting of the cells may be involved in reception of polarized light. The rhabdom is a closed or fused rhabdom in which the rhabdomeres from retinular cells touch each other. (modified from Wehner, 1976).

图 12-6 蜜蜂小眼的结构。蜜蜂的每个复眼约有 5 500 个小眼。每个小眼 8 个瘦长的视网膜细胞，第 9 个小眼短并限制了眼的基部。有半数感杆细胞为顺时针扭转，另一半逆时针扭转；细胞的扭曲可能牵涉到接受偏振光。感杆束为闭合型或融合型感杆束，其内视网膜细胞相互接触形成感杆（仿 Wehner,1976）。

attempted to explain insect phototactic behavior through behavioral, biochemical, physiological and molecular experiments. Based on the experimental findings, some hypotheses have been proposed. The light orientation hypothesis is based on the idea that insects understand starlight and moonlight as coordinates for flight or movement. Insect eyes always associate with a light source at a certain angle to maintain straight flight at night, and insects can confuse artificial light with starlight or moonlight. When they are too close to the light source, insects are unable to maintain a straight direction in relation to the light, and therefore rush toward or fly in spirals around the light.

The biological antenna hypothesis is based on the idea that insect antennae, in a manner similar to radio antennae, can sense the vibration of pheromone molecules released by the opposite sex. The oscillating waves of pheromone molecules belong to the far-IR spectrum, and insects feel these waves using their antenna when seeking a mate; therefore, some insects may be attracted to a light source that emits far-IR wavelengths.

The light interference hypothesis is concerned with the status of insects adapted to darkness at night. When insects approach the bright area around a light source, they rush towards the light because they are unable to promptly adapt to the bright environment. Some studies have suggested that this phototactic behavior occurs because the compound eyes of an insect adapted to low light levels are stimulated by bright light, thereby causing interference with normal behavior.

The light stress hypothesis addresses light stress occurring in insects exposed to light. Because nocturnal insects have become accustomed to a dark (or low-level light) environment over long periods, their ability to counteract stress caused by bright light may be diminished. Negative behavior towards light of a specific wavelength may indicate that the insects are trying to reduce the light stress. When insects are exposed to UV wavelengths, the activities of several antioxidant enzymes significantly increase, the acetylcholinesterase level decreases and the acetylcholine titer increases, so insects experience an excited state that can confuse the direction of flight (including rushing toward the light) or even result in death (insect corpses around the streetlight). Overall, the positive and negative phototropism of insects is a passive behavioral response of insects after physiological stress under light stress.

Although many studies have been conducted from various perspectives, a complete explanation for insect phototactic behavior still requires more research.

研究人员试图通过行为学、生物化学、生理学和分子学实验来解释昆虫的趋光行为。根据实验结果，人们提出了一些假说。光定向假说基于昆虫把星光和月光作为参照的概念。昆虫的眼睛与光源保持一定的角度，以便在夜间保持直线飞行，但昆虫会将人造光与星光或月光混淆。当它们离光源太近时，昆虫无法保持与光线有关的直线方向，因此会冲向或绕着光线螺旋式飞行。

生物天线假说是基于这个想法，即昆虫触角以一种类似于无线电天线的方式，感知异性释放的信息素分子的振动。信息素分子的振荡波属于远红外光谱，昆虫在寻找配偶时则用它们的触角感受频率范围内的波。因此，一些昆虫可能会被发出远红外波长的光源所吸引。

光干扰假说关注的是昆虫在夜间适应黑暗的状况。当昆虫接近光源周围的明亮区域时，它们会冲向光线，因为它们不能及时适应明亮的环境。一些研究认为，这种趋光行为的发生是因为适应了低光的昆虫的复眼受到强光的刺激，从而对正常行为造成干扰。

光胁迫假说涉及暴露在光下的昆虫发生的光胁迫。由于夜行性昆虫已经长期习惯于黑暗（或低水平的光）环境，它们对抗强光的能力可能被削弱。昆虫对特定波长光的负向趋光性可能表明，昆虫正试图减少光胁迫。当昆虫暴露在紫外线波长下时，几种抗氧化酶的活性明显增加，乙酰胆碱酯酶水平下降，乙酰胆碱滴度增加，因此昆虫会经历一种兴奋状态，这种状态会混淆昆虫的飞行方向（包括冲向光线），甚至会导致死亡（路灯周围的昆虫尸体）。总体来说，昆虫的正向和负向趋光行为是昆虫在光胁迫下产生生理应激后的被动行为反应。

尽管前人已经从不同的角度进行了研究，但对昆虫趋光性行为的完整解释仍需探索。

12.6　生物发光机制和光波通信

12.6　Bioluminescence Mechanisms and Light Wavelength Communication

Among terrestrial animals, the insects are the main group with luminescent species. About thirty distinct luciferin/luciferase systems producing distinct colors have been selected in species evolution. The color of insects themselves and emitted light are closely related to escape from pedastors, courtship and regulation of body temperatures. It plays a important role in communication between intraspecies and interspecies.

在陆生动物中，昆虫是拥有发冷光物种的主要群体。该物种进化过程中，大约有 30 种截然不同的荧光素 / 荧光素酶系统产生不同的色彩。昆虫自身的色彩和发出的光亮与避敌、求偶和体温调节密切相关。它对种内和种间的信息交流也起着重要作用。

12.6.1 Bioluminescence mechanisms

12.6.1　生物发光机制

12.6.1.1 Lightness emission insects and tissue structure

12.6.1.1　发光昆虫及其组织结构

The bioluminescent reactions mainly occurred in Coleoptera, including Drilidae, Homalisidae, Phengodidae and Cantharidea in Cantharoidea, and Elateridae in Buprestoidea.

All of eggs, larvae, pupae and adult in *Lampyridae* can emit light, but it is more important in adult. The light emission organs in *L.cruciata* is located under the transparent cuticle between the fifth and the sixth abdomen in male and the fifth abdomen in female. It is composed of a group of photogenic cells which derived from the mesodermal embryo. The reflecting cell contains urates in the inner laternal side. A large amount of organs and nerve axon innervated the light emission cell layers and provide oxygen. Numerous mitochondria and fluorescence particles distribute in the cells. Luciferin and luciferase fullfilled within the particle.

生物发光反应主要出现在鞘翅目昆虫，花萤总科的稚萤科、平萤科、光萤科和花萤科以及吉丁虫总科的叩甲科。

萤甲科的昆虫，卵、幼虫、蛹和成虫都能发光，但成虫期更重要。萤甲成虫的发光器官位于雄虫腹部第 5 ～ 6 节之间或雌虫的第 5 节透明的表皮下方。它由一群由中胚层发育而来的发光细胞组成。反光细胞内部侧面含尿酸盐。大量器官和神经的轴突支配发光细胞层并提供氧。细胞内分布大量线粒体与荧光颗粒。

12.6.1.2 The structures of luciferases

12.6.1.2　荧光素酶的结构

Luciferase from the firefly *Photinus pyralis* which folds into two distinct domains, a large N-terminal domain comprising residues 4-436, and a C-terminal domain formed from residues 440-544 (Fig. 12-7; Appendix color graphic Fig.12-7). The fold assumed by the luciferase polypeptide appears to be unique. The N-terminal domain consists of a β barrel and two β sheets flanked by α helices which form a five-layered αβαβα structure. The C-terminal domain, consisting of five β strands and three α helices, is folded into a compact structure that is connected to the N- terminal domain by a

北美萤火虫的荧光酶折叠成两个明显的区域，氨基端较大，含有第 4 ～ 436 个氨基酸残基；羧基端由第 440 ～ 544 个氨基酸形成（图 12-7；附彩图 12-7）。由荧光素酶多肽形成的折叠似乎是独特的。N- 端结构域由 β 桶状蛋白质和两个 β- 折叠片层组成，侧翼有 α 螺旋形成五层 αβαβα 结构。C- 端结构域由 5 个 β 折叠和 3 个 α 螺旋折叠成致密结构，通过无序环肽链（连接第 435 ～

disordered loop (connecting residues the 435th and 441st). There are three other disordered loops not visible in the electron density, one in the C-terminal domain connecting residues the 523rd and 529th, and two in the N-terminal domain connecting residues 198th-204th and residues 355-359 (Baldwin,1996).

441 个氨基酸残基）与氨基端相连。还有三个在电子密度中看不到的无序环，其中之一在羧基端连接第 523 ～ 529 个氨基酸残基，两个在氨基端连接第 198 ～ 204 个氨基酸残基和第 355 ～ 359 个氨基酸残基（Baldwin, 1996）。

12.6.1.3 Biochemical reactions of luminescence

The generally accepted mechanism of firefly bioluminescence is shown in Fig.12-8(Appendix color graphic Fig.12-8). This luciferase catalyzed reaction proceeds *via* the initial formation of an enzyme-bound luciferyl adenylate intermediate luciferase. They replaced luciferyl-AMP by the stable analogue 5′-*O*-[*N*-(dehydroluciferyl) sulfamoyl] adenosine (DLSA) and captured three key 'snapshots' of the reaction: ① luciferase bound to the reactants (ATP-Mg^{2+}); ② luciferase bound to the DLSA; ③ luciferase bound to the products oxyluciferin-AMP (OxyLH2-AMP); ④ the excited oxyluciferin decays to the ground state, emitting yellow-green light with very high efficiency. The wavelength of the emitted light depends on the polarity of the microenvironment at the phenol/phenolate terminal of the benzothiazole oxyluciferin (Navizet et al., 2010).

12.6.1.3 发光的生化反应

萤火虫生物发光的普遍接受的如图 12-8 所示（附彩图 12-8）。这个荧光素酶催化的反应通过酶结合荧光素酶中间体——荧光素腺苷酸的形成启动反应进程。稳定的类似物 5 ′ -*O*- [*N*-（脱氢荧光素）氨磺酰] 腺苷（DLSA）取代荧光素 - 腺苷酸，催化 3 个关键性的功能反应：①荧光素酶结合到反应体（ATP-Mg^{2+}）；②荧光素酶结合到 DLSA；③荧光素酶结合到产物氧化荧光素 - 腺苷磷酸（OxyLH2-AMP）。④激发的氧化荧光素衰减到基态，以非常高的效率发出黄绿色的光。发射光的波长取决于氧化荧光素内的苯并噻唑片段的酚/酚盐末端微环境的极性（Navizet等，2010）。

Fig.12-7 Ribbon diagram of the firefly luciferase structure.
图 12-7 荧光素酶的带状结构。

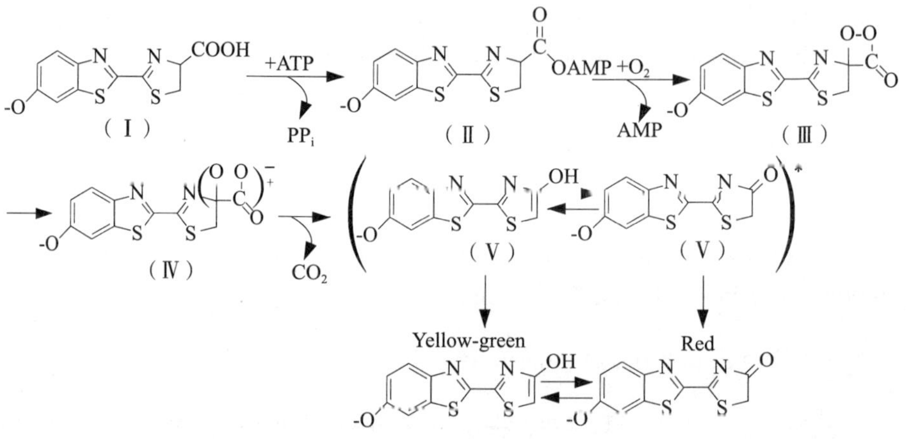

Fig.12-8 Proposed mechanism of the firefly bioluminescence reaction. The carboxylate group of firefly luciferin (Ⅰ) is activated by reaction with ATP to form the adenylated luciferin (Ⅱ). A proton is lost, allowing reaction with molecular oxygen to yield (Ⅲ). Cleavage of the dioxetanone ring (Ⅳ) yields the excited state of oxyluciferin (Ⅴ). (modified from Baldwin, 1996); oxyluciferin (Navizet et al., 2010).
12-8 萤火虫生物发光反应的建议机制。萤火虫荧光素（Ⅰ）羧基基团与 ATP 反应活化，产生荧光素腺苷酸（Ⅱ）；失去一个质子，与氧分子反应产生（Ⅲ）；裂解双氧环杂酮（Ⅳ）产生氧化荧光素的激发态（Ⅴ）（仿 Baldwin, 1996）；氧化荧光素（Navizet 等 , 2010）。

12.6.2 Biological function of bioluminscience

12.6.2.1 Reproduction

The best known examples of bioluminescence having a role in the propagation of the species is found among the beetles of the family Lampyridae (true fireflies and glow worms), although bioluminescence also occurs in several other beetle families. In glow worms only the female glows brightly with a steady light and by this attracts the male, and the same is the case in some firefly species. In other fireflies a sophisticated "light conversation" between males and females has evolved, with a different "language" for each species. Males send out an "interrogation" flash, and females respond. Species specificities are obtained both by the time course of the flash on a half-second time scale and on the time delay between "interrogation" and "answer."

12.6.2.2 Food acquisition

Female fireflies of the genus *Photuris* reply to the "interrogating flashes"from males of *Photinus* fireflies, lure them to approach, and eat them to obtain defensive compounds called lucibufagins. Without lucibufagans from a *Photinus* firefly, the *Photuris* firefly is attacked and eaten by a *Phidippus* jumping spider, but a *Photuris* firefly with lucibufagans can repel the spider. Most sly and cunning of them all are the females of *Photuris versicolor*, who know firefly languages sufficiently well to be able to prey on 11 different species. About a dozen of more than 3,000 species in the family have larvae which use bioluminescence in different ways to catch prey. Particularly famous are the larvae in the Te Anaau caves on the South Island of New Zealand. The gnat larvae sit on the roof and deploy luminescent and sticky "fish lines" to attract other insects, catch and devoured.

12.6.2.3 Sexual communication

Sexual communication between male and female fireflies involves the visual detection of species-specific bioluminescent signals. Firefly species vary spectrally in both their emitted light and in the sensitivity of the eye. Twilight-active species had similar pink screening pigments and produced yellow flash. But the visual pigment of *Photinus pyralis* peaked near 545 nm, while that of *P. scintillans* had a λ_{max} near 557 nm. The night-active *Photuris versicolor* had a yellow screening pigment that was uniquely

12.6.2 生物发光的生物学作用

12.6.2.1 生殖

生物发光在种系繁殖中起作用的最著名的例子是萤科甲虫（真萤火虫和发光幼虫），但是有其他几种甲虫也能生物发光。在萤火虫中，只有雌性会发出稳定的荧光，并以此吸引雄性，在一些萤火虫物种中也是如此。在其他萤火虫中，雌雄间进化出了复杂的“发光会谈”，每个种都有不同的“语言”。雄虫发出“询问”闪光，雌虫会做出回应。半秒时间内闪光的次数和“讯问”与“回答”之间的延缓时间具有种特异性。

12.6.2.2 捕获食物

Photinus 属的雌萤火虫响应阜提萤属雄萤火虫的“讯问闪光“，引诱并捕食它们以获得被称为荧光毒素的防御化合物。没有从 *Photinus* 萤火虫获得荧光毒素的 *Photuris* 萤火虫会被跳蛛 *Phidippus* 攻击并吃掉，但有荧光毒素的 *Photuris* 萤火虫能逃避蜘蛛的捕食。多数狡诈而奸猾的 *P. versicolor* 雌虫完全能够识别萤火虫语言，能猎食 11 种不同的萤火虫。在这个家族的 3 000 多个物种中，大约有十几个物种的幼虫以不同方式利用生物发光来捕捉猎物。最著名的是新西兰南岛上的蒂阿瑙洞穴内的幼虫。这种叮人幼虫稳坐穴顶，有效利用荧光和黏性“钓鱼线”，吸引其他昆虫进行捕获并吞食。

12.6.2.3 两性通信

雌雄萤火虫之间的两性通信牵涉到种特异性生物荧光信号的视觉侦测。光谱学上，萤火虫发出的光和眼睛感受的光有种间多样性。黄昏活跃的品种有相似的粉色滤光色素，发黄色荧光。但是 *P. pyralis* 的视觉色素峰值约为 545 nm，而 *P. scintillans* 最大视觉色素峰值约 557 nm。夜间活动的 *P. versicolor* 独具黄滤光色素而发绿光，其视觉色素与

localized, while it retained green emission and its visual pigment was similar to that of *P.pyralis* (Cronin et al., 2000). Firefly *L. substriata* emits single species-specific flash signal for courtship. The flash duration, intervals, flash intensity peak was different in the first response and the second response. When flying male land 5-10 cm near the female and continued to flash, but the signal frequency of male flash become lower with longer duration and intervals (Fu Xinhua et al., 2005).

P.pyralis 的相同（Cronin 等，2000）。萤火虫条背萤通过发出单脉冲周期性特异闪光信号进行求偶。闪光持续时间、间隔和闪光强度在第一次响应和第二次响应间不同。当雄虫降落到雌虫附近 5～10 cm 时继续发光，但发光频率下降，持续时间和间隔较长（Fu Xinhua 等，2005）。

12.6.2.4 Protection from predation

In most beetles of the family Lampyridae, almost all development stage of eggs, larvae, pupae and adults emitted light. By aposematic coloration, and easily recognized bright color patterns like the black-yellow of wasps, spots on ladybugs, and stripes on coral snakes, which warn a predator of nasty consequences of an attack, bioluminescence of firefly larvae serves a similar purpose.

12.6.2.4 防止被捕食

多数萤甲科昆虫各发育时期的卵、幼虫、蛹和成虫都可发光。通过警戒色和易于识别的明亮模式，如黄蜂的黑黄色、瓢虫身上的斑点和珊瑚虫身上的条纹警告捕食者攻击的危险后果，荧火虫幼虫的生物发光也有类似的目的。

12.7 信息感受与昼夜节律

12.7 Information Sensory and Circadian Rhythms

The essence of circadian rhythms involves the restriction of an activity to a particular time of day to coordinate biological processes with exogenous environmental cycles. This requires the regulation of genes involved in the output pathways by the circadian clock (Dunlap, 1999).

昼夜节律的本质是每天特定时间的限定的活动与外界环境周期变化相一致的生理过程。这需要昼夜节律钟对参与输出途径的基因进行调节（Dunlap，1999）。

12.7.1 Biological clock genes and the proteins

Circadian clocks control daily rhythms at the molecular, physiological, and behavioral levels. Like most organisms, the central circadian oscillator controlling the locomotor activity rhythm in *Drosophila* is proposed to consist of molecular feedback loops. The feedback loops comprise several core clock genes, such as *period (per), timeless* (*tim*), *Clock* (*Clk*), *cycle* (*cyc*), *double-time* (*Dbt*), *vrille* (*Vri*) and *cryptochrome* (*cry*). These genes encode PER, TIM, CLK, CYC, DBT, VRI and CRY proteins, respectively, and loss-of-function or dominant-negative mutations in these genes result in loss of normal circadian rhythmicity. CLK and CYC form a heterodimer to

12.7.1 生物钟基因及蛋白质

昼夜节律钟在分子水平、生理学和行为学上控制日常节律。像多数生物一样，果蝇的中枢昼夜节律振荡器控制着果蝇运动神经的活动节律，被认为由分子反馈回路组成。反馈回路包含几个核心生物钟基因：*period (per)*, *timeless (tim)*, *Clock (Clk)*, *cycle (cyc)*, *double-time (Dbt)*, *vrille (Vri)* 和 *cryptochrome* (*cry*)。这些基因分别编码 PER 蛋白、TIM 蛋白、CLK 蛋白、CYC 蛋白、DBT 蛋白、VRI 蛋白和 CRY 蛋白，这些基因的功能缺失或显性负突变体会导致失去正常

activate transcription of *per* and *tim*. PER and TIM form another heterodimer, are transferred into the nucleus, and then inactivate the transcriptional activity of the CLK/CYC heterodimer. The feedback loops are entrained and reset by the light induced degradation of TIM mediated by a photoreceptor pigment, CRY. In addition to the central oscillator, peripheral oscillators exist in many tissues, including the compound eyes, antennae, wings, legs, and Malpighian tubules in adults and the ring gland in pupae. These oscillators are self-sustained and directly light-entrainable even *in vitro*. In addition, the oscillators are clearly independent of the central circadian oscillator(s) in the brain. The peripheral oscillator shares the same genes with the central clock in the brain. However, there is a remarkable difference that, in the peripheral oscillator, CRY functions as a core component besides a photoreceptor. In the compound eyes, CRY is suggested to act as a repressor of the CLK/CYC transcriptional activity.

的昼夜节律。CLK 和 CYC 形成一个异源二聚体，激活 *per* 和 *tim* 的转录。PER 和 TIM 形成另一个异源二聚体，被转移到细胞核中，然后使 CLK/CYC 异源二聚体的转录失活。光感受色素，隐花色素蛋白介导光感应 TIM 的降解，使反馈回路节律点受控并被重置。除了中枢振荡器外，许多组织有外周振荡器，包括成虫的复眼、触角、翅膀、腿和成虫中的马氏管和蛹的环腺。即使在离体条件下，这些振荡器也能自我维护，节律点被光线直接控制。此外，这些振荡器明显独立于脑的中枢昼夜节律振荡器。外周振荡器与脑的中枢生物钟拥有共同基因，但外周振荡器有个明显的区别，CRY 除了作为光受体，还是功能上的核心成分。在复眼中，CRY 被认为是 CLK/CYC 蛋白转录活动的抑制因子。

12.7.2 Molecular mechanism of chronobiology and feedback

The *per* gene likely encodes a component of the *Drosophila* circadian clock. Circadian oscillations in the abundance of *per* messenger RNA and PER are thought to arise from negative feedback control of *per* gene transcription by PER. A second clock locus, *tim*, apparently regulates entry of PER into the nucleus. A restricted segment of TIM binds directly to a part of the PER dimerization domain PAS. The interaction between TIM and PER determines the timing of PER nuclear entry and therefore the duration of part of the circadian cycle (Weitz et al., 1995). In short, after transcription of *per* and *tim*, their proteins PER and TIM accumulate in the cytoplasm during darkness and then form a PER/TIM heterodimer that enters the nucleus to act as a negative regulator of the CYC/CLK dimer, while the negative regulator itselfis responsible for *per* and *tim* transcription.

The robustness of this cycling may be enhanced by another interlocked feedback loop, in which *per* positively regulates *clk* transcription, as well as by a number of post-transcriptional and post-translational mechanisms. For example, the stability of the PER protein is negatively regulated by phosphorylation by DBT, a protein kinase related to the mammalian casein kinase. Furthermore, the stability of the TIM protein may be independently

12.7.2　生物钟的分子机制及反馈

per 基因编码果蝇昼夜节律钟的组成。*per* 基因的 mRNA 和 PER 的丰度的昼夜震荡被认为是由 PER 对 *per* 基因转录的负反馈调控引起的。第二个时钟位点，*tim*，显然可调控 PER 进入细胞核。TIM 的一个限制性酶切片段直接与 PER 二聚体结构域 PAS 的一部分结合。TIM 和 PER 之间的相互作用决定了 PER 进入细胞核的时间，从而决定了部分昼夜节律周期的持续时间（Weitz 等，1995）。简而言之，在 *per* 和 *tim* 转录后，它们的蛋白质 PER 和 TIM 在黑暗中积聚在细胞质中，然后形成 PER/TIM 异源二聚体，进入细胞核作为 CYC/CLK 二聚体的负调节器，而 CYC/CLK 二聚体本身负责 *per* 和 *tim* 的转录。

这个循环的稳定性可被另一个连锁的反馈回路增强。*per* 基因与大量转录和翻译后机制一样可正向调节 *clk* 基因转录。例如，PER 的稳定性通过 *dbt* 磷酸化负调节，DBT 是一种与哺乳动物酪蛋白激酶相关的蛋白激酶。此外，TIM regulated by *cryptochrome*, a blue light photoreceptor that appears to act as the major photoreceptor of the *Drosophila* circadian system (Ceriani et al., 1999).

In the monarch butterfly *D.plexippus* (Zhu et al., 2005), some other Lepidoptera, the honey bee *A.mellifera*, the beetle *T.castaneum* and in mosquitoes, a second form of *cryptochrome* is also expressed which, unlike the type *CRY* 1 in *Drosophila*, which encodes the blue light photoreceptor, it seems to act as a transcriptional repressor of *CLK*/*CYC* in the feedback loop (Yuan et al., 2007). In *B. mori*, *CRY* gene are clearly dividec into two types:*CRY1* and *CRY2*. Bm*CRY1* have 12 exons and 11 introns, while Bm*CRY2* have 9 exons and 8 introns with whole gene length of 2,166 bp and 2,389 bp, respectively (Wang et al., 2011).

12.7.3 Photoperiodic mechanism

The insect photoperiodic response comprises a sequence of events from: photoreception, through measurement of night length (perhaps day length in some species) and the simultaneous accumulation of the covert effects of successive photoperiods by a 'counter' mechanism to the endocrine effectors that determine the subsequent diapause or nondiapause developmental pathways (Saunders, 2010).

Although low temperature and restricted diet may act as factors subordinate to photoperiod they may also induce diapause in their own right, often in areas such as the tropics where changes in day length or more significantly, night length are slight (Denlinger, 1974). Even where seasonal change in photoperiod is the predominate factor, there is considerable diversity in response. For example, some univoltine or long-lived insects such as beetles and dragonflies may regulate alternating phases of development and diapause using a circannual rhythm with an endogenous period of about 40-46 weeks synchronised to an exact 52 weeks by annual changes in photoperiod. Other species-particularly multivoltine insects producing successive nondiapause generations during the summer months-enter diapause in the autumn when night length exceeds a well-defined critical value. In some, there is now an abundance of evidence to suggest that night length measurement is a function of the circadian system. The most complex sequence occurs in some strains of the silk moth *B. mori* where diapause in the egg occurs as a result of the short night illumination of the eggs and young larvae of the female parent, almost a generation

的稳定性可被蓝光光感受器隐花色素单独调节，它在果蝇昼夜节律系统中是主要的光感受器（Ceriani 等，1999）。

在帝王蝶（Zhu 等，2005），还有其他一些鳞翅目，如意大利蜜蜂、甲虫和蚊虫体内，有第二种隐花色素表达，与果蝇编码蓝光感光细胞隐花色素 *CRY 1* 不同，似乎在反馈回路中起着抑制 *CLK*/*CYC* 转录的作用（Yuan 等，2007）。家蚕的 *CRY* 基因明显分成两类：*CRY1* 和 *CRY2*。Bm*CRY1* 有 12 个外显子，11 个内含子，而 Bm*CRY2* 有 9 个外显子，8 个内含子，全基因序列分别为 2 166 bp 和 2 389 bp（Wang 等，2011）。

12.7.3 光期机制

昆虫光期应答由一系列事件组成：感光作用，通过测定夜晚时长（有些种类可能是白天时长）和通过“计数器”机制，持续光期隐秘效果的同时累积到决定随后发育途径的滞育或非滞育内分泌效应器（Saunders, 2010）。

虽然低温和限制饮食可起着光期从属因子的作用，但它们本身可能诱发绝育，通常是在诸如热带地区，昼长或夜长的变化很轻微（Denlinger, 1974）。即使在光周期的季节性变化是主要因素的区域，其反应也有相当大的多样性。比如，有些一化性或长寿的昆虫，如甲虫和蜻蜓利用节律性节奏可交替调节发育和滞育期，其内生期为 40 ～ 46 周，与光期周年变化精确的 52 周协同运作。另外一些种类，特别是多化性昆虫在夏季数月相继产生非滞育后代，在秋季夜长超过规定的临界值时进入滞育。已有大量证据证明有些昆虫的昼夜节律系统具有夜长测定的功能。最为复杂的时序出现在家蚕的一些品系，由于母本卵和幼虫的短夜照，几乎在一代之前就会出现卵的滞育。在这些情况下，事件的长

before. In these cases, the long sequence of events also includes mechanisms for the transfer of photoperiodic "information" from one instar to another, or from one generation to the next.

The night-length timer and the photoperiodic counter operate hand-in-hand through the sensitive period, and that the combined "clock-counter" mechanism presents many properties consistent with circadian rhythmicity (Saunders, 2010). Truman (1984) proposed that the timing of eclosion in silkmoths is determined by the interactions between two circadian clocks: one located in the brain that produces a daily gate for release of eclosion hormone, and another outside the brain that produces a daily modulation of ecdysteroid levels. Circadian rhythms in ecdysteroid levels have been measured in the hemolymph of the blood sucking insect *R.prolixus* and the wax moth *G. mellonella*.

时序也包括将光期“信息”从一个龄期传到另一个龄期，或从一代传到另一代的机制。

夜长计时器和光周期计数器通过敏感期环环相扣，组合成“时钟计数器”机制，显示出昼夜节律节奏性的许多连续特性（Saunders, 2010）。Truman（1984）提出，家蚕羽化的时机由两个昼夜节律钟互作决定：一个位于脑，每天产生释放羽化激素的门控，而另一个在脑外，进行蜕皮甾类水平的日间调节。已在吸血昆虫长红猎蝽和蜡螟中测定过蜕皮甾类水平昼夜节律的节奏性。

12.7.4 Insect circadian rhythms related physiology and biochemistry processes

Recent studies revealed that the neuronal network controlling overt rhythms shows substantial similarity in various insect orders. However, there are variations in the molecular oscillatory mechanism, as well as input and output pathways among insects. Although similar clock genes have been found in most insects, the behavior of their protein products show considerable differences from that of *D. melanogaster*. In non-dipteran insects, mammalian-type cryptochrome (CRY2) seems to work as a transcriptional repressor in the feedback loop. Another difference is that PER and TIM, which translocate to cell nuclei in the fruit fly are predominantly cytoplasmic in neurons of other insects. However, PER was detected in cell nuclei in peripheral organs, such as larval midgut of *A. pernyi*, epithelial cells in the antenna of *M. sexta* and vas deferens epithelium in the *C. pomonella* and *S. littoralis*. In addition, oscillators necessary for adult eclosion, olfactory sensitivity and feeding rhythms have been identified in adult fruit flies. Reduced *per* mRNA levels affected the egg hatching rhythm in the silk moth *B. mori* and inhibited the locomotor activity rhythm and rhythmic electrical activity in the optic lobes in crickets *G. bimaculatus*. Furthermore, *per*RNAi inhibited sperm release rhythm in the moth *S. littoralis*. The circadian system located in the male reproductive tract of Lepidoptera is one of the best characterized peripheral oscillators in insects. In cotton leaf worm, *S. littoralis*, demonstrated that this oscillator controls the rhythm of

12.7.4 昆虫昼夜节律相关的生理生化过程

最近的研究显示，神经网络控制的明显节律在不同昆虫间表现出实质的相似性。但是在分子振荡机制方面与昆虫间输入和输出途径一样存在差异。虽然，发现多数昆虫有相似的生物钟基因，但其蛋白质产物状态与果蝇的蛋白质产物状态表现出明显差异。非双翅目昆虫，哺乳动物类隐花色素（CRY2）在反馈循环中似乎起着转录抑制的作用。另一个差别是 PER 和 TIM，在果蝇中转位到细胞核内，其他昆虫主要位于神经元细胞质内。然而，PER 在外周器官细胞核中可检测到，如柞蚕幼虫中肠、烟草天蛾的触角上皮细胞、苹果蠹蛾和斜纹夜蛾输精管上皮。此外，在果蝇成虫中还发现了成虫蜕变、嗅觉敏感和摄食节律所必需的振荡器。*per* mRNA 水平的降低影响了家蚕蛾卵的孵化节律，抑制了蟋蟀的运动活动节律和视叶的节律性电活动。此外，*per*RNAi 可抑制斜纹夜蛾精子的释放节奏。位于鳞翅目雄性生殖道的昼夜节律系统是昆虫中最具特征的外围振荡器之一。已证实棉花叶害虫棉贪夜蛾的这种振荡器控制睾

sperm release from the testis and coordinates sperm maturation in the upper vas deferens (Kotwica, 2011).

The Circadian clock can be entrained to light by the visual system and by nonvisual, dedicated circadian mechanisms. Upon light treatment, CRY is activated and transmits a signal that targets TIM for degradation by the proteasome. Although tyrosine kinase activity appears to be required for TIM degradation by light. Light alters the levels of a clock component, which resets the timing of all other events in the cycle. Interestingly, pulses of light delivered at night will reset the phase of the clock. The effect is different to depending upon the time of delivery: in the early night, a light pulse delays the clock (resetting to dusk) while in the late night it advances the clock (resetting to dawn). In molecular terms, a possible explanation may be provided by the levels of *tim* mRNA and the subcellular localization of PER and TIM. In the early night, the two proteins are cytoplasmic and mRNA levels are high and able to resynthesize the protein lost by degradation. Thus, the clock is delayed by the number of hours it takes to produce that amount of protein. In the late night, the PER-TIM complex is in the nucleus, repressing transcription. The protein cannot be replenished and the clock moves forward to the next cycle. So, post-translational mechanisms are critical for the entrainment of the clock to light. Free-running clock function relies upon regulated posttranslational events, even when *per* and *tim* mRNA are expressed with a robust rhythm (Zheng et al., 2008).

PDF neurons are involved in output pathways of the optic lobe pacemakers of the cricket and the cockroach and project to several brain areas in both species. The multiple functions of PDF or PDF positive neurons in crickets and cockroaches correspond with the observations in *Drosophila*, suggesting that PDF plays a central role in insect circadian timing systems. Cockroaches and crickets share great similarities in their circadian systems. Bilaterally located, mutually coupled, optic lobe pacemakers regulate rhythmicity in most behavioral and physiological processes. PDF neurons play a central role in their systems not only as pacemaker neurons, but also in the regulation of output, coupling and synchronization. Whereas coupling between the two optic lobe pacemakers is stronger in the cockroach than in the cricket, it appears to affect the free running period of locomotor rhythms in both species. For cockroaches, there is evidence that coupling

丸精子的释放节奏，并与上输精管精子的成熟相协调（Kotwica, 2011）。

生物钟可通过视觉系统和非视觉系统驯化成为光专用昼夜节律机制。在光照处理后，CRY 被激活并传递针对 TIM 的信号，使 TIM 被蛋白酶体降解。虽然酪氨酸激酶的活性显然是光降解 TIM 所必需的。光改变了生物钟组成的水平，从而重置周期计时器的其他所有环节。有趣的是，夜晚光发送的脉冲将重置生物钟的周期，其效果依赖于光发送的时间：在上半夜，光脉冲延缓生物钟（重置傍晚）而在后半夜使钟提前（重置黎明）。从分子方面解释，大概是 *tim* mRNA 的水平以及 PER 和 TIM 的亚细胞定位。在上半夜，这两种蛋白质在细胞质中的 mRNA 水平很高，能够重新合成因降解而损失的蛋白质。因此，生物钟延迟了产生该量蛋白质所需的时间。在下半夜，PER-TIM 复合体在细胞核中，抑制了转录。蛋白质无法得到补充，时钟就会进入下一个周期。因此，后转录机制对于将生物钟驯化成光敏感是至关重要的。即使在 *per* 和 *tim* mRNA 以固定的节律表达时，自由运转的钟功能也依赖于翻译后环节的调节（Zheng 等，2008）。

PDF 神经元参与蟋蟀和蜚蠊的视叶起搏点的输出途径，并投射到两个物件的几个脑区域。蟋蟀和蜚蠊的 PDF 或 PDF 阳性神经元的多种功能与在果蝇中的观察一致，表明 PDF 在昆虫昼夜节律计时系统中起着核心作用。蜚蠊和蟋蟀的昼夜节律系统有很大相似之处。双侧定位以及相互耦合的视叶起搏点调节大多数行为和生理过程的节律性。PDF 神经元在它们的系统中不仅作为起搏点神经元，而且还在输出、耦合和同步的调节中发挥着核心作用。虽然两个视叶起搏点神经元之间的耦合在蜚蠊中比在蟋蟀中更强，但它似乎会影响两个物种运动节律的自由运行期。对于蜚蠊来说，证据表明，位于两侧的起搏点之间

between the bilaterally located pacemakers contributes to photic entrainment (Vansteensel et al., 2008).

(Sang Wen　Qiu Baoli)

的耦合有助于光驯化。(Vansteensel 等，2008)。

（桑　文　邱宝利）

Exercises

1. What are the types of photoreceptors in insects? How is insect vision formed?
2. What is the structure of the compound eye in insects? What are the types of compound eyes in insects?
3. What is the biological role of bioluminescence?
4. How are circadian rhythms regulated in insects?

思考题

1. 昆虫的光感受器有哪几种类型？昆虫视觉是如何形成的？
2. 昆虫的复眼结构如何？昆虫的复眼有哪几种类型？
3. 生物发光有什么生物学作用？
4. 昆虫的昼夜节律是如何调控的？

第 13 章 | 信息化合物

Chapter 13 Pheromones

Key points

1.Pheromones are diversified in chemical nature and physiological functions.

2.In the sex pheromone detection processes, sex pheromone molecule (SPM), sex pheromone receptors (SPRs), co-receptor (Orco), pheromone binding protein (PBP), sensory neuron membrane protein (SNMP), and pheromone degradation enzyme (PDE) play individual and cooperative roles.

3.The insect pheromone components (*Z*, *E*)-10,12-hexadecadienol (bombykol) and (*Z*, *E*)- 10,12-hexadecadienal (bombykal) were characterized in the silkworm moth *B. mori*.

4.The semiochemical detection includes essential chemical, membrane potential, and physiological electric energy transformation, is under control of hormones.

知识点

1. 信息素在化学特性和生理功能上具有多样性。

2. 在性信息素检测过程中，性信息素分子（SPM）、性信息素受体（SPRs）、辅助受体（Orco）、信息素结合蛋白（PBP）、感觉神经元膜蛋白（SNMP）和信息素降解酶（PDE）发挥着个体和协同作用。

3. 昆虫信息素组分（*Z*，*E*）-10,12- 十六二烯醇（家蚕醇）和（*Z*，*E*）-10,12- 十六烷二烯醛（家蚕醛）以家蚕中的为典型。

4. 信息化合物的检测包括基本的化学、膜电位和生理电能转换，受激素控制。

Semiochemicals are signaling chemicals produced by an organism to send a message and elicit changes in the behavior or physiology of the receiving organism.

There are two main types of olfactory proteins involved in the sex pheromone perception, pheromone binding protein (PBP) and pheromone receptor (PR). PBPs belong to the odorant binding protein (OBP) family and abound in the lymph fluid of sex pheromone sensitive sensilla. PBPs are specifically or highly expressed in antennae of male moths, functioning to bind and transport the pheromone molecules onto the PR housed in the membrane of olfactory neurons.

Usually, SPM was released from female insects and then sensed

信息化学物质是指有机体为传递信息而产生的信号化学物质，引起接收的生物行为和生理上的变化。

有两种主要的嗅觉蛋白参与性信息素的感知，信息素结合蛋白（PBP）和信息素受体（PR）。PBPs 属于气味结合蛋白（OBP）家族，大量存在于性信息素敏感的感器淋巴液中。PBPs 在雄蛾的触角中特异或高度表达，其功能是将信息素分子结合并转运到位于嗅觉神经元膜中的信息素受体上。

通常，性信息素分子从雌性昆虫释放，然后被同种雄性昆虫的特定性信息素受体感应；同时，围绕性信息素受体

by specific SPR of conspecific male insects; meanwhile, the specific PBP surrounding the SPR transports the SPM to SPR and increases the magnitude of SPR response to SPM.

的特殊信息素结合蛋白将性信息素分子传输到性信息素受体，并增加性信息素受体对性信息素分子的响应程度。

13.1 信息化学物质的分类

13.1 Classification of Semichemical

13.1.1 Semichemicals

Semiochemicals have been defined as intraspecific agents influencing the physiology or behavior of members of the same species as that of the producer (pheromones), or as interspecific agents influencing a different species than that of the producer (allelochemicals). Nearly all sex pheromones, and many other types of pheromones, are blends of components, and the total blend generally is considered to be the pheromone. The blend or ratio of components allows more information to be encoded, and allows greater discrimination among closely related species, which often use the same, or some of the same components. Pheromones often serve multiple functions. The pheromone and plant-derived alkaloids are used by the female moth to evaluate the fitness of male moths, resulting in a model for sexual selection. Pheromones have been identified from more than 1,600 species of insects in more than 90 families from 9 orders.

13.1.1.1 Primer pheromone

A pheromone that has an effect on the physiology or biochemistry of an animal, such as suppression of ovary development in bees or stimulation of maturation in the desert locust, is called a primer pheromone because it requires a finite period of time to “prime the pump” in order to make it more effective. It causes relatively slow changes in the physiology, and usually behavior, of the animal.

13.1.1.2 Releaser pheromone

A sex pheromone that attracts a potential mate is called a releaser pheromone because the pheromone almost instantly releases some

13.1.1 信息化学物质

信息化学物质分为影响多数同种昆虫的生理或行为的种内特异性物质（信息素），或影响除产生者外不同种昆虫的种间特异性物质（异种化感物）。几乎所有的性信息素以及其他类型的信息素的成分都是混合物。总的混合物一般认为是信息素。混合物或成分比率允许编码更多信息，并允许在密切相关的种群间进行最大的区分，这些种群常用相同的或某些相同的混合物。信息素通常具有多种功能。雌蛾利用信息素和植物源生物碱来评价雄蛾是否合适，从而建立一种性选择模型。已鉴定出 9 个目、90 多个科、1 600 多个昆虫种的信息素。

13.1.1.1 原发性信息素

原发性信息素是指某种信息素对某种动物具有生理和生物化学效应，比如，对蜜蜂的卵巢发育具有抑制作用，对沙漠蝗的卵巢成熟具有刺激作用，即称之为原发性信息素。因为它需特定的时间段来“启动泵”，使其作用更为有效。它引起动物生理发生缓慢的变化，通常是行为的变化。

13.1.1.2 释放信息素

性信息素吸引潜在的伴侣称之为释放信息素，因为信息素几乎瞬间令行为

behavior, such as upwind searching behavior or attempted mating.

放松，如逆风搜寻或尝试交配。

13.1.1.3 Allelochemicals

Allomones, kairomones, and synomones influence the behavior or physiology of members of a different species than that of the producer. They are interspecific, semiochemicals, and sometimes are also called allelochemicals because they act between species.

13.1.1.3 异种化感物

利己素、利它素和互利素对其他种群行为或生理的影响，更甚于产生者种群。它们是种间信息化学物质。由于它们作用于种间，有时也称为异种化感物。

13.1.1.4 Allomones

Allomones are defensive secretions. Allomones benefit the producer. A kairomone often works to the detriment of the producer, and to the benefit of the receiver. The host odors that attract a phytophagous insect to its food plant are often called kairomones. Synomones are mutually beneficial to producer and receiver, and operate in cases of mutualism or commensalism.

Semiochemicals are used by insects in a variety of ways. Some well-known cases have been described in which a chemical can play more than one of the roles noted above. The sex pheromone of *D. fiontalis,* a bark beetle, not only attracts a potential mate, but attracts the clerid predator *T. dubius*. Thus, it functions as a pheromone to the beetles and as a kairomone to the predator. Many parasitoids and predators are evidently in an evolutionary race with their hosts to home-in on semiochemical components, while the hosts tend to modify the chemicals or blend they release.

13.1.1.4 利己素

利己素是防御性的分泌物。利己素对产生者有利。利它素起作用时对产生者造成伤害，而对接收者有利。宿主的气味引起植食性昆虫取食植物，常称为利它素。利己素对产生者和接受者双方都有利，运行互利共生或共栖机制。

信息化学物质被昆虫以不同的方式加以利用。一些熟悉的化学物质能起到上述不止一种的作用。树皮甲虫的性信息素不仅能吸引异性引起可能的交配，还可以吸引天敌郭公虫。因此，它对树皮甲虫起着信息素的作用，对天敌起着利它素的作用。很明显，许多拟寄生物和天敌在进化竞争中利用自动导向搜寻其寄主，而寄主则趋于修改其释放的化学物质或混合物。

13.1.2 Pheromone classification

Pheromones are often classified by the behavior elicited from the receiver. There are sex pheromones, aggregation pheromones, alarm pheromones, egg-laying pheromones, brood-tending pheromones, recruitment pheromones, trail-following pheromones, and territory-marking pheromones, to name a few. Either sex or both, can produce one or several pheromones.

For example, the mandibular glands of the stingless bee *M. solani* serve as the reservoir of the alarm pheromone. (*S*)-2-heptanol was the active compound involved in the alarm response of this species. Both male and female antennae of the Citrus Borer, *D. rotundicolle* respond to the natural enantiomer (*R*)-3-hydroxy-2-hexanone, suggesting that it may function as an aggregation-sex pheromone. The trail chemicals come from a gland, the Dufour's

13.1.2 信息素分类

信息素常常根据接收者引发的行为进行分类。有性信息素、聚集信息素、警报信息素、产卵信息素、孵蛋照料信息素、征募信息素、相随跟踪信息素和领地标记信息素，仅举几例。任意性别或两性都能产生一种或多种信息素。

例如，无刺蜜蜂的下颌腺体是报警信息素的贮存库。（*S*）-2- 庚醇是参与该物种报警响应的活性化合物。柑橘蛀虫的雄性和雌性触角都对天然对映体（*R*）-3- 羟基 -2- 己酮产生反应，表明它可能起到聚集性信息素的作用。示踪化学物质来自一个腺体，杜氏腺，在腹末附近。其他利用示踪信息素的昆虫

gland, near the tip of the abdomen. Other insects that use trail pheromones include termites, some caterpillars, and bumblebees. Trail pheromones are used in many contexts of colony life apart from bringing workers to a food source. They are used during house hunting, for recruitment to battlegrounds and escape, or to guide the building of tunnels. Trails may be formed of multiple pheromones with different properties, allowing colonies to form external memories or even "smell the past"(Czaczkes et al., 2015).

包括白蚁、一些毛虫和大黄蜂。踪迹信息素除了将工蚁带到食物来源外，还被用于群体生活的许多环境。它们被用于猎巢、征募到争斗场所和逃跑，或指导建造蚁穴隧道。踪迹信息素可能由具有不同属性的多个信息素组成，允许蚁群形成外部记忆，甚至“嗅到过去”(Czaczkes 等 , 2015)。

13.2 信息素与昆虫嗅觉感受器

13.2 Pheromones and Olfactory Sense in Insects

The chemical stimuli, including general odorants and pheromones, are dominantly detected by specialized olfactory neurons located in the antennal sensilla. To excite the olfactory neurons, the odorants should cross over the hydrophilic sensillum lymph. Considering the hydrophobicity of odorants, certain carriers are necessary to ferry the physiologically active odorants to the membrane of olfactory neurons. The soluble odorant binding proteins (OBPs) enriched in the aqueous sensillum lymph are thought to fulfill the mission. The antennae of many male Lepidoptera are plumose, with thousands of small hairs containing pheromone-sensitive sensory neurons (Fig.13-1; Appendix color graphic Fig.13-1). Although female Lepidoptera usually produce a sex pheromone that attracts males, many male Lepidoptera also release pheromonal components from hair.

In the hawkmoth the long trichoid sensillum is innervated by two pheromone-sensitive olfactory receptor neurons (ORN), one of the two responds to bombykal, the main pheromone component.

The male commercial silk moth, *B.mori*, is a good example of an insect with very high sensitivity to its sex pheromone, and was the first insect from which a sex pheromone component was chemically identified. The male has about 17,000 olfactory receptors on its antennae, and about 50% of them are tuned to detect the sex pheromone. The advantage of having a very large number of receptors tuned to the pheromone is that sensitivity to low concentrations in the air is greatly increased. These male

化学刺激包括普通的气味物质和信息素，主要由位于触角感器中的专门嗅觉神经元对其进行检测。为了刺激嗅觉神经元，气味物质应该穿过亲水感受器淋巴。考虑到气味物质的疏水性，需要某些载体将生理活性气味物质转运到嗅觉神经元的膜上。富含感器淋巴液的可溶性气味结合蛋白(OBPs)被认为完成了这项任务。许多鳞翅目雄性昆虫的触角是羽状的，数以千计的感觉毛包含对信息素敏感的感觉神经元(图 13-1 ; 附彩图 13-1)。虽然雌性鳞翅目通常产生性信息素引诱雄性，但很多雄性鳞翅目也从感觉毛释放信息素成分。

天蛾的长毛形感器分布有两个信息素敏感的嗅觉感受器神经元(ORN)，其中一个对信息素主要成分家蚕醛有应答。

家蚕雄蛾是信息素敏感度非常高的昆虫的范例，是第一个鉴定了性信息素化学组成的昆虫。雄蛾触角上有大约 17 000 个嗅觉感受器，其中约 50% 适用于检测性信息素。拥有大量对信息素敏感的受体的好处是对空气中低浓度信息素的敏感度大幅度增加。这些雄性信

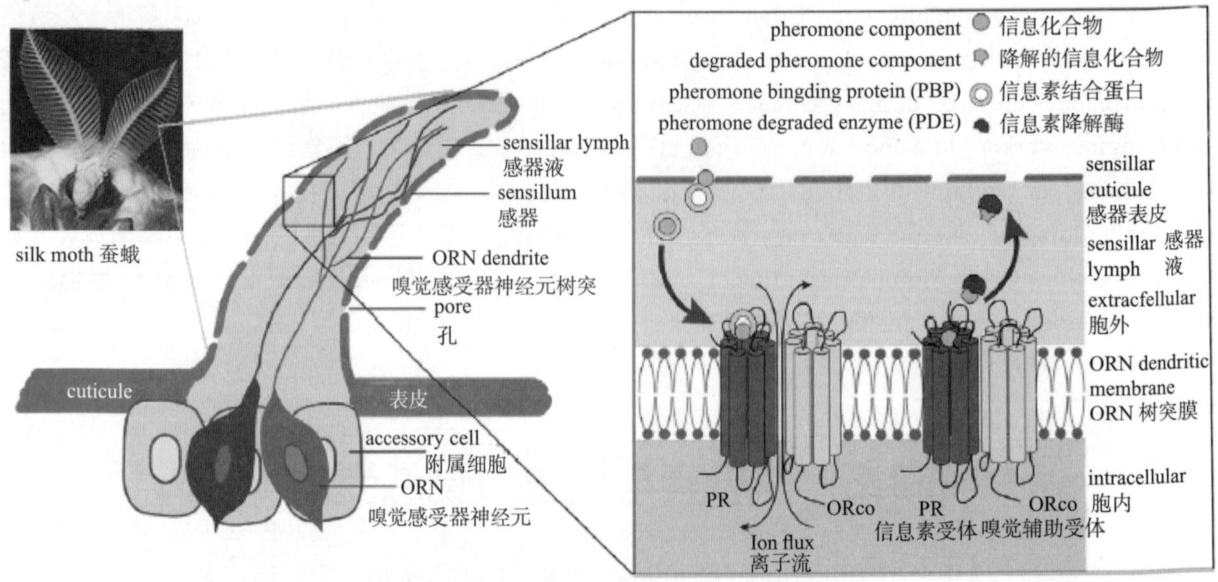

Fig.13-1 The plumose antenna of a male silk moth and the mechanism of receptors. There are thousands of pheromone receptors on the tiny hairs of the antennae (modified from Koutroumpa et al., 2014).

图 13-1 家蚕雄蛾的羽状触角感觉毛及其感受机制。触角小感觉毛上有数以千计的信息素受体（仿 Koutroumpa 等，2014）。

pheromone receptors fire spontaneously, with each antenna sending about 1,600 impulses per second into the brain even when not stimulated by pheromones.

Male moths have the ability to track a highly intermittent and fluctuating sex-pheromone plume, and successfully locate its source from relatively long distances. The odor-searching strategy of moths consists of two fundamental behaviors: upwind surge and crosswind casting.

The surge is a straight upwind flight, and is initiated shortly after the moth encounters the pheromone filaments. On the one hand, when tracking a pheromone plume with high-frequency dispersion, moths perform a reiterative upwind surge due to the high encounter rate of pheromone filaments, which results in a straighter flight. The function of the reiterative upwind surge is believed to be reorientation toward the pheromone source. On the other hand, the moth initiates cross-wind casting when it loses contact with the pheromone plume. The casting is a self-steering counterturning and is presumably an intrinsically programmed behavior, and its function is to recontact with the windborne odor. The fundamental strategy of flying moths can be observed in the odor-searching walk by the male silkmoths when they respond to conspecific female sex pheromones. Their walking behavior is a well-defined programmed behavior consisting of surge (straight-line walking) during pheromone reception, and zigzagging turns followed by

息素受体可自发传导，即使没有信息素的刺激，每个触角每秒钟内也能发送 1 600 个脉冲到脑部。

雄蛾有能力追踪高度间歇性和波动的性信息素缕流，并成功地从相对长的距离定位其来源。蛾的气味搜索策略包括两个基本行为：逆风涌动和侧风投射。

涌动是一种笔直的逆风飞行，在蛾遇到信息素缕流后不久开始。一方面，当跟踪具有高频色散的信息素缕流时，由于信息素缕流的高相遇率，飞蛾执行重复的逆风涌动会导致更直的飞行。反复逆风涌动的作用是朝向信息素源重新定向。另一方面，当蛾与信息素缕流失去接触时，它会启动侧风投射。投射是一种自动转向的反转，可能是一种固有的编程行为，其功能是与风传播的气味重新接触。在雄性蚕蛾对同种雌性信息素做出反应时，可以观察到飞蛾在气味搜索行走中的基本策略。它们的行走行为是一种清晰的编程行为，包括在信息素接收期间的涌动（直线行走）以及在信息素丢失后的曲折转弯后绕圈。通过

looping after the loss of the pheromone. The behavioral sequence is reset by an additional pheromone reception.

Therefore, their trajectories become straight when they repetitively encounter the pheromone. The male silk moth is an excellent model for odor-searching behavior because its neural basis has been well documented in behavioral, genetic, neurophysiological, and robotic studies (Ando et al., 2020).

Moths have long been a model system for insect olfaction. Females release species-specific volatile pheromones that males use to locate them. Female moth pheromones usually include a mix of several compounds. For some species, a single component from the mixture can be behaviorally attractive, while other species require a blend with appropriate ratios. OSNs that are sensitive to pheromones are housed in the long trichodea sensilla on male antennae and project to a special region in the antennal lobe called the macroglomerular complex (MGC). The MGC is male-specific and segregated from other glomeruli. Like many sexual signals, female pheromone blends and male preferences evolve rapidly (Zhao et al., 2020).

Furthermore, an upwind zigzag path could be generated by sensing loss of the plume's edge and turning back toward the plume's centerline. However, this is not the natural cause of zigzagging in moths. Instead, an internal metronome-like generator sets turns several times per second. An alternative to optomotor anemotaxis is a mechanoreceptive anemotaxis in which wind direction is assessed before takeoff.

The mechanism of insect upwind search cause of: ① airborne navigation to a distant odor source is accomplished by steering upwind when the odor is detected, and not by gradient sampling of the odor plume; ② wind direction is gauged by visually appraising how the interaction of wind heading modulates trajectory, an optomotor reaction. A front-to-rear image flow indicates a due upwind course; ③ gaps in the plume caused by turbulent diffusion and by shifts in wind direction complicate plume tracking, limiting the distance over which plumes can be traced and requiring maneuvers for re-establishing contact with a lost plume; ④ orientation in still air can utilize detection of plume distribution, at least over short distances (Cardé, 2021).

Electrophysiological recordings from the antenna indicate that when a single pheromone molecule strikes a receptor, there is a receptor response. When a total of about 200 molecules of the sex

额外的信息素接收来重置行为顺序。

因此，当它们重复性地遇到信息素时，它们的轨迹变得笔直。雄性蚕蛾是气味搜索行为的优秀模型，因为其神经基础已在行为、遗传、神经生理学和机器人研究中得到了充分的证明（Ando等，2020）。

长期以来，蛾一直是昆虫嗅觉的模型系统。雌蛾释放物种特异的挥发性信息素，雄蛾以此定位。雌蛾信息素通常是包含几种化合物的混合物。对于某些物种，混合物中的单一成分在行为上具有吸引力，而其他物种需要具有适当比例的混合物。对信息素敏感的OSNs位于雄蛾触角上的长毛形感器内，并投射到触角叶的一个特殊区域，称为巨型神经纤维球（MGC）。MGC是雄虫特有的，与其他神经纤维球分开。像许多性信号一样，雌蛾信息素是混合物，雄虫的偏好进化很快（Zhao等，2020）。

此外，逆风之字形路径，可以通过感知缕流边缘的散失，并转向缕流中心线来生成。然而，这并不是飞蛾飞之字形的自然原因。相反，一种内部节拍器式的发生器设定了每秒转动几次。趋风视觉运动的另一种选择是机械感受趋风性，可以在起飞前评估风向。

昆虫逆风搜寻的机制：①当检测到气味时，通过迎风转向，而不是通过气味缕流的梯度采样，实现对远距离气味源的空中导航；②风向是通过视觉评估其相互作用如何调节轨迹来测量的，这是一种视觉反应。从前到后的图像流表示正迎风方向；③由湍流扩散和风向变化引起的缕流间隙使缕流跟踪复杂化，限制了缕流可跟踪的距离，并需要策略以重新与丢失的缕流建立起联系；④静止空气中的定向可以至少在短距离上利用缕流分布探测（Cardé，2021）。

触角的电生理学记录证明，一个单个的信息素分子触碰一个受体，就有一个受体反应；当总数为200个性信息素

pheromone simultaneously strike the antenna of the male moth (i.e., 200 receptor neurons are activated), the male moth responds by searching upwind for the source of the chemical. The male of the commercial variety of silkworm, however, has lost the ability to fly because it is too heavy. When it detects the sex pheromone of the female, motor output to the wing muscles causes them to start vibrating, but it merely walks upwind in a zig-zag path towards the source of the pheromone.

分子同时触碰雄蛾的触角（即200个神经元被激活），雄蛾的反应就能逆风搜寻化学物质源。然而，家蚕蛾因为身体太重，失去了飞行能力。当检测到雌蛾的性信息素时，运动神经元发送到翅肌的指令驱使它们开始振翅，但很少能迎着信息素源，以逆风曲折路线行走。

Calling female moths attract their mates late at night with intermittent release of a species-specific sex pherosmone blend. Mean frequency of pheromone filaments encodes distance to the calling female. In their zigzagging upwind search male moths encounter turbulent pheromone blend filaments at highly variable concentrations and frequencies. The male moth antennae are delicately designed to detect and distinguish even traces of these sex pheromones amongst the abundance of other odors. Its olfactory receptor neurons sense even single pheromone molecules and track intermittent pheromone filaments of highly variable frequencies up to about 30 Hz over a wide concentration range. In the hawkmoth *M. sexta* brief, weak pheromone stimuli as encountered during flight are detected *via* a metabotropic PLC_β dependent signal transduction cascade which leads to transient changes in intracellular Ca^{2+} concentrations.

召唤中的雌蛾在后半夜间歇式释放种内性信息素引诱其配偶。信息素缕流的中间频率编码了其与召唤雌蛾的距离。在其弯曲逆风搜寻雄蛾时，遭遇到浓度和频率高度变化的信息素混合物缕流的涡流。雄蛾触角微妙的设计可在大量其他气味中探测和辨别出痕量的性信息素。其嗅觉受体神经元甚至感受得到单个信息素分子，并在宽幅浓度范围跟踪高达约30 Hz的高变频间歇式信息素缕流。烟草天蛾飞行过程中受到微弱的信息素刺激，通过代谢型磷脂酶C_β依赖性信号转导级联检测，引起胞内Ca^{2+}浓度瞬时变化。

Strong or long pheromone stimuli, which are possibly perceived in direct contact with the female, activate receptor- guanylyl cyclases causing long-term adaptation. In addition, depending on endogenous rhythms of the moth's physiological state, hormones such as the stress hormone octopamine modulate second messenger levels in sensory neurons. High octopamine levels during the activity phase maximize temporal resolution cAMP-dependently as a prerequisite to mate location. Thus, sliding adjustment of odor response threshold and kinetics is based upon relative concentration ratios of intracellular Ca^{2+} and cyclic nucleotide levels which gate different ion channels synergistically. In addition, a new hypothesis for the cyclic nucleotide-dependent ion channel formed by insect olfactory receptor/coreceptor complexes. Instead of being employed for an ionotropic mechanism of odor detection it is proposed to control subthreshold membrane potential oscillation of sensory neurons, as a basis for temporal encoding of odors.

可能感知到与雌蛾有直接接触，强烈而持久的刺激活化鸟苷酰环化酶受体引起长期的适应。此外，根据蛾生理状态的内源性节律，强化激素如章鱼胺调节感觉神经元的第二信使水平。在活跃阶段，高水平章鱼胺使cAMP依赖的瞬时溶解度最大化，作为交配场所的前提。因此，气味反应阈值和动力学的滑动调节是基于胞内Ca^{2+}的相对浓度比和环核苷酸水平，协作不同的离子通道门控。除此之外，环核苷酸依赖性离子通道新假设是由昆虫嗅觉受体/辅助受体复合体形成的。推测控制感觉神经元的阈下膜电位振荡作为气味瞬间编码的基准，而不是采用气味检测的离子促进机制。

13.3 信息素的性质和合成部位

13.3 Characteristics of Pheromones and their Synthesis Parts

13.3.1 Chemical characteristics of pheromones

In many species of insects, chemical communication is mediated by hydrocarbons found on the surface of the cuticle and rely on such chemical cues for kin recognition, aggregation, and to trigger alarm. For example, the characteristic female compounds 7,11-heptacosadiene and 7,11-nonacosadiene serve as attractants for males. In addition, the volatile sex-pheromone, 11-*cis*-vaccenyl acetate (cVA), synthesized exclusively by males, plays a dual role as an aggregation factor for females and males and an inhibitor of male courtship in *Drosophila*. In males, cVA co-localizes at the tip of the ejaculatory bulb with a second acetylated hydrocarbon(a new sex pheromone 3-*O*-acetyl-1,3-dihydroxy-28octacosa-11,19-diene), CH503 adds to the roster of identified pheromones involved in fly courtship behavior and may help to explain some of the long-term behavioral changes observed after mating.

The hundreds of different cuticular hydrocarbon components in insects can be divided into three major classes: *n*-alkanes, saturated methyl-branched components, and unsaturated hydrocarbons. There are examples of other types of hydrocarbons including unsaturated methyl-branched components, but these are rare and often are present in small amounts. Cuticular hydrocarbons (CHC) consist of complex mixtures of straight chain, unsaturated, and methyl-branched components with 21 to 40+ carbon atoms. They function both to restrict water loss to prevent a lethal rate of desiccation, and serve in chemical communication in many species. The major volatile insect pheromones consist of modified fatty acids and terpenoids. Many of the lepidopteran pheromones arise from fatty acid precursors, are modified with desaturases, and undergo limited chain shortening or elongation followed by modification of the carboxyl group to produce acetate esters, aldehydes, alcohols, and hydrocarbons.

13.3.1 信息素的化学性质

在许多种类的昆虫中，化学通信是通过在表皮表面发现的碳氢化合物介导的，并依赖这种化学信号进行亲缘识别、聚集和触发警报。例如，典型的雌性化合物 7,11- 七碳二烯和 7,11- 壬二烯用作雄性的引诱剂。此外，挥发性性信息素 11- 顺式醋酸乙烯酯（cVA）仅由雄性合成，在果蝇中作为雌性和雄性的聚集因子和雄性求偶的抑制剂发挥着双重作用。在雄性中，11- 顺式醋酸乙烯酯与第二种乙酰化碳氢化合物（一种新的性信息素，3-*O*- 乙酰基-1,3- 二羟基 - 二十八烷基 -11,19- 二烯 ）共同定位在射精囊的顶端，CH503 增加了参与果蝇求偶行为的已识别信息素候选名单，可能有助于解释交配后观察到的一些长期行为变化。

昆虫表皮上数百种不同的碳氢化合物成分可分为三大类：正烷烃、饱和甲基支链成分和不饱和碳氢化合物。有其他类型的碳氢化合物的例子，包括不饱和甲基支链组分，但这些是罕见的，通常以少量存在。表皮碳氢化合物（CHC）由具有 21 至 40 多个碳原子的直链、不饱和甲基化支链组分的复杂混合物组成。它们既可以限制水分流失，防止脱水致死率，又在许多物种中起着化学通信的作用。主要的挥发性昆虫信息素由改性脂肪酸和萜类组成。许多鳞翅目信息素来源于脂肪酸前体，用去饱和酶修饰，经过有限的链缩短或伸长，然后修饰羧基以产生醋酸酯、醛、醇和烃。

Many coleopteran pheromones are terpenoids, while still other insects use a variety of other compounds. The volatile, long-range pheromones produced by insects are often produced in specific glands, and pheromone glands on the abdomen of many lepidopterans produce 10-21 carbon atom pheromone components. In some coleopterans, midgut tissue produces terpenoid pheromones. In insects, behavior is modulated by cuticular hydrocarbons (CHC). In *Drosophila*, CHC pheromones are produced by oenocytes, and subsequently displayed on the cuticle. They play a significant role in social behaviors like aggregation, aggression, attraction, foraging, group behavior and courtship. Shifting the CHC to a more saturated mix is associated with increased life span in females, but shortened longevity in males. These changes in longevity are correlated with reduction in male mating behavior. Male longevity is shortened in response to sex-specific desaturase knockdown, regardless of mating status. Therefore, male longevity is directly correlated to male CHC profile (Blomquist et al., 2020).

To have sufficient volatility, airborne pheromones must be relatively small molecules. In general, the molecular weight must be less than 200 KDa, or not much over 200 KDa, to get the volatility needed. Not all phero mones are airborne; some insect pheromones are contact pheromones and some crustaceans have waterborne pheromones. No particular chemical structure is used exclusively by insects as a pheromone, but low- molecular-weight acids, esters, alcohols, aldehydes, ketones, epoxides, lactones, hydrocarbons, terpenes, and sesquiterpenes are common components in pheromones (Yu et al.,2007; Li et al., 2011). A few large molecules serve as pheromones, but they are predominantly contact pheromones. For example, (*Z*)-9-tricosene (a hydrocarbon composed of 23 carbons) is the sex pheromone of the housefly *M.domestica*. It has very low volatility. Male houseflies are stimulated to make contact with any small dark object, such as a knot in a black shoelace. They detect the pheromone, if it is present, after landing upon the object and then attempt to mate with it. Even larger cuticular hydrocarbons on the surface of female tsetse flies serve as contact pheromones. In general, a male tsetse fly cannot distinguish a female from a male until it contacts the body surface of the fly. The chemoreceptors may be located on the

许多鞘翅目信息素是萜类化合物，而其他昆虫则使用多种其他化合物。昆虫产生的挥发性、远距离信息素通常在特定的腺体中产生，许多鳞翅目昆虫腹部的信息素腺体产生 10 ～ 21 个碳原子的信息素成分。在一些鞘翅目动物中，中肠组织产生萜类信息素。在昆虫中，行为由表皮碳氢化合物（CHC）调节。在果蝇中，CHC 信息素由绛色细胞产生，随后出现在表皮上。它们在聚集、攻击、吸引、觅食、群体行为和求爱等社会行为中起着重要作用。将 CHC 转变为更饱和的混合物将延长雌性果蝇的寿命，但雄性寿命缩短。这些寿命的变化与雄性交配行为的减少有关。无论交配状态如何，雄性寿命都会因性别特异性除去饱和酶而缩短。因此，雄性寿命与雄性 CHC 分布直接相关（Blomquist 等，2020）。

为了拥有足够的挥发性，空气传播的信息素必须是相对较小的分子。一般而言，分子质量必须小于 200 KDa 或不大于 200 kDa，才能获得所需的挥发性。并非所有的信息素都是靠空气传播的，有些昆虫信息素是靠接触来传播的，而有些甲壳纲是利用水传播信息素。没有特定的化学结构可用于概括昆虫的特有信息素，但低分子质量的酸、酯、醇、醛、酮、环氧衍生物、内酯、羟、萜烯和倍半萜烯是信息素的普遍成分（Yu 等，2007 年；Li 等，2011 年）。一些大分子也起着信息素的作用，但主要是接触信息素。比如，（顺）-9- 诱虫烯（一种 23 个碳原子的碳氢化合物）是家蝇的性信息素，其挥发性极低。雄家蝇受到刺激后会与任何黑小物体接触，比如黑鞋带的结。家蝇侦测到可能出现的信息素，再飞落到该物体上，然后尝试交配。雌舌蝇表皮上的较大分子的表皮碳水化合物甚至也起着接触信息素的作用。通常，雄舌蝇在没有接触到舌蝇体表前，不能辨别雌雄。化学感受

tarsi of the male.

Pheromones often serve as species isolating mechanisms, and the receptors on the antennae of the opposite sex must be tuned to the pheromone of its species. There are several ways that species specificity is coded in pheromones. One of the most common adaptations is the use of a blend of two or more chemical components in the pheromone, as well as variations in different blend ratios when the same components are incorporated into the pheromone. Many of the sex pheromone components that have been identified from moths are acetates and alcohols with a backbone of 14 to 16 carbons with one or more double bonds in the molecule. The position of the double bond offers many variations on each backbone. Still further specificity can be encoded in (*E*) or (*Z*) configuration of the groups or atoms on the carbons at the double bond. Finally, chirality in molecules provides another way that insects can specify their pheromone signal.

Pheromone molecules to have more than one stereo center. Other atoms in addition to carbon may also be a source of chirality in a molecule, a situation not often relevant to pheromones. For chirality in a pheromone to provide specificity, a receptor has to recognize the difference in the two enantiomers. The opposite enantiomer of a chiral pheromone component may have no effect at the sensory neuron, may stimulate, or may inhibit the response to the natural enantiomer. Grain beetles in the family Cucujidae illustrate the probable utilization of pheromone chirality in evolution and isolation of species. The pheromonal compounds produced by males during feeding act as aggregation pheromones, they attract both sexes.

13.3.2 Pheromone gland and their location

The pheromone gland in most species of moths had been referred to either as a single layer of columnar cells located in the ventral region, as distinct areas in both ventral and dorsal regions or as a continuous ring. In *H. zea*, the pheromone gland was found in the intersegmental membrane between the 8th and 9th abdominal segments. Histology of this area showed a single layer of columnar cells that formed an almost continuous ring just below the cuticle. Both dorsal and ventral surfaces in this area had extensive convolutions that were more prominent in the 9th segment. At the ultrastructural level the pheromone gland cells showed characteristic apical microvilli. In pheromone producing females, these cells contained pockets of granular material and inter- and intracellular canals with distinct desmosomes. The cuticular hairs had a hollow core and many pore canals. In the diamondback

器可能位于雄舌蝇的跗骨上。

信息素常常起着物种隔离机制的作用，异性触角上的受体必须适应其物种信息素。种属特异性有几种形式是由信息素编码的。最普遍的适应之一就是利用信息素两种或多种化学成分混合，此外，当相同的成分掺入信息素时，其也有不同混合物比例的变化。从蛾中鉴定出的许多性信息素成分是分子骨架为 14 ～ 16 个碳原子的酯和醇，分子内有 1 个至多个双键。双键的位置在每个骨架上提供多个变异。更具特异性的是双键上的碳原子或基团能编码反式或顺式构象。最后，分子的手性为昆虫特化其信息素信号提供了另外一条途径。

信息素分子有多于 1 个的立体中心。除了碳原子，其他原子也可能作为分子手性的来源，这种情况通常与信息素无关。信息素的手性提供了特异性，受体不得不辨别两个对映体的差别。手性信息素成分相对应的对映体可能对感觉神经元没有作用，可能激发或抑制天然对映体的响应。扁甲科的谷甲证实了信息素手性分子在进化和种群分离中的可能用途。在取食过程中，产生的信息素成分起着聚集信息素的作用，吸引两性。

13.3.2 信息素腺及其位置

多数种群蛾信息素腺指的是位于腹部区域的单层柱形细胞，有别于腹部和背部区域，或指的是连续的环腺。发现玉米夜蛾的信息素腺位于第 8 腹部体节和第 9 腹部体节的节间膜上。这个区域的组织学显示出单层柱形细胞，形成的几乎连续的环腺位于表皮的正下方。这一区域的背腹表面有大范围的盘旋，在第 9 体节更为突出。信息素腺细胞在超微结构水平上展现出特征化的顶端微绒毛。在产生信息素的雌蛾中，这些细胞包含有颗粒状物质的袋口以及有清晰桥粒的胞内和胞间腺管。表皮毛为中空芯型，有很多孔管。

moth, the main sex pheromone gland was present in the 9th segment. Extensive basal invaginations were also reported in pheromone gland cells of *T. ampelophaga*. The pheromone gland occupies a rather narrow band, and to maximize the evaporation of the pheromone. It is smeared over a highly convoluted surface, during the rhythmic protrusion of the ovipositor, often referred to as calling.

The site of pheromone production is varied amongst the insects just as there are variable structures in the different orders (Jurenka,2004). The site of pheromone biosynthesis in the Blattodea, Coleoptera and Lepidoptera is varied but is usually located in the abdomen. In moths the pheromone gland is usually located as modified epidermal cells between the 8th and 9th abdominal segments. In *T. ampelophaga*, the gland is located on the dorsal part of abdominal segments the 3rd to 5th. The site of synthesis of hydrocarbon sex pheromones involved another group of cells called oenocytes that are usually associated with epidermal cells throughout the abdomen.

There is much variability among and within insect orders in the anatomical location of the cells or tissues involved in pheromone biosynthesis, accumulation, and release. In the female German cockroach, *B. germanica,* the volatile sex pheromone is produced in a gland located on the anterior of the last (the 10th) abdominal tergite called the pygidium. The contents of secretory vesicles from cells in the gland are transported through long ducts to the cuticular surface for release. The non-volatile contact pheromone present on the cuticle of female *B. germanica* is produced by epidermal cells. The sites of sex or aggregation pheromone synthesis, accumulation, and/or release have been examined in many species of Coleoptera, with the preponderance of studies revealing abdominal glands, such as *Trogoderma* spp., *D. maculates*, *D. ater*; *L. serricorne,* cowpea weevil, *B. atrolineatus*, western corn rootworm, *D. virgifera*, *S. lutus*, *A. obscurus*, *A. lineatus* (L.) (both Elateridae) and the sap beetle, *C. freemani*. The female sex pheromone gland of the cowpea weevil, *B. atrolineatus* localize to the dorsal and ventral intersegmental membranes that connect the 8th abdominal segment to the ovipositor. In female *A. obscurus*, and the lined click beetle, *A. lineatus*, the sex pheromone accumulates in opalescent, sacciform glands located in the 7th abdominal segment. This gland discharges geranyl hexanoate and geranyl octanoate posteriorly into the outer portion of the oviduct. In the sap beetle, *C. freemani*, it seems that males produce hydrocarbon aggregation pheromone in large disk-like abdominal oenocytes that occur within the body cavity.

These cells are connected by tracheae to the integument, with the pheromone secreted into tracheal-associated ductules eventually reaching the cuticular surface of the male through the spiracles. The recruitment of abdominal oenocytes for pheromone production

小菜蛾主要的性信息素腺出现在第9体节。葡萄蛾信息素腺也有外延的底膜凹陷。信息素腺占据颇窄的条带，这可使得信息素的散失最小化。信息素被涂抹在高度盘绕的表面，在产卵器节律性突出部，通常与召唤有关。

昆虫种间信息素产生的部位在昆虫之间是不同的，正如在不同目昆虫中，有不同的结构一样（Jurenka, 2004）。蜚蠊目、鞘翅目和鳞翅目昆虫的性信息素和聚集信息素合成部位不同，常位于腹部。蛾类信息素腺体常位于第8～9腹节之间，为修饰过的真皮细胞。但斑蛾科的腺体位于第3～5腹节的背部。羟类性信息素的合成涉及另一群细胞，称之为绛色细胞，常遍及腹部而与真皮细胞相连。

牵涉昆虫信息素的生物合成、累积和释放的细胞和组织在昆虫种间、种内有非常大的差异性。雌性德国蜚蠊挥发性的性信息素合成部位位于最后（第10）腹节背甲，其称为尾板，腺细胞分泌的物质通过长导管转运到表皮表面释放。非挥发性的信息素在雌蜚蠊真皮细胞合成和释放。对许多鞘翅目昆虫的性信息素或聚集信息素的合成、累积和/或释放部位进行了研究，深入研究的是腹腺，如斑皮蠹属、白腹皮蠹、钩纹皮蠹；烟草甲、绿豆虫（暗条豆象）、玉米根萤叶甲、宽背金叩甲和暗色叩甲、具条叩甲（都是叩甲科）和玉米液甲。暗条豆象雌性的性信息素合成腺体位于第8腹节和产卵器之间的腹、背面节间膜上。暗黑金针虫和具条叩甲的雌虫性信息素的积累在第7腹节的乳白色囊状腺体内。这个腺体释放辛酸香叶酯和己酸香叶酯到输卵管的外部。玉米液甲雄虫在腹部圆盘状绛色细胞内产生的聚集信息素为碳水化合物，位于体腔内。

这些细胞由气管和体壁相连，其性信息素被释放到腺管内，最终通过雄虫气门到达体表。家蝇和玉米液甲雄虫在腹部绛色细胞产生信息素作为补充，也

by both the common house fly, *M. domestica* and male *C. freemani* may reflect the biochemical similarities in the hydrocarbon pheromones. In other coleopteran taxa, abdominal glands are not the sites of sex or aggregation pheromone synthesis, accumulation, or release. A setiferous patch over exocrine glands was located in the prothoracic femora of the male red flour beetle, *T. castaneum* and the secretion from this patch was attractive to both sexes. a glandular organ in the apical (the 10th and 11th) antennal segments of the male antloving beetle, *B.oculatus* Aube (Pselaphidae), is involved in secreting a female attractant or other semiochemicals.

With the exception of the European elm bark beetle, *S. multistriatus* and the large elm bark beetle, *S. scolytus* where pheromone production, storage, and release correlate with accessory glands associated with vaginal palpi, accumulation of aggregation pheromone in scolytids has otherwise been generally localized to the alimentary canal, including the Malpighian tubules. Metathoracic flight muscle may be the site of biosynthesis of the pheromone precursor ipsenone by male *I. paraconfusus*. Because the major pheromone component (ipsenol) ultimately ends up in the hindgut, a transport mechanism from the metathorax might be involved. Fat body tissue isolated from *A. grandis* synthesized aggregation pheromone in culture.

Hydrocarbon pheromones of the model Diptera are synthesized in specialized subcuticular abdominal epidermal cells (oenocytes) and deposited onto the cuticular surface. The hydrocarbon pheromones synthesized in the abdominal oenocytes by fruit fly, *D. melanogaster*, are transported by lipophorin to epidermal cells for deposition on the cuticular surface.

The majority of lepidopteran females produce and release sex pheromone components from bulbous extrudable glands located between the 8th and 9th abdominal segments. These glands have secretory cells that are hypertrophied and modified epidermal cells that typically contain a well-developed endoplasmic reticulum involved in fatty acid metabolism. Indeed, in a survey of females of ten lepidopteran species, extracts of the ovipositor tips (which contain the glands) revealed unusual fatty acids that had the same carbon lengths, double-bond positions, and stereochemistries as the acetate, alcohol, or aldehyde pheromone components for the species. An exception to the normal lepidopteran gland morphology has been found with the spear-marked black moth, *R. hastata* (L.), in which the gland consists of a pair of internal tubular organs that extend from their common opening in the 9th abdominal segment anteriorly into the 7th abdominal segment. Similar paired tubular glands have been identified from the bog holomelina, *H. lamae*, while long, coiled tubular glands are present in the abdominal tip of another female arctiid, *Utetheisa*.

反映了两种昆虫在碳水化合物信息素的生物化学上具有相似性。其他鞘翅目昆虫，腹腺不是性信息素和聚集信息素合成、积累和释放的部位。赤拟谷盗雄虫的前胸腿节上的外分泌腺体覆有刚毛斑块，分泌物对两性均具有吸引力。蚁甲科昆虫的雄虫触角（第 10 ～ 11 节）的腺体器官参与分泌雌虫引诱剂和其他信息化学物质。

欧洲榆小蠹和欧洲大榆小蠹是个例外，其信息素的合成、贮存和释放与阴道须相连的附腺相关。聚集信息素的积聚位于食道，包括马氏管。加州五条小蠹的前胸翅肌可能是信息素前体的合成部位。因为信息素主要成分（齿小蠹烯醇）终止于后肠，可能牵涉到转运到前胸的机制。培养的棉铃象甲脂肪体也能分泌信息素。

双翅目模式昆虫碳水化合物信息素的合成位于腹部特化的真皮细胞中，即绛色细胞，并在体表沉积。果蝇的信息素由腹部绛色细胞合成，通过载脂蛋白转运到真皮细胞，并在表皮的表面沉积。

鳞翅目昆虫的性信息素合成位于第 8 腹节和第 9 腹节间的球根状突出腺体。这些腺体具有肥厚的分泌细胞，修饰过的真皮细胞含有发育良好的粗面内质网，与脂肪酸代谢有关。10 种鳞翅目昆虫产卵器尖（含腺体）抽提物显示出不同寻常的脂肪酸，碳链长度和双键位置相同，酮、醇和醛类化合物组分的立体化学也相同。鳞翅目昆虫正常腺体形态学上的一个例外是黑白汝尺蛾，其腺体由一成对的内部管状器管组成，从常见的第 9 腹节前端开口延伸到第 7 腹节。一种沼泽蛾类中也有相似的管状腺体，而另一种雌性灯蛾在腹部末端有长长的线绕式、螺旋状的腺体。

13.4 昆虫受体和检测过程

13.4 Insect Receptors and the Detection Process

Olfaction is essential in guiding insect behaviors such as foraging, host-seeking, and oviposition. In female mosquitoes a population of hair-like sensilla distributed over the surface of the antennae and maxillary palps act as a nose to detect chemical signals. The major proteins involved in the selectivity and sensitivity of the insect olfactory system are odorant-binding proteins (OBPs) and odorant receptors (ORs). OBPs are involved in the first step of odorant reception where they bind, solubilize and deliver odorant molecules to ORs. ORs are heterodimers comprised of highly variable odorant-binding subunits associated with one conserved coreceptor and localized on the dendrite membrane in the olfactory sensilla. They detect odorant compounds and transduce olfactory signals to the brain to mediate insect behaviors (Xu et al., 2010).

嗅觉是引导昆虫行为如搜寻食物、寻找寄主和产卵所必需的。雌蚊种群在触角表面和上颚骨触须上分布有毛形感器，用于嗅觉检测化学信号。昆虫嗅觉系统涉及选择性和敏感性的主要蛋白质是气味结合蛋白（OBPs）和嗅觉受体（ORs）。OBPs 牵涉到接收气味的第一步，此处气味分子结合、溶解并被发送到 ORs。ORs 是异源二聚体，由高度可变的气味结合蛋白亚基组成，与一个保守的辅助受体及位于嗅觉感器的树突膜连接。它们检测到气味化合物并将嗅觉信号转导到脑，介导昆虫行为（Xu 等，2010）。

Pheromones are generally detected by olfactory receptors located primarily on the antennae. Many male Lepidoptera have pheromonal receptors specialized for reception of the sex pheromone; these receptors are relatively insensitive to other chemicals. A sex pheromone receptor on the antenna typically consists of one or two nerve cells housed within a seta or fine "hair" on the antenna. The entire structure is called the sensillum. There often are many thousands of such sensilla on each antenna of a male moth. Each seta has microscopic pores along its length, through which the airborne molecules enter the sensillum and make contact with the sensory neuron.

信息素基本上是由位于触角上的嗅觉受体检测的。多数鳞翅目昆虫有特化的性信息素受体，这些受体相对其他化合物不敏感。触角上典型的性信息素受体由 1 ～ 2 个神经细胞封装在触角的 1 个刚毛或纤细的"毛"内。整个结构称为感受器。雄蛾触角上通常有数千个这样的感受器。每根刚毛节段有微孔，空气传播的分子由此进入感受器并与感觉神经元接触。

The following six steps occur as a part of the process of semiochemical detection (Kaissling 1987): ① absorption of all odor molecules by sensory hairs (setae) on the antennae; ② penetration of the molecule through pores in the setal wall; ③ receptor binding of the molecule and transport to the sensory nerve endings; ④ membrane alteration, probably opening of sodium channels; ⑤ receptor potential generation, a graded potential, followed by spikes in the axon hillock region; ⑥ inactivation of the odor molecule and removal (Inact- ivation clears the receptor so that it can respond again).

信息化学物质的检测过程通常有 6 步（Kaissling，1987）：①触角上感觉毛（刚毛）吸收一个气味分子；②分子通过刚毛壁上的孔穿透；③分子的受体结合并转运到感觉神经末端；④膜产生变化，可能打开 Na^+ 通道；⑤受体电位传导，在轴丘区域，神经轴传导分级电位，紧接着是峰电位；⑥气味分子的失活和移除（失活清除了受体使之可再次反应）。

13.4.1 Pheromone binding proteins

After a odour molecule enters a pore, it must crosses the sensillum liquor to each the dendritic nerve endings. Most pheromonal molecules are lipid soluble, and the sensillum liquor is an aqueous medium that does not readily dissolve lipids. Thus, the molecule must usually bind with a specific binding protein, which then transports it across or through the sensillum liquor. Pheromones bind to pheromone binding proteins (PBPs), a subset of a larger group of odorant binding proteins (OBPs) known from both vertebrates and invertebrates (Fig.13-2; Appendix color graphic Fig.13-2).

General OBPs (GOBPs, Fig. 13-3; Appendix color graphic Fig.13-3) of Lepidoptera bind a variety of odorant molecules associated with food, habitat, and oviposition substrates. GOBPs may occur in other insects, perhaps in males in some cases as well as in females.

PBPs are synthesized in male Lepidoptera just prior to adult emergence and are localized in the extracellular fluid of pheromone-responsive sensilla on the male antennae. Auxiliary

13.4.1　信息素结合蛋白

气味分子进入感受器的孔道，它必须通过感器液腔才到达树突的神经末梢。多数信息素分子是脂溶性的，感器液是水溶性介质，不能溶解脂类。因此，这些分子就必须与特异性结合蛋白结合，然后通过转运使其跨越或通过感器液。信息素与信息素结合蛋白（PBPs）结合，PBPs 是脊椎动物和无脊椎动物内一大类气味结合蛋白（OBPs）的亚类（图 13-2；附彩图 13-2）。

鳞翅目普通型气味结合蛋白（GOBPs，图 13-3；附彩图 13-3）与不同的气味分子结合，和食物、习性及产卵基质有关。GOBPs 也可能出现在其他昆虫中，在有些情况下雌雄中可能都有。

PBPs 在鳞翅目雄虫正要变为成虫前合成，位于雄虫触角信息素应答的感受器胞外液中。与受体细胞相关的辅助

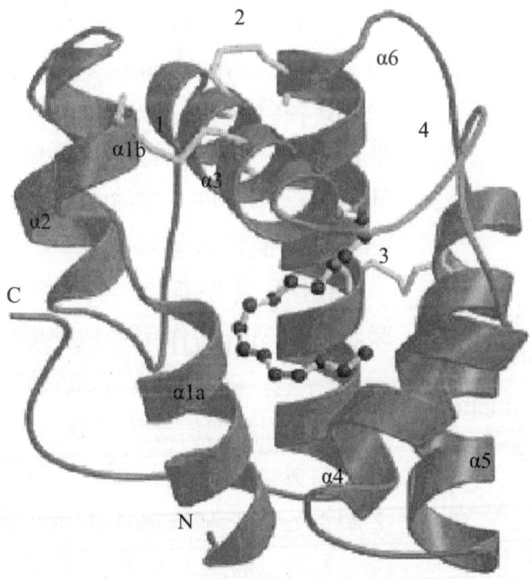

Fig.13-2 Overall structure of the *B. mori* PBP. Disulfide 1-3 bridges are shown between two α-helices, and the loop4 covered the binding pocket. Bombykol is shown in a ball-and-stick representation with double helix (modified from Sandler et al., 2000; Lee et al., 2002).

图 13-2 家蚕 PBP 总体结构。两条螺旋之间有 1 ~ 3 双硫键，环肽链 4 覆盖了结合袋区。蚕蛾性诱醇用双螺旋球链表示（仿 Sandler 等，2000; Lee 等，2002）。

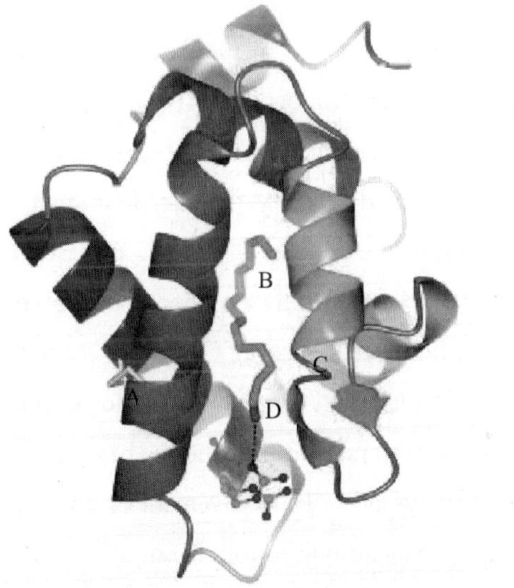

Fig.13-3 The cartoon structure of the silkworm moth GOBPs bound to an analogue of its sex pheromone, shows binding to the arginine amino acid (D) at the entry to the binding pocket (A-C) (modified from Gill, 2009).

图 13-3 蚕蛾 GOBPs 结合到其类似物性信息素的卡通结构，显示在结合袋区（A-C）入口结合到精氨酸（D）（仿 Gill，2009）。

cells associated with the receptor cell appear to be responsible for synthesis. By binding the hydrophobic pheromone molecule, the PBP aids in transport of the pheromone through the aqueous sensillum lymph so that it makes contact with specific receptor proteins in the dendritic nerve endings. One current hypothesis is that the PBPs with bound protein may attach directly to the dendritic membrane where subsequent activation of a G protein coupled cascade of events result in a receptor potential in the dendrite. Specific binding proteins may serve as filters that protect the dendritic endings from many other chemical molecules of the air that also enter sensilla pores. PBPs may also be involved in destruction of file pheromone after signal transduction.

The major component of the *A. polyphernus* sex pheromone, 6*E*, 11 *Z*-hexadecadienyl acetate, has only one binding site, but a second pheromonal component, 4*E*,9*Z*- tetradecadienyl acetate can bind in two slightly different ways (Du et al., 1994).

细胞承担合成任务。通过与疏水性的信息分子结合，PBP 协助信息素的转运，穿越水溶性感器液，以便与树突的神经末端内的特异性受体蛋白结合。最近的一个假设是 PBPs 和结合蛋白可直接到达树突膜，膜上 G 蛋白级联环节产生树突的受体电位。特异性结合蛋白作为一个滤器保护树突末端，免受空气中其他也进入到感受器孔道的化学分子的损害。PBPs 也可能参与信号传导后的一系列信息素的破坏。

多音天蚕主要信息素成分是 6 顺，11 反 - 十六碳二烯酯，它只有一个结合位点，但第二个信息素成分，4 顺，9 反 - 四碳二烯酯能与其以略有不同的两条结合途径（Du 等，1994）。

13.4.2 Signal transduction and receptor response

At the dendritic ending, the pheromone probably combines with a receptor protein in the dendritic membrane. G proteins, cAMP, and inositol triphosphate (IP_3) are involved in the odor transduction process in several vertebrates, and some or all of these may mediate and amplify the pheromone signal at the dendritic ending in insects.

PBAN acts firstly on pheromone gland cells of mature females by binding to a specific G-protein coupled membrane receptor, then one of three different pathways for intracellular signal transduction of PBAN is opened according to insect species. The three pathways are: ① involvement of cAMP; ② involvement of cAMP and phosphatidylinositol; ③ only involvement of Ca^{2+} as the second messenger (Wang et al., 2006).

Early research revealed the presence of G-protein, belonging to the Gαq family, in antennal preparations (especially the pheromone-sensitive sensilla trichodea) of *B.mori* and *Antheraea pernyi*, implied a participation of G-protein of the Gαq family in the signal transduction of OR cells in moths. The bombykol stimulation of *Xenopus laevis* oocytes expressing BmOR1 and BmGαq elicited robust dose-dependent inward Ca^{2+}-dependent

13.4.2 信号转导与受体反应

在树突末端，信息素分子可能在树突膜上与受体蛋白结合。G 蛋白，cAMP 和三磷酸肌醇（IP_3）涉及几种脊椎动物的气味转导过程。其中一些或全部第二信使也许介导和应用到昆虫的树突末端的信息素信号中。

PBAN 首先通过与特定的 G 蛋白偶联膜受体结合而作用于成熟雌蛾的信息素腺细胞，随后依据昆虫种类的不同，PBAN 在细胞内信号转导以三种不同途径之一打开。三条途径是：①涉及 cAMP；②牵涉 cAMP 和磷脂酰肌醇；③仅涉及 Ca^{2+} 为第二信使（王等，2006）。

早期研究表明，在桑蚕和柞蚕的触角样本（尤其是信息素敏感的毛形感器）中存在属于 Gαq 家族的 G 蛋白，这意味着 Gαq 族的 G 蛋白参与了蛾类 OR 细胞的信号转导。家蚕醇刺激非洲爪蟾卵母细胞表达 BmOR1 和 BmGαq 在双电极电压钳记录上引发强烈的剂量依赖内向 Ca^{2+} 依赖性 Cl^- 电流，表明家

Cl^- currents on two-electrode voltage-clamp recordings, demonstrating that the binding of bombykol to BmOR1 leads to the activation of a BmGαq-mediated signaling cascade. The GPCRs usually direct the recruitment, activation, and scaffolding of the cytoplasmic signaling complexes *via* two multifunctional adaptor and transducer molecules, *β*-arrestins 1 and 2, and arrestins also function to activate signaling cascades independently of G-protein activation or mediate receptor desensitization. The knockdown of *β*-arrestins significantly reduced bombykol-induced calcium influx through BmOR3-BmOR2, which was accompanied by the collapse of the receptor complex, suggesting that the *α*-arrestins mediate Ca^{2+} response mainly by regulating the structural and functional integrity of the BmOR3-BmOR2 complex. The summarized researches show that insect pheromone receptors may both have G-protein and arrestin downstream pathways.

蚕性诱醇与 BmOR1 的结合导致 BmGαq 介导的信号级联的激活。GPCRs 通常通过两个多功能衔接子和传感器分子 *β*-抑制蛋白 1 和 2 来引导细胞质信号复合物的募集、激活和细胞质信号复合物的折叠，抑制蛋白还激活 G 蛋白非依赖性信号的级联，或介导受体的脱敏。*β*-抑制蛋白的敲除显著减少了家蚕性诱醇诱导的通过 BmOR3-BmOR2 的 Ca^{2+} 内流，这伴随着受体复合物的分解，表明 *α*- 阻遏蛋白主要通过调节 BmOR3-BmOR2 复合物的结构和功能完整性来介导 Ca^{2+} 反应。综上所述，昆虫信息素受体可能同时具有 G 蛋白和抑制蛋白这两种下游通路。

BmOR1 and BmOR3 of *B. mori* are mutually exclusively expressed in a pair of adjacent pheromone-sensitive neurons of male antennae, and both of which are coexpressed with the highly conserved insect Orco, operate as receptors for the sex pheromones bombykol (*E*10, *Z*12-hexadecadienol) and bombykal (*E*10, *Z*12-hexadecadienal), respectively. The BmOR3-BmOR2 combination elicited a response to bombykal and showed similar channel properties, and the coupling of BmOR3 and BmOR2 forms a cation channel with the detection of calcium influx. From the view of biophysics, there was also physical interaction between BmOR3 and its Orco BmOR2. On bombykal stimulation, the cytoplasmic parts intracellular loops 1 (ICL1), ICL2, and ICL3 moved away from the N-terminus, while the C-terminal helical kink moved close to the N-terminus of BmOR3. On the contrary, the lower part of loop7a-7b moved away from the N-terminus in BmOR2. ICL1, ICL2, and ICL3 also moved away from the N-terminus of BmOR2. The substitution of transmembrane 7 in the two receptors confirmed its indispensable role in the ionotropic function in BmOR3- BmOrco coupling.

家蚕的 BmOR1 和 BmOR3 在雄蛾触角的一对相邻的信息素敏感神经元中相互排他性表达，两者均与高度保守的昆虫 Orco 共表达，分别作为性信息素家蚕性诱醇（*E*10, *Z*12- 十六二烯醇）和家蚕醛（*E*11, *Z*12- 十八二烯醛）的受体。BmOR3-BmOR2 组合引发了对家蚕醛的反应，并显示出类似的通道特性，BmOR3 和 BmOR2 的耦联形成了一个检测 Ca^{2+} 内流的阳离子通道。从生物物理学的角度来看，BmOR3 与其 Orco BmOR2 之间也存在物理相互作用。在家蚕醛刺激下，细胞内环肽链 1（ICL1）、ICL2 和 ICL3 的细胞质部分远离 N- 端，而 C- 端螺旋扭结靠近 BmOR3 的 N- 端。相反，环肽链 7a-7b 的下部远离 BmOR2 的 N- 端，ICL3 也从 BmOR2 的 N- 端远离。两种受体中的 7 跨膜的替换证实了其在 BmOR3-BmOrco 偶联中对具有不可缺少的促离子作用。

Senthilkumar et al. (2019) characterized the pheromone biosynthesis-activating neuropeptide (DH-PBAN) and pheromone /odorant binding proteins in *S. recurvalis*. The open reading frame of 600 base pairs encodes a 200-amino acid protein possessing five neuropeptide motifs (DH, PBAN, *α*-, *β*-, and *γ*- subesophageal ganglion neuropeptides) and shares a characteristic conserved

Senthilkumar 等（2019）描述了信息素生物合成激活神经肽（DH-PBAN）和信息素 / 气味结合蛋白在斜纹夜蛾中的特征。600 个碱基对的开放阅读框编码一个 200 个氨基酸的蛋白质，具有五个神经肽基序（DH、PBAN、*α*、*β* 和 *γ*- 食管下神经节神经肽），并共享一个

C-terminal pentapeptide fragment FXPRL. The full-length genome of Spre- DH-PBAN was 4,295 bp in length and comprised of six exons interspersed by five introns. It is the first report on the molecular analysis of PBAN gene and binding proteins confirmed their role in recognition and chemoreception of sex pheromone components, and they were distinct, tissue- and sex-specific.

特征性保守的 C- 端五肽片段 FXPRL。斜纹夜蛾 DH-PBAN 的全长基因组长度为 4 295 bp，由 6 个外显子夹杂着 5 个内含子组成。这是对 PBAN 基因和结合蛋白的分子分析的首次报道，证实了它们在性信息素成分的识别和化学接收中的作用，并且它们具有独特的组织特异性和性别特异性。

Axons from sex pheromone receptor neurons located peripherally project into the antennal lobe of the deutocerebrum as labeled-lines (pheromone-specific neurons), while general food, host odors, and environmental odors are transmitted by across-fiber patterning. It currently appears that an odor is defined in the deutocerebrum by an across-glomeruli pattern based on input that the glomeruli receive from peripheral receptors.

信息素受体神经元发出的轴突位于外周，伸入中脑触角叶，作为标记线（信息素特化神经元），而一般的食物、寄主气味和环境气味通过跨纤维模式传送。现在看来，基于神经纤维球接收的外周受体的传入，在中脑由横跨神经纤维球模式阐明了气味。

Pheromone receptors on the antennae of perhaps on most insects and perhaps on most insects are spontaneously active and display a constant low level of firing contact with a pheromone component to which a receptor is sensitive usually results in an increase in firing. In some cases, a receptor is differentially sensitive to a particular component of a pheromone blend, and it typically responds to low concentrations of that component by a large increase in rate of firing. Exposure of a specialist receptor to large amounts of other blend components may cause an increase in firing to nonspecific components, but usually at a much lower rate (i.e., it is less sensitive to the other components).

也许多数昆虫触角上的信息素受体是自发起作用的，显示与信息素成分恒定的低水平触发，其受体通常是敏感的，引起触发的增加。有些情况下，受体对信息素混合物的某个特定成分的敏感性有差别，一般在低浓度下对那个成分有反应，触发速率大大增加。专化型受体暴露在大量其他混合物成分中可能会引起非特异性成分触发的增加，但其速率通常极低（即对其他成分敏感度低）。

Most insects with pheromone receptors appear to have more than one type of receptor, each being sensitive to one or more of the blend components of the pheromone. It may be that when a responding insect flies upwind in a plume of pheromone composed of several components, each of its several receptor types responds to one component while ignoring other components in the blend. Thus, the nerve activity going into the deutocerebrum and other parts of the brain is *via* labeled lines, and the brain presumably has to integrate the input as the relative activities of different receptor neurons.

多数昆虫的信息素受体多于一个类型，每一个受体对信息素混合物成分之一或几个敏感。当响应昆虫逆风飞向缕流内含有几种成分组成的信息素时，其几个类型的受体每一个均只对一个成分有反应，而忽略混合物的其他成分。因此，进入中脑和其他部分脑的神经活动通过标记线进行的，大脑很可能要将输入整合为不同受体神经元的相对活动。

13.4.3 Pheromone inactivation and clearing of the receptor

13.4.3 信息素的失活和受体清除

When pheromone molecules have made contact with the dendritic endings, it is important to destroy or inactivate them in order to clear the receptor active site and allow the receptor to be sensitive to incoming pheromone. An esterase in *A. polyphemus*

当信息素分子已经与树突终板接触后，重要的是破坏或使之失活，以便清除受体活性位点，并使受体对进入的信息素敏感。已分离到美洲野蚕酯酶和

and an aldehyde oxidase from *M. sexta* have been identified. In less than 0.5 s the esterase in *M. sexta* antennae can destroy a million pheromone molecules, and this seems consistent with the rapid changes in upwind or casting behavior that males make in a changing pheromone plume. The pheromone-destroying enzymes may also aid an insect by destroying small amounts of pheromone slowly leaking into the pores after adsorption on the antennal surface. Such a slow, persistent pheromone leak might create high background noise in the system.

烟草天蛾的醛脱氢酶。在少于 0.5 s 内，烟草天蛾触角内的酯酶能破坏掉成百万的信息素分子，这似乎与逆风搜寻的快速变化或与雄蛾在不断变化的信息素缕流中的投射行为一致。在触角表面吸收信息素分子之后，信息素降解酶有助于昆虫通过破坏少量信息素，在触角表皮吸收了信息素之后减缓漏进孔内的信息素。这种缓慢而持续的信息素泄漏会产生系统的高背影干扰。

The degradation process of sex pheromones might be related to aldehyde oxidases (AOXs), aldehyde reductases (ARs), and carboxylesterases (COEs). AOX and AR can convert aldehyde sex pheromone components to the corresponding acids and alcohols, respectively. The female sex pheromones of *A. pernyi* have been identified as a blend of *E*6, *Z*11-16：Ald, *E*6, *Z*11-16：Oac, and its chain shortened form, *E*4, *Z*9-14:OAc with a ratio of 100：20：40. 141 candidate genes that are putatively related to pheromone biosynthesis, degradation, and chemoreception in *A. pernyi*.

性信息素的降解过程可能与醛氧化酶（AOXs）、醛还原酶（ARs）和羧酸酯酶（COEs）有关。AOX 和 AR 可分别将醛性信息素组分转化为相应的酸和醇。柞蚕雌性信息素已鉴定为 *E*6、*Z*11-16:Ald、*E*6、*Z*11-16：Oac 及其链缩短形式。*E*4、*Z*9-14：Oac 的混合物的比例为 100:20:40。柞蚕的 141 个候选基因被推测与信息素生物合成、降解和化学接收有关。

13.4.4 Pheromone signal processing

Signal processing has been studied intensively in only a few insects. Males of *M. sexta* have approximately 10^5 sensilla with single walls and pores that house about 3×10^5 receptor neurons. Each sensillum typically contains two sensory neurons, one that is sensitive to (*E*,*Z*)- 10,12-hexadecadienal (*E*,*Z*-10,12)-16：AL and the other sensitive to (*E*,*E*,*Z*)-10,12,14-hexadecatrienal (*E*,*E*,*Z*)-10,12, 14-16：AL, the two main components of the eight C16 aldehydes in the *M. sexta* female-produced pheromone. All eight components are important and give the best results in the field, but detailed neurophysiological data are only available for the two main components. The antennal receptor neurons respond to pheromone aldehydes by opening Na^+, K^+, and Ca^{2+} channels, which are not ligand (pheromone) gated, but mobilized through the second messenger system of G-proteins. Axons from olfactory receptors on the antennae pass through the antennal nerve and enter the large antennal lobe (AL) of the deutocerebrum.

In moths, the antennal nerve breaks into two branches as it enters the AL. One branch carries axons from non- pheromone olfactory

13.4.4 信息素的信号处理

仅几种信息素的信号进程有详细的研究。烟草天蛾的雄虫中约有 10^5 个单壁和单孔感器，包被约 3×10^5 个受体神经元。每个典型的感器包括两个受体神经元，一个对（*E*,*Z*）-10,12-（即反 10，顺 12-）十六碳二烯醛敏感，另一个对（*E*,*E*,*Z*）-10,12,14-十六碳三烯醛敏感。这是烟草天蛾雌虫产生的 8 种 16 碳醛类信息素的两种主要成分。所有 8 种成分都很重要，并在田间产生最佳效果。但只有两种有详细的神经生理学资料。触角受体神经元通过打开 Na^+、K^+ 和 Ca^{2+} 通道对醛类信息素做出反应。它们不是通过配体（信息素）门控的，而是通过 G– 蛋白的第二信使系统调用。从触角上的嗅觉受体来的轴突，通过触角神经并进入中脑大触角叶（AL）。

蛾类的触角神经在进入 AL 时被分成两个分支。一个分支的轴突从非信

receptors to glomeruli in a mechano- and taste-sensitive region of the AL, while the second branch carries axons from pheromone receptors to glomeruli in the macroglomerular complex (MGC). In the MGC, incoming axons synapse with antennal lobe interneurons that interconnect parts of the AL (local interneurons) or pass to other parts of the brain, including the protocerebrum (projection interneurons). Output neurons from this region project along three parallel pathways, the medial, mediolateral, and lateral tracts.

All lateral-tract MGC neurons targeted the column, a small region within the superior intermediate neuropil as a unique converging site for MGC lateral-tract neurons responsive to pheromones, as well as a dense congregating site for plant odor information since a substantial number of lateral-tract neurons from ordinary glomeruli also terminates in this region (Chu et al., 2020). In male moths, the glomeruli dedicated to sex pheromone signal integration have a larger volume than ordinary glomeruli and gather in the MGC. The size of glomeruli within the AL is, however, subjected to plasticity as a function of for example age, rearing conditions and experience. In *S. littoralis*, inactivation of the receptor specifically affected the responses of the corresponding antennal neurons did not impact the number of MGC in the antennal lobe but reduced the size of the functional MGC (Koutroumpa et al., 2022).

In all Lepidoptera that have been investigated, the MGC is a sex-specific region found only in the AL of males. In male *M. sexta* the large globular glomerulus near the point at which the antennal nerve enters the AL is called the cumulus (because of its resemblance to the cumulus cloud shape), and the ring or donut-shaped area beneath it is called the toroid. Similar glomeruli varying in shape and size are also known in a number of other male moths, including *B. mori*, *A. polyphemus*, *A. segetum*, *T. ni*, *S. littoralis*, *H. zea*, and *H. virescens*. All have one large glomerulus at the entrance of the antennal nerve into the AL and several smaller satellite glomeruli beneath the large one.

Glial cells invest the glomeruli and provide protection. In addition to glomeruli, the AL contains lateral and medial groups of cell bodies of the interneuron associated with the glomeruli. The cell bodies of the primary olfactory neurons are located peripherally in the antennae, as is characteristic of sensory neurons in insects.

息素嗅觉感受器到触角的机械和味觉敏感区的神经纤维球，而第二个分支的轴突从信息素受体到巨型神经纤维球（MGC）内的神经纤维球。在MGC内，传入轴突与触角叶中间神经元突触连接，使之与触角叶的部分相连（局部联络神经元）或传到脑的其他部分，包括前脑（投射联络神经元）。该区域的输出神经元沿三条平行路径投射，即内侧、中侧和外侧束。

所有侧束MGC神经元都靶标于上中间神经毡内的一个小柱形小区，是MGC侧束神经元响应信息素的唯一汇聚点，也是植物气味信息的密集聚集点，因为来自普通神经纤维球的大量侧束神经元也终止于该区域（Chu 等，2020）。在雄蛾中，专门用于性信息素信号整合的神经纤维球球体积大于普通神经纤维球，并聚集在MGC的特定AL区域。然而，触角叶内神经纤维球的大小由于虫龄、饲养条件和经验等因素而具有可塑性。在斜纹夜蛾中，受体的失活特异性地影响相应触角神经元的反应，不影响触角叶中MGC的数量，但减少了功能性MGC的大小（Koutroumpa 等，2022）。

对所有鳞翅目昆虫的研究发现，只有雄虫的触角叶内的MGC是性专一的。烟草天蛾雄虫的大球状神经纤维球靠近触角神经进入触角叶的位点称为“积云”（因为它的装配形状像积云），其下面是环状或同心圆环状的区域，称为圆环面。其他一些蛾类也有相似的神经纤维球，形状和大小有变化，如家蚕、美洲野蚕、黄地老虎、粉纹夜蛾、斜纹夜蛾、美洲棉铃虫和烟芽夜蛾。在触角神经进入AL都有一个大神经纤维球，它的下面有几个稍小的卫星神经纤维球。

胶细胞包埋了神经纤维球并提供保护。除了神经纤维球，触角叶包含与之相连的联络神经元的侧神经和中神经细胞体。原发性嗅觉神经元的细胞体位于触角外周，是昆虫感觉神经元的特征。

M. sexta males can detect the pheromone quality, quantity of pheromone, and the frequency of pulses in pheromone plumes. Some pheromone receptor neurons in the male act as pheromone generalists that respond to either of the two aldehyde components or to the total pheromone blend, while other pheromone specialists discriminate between the two aldehydes and respond differently to them. Thus, the response to either aldehyde or to a blend of both is sent to the MGC as information about the blend (i.e., its quality).

烟草天蛾的雄虫能检测信息素的质量和数量以及信息缕流的脉冲频率。有些信息素受体神经元起着"普通信息素"的作用，对两种醛类组分或全部信息素混合物反应，而另一些"专化信息素"可识别两种醛类，并产生不同的反应。这样对醛的反应或对混合物反应都作为信息输送到 MGC（即其质量）。

A subset of pheromone-specialist neurons provides further discrimination by responding in opposite ways to the two aldehydes. For example, some of the neurons are stimulated by (*E*, *Z*)-10,12-hexadecadienal, resulting in excitation, exposure to (*E*, *E*, *Z*)-10,12,14-hexadecatrienal inhibits these same neurons. The opposite scenario can also occur, that is, (*E*, *E*, *Z*)-10,12,14- hexadecatrienal may stimulate some neurons while (*E*, *Z*)-10, 12-hexadecadienal inhibits them. A blend of aldehyde components results in a unique mixture of inhibitory and excitatory responses, depending on the mixing of these two input channels. These pheromone specialists respond and recover rapidly enough to detect the natural intermittent pheromone release by the female at frequencies of about ten pheromone plumes or pulsed releases per second.

"专化信息素"的亚类神经元通过对两种醛类相反的反应途径进一步的区分。比如，有些神经元受到（*E*, *Z*）-10,12-十六碳二烯醛刺激，导致兴奋，浸泡在（*E*, *E*, *Z*）-10,12,14- 十六碳三烯醛抑制了相同的神经元。相反的局面也可能出现，即（*E*, *E*, *Z*）-10,12,14- 十六碳三烯醛刺激某些神经元，而（*E*, *Z*）-10,12- 十六碳二烯醛抑制它们。醛类组分的混合物依赖于这两种化合物输入通道的混合，导致抑制和激活反应的特有的混合。这些"专化信息素"对雌虫自然间歇性释放的信息素能够快速反应和恢复，其频率约为每秒 10 个信息素缕流或脉冲。

In some moth species (*A. segetum*, *S. littoralis*, and *M. sexta*) receptor neurons responding to different pheromone components synapse in different glomeruli in the MGC. For example, receptor neurons of male *M. sexta* that respond when one of the sex pheromone components [(*E*, *Z*)-10.12-hexadecadienal] is blown onto the antenna have terminal arborizations in the toroid, while receptors responding to (*E*, *E*, *Z*)-10,12,14-hexadecatrienal, a second component, terminate in the cumulus (Hansson et al., 1991). In some other species (*H. virescens* and *A. polyphemus*) receptor neurons tuned to different pheromone components synapse in the same MGC glomeruli, with some neurons also possibly projecting to a second glomerulus. A number of examples are known of male moths that have receptor neurons on the antennae sensitive to one or more components that inhibit pheromone response.

据报道，有些蛾类品种（如黄地老虎、斜纹夜蛾和烟草天蛾）受体神经元在 MGC 内对不同的信息素组分的反应有不同的神经纤维球突触连接。比如，烟草天蛾雄蛾的受体神经元，对性信息素组分（*E*, *Z*）-10,12- 十六碳二烯醛的终端树突位于环形室，而对第二个组分，（*E*, *E*, *Z*）-10,12,14- 十六碳三烯醛的反应受体终止于"积云"（Hansson 等，1991）。其他一些昆虫品种（美国野蚕和烟蚜夜蛾）的受体神经元调整为相同 MGC 内的神经纤维球对不同的信息素组分突触连接，有些神经元也可能伸出到第二个神经纤维球。在雄蛾触角上，有些受体神经元对一种或几种抑制信息素反应的组分敏感。

These inhibitory components may serve as isolating mechanisms for closely related species that share blend components. In all cases known, the receptor neurons that detect inhibitory compounds synapse in a glomerulus that does not receive input from pheromone components. Male and female insects have dimorphic

这些抑制性组分可在共享混合物成分的邻近种间起着隔离机制。在所有已知情况下，神经纤维球内检测抑制性化合物突触的受体神经元不接收信息素组

numbers of receptors sensitive to sex pheromone on the antennae. Moths and cockroaches, in particular, have large numbers of receptors on the male antennae that respond mainly to the sex pheromone produced by females. Sex-specific sex pheromone receptors do not occur (or do not occur in large numbers) on the antennae of female moths and cockroaches. In mosquitoes, only female adult need a protein-rich blood meal to acquire nutrients necessary for eggs maturation after mating, while males do not feed on blood. Therefore, female specific or female enriched OBPs may be more specifically involved in host-seeking behavior.

分的输入。雌性昆虫和雄性昆虫的触角上对性信息素敏感的受体数目具有双向性，特别是蛾和蜚蠊的雄虫触角上有大量受体，主要对雌虫产生的性信息素响应。它们的触角上没有（或没有大量）性别特异性信息素受体。蚊子只有雌性成虫才需要取食蛋白质丰富的血液，以获得交配后卵成熟所需的营养，而雄虫不需要取食血粉。因此，雌特异性或雌富含的 OBPs 可能对寄主搜寻行为更为特殊。

13.5 信息素合成和释放的激素调控

13.5 Hormonal Control of Pheromone Synthesis and Release

13.5.1 Hormonal control of pheromone synthesis and release

Pheromone production in some insects, and perhaps in most if not all Lepidoptera, is under hormonal control. Early evidence for possible hormonal control of pheromone production and/or release came from studies on the cockroaches *B. fumigata* and *P. surinamensis*. Allatectomized cockroaches do not produce sex pheromone, leading to the hypothesis that JH from the corpora allata is probably a pheromone-tropic hormone in cockroaches. A polypeptide hormone that controls the synthesis of the sex pheromone in moths, called PBAN (pheromone biosynthesis activating neuropeptide) (Raina et al., 1984), was first isolated from the subesophageal ganglion of the moth *H. zea*, and determined to consist of 33 amino acids with M.W. of 3,900 kDa. The PBANs belong to a class of peptides called pyrokinins and those isolated thus far have 33 to 34 amino acid residues. All PBANs share, with other pyrokinins, a common C-terminal sequence of five amino acids-Phe-X-Pro-Arg-Leu-NH_2 (X can be Gly, Ser, Thr, or Val) that is required for biological activity. Substitution in file X position is extremely critical to pheromonotropic activity; putting glycine in the X position causes loss of activity, whereas a molecule

13.5.1 信息素合成和释放的激素控制

有些昆虫，如不是全部，也许是多数鳞翅目昆虫的信息素分泌都受到激素的调控。早期对蜚蠊，古巴挖地蟑螂和 / 或蔗蠊的研究证实了信息素的产生和释放可能受到激素调控。切除咽侧体的蜚蠊不产生性信息素，导致咽侧体产生的 JH 可能是蜚蠊促信息素激素的假设。控制蛾类性信息素合成的多肽激素，称为 PBAN（信息素生物合成激活神经肽）（Raina 等，1984），最先是从美洲棉铃虫的咽下神经节分离出来的，有 33 个氨基酸，分子质量为 3 900 kDa。PBANs 属于一类焦激肽，截至目前，分离到的有 33 ～ 34 个氨基酸残基。所有 PBANs 和其他焦激肽的共享序列是 C- 末端的 5 个氨基酸—苯丙氨酸 -X- 脯氨酸 - 精氨酸 - 亮氨酸 -NH_2（X 可以是甘氨酸、丝氨酸、苏氨酸或缬氨酸）是生物学活性必需的。X 位点的取代对激素活性极为关键；把甘氨酸代入 X 可引起活性的丧失，而分子的 X 位

containing threonine in position X is active.

Pyrokinin-type peptides have several different functions in insects. For example, one has myotropic activity in an *in vitro* cockroach hindgut assay and another (Bom-DH) functions as an egg diapause hormone in *B. mori.* Some degree of cross-reactivity in the different bioassays is common, and is probably due to the characteristic C-terminal peptide sequence. Some of the cockroach and locust pyrokinins also stimulate sex pheromone synthesis in *B. mori* females.

There may be a number of neuropeptides involved with pheromone production, and that some of the neuropeptides may have other physiological functions as well. Genes encoding for PBAN in *H. zea* and in *B. mori* encode for propheromones. In *B. mori*, the propheromones give rise to Bom-PBAN, Bom-DH, a peptide with homology to Pss-Pt (a pheromonotropic peptide from the armyworm *P. separata*), and other peptides.

置含有苏氨酸可增强其活性。

昆虫的焦激肽类的多肽具有几种不同的功能。例如，一种在蜚蠊后肠分析中具有体外肌原活性，而另一种（Bom-DH）在家蚕中起着卵滞育激素作用。可能是由于C端序列的特性，不同的生物学分析交互反应的某些程度是相同的。有些蜚蠊和蝗虫的焦激肽也能刺激家蚕雌蛾合成性信息素。

信息素产生可能牵涉一些神经肽，有些神经肽还有其他生理学功能。美洲棉铃虫和家蚕编码PBAN的基因都编码信息素原。家蚕信息素原产生Bom-PBAN、Bom-DH、一个与Pss-Pt同源的肽（来自黏虫的一个促信息素肽）和其他肽。

13.5.2 Mode of action of PBAN

In the red banded leaf roller, *A. velutinana*, PBAN regulates pheromone biosynthesis by increasing the supply of octadecanoyl and hexadecanoyl fatty acids needed for pheromone biosynthesis. PBAN, regulates Δ^{11} desaturase, an enzyme widely distributed in Lepidoptera. The enzyme introduces a double bond into the pheromone precursor in *M. brassicae* and *C. chalcitex*. In two lepidopterans, *B. mori* and *S. littoralis*, PBAN influences the reduction of the fatty acid to the alc-precursor of the pheromone.

PBAN may control pheromone biosynthesis through regulation of fatty acid biosynthesis in *H.zea*, or it may control a step prior to fatty acid synthesis (Fig.13-4).The sex pheromone of *H.zea* is secreted from glandular cells in the intersegmental membrane located between the 8th and 9th abdominal segments. The major component (up to 92%) of the pheromone is (*Z*)-11-hexadecenal, derived from fatty acid metabolism. The females normal synthesize and release the pheromone during the scoto- phase period, and only then can it be detected in the glandular tissue; little or no pheromone occurs in the gland during the photo phase. The hormone is detectable in the hemolymph only during the time when the pheromone is being produced. Injection of exogenotic PBAN into a female moth can cause pheromone synthesis independent of the photoperiod. As little as 0.06 pmol of Hez-PBAN injected

13.5.2 PBAN 的作用模式

在红斑卷叶虫和红带卷蛾中，PBAN通过增加信息素生物合成所需的十八酰基和十六酰基脂肪酸的供应来调节。PBAN调节Δ^{11}去饱和酶，该酶在鳞翅目昆虫中广为分布。在甘蓝菜蛾和铗纹夜蛾中，这种酶在信息素前体分子引入一个双键。在两种鳞翅目昆虫中，家蚕和灰翅夜蛾，PBAN会影响脂肪酸还原成信息素的前体醇。

谷实夜蛾的PBAN通过调节脂肪酸的生物合成，或控制脂肪酸合成的前一步，控制信息素的生物合成（图13-4）。其性信息素是由第8到第9腹节的节间膜上的球形细胞分泌的。信息素的主要成分（～92%）是脂肪酸代谢衍生的顺11-十六碳醛。雌虫正常合成和分泌信息素是在暗期，只有这时能在球形组织中检测到。在光期，腺体中有少量或检测不到信息素。血液中的信息素也只能在合成时间内检测到。对雌蛾施用外源的PBAN可使信息素的合成不依赖于光期。刺激信息素的生物合成仅需要0.06 pmol的Hez-PBAN注射到雌

into a female moth stimulated pheromone biosynthesis, and 2 pmol stimulated maximum pheromone synthesis; higher doses produced no additional increase in synthesis.

蛾；2 pmol 的刺激达到合成的最大值；更高的剂量不产生合成的额外增加。

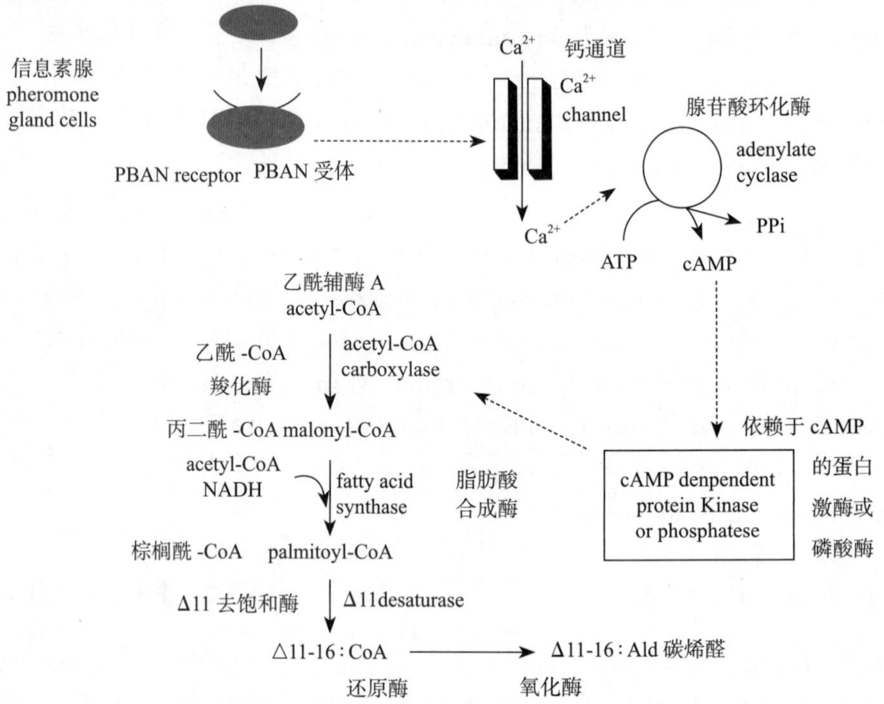

Fig.13-4 Proposed model for PBAN-mediated stimu lation of pheromone bio synthesis at the pheromone gland in the female corn earworm, *H. zea* (modified from Tillman et al., 1999).

图 13-4 谷实夜蛾信息素腺 PABN 对信息素生物合成的介导促进作用（仿 Tillman 等，1999）。

13.5.3 The originate, transport and release of PBAN

PBAN is produced in the subesophageal ganglion (SEG) of *H. zea*, *B. mori* and *H. assulta* (Xu et al., 2000). In studies with *H. zea*, PBAN is released from the SEG and transported by the hemolymph to the pheromone gland, its target. Teal et al. (1989) have evidence that PBAN may be transported through the ventral nerve cord to the terminal abdominal ganglion (TAG), where it causes the release of some second messenger that acts on the pheromone gland cells.They propose that the target for PBAN is the terminal abdominal ganglion (TAG), and that nerves from the TAG must be intact to the pheromone gland for extracts of brain-SEG applied to the TAG, to ultimately elicit pheromone synthesis. In the female European corn borer, *O. nubilalis*, that PBAN was synthesized in three sets of NSC in the SEG and released from the corpora cardiaca. Although PBAN immunoreactivity was present through- out the ventral nerve cord, complete removal of the nerve

13.5.3 PBAN 起源、转运和释放机制

在美洲棉铃虫、家蚕和烟实夜蛾（Xu 等，2000）中，PBAN 是由咽下神经节产生。美洲棉铃虫的研究中通过血液循环被转运至其靶器官信息素腺体。Teal 等（1989）认为，PBAN 可通过神经索转运到腹末神经节（TAG），导致它释放出第二信使并作用于信息素腺体细胞。他们认为 PBAN 的靶器官是 TAG，那么从 TAG 来的神经对信息素腺体而言必须是未经活化的，脑 – 咽下神经节抽提物施用到 TAG，最大限度地激发信息素的合成。欧洲玉米螟雌虫的 PBAN 由咽下神经节的 3 套神经分泌细胞产生，并通过心侧体释放。虽然 PBAN 的免疫反应出现在整个腹神经索，但完全去除腹神经索也不能改

cord did not alter female response to exogenous PBAN. In the gypsy moth, *L. dispar*, PBAN-immune- reactive material can be detected in the SEG and cells in each segmental ganglion, and transection of the ventral nerve cord disrupts pheromone production in females. All life stages of *P. xylostella* have similar PBAN immunoreactivity patterns in the central nervous system, and the accumulation of PBAN is similar in all life stages except adult males. PBAN immunoreactive signals were observed in the brain and SEG, and fluorescent signals from SEG extended to the entire length of the ventral nerve cord and terminated at the terminal ventral ganglion (Chen et al., 2019). It thus appears that the origin, transport, and release mechanism of PBAN vary with species.

变雌蛾对外源性 PBAN 的响应。吉卜赛蛾的 PBAN 免疫化学物质可在咽下神经节和每一个体神经节细胞中被检测到，横切腹神经索即可打断雌蛾信息素的产生。小菜蛾的所有生命阶段在中枢神经系统中具有相似的 PBAN 免疫反应模式，除成年雄性外，所有生命阶段中 PBAN 的积累相似。在脑和 SEG 中观察到 PBAN 免疫反应信号，来自 SEG 的荧光信号延伸到腹侧神经索的整个长度，终止于腹末神经节（Chen 等，2019）。因此，PBAN 的起源、转运和释放机制因种而异。

13.5.4 Biosynthesis of pheromones

Most insects that have been studied synthesize their pheromonal components from small metabolic pool precursor molecules. Some insects modify precursors obtained in the food. For example, *scolytid* bark beetles use one or more of the terpenes in the host tree they feed on as a pheromone precursor. Males of *B. dorsalis* fruit flies are strongly attracted to a Hawaiian lei flower, feed on phenylpropanoid compounds on the petals, and use the compounds to make a pheromone that attracts female flies. Many moths use mono-unsaturated alcohols, acetates, or aldehydes with 12 to 16 carbons as pheromonal components. The starting material for synthesis of these pheromonal components is a saturated fatty acid synthesized from the acetate pool. Some moths use diunsaturated pheromonal components, and two double bonds can be introduced by the Δ^{11} desaturase acting before and after chain-shortening. Deuterium-labeled palmitic and myristic acids were used to demonstrate production of (*E*)-11-14 acetate and (*E*, *E*)-9,11 acetate, major components of the pheromone of *E. postvittana* (Fig.13-5; Appendix color graphic Fig.13-5).

In addition to the Δ^{11} desaturase, a Δ^{10} desaturase has been identified in the New Zealand leaf roller *P. excessana*, and a Δ^{9} desaturase in the brown-headed leaf roller *C. obliquana*. It is speculated that in the early evolutionary stages of these Tortricidae, the β-oxidation step is used to shorten oleic and/or palmitoleic acids from which shorter-chain pheromone components were synthesized. Later, evolution of the Δ^{10} and Δ^{11} desaturases make it possible for the moths to biosynthesize a wide range of unique pheromonal components and may have paved the way for evolution of many different species-specific blends.

13.5.4 信息素的生物合成

已经研究了多数昆虫从小代谢池前体分子合成信息素成分。有些昆虫从食物中获取修饰前体。例如，刺胫小蠹皮甲利用其取食的树上的一种或多种萜烯作为信息素前体物。雄性东方果实蝇被夏威夷花环花深深地吸引住，取食花瓣上的苯基类丙烷成分，利用这一成分产生信息素吸引雌蝇。许多蛾类利用 12 ～ 16 个碳原子的单不饱和醇、乙酸酯或醛作为信息素成分。合成这些信息素成分的起始原料是从乙酸酯合成的饱和脂肪酸。有些蛾用二不饱和信息素成分，在碳链缩短前后通过 Δ^{11} 去饱和酶的作用引入两个双键。氘标记棕榈酸和肉豆蔻酸被用于论证苹果褐卷叶蛾信息素的主要成分，(*E*) -11-14 醋酸酯和 (*E*, *E*) -9,11 醋酸酯的产生（图 13-5；附彩图 13-5）。

除了 Δ^{11} 去饱和酶，在新西兰卷叶蛾中分离到 Δ^{10} 去饱和酶，在褐头卷叶蛾中分离到 Δ^{9} 去饱和酶。据此猜测，在卷叶蛾进化早期，β- 氧化步骤用于缩短油酸和 / 或棕榈酸，因而合成出较短链的信息素化合物。其后，Δ^{10} 和 Δ^{11} 去饱和酶进化，使得蛾类能生物合成大范围的独特信息素成分，为许多种间混合物的进化铺平了道路。

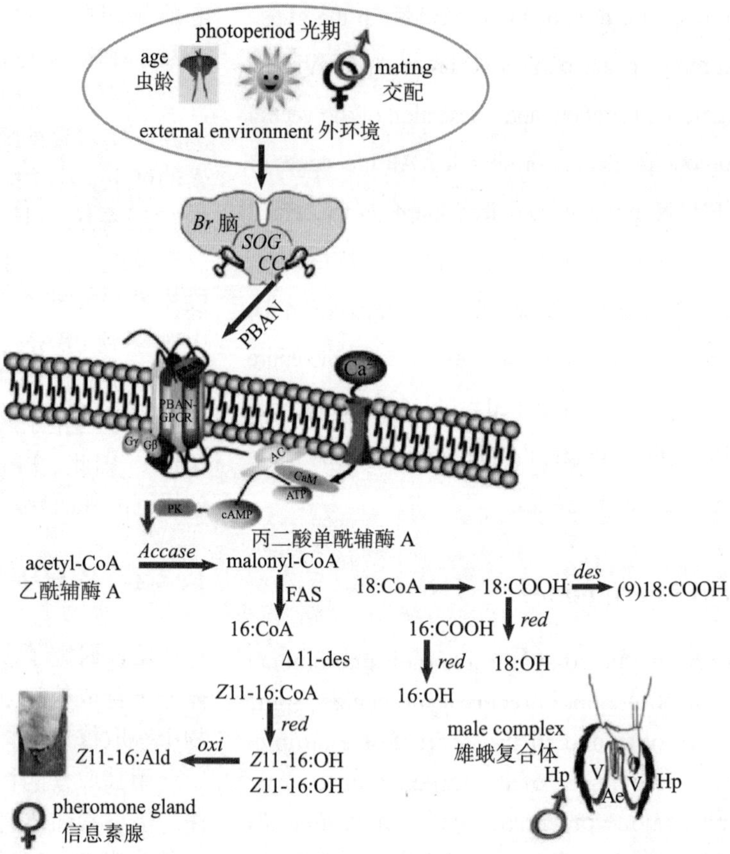

Fig.13-5 A diagrammatic representation of sex pheromone biosynthesis resulting from PBAN release into the hemolymphand its interaction with the PBAN GPCR in female pheromone glands (PG) and male hair pencil- aedaegus-complexes of *H. armigera*. AC. adenylate cyclase; ACCase. acetyl-CoA carboxylase; Br. brain; CaM. calcium-calmodulin; cAMP. cyclic-AMP; CC. corpora cardiaca; Δ^{11}-des. Δ^{11} desaturase; FAS. fatty acids ynthetase; GPCR. G-protein coupled receptor; oxi. oxidase; PBAN. pheromone biosynthesis activating neuropeptide; PK. proteinkinaseA; red. reductase; SOG. subesophageal ganglion. V. valves; HP. hairpencil; Ae. aedaegus(modified from Jurenka et al., 2011).

图 13-5 棉铃虫释放 PBAN 进入血液并与雌性信息腺（PG）和雄性笔状阳茎复合体的 PBAN GPCR 相互作用，引起信息素生物合成的图解描述。AC. 腺苷酸环化酶；ACCase. 乙酰辅酶 A 羧化酶；Br. 脑；CaM. 钙 - 钙调蛋白；cAMP. 环腺苷酸；CC. 心侧体；Δ11-des. Δ11 去饱和酶；FAS. 脂肪酸合成酶；GPCR. G- 蛋白耦联受体；Oxi. 氧化酶；PBAN. 信息素生物合成激发神经肽；PK. 蛋白激酶 A；red. 还原酶；SOG. 咽下神经节；V. 瓣膜；HP. 毛鬃；Ae. 阳茎（仿 Jurenka 等，2011）。

(Li Wenchu)

（李文楚）

Exercises

1.What is pheromone? How to classify?

2.How does the silkworm moth detect the sex pheromone released by the seduction gland of the female moth and initiate the zigzag upwind search behavior?

3. How is sex pheromone synthesis regulated by PBAN?

思考题

1. 什么是信息素？如何分类？

2. 家蚕蛾是如何检测雌蛾诱惑腺释放的性信息素，启动之字形逆风搜寻行为的？

3. 性信息素合成如何受 PBAN 调控？

第 14 章 昆虫丝腺及丝蛋白质合成

Chapter 14 Insect Silk Gland and Silk Protein Synthesis

Key points

1.Not only Lepidoptera insects can spin silk, but also others, such as Diptera, Coleoptera Trichoptera and Hymenoptera insects can spin silk and cocoon as well, which is considered as one of the excretion behaviors of these insects. The silk matcrial of silkworm moths has had a profound impact on human development.

2.Due to the special structural features of silk glands, each silk gland division has special functions in the biosynthesis of silk fibroin and sericin.

3.Silk fibroin protein is composed of light chain, heavy chain and P25, antiparallel β- sheet. The content is up to 50%, and its gene structure has been clarified, providing an important material for molecular biomimetic.

4 Silkworm and other lepidopteran insects have colorful cocoons, which reflect that heredities, that and are mainly regulated by genes.

5.Technical measures should be taken in cocoon production to avoid cocoon free silkworm. The progress of science and technology promotes more extensive utilizations.

知识点

1. 不仅鳞翅目昆虫能吐丝，而且其他如双翅目、鞘翅目、毛翅目和膜翅目昆虫也能吐丝和营茧，这被认为是这些昆虫的排泄行为之一。蚕的蚕丝原料对人类的发展产生了深远的影响。

2. 由于丝腺的特殊结构特征，各丝腺分区在丝心蛋白和丝胶生物合成过程中具有特殊的功能。

3. 家蚕丝心蛋白由轻链和重链以及 P25 组成，反平行 β- 折叠片含量达 50%，其基因结构已明确，这为分子生物仿生学提供了重要材料。

4. 家蚕和其他鳞翅目昆虫的茧形多姿多彩，颜色丰富，体现了遗传特性主要受到基因的调控。

5. 蚕茧生产上应采取技术措施，可避免不结蚕茧。科学技术的进步拓宽了茧丝更为广泛的用途。

Silk fibers are fine filaments produced by the silkworm, *B. mori*, and other arthropods, including mites, spiders, moths, butterflies, ants, scorpions, fleas, wasps and bees, kinds of lustrous fibrous protein polymers. Most silk producing arthropods have four different stages in their life cycle. In the final stage of the larva, it begins to secrete silk fibers from its mouth and rotate them around its body to form a protective structure called cocoon.

丝纤维是由家蚕和其他节肢动物，包括螨、蜘蛛、蛾、蝴蝶、蚂蚁、蝎子、跳蚤、黄蜂和蜜蜂等产生的细丝，是一种有光泽的纤丝蛋白聚合物。大多数产丝节肢动物在其生命周期中有四个不同阶段。在幼虫的最后阶段，它开始从口器分泌丝纤维，并将其绕着身体旋转，形成一种称为茧的保护结构。

Except those Lepidoptrean in insecta spinning silk, there are still

在昆虫纲内，除了那些鳞翅目吐

other insects also spinning, mainly Diptera (*Simulium vittatum* and *Simulium. tribulatum*), Hymenoptera, coleoptera, and some aquatic insects in Trichoptera, such as limnephilid larvae of *Pycnopsyche guttifer* (Walker) and *Neophylax concinnus*, *Stenopsyche marmorata* McL., *Pycnopsyche designatus* Walker (Marshell, 1915; Hatano et al., 2014), caddis worm (*H. nigrovittatus*) (Kim et al. 2020), *R. obliterate*; *H. angustipennis* (Yonemura et al. 2009) (Trichoptera: Limnephilidae) (Engster M.S.,1976); *Polyrhachis dives* (Fountain et al., 2011) , etc. They use their mouthparts, foreleg, digestive tracts, and skin secreting glands to produce sick.

The members of family Saturniidae are one of the largest and most spectacular of the lepidopteran insects and estimated 1,300-1,500 different species distributed worldwide. The *Bombyx mandarina* Leech, and wild saturniid silkworm *Antheraea mylitta* are also holometabolous insects that complete their life cycle else in four stages. The non-mulberry silkworms comprise the tropical tasar (*A.mylitta*), the Chinese oak tasar (*A. pernyi*), the muga (*A.assamensis*) and the eric silkworm (*P.ricini*),etc., and still other lepidopteran silk spinning insects such as ghost moth (*H. californicus)*, *Corcyra cephalonica*, *Ephestia kuehniella*, etc.

B. mori is a lepidopteran insect, which is often used as a model animal for entomological and zoological research. Domestic silk materials are widely used, not only for deep processing into raw silk, silk and satin clothing products, but also as biomaterials in beauty and chemical industry, medical drug release, surgical artificial tissue reconstruction surgery, and even military industry.

丝，还有其他昆虫，主要是双翅目（带蚋、*S. tribulatum*）、一些膜翅目、膜翅目和水生毛翅目昆虫也吐丝，如沼石蛾科的幼虫 *P. guttifer*（Walker）、*N. concinnus* McL.、石蛾（Marshell，1915；Hatano et al., 2014）、*P.designatus*（Marshell，1915）石蚕（*H. nigrovitt-atus*）（Kim 等，2020）、齿肢原石蛾；石蚕蛾（Yonemura 等，2009）的幼虫（毛翅目：狐蝠科）（Engster M.S., 1976）和黑棘蚁（Fountain 等，2011）等。

大蚕蛾科昆虫是鳞翅目昆虫中最大、最壮观的一种，估计有 1 300 ～ 1 500 个种类分布于世界各地。野蚕和野生天蚕也是完全变态昆虫，其生命周期分四个阶段完成。非桑蚕包括热带柞蚕、中国柞蚕、琥珀蚕和蓖麻蚕等，还有其他鳞翅目吐丝昆虫，如鬼蛾（虫草蝠蛾）、米蛾和地中海粉螟等。

家蚕是一种鳞翅目昆虫，常被用作昆虫学研究和动物学研究的模式动物。家蚕丝物质的用途非常广泛，不仅用于深加工成生丝和绸缎服装等产品，也作为生物材料用于美容化工、内科药物释放、外科人工组织重构手术，甚至军工等领域。它已成为一种被研究得非常透彻的天然丝蛋白质。

14.1 昆虫丝物质简述

14.1 Brief Introduction of Insect Silk Substances

Silk of *Bombyx* is composed of 70%-75% fibroin and 20%-25% sericin, both of which are mainly protein (Ho et al., 1944). A silk cocoon is a natural polymer composite. A considerable part of the

家蚕的蚕丝由 70% ～ 75% 的丝心蛋白和 20% ～ 25% 的丝胶组成，两者主要成分是蛋白质（Ho 等，1944）。蚕

mass and energy obtained by silkworm from larval stage is used to produce silk for cocooning, and then spend most of its life in cocoon as pupa before eclosion into moth.

Wild cocoons are considered to protect pupae from the environment, parasites and predators. Domestic silkworms are species cultivated by human beings and have provided silk fibers for the textile industry for more than 5,000 years.

The caterpillar has a flexible adaptive response to the structural features of its environment. Usually the larva starts by constructing a macrostructure of a widely spaced supporting network of parallel strands, and stretching and turning its segmented body to survey potential attaching substrates. Twigs or leaves can be pulled around for spinning peduncles or attachments. Within this framework the outermost cover of the cocoon is then built to form a characteristic shape. The larva turns it body around by swinging its head in a figure of "eight" motion to overlap silk fibers, making layers of silk sheets producing a cocoon with uniform thickness and symmetrical shape.

Cocoons can be suspended from the twigs of host plants through a silk attachment called cocoon stalk, attached along the twigs or other supports, or build cocoons between the leaves of host plants that do not attach twigs. The construction strategies depend on both the spinning insectl's species and the environments. Silkworm cocoons in terms of relative size, robustness, structure, and composition of cocoon wall, shape, color, attachment, and orientation, might relate them to these properties to the biological functions of the cocoon.

The composition, structure and properties of silk vary according to its specific source and function. Spiders and insects secrete glycine rich silk, which is characterized by unique synthetic and processing properties, strength and ductility.

茧是一种天然聚合物复合材料。蚕从幼虫期获得的质量和能量中有相当大的一部分用于产丝结茧，然后在羽化成蛾前以蛹的形式在茧中度过其一生的大部分时间。

野生蚕茧被认为可保护蛹免受环境、寄生和捕食者的侵害，而家蚕是人类培育的物种，为纺织工业提供丝绸纤维已有 5 000 多年的历史。

毛虫对其环境的结构特征具有灵活的自适应性反应。在通常情况下，幼虫首先构建一个由平行链组成的大空间支撑网络的宏观结构，然后拉伸和转动其分节的身体，以勘测潜在的附着基质。树枝或树叶可以用来作旋转的茧梗或附着物。在这个框架内，蚕茧的最外层被构建成一个独特的形状。幼虫以“8”字形摆动头部，使其身体转动，以重叠丝纤维，并使丝片层层堆叠，形成厚度均匀、形状对称的茧。

茧可以通过一种被称为茧梗的丝附着物从寄主植物的小枝上悬挂下来，沿着树枝或其他支撑物附着，或者在不附着小枝的寄主植物的叶片之间构建茧。构建策略取决于吐丝昆虫的品种和环境。蚕茧的相对大小、坚固性、茧壁结构、组成、形状、颜色、附着和方向，这些茧的特性可能与生物功能相关。

丝的成分、结构和性能因其特定来源和功能而异。蜘蛛和昆虫分泌富含甘氨酸的丝，其特点是具有独特的合成和加工特性、强度和延展性。

14.2 丝腺的结构

14.2 The Structure of Silk Gland

The silk gland of silkworms composed of fusulus, anterior silk gland (ASG), middle silk gland (MSG), and posterior silk gland (PSG) (Fig.14-1; Appendix color graphic Fig.14-1).

蚕的丝腺由吐丝器、前部丝腺（ASG）、中部丝腺（MSG）和后部丝腺（PSG）组成（图 14-1；附彩图 14-1）。

14.2.1 Fusulus

It is located at the head and is a thin tube, which is opened with a silk spinning hole. According to their different forms and skills, they can be divided into three parts: spinning tube, extracting area, common tube and Filippi's glands (Fig.14-2; Appendix color graphic Fig.14-2).

Spinning tube also known as front area. It is a thin tube, which refers to the part protruding out of the lower lip and opening at the apex of the silk tube (the silk hole), which is the channel for cocoon silk to spit out.

14.2.1 吐丝器

吐丝器位于头部，呈细管状，以吐丝孔开口。根据其形态和功能的不同，又可分为吐丝管、榨丝区和共通管三部分以及菲氏腺（图 14-2；附彩图 14-2）。

吐丝管也称前区呈细管状，是指突出于下唇之外的部分，开口于吐丝管的前端（吐丝孔），是茧丝吐出的通道。

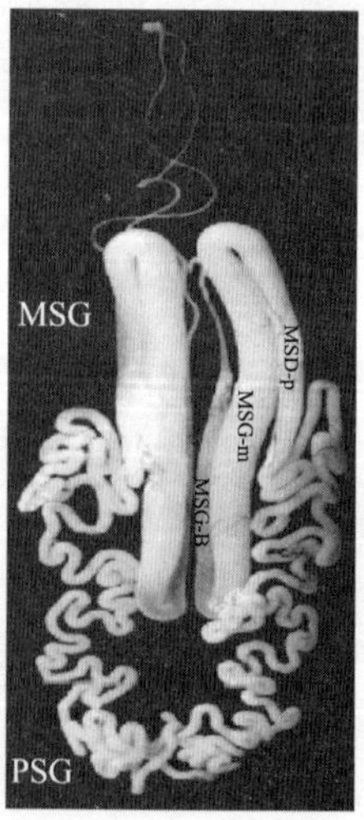

Fig.14-1 Appearance structure of silk gland.
图 14-1 丝腺的外观结构。fusulus. 吐丝器；ASG (anterior silk gland). 前部丝腺；MSG（middle silk gland）. 中部丝腺；PSG（posterior silk gland）. 后部丝腺；MSG-a. 中部丝腺前部；MSG-m. 中部丝腺中部；MSG-p. 中部丝腺后部。

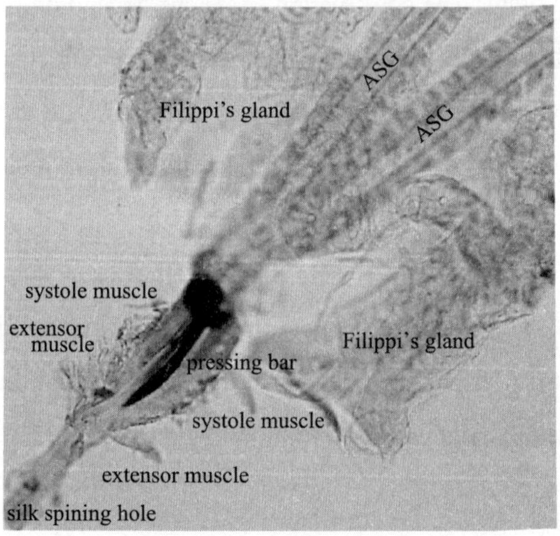

Fig.14-2 Structure of fusulus.
图 14-2 吐丝器的结构。silk spinning hole. 吐丝孔；extensor muscle. 舒展肌；systole muscle. 压缩肌；pressing bar. 压杆；Filippi's gland. 菲氏腺；ASG. 前部丝腺。

Extracting area located in the head. The tube shape is relatively expanded, which is the thickest part of the spinning part, but the inner cavity of the tube is the thinnest, showing a "half moon shape". In the center of the back of the lumen, there is a sword shaped dark brown bone rod, called pressing bar. On the back of the stick, six muscle bundles are symbiotic upward, sideways and downward. The other end of these muscles is attached to the body wall, which are called systole muscle of the lower lip, and extensor muscle of extracting area, respectively. On the ventral surface of the tube, there is a semi-circular dark brown sclerotin pressing bar. Functions of silk extracting area might be: ① control the flow of coarse and fine cocoon silk. ② compresse the silk fibers from the two silk glands into one filament. ③ in the process of mounting, when the silkworm is suddenly vibrated or clings to the mounting tool, the silk extracting area can shrink rapidly and clamp the silk, so that the silkworm can be hung in the air or mounting tool space by the silk. ④ extracting area can crush and break the silk. Common area also known as common tube or rear area where two anterior silk glands meet.

A pair of glands like grape located on both sides of the dorsal junction between the common tube and the two anterior silk glands, which open on both sides of the dorsal posterior segment of the common tube, which are called filippi's glands. The function of Filippi's glands has not been fully proved. Its secretion is white wax with a pH value of about 2.8. Its function is generally considered as follows: ① dehydrate the liquid silk and promote the fibrosis of the liquid silk; ② secrete substances that lubricate the common tube and glue the two monofilaments.

14.2.2 ASG

It is connected to the common area of the spinning part in the front, and then to the third abdominal segment in the middle silk gland. Its length is about 45 mm and its diameter is about 0.2-0.3 mm in matured silkworm. ASG is the thinnest part among the anterior, middle and posterior silk glands. It does not synthesize and secrete silk substances, and serves as the conduit through which silk substances flow.

14.2.3 MSG

MSG of adult silk moth *B. mori* is connected to the trailing end of the front silk gland with 85 mm long and 4.0-5.0 mm in diameter in matured silkworm. It is the thickest part of the whole silk gland. MSG is usually divided into three segments due to two flexion regions, which are called anterior, middle and posterior region.

榨丝区位于头壳内，管形较膨大，是吐丝部最粗的一部分，但其管内腔最细呈“半月形”。在管腔背面中央，生有一剑状的黑褐色骨质棒，称为压杆。在该棒的背面，向上、向侧、向下共生出 6 条肌肉束，这些肌肉的另一端附着于体壁上，分别是下唇收缩肌及榨丝区舒展肌。在管的腹面，有一个半环形的黑褐色骨质压片。榨丝区的功能：①控制丝物质的流量，调节茧丝的粗细。②将两条丝腺的丝纤维压合为一。③在上簇过程中，当蚕突然受振的瞬间或攀附簇具时，榨丝区可迅速收缩将丝卡住，使蚕靠丝悬挂于空中或簇具空间中。④榨丝区可将丝斩断。共通区也称为共同管或后区，是两条前部丝腺汇聚之处。

菲氏腺是一对葡萄状腺体，位于共通管和两个前部丝腺之间的背侧连接的两侧，在共通管的后背段两侧开放。菲氏腺的作用尚未完全探明，其分泌液为白色蜡状物，pH 为 2.8 左右。其作用一般为：①使液状丝脱水，促进液状丝纤维化；②分泌有润滑共通管作用和使两条单丝胶粘作用的物质。

14.2.2　前部丝腺

前部丝腺前与吐丝部共通区相连，后与中部丝腺相连于第 3 腹节。熟蚕时，该部长约为 45 mm，直径为 0.2 ～ 0.3 mm。前部丝腺是前、中和后部丝腺中最细的部分，没有合成和分泌丝物质的功能，是丝物质流经的导管。

14.2.3　中部丝腺

成熟家蚕的中部丝腺尾端连接前部丝腺，长为 85 mm，直径为 4.0 ～ 5.0 mm。它是整个丝腺中最粗的部分。中部丝腺通常因为两个屈曲分为三节，称为中部丝腺前段、中部丝腺中段和中部丝腺后段。

14.2.4 PSG

The posterior silk gland starts from the fourth abdominal segment, moves forward to the front edge of the third abdominal segment, and turns back to 6-8 abdominal segments, ending with a blind tube, forming many distortions. The length of the posterior silk gland is 170-200 mm and the diameter is about 1-1.4 mm in matured silkworm. There are muscles distributed on the silk gland. The main function of muscle is to fix the position of silk gland. Trachea is also distributed on the middle silk gland and the rear silk gland, which can not only supply oxygen, but also fix the position.

14.2.5 Cells of silk gland

Four different cell types were identified in the silk gland of *B. mori.* The larval silk glands have about 200 cells in the anterior, 250 in the middle and 525 in the posterior segments. The silk glands grow to a size of 25-30 cm by the 5th instar of larval development without an increase in the cell numbers. There are 10 distinct clusters in the silk gland of silkworm, which perform the functions of silk fibroin synthesis, sericin synthesis, liquid silk fibrosis, cell reconstruction, silk protein metabolism, chitin metabolism and so on. The cell clusters are categories into fibroin protein synthesis cells, death and remodeling regulation cells, sericin protein synthesis cells (SPSs), liquid silk fibrosis cells, epithelial cell remodeling cells, endoplasmic reticulum stress signaling cells (ERSSs), fibroin protein catabolism cells, chitin metabolism cells, traction forces cells, and sericin protein catabolism cells. Interestingly, the silk protein gene is highly expressed not only in the mature silk gland cells, but also in the dormant stage; Gene expression is very active in silk gland cells during dormancy. Some signal transduction pathways, such as ecdysone signal and Hippo signal, play an important role in regulating silk gland development and silk protein synthesis (Ma et al., 2022).

It is believed that the sericins secreted in the MSG does not mix with the fibroin and exist as a lubricating layer around the fibroin as the gland content is transported to the ASG (Magoshi et al., 1993). The presence of three differentially stained layers in the gland lumen. Cell type IV is responsible for producing carbonic anhydrase. At the funnel, where cell type IV begins, a cuticular intima appears at the apical cell membrane and is present all along the ASG, similar to the cuticular intima that has been observed

14.2.4 后部丝腺

后部丝腺从第 4 腹节开始，前行至第三腹节前缘，又向后折回至 6 ～ 8 腹节以盲管终了，形成许多扭曲。在熟蚕时后部丝腺长达 170 ～ 200 mm，直径为 1 ～ 1.4 mm。丝腺上有肌肉分布。肌肉的主要作用是固定丝腺的位置。中部丝腺和后部丝腺上还有气管分布，除供给氧气外，也有固定位置的作用。

14.2.5 丝腺细胞

在桑蚕丝腺中鉴定出四种不同的细胞类型。幼虫丝腺有 200 个前部丝腺细胞，250 个中部丝腺细胞，525 个后部丝腺细胞。幼虫发育到第 5 龄，丝腺长到 25 ～ 30 cm，细胞数量没有增加。家蚕丝腺中有 10 种主要细胞类群，它们分别执行丝心蛋白合成、丝胶蛋白合成、液态丝纤维化、细胞重建、丝蛋白质代谢和几丁质代谢等功能。这些细胞群归类为丝心蛋白合成细胞、死亡和重塑调节细胞、丝胶蛋白合成细胞（SPSs）、液体丝纤维化细胞、上皮细胞重塑细胞、内质网应激信号细胞（ERSSs）、丝心蛋白分解代谢细胞、几丁质代谢细胞、牵引力细胞和丝胶蛋白质分解代谢细胞。有趣的是，丝蛋白质基因不但在龄期丝腺细胞高表达，在眠期也有不同程度的表达。眠期丝腺细胞中基因表达非常活跃，一些信号转导通路，如蜕皮激素信号以及 Hippo 信号等参与其中，发挥着调控丝腺发育和丝蛋白质合成的重要作用（Ma 等，2022）。

人们认为，中部丝腺分泌的丝胶不会与丝心蛋白混合，而是作为丝心蛋白周围的润滑层存在，因为腺体内容物被输送到前部丝腺（Magoshi 等，1993）。腺体腔中存在三个不同的染色层。IV 型细胞负责产生碳酸酐酶。在压丝区，即 IV 型细胞开始的地方，角质层内膜出现在顶端细胞膜上，并沿前部丝腺分布，

along the S-shaped duct in spider silk glands (Andersson et al., 2013). The cuticle acts to resist unwanted deformation of the duct epithelium (Davies et al.,2013). The giant ramified nuclei present in all four cell types are characteristic of polyploidation, and have previously been identified in a number of lepidopteran species, including *B. mori* (Henderson et al., 1992; Ichimura et al., 1985).

类似于沿蜘蛛丝腺 S 形导管观察到的表皮内膜（Andersson 等，2013）。表皮的作用是抵抗吐丝管上皮的不必要变形（Davies 等，2013）。存在于所有四种细胞类型中的巨大分枝核都具有多倍体的特征，并且先前已在包括家蚕在内的许多鳞翅目物种中被鉴定出来（Henderson 等，1992；Ichimura 等，1985）。

The basic tissue component of silk gland is basement membrane, which has no cell structure and is composed of uniform membrane of each part. An intima composed of chitin. Silk substances can be secreted through micropores. The silk gland cells are hexagonal, and the left and right rows of cells form a tube (Fig.14-3; Appendix color graphic Fig.14-3). The gland cells never divide after the period of late embryonic development, although they increase in weight more than 50,000 times during the larval life (Akai, 1983).

丝腺的基本组织成分是基膜，无细胞结构，由各部分均匀的薄膜组成。由几丁质组成内膜。丝质物质可以通过微孔分泌。丝腺细胞呈六角形，左右两排细胞形成管状（图 14-3；附彩图 14-3）。在胚胎发育后期，腺体细胞不再分裂，尽管在幼虫期，腺体细胞的重量增加了 50 000 倍以上 (Akai, 1983)。

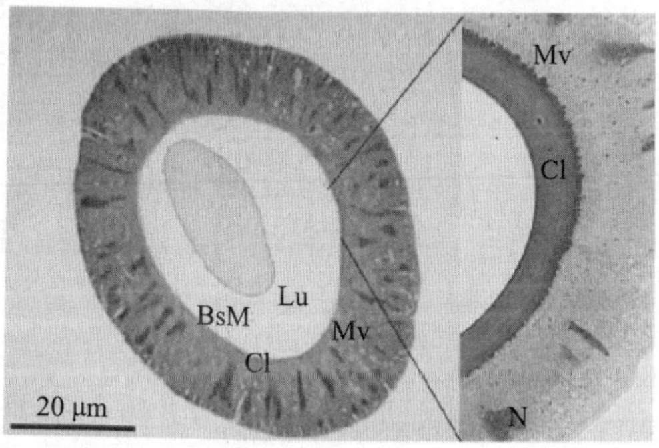

Fig.14-3 Cross sectional structure of silk gland. BsM. basal membrane; CI. cuticular Intima; Lu. lumen; Mv. microvilli; N. nucleus.

图 14-3 丝腺的横截结构。BsM. 底膜；CI. 表皮质内膜；Lu. 丝腺腔；Mv. 微绒毛；N. 腺细胞核。

14.3 丝物质的生物合成与分泌

14.3 Biosynthesis and Secretion of Silk Substances

14.3.1 Secretion of Silk Substances

14.3.1.1 Secretion of silk substances in every parts of SG

PSG synthesizes the silk structural proteins, the fibroin L and H chains and fibrohexamerin (formerly known as P25), whereas the MSG synthesizes the four types of glue proteins. The anterior division is a mere duct lined with a thick cuticular intima, and does not contribute to the secretion of silk materials.

Sericin can be divided into three layers: inner layer、middle

14.3.1 丝物质的分泌

14.3.1.1 丝腺各部位丝质物质的分泌

后部丝腺合成丝结构蛋白、丝心蛋白 L 和 H 链以及丝心蛋白六聚体（以前称为 P25），而中部丝腺合成四种丝胶蛋白。前部丝腺只是一条导管，内衬厚厚的表皮内膜，无助于丝物质的分泌。

丝胶可分为三层：内层、中层和外

layer and outer layer. According to its solubility and resolve speed, sericin can be divided into sericins Ⅰ、 Ⅱ、 Ⅲ、 Ⅳ, sericin Ⅳ is the most difficulty to be dissolved. It forms a layer of thin membrane on the surface of fibroin to protect mix drifting of fibroin and sericin.

14.3.1.2 Secretion of silk substances

In posterior silk gland, DNA synthesis continued at a fairly high level throughout the 4th instar and at a very high level in the early 5th instar, but then gradually decreased until it became undetectable on the 6th day. Electron microscopic autoradiography shows the position of particles on dispersed chromatin rather than condensed chromatin. At the active stage of RNA synthesis at the 5th instar, the nuclear material was close to the envelope, indicating that the material was transferred from the nucleus to the cytoplasm. Extrusion of some nuclear material into the cytoplasm, associated with a partial disappearance of the nuclear membrane. Several intracellular organelles, such as rough-surfaced endoplasmic reticulum (rough ER), free ribosomes, Golgi complexes, fibroin globules, mitochondria, microvilli, bandles of microfilaments and microtubules are distributed in the cytoplasm of cells in the posterior gland division (Fig.14-4; Appendix color graphic Fig.14-4). It was demonstrated that rough ER is the site of fibroin synthesis. In an early period of the 5th instar only a small number of tubular rough ER is scattered in the cytoplasm which contains a large amount of widely distributed free ribosomes. The rough ER develops rapidly during the first half of the 5th instar and in spinning larvae it occupies the entire cytoplasm while the free ribosomes disappear

Mature Golgi vacuoles containing a high concentration of the elementary fibroin fibers leave the Golgi complex as fibroin globules and move to the apical cytoplasmic region where they are discharged as spherical masses into the silk layer (Fig.14-4; Appendix color graphic Fig.14-4). The latter is separated from the gland lumen by a discontinuous bordering membrane. The spherical fibroin masses pass through spaces in this membrane and accumulate as columnar fibroin in the gland lumen. Bundles of microfilaments are located in the apical region of the gland cells. A large number of microtubules, which are distributed throughout the cytoplasm, apparently plays a role in the intracellular transport of fibroin globules. Discharge of fibroin globules is probably an energy dependent process because reaction products of an ATPase test were found concentrated on the microvilli of the apical cell

层。根据其溶解性和溶解速率，丝胶可分为丝胶Ⅰ、丝胶Ⅱ、丝胶Ⅲ、丝胶Ⅳ，其中丝胶Ⅳ最难溶解。它在丝心蛋白表面形成一层薄膜，以保护丝心蛋白和丝胶的混合蠕动。

14.3.1.2 丝物质的分泌

在后部丝腺，DNA 合成在整个 4 龄期间持续在相当高的水平上，在 5 龄早期达到非常高的水平，但随后逐渐减少，直到第 6 天变得不可检测。电子显微放射自显影显示颗粒位置在分散的染色质上，而不是在凝聚的染色质上。在 5 龄 RNA 合成的活跃期，细胞核物质接近包膜，表明物质从细胞核转移到细胞质中。某些核物质被挤出，进入细胞质中，与核膜部分消失有关。在后部丝腺分裂的细胞质中分布着几种胞内细胞器，如粗面内质网、游离核糖体、高尔基体复合体、丝心蛋白球、线粒体、微绒毛、微丝束和微管（图 14-4；附彩图 14-4）。已证实粗面内质网是丝心蛋白合成的部位。在 5 龄早期，只有少量管状粗面内质网被分散在细胞质中，含有大量广泛分布的游离核糖体。粗面内质网在整个过程中迅速发育，在 5 龄前半期和吐丝幼虫中，它占据整个细胞质，而游离核糖体消失。

含有高浓度基本丝心蛋白纤维的成熟高尔基体液泡离开高尔基体，形成丝心蛋白球并移动到顶端细胞质区，在那里它们作为球状物质排入丝层（图 14-4；附彩图 14-4）。后者通过不连续的边缘膜与腺腔分离。球形丝心蛋白团穿过膜中的间隙，并在腺体腔中积聚为柱状丝心蛋白。微丝束位于腺体细胞的顶端区域。大量微管分布在细胞质中，显然在丝心蛋白球的细胞内运输中起作用。丝心蛋白球的排出可能是能量依赖性的，因为 ATP 酶测试的反应产物集中在顶端细胞表面的微绒毛上。丝心蛋

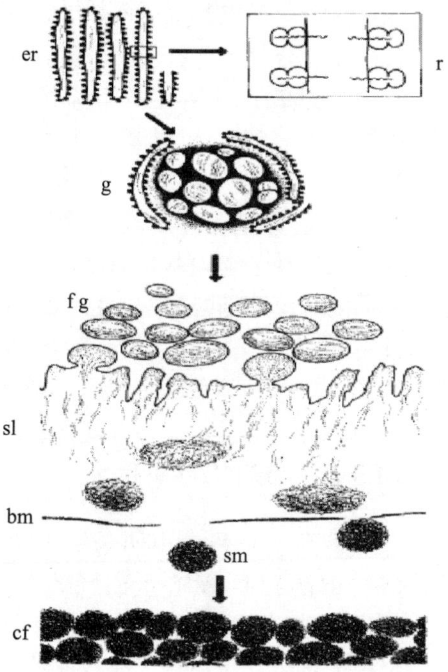

Fig.14-4 Schematic drawing of the process of fibroin secretion in posterior division. bm. Boundary membrane of cuticular intima between the silk layer and the lumen; cf. columnar fibroin accumulated with spherical masses of fibroin fiber in the lumen; er. rough endoplasmic reticulum; fg. fibroin globules; g. Golgi complex; r. ribosome; sl. silk layer; sm. spherical masses of fibroin fiber (Akai, 1983).

图 14-4 后区丝心蛋白分泌过程。bm. 丝层和管腔之间的表皮内膜边界膜；cf. 柱状丝心蛋白在管腔内堆积成球状丝心蛋白纤维团；er. 粗面内质网；fg. 纤维蛋白球；G. 高尔基复合体；r. 核糖体；sl. 丝层；sm. 丝心蛋白纤维的球形团块（Akai，1983）。

surface. Fibroin globules move to the surface of apical cells and release their contents to the silk layer. The discontinuous membrane separates the filament layer from the lumen of the gland. Sericin is pushed through the holes in the membrane and accumulates on the moving columnar silk fibroin as an inner sericin layer downward from the posterior gland.

Middle layer sericin is secreted by the cells in the posterior section of the middle part of the middle gland division. Sericin fibers floating in the globules are released into the silk layer and later pass through the bordering discontinuous membranes onto the surface of the previously deposited inner layer sericin.

Outer layer sericin is secreted in the anterior part of the middle gland division. These sericins also pass through the silk layer and then accumulate on the middle layer sericin to form the outer sericin layer of the silk.

The non-protein components in the cocoons include carbohydrate, salt wax and other chemicals. The active compounds traced in the cocoons are presumed to be sequestered from their host plants. For example, biophysical analysis of *A. mylitta* methanolic cocoon extraction revealed that the active compounds are fatty acids, alkane hydrocarbons, esters, terpene alcohols, aldehydes, aromatic carboxylic acids, steroidal compounds, and antimicrobial compounds of 1,3,5-triazine-2,4-diamine, nonadecanoic acid, oleic acid, cyclopropanepentanoic acid, 2-undecyl methylester, bis (2-ethylhexyl) maleate, decanal, 1,2-benzenedicarboxylic acid, cholestane, ethyl iso-allocholate

白小球移动到顶端细胞表面并将其内容物释放到丝层。不连续的膜将丝层与腺体的内腔分开。丝胶被推过膜上的孔，并在移动的柱状丝心蛋白上积聚，形成从后部丝腺向下的内层丝胶。

中层丝胶由位于中部丝腺分区的中后部细胞分泌。漂浮在球状物中的丝胶纤维被释放到丝层中，然后通过相邻的不连续膜到达之前沉积的内层丝胶表面。

外层丝胶在中部丝腺的前部分泌。这些丝胶也通过丝层，然后积聚在中间层丝胶上，形成丝的外丝胶层。

茧中的非蛋白质成分包括碳水化合物、盐蜡和其他化学物质。蚕茧中追踪到的活性化合物被认为是从寄主植物中接收而来的。例如，对野生天蚕甲醇蚕茧提取物的生物物理分析表明，活性化合物为脂肪酸、烷烃、酯、萜烯醇、醛、芳香羧酸、甾体化合物以及抗菌化合物 1,3,5- 三嗪 -2,4- 二胺、壬癸酸、油酸、环丙烷戊酸、2- 十一烷基 - 甲酯、顺丁烯二酸双（2- 乙基己基）、癸醛、1,2- 苯二甲酸、胆甾烷、异别胆酸乙酯和细胞毒性诱导的三氯乙酸十六

and cytotoxicity inducing trichloroacetic acid hexadecyl ester. The methanolic extract of *B.mori* cocoons identified eight non-protein compounds of cyclopentane-undecanoic acid, pterin-6-carboxylic acid, Z-1, 9- hexadecadiene, and five new phytoconstituents, hexanal, *n*-pentanyl-3,4-(dioxy-methylene) benzene, *n*-pentanyl-2,3- dioxymethy-lene-4- hydroxy benzene, n-decanyl-3,4- dioxymethy-lene-4-hydroxylbenzene, dihydrocitronellyl glucoside and dihydrocitronelloyl glucuronosidic ester, which are considered as medicinally significant compounds (Ahamad et al., 2020).

酯。桑蚕茧的甲醇提取物鉴定出八种非蛋白质化合物，包括环戊二烯二酸、蝶呤-6-羧酸、Z-1,9-十六二烯和五种新植物成分己醛、正戊基-3,4-（二氧甲基）苯、正戊基-2,3-二氧甲基-4-羟基苯、正癸基-3,4-二氧甲基-4-羟基苯、二氢香茅烯丙基葡萄糖苷和二氢香茅烯丙基葡萄糖醛酸酯，它们被认为是具有医学意义的化合物（Ahamad 等，2020）。

14.3.2 Chemical structure of silk

Fibroin from the silkworm *B. mori* is composed of two chains (H-fibroin heavy chain and L-fibroin light chain) linked by a disulfide bond. Another component of the protein is P25, which has been suggested to act as a chaperone to assist in the transport and secretion of H-fibroin. During the formation of fibroin complexes, P25 is speculated to facilitate both the efficient secretion of the vast amount of fibroin from PSG cells into the lumen by reverse pinocytosis and maintaining fibroin solubility during luminal transport. However, some researchers suggested that P25 is dispensable and its loss has only a small effect on the quantity, quality, and composition of the silk (Zabelina et al., 2021). H-fibroin, L-fibroin and P25 are thought to be assembled in the ratio 6∶6∶1. This assembly appears to be standard in silk secreted by members of the lepidoptera order, with the exception of the Saturniidae family, a more basal lepidopteran clade, Hepialioidea (Collin et al., 2010), where L-fibroin and P25 are absent. In addition, H-L linkage could contribute to the maintenance of the elementary unit but the formation of the H-L heterodimer is not essential for secretion of the H-chain into the PSG lumen; and the H-chain and the H-chain fusion protein could be secreted directly into the PSG lumen and spun into cocoons without forming the H-L heterodimer, especially for the small molecules and hydrophilic H-chain fusion proteins (Long et al. 2015).

Silk fibroin can exist in a number of different conformational states, silk Ⅰ, silk Ⅱ and silk Ⅲ. Silk I is the structure of *B. mori* silk fibroin in the solid state obtained from the middle silk gland after drying. The backbone structure of the silk Ⅰ conformation

14.3.2 丝的化学结构

桑蚕丝心蛋白由两条链（H-丝心蛋白重链和L-丝心蛋白轻链）组成，并通过二硫键连接。该蛋白质的另一个组成部分是P25，它被认为是一种分子伴侣，有助于H-丝心蛋白的运输和分泌。在丝心蛋白复合物的形成过程中，推测P25能促进两者高效分泌大量的丝心蛋白，进一步通过反向胞饮作用使PSG细胞进入管腔，并维持丝心蛋白在管腔转运中的溶解度。然而，有些研究者认为P25可有可无，它的缺失对生丝的数量、质量和成分只有很小的影响（Zabelina 等，2021 年）。H-丝心蛋白、L-丝心蛋白和P25被认为以6∶6∶1的比例组装。这种组装似乎是鳞翅目成员分泌丝的标准，但大蚕蛾科、鳞翅目基部的一个分支及蝙蝠蛾总科除外（Collin 等，2010），这些科缺少L-丝心蛋白和P25。此外，H-L连接可能有助于维持基本单元，但H-L异二聚体的形成对于H-链分泌到PSG管腔不是必需的。H-链和H-链融合蛋白质可以直接分泌到PSG内腔并吐丝成茧而不形成H-L异二聚体，特别是对小分子和亲水性H-链融合蛋白质（Long 等，2015）。

丝心蛋白能以多种不同的构象状态存在，如丝心蛋白Ⅰ、丝心蛋白Ⅱ和丝心蛋白Ⅲ。丝心蛋白Ⅰ是干燥后从中部丝腺获得的固态桑蚕丝心蛋白结构。$(AG)_n$的丝心蛋白Ⅰ构象的主链结构

for $(AG)_n$ was determined to be a repeated type Ⅱ β-turn structure. The conformation of silk Ⅰ is often described as a range of conformations, from lacking secondary structure, through partially ordered to well-formed crystals. Silk Ⅱ is the solid fibroin that can be found in the spun silk thread, while silk I is the dissolved, meta-stable form stored in the silk gland. The silk Ⅱ structure formed after silk spinning was proposed as a regular array of antiparallel β-sheet. However, some intrinsic structural disordered exist in the silk Ⅱ structure. Silk Ⅲ, a structure stabilized at interfaces that optimizes the surfactancy of the silk repetitive motif.

被确定为重复的Ⅱ型 β- 转折结构。丝心蛋白 I 的构象通常被描述为一个构象范畴，从缺乏二级结构，经部分有序到形成良好的晶体。丝心蛋白Ⅱ是在丝束中发现的固体丝心蛋白，而丝心蛋白 I 是贮存在丝腺中的可溶性亚稳态形式。吐丝后形成的丝心蛋白Ⅱ结构为规则排列的反平行 β 片结构。但是丝心蛋白Ⅱ结构中存在一些内在结构紊乱。丝心蛋白Ⅲ是丝心蛋白重复单元表面活性优化在界面的稳定结构。

According to researches, the pH, Ca^{2+}, K^+, and Cu^{2+} concentrations, and water content are thought to change as the silk flows through the secretory pathway. For example, pH values of silkworm at ASG, MSG-a,m,p, and PSG are 4.8, 5.0,5.2,5.6 and 6.9, respectively. These changes in pH and charge may contribute to the change in the physical state of the protein from an isotropic phase, micelle, gel and liquid crystal, which finally undergo strain induced β-sheets formation. The predicted pI in all the lepidopteran insects, *A. mylitta, A. pernyi,A. yamamai, G. mellonella, B. mori* , and *B.mandarina* heavy chain silk fibroin sequences examined is low and remarkably constant ranging from 4.37 to 5.05. However, the pI of the N-terminal sequences varied from 4.6 to 5.8, This reduction in charge on acidification may allow hydrophobic and hydrophilic interactions to occur initiating both molecular aggregation and changes in secondary structure. The pI of the C-terminal sequences is well above neutrality, pI 7.8 to 11.5. The pI of the light chain and P25 of silkworm is 5.06, 6.69, respectively, which is also still negatively charged in the posterior division of the silk gland. Further, the charge on the hydrophilic core spacers that lie between the hydrophobic blocks in silk proteins is more acidic, only 4.03. Therefore, the heavy and the light chains might linked by an intermolecular disulfide bond through the first of three cysteines (CysC20) at the C-terminus of the heavy chain and Cys-172 of the light chain (Foo et al., 2006).

根据研究，随着丝物质流经分泌途径，pH、Ca^{2+}、K^+ 和 Cu^{2+} 的浓度以及水分含量都会发生变化。例如，家蚕的 ASG、MSG-a、MSG-m、MSG-p 和 PSG 的 pH 分别为 4.8、5.0、5.2、5.6 和 6.9。pH 和电荷的这些变化可能有助于蛋白质物理状态的变化，从均质态、丝胶束、丝胶和液晶化，最终经扭曲诱导 β 片形成。所有鳞翅目昆虫，如印度柞蚕、中国柞蚕、天蚕、家蚕和野蚕重链丝心蛋白序列预测的 pI 都很低，在 pH 4.37 ～ 5.05 范围内非常稳定。然而，N- 末端序列的 pI 在 4.6 ～ 5.8 之间发生变化，这种酸化电荷的减少可能会导致疏水和亲水相互作用的发生，从而引发分子聚集和二级结构的变化。C 端序列的 pI 远高于中性，为 7.8 ～ 11.5。家蚕的轻链和 P25 的 pI 分别为 5.06 和 6.69，在丝腺的后部仍然带负电。此外，位于丝蛋白质中疏水块之间的亲水内核空间的电荷酸性更强，仅 4.03。因此，重链和轻链能通过重链 C 端的三个半胱氨酸（CysC20）和轻链的 Cys-172 形成分子间二硫键进行连接（Foo 等，2006）。

14.3.2.1 Primary and secondary structure of silk proteins

14.3.2.1 丝蛋白质的一级和二级结构

Silk proteins have a highly repetitive primary structure, which gives rise to very regular conformations at the primary level. The repeating units are mostly poly glycine-alanine blocks in fibroin. High glycine content is thought to allow greater conformational

丝蛋白质具有高度重复的一级结构，在一级水平上形成非常规则的构象。重复单元主要是丝心蛋白中的聚甘氨酸 - 丙氨酸嵌段。高甘氨酸含量被认

variability. The central region of the protein is mostly hydrophobic and repetitive. The N-terminal and the C-terminal domains are non-repetitive and more hydrophilic. In fibroin, the hydrophilic spacer units are more complex and smaller than the hydrophobic repeating units. They also contain a much higher proportion of charge-carrying amino acids.

The amino acid composition of the H-chain showed the predominance of four amino acids: Gly (42.9%), Ala (30.0%), Ser (12.2%), and Tyr (4.8%). Detailed primary structure is unusually repeated sequence in the H-chain of *B. mori* silk fibroin. An analysis of the primary structure of H-chain indicated that the sequence may be roughly divided into four modular motifs: ① contains a highly repetitive GAGAGS sequence and comprises the crystalline regions; ② module Ⅱ contains relatively less repetitive sequences with hydrophobic and/or aromatic residues –GAGAGY, GAGAGV, and GAGAGVGY – and make up the semicrystalline regions; ③ module Ⅲ is very similar to module Ⅰ except for the presence of an AAS motif; ④ module Ⅳconstitutes the amorphous regions containing negatively charged, polar, bulky hydrophobic, and/or aromatic residues. e.g., TGSSGFG-PYVANGGYSGYEY AWSSESDFGT.

In fibroin, which is composed of L-fibroin and P25, L-fibroin is mostly hydrophobic, but has two disulfide bridges (one internal and one linking it to the heavy chain) which might be responsible for 'exposing' the hydrophilic N-terminal through a specific conformation. P25 consists of alternating hydrophobic and hydrophilic regions and many charged residues (Fig.14-5; Appendix color graphic Fig.14-5). The heavy and light chains are linked by a disulfide bond, and P25 associates with disulfide-linked heavy and light chains by noncovalent interactions (Inoue et al. 2000).

P25 is also known as fibrohexamerin (Fhx), a large polypeptide with 350 kDa that mostly consists of glycine- and alanine- rich repeats in 220 amino acids.

Compared with silk of *B.mori, S.cynthia* and *N. clavata*, the proportion of each secondary structure varies from one silk fiber to another. The silk fibers exhibit a common structural pattern characterized by mixtures in different amounts of crystallites, amorphous β-sheets, and other more flexible structures, such as

为可允许产生更大的构象变异性。蛋白质的中心区域大部分是疏水性和重复性的。N- 端和 C- 端结构域是非重复的且亲水性更强。在丝心蛋白中，亲水性间隔单元比疏水性重复单元更复杂且更小。它们还含有更高比例的带电荷氨基酸。

H- 链的氨基酸组成以 4 种氨基酸为主：甘氨酸（42.9%）、丙氨酸（30.0%）、丝氨酸（12.2%）和酪氨酸（4.8%）。详细的一级结构是桑蚕丝心蛋白 H- 链中异常重复的序列。对 H- 链一级结构的分析表明，该序列可大致分为 4 个模块化基序：①模块Ⅰ包含高度重复的甘-丙-甘-丙序列，并组成结晶区；②模块Ⅱ含有相对较少的重复序列，带有疏水和 / 或芳香残基甘 - 丙 - 甘 - 丙 - 酪、甘 - 丙 - 甘 - 丙 - 缬和甘 - 丙 - 丙 - 缬 - 甘 - 酪，并构成半结晶区域；③模块Ⅲ与模块Ⅰ除了出现丙 - 丙 - 丝基序外非常相似；④模块Ⅳ构成非晶形区域，含有带负电、极性、大体积疏水性的、和 / 或芳香残基。例如，苏-甘-丝-丝-甘-苯丙-甘-脯-酪-缬-丙-天冬酰-甘-甘-酪-丝-甘-酪-丝-甘-酪-色-丝-丝-谷-丝-天冬-苯丙-甘-苏。

在由 L- 丝心蛋白和 P25 组成的丝心蛋白中，L- 丝心蛋白主要是疏水性的，但有 2 个二硫桥（一个在内部，一个连接到重链），这可能负责通过特定构象来“暴露”亲水性 N 末端。P25 由交替的疏水和亲水区域以及许多带电残基组成（图 14-5；附彩图 14-5）。重链和轻链通过二硫键连接，P25 通过非共价相互作用与二硫键连接的重链和轻链结合 (Inoue 等，2000)。

P25 也被称为纤维六聚体（Fhx），是一个 350 kDa 的大多肽，主要由 220 个氨基酸中富含甘氨酸和丙氨酸的重复序列组成。

比较家蚕、蓖麻蚕和蜘蛛（棒络新妇）丝可知，每个二级结构的比例因丝纤维而异。丝纤维表现出一种共同的结构模式。其特征是以不同数量的微晶、无序 β 片和其他更灵活的结

β-turns, 3_1-helices, and random structures that are organized with specific levels of orientation. For *B. mori*, the β-sheet content is 50%, which matches the proportion of residues that form the GAGAGS fibroin motifs. For the *N. dragline* and *S.cynthia* cocoon, the *β*-sheet content (36%-37% and 45%, respectively) is higher than the proportion of residues that belong to polyalanine blocks (18% and 42%, respectively), showing that adjacent GGA motifs are incorporated into the β-sheets (Lefèvre et al., 2007).

构，如 β 弯、3_1 螺旋和随机结构，其随机结构以特定层次取向组装。对于家蚕而言，β - 折叠含量为 50%，这与形成 GAGAGS 丝心蛋白基序的残基比例相匹配。在蜘蛛丝和蓖麻蚕茧中，β- 片含量分别为 36% ～ 37% 和 45%，高于多聚丙氨酸嵌段的残基比例（分别为 18% 和 42%），表明相邻 GGA 基序被纳入 β- 片（Lefèvre 等，2007）。

B.mori	MKP IFLWLLVAT SAYAAPSWTINQY—SDNE IPRDIDDGK—ASSVISRRWDYVDDTDKS IAILNVQE ILKDMASQGDYAS	77
B. mandarina	MKP IFLWLLVAT SAYAAPSWTINQY—SDNE IPRD IDDGK—ASSVISRAWDYVDDTDKS IAILNVQEILKDMASQGDYAS	77
M.sexta	MKPIVLVLFVAAGAFAAPSVVLNEY—SENELPR————SNSL TERAFGL IDGGDMN IHILNLQAVLKDL-SERDPRS	70
G. mellonella	MLP FVL VLL VATSAL AAP SVVISQD—NINN IAPRVGNGRpISSAL IDRAFEIVDGGDTN IYILTIQQ ILNDL ADQPDGLS	79
V. tameamea	MLP IVL VLFVAQSAFAAP SVVGVQFkNFNEVAPVSDNGKIVSSYITDGSIAYIDGGDTPLYAQMIIQSINDMADTSDTDS	80
H. occidentalis	——MALSLLIGALLAIQGASFVASSQ——————ISASLLEETWNLVDQGEVEPFTLLLKKEV—VATGGVY—	59

B.mori	QASAVAQTAGIIAHL SAGIPGDACAAANVINSYTDGVRSGNFAGFRQSLGP—FFGHVGQNLNL INQLVINPGQLRYSVGP	156
B. mandarina	QASAVAQTAGIIAHL SAGIPGD ACAAANVINSYIDGVRSGNFAGFRQSLGP—FFGHVGQ NLNL INQL VINPGQLRYSVGP	156
M.sexta	QAL AFGQILGVL VDLATGIPGDSCASAALINAYAAGVKSGNFSGFRSVLYN—FVRQLASNLKL IVKL ASNPASVRFSTGP	149
G. mellonella	QSLAVTQAVAALGELATGVPGNSCEAAAVIDAYANSVRTGDNSALSIAVAV—YINRLSSNIGLISQLASNPDSLRYSSGP	158
V. tameamea	WANAVIQTIATVGELANGVPGDACEAAAL INAYAYSVSTGNSAGLRAALPN—FVQRIAGYVDTIVQLVVNPNGVRYSSGP	159
H. occidentalis	———GLGATL TGVGEL AWPRPASGCGHSKL INANVAL NDGTLAWGELEDAVDSYAVVLAQAVDNLRILG—LSCIIPAPWP	134

B.mori	AL—GCAGGGRIYDFEAAWDAILASSD——SSFLNEEYCIVKRL YNSRNSQSNNIAAYI TAHLLPPVAQVFHQSAGSI—	229
B. mandarina	AL—GCAGGGRIYDFEAAWDAILASSD——SGFLNEEYCIVKRLYNSRNSQSNNIAAYITAHLLPPVAQVFHQSAGSI—	229
M.sexta	SG—GCLGSGRSYQFEAAWDAILSSSS——SSLVHEQYCTAKRLYYAFNSRSNNVGATITALSLPPVAQVLEAAGEST—	222
G. mellonella	AG—NCAGGGRSYQFEAAWDAVLNNANpyQIGLINEEYCAARRLYNAFNSRSNNVGAAITAGAVVAQTQAAQIILPSL—	234
V. tameamea	RG—KCLGGGRSYQFEEAWDGILDKANayQIGLLNEEYCAAKRLYNAFNIRSNNVAAGVTAVTTRPVNQLVQLALPQLaq	237
H. occidentalis	TLenSCGDWGRIYDFENSWD—————LSNVNNGVVCAARRLYTAFGARANNVGAAATSAATDAAITIISDVEDEL—	203

B.mori	TDLLRGVGNGNDATGLVANAQRYIAQAASQVHV	262
B. mandarina	TDLLRGVGNGNDATGLVANAQRYIAQAASQVHV	262
M.sexta	SNLLRAVASGYDASAAAASANKALLRAVERIQL	255
G. mellonella	VNVLSAVAAGGNVAGAAAQAGQALANAAANVQL	267
V. tameamea	VSFLRVVANGGNPAQAGANAKNALLQSLPKIQY	270
H. occidentalis	VSYLEAVLSKSAGPGCKSKQKQLRTLAGSLKAS	253

Fig. 14-5 Primary structure of silk L-fibroin of *Bombyx mori, Bombyx mandarina, Manduca sexta, Vanessa tameamea* and the *caddisflies Hesperophylax occidentalis* (Lowercase letters represent nonhomologous amino acid regions）.
图 14-5 家蚕、野蚕、烟草天蛾、夏威夷红蛱蝶和石蛾的丝心蛋白轻链的一级结构比较（小写字母表示非同源氨基酸区域）。

Generally, silk proteins are thought to be two phased semi-crystalline polymers containing both crystalline and non-crystalline regions. The crystalline regions have been identified as rigid, tightly packed anti-parallel β-pleated sheets. The β-sheets are either poly-Ala, or poly-Gly-Ala, interlocking with adjacent chain *via* hydrogen bonds. The non-crystalline domains are often described as poorly orientated, randomly coiled sections of the peptide, or 'amorphous'. The ratio of crystalline to non-crystalline domains is known to vary between spider and silkworm silk, with spider silk containing less β-sheet and more helical regions. This is thought to be important for assembly and mechanical features.

丝蛋白质通常被认为是包含结晶和非结晶区域的两相半结晶聚合物。结晶区被确定为刚性、紧密排列的反平行 β-折叠片。β- 片为聚丙氨酸或聚 - 甘氨酸 - 丙氨酸，通过氢键与相邻链连接。非晶体结构域通常被描述为肽的取向紊乱、肽的随机卷曲部分或“非结晶”。已知蜘蛛丝和蚕丝的结晶域和非结晶域的比例不同，蜘蛛丝含有较少的 β- 折叠区域和更多的螺旋区域。这被认为对装配和机械特性很重要。

14.3.2.2 Tertiary structure

The combination of extensive hydrogen bonding, high levels of crystallinity, and general hydrophobic nature are all thought to stabilize tertiary structure. The hydrophobic nature of silk is essential to exclude water and produce the high packing density and β-crystallinity necessary for its mechanical function. Silk proteins are therefore insoluble in most solvents, including water and dilute acid and alkali (Fig.14-6; Appendix color graphic Fig.14-6).

14.3.2.2 三级结构

大量的氢键、高水平的结晶度和普遍的疏水性质都被认为可以稳定三级结构。丝的疏水性对于排除水分、产生其机械功能所需的高堆积密度和β-结晶度至关重要。因此，丝蛋白质不溶于大多数溶剂，包括水、稀酸和稀碱（图 14-6；附彩图 14-6）。

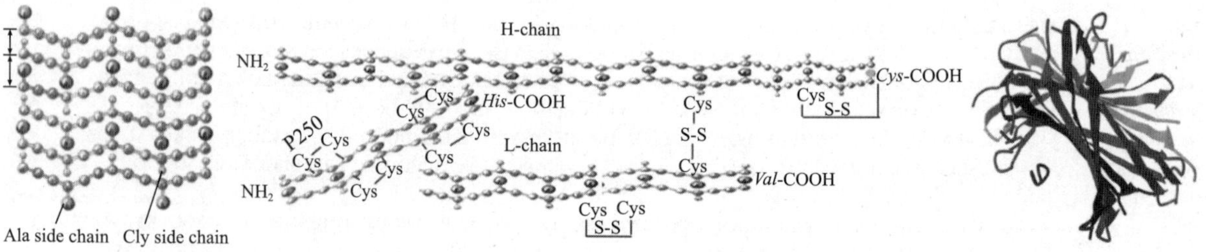

Fig.14-6 Secondary and tertiary structure of fibroin (modified from Tanaka et al., 1999; He et al., 2012).
图 14-6 丝心蛋白的二级和三级结构（仿 Tanaka 等，1999；He 等，2012)。

14.3.3 Genes of silk

At least three genes are responsible for sericin production. *Ser1* was discovered earlier by Okamoto et al. (1982) and almost fully sequenced by Garel et al. (1997). It was determined that the gene consists of 9 exons and generates 4 mRNAs (2.8 kb, 4.0 kb, 9.0 kb and 10.5 kb) by alternative splicing (Garel et al., 1997). The *Ser2* gene was also discovered earlier and mapped with restriction enzymes. Its transcription start, part of the final exon, and the major repetitive motif were identified. Two *Ser2* mRNA variants, 3.1 kb and 5.0~6.4 kb were proposed to arise *via* alternative splicing mechanism (Michaille et al., 1990). Takasu et al. (2007) described the structure and sequence of the third gene, *Ser3*, which shown to generate a single transcript of 4.5 kb.

As well known, silk fibroin contains 350 kDa H-chain, 25 kDa L-chain polypeptide with a disulfide bond at the C-terminus of the two subunits, and another *P25* gene encoding a 25 kDa polypeptide associates with the H-L complexes primary by hydrophobic interaction.

Cloning of silk fibroin H-chain consists of one intron and two exons. The intron contains a truncated sequences of *B.mori*

14.3.3 丝基因

至少有三个基因负责丝胶的产生。*Ser1* 由 Okamoto 等（1982）发现，Garel 等（1997）几乎完全测序。确定该基因由 9 个外显子组成，并通过选择性剪切产生 4 个 mRNA（2.8 kb、4.0 kb、9.0 kb 和 10.5 kb）(Garel 等，1997）。*Ser2* 基因也更早地被发现，并用限制性内切酶进行了定位。确定了其转录起始、最终外显子的一部分和主要的重复基序。两种 *Ser2* mRNA 变体，3.1 kb 和 5.0 ～ 6.4 kb 被认为是通过选择性剪切机制产生的（Michaille 等，1990）。Takasu 等（2007）对第三个基因 *Ser3* 的结构和序列进行了描述，并显示其产生了 4.5 kb 的单个转录子。

众所周知，丝心蛋白含有 350 kDa H 链、25 kDa L- 链多肽，在两个亚单位的 C- 端有一个二硫键，另一个 *P25* 基因编码一个 25 kDa 多肽，通过疏水相互作用与 H-L 复合物相结合。

丝心蛋白 H- 链由一个内含子和两个外显子组成。内含子包含一个位于 +565 ～

repetitive element at +565~+572 and multiple octamer-like AT-rich fibroin-modulator for the transcriptional regulation. The exon 1 is only 67bp long which containing a 25bp untranslated domain (+1~25) and a coding region for 14 amino acid residue (+26~67). The exon 2 locates between +1,039~16,788 comprising the majority of fibroin gene. The repetitive core of the central repetitive region (+1,449~+16,610) is composed of 12 repetitive domains separated by 11 amorphous region. The distinct property of repetitive domains is their high content of GC. What relatively well conserved is the amorphous domains and their upstream flanking regions in variable lengths and different composition of repetitive domains. H-chain of fibroin has five Cys residues located near the N-and C-terminus (Zhou et al.,2000).

+572 的家蚕重复元件的截短序列以及用于转录调控的多个八聚体类富 -AT 丝心蛋白调节基因。外显子 1 只有 67 bp 长，包含一个 25 bp 的非翻译结构域（+1 ～ 25）和一个 14 个氨基酸残基（+26 ～ 67）的编码区。外显子 2 位于 +1 039 ～ 16 788 之间，包含大部分丝心蛋白基因。中间重复区（+1449 ～ +16610）的重复核心由 11 个非晶区分隔的 12 个重复域组成。重复结构域的显著特点是其 GC 含量高。相对保守的是可变长度的非晶态结构域及其上游侧翼区域和不同组成的重复结构域。丝心蛋白的 H- 链在 N- 端和 C- 端附近有五个 Cys 残基（Zhou 等，2000）。

The molecular mass of fibrohexamerin (P25) contains the three N-linked oligosaccharide chains. The total length of *P25* mRNA was estimated to be (1,035 ± 54) bps containing 5 exons and 4 introns. The size of 4 exons were estimated to be (80 ± 10) bps, (212 ± 16) bps, (148 ± 23) bps and (241 ± 18) bps, respectively. Given the fact that the non-annealed portion of the mRNA on the 3′ side is (405 ± 49) nucleotides, and that the mean size of the poly-A tail is 150 bps, then the 3′ region of *P25* gene would extend to 250 bps. The ATG initiation codon lies in position +23(Couble et al.,1985).

纤维六聚体（P25）的分子团包含三条 N- 连接的低聚糖链。*P25* mRNA 的全长估计为（1 035 ± 54）bps，包含 5 个外显子和 4 个内含子。估计 4 个外显子的大小分别为（80 ± 10）bps、（212 ± 16）bps、（148 ± 23）bps 和（241 ± 18）bps。考虑到 3′ 端 mRNA 的未退火部分为（405 ± 49）个核苷酸，且 poly-A 尾部的平均大小为 150 bps，那么 *P25* 基因的 3′ 区将延伸至 250 bps, 起始密码子 ATG 位于 +23（Couble 等，1985）。

The disulfide linkage between Cys-172 of L-chain and Cys C20 (20th residue from the C-terminus) of H-chain has been demonstrated to be essential for the secretion of a large amount of fibroin from PSG cells. The major role of L-chain is to prevent the retention of H-chain in ER by forming the disulfide linkage with Cys C20 of H-chain. The major role of P25 seems to assemble and hold the elementary unit containing six sets of the H-L dimer.

L- 链的 Cys-172 和 H- 链的 Cys C20（C- 端的第 20 个残基）之间的二硫键已被证明是 PSG 细胞分泌大量丝心蛋白所必需的。L- 链的主要作用是通过与 H- 链的 Cys C20 形成二硫键来防止 H- 链在 ER 中的滞留。P25 的主要作用似乎是组装和保持包含六组 H-L 二聚体的基本单元。

14.3.4 The structure of *B. mori* silk thread

B. mori silk fibre has been shown to be composed of two protein-monofilaments (named brins) embedded in a gluelike sericin coating. The brins are fibroin filaments made up of bundles of nanofibrils, approx 5 nm in diameter, with a bundle diameter of around 100 nm. The nanofibrils are oriented parallel to the axis of the

14.3.4 桑蚕丝束的结构

桑蚕丝纤维由两条蛋白质单丝（命名为 brins）组成，嵌在胶状丝胶包被层中。brins 是由纳米纤维束组成的丝心蛋白细丝，直径约为 5 nm，丝心蛋白束直径约为 100 nm。纳米纤维的方向平行于纤维的轴线，并且被认为彼

fibre, and are thought to interact strongly with each other. Silk threads comprise approximately 75% insoluble fibrous proteins in the inner layer and 25% hydrophilic glue sericin proteins in the outer layer (Fig.14-7; Appendix color graphic Fig.14-7). The sericin coating helps in the formation of silk cocoon by gluing the fibres together. It is made of sericins, gum-like proteins that vary in their molecular weight between 10-300 kDa, and are rich in the amino acid serine. It is also thought to contain carotenoids, which are responsible for cocoon pigmentation .

Sericin has several extraordinary properties: they resist oxidation, are antibacterial, UV resistant, and can absorb and release moisture easily. These properties are valuable in the protection of silk from microbial degradation, animal digestion, and other damaging processes. However, sericin is still considered an obstacle for *B. mori* silk biocompatibility, despite the coating being soluble in hot alkaline water and therefore removable. Insufficient sericin removal is still an issue in many biomedical applications.

此之间有强烈的相互作用。丝心蛋白束内层含有约 75% 的不溶性纤维蛋白质，外层含有 25% 的亲水性胶状丝胶蛋白（图 14-7；附彩图 14-7）。丝胶包被层将丝纤维黏合在一起有助于形成蚕茧。它由丝胶蛋白组成，是一种分子质量在 10 ～ 300 kDa 之间变化的胶状蛋白质，富含氨基酸丝氨酸。它还被认为含有类胡萝卜素，负责蚕茧色素的沉着。

丝胶有几个特殊的特性：它们能抗氧化、抗菌和抗紫外线，并且能很容易地吸收和释放水分。这些特性在保护蚕丝免受微生物降解、动物消化和其他破坏过程中具有重要价值。然而，丝胶仍然被认为是桑蚕丝生物相容性的障碍，尽管该包被层可溶于热碱性水，因而可被去除。但丝胶去除不足仍然是许多生物医学应用中的一个问题。

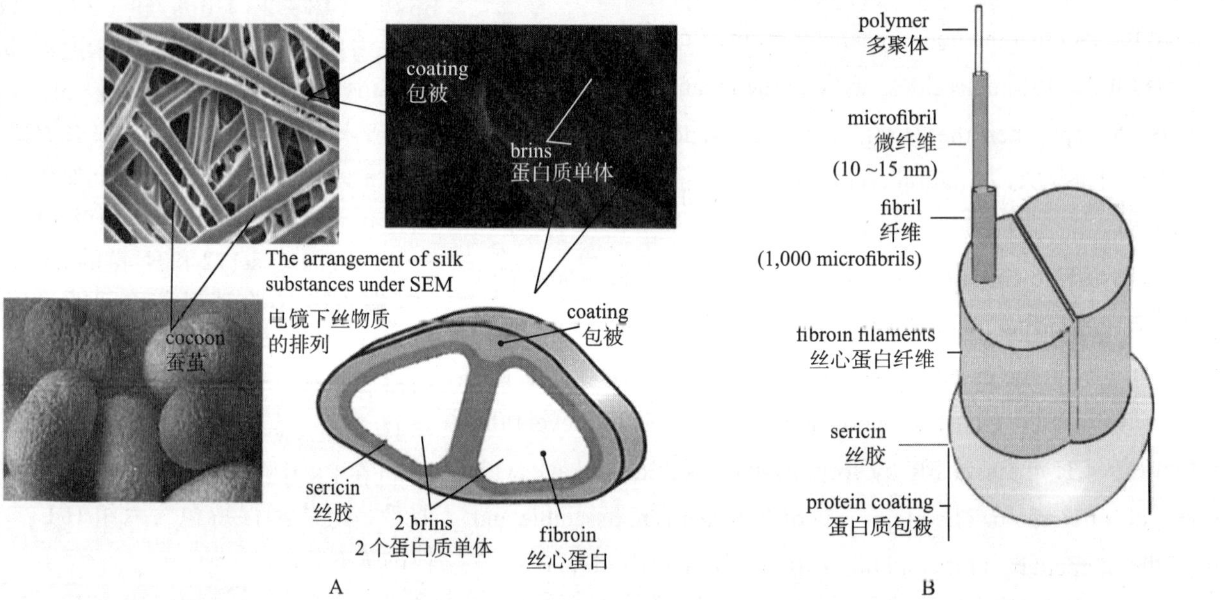

Fig.14-7 Structure of silk thread from macro to micro view. A. The cocoon is composed of silk thread, which consists of coated protein and monofilament. Each monofilament is composed of sericin and fibroin; B. Cross sectional view of monofilament.

图 14-7 丝束宏观到微观结构。A. 蚕茧由丝束组成，丝束有蛋白质包被和单纤丝，每根单纤丝由丝胶和丝心蛋白组成；B. 单纤丝横截面图。

14.4 茧的形态与结构

14.4 Morphology and Structure of Cocoons

14.4.1 Shape and layers of cocoon

The cocoon is usually in different shapes such as long oval, oval waist, spherical or spindle shape, or slightly constricted in the middle. The cocoon has white, yellow, light green, meat red and other colors, about 3-4 cm long, 1.7-2.1 cm in diameter, white surface, irregular wrinkles, attached silk and villous shape. The silk lines on its inner wall are very regular. Light and tough, not easy to tear. The cocoon without eclosion contains a yellow brown silkworm pupa and a light brown and wrinkled silkworm skin removed before pupation. The protective layer includes cocoon coat, cocoon layer and pupa lining. The cocoon layer can be reeled, and the cocoon clothes and the reeled waste silk can be used as raw materials for silk, cotton and silk spinning.

14.4.1 茧的形状和茧层

蚕茧通常呈长椭圆形、椭圆束腰形、球形或纺锤形等不同形状或中部稍缢缩，茧有白、黄、淡绿和肉红等颜色，长为 3 ～ 4 cm，直径为 1.7 ～ 2.1 cm，表面白色，有不规则皱纹，并有附着的蚕丝，呈绒毛状。其内壁的丝纹很有规律。质轻而韧，不易撕破。未经羽化的蚕茧，内有黄棕色的蚕蛹一枚以及成蛹前脱下的淡棕色、皱缩的蚕皮。保护层包括茧衣、茧层和蛹衬等部分。茧层可以缫丝，茧衣及缫制后的废丝可作丝棉和绢纺原料。

14.4.2 Cocoon color

Cocoon color can be divided into white cocoons and colored cocoons, and colored cocoons can be divided into yellow cocoons and green cocoons. Yellow cocoons have golden, light yellow, ligusticum color, flesh color, light red and other cocoon colors. Generally speaking, the silkworm with yellow cocoon has yellow blood and the silkworm with white cocoon has colorless blood. However, it is found that the silkworm with yellow blood and white cocoon is controlled by different genes. Green cocoons and light bamboo cocoons with various concentration differences. *Bilian* and *Dazao*, which are native to China, are typical green cocoons. The offspring of Chinese white cocoon species and Japanese white cocoon species often produce light bamboo colored cocoons.

14.4.2 茧色

茧色分白色茧和着色茧两大类，着色茧又分黄茧和绿茧两类。黄茧类有金黄、浅黄、藁色、肉色和淡红等茧色。一般情况，黄茧的蚕是黄血，白茧的蚕血是无色的，然而，人们发现黄血蚕和白茧蚕是由不同基因控制的。有各种浓度差异的绿色茧和淡竹色茧。中国地方种的碧莲及大造是典型的绿茧。中国白茧种和日本白茧种杂交的子代往往产生淡竹色茧。

14.4.3 Mechanism of cocoon color formation

The yellow and flesh cocoons depend on carotenoid pigments, green cocoons are determined by flavonoid pigments. The cocoon's color is affected by the genes controlling penetration process from midgut to coelom and silk gland. The golden cocoon gene must

14.4.3 茧色形成的机制

黄茧和肉茧是由类胡萝卜素色素决定的，而绿茧取决于类黄酮色素。茧的颜色受到中肠至体腔和丝腺渗透过程基因控制的影响。黄茧和黄血是由不同的

be complementary to the yellow hemolymph gene to show the outer golden yellow. Light bamboo cocoon has two genes, G^a and G^b, which appear white when they exist alone. Through hybridization, the two genes complement each other to produce light bamboo color.

Insects such as silkworm, *B.mori, A. yamamai* and *A. pernyi* obtain pigments by feeding on plants. Under gene regulation, they are digested and absorbed through the midgut, enter the hemolymph circulation, and the pigments enter the silk gland cavity, and incorporated into cocoons.

Carotenoids cannot be synthesized by animals or insects and must be acquired from exogenous sources by plants, bacteria, and fungi. Carotenoids can be classified into two groups: nonpolar hydrocarbons carotenes, such as β-carotene and lycopene; and polar, oxygen-containing hydrocarbons xanthophylls. Xanthophylls include zeaxanthin, lutein, cryptoxanthin and canthaxanthin. Carotenoids are absorbed in the intestine of the caterpillar, transferred to the hemolymph lipoprotein, lipophorin, and ultimately, accumulate in the middle silk gland, resulting in the formation of golden yellow cocoons containing both lutein and β-carotene.

The lutein-binding protein is present in the midgut, fat body, and testis tissues, and in ovary and silk gland tissues, but not found in the hemolymph, hormonal sac, or eggs. equally distributed throughout the length of the midgut of all larval instars have the lutein-binding protein in equal amounts, suggesting that this protein it is involved in lutein uptake throughout the larval development stages.

Lipids are water insoluble, and their transport to various tissues via aqueous hemolymph is intimately linked to lipoprotein metabolism. In insects, two major lipoproteins exist in the hemolymph, lipophorin and lipid transfer particle. Scavenger receptors at target tissues recognize these lipoproteins and facilitate the movement of carotenoids into the cell, such as carotenoid binding protein (CBP) and lutein binding protein. In yellow-cocoon strains, CBP is expressed in the midguts and MSGs, and therefore the larval hemolymph, silk glands, and cocoons all appear yellow color. In $+^Y$ allele strains, due to the absence of the CBP, the midgut epithelium cannot absorb and transport carotenoids into hemolymph or silk glands, resulting in the olorless hemolymph, colorless silk glands, and white cocoons (Yuan et al., 2021). Lipophorin-mediated lutein and β-carotene transport to the middle silk gland are controlled by the Yellow cocoon (*Y*) gene and the flesh (*F*) gene, respectively. The middle silk gland of the *Y* mutant, homozygous for the recessive $+^F$

基因控制的。金黄茧基因必须与黄血基因互补，才表现外层金黄色。淡竹色茧有 G^a 和 G^b 两种基因，单独存在时各为白色，通过杂交，两基因互补，产生淡竹色。

家蚕、天蚕和柞蚕等昆虫通过取食植物获取色素。在基因调控下，经中肠消化和吸收，进入血液循环，色素进入丝腺腔并掺入茧。

类胡萝卜素不能由动物或昆虫合成，必须从植物、细菌和真菌外源获得。类胡萝卜素可分为两类：非极性碳氢化合物类胡萝卜素，如β-胡萝卜素和番茄红素以及极性含氧碳氢化合物叶黄质。叶黄质包括玉米黄质、叶黄素、隐黄素和角黄素。类胡萝卜素在虫体肠道中被吸收，转移到血淋巴脂蛋白，载脂蛋白，最终积累在中部丝腺，导致含有叶黄素和β-胡萝卜素的黄茧形成。

叶黄素结合蛋白存在于中肠、脂肪体、睾丸组织、卵巢和丝腺组织中，而在血淋巴、激素囊或卵子中未被发现。叶黄素结合蛋白在所有幼虫龄期的整个中肠上均匀分布且数量相等，表明该蛋白质在整个幼虫发育阶段参与叶黄素的吸收。

脂质是不溶于水的，它们通过水相血淋巴运输到各种组织，与脂蛋白代谢密切相关。在昆虫中，血淋巴中存在两种主要的脂蛋白，即载脂蛋白和脂质转移颗粒。靶组织的清道夫受体，如类胡萝卜素结合蛋白（CBP）和叶黄素结合蛋白，识别这些脂蛋白并促进类胡萝卜素进入细胞。在黄色茧品系中，CBP在中肠和MSGs中表达，因此幼虫血淋巴、丝腺和茧都呈现黄色。$+^Y$ 等位基因品种由于缺乏CBP，中肠上皮无法吸收类胡萝卜素并将其运输到血淋巴或丝腺中，从而形成无色血淋巴、无色丝腺和白色茧 (Yuan 等 , 2021)。载脂蛋白介导叶黄素和β-胡萝卜素向中部丝腺的转运，分别由黄茧（*Y*）基因和肉色（*F*）基因控制。*Y* 突变体的中部丝腺是隐性 $+^F$ 等位基因的纯合子，在细胞摄

allele, has a defect in the cellular uptake of lutein, resulting in the formation of cocoons of creamier yellow, containing mainly β-carotene. Conversely, the middle silk gland of the *F* mutant, homozygous for the recessive $+^F$ allele, has a defect in the cellular uptake of β-carotene, resulting in the formation of light yellow cocoons containing mainly lutein.

In silkworm genetic linkage map, more than 10 genes (loci) that play roles in the uptake and transport of carotenoids have been identified, such as Yellow hemolymph (*Y*), Yellow cocoon (*C*), Flesh cocoon (*F*), the Yellow hemolymph inhibitor (*I*), Rusty cocoon (*Rc*), Pink cocoon (*Pk*), Straw-colored cocoon (*Cst*). The *Y* gene encodes a CBP which controlled the uptake of carotenoids into the intestinal mucosa and the silk glands; *I* gene prevents carotenoids from entering body fluid from midgut epithelial cells, while the recessive $+^I$ can make carotenoids enter body fluid; The *C* gene is mainly responsible for the transport of lutein from the hemolymphto the silk gland, while the *F* gene determines selective uptake of β-carotene in the middle silk glands. Only larvae with the phenotype $Y+^IC$ make yellow cocoons (Chai et al., 2014; Fig.14-8; Appendix color graphic Fig.14-8).

取叶黄素方面存在缺陷，导致形成乳黄色的茧，主要含有 β- 胡萝卜素。相反，*F* 突变体的中部丝腺是隐性 $+^F$ 等位基因的纯合子，在细胞摄取 β- 胡萝卜素方面存在的缺陷导致形成主要含有叶黄素的淡黄色茧。

在家蚕遗传连锁图谱中，已鉴定出 10 多个与类胡萝卜素吸收和运输有关的基因（位点），如黄血淋巴（*Y*）、黄茧（*C*）、肉茧（*F*）、黄血淋巴抑制基因（*I*）、锈茧（*Rc*）、粉红茧（*Pk*）和稻草色茧（*Cst*）。*Y* 基因 CBP，该蛋白质控制类胡萝卜素进入肠黏膜和丝腺的摄取；*I* 基因阻止类胡萝卜素从中肠上皮细胞进入体液，而隐性等位基因 $+^I$ 可以使类胡萝卜素进入体液。*C* 基因主要作用于叶黄素从血液向丝腺的运输，*F* 基因决定中丝腺中 β- 胡萝卜素的选择性摄取。只有基因型 $Y+^IC$ 的幼虫才可以产生黄茧（Chai 等，2014 年；图 14-8；附彩图 14-8）。

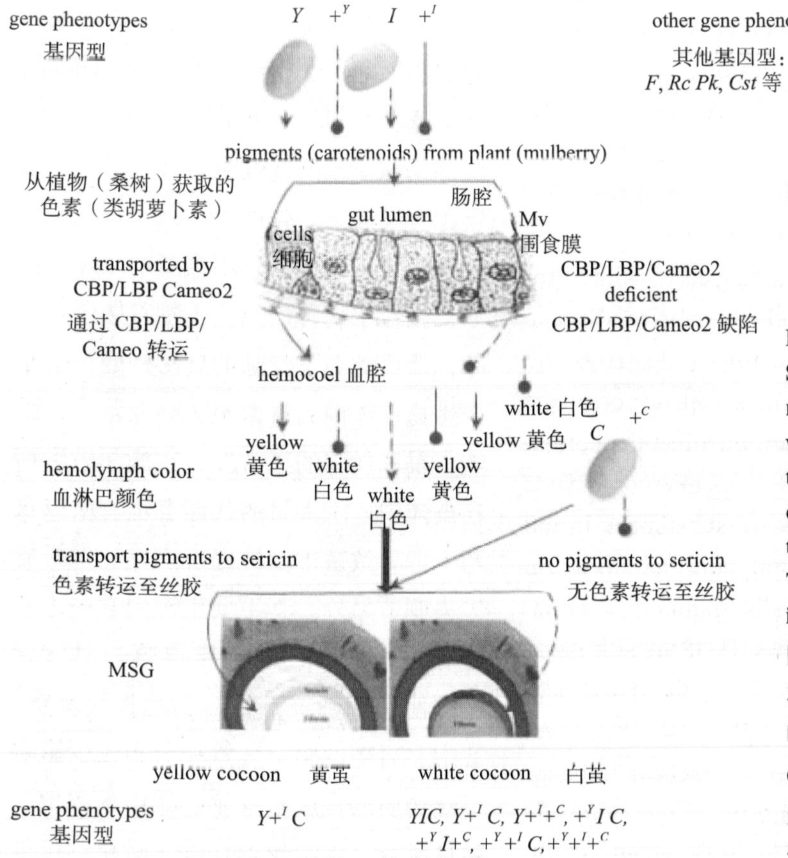

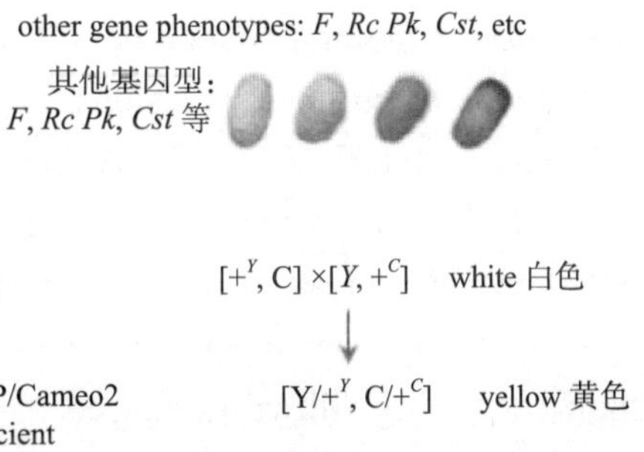

Fig.14-8 Mechanism of cocoon color formation. Silkworm and other insects uptake pigments from mulberry trees or other plants, such as carotenoids, which are absorbed in midgut, and transported into the hemocoel through CBP encoded by *Y* gene to color the silkworm hemolymph, which circulates to the silk gland and infiltrates into the silk gland cells. The pigments accumulated in MSG and incorporated into sericin layer, resulting in color cocoons.

图 14-8 茧色形成机制。家蚕等昆虫取食桑树或其他植物来的色素，如类胡萝卜素等在中肠被吸收，通过 *Y* 基因编码的类胡萝卜素结合蛋白 CBP 转运进入血腔，蚕血液着色；血液循环至丝腺并渗入丝腺细胞，在中部丝腺累积的色素进入丝胶层，从而产生色茧。

At the end of the fifth instar, the midgut loses its bright yellow color and takes on a faint yellow color, whereas the silk gland, specifically the middle part, turns bright yellow in color as a result of the mobilization of lutein from the midgut to the silk gland. In addition, when the insect stops feeding and during fibroin and sericin synthesis, uptake of dietary carotenoids in the midgut and silk gland is halted. At the end of the spinning period, the amount of carotenoids in the silk gland decreases sharply because of its utilization in the cocoon.

在5龄结束时，中肠失去了明亮的黄色呈现出微弱的黄色，而丝腺，特别是中部，由于叶黄素从中肠向丝腺的移动而变成明亮的黄色。此外，当昆虫停止取食，在丝心蛋白和丝胶合成期间，中肠和丝腺对食物类胡萝卜素的吸收也停止。在吐丝期结束时，由于类胡萝卜素用于结茧，丝腺中的含量急剧减少。

14.5 不结茧蚕发生的生理因素

14.5 Physiological Factors of Non-cocooning Silkworms

After the cocooning peak, the body color of some silkworms is still as green as that of in full feeding period. After mulberry feeding, only a small amount of leaves can be eaten. After 2~3 days, the body color appearances mature. It is dormant in the cocooning tool or creeps slowly, but does not form cocoons. The worm will dies a week later, and there is no abnormality on the body surface. Dissect the non-cocooning silkworm body and observe the silk gland has no obvious lesions, but the MSG is slightly turbid. Two days later, dissect the ASG of the non-cocooning silkworm, and some of them appear rusty red. After two days more, the ASG shows an increasing rusty red, and the MSG feels hard and has poor elasticity when pinched by hand.

In addition to non-cocooning silkworm caused by silkworm disease, pathogenic microorganisms such as pebrine disease directly invade the silk gland, which seriously destroys the spinning function of the silk gland; Mass reproduction of microorganisms in the body and capture nutrition affect the normal development of silkworm and silk gland; Besides microorganisms producing a large number of metabolic toxic substances in the silkworm body or pesticide poisoning, paralyzing the silkworm central nerve and losing the regulation of silk spinning function. The main physiological factors arc as follows: ① if the silk gland is too hypertrophic, it will compress the first pair of tracheal plexus, resulting in insufficient oxygen supply to the head and nerve paralysis. ② due to the less distribution of tracheal branches on the middle and posterior silk glands, it affects the synthesis and secretion of silk substances. ③ caused by poor growth of silk

上蔟峰期后，部分蚕的体色仍像盛食蚕一样体色较青，给桑后仅少量食叶，再过2～3 d体色呈熟状，蛰伏簇中或缓慢蠕动，但不结茧，一周后死亡，尸体表面没有异常。解剖不结茧蚕体，观察其丝腺无明显病变，但中部丝腺轻度浑浊。2 d后，解剖不结茧蚕的前部丝腺，部分出现锈红色，再过2 d，前部丝腺呈现的锈红色增多，手捏中部丝腺有发硬感，弹性差。

除了蚕病会引起不结茧蚕以外，比如，微粒子病等病原微生物直接侵入丝腺，严重破坏了丝腺的分泌机能；病原微生物在体内大量繁殖，夺取营养，影响了蚕和丝腺的正常发育；病原微生物在蚕体内产生大量的代谢有毒物质或农药中毒导致蚕中枢神经麻痹，吐丝机能失去调节之外。其主要生理因素：①若丝腺过于肥大，则会压迫第一对气管丛，使头部供氧不足，引起神经麻痹。②中、后部丝腺上气管枝分布较少影响了丝物质的合成与分泌。③由丝腺生长不良所致。正常丝腺长度从孵化到极度

gland. The length of normal silk gland increases about 50 times from hatching to extreme growth, while that of non-silkworms increases only 10 ~ 20 times. ④ abnormal or blocked spinning tube hinders spinning. ⑤ the central silk gland is thick, the PSG shrinks, or the whole silk gland shrinks, which affects the synthesis and secretion of silk substances. ⑥ due to the degeneration of local gland cells, one or several places of the MSG are constricted or distorted that affecting the synthesis and secretion of silk substances. ⑦ the nucleus of gland is underdeveloped. The nucleus of normal silkworm has many branches at the 5th instar, accounting for most of the gland cells, while the nucleus of non-cocooning silkworm has few branches.

成长约增加 50 倍，而不吐丝蚕仅增加 10 ～ 20 倍。④吐丝管畸形或闭塞妨碍吐丝。⑤中部丝腺粗大、后部丝腺萎缩或整个丝腺萎缩，影响了丝物质的合成与分泌。⑥局部腺细胞退化引起中部丝腺一处或数处缢缩或扭曲，影响丝物质的合成与分泌。⑦腺细胞核不发达。正常蚕的细胞核在 5 龄期分支很多，占腺细胞的大部分，而不结茧蚕的腺细胞核分枝很少。

Except the factors previously describe, non-cocooning silkworm also can caused by undesirable environment, such as the feeding room temperature of the 5th instar was too high, the leaf quality was too tender, and the endocrine disorder caused the abnormal development of silk gland. When the mounting room under high temperature and wet, or low temperature and dry, the number of non-cocooning silkworms will also increase greatly.

除这些因素外，不结茧的家蚕也可由不良环境引起，如 5 龄饲养室温度过高、叶质太嫩和内分泌失调导致丝腺发育异常等。当上蔟室处于高温多湿或低温干燥条件下时，不结茧蚕的数量也会大大增加。

14.6　丝物质的应用

14.6　Applications of Silk Substances

Silk extracted from silkworm has attracted much attention because of its natural large-scale production, excellent biocompatibility, unique mechanical properties, controllable degradation and processing into various material forms. For thousands of years, silk has been widely used as textiles because of its luster, flexibility and light weight. In addition, it has been used as a surgical suture. Due to its attractive biological applications, people have increased interest in the morphology and structure of silk fibroin. Because of the biocompatibility between silk fibroin fiber and human living tissue, silk fibroin fiber has been used as suture for a long time in the field of surgery. In addition, it has been proved that silk can be used as a substrate for enzyme immobilization in biosensors.

从桑蚕中提取的蚕丝因其天然的规模化生产、优良的生物相容性、独特的机械性能、可控的降解以及可加工成各种材料形式而备受关注。几千年来，蚕丝因其光泽、柔韧性和重量轻而被广泛用作纺织品。此外，它已被用作外科缝线。由于其在生物相关应用方面的吸引力，人们对丝心蛋白形态和结构的兴趣增加。由于丝心蛋白纤维与人体活体组织具有良好的生物相容性，因此丝心蛋白纤维在外科领域被长期用作缝合线。此外，已经证明丝可以用作生物传感器中酶固定化的底物。

The striking and charming characteristics of silk proteins from a point of view for industrial materials can be summarized as follows with the comparison of other naturally occurring biopolymers. Silk

从工业材料的角度来看，与其他天然生物高聚物相比，丝蛋白质的特点和物质特性可概括如下。桑蚕丝纤维可以在

fibers from *B. mori* silkworms can be dissolved in concentrated neutral salt solutions such as calcium chloride at elevated temperatures without using any harmful and reducing agents. After dialyzing against water a pure silk fibroin solution can be obtained.

Various forms of silk proteins, such as silk powders, silk gels and silk films can be produced controlling the cast-dry speed of the silk fibroin solution. Silk protein can be made insoluble by immersing it in alcohol solution.

A biomaterial is a non-living material used in a medical device, intended to interact with a biological system. The chief function of biomaterials is to repair, replace or augment a tissue. The ability of a biomaterial to perform with the appropriate host response in a specific situation is known as biocompatibility.

New applications are using silk nanoparticles for drug delivery; making dissolvable plates and screws for orthopedic repair devices, and providing the scaffolding for growing tissues and organs. Some researchers now working on implantable silk biosensors. The outstanding mechanical properties of silk have been further enhanced by Chinese chemists who fed larval silkworms mulberry leaves sprayed with single-walled carbon nanotubes and graphene (Patel 2016; wang et al., 2021). Parts of the fed carbon nanomaterials were incorporated into the spun silk fibers and, compared to regular silk, the carbon-enhanced silks were found to be twice as tough and could withstand half again higher stress before breaking. Researchers have also fed titanium dioxide nanoparticles to silkworms to create ultra-strong silk that resists ultraviolet degradation and conducts electricity, making it usable for sensors embedded in textiles and possibly useful in medical applications related to reading nerve signals. Ethanol treatment could induce the formation of β-sheet of H-fibroin in composite nanofibers, thus improving the mechanical properties of PCL(poly ε-caprolactone) /H-fibroin nanofiber scaffold. PCL/H-fibroin nanofiber scaffold was more conducive to cell adhesion and proliferation with the increment of H-fibroin. PCL/H-fibroin core–shell nanofibers could effectively reduce the prophase burst of drug molecules and show the sustained drug release (Wang et al., 2021).

In general, silk has oxygen permeability, can be used as artificial skin, substrates for enzyme immobilization, substrates for diseases detecting biosensors, soft tissue compatible polymers, silk fibroin fibers as artificial tendons, composite materials made of silk fibroin and ethylene glycol, drug delivery system, grafting and chemical modification, biomimetic processing of silk protein, new

高温下溶解在浓缩的中性盐溶液中，如氯化钙，而不使用任何有害的还原剂。用水透析后，可获得纯丝心蛋白溶液。

通过控制丝心蛋白溶液的浇铸干燥速率，可以生产各种形式的丝蛋白质，如丝粉、丝凝胶和丝膜。将丝蛋白质浸泡在酒精溶液中可使其不溶。

生物材料是用于医疗设备的非生物材料，旨在与生物系统相互作用。生物材料的主要功能是修复、替换或增强组织。生物材料在特定情况下具有适当与宿主反应的能力，这种能力称为生物相容性。

新的应用是使用丝纳米颗粒进行药物输送；为骨科修复设备制造可溶解的钢板和螺钉，以及为生长的组织和器官提供支架。研究人员正在研究植入式丝生物传感器。中国化学家用喷涂有单壁碳纳米管和石墨烯的桑叶喂养幼虫，进一步增强了蚕丝卓越的机械性能（Patel 2016；Wang 等，2021）。部分饲喂的碳纳米材料被加入绢丝纤维，与普通丝相比，碳增强丝的韧性是普通丝的两倍，并且在断裂前可以承受一半以上的应力。研究人员还向蚕喂入二氧化钛纳米颗粒，以制造出抗紫外线降解和导电的超强丝，使其成为嵌入纺织品中的传感器，并可能用于与读取神经信号相关的医疗应用。乙醇处理可以诱导复合纳米纤维中 H- 丝心蛋白 β- 片的形成，从而提高聚 PCL（ε- 己内酯）/H- 丝心蛋白纳米纤维支架的力学性能。随着 H- 丝心蛋白含量的增加，PCL/H- 丝心蛋白纳米纤维支架更有利于细胞的黏附和增殖。PCL/H- 丝心蛋白核壳纳米纤维可以有效减少药物分子的前期猝发，并表现出药物的持续释放（Wang 等，2021）。

概括来讲，蚕丝具有透氧性，可用作人造皮肤、酶固定化基质、疾病检测生物传感器基质、软组织相容性聚合物、人工肌腱的丝心蛋白纤维、丝心蛋白和乙二醇复合材料、药物输送系统、移植和化学改性、丝蛋白质的仿生加工

silk protein-modification of silk protein by gene engineering for production of biomaterials.

(Li Wenchu)

以及通过基因工程对丝蛋白质进行新的丝蛋白质改性以生产生物材料。

（李文楚）

Exercises

1.What's detailed composition and structure of silk gland?

2.How insect sysnthesis silk substances?

3.Do you really understand the molecular mechanisms of colorful cocoon?

4.How to avoid nonspining silkworm in practical cocoon production?

思考题

1. 丝腺的详细组成和结构是什么？

2. 昆虫如何合成丝物质？

3. 你真的了解彩色茧的分子机制吗？

4. 在实际蚕茧生产中，如何避免不结茧蚕？

第 15 章 肌肉和运动

Chapter 15 Muscle and Movement

Key points

1.A muscle can be subdivided into fibers, and a fiber into myofibrils. Myofibrils are composed of sarcomeres, each of which contains actin, myosin, and other proteins involved in the contraction mechanism. In addition, muscle contains sarcoplasmic reticulum.

2.The major proteins in muscle fibers are actin, tropomyosin, troponin, and myosin.

3.The main functions of actin and myosin and the contraction mechanism of insect muscle filament sliding.

4.Stimulatory muscle sarcolemma causes contraction, and the process of muscle and movement.

知识点

1. 肌肉可以细分为纤维，纤维可以细分为肌原纤维。肌原纤维由肌节组成，每个肌节都含有肌动蛋白、肌球蛋白和其他参与收缩机制的蛋白质。此外，肌肉含有肌浆网。

2. 肌肉纤维中的主要蛋白质是肌动蛋白、原肌球蛋白、肌钙蛋白和肌球蛋白。

3. 肌动蛋白和肌球蛋白的主要功能以及昆虫肌丝滑动的收缩机制。

4. 刺激肌肉肌膜引起收缩以及肌肉运动的过程。

Insect muscles are derived from ectodermal tissues in embryo stages. The most muscles belong to skeletal muscles, or called somatic muscles. Few of them are visceral muscles. Skeletal muscles are attached to the cuticle, typically by tonofibrillae that pass through the endo- and exocuticle and attach to the inner layer of the epicuticle. Mitochondria, sometimes called sarcosomes because of their large and irregular size, are the power-houses for muscles function.

Muscles are usually divided into broad categories such as skeletal, flight, heart, alary, and gut muscles based on their location within the body, structure, and function. Skeletal muscles are a mixture of fast and slow contract ing muscles. Fibrillar muscles are exclusively fast con tracting muscles. Heart, alary, and gut muscle are slower contracting muscles.

昆虫肌肉是由中胚层特化而成的组织。大多数肌肉属于骨骼肌或称体壁肌，少数为脏肌。骨骼肌与表皮相连，通常是通过贯穿内外表皮的皮肌纤维和上表皮的内层相连。线粒体因为大而无规则，称为肌粒（即肌细胞线粒体），是肌肉发挥功能的发电站。

肌肉通常根据其在体内的位置、结构和功能广义地划分为骨骼肌、飞行肌、心肌、翼肌和肠道肌。骨骼肌是快收缩肌和慢收缩肌的混合体。纤丝状肌是唯一的快收缩肌。心肌、翼肌和肠道肌都是慢收缩肌。

15.1　昆虫肌肉的类型和结构

15.1　The Types and Structure of Insect Muscle

15.1.1 The composition of muscle

A muscle can be subdivided into fibers, and a fiber into myofibrils. Myofibrils are composed of sarcomeres, each of which contains actin, myosin, and other proteins involved in the contraction mechanism (Fig. 15-1; Appendix color graphic Fig. 15-1). In addition, muscle contains sarcoplasmic reticulum (SR). The SR is an extensive network of internal membranes broken into vesicles that run longitudinally on the surface of the muscle fibers. The SR plays a major role in the contraction process as a storehouse of calcium ions. Transverse (T) tubules penetrate the muscle from the outside, originating usually; but not always, at the Z bands. The network of T tubules and SR membranes do not open to each other, but they do intersect at closed junctions believed to be the major sites of calcium storage. These junctions are called dyad or triad junctions (Fig. 15-2; Appendix color graphic Fig.15-2), depending on whether a T tubule intersects with one or two SR vesicles, respectively.

The T tubules carry the electrical wave of excitation arriving at file surface of a muscle (*via* a nerve) inward where it also spreads to the SR and releases bound calcium as the free ions necessary for contraction to occur.

The myofibrils are made up of repeating sarcomere units. A sarcomere, the region between two Z bands (also called Z line, Z disc), is typically 2-3 μm long in a muscle at rest, but sarcomeres up to 10 μm long occur in some very slow muscles. Sarcomere length is shorter in fast-contracting muscles. The Z line is a plate-like sheet of protein to which actin and some other muscle proteins are attached. The thin filaments extend on either side of the Z line about two thirds to nearly the mid-point of a sarcomere. Thick filaments of myosin, about 20 nm in diameter, lie between the thin filaments. The thick filaments extend across the middle of a sarcomere but usually do not extend to the Z line. The various overlapping regions of thick and thin filaments give muscle in thin

15.1.1　肌肉的组成

肌肉可分为肌纤维，肌纤维再分为肌原纤维。肌原纤维由肌节组成，每个肌节包含肌动蛋白、肌球蛋白和其他牵涉到收缩机制的蛋白质（图 15-1；附彩图 15-1）。此外，肌肉包含肌质网（SR）。SR 是内膜的网状延伸，被沿着肌纤维表面纵向分布的小囊泡分隔。SR 在肌肉收缩过程中作为钙离子贮存库起着主要作用。横管（T 管）从肌肉外部穿过，通常而不是始终起源于 Z 带。T 管和 SR 膜相互之间并不开放，但它们确实在封闭连接处交叉，被认为是钙库的主要位置。这些连接根据 T 管与一个肌质网池或两个肌质网池连接分别称为二联体或三联体连接（图 15-2；附彩图 15-2）。

T 管（通过神经）输送兴奋性电波到达肌肉纵列表面，向内也传送到 SR，将结合态钙作为游离的离子释放到产生收缩需要的地方。

肌原纤维由重复的肌节单位组成。肌节是 Z 带（也称为 Z 线、Z 盘）之间的区域，通常在肌肉处于静息时长为 2 ～ 3 μm，但在一些非常慢的肌肉中长达 10 μm。快速收缩肌较短。Z 线是一板状薄层，附着肌动蛋白和一些其他蛋白质。细肌丝在 Z 线一侧延伸约 2/3，接近肌节的中点。肌球蛋白的粗肌丝直径约 20 nm，位于细肌丝之间。粗肌丝通常越过肌节的中线延伸，但一般不延伸到 Z 线。粗肌丝和细肌丝各种重叠区域使肌肉的薄组织切片在光线

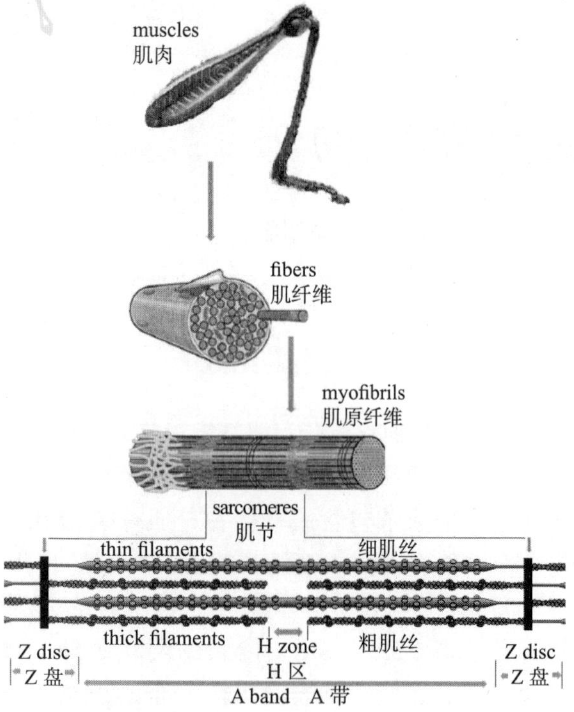

Fig.15-1 An illustration of smaller units that compose muscles (modified from Herrid, 1977).
图 15-1 肌肉组成的较小单元图解（仿 Herrid, 1977）。

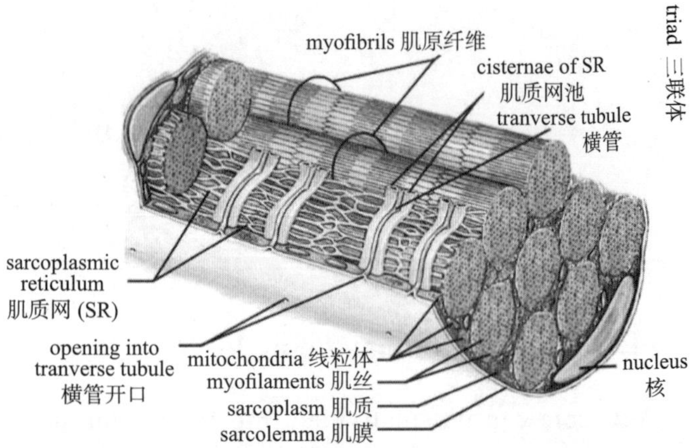

Fig.15-2 Diagrammatic illustration of the sarcoplasmic reticulum (SR) and transverse (T) tubules. Sarcolemma and mitochondria-abundant sarcoplasm enveloped myofibrils. There are openings into transverse tubules on the surface of the sarcolemma. The linkage of T-tubule intersecting with two SR vesicles (or cisternae) are called triad junction (modified from Mic-Graw Hill Company).

图 15-2 肌质网（SR）和横管（T）概要图解。肌原纤维外围是肌膜和富含大型线粒体的肌质。横管在肌膜表面有开口，与两个肌质网池相连时称为三联体连接（仿 Mic-Graw Hill 公司）。

histological sections a banded appearance as light passes through regions of different density. The A band appears dark because light must pass through the overlapping regions of the actin and myosin filaments, while the H zone in the middle of the sarcomere and the I band near the Z line transmit lighter because those regions contain only myosin or actin filaments, respectively. The M line across the middle of the H zone is created by cross-links between myosin filaments that help hold the myosin filaments in place. In general, in skeletal and flight muscles the striations or bands of adjacent myofibrils are aligned side by side. Thus, the light and dark bands appear evenly lined up in a large section of muscle (Fig.15-3A; Appendix color graphic Fig.15-3).

Each thick filament surrounded by six thin filaments, with the thin filaments positioned about equally between two thick filaments (Fig.15-3B) to give a ratio of three thin filaments to one thick filament. Other ratios are found in some muscles, including up to 12 thin filaments arranged in such a way as to present a 6 thin: 1 thick filament arrangement in intersegmental muscles of the cockroach *P. americana* and *R. prolixus*, and in wing muscles of large saturniid silk moths that have slow wing beats of five to six beats per second (Fig.15-3C).

穿过不同密度的区域时呈现带状。A 带呈黑色，因为光线必须透过肌动蛋白和肌球蛋白的重叠区域，而肌节中线的 H 区和靠近 Z 线的 I 带透过更多的光线，因为这些区域分别只含有肌球蛋白或肌动蛋白。M 线穿过 H 区的中间，由肌球蛋白纤丝间交叉连接引起，以协助固定肌球蛋白纤丝的位置。总之，骨骼肌和飞行肌相邻肌原纤维的条纹或带排列齐整。所以在大肌肉切片中，明带和暗带均等排列（图 15-3 A；附彩图 15-3）。

每根粗肌丝被 6 根细肌丝包围，细肌丝在两根粗肌丝间的位置大致相等（图 15-3B），其比例为 3 根细肌丝：1 根粗肌丝。在有些肌肉中发现其他比例，包括在美洲大蠊、长红猎蝽的节间肌肉，以及翅膀缓慢拍打 5～6 次 /s 的大蚕蛾的翅膀肌肉中多达 12 条细丝呈现 6 细 1 粗的排列方式（图 15-3C）。

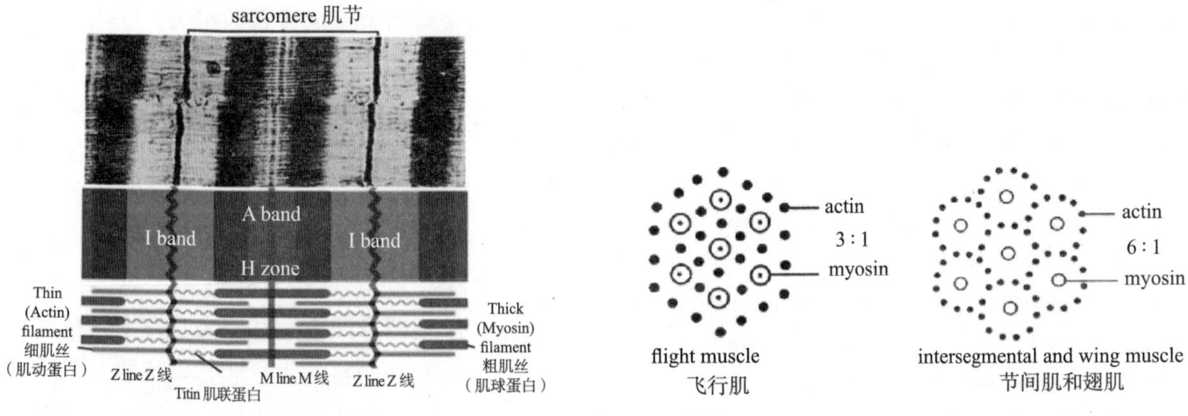

Fig.15-3 Electron microscopic division of sarcomere units and the arrangement of actin and myosin filaments in different muscles. A. Composition of sarcomere; B. In the flight muscle of most insects, producing a 3:1 ratio of thin to thick filaments; C. An arrangement in which 12 actin filaments are arranged between myosin filaments to give a 6∶1 ratio. (modified from Toselli et al., 1968.)

图 15-3 肌节单位的电镜学划分和不同肌肉中的肌动蛋白和肌球蛋白丝的排列。A. 肌节的组成；B. 飞行肌细肌丝和粗肌丝的比例为 3∶1；C. 肌球蛋白丝间 12 根肌动蛋白丝的排列产生 6∶1 的比例（仿 Toselli 等，1968）。

15.1.2 Synchronous and asynchronous muscles

Muscles of insects can be divided into two groups based on the contraction rate per nerve input. Contraction frequency in synchronous muscles is directly controlled in a 1:1 manner by the output of nerve impulses from the central nervous system. Asynchronous muscles oscillate and produce several to many contractions in response to a single neuronal stimulus. Synchronous muscles typically have a well-developed SR and T systems with regularly occurring triad junctions.

15.1.3 The types of skeletal muscles

Skeletal muscle could be divided into tubular muscle, close-packed muscle and fibrillar muscle based on the shape and arrangement way of myofibrils (Fig. 15-4).

The fibers of tubular muscle are multinucleate, with nuclei located along the center of the fiber and slab-like myofibrils typically radiating from the center like spokes of a wheel. These slab-like myofibrils run the length of the muscle and shorten in the plane of the long axis of the muscle. Numerous large, somewhat elongated mitochon- dria are arranged radially between myofibrils. Both longitudinal and transverse tubules of the sarcoplasmic system are well developed in tubular muscles. Tubular muscles are found in the flight musculature of many insect groups, including both direct and indirect flight muscles in some insects. These flight muscles, which are the typical flight musculature in the lower

15.1.2 同步肌和异步肌

昆虫肌肉根据每次神经传入收缩的速率可分为两类。同步肌的收缩频率由中枢神经系统而来的神经脉冲输出直接以 1∶1 方式控制。异步肌对单一神经刺激的反应是振荡并产生几个到多个收缩。同步肌一般有发达的 SR 和 T 管系统，伴随着有规律的三联体连接。

15.1.3 骨骼肌的类型

骨骼肌按肌原纤维形状和排列方式，分为管状肌、束状肌和纤维状肌（图 15-4）。

管状肌纤维是多核的，胞核位于纤维的中心，厚片状肌原纤维常像车轮辐条一般从中心辐射。这些厚片状肌原纤维沿肌肉的长度延伸，在肌肉长轴平面收缩。无数大型、稍微加长的线粒体放射状排列在肌原纤维间。在管状肌内，肌质的纵管和横管很发达。管状肌被发现于许多昆虫的飞行肌组织中，包括有些昆虫的直接飞行肌和间接飞行肌。这些飞行肌是低等直翅目（如蜚蠊科）和

Orthoptera (such as the Blattidae) and in Odonata, may be most similar to the evolutionarily primitive flight muscles (Fig. 15-4A).

Other synchronous muscles in the flight musculature of many groups of insects are called close-packed fibers. The muscle fibers are typically 10-100 μm in diameter. Nuclei are numerous, peripherally located, and somewhat flattened. Small myofibrils of 0.5-1 μm in diameter are interspersed with columns of large mitochondria. The muscle fibers may be circular in shape or have a polygon shape with many angles. The flight muscles of some Orthoptera, Trichoptera, and Lepidoptera are close-packed muscle. Both longitudinal and transverse tubules of the sarcoplasmic reticulum are extensively developed. Close-packed muscle could be divided into thin SR and thick SR as well. The former contains seldom SR. The muscle cell almost fulfilled with myofibrils. The nuclei directly attached between the membrane and myofibrils; The later have more SR, among which the nuclei are embedded (Fig. 15-4B).

Asynchronous muscles, also called fibrillar muscles (Fig. 15-4C), and the majority of insects that fly do so with asynchronous muscles. Asynchronous muscles are more efficient than synchronous muscles because they reduce the repetitive cycling of Ca^{2+}, and thus minimize the energy cost of such cycling. Moreover, asynchronous muscles likely have been selected during evolution of flying insects. Asynchronous flight muscles occur in both direct

蜻蜓目典型的飞行肌组织，可能类似于进化上最原始的飞行肌（图 15-4A）。

多种昆虫飞行肌组织的其他同步肌被称为束状肌。肌纤维直径通常为 10 ～ 100 μm。核的数量巨大，位于外周，稍微扁平。小肌原纤维直径为 0.5 ～ 1 μm，点缀着柱形大线粒体。肌纤维呈圆形或多角体。有些直翅目、毛翅目和鳞翅目的飞行肌是束状肌。肌质网的纵管和横管都很发达。束状肌还可以分为薄 SR 和厚 SR。前者 SR 极少，肌原纤维几乎充满整个肌细胞，细胞核直接附着在细胞膜和肌原纤维之间；后者具有较多的 SR，细胞核包埋其中（图 15-4B）。

异步肌也称纤维状肌（图 15-4C），多数昆虫飞行时用的是异步肌。异步肌比同步肌更为有效，因为其减少了 Ca^{2+} 的重复循环，而使这种循环的能量成本降到最低。此外，异步肌可能是飞行昆虫进化过程中的选择。异步飞行肌出现在双翅目、膜翅目、鞘翅目、一些

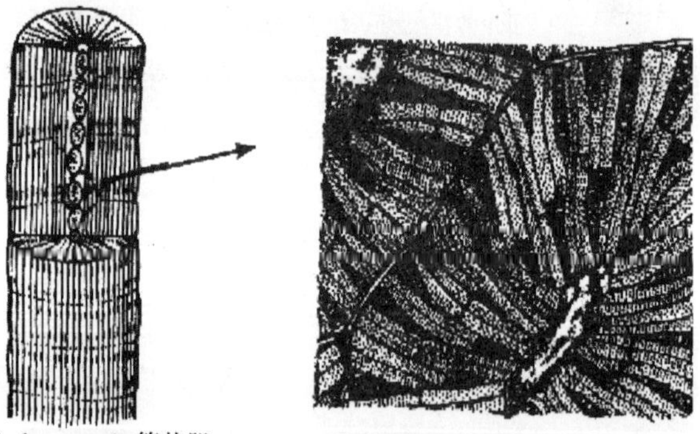

tubular muscle 管状肌

dragonfly fight muscles 蜻蜓飞行肌

A

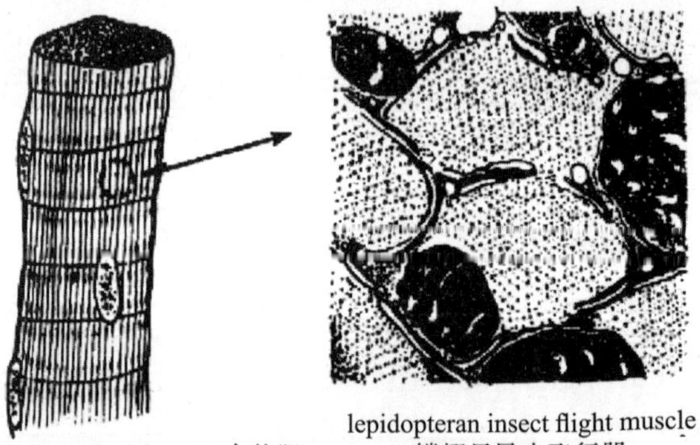

close-packed muscle 束状肌

lepidopteran insect flight muscle 鳞翅目昆虫飞行肌

B

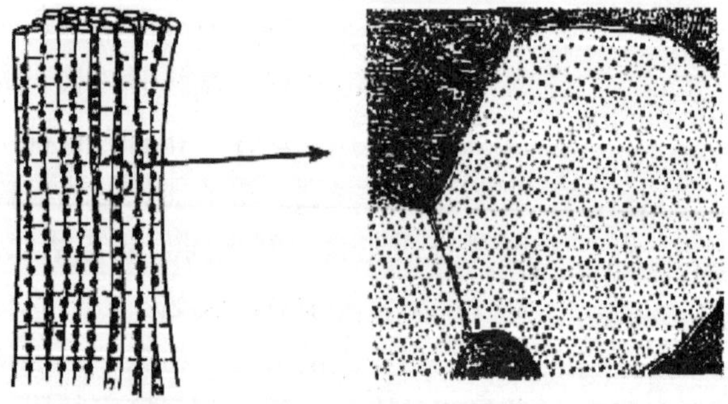

fibrillar muscle 纤维状肌

wasp flight muscle 胡蜂飞行肌

C

Fig.15-4 The types of insect skeletal muscle (modified from Snodgrass,1925; Smith, 1984).
图 15-4 昆虫骨骼肌的类型（仿 Snodgrass, 1925; Smith，1984）。

and indirect muscles associated with wing move ments in Diptera, Hymenoptera, Coleoptera, some Hemiptera, and in a number of other insect groups. Wing beat frequencies greater than 100 Hz are common in insects with asynchronous flight muscles. One very small midge (a dipteran) may beat the wings up to 1,000 times per second (Sotavalta, 1953). The muscle fibers are cylin drical in shape and multinucleate, with nuclei usually lying peripherally in neat rows. Sarcomeres vary from 1.7-2.5μm in length. The A band covers about 90% of the sarcomere length, and I bands are very narrow. The sarcoplasmic reticulum of myofibrils is poorly developed and junctions between the SR and T system of tubules are reduced to dyad junctions. Six thin filaments (actin) surround the thick filament (myosin) in a very regular order so that the ratio of thick to thin filaments is 1 : 3.

半翅目和许多类其他昆虫的与翅膀运动直接或间接相关的肌肉中。在异步飞行肌昆虫中翅膀的拍击频率普遍大于 100 Hz。一种非常小的摇蚊（双翅目）拍击翅膀的频率达到每秒 1 000 次（Sotavalta, 1953）。肌纤维呈柱状，多核，胞核常常位于匀称成排的纤维外周。肌节长为 1.7 ~ 2.5 μm。A 带覆盖整个肌节长度的 90%，I 带很窄。纤维状肌的肌质网不发达，和横管间的连接简化为二联体连接。6 根细肌丝（肌动蛋白）很有规律地围绕着粗肌丝（肌球蛋白），因而粗肌丝和细肌丝比为 1 : 3。

15.1.4 Muscle attachments to the exoskeleton

One end of a skeletal and wing muscle is anchored to a relatively non-moveable part of the exoskeleton. This is called the origin of the muscle, and the opposite end that is attached to a movable part of the body, such as a wing hinge, an appendage, or a part of the exoskeleton is called the insertion. Skeletal muscles are usually anchored to the epicuticle layer of the exoskeleton (Fig. 15-5). A large 2.5 kDa extracellular matrix protein (Dumpy protein) encoded by the *dumpy* (*dp*) gene in *D. melanogaster* connects the outer membrane of muscle fibers to the basal membrane of epidermal cells through a crosslinking zona pellucida domain and a transmem brane anchoring sequence. Dumpy protein provides a

15.1.4 肌肉与外骨骼的连接

骨骼肌和翅肌的一端固定在相对不能移动的外骨骼上，这称为肌肉的起点；另一端与虫体可移动部分相连，如翅和附肢的关节或外骨骼的一部分，称为嵌入。骨骼肌常常与外骨骼的上皮层锚定（图 15-5）。果蝇中有一个分子质量为 2.5 kDa 的胞外基质蛋白，即粗短蛋白由 dump（dp）基因编码，通过透明带区域与跨膜固定序列的交联，把肌纤维的外膜和真皮细胞的底膜连接起来。粗短蛋白为肌肉和细胞及表皮提供

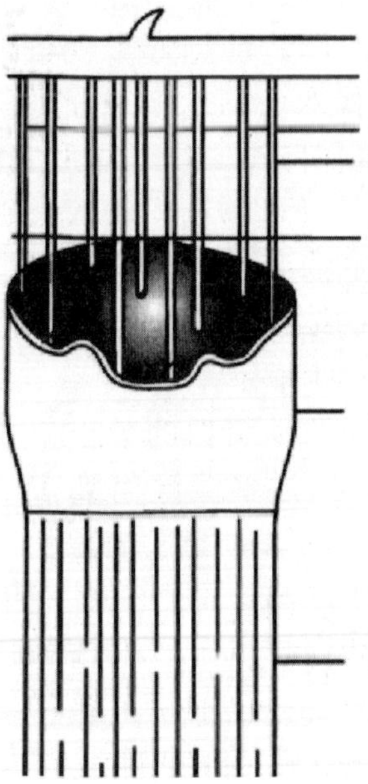

Fig.15-5 A schematic illustration of tonofibrillae connecting muscle fibers with the epicuticle (modified from Elder, 1975).
图 15-5 张力原纤维通过肌纤维与上皮连接的图解说明（仿 Elder，1975）。

strong attachment for muscles to cells and cuticle, allowing high mechanical tension and preventing the tearing of muscle away from the cuticle. Bundles of intracellular microtubules originate at the junction with the epidermal cells, pass through the epidermal cells, and, at the cuticle, are cemented to fibers of chitin (tonofibrillae) that are formed extracellularly and embedded in the cuticle. Tonofibrillae are very resistant to the action of molting fluid, and allow muscles to remain attached to the exoskeleton after apolysis and secretion of new cuticle begins. In some insects, apolysis occurs hours and even days before ecdysis, and movement during the interval between apolysis and ecdysis may be critical to feeding and predator escape. The final factors that enter into the breaking of the tonofibrillae to allow the old cuticle to be ecdysed have not been elucidated.

牢固的连接，允许强烈的机械张力，防止肌肉从表皮上撕裂。胞内微管束起源于真皮细胞与肌肉连接处，通过真皮细胞，在胞外表皮层和几丁质纤维（张力原纤维）胶合，并包埋在表皮中。张力原纤维对蜕皮液的作用具有很强的抗性，并允许肌肉在皮层溶离后和新表皮层形成之前仍附着外骨骼。有些昆虫皮层溶离后的几小时，甚至一天才蜕皮皮层溶离和蜕皮间隔期内的运动对昆虫觅食和逃避天敌非常关键。进入裂解中的因子将张力原纤维分解，以允许肌肉与旧表皮脱离。

15.2 肌纤丝的结构和功能

15.2 The Structure and Function of Muscle Filament

15.2.1 Muscle proteins

The major proteins in muscle fibers are actin, tropomyosin, troponin (all part of the thin filaments), and myosin (the thick filaments). Shortening of a muscle occurs by the sliding of actin and myosin filaments over each other, pulling the Z bands closer together, and ultimately shortening the muscle toward the point of its fixed attachment, its origin. A single myosin molecule has the shape of a double-headed golf club (Fig. 15-6A). Myosin molecules are arranged with their long tails forming the core of the thick filament with heads projecting from the filament. The globular heads have calcium-dependent ATPase activity and contain the binding sites for attachment to actin during contraction. The myosin heads form the structures called cross-bridges.

The smallest unit of actin is G actin. G actin subunits are linked together by polypeptide bonds to form a long chain of filament actin (F actin) (Fig.15-6B、Fig.15-6C) and a thin filament consists of two chains of F actin twisted around each other in an α-helix. Each G actin subunit in the chain has an active site where a myosin

15.2.1 肌肉蛋白

肌纤维的主要蛋白质是肌动蛋白、肌球蛋白和肌钙蛋白（都是细肌丝的部分）。肌肉收缩通过肌动蛋白和肌球蛋白纤丝相互滑动而发生的，把 Z 盘拉得更近，最大限度地将肌肉向其附着点，即其起点收缩。单肌球蛋白分子形状像双头高尔夫球杆（图 15-6A）。肌球蛋白分子以其长尾形成粗肌丝核，头部从纤丝伸出。球状头部有钙依赖性 ATP 酶活性，包含收缩时附着于肌动蛋白的结合位点。肌球蛋白头部形成的结构称为横桥。

肌动蛋白的最小单位是肌动蛋白 G。通过多肽键连接在一起形成一条长长的肌动蛋白纤丝链（F- 肌动蛋白）（图 15-6B、图 15-6C）以及由两 F- 肌动蛋白链相互，用 α 螺旋缠绕形成的细肌丝。链上的每个肌动蛋白 G 亚基都有一个活跃的肌球蛋白头部附着位点。

head can attach. Another protein, tropomyosin, is associated with the thin filaments, and it consists of a long filamentous chain running along each of the two grooves created by the F actin helix. Each tropomyosin chain is composed of two α-helical units in a coiled-coil pattern, and the coiled helix. Tropomyosin filaments cover the active along the groove created by the F actin where a sites chain runs myosin head can attach. Troponin is a globular protein composed of three subunits: TnT (tropomyosin binding), TnC (calcium binding), and TnI (actin binding) (Fig.15-6C, Fig.15-6 D). It functions during nerve stimulation to the muscle by changing shape and pulling tropomysin away from file myosin binding sites. A troponin unit is associated with each G actin unit. All muscles contain thick and thin filaments. Thick filaments are composed of myosin. Thin filaments are a double helix of polymerized actin monomers.

另一个蛋白质，原肌球蛋白与细肌丝有关，沿 F 肌动蛋白螺旋产生的链组中的任一条延伸。每条原肌球蛋白链由线绕式两个 α 螺旋单元组成，两条槽相连，它由一条长长的纤丝化链，且以线绕式沿 F 肌动蛋白产生的槽分布。原肌球蛋白纤丝覆盖了肌球蛋白头部附着的活跃位点。肌钙蛋白是一种球状蛋白质，由三个亚基组成：TnT（原肌球蛋白结合）、TnC 钙结合和 TnI（肌动蛋白结合）（图 15-6C、图 15-6D）。在神经激发到肌肉过程中通过改变形状将原肌球蛋白从肌球蛋白结合位点推开。肌钙蛋白单体与肌动蛋白 G 单体相连。所有肌肉都包含粗肌丝和细肌丝。粗肌丝由肌球蛋白组成，而细肌丝是肌动蛋白单体聚合而成的双螺旋结构。

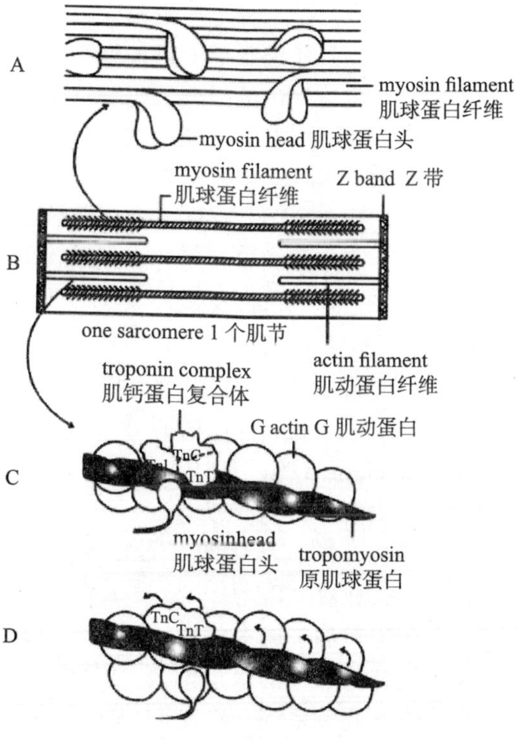

Fig.15-6 The arrangement of the major proteins involved with contraction in each sarcomere of a myofibril. A. A diagram of myosin molecules to show the double heads of each myosin filament; B. A sarcomere unit illustrating actin and myosin filaments;C. Helically entwined strands of F actin with a helical coil of tropomyosin lying in the groove of the F actin molecules. Troponin, consisting of three polypeptide units, is bonded through one polypeptide to a G actin molecule and by another subunit to tropomyosin, leaving one subunit free to bind to calcium ions; D. When the TnC part of troponin binds calcium ions, a change in shape of troponin occurs and tropomyosin is pulled away from the active site on G actin, allowing the myosin heads to bind to actin and initiate the power stroke of contraction (modified from Nation, 2022)

图 15-6 肌原纤维的每个肌节内涉及收缩的主要蛋白质排列。A. 肌球蛋白图解显示每条肌球蛋白纤丝双头部；B.1 个肌节单位阐明肌动蛋白和肌球蛋白纤丝；C.F- 肌动蛋白线绕式螺旋缠绕的双链，位于 F- 肌动蛋白分子的凹槽内。肌钙蛋白由三个多肽单位组成，通过一条多肽链绑定到 G 肌动蛋白，另一个亚基结合到原肌球蛋白，留下一游离亚基与 Ca^{2+} 结合；D. 肌钙蛋白的 TnC 部分与 Ca^{2+} 结合，肌钙蛋白形状发生了变化，原肌球蛋白被从 G- 肌动蛋白活跃位点拉开，并允许肌球蛋白头部结合到肌动蛋白，启动肌肉收缩的动力过程（仿 Nation, 2022）。

15.2.1.1 Myosin

Muscle myosin is a large, asymmetric, multidomain molecule composed of six polypeptides, two heavy chains of molecular mass 200-220 kDa and two pairs of light chains of molecular masses 17-23 kDa. The carboxy-terminal portion of the myosin heavy

15.2.1.1 肌球蛋白

肌肉肌球蛋白是一大型、非对称、多功能域分子，由 6 个多肽组成，两条重链分子质量为 200 ～ 220 kDa 和两对轻链分子质量为 17 ～ 23 kDa。肌球

chains associate to form an α-helical coiled-coil rod. The amino-terminal portion of each heavy chain associates with two different light chains to form an elongated globular domain; this structure contains the sites for hydrolysis of ATP and for binding actin. Myosin can be fragmented by limited proteolysis with papain, trypsin, and chymotrypsin into several structural and functional domains. The globular, actin binding head domain (Sl) is sufficient to cause sliding movement of actin filament (Winkelmann, et al., 1991). The lever arm consists of a small folded domain, dubbed the converter, a long α-helical peptide that binds the two light chains, an essential light chain, and a regulatory light chain.

蛋白重链碳末端部分联合形成线绕式α-螺旋杆。每条重链的氨基端与两条不同的轻链结合形成一个延伸的球状域，这个结构包括ATP的水解和结合位点。肌球蛋白可被木瓜蛋白酶、胰岛素和糜蛋白酶限制性水解成几个结构和功能域片段。球状肌动蛋白结合的头部区域（S1）足以引起细肌丝的滑动（Winkelmann等，1991）。杠杆臂由小折叠区域，被称为转换器的一条长α-双螺旋状肽链组成，肽链与两条轻链结合，一条必需轻链，一条调节轻链。

Myosin has three important characters: ① polymerize into thick filament; ② binding actin to form actomyosin with junction of crossbridge; myosin always regulated by phosphorylation of myosin light chain; ③ hydrolysis ATP that convert chemical energy into mechanical energy, and provide energy for molecules conformation at cross-bridge.

肌球蛋白具有3个重要性质：①聚合成粗肌丝；②与肌动蛋白结合，形成以横桥连接的肌动球蛋白，肌动球蛋白常通过肌球蛋白轻链的磷酸化进行调节；③水解ATP，将化学能转换成机械能，为横桥处分子变构提供能量。

Myosin is a motor protein that is essential for a variety of processes ranging from intracellular transport to muscle contraction. Folding and assembly of myosin relies on a specific chaperone, UNC-45. The role of UNC-45 as a myosin-specific chaperone, promoting the folding of the myosin ATPase domain as well as coordinating the assembly of thick filaments during muscle development. The chaperone can form a linear protein chain, which constitutes a myosin assembly line licensing HSP70 and HSP90 to act in a defined periodicity on myosin heads protruding from the myofilaments (Hellerschmied, et al., 2019).

肌球蛋白是一种马达蛋白质，对从细胞内运输到肌肉收缩的各种过程都至关重要。肌球蛋白的折叠和组装依赖于特定的分子伴侣UNC-45。UNC-45作为肌球蛋白特异性分子伴侣的作用主要有，促进肌球蛋白ATP酶结构域的折叠，并协调肌肉发育过程中粗肌丝的组装。分子伴侣可以形成一条线性蛋白质链，该蛋白质链构成一条肌球蛋白装配线，允许HSP70和HSP90以规定的周期作用于从肌丝伸出的肌球蛋白头部（Hellerschmied等，2019）。

Insect actomyosin's activation by and hydrolysis of ATP, that its activation requires both Mg^{2+} and Ca^{2+}, and the effect of ADP on ATPase activity or muscle fiber tension. Insect actomyosin is both thin and thick filament dual regulated. An unusual aspect of the thin filament regulatory system in asynchronous flight muscle is that both the typical Ca^{2+}-activated and a stretch-activated troponin. In addition to activation *via* direct Ca^{2+} binding, insect actomyosin also has a third regulatory pathway involving myosin light chain phosphorylation. A calmodulin-based myosin light chain phosphorylation system regulates Locusta oviduct contraction (Hooper et al., 2009).

昆虫肌动球蛋白通过ATP和ATP水解活化，即其活化同时需要Mg^{2+}和Ca^{2+}，ADP对ATP酶的活性或肌纤维张力产生影响。昆虫的肌动球蛋白受到细肌丝和粗肌丝的双重调节。异步飞行肌细肌丝调节系统一个不同寻常的两个方面是典型的Ca^{2+}活化和张力活化肌钙蛋白。除了通过直接Ca^{2+}结合活化，昆虫肌动球蛋白还有涉及肌球蛋白轻链磷酸化的第三条调节途径。基于钙调蛋白的肌球蛋白轻链磷酸化系统调节蝗虫产卵器的收缩（Hooper等，2009）。

15.2.1.2 Actin

Insect flight muscle thin filaments are very similar to vertebrate thin filaments, with the double helix repeating once every 35-40nm in 13-15 monomers. In this muscle the actin repeat distance is clearly about 38 nm. To be main composition of thin filament, actin exist two forms: globular protein monomer with molecular weight about 42 kDa, which is called G-actin. Filament actin or F-actin shows bundle protein in shape which is polymerized polypeptide that present strands of bead. The different actins are functionally nonequivalent. *Drosophila* indirect flight muscle actin requires posttranslational modification for normal polymerization. In Lepidoptera (*B. mori*) three actin genes have been identified, one cytoplasmic and two muscle, one of which is expressed only in adult muscle and the other in both larval and adult muscles.

Thin filament is a double strand of α-helix composed of actin filament. The same end polar of actin monomer polymerized with head to tail style in double strand helix. Actin has at least three functions as the follow.

① The primary role of actin is to transmit force generated by myosin, but evidence that, upon activation, actin undergoes a major structural change in its own right is accumulating (Iwamoto, 2009).

② The binding sites for myosin located on the surface of globular monomer of actin that can form actomyosin cross-bridge with myosin molecular head.

③ Actin filament may improve ATPase activity of myosin head. Once the catalytic domain has attached, the head can sequentially release hydrolysis products Pi and ADP. Because release is relatively slow, although myosin itself hydrolysis is not slow. While actin binding with ADP and Pi, it increases releasing rate. When muscle relaxed, actin bind ATP again which results in detaching of actin and actomyosin, and the second contraction reaction may go on.

15.2.1.3 Titin

Titin, a protein of molecular mass of more than 30,000 kDa, is the third most abundant protein in muscle after actin and myosin and is responsible for much of the elasticity of myofibrils in various

15.2.1.2 肌动蛋白

昆虫飞行肌的细肌丝与脊椎动物细肌丝非常相似，有双股螺旋，每 35 ～ 40 nm 重复 13 ～ 15 个单体。这种肌肉的重复距离很清楚，约 38 nm。作为细肌丝主要成分的肌动蛋白存在两种形式：分子质量约 42 kDa 的球形蛋白单体，称为 G- 肌动蛋白。纤维状肌动蛋白或 F- 肌动蛋白表现为捆绑形蛋白，其聚合多肽呈串珠状。不同的肌动蛋白功能上不等效。果蝇间接飞行肌肌动蛋白正常的聚合需要翻译后加工修饰。鳞翅目（家蚕）的三个肌动蛋白基因被鉴定出一个在细胞质，两个在肌肉，其中之一仅在成虫肌肉中表达，另一个在幼虫和成虫期都有表达。

细肌丝是由两条肌动蛋白丝形成的双股 α- 螺旋。在双股螺旋中，每个肌动蛋白单体的极性相同，都以头尾相接的方式进行聚合。肌动蛋白至少有三种功能。

①肌动蛋白的基本功能是传递由肌球蛋白产生的动能，但越来越多的证据表明，在激活后，肌动蛋白自身会发生重大的结构变化 (Iwamoto, 2009)。

②在球形单体的表面有肌球蛋白的结合位点，可与肌球蛋白的头部结合形成肌动球蛋白横桥。

③肌动蛋白丝可以提高肌球蛋白头部的 ATP 酶的活性。一旦催化部位附着，头部依次释放出水解产物 Pi 和 ADP。因为肌球蛋白虽然自身水解 ATP 速率不慢，但释放相当慢，而肌动蛋白与 ADP 及 Pi 结合增加了它们的释放速率。当肌肉放松后，肌动蛋白再与 ATP 结合，导致肌动蛋白和肌动球蛋白脱离，可持续第二次收缩反应。

15.2.1.3 肌联蛋白

肌联蛋白是一个分子质量大于 30 000 kDa 的蛋白质，为肌肉中仅次于肌动蛋白和肌球蛋白的第三丰度蛋白

organisms. Titin molecules are associated with each thick filament in skeletal muscle, where they interact with myosin and with a variety of sarcomere proteins. Each titin molecule extends from the M-band to the Z-line (Nave, et al., 1990). As the sarcomere increases in length, titin molecules stretch. Titin is responsible for most of the elastic force that returns stretched muscle to its resting length. When the muscle is stretched past overlap, titin is responsible for the thin filaments returning to interdigitate with thick filaments. Further, titin may be function physiologically, together with Skeletor, Megator (Qi, et al., 2004) and Chromator, as part of a spindle matrix (Fabian, et al., 2007) which is thought to be involved in chromosomal elasticity.

质，在各种生物体内是肌原纤维弹性的起因。肌联蛋白分子在骨骼肌中与各粗肌丝相连，在此处它们和肌球蛋白及各种肌节蛋白相互作用。每个肌联蛋白分子从M带延伸到Z线（Nave等，1990）。随着肌节长度增加，肌联蛋白分子伸展。肌联蛋白是多数伸展肌借弹力回到静息长度的原因。当肌肉交替伸展时，肌联蛋白负责细肌丝恢复到与粗肌丝的相间交叉。此外，肌联蛋白在生理功能上还和Skeletor、Megator（Qi等，2004）和Chromator一起成为纺锤体基质（Fabian等，2007）的一部分，被认为涉及染色体伸缩性。

15.2.1.4 Tropomyosin

Tropomyosin is an important component of immune and allergic reactions. *Drosophila* has two muscle tropomyosin genes. Each produces four or five tissue or developmentally specific isoforms (including nonmusical forms) by a combination of alternative splicing and multiple promoters. Beetles and centipedes have three muscle tropomyosins, which show some muscle specificity. Locust has multiple isoforms with muscle specific expression. Tropomyosin is a filamentous protein which inlayed in the grooves of the twister actin with double strands α-helix. Every seven actin molecules combine with a tropomyosin. The molecular weight of tropomyosin in *Phormia* adult is 65.6 kDa, but that in larvae is 84.4 kDa. The function of tropomyosin is that it can ceases the formation of crossbridge of actin when it covers the active site of myosin head.

15.2.1.4 原肌球蛋白

原肌球蛋白是免疫和过敏反应的重要组分。果蝇有两个肌肉原肌球蛋白基因。每个基因通过可变的剪切组合和多个启动子产生4个或5个组织特异性或发育特异性异构体（包括非肌肉形式）。蜜蜂和蜈蚣有三个原肌球蛋白基因，表现出一些肌肉特异性。蝗虫有多个肌肉特异表达的异构体。原肌球蛋白是一种纤丝状蛋白质，以双股*α*-螺旋镶嵌在肌动蛋白缠绕的凹槽内，每7个分子的肌动蛋白球结合一个分子的原肌球蛋白。伏蝇成虫的原肌球蛋白分子质量为65.6 kDa，但幼虫的分子质量84.4 kDa。其功能是当它覆盖肌球蛋白头部的活性部位时，能阻止肌动球蛋白横桥的形成。

15.2.1.5 Troponin

The regulation of the sliding of thin over thick filaments during muscle contraction depends to a great extent on the proper activity of the troponin complex. This is a tetrameric protein ensemble composed of three troponins: I, C and T, and tropomyosin. In thin filament regulated muscle contraction, troponin I (TnI) inhibits the interaction between actin and the myosin head at rest. Troponin C (TnC) is the calcium binding component when this ion is released from the storage compartments at the muscle activation point.

15.2.1.5 肌钙蛋白

在肌肉收缩时，细肌丝在粗肌丝上滑动的调节作用很大程度上依赖于合适的肌钙蛋白复合体的特性。这是一种四聚体蛋白质组合，由3个肌钙蛋白I、肌钙蛋白C、肌钙蛋白T和原肌球蛋白组成。在细肌丝调节的肌肉收缩中，肌钙蛋白I（TnI）静息状态下抑制肌动蛋白和肌球蛋白间的相互作用。当Ca^{2+}从肌肉活化点的贮存区室被释放出来

Troponin T (TnT) anchors the complex to tropomyosin providing a regular spacing pattern along the actin based thin filament. TnT in insects contains a highly diverged structure consisting of a long glutamic acid–rich C-terminal extension of ~70 residues, which functions as a myofilament Ca^{2+} buffer/reservoir and is potentially critical to the high-frequency asynchronous contraction of flight muscles (Cao et al., 2020). In addition, tropomyosin along with tropomodulin is known to play a role in thin filament capping by preventing actin monomer depolymerization at the pointed ends. A fourth troponin, troponin H (tropomyosin isoform), which is a fusion product of tropomyosin and a hydrophobic proline-rich sequence, is alternatively spliced from one of the tropomyosin genes, at least partly replaces troponin I, and exists in all indirect (asynchronous) and some direct (synchronous) flight muscles.

The most characteristic role of TnI is its inhibitory effect of muscle contraction by preventing the actin-myosin head interaction, that is, the actomyosin Mg^{2+}-ATPase enzymatic activity. However, the binding of TnI to actin causes a conformational change in the latter that result in allosteric effects at additional sites of the actin monomer.

TnC is a Ca^{2+}-binding protein about 155 amino acids long that represents a specialized form of calmodulin, to which it is evolutionary related. Insect TnC isoforms differ in the α-helix separating the two globular domains. Single amino acid insertions in this helix are frequent in certain isoforms of *Drosophila, Anopheles*, *Apis* and *Lethoceru*. These structural peculiarities are likely to sustain specialized functional properties, such as stretch activation.

Asynchronous flight muscle related thick filament protein, projectin, have kinase activities which can phosphorylate troponins that increases Ca^{2+} sensitivity. Projectin is an 800-1,000 kDa protein. In the indirect flight muscle (IFM), projectin molecules are anchored at the Z-disk, extend over the I-band region, and associate with myosin at the A-band. Mutant forms of projectin have been shown to alter insect flight dynamics

后，肌钙蛋白 C（TnC）是其结合组分。肌钙蛋白 T（TnT）将复合体锚定在原肌球蛋白上，提供沿肌动蛋白组成的细肌丝一个固定的空间形式。昆虫体内的 TnT 含有高度多样化结构，由约 70 个残基的富含谷氨酸的长 C- 端延伸组成，其功能是作为肌丝 Ca^{2+} 缓冲 / 储库发挥作用，对飞行肌肉的高频异步收缩可能至关重要（Cao 等，2020）。此外，已知原肌球蛋白与原肌球蛋白调节蛋白一起在细肌丝帽化中发挥作用，防止肌动蛋白单体在突出端解聚。第四个肌钙蛋白是肌钙蛋白 H（原肌球蛋白异构体），是原肌球蛋白和疏水性富脯氨酸序列的融合产物，是从原肌球蛋白的一个基因交替剪切的，至少部分地替代 TnI，在所有间接飞行肌（异步肌）和有些直接飞行肌（同步肌）中存在。

TnI 最具特色的作用是其通过阻止肌动蛋白 - 肌球蛋白头部的相互作用对肌肉收缩产生抑制作用，即肌动球蛋白 Mg^{2+}- ATP 酶的酶活性。但是 TnI 结合到肌动蛋白后阻止了构型的变化，导致肌动蛋白单体在其他位点的变构效应。

TnC 是一种 Ca^{2+} 结合蛋白，长约 155 个氨基酸，代表了钙调蛋白的特殊形式，并与进化相关。昆虫的 TnC 异构体在 α- 螺旋拆分两个球状域上有差异。果蝇、疟蚊、蜜蜂和负子蝽的某些异构体常常在这个螺旋上有单个氨基酸的插入。这些结构特性很可能用于保持特异功能属性，如拉伸激活。

异步飞行肌的粗肌丝关联蛋白、凸出蛋白拥有激酶活性，可将肌钙蛋白磷酸化增加 Ca^{2+} 敏感性。凸出蛋白是分子质量为 800 ～ 1 000 kDa 的蛋白质。在间接飞行肌（IFM）中，凸出蛋白分子锚定在 Z 盘，延伸覆盖 I 带区域，在 A 带与肌球蛋白相连。凸出蛋白突变体已被证实能改变昆虫的飞行动力学。

15.2.1.6 Paramyosin

Sequence analysis of paramyosin isoforms isolated from different species reveal a rod-like molecule with a central α-helical region and two nonhelical terminal domains that can dimerize into a coiled-coil structure. Like the myosin tail, amino acids of the paramyosin helical region have a periodic 28-residue positive and negative charge repeat distribution.

Paramyosin is a major structural component of thick filaments in invertebrate muscle. The paramyosin to myosin ratio, which positively correlates with thick filament length and maximum active tension development, varies widely between different invertebrate phyla and even between different muscles within the same animal. For instance, this ratio is 1 : 34 in the fibrillar flight muscle of *D. melanogaster* but is 1 : 6 in larval muscles. This charge arrangement on the surfaces of paramyosin and the myosin tail may regulate the packing of these two proteins in the thick filament. Paramyosin is important for thick filament formation and for filament assembly into myofibrils. *Drosophila* paramyosin can be phosphorylated in its N-terminal nonhelical domain and that this post-translational modification is not critical for normal indirect flight muscle (IFM) assembly but is important for flight muscle power output and normal flight ability. Paramyosin also may reinforce sarcomere stiffness by stabilizing the connection between thick filaments and the M-line (Liu et al.,2005).

Miniparamyosin contains the same C-terminal region as paramyosin, permitting the coiled-coil formation of heterodimers. Thus, it is possible that the phosphorylated N-terminal of paramyosin initiate or stabilize an interaction between paramyosin and miniparamyosin. Because paramyosin is differentially phosphorylated, phosphor-variants may be differentially localized along the length of the thick filament. Phosphovariant-specific localization directed at the M-line may be one mechanism by which an interaction of paramyosin with miniparamyosin is enhanced.

In addition, paramyosin may stabilize thick filament interactions with proteins such as projectin or kettin. Kettin is a high molecular mass protein of insect muscle that in the sarcomeres binds to actin and actinin which is attached not only to actin but also to

15.2.1.6 副肌球蛋白

从不同物种纯化到的副肌球蛋白异构体序列分析显示其为一个棒状分子，有一个中央 α- 螺旋区域和两个能二聚成线绕式结构的非螺旋终端区域。像肌球蛋白尾一样，副肌球蛋白螺旋区的氨基酸有周期性的 28 个带正电和负电残基的重复分布。

副肌球蛋白是无脊椎动物粗肌丝的主要结构成分。副肌球蛋白对肌球蛋白的比率与粗肌丝长度和发生的最大活性张力正相关，不同的无脊椎动物种群间变异很大，甚至同种动物不同的肌肉间也有很大差异。例如，这个比例在果蝇飞行肌纤维中是 1 : 34，但在幼虫肌肉中是 1 : 6。副肌球蛋白表面和肌球蛋白尾上的这种电荷排列可能调控粗肌丝的这两个蛋白质的包装。副肌球蛋白对粗肌丝的形成和将纤维装配成肌原纤维很重要。果蝇副肌球蛋白在其 N- 端非螺旋区被磷酸化，这种翻译后加工修饰对正常的间接飞行肌（IFM）并非关键性的，但对飞行肌动能输出和正常的飞行能力很重要。副肌球蛋白通过稳定粗肌丝和 M- 线之间的联系也可加强肌节的刚性（Liu 等，2005）。

小副肌球蛋白含有与副肌球蛋白相同的 C- 端区域，允许线绕式异源多聚体的形成。因此，副肌球蛋白 N- 端的磷酸化可能启动或稳定副肌球蛋白和小副肌球蛋白间的相互作用。因为副肌球蛋白是分别磷酸化的，磷酸化突变体也就可能沿粗肌丝长度不同地定位。磷酸化突变体特异性在 M- 线上的引导性定位可能是副肌球蛋白和小副肌球蛋白相互作用增强的机制之一。

此外，副肌球蛋白可稳定粗肌丝和蛋白质，如凸出蛋白或 Kettin 的相互作用。Kettin 是昆虫肌肉中的高分子质量蛋白质，在肌节中与肌动蛋白和辅

the end of the thick filament. In *B.mori*, a kettin homologue on Z chromosome to that of *Drosophila* with molecular weight of 380 kDa. Southern blot analysis revealed that *Bm Kettin* mRNA was 9.1kb in length, and that the level of mRNA in males was two times greater than that of females. Kettin along with projectin, are members of titin proteins superfamily, mainly responsible for the high passive stiffness of insect indirect flight muscles, which is needed to generate oscillatory work during flight. Possibly, the two proteins modulate myofibrillar stiffness by expressing different size isoforms (Lakey, et al., 1993). Kettin is in the I-band, and projectin is in the A-band, except in IFM, in which a large part of the molecule is in the short I-band. Kettin is an alternatively spliced product of the *Drosophila sls* gene. In contrast to projectin, kettin is a 540 kDa protein. In the IFM, the molecule is anchored to the Z-disk, extends over the I-band running along the actin filament, and then attaches to the thick filaments. The distribution of varies proteins are indicated in Fig.15-7(Appendix color graphic Fig.15-7).

肌动蛋白结合，它不仅锚定在肌动蛋白上，也锚定在粗肌丝尾端。家蚕 Z 染色体上的 Kettin 与果蝇同源，分子质量为 380 kDa。Southern 印迹分析显示，家蚕 *Kettin* 的 mRNA 长 9.1 kb，雄蛾体内 mRNA 水平是雌蛾的两倍。Kettin 和凸出蛋白都是粗肌丝肌联蛋白超家族的成员，主要负责昆虫 IFM 的高被动刚性，这是飞行中产生振荡作用所必需的。两种蛋白质可能通过表达不同大小异构体来调节肌纤维的刚性（Lakey 等，1993）。Kettin 在 I 带，凸出蛋白在 A 带，除了 IFM，其分子的大部分在短 I 带内。Kettin 是果蝇 *sls* 基因交替剪切的产物。与凸出蛋白对比，其分子质量为 540 kDa。在 IFM 内，分子锚定在 Z 盘，延伸覆盖 I 带，沿肌动蛋白纤丝游走，之后锚定在粗肌丝。各种蛋白质的分布如图 15-7 所示（附彩图 15-7）。

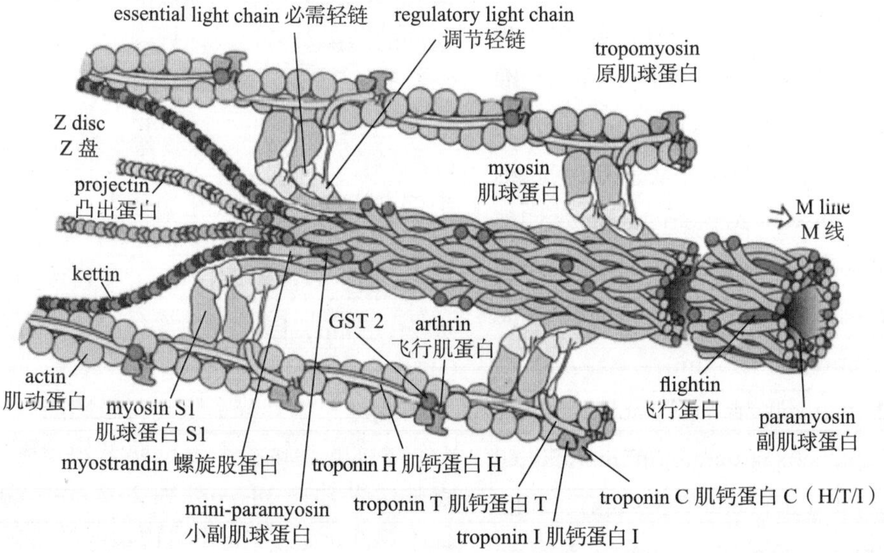

Fig.15-7 The position of varies proteins on muscle thin and thick filament in adult indirect flight muscle (modified from Vigreaux, 2005).

图 15-7 成虫间接飞行肌各种蛋白质在细肌丝和粗肌纤丝上的位置（仿 Vigreaux, 2005）。

15.2.2 Thin filament (actin filament)

Although there was early controversy over repeat distances and numbers of actin monomers per repeat, biochemical, X-ray, and electron microscopy work indicates that insect flight muscle thin filaments are very similar to vertebrate thin filaments, with the double helix repeating once every 35-40 nm in 13-15 monomers.

15.2.2 细肌丝（肌动蛋白丝）

尽管早期肌节内肌动蛋白单体的重复距离和数量还存在争议，但生物化学、X 射线和电镜学研究表明，昆虫飞行肌细肌丝和脊椎动物细肌丝非常相似，其双螺旋在 13 ～ 15 个单体中每 35 ～ 40 nm 重复一次。昆虫飞行肌已

Insect flight muscle has been extensively studied, and in this muscle the actin repeat distance is clearly about 38 nm with the small differences in reported values being due to muscle state variations, i.e., Ca^{2+} concentration. Insect thin filaments contain a ubiquinated actin monomer, Arthrin, every seventh subunit; not all six *Drosophila* actins substitute for each other equally well.

被广泛研究过，其肌动蛋白的重复距离很明确是 38 nm，报道的数据因肌肉的状态，即 Ca^{2+} 浓度而有微小差异。昆虫肌动蛋白每 7 个亚基包含一个泛素化肌动蛋白单体 Arthrin；果蝇的 6 个肌动蛋白并不能相互替代。

15.2.3 Thick filament (myosin filament)

Muscle thick filaments are composed of myosin. All known invertebrate muscle is end polarization, in which the myosin molecules are oriented in opposite directions at each end of the thick filament. This orientation results in a cylindrical filament (diameter 14-16 nm) with a central region without heads and two peripheral regions out of which heads protrude. Some thick filaments have solid cores, and others, particularly in insect flight muscle being hollow tubes (Craig, et al., 2001).

The two heads arising from one myosin molecule actually interact with each other, with one head being free and the other bound to the free head's motor domain and essential light chain. The continuous strands result from the motor domain of the free head of one crown interacting with the essential light chain of the bound head on the crown above. These filaments are 20 - 35 nm in diameter. Every 14.5 nm four evenly spaced pairs of myosin heads emerge from the thick filament. The crowns rotate right-handed 30° every 14.5 nm, axial repeat distance 43.5 nm for the right handed helix and 21.75 nm for the left handed, and thus the heads come back into angular register every 3 crowns (43.5 nm).

Lethocerus flight muscle thick filaments are now known to have a four folds rotational symmetry and helix structure (off meriodonal reflections 38.5 nm and 23.3 nm) and diameter (23.5 nm) similar to those of chelicerate thick filaments. The altered crown rotation results in this filament's heads coming back into register every 8 crowns (116 nm) instead of every 3 (43.5 nm). Instead of strong strands like the other filaments, this thick filament thus instead has rings around it.

15.2.3 粗肌丝（肌球蛋白丝）

肌肉粗肌丝由肌球蛋白组成。所有已知的无脊椎动物肌肉都是末端极化的，其中肌球蛋白分子在粗肌丝每个末端的取向是相对的。这种取向产生圆柱形纤丝（直径为 14 ～ 16 nm），有个中心区域而没有头部，两个外周区域可使头部突出。有些粗肌丝有一个实心的核，但其他的，特别是昆虫的飞行肌变成了中空管（Craig 等，2001 年）。

产生于同一肌球蛋白分子的两个头部确实是相互作用的，一个头部处于游离状态，另一个则与游离头部的运动区域和必需轻链结合。连续的链即是一个冠部的游离头部运动区域与上一个冠部结合态头部必需轻链相互作用的结果。这些肌丝直径 20 ～ 35 nm。每隔 14.5 nm 就有 4 个空间均等的成对肌球蛋白头部有规律地出现在粗肌丝上。冠部每隔 14.5 nm 就右手转动 30°，轴向重复距离右手转动为 43.5 nm，左手转动为 21.75 nm，因此，头部每隔 3 个冠部（43.5 nm）回到原来的角度。

现在已知，负子蝽飞行肌粗肌丝有 4 个对称的旋转和螺旋折叠结构（离起点映像纵轴长 38.5 nm 和 23.3 nm），直径（23.5 nm）与螯肢动物粗肌丝类似。可变的冠部转动导致这种粗肌丝每 8 个冠部（116 nm）才回到原来的角度，而不是之前的 3 个冠部（43.5 nm）。与其他粗肌丝的强劲链不同，这种粗肌丝有环围绕。

15.3 肌肉收缩

15.3 Muscle Contraction

15.3.1 A sliding filament model for muscle contraction

The widely accepted theory about insect muscle contraction mechanism is that proposed by Hanson and Huxley. According to the sliding filament theory when calcium ion concentration rises, binding sites on actin filaments open, and cross-bridges attach. Upon binding to actin, cross-bridges spring from the cocked position and pull on actin filaments. ATP binds to the cross-bridge, causing myosin head to release from the actin filament. ATP breakdown provides energy to cock the unattached myosin cross-bridge. As long as ATP and calcium ions are present, the cycle continues. When calcium ion concentration is low, the muscle remains relaxed.

In most insects each contraction is triggered not by a motor neuron spike but by mechanical stretch imposed by antagonistic muscles. Whereas 'stretch activation' and its reciprocal phenomenon 'shortening deactivation' are observed to varying extents in all striated muscles, both are particularly prominent in the indirect flight muscles of insects (Josephson et al., 2000).

15.3.1 肌纤丝滑动学说

昆虫肌肉的收缩机制被广泛接受的是 Hanson 和 Huxley 提出的肌纤丝滑动学说。根据纤丝滑动学说，当 Ca^{2+} 浓度上升，肌动蛋白纤丝结合位点被打开，横桥附着。一旦结合到肌动蛋白，横桥就从压簧杆位置松开并拉动肌动蛋白纤丝。ATP 结合到横桥，使肌球蛋白头部从肌动蛋白纤丝释放。ATP 降解提供能量使未结合的肌球蛋白横桥竖起。只要有 ATP 和 Ca^{2+} 存在，周期就继续。Ca^{2+} 浓度低，肌肉则保持松弛。

在大多数昆虫中，每一次收缩都不是由运动神经元峰电位触发的，而是由拮抗肌肉施加的机械拉伸触发的。尽管在所有横纹肌中都不同程度地观察到"拉伸激活"及其相互作用的"缩短失活"现象，但在昆虫的间接飞行肌中都特别突出（Josephson 等 , 2000）。

15.3.2 Contraction of myofibril and driving force

15.3.2.1 Binding of myosin heads to actin and the sliding of filaments

The signal for initiating a contraction is the arrival of a nerve impulse at the junctional ending on the muscle surface, resulting in the release of neurotransmitter at the proneuronal neuromuscular junctions. The nerve impulse sets in motion a cascade of reactions mediated by G proteins and results in the release of Ca^{2+} from binding sites in the SR at the triad (synchronous muscle) or dyad (asynchronous muscle) junctions in the muscle fiber. The threshold of free Ca^{2+} ions for activation of the contraction process is about

15.3.2 肌纤丝的收缩及收缩动力

15.3.2.1 肌球蛋白头部与肌动蛋白的结合及肌丝的滑动

神经脉冲到达其与肌肉表面连接末端处是肌肉收缩的起始信号，导致神经肌肉连接处多神经元的递质释放。神经脉冲决定 G 蛋白介导的级联反应处于兴奋状态，并导致肌纤丝内肌质网上的三联体（出现在同步肌）和二联体（出现在异步肌）连接处结合位点的 Ca^{2+} 释放。激活肌肉收缩过程的游离 Ca^{2+}

10^{-7} mol/L, with maximum contraction occurring at about 10^{-5} mol/L. The free Ca^{2+} ions in the sarcoplasm and availability of ATP create an "active state" in each myofibril. Once initiated, the active state can persist as long as free Ca^{2+} ions and ATP are available. When the concentration of free Ca^{2+} ions fall below about 10^{-7} mol/L in the sarcoplasm because of the sequestering of Ca^{2+} by the sarcoplasmic reticulum, the active state is abolished, and a further contraction is not possible until a new volley of nerve impulses releases more Ca^{2+}. The active state persists much longer in asynchronous muscle than in synchronous muscle because the poorly developed SR in asynchronous muscle sequesters Ca^{2+} ions only slowly.

During the active state, free Ca^{2+} ions bind to the TnC subunit of troponin. This promotes a change in shape of the troponin complex, causing the TnT subunit to pull tropomyosin away from the active site where a myosin head can attach to actin. Attachment sites covered by troponin occur on each of the globular or G actin subunits, spaced about 38.5 nm apart. This allows myosin heads to bind to actin at many sites during the active state. When myosin binds to actin, a conformational change occurs in the myosin molecule in a "hinge-like" region of the myosin head and at another hinge site in the arm to which the head is attached, causing the arm to bend similar to the bending of a person's arm. These shape changes create the power stroke that causes sliding of actin over myosin.

15.3.2.2 Release of myosin heads from actin

The myosin head must release from actin in order to return to its original state and prepare for another power stroke. The conformational change in the myosin molecule associated with the power stroke exposes a site on the head where ATP can bind. ATP binds to this head site, and myosin, which is an ATPase as well as a structural protein, splits the bound ATP into ADP and PO_4. The energy liberated enables the myosin head to release from actin and to return to its original shape, ending the power stroke. ADP and inorganic phosphate remain attached to the head until a new attachment to actin occurs, at which time they are also released from the head and a new power stoke begins. The structure of inhibited myosin heads folded back on the thick filament in relaxed *Lethocerus* fibers has not been seen in

临界值大约为 10^{-7} mol/L，其收缩最大浓度为 10^{-5} mol/L。肌质内的游离钙和可获得的 ATP 在每条肌纤维内产生一个“活动状态”。一旦启动，只要有游离钙和 ATP 就能维持活跃状态。由于肌质网的螯合，当肌质上的游离 Ca^{2+} 浓度低于 10^{-7} mol/L 时，活动状态即废除。在新的神经脉冲释放更多的 Ca^{2+} 之前，不可能进一步收缩。因为异步肌的肌质网不发达，螯合 Ca^{2+} 只能缓慢进行，其活动状态的维持时间比同步肌更长。

在活动状态期，游离 Ca^{2+} 与肌钙蛋白 TnC 亚基结合。这使得肌钙蛋白复合体形状发生变化，导致 TnT 亚基把原肌球蛋白推离活化位点，此位点即是肌球蛋白头部和肌动蛋白的结合位点。肌钙蛋白覆盖的连接点出现在每一个球状亚基或 G- 肌动蛋白上，空间距离约 38.5 nm。这就允许肌球蛋白的头部在活动状态时能与肌动蛋白的很多位点结合。当肌球蛋白与肌动蛋白结合时，肌球蛋白头部“铰链样”区域的分子构象发生变化，臂上另一个铰链位点与头部连接引起臂的弯曲。这种形状的变化产生了动力冲程，导致肌动蛋白在肌球蛋白上滑动。

15.3.2.2 肌球蛋白头部从肌动蛋白的释放

肌球蛋白的头部必须从肌动蛋白中释放出来，回到其原来的状态，并准备另一个动力冲程。与动力冲程相关的肌球蛋白分子构象变化使头部 ATP 结合位点暴露出来。ATP 结合到这个头部，肌球蛋白除了是一个结构蛋白外，同时也是一个 ATP 酶。ATP 和头部位点结合后，把结合态 ATP 分解成 ADP 和 PO_4。产生的能量使肌球蛋白头部也从肌动蛋白释放出来，恢复到其原始状态，动力冲程终止。在与新的肌动蛋白结合之前，ADP 和无机磷保持附着状态；结合时，它们也要从头部释放出来，从而开始另一个新的动力冲程。大负子蝽被抑制的肌球蛋白头部折叠在松

other species and may be an adaptation to the rapid contractions of flight muscle (Bullard et al., 2019).

The binding of ATP, detachment of the myosin head from actin, and reattachment occur extremely rapidly, and are estimated to require only about 0.1ms. If calcium ions are available to combine with TnC and pull tropomyosin away from the active sites, myosin call bind, accomplish its power stroke, detach, and bind again many times per second. The movement caused by a power stroke has been estimated to be 5 to 10 nm, and thousands of such movements along the length of the muscle fiber create total fiber shortening. After death of an insect, the myosin heads bind to actin and contraction occurs, but the lack of ATP prevents them from detaching from actin, resulting in the condition of rigor mortis, as in other animals. ATP must be available at all times for normal muscle functions because it is necessary for the SR to actively sequester calcium, and for binding to the myosin heads so that the heads detach from actin.

弛的肌纤维的粗肌丝上，这种结构在其他物种中未见，可能是对飞行肌肉快速收缩的适应（Bullard 等，2019）。

ATP 的结合、肌球蛋白头部从肌动蛋白上分开和再附着过程发生极快，据估算仅需要 0.1 ms。如果有 Ca^{2+} 结合到 TnC，拉动原肌球蛋白离开活化点，肌球蛋白寻求结合、完成动力冲程、解离并再结合，每秒钟发生很多次。每个动力冲程移动距离估计是 5 ～ 10 nm，肌纤维上可发生上千次这样的移动，最终产生肌纤丝的收缩。昆虫死亡后，肌球蛋白头部与肌动蛋白结合并产生收缩，但缺少 ATP，使之不能从肌动蛋白解离出来。像其他动物一样，这导致了尸体的僵直。正常肌肉的运动必须随时有 ATP 的供应，因为它是肌质网活跃地螯合钙，和肌球蛋白头部的结合以及从肌动蛋白分开都是必需的。

15.3.2.3 Cross-bridge and cycle regulation

Cross-bridges are the heads of myosin molecules, reaching from thick toward thin myofilaments, and known to be the molecular motors that produce force and shortening in muscle by ATP driven cycles of mechanochemical partnership with the actin of thin filaments (Huxley, 1969; Lymn et al., 1971). X-ray modeling of ATP relaxed *Lethocerus* IFM shows that a myosin head conformation similar to "prestroke" crystal structures, is arrayed in the 14.5 nm periodic "shelves" along thick filaments such that only one head of each molecule is well-positioned, as if poised to bind to actin upon activation, while the other head curves around the thick filament shaft (Tregeal et al., 2004).

Initial binding of the myosin head to the thin filament is weak with the head having a 45° angle relative to the thin filament long axis. Myosin head binding throughout the 38.7 nm repeat is indicative of a myosin head relay mechanism that occurs over small changes in sarcomere length (Wu et al., 2012). As binding proceeds, the portion of the myosin heavy chain engaged with the actin retains its position and shape, but the region closer to the thick filament rotates toward the Z-line, which produces an M-line directed force on the thin filament. Force is thus not generated

15.3.2.3 横桥及其周期调节

横桥是肌球蛋白分子的头部，从粗肌丝到达细肌丝，已知是通过 ATP 驱动细肌丝肌动蛋白的机械化学组合周期性地产生力量并收缩肌肉的分子马达（Huxley，1969；Lymn 等，1971）。负子蝽松驰的间接飞行肌 ATP X- 射线模型表明，肌球蛋白构象与“预冲前”晶状结构相似，沿粗肌丝排成 14.5 nm 的周期性“架子”，这样，每个分子只有一个头部有很好的位点，一旦激活，就要结合到肌动蛋白，而其他头部围绕粗肌丝轴（Tregeal 等，2004）。

在起始状态下，肌球蛋白头部与细肌丝的结合较弱，相对于细肌丝长轴，头部的角度为 45° 。肌球蛋白头部的结合始终以 38.7 nm 重复，是发生在肌节长度内小变化的肌球蛋白头部延缓机制的象征（Wu 等，2012）。随着进一步的结合，肌球蛋白的重链使肌动蛋白保持它的位置和形状，但与粗肌丝较接近的部分向 Z- 线转动。这在细肌丝上产生了一个 M- 线定向的力量。因此，动力不是由

by rotation of the entire myosin head, but instead in a lever-like manner in which rotation of a more distant portion of the head uses the actin-binding portion to transfer force to the thin filament. At the end of the stroke the lever arm has a ~135° angle relative to the thin filament long axis and points to the Z-line. If this were the end of the process, the muscle would not contract further, and, furthermore, would become rigid, since the tight binding of the myosin head to the actin, and the inability of the myosin head to rotate back to its original angle, would lock the thin and thick filaments into an unyielding conformation (this is the basis of rigor mortis). However, myosin head rotation is accompanied by Pi unbinding and then ADP unbinding. ATP can then bind to the myosin head, which causes the head to detach from the thin filament. The ATP is then dephosphorylated, at which time it can again bind the thin filament. This cross-bridge cycling is the fundamental mechanism for generating force in all muscles (Hooper et al., 2008). Behaviorally relevant muscle contraction and relaxation requires that cross-bridge cycling be regulated. Three types of regulation as the follow.

① Actin based and depends on tropomyosin and troponin. Tropomyosin binds along the thin filament and troponin controls tropomyosin's position on it. When Ca^{2+} binds to troponin the tropomyosin filament is displaced and the myosin heads can attach to the thin filament.

The three components of troponin work together to regulate actomyosin activity. Troponin I binding to the actin/tropomyosin chains results in the site on the actin with which the myosin heads interact becoming blocked. Adding troponin C, with or without Ca^{2+}, induces cross-bridge cycling due to troponin I binding to a high-affinity site on troponin C. In the absence of Ca^{2+}, troponin T prevents troponin C's activating effect by weakening troponin I's affinity for troponin C. Ca^{2+} activates cross-bridge cycling by binding to troponin C and changing its conformation so that troponin T no longer masks troponin C's high affinity site for troponin I.

② Mechanism for regulating cross-bridge cycling is myosin based. All activation of actin and depolarization of muscles are involved Ca^{2+} to be median. In most species, there are two ways that include activation through Troponin-tropomyosin/actin and activate thick filament by combination of Ca^{2+} and light chain of myosin.

③ Regulatory mechanism based on troponin. TnC and Ca^{2+} combination has systematic diversity. Asynchronous insect

肌球蛋白头部转动产生的，而是由类杠杆的方式产生的。头部较远部分的转动利用肌动蛋白的结合点，将动力传给细肌丝。到动力过程结束，杆臂相对于细肌丝和Z线尖端的角度变成135°。如果这是过程的终结，那么肌肉就不能进一步收缩，甚至肌肉会变得僵硬。因为肌球蛋白头部与肌动蛋白牢固地结合，使之不能回到原来的角度，把细肌丝和粗肌丝固定在了不能弯曲的构象。然而，肌球蛋白头部的转动伴随着Pi的释放，接着是ADP的释放。之后，ATP就能结合到肌球蛋白头部，导致头部从细肌丝上分开。ATP接着去磷酸化，又能再与细肌丝结合。这一横桥周期性是所有肌肉产生动力的基本机制 (Hooper等，2008)。动作相关的肌肉伸缩要求横桥的周期性可调。调节的类型有三种。

①基于肌动蛋白并依赖于原肌球蛋白和肌钙蛋白的机制。原肌球蛋白沿细肌丝结合，结合位点即由肌钙蛋白控制。当Ca^{2+}与肌钙蛋白结合，原肌球蛋白丝被移走，肌球蛋白就可以和细肌丝结合。

肌钙蛋白的三个亚基的共同作用，可调节肌动球蛋白的活性。TnI与肌动蛋白/原肌球蛋白链的结合，导致肌动蛋白上的肌球蛋白头部结合位点被阻断；加上TnC，无论有钙无钙，都能引起横桥周期，因为TnI在TnC上有一个高亲和力的结合位点。无Ca^{2+}时，TnT通过弱化TnI对TnC的亲和力来阻止TnC的活化效果。Ca^{2+}通过与TnC的结合活化横桥周期，并改变构象，所以TnT不再遮盖TnC上对TnI具有高亲和力的位点。

②基于肌球蛋白调节横桥周期的机制。所有的肌动球蛋白的活化和肌肉去极化都牵涉到Ca^{2+}作为中介；在多数种内，有两种方式即肌钙蛋白-原肌球蛋白/肌动蛋白的活化作用和通过Ca^{2+}与肌球蛋白轻链的结合激活粗肌丝。

③基于肌钙蛋白的调节机制。TnC和Ca^{2+}结合具有系统多样性。具异步肌的昆虫分成两类Tn簇：第一类是在

classified into two types of manifold Tn: the first is expressed in asynchronous muscles that combined with only one Ca^{2+}; the second is expressed in both synchronous and asynchronous muscle which combined with two Ca^{2+}.

In relaxed muscle, myosin heads bind to each other and to the thick filament backbone in an inhibited head motif (IHM) that maintains the filaments in an OFF state. The OFF state is characterised by low ATPase activity and is also known as the super relaxed state. The majority of myosin heads are inhibited from binding to actin until thick filaments experience a force and heads in the IHM are released to the ON state.

异步肌表达，只和一个 Ca^{2+} 结合；第二类在同步肌和异步肌中都有表达，和两个 Ca^{2+} 结合。

在松弛的肌肉中，肌球蛋白头部相互结合，并结合到抑制性头部基序（IHM）内粗肌丝骨架，使肌纤维处于 OFF 状态。OFF 状态以低 ATP 酶活性为特征，也称为超松弛状态。大多数肌球蛋白头部与肌动蛋白的结合被抑制，直到粗肌丝感受到一股力，IHM 中的头部才被释放到 ON 状态。

15.4 肌肉兴奋的调控机制

15.4 Regulatory Mechanism of Muscular Excitation

15.4.1 Polyneuronal innervation and multiterminal nerve contact

Because of the small size of insects, only a few neurons are allocated to control each muscle. Typically, the leg muscles and many other muscles, receive two stimulatory neurons, commonly described as the slow neuron and the fast neuron. At the muscle surface, the two nerves usually share a glial sheath and lie in a shallow groove on the muscle fiber surface, or in some cases they are invaginated within the muscle fiber outer membrane. Both fast and slow axons break into a number of arborizations, with arborizations making multiple contacts a few micrometers apart on muscle fibers. This is called multiterminal innervation. The distance between the multiterminal contacts on a muscle fiber varies with different species, but the contacts are always close together.

Insect skeletal muscle usually does not develop a propagated action potential. Instead, graded potentials are produced around each of the junctional endings, and local contraction of the muscle fiber occurs around each ending. Because junctional terminals are very close together, the net result is a nearly simultaneous contraction of the whole fiber without a propagated muscle potential. Thus, excitation of insect muscle fibers is integrated with local excitation formed by many multiterminal which is distributed on sarcolemma. The synapsis between the ending of motor neuron and muscle conjunction always called junctional terminals.

15.4.1 多神经元分布和多终端神经接触

昆虫体小，仅有少数神经元控制肌肉。通常其腿肌和其他肌肉接收两类神经元的刺激，常称为慢神经和快神经。在肌肉表面，两个神经常共用一个胶细胞鞘，位于肌纤维表面的浅槽内，或者在某些情况下凹陷入肌纤维外膜内。快神经和慢神经的轴突分成许多树突，通过树突和几微米远的肌纤维接触，这称为多终端神经分布。多终板与肌纤维之间接触的距离在不同种间呈现出多变性，但其接触常常是密切的。

昆虫骨骼肌通常不产生动作电位，取而代之的是每个终板周围产生的分级电位，肌纤维的局部收缩发生在每个终板周围。因为终板彼此非常近，最终结果是整条肌纤维几乎是同时收缩的而没有肌电位的传播。因此，昆虫肌纤维的兴奋由分布在肌膜上的许多终板形成的局部兴奋综合而成。运动神经末梢与肌肉联结的突触常称为终板。

The fast neuron, produces a fast, twitch-type response in the muscle, while the slow axon typically produces a much slower, graded response. The slow neuron junctional contacts at the muscle show facilitation, and a characteristic frequency of nerve impulses must arrive at the junction before the muscle contracts. Typically, a muscle innervated by a fast axon contract rapidly in response to each arriving nerve impulse. The designations "fast" and "slow" indicate the speed of the muscle contraction, not the rate of nerve impulse conduction. This dual (or triple if an inhibitory neuron is present) innervation of a muscle in insects is referred to as polyneuronal innervation. A third neuron, an inhibitory one that causes hyperpolarization.

快神经在肌肉中产生快速、颤动类的反应，而慢神经一般产生非常慢的分级电位。肌肉慢神经的终板连接表现出易化作用，肌肉收缩前，必须有一个特定频率的神经脉冲到达终板。通常，受快神经控制的肌肉在每一个神经脉冲到达时能快速收缩。“快”和“慢”的名称指肌肉收缩的速率，而不是神经脉冲传导的速率。昆虫这种肌肉的双神经（如果还有1条抑制神经即是三神经）分布被称为多神经元分布。第三个神经元是抑制性的，引起超极化。

Only a few large muscles get more than two stimulatory neurons. For example, seven to nine axons to the basalar flight muscle of large beetles. The locust dorsal longitudinal muscles (DLM), which are large, powerful muscles involved in flight, receive five fast motoneurons. Thus, although the flight muscles of some insects receive multiple axons, and skeletal muscles may receive both slow and fast axons, indirect flight muscles of Hymenoptera and Diptera receive only fast axons. Slow flapping of the wings may be unnecessary and even impossible in some insects. The tymbal muscles of cicadas also receive only fast axons.

仅少数大肌肉有多于2条兴奋神经。例如，大甲壳虫飞行肌前上侧片有7～9条轴突，蝗虫的背纵肌（DLM）是牵涉到飞行的大而有力的肌肉，有5条运动快神经元。因此，尽管有些昆虫的飞行肌接收多个轴突，骨骼肌也接收快神经和慢神经，但膜翅目和双翅目昆虫只能接收快神经。有些昆虫飞行时翅的轻拍不需要慢神经，而蝉的鼓膜肌也只有快神经。

Facilitation and summation commonly occur at the junctional endings when the slow axon is activated. In the leg, typically about 15 neuronal impulses per second must arrive at the junctional endings of the slow axon to produce contraction. Temporal facilitation and summation can occur when several small volleys of nerve impulses arrive within one second. Thus, an insect that has some combination of fast, slow, and inhibitory axons to the same muscle may realize a number of degrees of graded muscle responses from the neuronal system or systems utilized, and response may be further modified by neuropeptides secreted at some nerve endings.

当终板被慢神经激活时，易化与总和普通发生在终板。通常，腿部的慢神经终板每秒接收15个神经脉冲才能产生收缩。当1 s内有几个小的神经脉冲到达时，会发生时间的易化与总和。因此，昆虫相同的肌肉有一些快神经、慢神经和抑制性的神经轴突的组合，可实现一些神经系统或系统利用的分级肌肉反应，其反应可被神经末端分泌的神经肽进一步调整。

15.4.2 The transmitter chemical at lamina terminals

15.4.2 终板的化学递质

The synaptic cleft between nerve ending and muscle membrane is about 30 nm wide, and a transmitter chemical must diffuse across this gap from axonal ending to muscle sarcolemma. The stimulatory transmitter chemical at fast and slow nerve-muscle junctions is *L*-glutamic acid and possibly *L*-aspartate, also. At inhibitory junctions, it is γ-aminobutyric acid (GABA). The transmitter chemicals are contained in large numbers of vesicles

神经末端和肌膜间的突触间隙宽约30 nm。化学递质必须通过扩散跨过轴突终端与肌肉肌膜的间隙。快神经与慢神经的终板兴奋性化学递质是*L*-谷氨酸，也可能是*L*-天冬氨酸。而抑制性化学递质即是γ-氨基丁酸（GABA）。化学递质

that are from 20-60 nm in diameter.

Several types of glutamate receptors (GluR) have been characterized in insect muscles and also in the insect CNS, based on binding to indicator compounds including quisqualate (quisqualate glutamate receptor, qGluR), ibotenate (iGluR), and aspartate (aspGluR). Multiple receptors may be involved in different degrees of graded responses from the muscle. In *S. gregaria* and *T. molitor*, where the most detailed studies have been done, populations of these receptors are present at nerve-muscle junctions, as well as at extra-junctional sites.

γ-GABA can activates chlorine ion channel, which inhibits muscle excitation or decreases it (hyperpolarization). Furthermore, insect muscle is innovated by visceral nerve system. For example, median nerve in locust dorsal can secrete octopamine as a chemo-modulator which stimulates cAMP in muscle sarcolemma that increases metabolism of cAMP and supply of ATP; therefore, it increases the speed and physical strength of muscle contraction, and come back rapidly relaxation after contraction.

被直径 20 ～ 60 nm 的无数囊泡包裹。

几种类型的谷氨酸受体（GluR）已经在昆虫的肌肉和中枢神经系统中被表征，根据结合的指示化合物，包括使君子氨酸（使君子氨酸受体）、鹅膏蕈氨酸盐受体（iGluR）和天冬氨酸盐受体（aspGluR）。多种受体可能不同程度地参与了肌肉的分级反应。在沙漠蝗和黄粉甲中，这些受体群不但出现在神经肌肉终板，也出现在外部接点。

γ- 氨基丁酸能激活 Cl^- 通道，使肌肉不产生兴奋或降低其兴奋性（超极化）。此外，昆虫肌肉被交感神经支配。如蝗虫背部中神经元能分泌章鱼胺作为化学调质，刺激肌膜上的腺苷酸环化酶，增加 cAMP，刺激代谢，加快 ATP 供应，从而加强神经元的刺激作用，增加肌肉兴奋收缩的速率和力量，并在收缩后加速恢复到松弛状态。

15.4.3 The other regulatory factors

Regulation of stimulatory muscle is affected by hormone, hemolymph chemical facts and mechanical stretch as well as nerve transmitters and modulators. They influence not only on self-generative activity muscle, but on muscles which innervated by nerve. The other factors also influence on muscle contraction, such as proctolin, Ach, neurohormone, 5-hydroxytryptamine, crustacean cardioactive peptide (CCAP) and FMRFamide etc. stimulate spontaneous contraction activity of muscle contraction. Satiation may improve muscle's contraction. Myosuppressins reduce the frequency of insect muscle contractions.

Proctolin in insect muscles activates the phospholipase C (PLC) pathway leading to production of the second messengers 1,4,5-inositol-trisphosphate (IP_3) and diacylglycerol (DAG). Proctolin also increases muscle membrane resistance through a reduction of the resting K^+ conductance. The action of proctolin is dependent on the presence of extracellular Ca^{2+}, and comprises an influx of Ca^{2+} into the muscle. An involvement of both voltage dependent and non-voltage-dependent Ca^{2+} channels during proctolin action has been proposed for the oviduct of *L. migratoria*

15.4.3 其他调控因子

肌肉的兴奋性既受到激素、血淋巴化学因子和机械拉伸的影响，也受神经递质和调质的影响。它们不仅影响自发活动的肌肉，还影响神经支配的肌肉。其他影响肌肉收缩的因子，如直肠肽、乙酰胆碱、神经激素、5- 羟色胺、甲壳纲心激肽（CCAP）和酰胺化 FMRF 等同时刺激肌肉收缩活动。饱食可促进肌肉收缩。肌球蛋白抑制素降低昆虫收缩的频率。

昆虫肌肉直肠肽激活磷脂酶 C（PLC）途径，导致第二信使三磷酸肌醇（IP_3）和二酰甘油（DAG）的产生。直肠肽也通过静息 K^+ 电导的还原来增加肌肉膜的抗性。直肠肽的作用包括依赖于胞外的 Ca^{2+} 以及进入肌肉的钙电流。蝗虫产卵器和马德拉蜚蠊后肠的直肠肽作用过程中涉及电压门控和非

and the hindgut of the cockroach *L. maderae*. Periviscerokinin, are abundant myotropic neuropeptides in the abdominal neurosecretory system of the cockroach *P. americana* and are stored in the abdominal perisympathetic organs. The two known PVKs (PVK-1 : GASGLIPVMRNamide, and PVK-2 : GSSS- GLISMPRVamide) both have myotropic effects on the heart, segmental vessels, and the hyperneural muscle of *P.americana*, both peptides act postsynaptically at the muscle membrane. The mode of action of both peptides comprises additional mechanisms besides activation of voltage-dependent Ca^{2+} channels (Wegener et al., 2000).

电压门控 Ca^{2+} 通道。外周交感激肽是美洲大蠊腹神经分泌系统含量丰富的影响肌肉的神经肽，贮存于腹部副交感神经器官。两种已知的外周交感激肽（PVK-1 ：GASGLIPVMRNamide 和 PVK-2 : GSSS- GLISMPRVamide）都会影响美洲大蠊的心肌、节间血管和亢奋神经肌肉，两种肽都在肌肉突触后膜起作用。其作用模式包括除电压门控 Ca^{2+} 通道外的额外机制（Wegener 等，2000）。

15.5 肌膜兴奋与肌肉收缩

15.5 Sarcolemma Excitement and Muscle Contraction

Stimulatory muscle sarcolemma causes contraction, the process usually includes sarcolemma depolarization, transversal tubular system conduct excitation, SR system releases Ca^{2+}, Ca^{2+} then activates contractile proteins system, and actomyosin utilize ATP undergoing a series of contraction.

肌膜兴奋导致肌肉收缩，通常包括：肌膜的去极化、横管系统传导兴奋、内质网系统释放 Ca^{2+}、Ca^{2+} 激活收缩蛋白系统以及肌动球蛋白利用 ATP 进行收缩等一系列过程。

15.5.1 Sarcolemma depolarization and excitement transmission

The sarcolemma of insect myofibrils as well as nerve fibers is nearly impermeable to varies of ions. To transfer ions into and out of the sarcolemma, the membrane provides two structures: ion pump and ion channels. The limiting membrane of the striated muscle fiber, the sarcolemma, consists of several layers and varies with species and fiber type.

15.5.1 肌膜的去极化与兴奋传导

与神经纤维的一样，昆虫肌纤维的肌膜对不同离子几乎没有透过性。为了让离子转入和转出肌膜，细胞膜提供了两种结构：离子泵和离子通道。横纹肌纤维的界膜，即肌膜由几层组成，并随物种的纤维类型而变化。

15.5.2 Muscle excitation-contraction coupling mechanism

The arrival of an impulse in the nerve fiber supplying a portion of the indirect musculature sets up an excitation process in the muscle fibers which produces an electrical spike. To produce an active contraction of the myofibrils it is evidently necessary both that the system should be in an excited state and that the mechanical conditions should be correct (Pringle, 1949). The series of events that allows the transduction of the sarcolemma depolarization into Ca^{2+} release is called excitation-contraction

15.5.2 肌肉兴奋 - 收缩耦联机制

神经纤维脉冲到达时，供给部分间接肌肉组织引起肌纤维的兴奋过程，产生峰电位。肌原纤维要产生活跃的收缩，很明显，肌肉系统必须处于兴奋状态，而且体能条件也要合适（Pringle，1949）。一系列环节允许肌膜去极化的转导使 Ca^{2+} 释放，称之为兴奋 – 收缩耦联。两个膜系统：肌膜和肌质网，非

coupling. The two membrane systems, plasmalemma and SR, communicate very efficiently with each other at specific structures called junctions. In fact, excitation- contraction coupling in cardiac muscle fails in the absence of extracellular Ca^{2+}.In skeletal muscle, however, excitation- contraction coupling continues in the absence of extracellular Ca^{2+}. In the dragonfly flight muscles, L-type Ca^{2+} channels of body muscles are located in close proximity to the SR feet. This location support of the hypothesis that Ca^{2+}—induced Ca^{2+} release is a possible mechanism of excitation-contraction coupling in the muscles (Takekura et al., 2002).

常高效地在一个被称为汇聚的特殊结构内相互联系。事实上，脏肌内的兴奋－收缩耦联在胞外无 Ca^{2+} 时衰减。然而，骨骼肌的兴奋 - 收缩耦合在胞外无 Ca^{2+} 时也能持续。蜻蜓飞行肌内，体肌 L-型 Ca^{2+} 通道位于接近 SR 基部。这一定位支持了肌肉兴奋 - 收缩偶联机制中关于 Ca^{2+}——诱导 Ca^{2+} 释放的假说（Takekura 等，2002）。

15.5.3 Slow muscles and fast muscles

The muscles are classified into slow and fast muscle according to the structure of muscle cell and sarcolemma. Slow muscle also called tonic contraction muscles. The tonic contraction is characterized by a prolonged maintenance of tension, which is similar to the "catch" contraction. There are two hypotheses for the "catch" mechanism. One is the "parallel hypothesis" and proposes that the tension maintenance during the tonic contraction results from interaction between paramyosin molecules without involvement of the contractile actomyosin system; the other, the "cross-bridges hypothesis", proposed that the actomyosin cross cross-bridges break at a very slow rate during the tonic contraction. The tension of the tonic contraction elicited by repetitive stimulation increased gradually and finally reached a plateau, on which the phasic contraction appeared in response to each stimulus. As the intensity of current pulses was increased, the summed tonic tension increased several times while the phasic tension did not increase so markedly. After the cessation of stimulation the tonic contraction decayed very slowly with a half-time of relaxation. Longer duration or higher intensity of current pulse is needed to elicit the tonic contraction compared with the phasic contraction. The phasic contraction might be generated mainly by calcium ions released from the membrane, and the tonic contraction might be caused mainly by calcium ions released from the sarcoplasmic reticulum. The most of flight muscles and visceral muscles are tonic contractile muscles. Phasic contraction muscle, also called fast muscle, is sensitive to single nerve pulse and responses rapidly with non-or-all. It contracts fast and recovers rapidly. Synchronous flight muscles and jumping muscles with consistence between contraction frequency and motor nerve pulse belong to this type.

15.5.3 慢肌与快肌

根据肌细胞与肌膜的结构，分为慢肌与快肌。慢肌又称紧张性收缩肌。紧张性收缩定性为张力的持续维持，类似于“闭锁”收缩。“闭锁”机制有两个假说：一个是“平行假说”，紧张性收缩过程中张力的维持是副肌球蛋白分子间的相互作用，而没有涉及肌动球蛋白系统的收缩；另一个是“横桥假说”，认为肌动球蛋白跨越横桥，在紧张性收缩时以非常慢的速率中断。紧张性收缩的张力通过逐步增加的重复刺激而激发，最终到达平稳状态，此时，每次刺激可产生时相性收缩。当脉冲电流强度增加，紧张性张力的总和增加数倍，而时相性张力的增加则没这么明显。紧张性刺激停止后，在松弛的一半时间内非常缓慢地衰减。与时相性收缩相比，引发紧张性收缩需要更长时间或更高的脉冲电流。时相性收缩可能主要通过膜释放 Ca^{2+} 产生，而紧张性收缩可能主要由肌质网释放的 Ca^{2+} 产生。大多数飞行肌和脏肌都属于紧张性收缩肌。时相性收缩肌也称快肌，对单个神经脉冲敏感，并迅速做出全无反应或全有反应。收缩快、恢复快。收缩频率与运动神经脉冲一致的同步飞行肌和跳跃肌都属于这类。

15.5.4 Simple contraction and compound contraction

A single nerve impulse effectively stimuli muscle which causes a muscle contraction, called simple contraction. Single stimuli usually release enough acetylcholine in the neuromuscular junctions of the motor unit to produce action potentials in the sarcolemma. This will cause the muscle to contract after a short delay. A simple contraction gives only 20%-30% of the maximum tension possible. If a second stimulus is given before a muscle relaxes the muscle will shorten further, building up more tension than a simple contraction. The continuous contracting processes is called compound contraction, or called summation. If many stimuli are given very close together the muscle will go into a smooth continuous contraction called tetanus. Each single contraction experienced three stages: instant of stimuli to the latent period; contraction period when muscle starts contracting to the shortest length; and relaxed period as muscle begins relaxation to recover to the original state.

In fact, compound contraction is an overlapping process of a series of single contractions. If high stimuli frequencies are given, the next stimuli are accepted before the latent period of a single contraction or initial stage of contraction, muscle then undergoes constant contraction not experience relaxation stage. It is called complete tonic contraction. In contrast, if the later stimuli cause muscle contraction in relaxation period of the former single contraction, a little relaxation occurred between two contractions (still not recover to the original state), called incomplete tonic contraction.

15.5.4 单收缩和复收缩

肌肉接受一次有效的刺激所引起的收缩过程称为单收缩。单刺激常常在运动单元的神经肌肉连接处释放出足够的乙酰胆碱，在肌膜产生动作电位。这将导致肌肉在短延缓后收缩。单收缩只引起最大可能张力的 20% ～ 30%。在肌肉松弛之前，如果给予第二次刺激，肌肉将进一步收缩，构建起比单收缩更强的张力。产生连续收缩的过程称为复收缩或总和。如果给以非常多、非常密集的刺激，肌肉将进行连续平稳的收缩，称为强直收缩。每个单收缩分 3 个时期：从刺激的瞬间到开始收缩的潜伏期；从开始收缩到肌肉收缩到最短的收缩期；从开始松弛到恢复原状的松弛期。

复收缩实际上是一连串单收缩的重叠过程。如果刺激频率高，在前一个单收缩的潜伏期或收缩期的初期就接受了下一个刺激，肌肉不经松弛期而进行持续的收缩，称为完全紧张性收缩。反之，后一次刺激引起的收缩发生在前一次单收缩的松弛期，肌肉两次收缩之间稍有松弛（没有恢复到原状），称为不完全紧张性收缩。

15.5.5 Isometric contraction and isotonic contraction

15.5.5.1 Isometric contraction

An isometric contraction of a muscle generates force without changing length. The force generated during an isometric contraction is completely dependent on the length of the muscle contracting. Tension varies with the contraction processes which accumulate potential energy of muscle's movement.

15.5.5.2 Isotonic contraction

Isotonic contraction is a type of muscle contraction in which the tension remains unchanged and the distance between the muscle's origin and insertion becomes lessened. In an isotonic contraction the muscle shortens, keeping a constant tension. In an isometric contraction the muscle does not shorten and tension builds up;

15.5.5 等长收缩与等张收缩

15.5.5.1 等长收缩

等长收缩的肌肉产生力量时，其长度不变。等长收缩产生的张力完全依赖于肌肉收缩的长度。张力随收缩过程而发生变化，收缩过程积累了肌肉运动的势能。

15.5.5.2 等张收缩

等张收缩是一类张力不变，而肌肉起点与附着点之间距离变小的肌肉收缩。等张收缩肌肉变短，保持持续不断的张力。等长收缩肌肉不变短而产生

most real contractions are mixtures of the two types.

Isometric contraction of IFM at high Ca^{2+} concentration is required for preflight warm-up in large insects. Usually, energy consumed at the beginning of muscle contraction is called activation heat; acting on traction load of muscle is called maintenance heat; and heat produced by energy consumed during muscle shorten itself is relaxation heat.

张力。

高 Ca^{2+} 浓度下 IFM 的等长收缩是大型昆虫起飞前预热所必需的。一般将起始消耗的能量称为激活热；牵动负荷做功消耗的能量称为维持热；肌肉缩短时本身消耗的能量产生的热称为舒张热。

15.5.6 Concentric contraction

A concentric contraction is a type of muscle contraction in which the muscles shorten while generating force. During a concentric contraction, a muscle is stimulated to contract according to the sliding filament mechanism. This occurs throughout the length of the muscle, generating force at the musculo-tendinous junction, causing the muscle to shorten and changing the angle of the joint.

15.5.6 向心收缩

肌肉缩短的同时产生力量的一类肌肉收缩被称为向心收缩。向心收缩过程中，受到刺激时肌肉根据纤丝滑动机制收缩。这发生在整个肌肉长度上，产生的力量位于肌肉 - 肌腱连接处，引起肌肉缩短，并改变结合处的角度。

15.6 肌肉与运动的关系

15.6 The Relationship between Muscle and Movement

15.6.1 Muscle groups involved in flight

Two main categories of insect muscles involved in flight; the one is operator muscles which regulate stroke direction and twist form of wing; the other is flight muscles which responses to move up and down. Among them, muscles directly insert the basal wing are called direct muscle including elevator and depressor; while elevators of the most insects are dorsoventral muscle, indirect flight muscle (IFM).

At least ten pairs of muscles are involved in flight, wing orientation, and steering. The IFM power wing movements by changing the shape of the elastic thorax. These are power muscles and include the dorsal longitudinal muscles (the wing depressors) that arch the tergum and the dorsoventral and oblique dorsal muscles (the wing elevators). The direct wing muscles insert directly on the wing hinge sclerites, or on axillary sclerites or movable sclerites of the pleuron of the thorax. These muscles are the basalar (wing depressors), subalar (wing depressors and twisting or supinators of the wing), and third axillary (wing-folding) muscles. A third group of muscles in the thorax are accessory indirect muscles consisting of the pleurosternal, anterior

15.6.1 参与飞行的肌肉群

参与昆虫飞行的肌肉主要分两大类：一类调节翅的拍击方向和扭转形式，称为操作肌；另一类负责翅的上下运动，称为飞行肌。其中与翅基直接相连的称为直接翅肌，包括升肌和降肌。而大多数昆虫的升肌是背腹肌，间接翅肌（IFM）。

参与飞行、翅的定向和转向的肌肉至少有 10 对。IFM 通过改变弹性胸部的形状为翅膀运动提供动力。这些是力量肌，包括拱起背板的背纵肌（翅的降肌）、背腹肌和背斜肌（升肌）。直接翅肌直接嵌入在胸侧板的翅铰链或腋窝骨片或活动骨片上，这些肌肉是前上侧肌（降肌）、翅下肌（降肌和扭转肌或旋后肌）和第三腋窝肌（翅的折叠）。胸部的第三群肌肉是间接肌的辅助肌，由侧

tergopleural, posterior tergopleural, and intersegmental muscles (Fig.15-8; Appendix color graphic Fig.15-8). They modify the way in which the power-producing muscles move the wings and the angle of attack. For example, contraction of a small muscle, the pleuroalar (also called the pleumaxillary) muscle, in the locust *L. migratoria* reduces pronation of the forewings during the downstroke and supination during the upstroke, thus helping to control the angle of attack and the development of additional lift forces (Wolf, 1990). The muscle inserts on the third pleuroaxillary sclerite and has a broad, fan-shaped anchor on the pleural wing process. The angle of attack is important to generation of lift forces and to speed of flight and hovering.

The frontal edge of the wing leads in both the upstroke and the downstroke. Airflow hits the lower side of the wing during the downstroke, which generates the principal lift forces, and hits the upper side during the upstroke, which can also produce lift forces. The wing downstroke is produced by contraction of dorsal longitudinal muscles (in all insects except dragonflies and damselflies, in which the mechanism is different). The dorsal longitudinal muscles are indirect muscles that do not attach directly to the wing hinge sclerites. These large, powerful muscles are attached to the phragma, which are invaginated and hardened cuticular processes at the anterior and posterior of each of the meso- and metathoracic segments. When they contract, they shorten the thoracic segments by arching the tergum, slightly lifting the attachment base of the wings at the tergopleural junction,

腹板肌、前背侧肌、后背侧肌和节间肌组成（图 15-8；附彩图 15-8）。它们改变了产生力量的肌肉移动翅膀的方式和冲(迎)角。例如，飞蝗胸膜(也称胸腋)肌，一块小肌肉的收缩在俯冲时降低前翅的前旋，向上飞时减少后旋，因而有助于控制冲（迎）角，并产生额外的上举力（Wolf, 1990）。嵌入第三腋窝骨片的肌肉宽大呈扇形锚定在侧翅突上。冲（迎）角对上产生的举力和加速飞行及悬停是重要的。

翅的前缘引导上升冲程和下行冲程。下行冲程时气流撞击翅的下边，产生主要的上举力；上升冲程时气流撞击翅的上边，也能产生上举力。翅的下行冲程是由背纵肌收缩产生的（除蜻蜓和蜻蛉外的所有昆虫，其机制不同）。背纵肌是间接飞行肌，与翅膀的铰链骨片不直接相连。这些大而有力的肌肉连接在悬骨上。悬骨是每个中胸和后胸体节的前部和后部的凹入硬化的表皮。当它们收缩时，通过把背板弯成拱形来缩短胸节，使背侧板连接处的翅基轻微上举，促使翅膀向下。翅膀到达不稳定的水平位置时，它们在侧翅突上突然以向下为枢

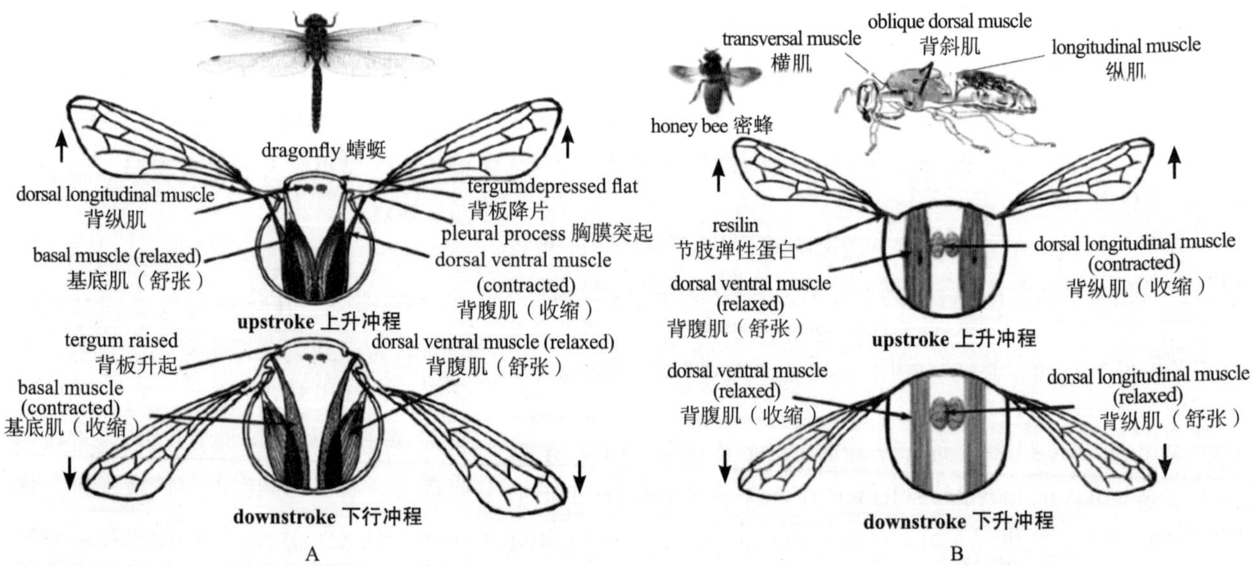

Fig.15-8 The action model of insect direct and indirect muscles during flight. A. The direct muscle of dragonfly; B. Indirect muscle function of bee (modified from Chari et al., 2021).

图 15-8 昆虫间接肌和直接肌在飞行中的作用模式。A. 蜻蜓直接肌的作用；B. 蜜蜂的间接肌的作用（仿 Chari 等，2021）。

and forcing the wings downward. As the wings approach the unstable horizontal position, they suddenly pivot downward over the pleural wing process. This releases the load on the dorsal longitudinal muscles and they cease to shorten, but the new position of the wings introduces the load to the antagonistic set of muscles, the dorsoventral muscles. Contraction of one pair of muscles stretches the antagonistic pair, and serves as a stimulus for a repeated contraction. Thus, several contractions may occur for each nerve impulse received.

There are two sets of IFMs in the thorax, the dorsal longitudinal muscle (DLM) and the dorsal ventral muscle (DVM). The two sets of muscles are arranged in cross orientations and work together to drive oscillated thoracic distortions and wing beats. When taking off, a nerve impulse triggers the DVM to contract, which deforms the thorax, causing a stretch of DLM to induce a delayed stretch activation. The delayed contraction of the DLM in turn causes the DVM to extend and is stretch activated. This cycle of stretching and contracting is repeated at high frequency with low frequency stimulating impulses from motor nerves, by which an intermediate rise in cytosolic Ca^{2+} will sustain continuous oscillations of contraction and relaxation of insect asynchronous IFM (Cao, et al., 2020).

In insect flight muscle, myosin heads rotate when muscle stretches, and this enables them to bind more strongly. In other words, stretch activation is a fundamental consequence of the interaction between actin and myosin in insect flight muscle. In both *Lethocerus* and *Drosophila* IFM, two isoforms of TnC exist, one regulated by stretch, the other by Ca^{2+}. Both are assumed to relieve the inhibitory effect of TnI on crossbridge formation by altering the position of tropomyosin to uncover actin-binding sites. The myosin heads that attach to actin during lengthening will be pulled towards the Z-disk, opposite to their direction of motion during power strokes. These strained crossbridges are therefore likely candidates for the 'sensors' or sensorimotor (Mischiati, et al.,2015) that activate the stretch-dependent TnC isoform, leading to the rapid binding of myosin heads at target zones (Dickinson et al., 2005).

纽。这释放了背纵肌的负荷，收缩停止，但翅膀的新位置引入一组负载拮抗性肌肉，即背腹肌。一对肌肉的收缩拉伸一对拮抗肌，起到刺激反复收缩的作用。因此，接收到的每一个神经脉冲都可能发生多次收缩。

胸部有两组 IFM，背纵肌（DLM）和背腹肌（DVM）。这两组肌肉呈交叉方向排列，共同驱动摆动的胸部扭曲和翅膀拍击。起飞时，神经脉冲触发 DVM 收缩，使胸部变形，导致 DLM 拉伸，产生延迟拉伸激活。DLM 的延迟收缩进而导致 DVM 伸展并被伸展激活。这种伸展和收缩的循环在运动神经的低频刺激脉冲下以高频重复，胞质间 Ca^{2+} 升高将维持昆虫异步 IFM 收缩和松驰的连续振荡 (Cao 等 , 2020)。

在昆虫飞行肌肉中，肌球蛋白头部在肌肉拉伸时旋转，这使它们能够更牢固地结合。换句话说，伸展激活是昆虫飞行肌肉中肌动蛋白和肌球蛋白相互作用的根本结果。在大负子蝽和果蝇的 IFM 中，存在两种肌钙蛋白 C 亚型，一种受拉伸调节，另一种受 Ca^{2+} 调节。这两种方法都可以通过改变原肌球蛋白的位置来暴露肌动蛋白结合位点，从而减轻肌钙蛋白 I 对横桥形成的抑制作用。在拉长过程中附着在肌动蛋白上的肌球蛋白头将被拉向 Z 盘，与它们在动力冲程中的运动方向相反。因此，这些应变横桥可能是“传感器”或感觉运动器（Mischiati 等，2015）的候选，用于激活伸展依赖性肌钙蛋白 C 亚型，导致肌球蛋白头在靶区的快速结合（Dickinson 等 , 2005 ）。

15.6.2 Wing source theory

15.6.2 翅源学说

15.6.2.1 Tracheal gill origin theory

15.6.2.1 气管鳃起源理论

In 1870, Carl Gegenbaur proposed that insect wings evolved from tracheal gills similar to those present on modern-day aquatic insect larvae. The original form of Gegenbaur's "gill theory" of

1870 年，Carl Gegenbaur 提出，昆虫翅膀是从气管鳃进化而来的，类似于当代水生昆虫幼虫的鳃。Gegenbaur 关

wing evolution involved gills enlarging for improved gas exchange, then a partial transition to terrestrial life, in which gills might have aided gliding, or steering during leaps, and then with improvements in musculature and articulations, they became wings that could flap for powered flight.

于翅膀进化的"鳃理论"的最初形式涉及鳃的扩大以改善气体交换，然后部分过渡到陆地生物，鳃可能有助于滑翔或跳跃时的转向，再随着肌肉组织和关节的改善，它们变成了可以为动力飞行而拍打的翅膀。

15.6.2.2 Paranotal lobe theory

Fritz Müller observed lateral tergal lobes on the thorax of certain termite nymphs, and considered that these lobes were incipient wings. Since the lobes did not contain obvious tracheae, he rejected Gegenbaur's contention that wings arose from tracheal gills. Crampton (1916) seems to have coined the term "paranotal lobes" for Müller's tergal expansions, and the theory has been known as the "paranotal lobe theory" ever since. Crampton produced a detailed list of the evidence in favor of both the gill theory and the paranotal lobe theory. He concluded that the weight of evidence, e.g., wings not being serially homologous with abdominal tracheal gills, and strong evidence for aerial respiration being primitive in insects (even aquatic ones) was against the gill theory. He also described evidence favoring the paranotal theory widespread occurrence of leaping ability and of paranota on the prothorax of extant insects so he argued that the paranotal theory was a better fit to the available evidence. The lateral tergal margin, which is homologous between apterygote and pterygote insects, comprises a growth organizer to expand the body wall to form adult wing blades in *Gryllus*. Wings evolved from the pre-existing lateral terga of a wingless insect ancestor, and the reactivation or redeployment of Wnt/Fat-Dachsous/Hippo-mediated feed-forward circuit might have expanded the lateral terga (Ohde T et al., 2022).

15.6.2.2 侧背板翅源说

Fritz Müller 观察到某些白蚁若虫胸部的外侧背板，并认为这些背板就是早期的翅膀。由于背板中没有明显的气管，他驳斥了 Gegenbaur 关于翅膀起源于气管鳃的论点。Crampton（1916）似乎为 Müller 的背板扩展创造了"侧背板"一词，自那以后，该理论一直被称为"侧背板理论"。Crampton 提出了一份支持鳃理论和侧背板理论的详细证据清单。他得出结论的重要证据，例如翅膀与腹部气管鳃不连续同源，有力的证据表明昆虫（甚至水生昆虫）的空中呼吸是原始的，这与鳃理论相悖。他还描述了支持侧背板理论的证据，即跳跃能力和侧背板在现存昆虫的前胸上广泛存在，因此他认为侧背板理论更适合现有证据。侧背板缘在无翅亚纲昆虫和有翅昆虫之间是同源的，它包括一个生长组织者，用于扩展体壁，以形成蟋蟀的成虫翅膀叶片。翅膀是从无翅昆虫祖先已有的侧背板进化而来的，Wnt/Fat-Dachsous/Hippo 介导的前馈回路的重新激活或重新部署可能扩大了侧背板（Ohde T 等，2022）。

15.6.2.3 The pleural-appendage theory

Wigglesworth (1976) described mayfly larvae that use some of their modified gills for both covers and ventilation, or even as swimming paddles. He suggested that both abdominal gills and wings were modified coxal exites, and proposed that thoracic gill covers might have evolved into paddles. Such paddles then might have been used aerodynamically by semi-aquatic insects stranded by drying ponds and rivers, and blown into the air by winds and updrafts. In this way, those insects which managed to control their flightpath and return to water would have had an advantage,

15.6.2.3 胸腔附属学说

Wigglesworth（1976）描述了蜉蝣的幼虫使用一些经过改造的鳃来覆盖和通风，甚至用作游泳桨。他认为腹鳃和翅膀都是改造的基节外叶，并提出胸鳃盖可能已经进化为桨。这种桨可能被那些困在干涸的池塘和河流中的半水生昆虫在空气动力学上使用，并被风和上升气流吹向空中。用这个方法，那些能够控制飞行路径并返回水中的昆虫就有了优势，

selecting for better aerial control, and eventually flapping.

选择更好的空中控制，并最终拍打。

15.6.2.4 Pleuron theory

Kukalová-Peck (2008) used her interpretations of various fossils to theorize on the origin of wings. Based on fossil specimens and geological data, she believes that wings originated from the epicoxa of the foot and the coxal exites. The epicoxa combined with the body wall and differentiated into dorsal and abdominal joint segments. The epicoxa widened to form protowing, and the protowing finally separated from the side plate composed of the epicoxa and subcoxa, and became movable under the traction of the foot muscle of the epicoxa. The protowing continued to grow and eventually evolved into a wing.

15.6.2.4 侧板翅源说

Kukalová-Peck（2008）利用她对各种化石的解释来推测翅膀的起源。她主要依据化石标本和地质资料，认为翅起源于足的上基节及其基节外叶。上基节与体壁结合，分化成背、腹关节片。其基节外叶变宽，形成原翅，而原翅最终与上基节和亚基节组成的侧板分离，在上基节足肌的牵拉下，变为可动。原翅继续变大，最终演化为翅。

15.6.2.5 Surface skimming theory

Focusing on behavior rather than models, Marden and Kramer (1994) observed surface-skimming behavior in extant stoneflies, where adults of some species are too weak to fly, but use flapping wings as propellers to skate over the surface of water. Even with shortened wings, the stoneflies could still locomote over the water surface. They suggested that surface skimming might be an intermediate stage in the evolution of flight. If an aquatic insect with mobile gill covers evolved an air-breathing adult stage, it might be able to skim across the water surface even with small, weak winglets. Selection for faster skimming would lead to longer wings and faster, stronger muscles. These improvements might then allow hops off the water at first, leading eventually to powered flight.

Others suggested that wings arose from hypertrophied spiracles or wings arose from legs by suggesting that wings and legs had a common origin, both being variations on dermal outgrowths that believed were originally respiratory, and that gave rise to all arthropod appendages (Alexander, 2018).

15.6.2.5 水面掠过学说

Marden 和 Kramer（1994）将重点放在行为而非模型上，观察了现存石蝇的水面掠过行为，其中一些物种的成虫身体太弱，无法飞行，但使用扇动的翅膀作为推进器在水面上滑行。即使翅膀变短，石蝇仍然可以在水面上活动。他们认为，表面掠过可能是飞行进化的中间阶段。如果一种可移动鳃盖的水生昆虫进化成一个呼吸空气的成年阶段，那么它甚至可以用小而脆弱的小翼掠过水面。选择更快的掠过速率将导致更长的翅膀和更快、更强的肌肉。这些改进可能允许它们首先快速滑行并离开水面，最终实现动力飞行。

其他学者则认为翅膀是由过度增大的气门形成的或者说翅膀起源于腿，表明翅膀和腿有一个共同的起源，都是最初起呼吸作用的真皮外生长的变化，且产生了所有节肢动物的附属物（Alexander，2018）。

15.6.3 The aerodynamics of lift forces produced by wings

To remain airborne, an insect must generate lift forces at least equal to its weight; and to move forward, the horizontal thrust vector must exceed the drag of air resisting forward motion (Pringle,1983). The dynamics of flight are complex, and the lift force needed is related to factors such as body weight, wing size, wing shape, speed of air movement over the wings (i.e., wing beat

15.6.3 翅膀产生上举力的空气动力学

为了逗留在空中，昆虫必须产生至少等同于其体重的上举力。要向前移动，水平推动的动力必须超过向前移动的空气阻力（Pringle, 1983）。飞行的动力学很复杂，所需的上举力与体重、翅

frequency), and angle of attack of the wings.

Weis-Fogh (1973) initially described an example of an unsteady lift condition that occurs in very small chalcid wasps as a "clap" and "fling" wing motion. In this small insect, the wings "clap together" at the top of the upstroke, and then twisting motions controlled by some of the small muscles fling them apart at the start of the downstroke. The rapid flinging motion sets up air movements above the wings that increase the lift force of the downstroke, and may aid very small insects in generating enough lift for flight despite having little airfoil surface on their tiny wings. The clap and fling mechanism seems to work mostly in small insects, although *D. melanogaster* also uses it at the start of the downstroke.

Using a dynamically scaled model of *D. melanogaster* with built-in sensors, Dickinson et al. (1999) describe three interacting mechanisms: ① the upstroke and downstroke of the wings with a high angle of attack (delayed stall); ② rotational circulation of air eddies above the wings; ③ wake capture, with the latter two mechanisms promoted by the pronation and supination of the wings as they rapidly rotate and change direction at the end of each half-stroke. Larger insects, such as sphingid moths, generate lift forces from the development of a leading-edge vortex that produces transient aerodynamic lift forces to keep the insect in the air. Although the turbulence or wake following a moving object in a fluid medium typically produces drag, by rotating the wings at the end of each half-stroke, the wing encounters its own wake in such a way to generate momentary positive lift (Dudley,1999). The magnitude of the lift generated by wake capture varies from positive to negative, depending on exactly when in time the rotation occurs relative to the beginning of the next half-stroke; wing rotation that is delayed until the start of a new stroke direction produces negative lift.

Dragonfly wings act as ultra-light aero foils during gliding flight. In a cross-sectional view, the wings have well-defined corrugations in which rotating vortices develop. The corrugations might be expected to lead to high drag values; but, in fact, the drag from air flowing over the wings is very low. The vortices filling the profile valleys smooth out the profile geometry, resulting in low drag similar to that of air flowing over flat plates (Kesel, 2000), while lift forces are much higher than expected from flat plates.

的大小、翅的形状、掠过翅膀的空气流动速率（即翅膀拍击的频率）和翅膀的冲（迎）角等因素都有关。

Weis-Fogh（1973）最初描述的一个非稳定态举力的范例是非常小的小黄蜂，把翅的运动分为“合拢”和“打开”。这种小昆虫的翅膀在上行冲程的顶点“一起合拢”，然后，有些小肌肉控制的扭曲运动使之打开离开下行冲程起点。快速的打开动作在翅膀上引起空气流动，增加了下行冲程的上举力。在它们的微型翅膀上有小气翼表面，这有助于非常小的昆虫产生足够的上举力飞行。合拢和打开机制多数在小昆虫中起作用，果蝇在下行冲程开始时也用它。

Dickinson（1999）将果蝇的动力学比例模型建立在感受器上，描述了果蝇飞行中 3 类提供飞行上举力的互作机制：①大冲（迎）角翅膀下行冲程和下行冲程（延时悬停）；②翅膀上空气涡流的旋转循环；③尾流捕获，后两种机制由翅膀的内旋和后旋推动，因为它们在每半个合拢结束时快速旋转和改变方向。大昆虫，如天蛾，从翅前缘涡流引发的上举力产生瞬时空气动力，其上举力将昆虫滞留在空中。在流体介质中，虽然湍流和尾流紧随着移动的物体，通常产生阻力，但通过翅膀每半次拍击结束时旋转，翅膀遇到其本身的尾流，这个途径可产生暂时性的正向举力（Dudley, 1999）。俘获尾流产生举力的量级从正向到负向变化完全根据相对于下一个半次拍击、旋转发生的快慢而定；翅膀的旋转延缓到新的拍击方向开始，产生负向举力。

蜻蜓的翅膀在滑翔过程中起着超轻型风板的作用。横切面观察，翅膀有清晰的波纹产生旋转涡流。波纹可能被认为会导致高阻力值；实际上，空气掠过翅面的阻力很低。充填进侧面翅谷的涡流平缓地掠出几何侧面，产生的低阻力与吹过平板的阻力相似（Kesel, 2000），而上举力远高于平板的预期阻力。

15.6.4 Control of pitch and twisting of wings

Pitch and twisting of the leading and trailing edges of the wings during a wing stroke are important in generating lift forces. The basalar and subalar muscles inserted directly on the wing hinges are mainly involved in controlling the wing angle during a wing stroke. The basalar muscles attached to hinge points in front of the pleural process pull down (pronate) the leading edge of the wings in a downstroke, while the subalar muscles attached to hinge points behind the pleural process supinate or pull down the posterior edge of the wings and cause the leading edge to tilt upward during an upstroke. These movements cause air to flow faster over the upper surface of the wing than over the lower surface and a lift force is produced. Thrust forces push the insect forward through the air, and the wing tip tends to describe a figure "eight" during flight.

In flies, bees, and wasps, a different mechanism for pitch and twist of the wings occurs because of the way in which the thorax is constructed. The tergum of the thorax is divided by the scutal cleft into two plates, the scutum and the scutellum. Sclerites at the posterior of the wing base join to the posterior plate (the scutellum). When the scutellum is pulled slightly forward by the contraction of the dorsal longitudinal muscles, the wing tilts slightly forward, or pronates, as it pivots down over the pleural process. Relaxation of the dorsal longitudinal muscles at the end of the downstroke allows the tergal plates to move further apart, and the wing tilts slightly backward (supinates) as it is forced upward by the contraction of the dorsoventral muscles.

Prey interception is a behaviour that is common to many insects, and is functionally similar to reaching. Because prey make unexpected turns, the predator's interception course cannot be pre-planned, and the control of steering has been viewed as purely reactive. Dragonflies in particular are thought to use a classic interception strategy known as parallel navigation. In parallel navigation, the pursuer steers so as to steadily reduce the length of the range vector from pursuer to prey while holding its direction constant (Mischiati et al.,2015).

15.6.5 Leg muscles and jumping movement

The main muscle in the femur or jumping leg of a grasshopper or locust is composed of many short muscle fibers with origins on the epicuticle of the femur. The muscle fibers insert on the tendon

15.6.4 翅的俯仰角和扭转的控制

在翅的拍击过程中，其前后缘的俯仰角和扭转的控制对上举力的产生是重要的。前上侧肌和翅下肌直接插入翅枢纽，在拍击过程中主要参与翅的角度控制。前上侧肌在侧翅突前方与支点相连，翅下行冲程使翅的前缘向下（前旋），而翅下肌与侧翅突后方支点相连，使翅的后缘向后旋或向下，导致上升冲程中翅前缘向上倾斜。这些运动在翅面产生的气流比翅腹面的气流更快，上举力由此而生。爆发的力量通过空气推动昆虫向前，飞行时翅尖的运动轨迹就像“8”字形。

在蝇类、蜜蜂和黄蜂中，存在一种不同俯仰角和扭转控制机制。这种途径是由胸的构造所决定的。胸背板被盾沟分成两片，盾片和盾盖。翅基后部的骨片与后部的盾板（盾盖）连接。当背纵肌收缩时，盾盖被轻微的向前推动，使翅的倾斜度轻微向前或前旋。此时，以侧翅突为支点向下飞。向下冲程结束，背纵肌舒展时，侧板进一步移动分开，翅的倾斜度轻微向后（后旋），此时背腹肌收缩促使昆虫上飞。

拦截猎物是许多昆虫常见的行为，在功能上类似于到达。因为猎物会做出意外的转弯，捕食者的拦截路线无法预先计划，而转向控制被视为纯粹的反应。蜻蜓尤其被认为使用了一种称为平行导航的经典拦截策略。在平行导航中，追踪者进行转向，以便在保持方向不变的情况下，稳步缩短从追击者到猎物的距离矢量长度（Mischiati 等，2015）。

15.6.5 足肌与跳跃运动

蚱蜢和蝗虫的腿节或跳跃足的主要肌肉由很多短的肌纤维组成，起源于腿节的上表皮。肌纤维从胫节插入肌腱或

or apodeme from the tibia that runs the length of the femur. Thus, each muscle fiber shortens very little, but the many fibers along the length of the long femur apply tension that straightens the leg and propels the locust into the air in a jump. Furthermore, the long femur and tibia of the metathoracic leg give the locust something of a "pole-vaulting" advantage.

表皮内突，贯穿整个腿节。这样，每个肌纤维收缩一点点，但长腿节上的许多肌纤维的张力使腿伸直，促使蝗虫跳入空中。此外，后胸的长腿节和胫节给予蝗虫一些"撑竿跳"的优势。

Very massive muscles whose fibers must shorten only a small percentage of the muscle length are a common adaptation in fleas that make extraordinary jumps. In preparation for a jump, a flea utilizes a cocking mechanism in which the femur is pulled up to overlap the large coxa. In this cocked position, both coxa and femur are perpendicular to the body axis and to the substrate. The cocking action compresses an arch of resilin in the pleural arch near where the coxa is attached to the thorax. In addition, the cocking action clamps the three thoracic segments together by engagement of "catch" mechanisms between the three thoracic segments. The sternum of the mesothorax, in particular, is rigidly held against the metathorax, and this allows the large muscles involved in raising the femur and locking the catches to relax, resisting tiring and reducing energy requirements. In this crouched position, a flea is poised for its leap with trochanter and tibio-tarsal joints and most of the tarsi resting on the substrate. The flea propels itself forcefully into the air by relaxing the elevator muscle holding the femur in the perpendicular position and relaxing the ventral longitudinal muscles holding the catches in the cocked position. The compressed resilin, similar to a compressed spring, forcefully drives the trochanter against the substrate to provide the initial leverage off the substrate. Powerful muscles extend the joint between the trochanter and the femur, and extend the leg, providing further thrust as the terminal tarsal segments and tarsal claws press against the substrate.

跳蚤共同的适应是非常多的肌肉中仅小部分肌肉必须收缩，使得它们做出非同寻常的跳跃。准备跳跃时，跳蚤使用压簧杆机制，其腿节拉起与基节重叠。在这个压簧杆位置，基节和腿节都和体轴及基质成直角。压簧杆动作压缩了基节与胸部连接处附近的侧板弓内的节肢弹性蛋白。此外，压簧杆动作通过三个胸节之间的"闭锁"机制将三个胸节夹紧在一起。中胸的胸片很特别，牢牢地控制着后胸，这允许参与腿节上举的大肌肉和闭锁的锁定状态放松，抵抗疲累和降低能耗。在这个蹲踞姿势，跳蚤用其转节和胫跗关节以及基质上静息的多数跗节使之处于跳跃的准备状态。通过舒展垂直状态的腿节上的提肌和处于压簧杆状态、保持闭锁的腹纵肌，跳蚤强有力地跳入空中。压缩的节支蛋白就像压缩弹簧，强力驱动转节抵住基质，并提供离开基质的初始动能。延伸到转节和腿节间的强壮肌肉和伸展的腿，在跗节与跗爪压向基质末端提供进一步的爆发力。

15.6.6 Tymbal and stridulatory muscles

15.6.6 鼓室和摩擦发音的肌肉

Tymbals are thin, often ribbed patches of cuticle in male cicadas (Homoptera), some other homopterans, some Hemiptera, and some Lepidoptera that are used for sound production and communication. The large tymbal muscle, the sound-producing muscle, is anchored on the sternal cuticle and inserted dorsolaterally on the tymbal membrane, sometimes by means of a thickened rod of chitin. Contraction of the muscle is initially isometric against the resistance of the convex tymbal, but suddenly the tymbal membrane buckles inward into an unstable, concave shape with a loud clicking sound. The inward buckling of the tymbal releases the load on the tymbal muscle and it stops developing tension and shortening. The sudden reduction of muscle tension allows the

鼓室是雄性蝉（膜翅目）及其他一些膜翅目昆虫、半翅目、鳞翅目昆虫用于发声和通信的表皮质薄片。大的鼓膜肌是产生声音的肌肉，锚定在胸骨表面，另一端有时借助于加粗的几丁质束从背侧面插入鼓膜。肌肉的收缩相对于凸起的鼓室而言，起初是等长度的，但突然鼓膜内陷成不稳定的凹形，发出喀喇音。内陷释放了鼓膜肌的负荷，使紧绷的张力和收缩停止。肌肉张力的突然减弱，鼓膜表皮的自然弹性使之回复到

natural elasticity of the tymbal cuticle to click back into the resting convex shape. Sounds may be produced by either, or both, inward and outward movement of the tymbal membrane. Some species have ribs in the convex tymbal surface that buckle progressively inward from posterior to anterior, producing a series of sound pulses as the successive ribs buckle, although only a single muscle twitch is involved.

Cicada tymbal muscle is innervated by a single motor axon with multiterminal branches to all the muscle fibers in the muscle. This sound is produced by a pair of tymbals, each of which is buckled by a large tymbal muscle. The frequency of these oscillations is strongly temperature dependent, and at higher temperatures approaches the normal operating frequency of the muscle. In part, the rapid contractions are enabled by very short twitch duration times in the fastest synchronous muscles. However, a contraction rate of 550 Hz in a synchronous muscle means that the nerve must repolarize and be ready to fire again in less than 2ms. The muscle may be delicately balanced on the edge of oscillatory instability, and may possibly go into oscillatory behavior (Josephson et al., 1985). The inward click of the tymbal cuticle releases the load against which the muscle works, and the natural cuticular elasticity that causes it to click back into the convex shape could then stretch and reintroduce the load to the activated muscle, inducing another contraction if the concentration of free calcium ions is still high enough in the cytoplasm. Thus, these tymbal muscles may work like the asynchronous wing muscles.

Another sound-producing mechanism, stridulation, involves the rubbing together of two parts of the body. Stridulation is widespread among insect orders, with examples known from Odonata, Orthoptera, Coleoptera, Mecoptera, Lepidoptera, Siphonaptera, and Hymenoptera. As in the tymbal muscles of some cicadas, synchronous muscles involved in stridulation by some katydids (Orthoptera: Tettigoniidae) show unusually high contraction frequencies. Male katydids stridulate to attract females by rubbing their forewings (mesothoracic wings) together. In the mesothoracic segment, the same muscles used to power the downstroke of the wings during flight, the mesothoracic dorsal longitudinal muscles (DLM), are used in stridulation. The wing-stroke frequency of the DLM in the mesothoracic segment during stridulation of a katydid, *N. robustus* (orthoptera Tettigoniidae), has been measured at 200 Hz. The DLM in both segments are used in flight, but then the wing-stroke frequency is 20 Hz. The tropical bush cricket *H. unicolor* (orthoptera: conocephalidae) produces a song with a stridulatory frequency of 320-415 Hz using synchronous muscles.

(Li Wenchu)

静息时的凸面。声音即是鼓膜的内外运动之一或同时运动产生的。有些种群在鼓膜凸面上有棱纹，从后向前逐步朝内弯曲，尽管仅牵涉单个肌肉的震颤，当连续的棱纹弯曲时也会产生一连串声音脉冲。

蝉的鼓膜肌受一个简单的运动神经轴控制，肌肉中所有肌纤维都分布有多终板神经分支。声音是由一对鼓膜发出的，每个鼓膜都由一大型鼓膜肌弯曲。这些振荡频率强烈地依赖温度，较高温才使肌肉到达正常的运行频率。最快的同步肌通过极短的颤动持续时间能快速收缩。但是同步肌中 550 Hz 的收缩频率意味着神经必须在低于 2 ms 内复极化并准备再传导。振荡不稳定性在肌肉边缘可能很好地被平衡了，并可能进入振荡行为（Josephson 等，1985）。鼓膜表皮向内的吸气声释放了对肌肉运作的负载，导致鼓膜倒吸气回到凸形，从而弹出并再引入负载到活跃肌，如果肌质中 Ca^{2+} 浓度仍然很高即产生另一个收缩。因此，这些鼓膜肌可能像异步翅肌一样工作。

摩擦发声是产生声音的另一种机制，牵涉身体的两个部分同时摩擦。摩擦发声在昆虫目间普遍存在，已知的例子来自蜻蜓目、直翅目、鞘翅目、长翅目、鳞翅目隐翅目和膜翅目。就像有些蝉的鼓膜肌一样，有些蛸蛉（直翅目：螽斯科）的同步肌参与摩擦发声，不同寻常地表现出高收缩频率。雄性蛸蛉通过一起摩擦前翅（中胸的翅膀）吸引雌性。在中胸体节中，在飞行时用于发动翅膀下行冲程的相同的背纵肌（DLM）被用于摩擦发声。在蛸蛉摩擦发声过程中，中胸体节背纵肌的翅拍击测得的频率为 200 Hz。体节的背纵肌用于飞行，但其拍击的频率是 20 Hz。素色拟织螽（直翅目阜螽科）用同步肌可产生摩擦频率 320 ～ 415 Hz 的歌声。

（李文楚）

Exercises

1.How many types of insect muscle?

2.What are the chemical components of insect muscle?

3.What are functions of actin and actomyosin in muscle contraction?

4.What do you think about “Calcium and ATP are the source of life?”

5.Discuss the wing source theory, which one you feel the best fit to the truth?

思考题

1. 昆虫肌肉有多少种？

2. 昆虫肌肉的化学成分有哪些？

3. 肌动蛋白和肌球蛋白在肌肉收缩中的作用是什么？

4. 你如何看待“钙和 ATP 是生命之源？”

5. 讨论翅膀起源学说，你觉得哪一个最贴合真实？

第 16 章 生殖及其生物化学

Chapter 16 Reproduction and Its Biochemistry

Key points

1.The unique structure of the internal organs of the insect reproductive system provides a place for cell division and nutrients for the germ cells.

2.Ovary development and synthesis of egg proteins arc controlled by hormones JH and ecdysone, together with other peptides.

3.Sequestering of Vgs and YPs by oocytes is the essential for eggs development, as well as important of spermatogenesis and maturation of sperm.

4.The diversity of insect reproduction enables insects to evolve into the largest animal population on the earth.

知识点

1. 昆虫生殖系统内部器官结构的独特性为生殖细胞提供了细胞分裂场所和营养物质。

2. 卵巢发育和卵蛋白的合成受激素 JH 和蜕皮激素以及其他肽的控制。

3. 卵母细胞螯合 Vgs 和 YPs 是卵发育所必需，对精子发生和精子成熟同样重要。

4. 昆虫繁殖的多样性使昆虫进化成地球上最大的动物群体。

Insects display great diversity in modes of reproduction. Most insects reproduce in the adult stage by laying eggs, but a few produce gametes and reproduce during an immature stage, aprocess known as paedogenesis. Some insects such as aphids and flies give birth to live young. Although sex reproduction by union of male and female gametes is typical, certain insects reproduce some or all of the time by laying unfertilized eggs, called parthenogenesis.

With the diversity in insect life histories and ecology, it represents in the details of mating behavior, endocrine regulation of egg development and pheromone production, and the physical structures associated with reproduction as well. For example, the external genitalia of the silkworm male moth is derived from the 9th and 10th ventral segments of the larva, and the external genitalia of the female moth is derived from the 8th, 9th and 10th ventral segments. During mating, female moths stretch out the ovipositor, and the sex pheromones (silkworm alcohol and silkworm aldehyde)

昆虫的生殖模式表现出很大的多样性。在成虫期，大多数昆虫的生殖是产卵，但少数昆虫产生配子，在未成熟阶段生殖，常称为幼体生殖。有些昆虫如蚜虫和苍蝇为胎生。尽管一般是雌雄个体配子的两性生殖，但某些昆虫产生部分或全程不产生受精卵称孤雌生殖。

昆虫的生活史和生态的多样性还表现在交配行为、卵发育的内分泌调控和信息素产物以及生殖相关的生理结构。如家蚕雄蛾外生殖器由幼虫的第 9 腹节、第 10 腹节变成，雌蛾的外生殖器由第 8 腹节、第 9 腹节、第 10 腹节衍生而来。交配时，雌蛾伸出产卵器，由诱惑腺释放出性信息激素（桑蚕醇和桑

released from the temptation gland lure male moths. Fertilized eggs can be produced within 1.5 to 2 hours of mating. A female moth lays about 400-700 eggs, most of which are laid on the day of eclosion and end on the third day.

蚕醛）引诱雄蛾，交配 1.5 ～ 2 h 即可产生受精卵。一头雌蛾产卵 400 ～ 700 粒，其中绝大多数在羽化当日产下，至第 3 天结束。

16.1 雌性生殖系统

16.1 Female Reproductive System

16.1.1 Structure of ovaries

The general anatomy of the female internal organs could be consisting of the paired ovarioles, lateral oviducts, common oviduct, spermatheca, and accessory glands. The internal reproductive structures of both sexes are located dorsally to the alimentary tract. Each ovary consists of one to many ovarioles, with each ovariole containing a string of egg-shaped chambers called follicles. A follicle is separated from the preceding one by a constriction and a bit of interfollicular tissue. Each ovariole is typically enclosed in an epithelial sheath of variable structure in different insects (Fig.16-1).

The number of ovarioles per ovary varies in different species, and even to some extent within a species. Viviparous Diptera have one or two; the American cockroach *P. americana* and Lepidoptera has eight; *D. manenogaster* has from 10 to 30; the blowfly *C. ery throcephala* has about 100; and termite queens may have up to 2,000. Smaller *A. aegypti* have only 30, while bigger one up to 175. Just one ovary containing one ovariole occurs in some aphids, and Collembola have sac-like ovaries that do not contain ovarioles.

Insect ovaries and ovarioles depend on whether there are nurse cells associated with the developing egg (meroistic ovarioles) or no nurse cells (panoistic ovarioles). Meroistic ovarioles can also be divided into two types: polytrophic and telotrophic. Panoistic ovarioles are considered to be the earliest type to evolve, and occur in present-day Thysanura, Odonata, Plecoptera, Dictyoptera, and Isoptera.

Meroistic ovarioles occur in most of the Holometabola (except Siphonaptera), and in Hemiptera, Dermaptera, Psocoptera, Anoplura, and Mallophaga in Hemimetabola. Meroistic ovaries

16.1.1 卵巢的结构

雌性内部生殖器官在解剖学上主要由成对的卵巢管、侧输卵管、共通区、受精囊和附腺等部分组成。两性生殖器官均位于背部延伸至消化道。每个卵巢有 1 至数根卵巢管，每根卵巢管含有一串卵形的小室称为滤泡。滤泡细胞通过括约肌和小束滤泡间组织与前一个分开。不同的昆虫的每个卵巢通常由不同的上皮鞘结构包裹（图 16-1）。

每个卵巢的卵巢管数量因品种而异，甚至种内也有一定程度的变化。胎生双翅目昆虫只有 1 根或 2 根；美洲大蠊和鳞翅目昆虫有 8 根；黑腹果蝇有 10 ～ 30 根；丽蝇有约 100 根，白蚁蚁后（等翅目）多达 2 000 根。伊蚊个体小的仅 30 根，个体大的多达 175 根。有些蚜虫一个卵巢刚好有一根卵巢管，弹尾目有类似气囊的卵巢，无卵巢管。

昆虫卵巢和卵巢管根据是否有滋养细胞与发育中的卵相连可分为有滋式或无滋式。有滋式卵巢管也可以分为两类：多滋式和端滋式。无滋式卵巢管被认为是进化的早期类型，出现在现代缨尾目、蜻蜓目、襀翅目、脉翅目和等翅目中。

有滋式卵巢管出现在大多数完全变态昆虫（蚤目例外）和半翅目、革翅目、啮虫目、虱目以及不完全变态昆虫的食毛目中。有滋式卵巢管的滋养细胞排列

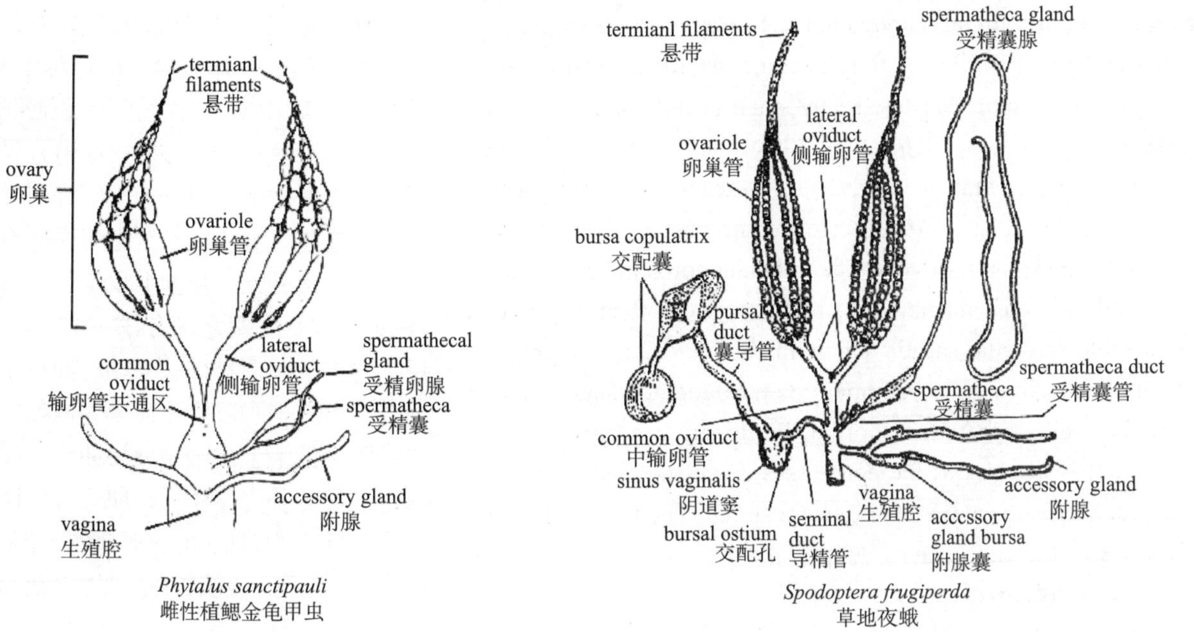

Fig.16-1 Diagram of female internal reproductive organs of *Phytalus sanctipauli* (Coleoptera, Scarabaeidae) and *Spodoptera frugiperda* (Lepidoptera, Noctuidae) (modified from other authors).
图 16-1 雌性植鳃金龟甲虫（鞘翅目 金桂科）、草地夜蛾（鳞翅目 夜蛾科）内部生殖器官图解（仿其他作者）。

contain nurse cells arranged in one of two ways: ① in polytrohic ovarioles. Each oocyte is closely associated with nurse cells in its follicle. The nurse cells may be present in the follicle containing the developing oocyte as in higher Diptera, or they occupy a separate follicle adjacent to that of the developing oocyte. ② in telotrophic ovarioles. The nurse cells are located at the distal apex of the ovariole, in the germarial region, and long connecting nutritive chords extend from the nurse cells to each developing oocyte. Each ovariole terminates distally in a thin, slightly elastic filament that attaches to the dorsal diaphragm or dorsal cuticle. Frequently, terminal filaments from all ovarioles fuse into a suspensory ligament that is likewise attached. Proximally each ovariole connects with the lateral oviduct, and the two lateral oviducts join the common or medial oviduct as a passage for the eggs to the outside.

方式主要有两种：①多滋式卵巢管。每个卵母细胞在滤泡内与滋养细胞密切相连。在高等双翅目昆虫中，滤泡中的滋养细胞含有发育中的卵母细胞或单独占据滤泡与发育中的卵母细胞连接。②端滋式卵巢管的滋养细胞位于卵巢管顶部的原卵区内。从滋养细胞伸出一条长长的营养索与每个发育的卵母细胞相连。每根卵巢管的顶端终止于一薄而略有弹性的纤丝与背膈或背部表皮连接。通常，所有从卵巢管而来的终纤丝融合成一根悬韧带。每根卵巢管就近与侧输卵管相连，两根侧输卵管进入共通区或中输卵管，成为卵产出的通道。

16.1.1.1 Panoistic ovarioles

Panoistic ovarioles do not have nurse cells. The paniostic ovary in *T. domestica* is a well-studied example of a panoistic ovary. Each ovary is composed of only five ovarioles, with each ovariole enclosed in a thin, largely membranous, epithelial sheath containing occasional nuclei. The distal germarial region contains oogonia that can undergo mitosis to produce additional oogonia. Proximal to the germarium, and in the area nearest the

16.1.1.1　无滋式卵巢管

无滋式卵巢管没有滋养细胞，缨尾虫是研究较为透彻的昆虫。其每个卵巢仅有 5 根卵巢管，每根卵巢管被包裹在薄而大的膜状上皮鞘细胞内，偶有细胞核。顶部原卵区含可进行有丝分裂产生另外的卵原细胞。最靠近原卵区和原卵区内附近区域的是还没有单串排列的未

germarium, are young oocytes not yet arranged in a single file, although more proximally they become arranged in to a single string of developing oocytes. Many small prefollicular nuclei are present in a common cytoplasm, but later these nuclei acquire cell boundaries and arrange themselves around each developing oocyte as a follicular epithelium. Interfollicular tissue separates each follicle from the next above it. The follicular epithelial cells secrete a noncellular, thick membranous tunica between them selves and the outside epithelial sheath. The terminal oocyte, sequesters yolk and increases in size to become a mature oocyte. The follicular epithelial cells secrete the chorion. Fertilization occurs after the eggs shell is in place as the egg passes down the median oviduct and past spermatheca where sperm are stored. The final maturation divisions of the egg nucleus do not take place in most eggs until the egg has been ovulates.

成熟卵母细胞，尽管在更近的地方，它们已被排列成一串发育中的卵母细胞。共同的胞质中有许多小的原滤泡细胞核，但随后这些细胞核获得细胞分界并围绕在发育中的卵母细胞周围排列形成滤泡上皮层。滤泡间组织把每个滤泡从前一个分离开来。滤泡上皮细胞在其自身和外部上皮鞘之间分泌一种无细胞层、厚厚的膜质被膜。终端的卵母细胞螯合卵黄并增加其大小而成为成熟的卵母细胞。滤泡上皮细胞分泌卵壳。受精发生在卵的外壳就位后，卵穿过中间输卵管和经过储存精子的受精囊时。最后的卵核成熟分裂，大多数在产卵后才进行。

16.1.1.2 Telotrophic ovariloes

There are nurse cells located in the germarial region of telotrophic ovarioles, and long cytoplasmic cords or strands extend from the germarium to the developing oocytes. Nutrients and maternal gene products pass down the cytoplasmic cords. The milkweed bug *O. fasciatus* has been intensively studied as a model of an insect with a telotrophic ovary. The germarial region at the distal end of each of the seven ovarioles in each ovary is primary occupied by the apical trophic tissue. These cells send long cytoplasmic strands to each developing oocyte through which nutrients and genetic messages pass to the oocytes. Proximal to the apical trophic tissue are young oocytes. All of which are produced during the nymphal stage. No additional oocytes are produced in the milkweed bug during the adult stage.

Each ovariole is covered by an inner envelop composed of a single layer of cells and an outer thin syncytial epithelial sheath. As the terminal oocyte mature, these two outer layers become stretched very thin. A single layer of follicular epithelium is arranged around each oocyte, and as the oocyte enlarges and grows toward maturity, the follicular epithelial cells first become large, binucleate rounded cells, and finally become squamous. Prior to the egg moving down the oviduct, the chorion is secreted by the follicular epithelium.

16.1.1.2 端滋式卵巢管

端滋式卵巢管的原卵区有滋养细胞，原卵区和发育中的卵母细胞通过长细胞质索相连。营养和母体基因产物沿细胞质索传送。乳草虫是研究较为透彻的端滋式模式昆虫。每个卵巢 7 根卵巢管的各顶端的原卵区起初充满顶端营养组织。这些细胞发出的长细胞质索通向每个发育中的卵母细胞，输送营养和基因信息。靠近顶端营养组织的是未发育的卵母细胞，这些卵母细胞在若虫期进行发育。乳草虫在成虫期不产生其他的卵母细胞。

每根卵巢管都被一层由单细胞和多核鞘细胞组成的内层包膜所包裹。当末端卵母细胞成熟时，这两个外层伸展得非常薄。单层滤泡上皮围绕每个卵母细胞，随着卵母细胞的增大和逐渐成熟，滤泡上皮细胞起初变成大的双核圆形细胞，最后变成鳞状。卵在移向输卵管之前，其滤泡上皮细胞分泌卵壳。

16.1.1.3 Polytrophic ovarioles

In polytrophic ovarioles, the nurse cells are located in the egg chambers, either in the same egg chamber or in an adjacent chamber. The earwig *A. maritime* has five ovarioles in each ovary, with each ovariole enclosed in a syncytial sheath. Each

16.1.1.3 多滋式卵巢管

多滋式卵巢管的滋养细胞位于卵室，或者与卵母细胞位于同一个卵室，或者位于相连的区室。蠼螋的每个卵巢有 5 根卵巢管，每根卵巢管包被在多核

ovariole consist of a terminal filament, a germarium, and string of egg chambers. Each follicle contains one developing oocyte and one trophocyte or nurse cell. Initially, the trophocyte is the larger of the two cells in a follicle, but this change as the oocyte mature. The cells of the follicular epithelium divided by mitosis to accommodate the need for more cells to surround the growing oocyte during the previtellogenesis growth; but later during vitellogenesis, the follicular epithelial cells mainly grow and change shape, becoming more thin and squamous as they stretch to cover the enlarging oocyte.

Many insects have multiple trophocytes per oocyte: Lepidoptera typically have 5 nurse cells per oocyte, the diving beetle *D. marginalis* has 15, the gyrinid beetle *D. nigrior* has 7, *D. melangater* and higher Diptera have 15, and the honeybee A. *mellifera* has 48 nurse cells per oocyte (Fig.16-2).

16.1.2 Nutrients from oogenesis

The availability of nutrients during oogenesis is a major limiting factor in the ability of an insect to successfully reproduce. In addition, mating and physical activity, such as flight, influence the physiological availability of nutrients in many insects. Mating is

的外层上皮鞘细胞内。每根卵巢管由终端纤丝、原卵区和一串卵室组成。每个滤泡中含有 1 个发育中的卵母细胞和 1 个滋养细胞。起初，滋养细胞是卵室内两个细胞中较大的，但随着卵母细胞的成熟发生了变化。滤泡上皮的细胞通过减数分裂，在卵黄发生前期累积更多的细胞，包围在成长中的卵细胞周围；但在其后的卵黄发生期，滤泡上皮细胞大部分增长和改变形状，当它们伸展以覆盖变大的卵细胞时，变得更薄和形成鳞状。

许多昆虫每个卵母细胞有多个滋养细胞：通常鳞翅目昆虫每个卵细胞有 5 个滋养细胞，龙虱有 15 个滋养细胞，甲虫有 7 个滋养细胞，果蝇和高等双翅目昆虫有 15 个滋养细胞，而蜜蜂每个卵母细胞有 48 个滋养细胞（图 16-2）。

16.1.2 卵黄发生的营养物质

昆虫成功繁殖能力的主要限制因素是卵黄发生过程中可获取的营养。此外，很多昆虫交配和生理活动，如飞行会影响营养的生理获取。对于有些雌性昆虫

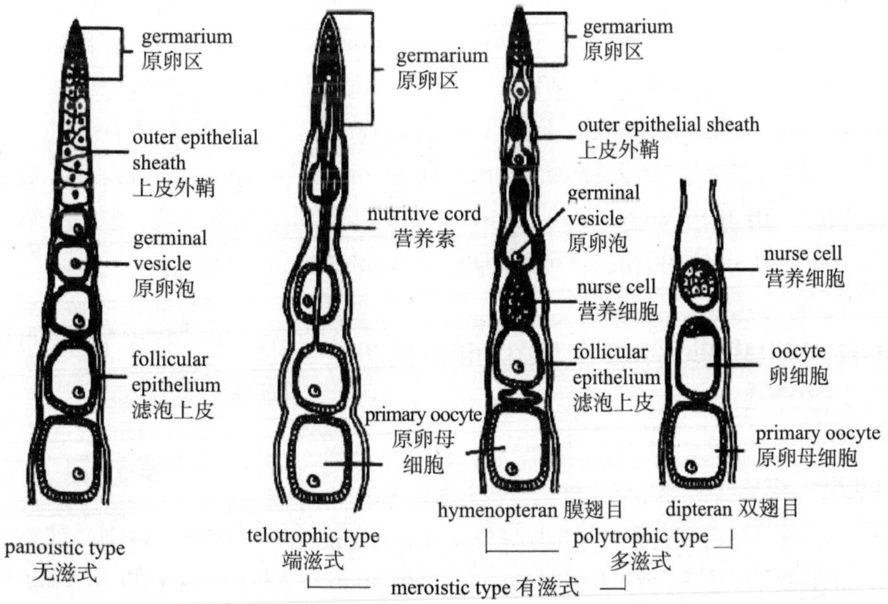

Fig.16-2 Major types of ovary structure in insects (modified from Nation, 2022) .
图 16-2 昆虫卵巢结构的主要类型（仿 Nation, 2022）。

a stimulus that induces mobilization of reserves in some females. The male may make nutrient contributions to the female during courtship and mating by offering nuptial gifts of food, and nutrients may obtained from seminal fluid and a spermatophore.

Among Diptera, the terms of autogenous, the ability to develop a first set of eggs without an exogenous nitrogen source, and anautogenous, the need for a protein or nitrogen source as an adult to develop eggs, are used to describe the requirements for a blood meal to provide nutrients for egg development. Some individuals in all populations are likely to show autogeny. Autogenous individuals have been found in populations of two anautogenous higher dipterans *S. bullata* and *M. domestica*.Some species of mosquitoes are autogenous while others are anautogenous. The autogenous species can mature one set of eggs without a nitrogen source as an adult, subsequent, egg development depends on taking a blood meal. Anautogenous species of mosquitoes need a blood meal in order to develop the first and each subsequent set of eggs. Some parasitic insects produce small, almost yolkless eggs that are deposited in a host where the developing embryo absorbs nutrients from the host through a thin chorion shell.

Physical activity, and specially flight, which demands so much energy, can compete with the ovary for nutrients. Oogenesis generally is inhibited during migratory flights or when lengthy flights must be taken to locate a new host or suitable oviposition site. A new brood of scolytid bark beetle emerging from the log or tree in which they have developed often must fly some distance in seeking a new suitable tree to colonize. When a suitable one has been selected, the wing muscle of some females degenerate, making nutrients available quickly for oogenesis. After the mating flight, queen ants break off their wings, and the wing muscles degenerate to make nutrients available for the first oogenesis cycle.

而言，交配是诱导贮存物代谢的刺激因素。雄虫在求偶和交配阶段可作为雌虫的营养贡献者，通过提供食物给雌虫作婚姻礼物，可从精液和精包中获取营养。

双翅目中的术语“自体发生”是指没有外源性营养来源发育第一批卵的能力；“非自体发生”是指成虫卵的发育必须有蛋白质或营养来源用于描述为卵子发育提供营养的血粉需求。所有群体中有些个体显示出单体生殖。自体发生的个体在两个非自体发生的群体高等双翅目昆虫麻蝇和家蝇中被发现。蚊虫的有些种是自体发生的，而另一些则是非自体发生的。自体发生的品种在成虫期没有外源营养时能成熟一批卵，之后，卵的发育靠吃血粉。非自体发生的蚊虫品种在第一批卵和其后的每一批卵都需要血粉。有些寄生性昆虫产生小而几乎没有卵黄的卵，沉积在宿主体内，当胚胎发育时，通过薄薄的卵壳吸收寄主的营养。

生理活动，特别是飞行，需要非常多的能量与卵发育的营养需求形成竞争。在迁飞过程中，卵黄的发生通常被抑制或漫长的飞行时，必须寻找新寄主或合适的产卵场所。蛀虫树皮甲在它们发育的原木或树上出现后必须飞行一段距离，从而寻找新的树进行繁殖。选择好合适的树之后，有些雌虫的翅肌退化，迅速获取营养进行卵黄的发生。交配飞行完成后，蚁后折断翅膀，翅肌退化，获取第一个卵黄的发生周期所需的营养物质。

16.1.3 Hormonal regulation of ovary development and synthesis of egg proteins

16.1.3.1 Regulation of JH

In Dictyoptera, JH has numerous pleiotropic actions on development and reproduction of cockroaches, including maturation of gonads, production of attraction and courtship pheromones, “calling” behavior and pheromone release, and sexual receptivity. Mating, high-quality nutrition, social interaction, and the presence

16.1.3 卵发育和卵蛋白质合成的激素调控

16.1.3.1 保幼激素的调控

在网翅目中，蜚蠊保幼激素对发育和生殖有数量众多的多效作用，包括性腺的成熟、引诱和求偶信息素的产生、“召唤”行为和信息素释放以及性的接受度。交配、高品质营养、社会性

of vitellogenic ovaries influence JH synthesis by the corpora allata. In Orthoptera, JH is principle hormone that stimulates fat body cells and the ovary, and the adipokinetic hormone from the corpora cardiaca inhibits Vgs mRNA translation at the end of an egg production cycle. JH controls the Vg gene because JH analogs promote transcription of a gene coding for Vg in the fat body, and the mature ovary appears to have feedback to the brain and/or corpora, which regulates down the production of JH until the primary set of eggs is laid. In Diptera, exhibit complex control of reproduction involving multiple hormones. Some mosquitoes, such as *A. aegypti* and *C. pipiens*, are anautogenous and must have a blood meal in order to mature the first set of eggs. Other mosquitoes, including *A. taeniorhynchus*, *A. atropaplus*, and *C. pipiens*, are autogenous. Hormone controls appear to be the similar in both kinds of mosquitoes. Autogeny and anautogeny occurs also in the cyclorrhaphous Diptera.

行为和卵巢卵黄的出现影响到咽侧体合成JH。直翅目JH是刺激脂肪体细胞和卵巢的主要激素，心侧体释放的脂动激素在产卵循环的末期抑制Vgs mRNA的翻译。JH控制了Vg基因，因为JH同系物启动脂肪体内编码卵黄发生的基因。JH也调节发育中的卵母细胞吸收Vg。成熟卵对脑和心侧体有反馈调节作用，在第一批卵产出前下调JH产物。双翅目昆虫表现出对生殖的复杂控制，牵涉到多种激素。有些蚊虫，如埃及伊蚊和尖音库蚊是非自体发生的，为了成熟第一批卵必须吃血粉。其他蚊虫，包括带喙伊蚊和尖音库蚊都是自体发生的。激素对两种蚊虫的控制似乎相同。自体发生和非自体发生殖也出现在环裂双翅亚目中。

The receptor for JH is the basic helix-loop-helix Per/Arnt/Sim (bHLH-PAS) protein Methoprene tolerant (Met) that binds JH at physiological (nanomolar) levels. As a bHLH-PAS domain protein, Met requires dimerization with another bHLH-PAS protein to form a functional transcription factor. The best-studied Met binding partner is the insect homolog of vertebrate steroid receptor coactivator (SRC), variably termed Taiman in *D. melanogaster*, FISC in *A. aegypti*, and SRC in some other insects. A key step in the JH signaling pathway is the nuclear import of its receptor, Met. Ligand bound Met interacts with a chaperone, heat-shock protein 83, which, in turn, interacts with the tetratricopeptide repeat domain at the N-terminal end of the nucleoporin Nup358, a component of the nuclear pore complex, facilitating Met nuclear translocation. In the nucleus, Met and Tai form a JH receptor complex that interacts with target gene promoter DNA elements, termed JH response elements (JHREs; Fig.16-3; Appendix color graphic Fig.16-3). The presence of a JH signaling branch that operates through a membrane receptor, functions by means of phospholipase C, resulting in rapid increases in the levels of inositol triphosphate, diacylglycerol, and intracellular calcium, which, in turn, activate protein kinase C and calcium/calmodulin-dependent protein kinase Ⅱ. These events culminate in the phosphorylation of Met and Tai, necessary for the JH-mediated induction of target gene expression (Roy et al., 2018).

JH受体是碱性螺旋-环-螺旋Per/Arnt/Sim（bHLH-PAS）蛋白质Methoprene tolerant（Met），在生理学（nmol）水平上结合JH。作为bHLH-PAS结构域蛋白，Met需要与另一bHLH-PAS蛋白二聚化形成功能转录因子。研究得最好的Met结合搭挡是脊椎动物类固醇受体辅激活因子（SRC）的昆虫同源物，可变命名在果蝇中为Taiman，在埃及伊蚊中称为FISC，其他一些昆虫中为SRC。JH信号通路中的一个关键步骤是其受体Met的核导入。配体结合的Met与分子伴侣HSP 83相互作用，进而与核孔复合体成分Nup358的N-末端的四肽重复结构域相互作用，促进Met核移位。在细胞核中，Met和Tai形成一个JH受体复合体，与靶基因启动子DNA元件相互作用（称为JHREs；图16-3；附彩图16-3）。JH信号支路通过膜受体，磷脂酶C的功能发挥作用，导致三磷酸肌醇、二酰甘油和细胞内钙水平迅速升高，进而激活蛋白激酶C和钙/钙调蛋白依赖性蛋白激酶Ⅱ。这些事件最终导致Met和Tai的磷酸化，这是JH介导的靶基因表达诱导所必需的（Roy等，2018）。

JH has at least three actions: ① It makes females receptive to mating; ② It stimulus previtellogenic growth of ovaries; ③ It

JH至少有3个作用：①使雌虫接受交配；②刺激卵巢在卵黄发生前期的

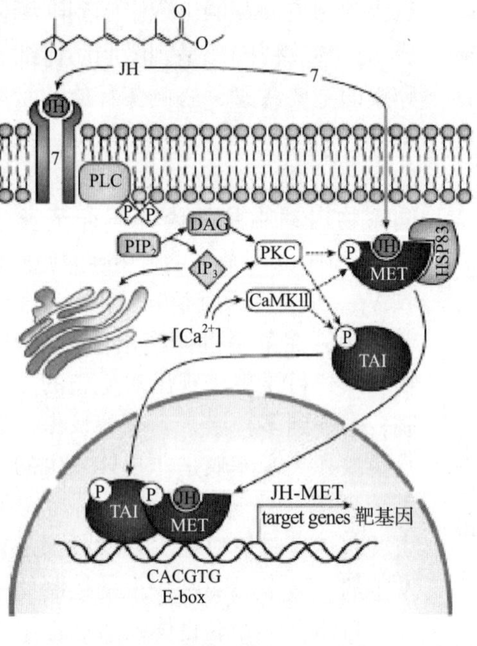

molecular mechanism and signal pathway of JH action
JH 作用的分子机制和信号途径

Fig.16-3 The JH receptor Methoprene tolerant (Met) and its partner Taiman (Tai) are central to JH signaling. In the *Aedes aegypti* mosquito, an additional membrane receptor, working through inositol triphosphate (IP_3), diacylglycerol (DAG), and intracellular calcium ([Ca^{2+}]), activates protein kinase C (PKC) and calcium/calmodulin-dependent protein kinase II (CaMKII), leading to the phosphorylation of Met and Tai. The JH receptor complex binds to E-box elements in the promoter of JH-Met target genes, inducing their expression (modified from Roy et al., 2018).

图 16-3 JH 受体 Methoprene tolerant（Met）及其搭档蛋白 Taiman（Tai）是 JH 信号传导的核心。在埃及伊蚊中，一个额外的膜受体通过三磷酸肌醇（IP_3）、二酰甘油（DAG）和细胞内钙［（Ca^{2+}）］发挥作用，激活蛋白激酶 C（PKC）和钙/钙调蛋白依赖性蛋白激酶 II（CaMKII），导致 Met 和 Tai 的磷酸化。JH 受体复合物与 JH-Met 靶基因启动子中的 E-box 元件结合，诱导其表达（仿 Roy 等，2018）。

prepares the fat body so that it is competent for responding to later hormones and secretion of vitello genins.

成长；③脂肪体准备好与稍后分泌的激素进行竞争并分泌卵黄原蛋白。

16.1.3.2 Regulation of ecdysone

Ecdysteroids, including ecdysone and 20E are also critical developmental hormones in insects that in larvae/nymphs are produced by the prothoracic glands (PGs) and in the absence of PGs in adults, by gonads, epidermis, and oenocytes. Follicles begin to separate from the growing pre vitellogenic ovary, possibly under influence of residual 20E remaining from the pupal to adult transformation. Soon after emerging, females seek a blood meal, which provides proteins, other nutrients, and initiate the vitellogenic stage. With availability of nutrients, the ovary, stimulated by JH, release CCSF (Corpora cardiaca stimulating factor). The target cells for CCSF are in the CC, which release EDNH (egg development neurohormone), recently name ovarian ecdysteroidogenic hormone Ⅰ (OEH). OEH is a polypeptide comprising 86 amino acids, M.W. 8,803 Da. OEH is synthesized in brain medial neurosecretoy cells and stored in the CC. Both humoral and nervous stimuli are important in causing the release of OEH from the CC. Follicular epithelium cells in the ovary respond to OEH by producing and releasing ecdysone into the circulating hemolymph. Ecdysone is converted rapidly

16.1.3.2 蜕皮激素的调控

蜕皮甾类包括蜕皮激素和 20E 也是昆虫的关键发育激素，在幼虫/若虫中，它是由前胸腺（PGs）产生的，而在成虫中缺少 PG，则是由性腺、真皮和绛色细胞产生的。滤泡细胞开始从发育中的卵黄发生前期的卵巢中分离出来，可能受到从蛹期到成虫变态时遗留的 20E 残留量的影响。变为成虫后，雌虫立即搜寻血粉以提供蛋白质及其他营养物质，并启动卵黄发生期。获取营养后，卵巢受到 JH 刺激，释放心侧体刺激因子（CCSF）。CCSF 的靶细胞位于心侧体内，释放卵发育神经激素（EDNH），最近命名卵巢Ⅰ型蜕皮甾类生成激素（OEH）。OEH 是由 86 个氨基酸组成的多肽，分子质量为 8 803 Da。OEH 由脑的中神经分泌细胞分泌，贮存在 CC 中。从 CC 释放 OEH，体液和神经的刺激都很重要。卵巢的滤泡上皮细胞对 OEH 的反应就

to 20E by many types of cells throughout the body, including the target fat body cells. Fat body cells respond to 20E by synthesizing vitellogenins. A receptor for 20E is expressed in fat body and ovary. JH also stimulates Vgs synthesis.

The postvitellogenic stage terminates Vgs production in the fat body when the primary oocytes complement is approaching or has reached maturity. The ovaries release an oostatic hormone, called TMOF in *A. aegypti*, and simply OSH. The OSH stops the uptake of yolk by secondary oocytes, thus stopping their growth until the primary set of eggs is laid. The function of TMOF is to stop the synthesis of late trypsin enzyme in midgut cells, interrupting blood meal digestion and denying secondary oocytes nutrients. The overall function of the OSH is to keep the ovary and abdomen from becoming over distended by too many eggs growing to maturity at the same time. Females now seek an appropriate site for oviposition. Hormones inhibiting host-seeking behavior and stimulating preovipostion seeking behavior are activated by an ovarian releasing factor.

20E acts through its cognate receptor, but its functional receptor is a heterodimer composed of two nuclear receptors, the ecdysone (E) receptor (EcR) and ultraspiracle protein (USP). The 20E-bound EcR-USP heterodimer interacts with the E response element (EcRE), consisting of two hexamers (AGGTCA) separated by a variable number of nucleotides that can form an inverted or direct palindrome. The EcRE-bound 20E-EcR-USP complex activates the early genes *Broad complex* (*BrC*), *E74*, and *E75*, the protein products of which regulate the expression of target genes (Fig.16-4; Appendix color graphic Fig.16-4). *BrC* encodes a family of C2H2-type zinc-finger DNA-binding transcription factors critical for the 20E regulatory hierarchy and required for expression of the other early genes, *E74* and *E75*. The *E74* gene consists of two transcription units, *E74A* and *E74B*. As a result, there are two isoforms of this early gene that belong to the ETS family of transcription factors, and these are involved in regulating several downstream genes in stage-and tissue-specific manners. The nuclear receptor *E75*, also presented as multiple isoforms, has essential roles in the 20E-mediated regulation of development and reproduction. The early-late gene

是产生和释放蜕皮激素进入血液循环。蜕皮激素被遍布虫体的许多细胞快速转换成 20E，包括脂肪体靶细胞。脂肪体细胞对 20E 的响应是合成卵黄原蛋白。20E 的一个受体在脂肪体和卵巢中表达，JH 也刺激 Vgs 的合成。

当初级卵母细胞得到补充或达到成熟时，在脂肪体中卵黄发生的末期终止于 Vgs 产物。卵巢释放一种抑卵激素，在埃及伊蚊中被称为酪氨酸调节抑卵因子（TMOF）或简称为抑卵激素（OSH）。OSH 通过二级卵母细胞停止对卵黄的吸收，因而直到第一批卵产下前停止卵的生长。TMOF 的功能是在中肠细胞停止合成后期的酪氨酸酶，中断取食血粉的消化并阻断二级卵母细胞的营养。OSH 的总体功能是在很多卵同时成长为成熟卵时防止卵巢和腹部避免过度膨胀。此时的雌虫寻找合适的产卵场所。激素抑制宿主搜寻行为和刺激产卵前搜寻行为是由一种卵巢释放因子激活的。

20E 通过其同源受体发挥作用，但其功能受体是由两个核受体组成的异二聚体，即蜕皮激素（E）受体（EcR）和超气门蛋白（USP）。20E 结合的 EcR-USP 异二聚体与 E 反应元件（EcRE）相互作用。该元件由两个六聚体（AGGTCA）组成，由不同数量的核苷酸分开，可形成反向或直接的回文。EcRE 结合到 20E-EcR-USP 复合体激活早期基因 *Broad* 复合体（*BrC*）、*E74* 和 *E75*，其蛋白质产物调节靶基因的表达（图 16-4；附彩图 16-4）。*BrC* 编码一个 C2H2 型锌指 DNA 结合转录因子家族，该转录因子对 20E 调控序位体系至关重要，且是其他早期基因 *E74* 和 *E75* 表达所必需的。*E74* 基因由两个转录单位 *E74A* 和 *E74B* 组成。因此，该早期基因有两种亚型属于 ETS 转录因子家族，它们以阶段和组织特异性方式参与调节几个下游基因。核受体 *E75* 也表现为多种亚型，在 20E 介导的发育和生殖调控中具有重要作用。已知早 - 晚

HR3 is known to activate the competence factor *β*FTZ-F1, which is necessary for a stage-specific response to a subsequent 20E pulse. HR3, an orphan nuclear receptor that has a central role in regulating developmental transitions during the termination stage of vitellogenesis, is expressed during the declining 20E titer at 24 to 36 h post–blood meal and downregulates *Vg* expression.HR3 binding to its cognate recognition site in the *Vg* gene promoter determines the negative effect. A timely cessation of expression

期基因 *HR3* 激活感受态因子 *β*FTZ-F1，这是对随后 20E 脉冲作出阶段特异性反应所必需的。HR3 是一种孤核受体，在卵黄形成终止阶段的发育转变中具有核心的调节作用，在进食血粉后 24 ～ 36 h 20E 滴度下降期间表达，并下调 *Vg* 表达。HR3 与其 *Vg* 基因启动子的同源识别位点结合决定了负效应。及时停止 *Vg* 和其他 *YPP* 基因的表达对

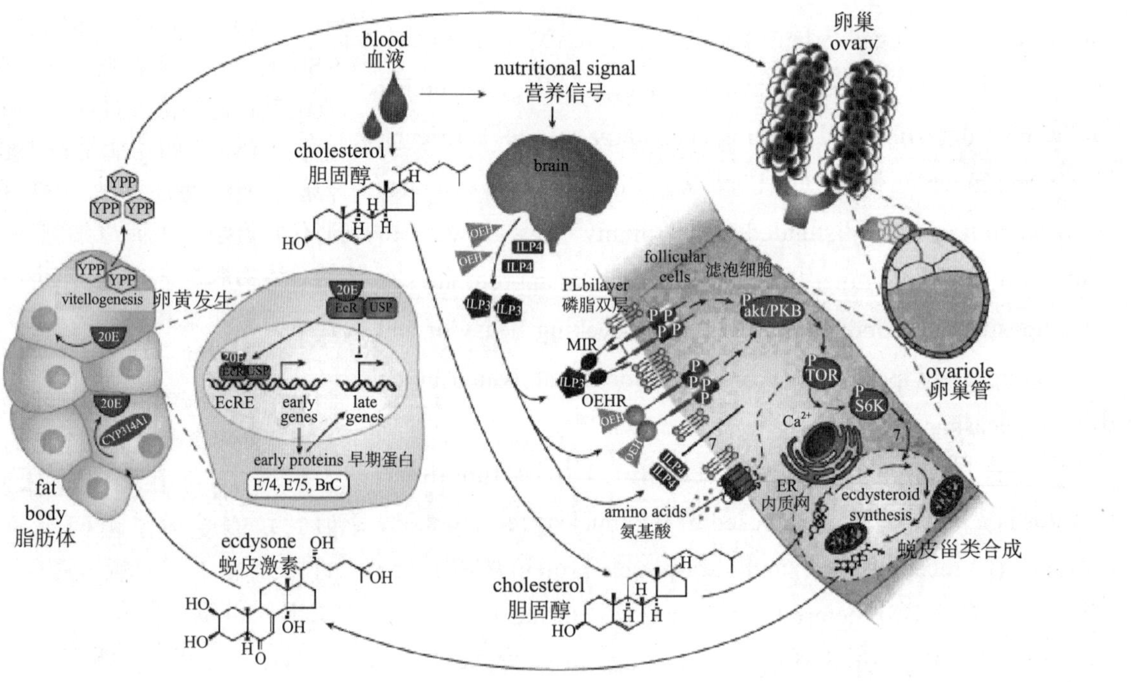

Fig. 16-4 Blood meal–activated nutritional signals trigger the release of the neuropeptide OEH and ILP3 and ILP4 from the brain of *Aedes aegypti* females. OEH binds OEHR; ILP3 binds MIR, and ILP4 binds to an unknown receptor. Phosphorylated OEHR and MIR, along with amino acid signaling, activate insulin and AA/TOR signaling pathways, manifested by phosphorylation of Akt (PKB), TOR, and S6 kinases. These pathways convert cholesterol to ecdysteroids through a multistep process in the ER and mitochondria of the follicle cells. The role of Ca^{2+} flux in ecdysteroidogenesis is important. Ecdysone secreted into the hemolymph is converted by the cytochrome P450 enzyme CYP314A1 to its active form of 20E in the peripheral tissues. Once released into the hemolymph, 20E binds the heterodimeric receptor complex of EcR and USP. Receptor-bound 20E activates the transcription of the early genes, *BrC*, *E74*, and *E75*. The early proteins are involved in the transcriptional regulation of genes encoding for *YPP*, such as Vg, in the fat body. The secreted YPPs are subsequently taken up by oocytes during maturation (modified from Roy et al., 2018).

图 16-4 雌性埃及伊蚊血粉激活的营养信号触发大脑释放神经肽 OEH、ILP3 和 ILP4。OEH 结合 OEHR；ILP3 结合 MIR；ILP4 与未知受体结合。磷酸化 OEHR 和 MIR 与氨基酸信号一起激活胰岛素和 AA/TOR 信号通路，表现为 Akt（PKB）、TOR 和 S6 激酶的磷酸化。这些途径通过卵泡细胞内质网和线粒体的多步骤过程将胆固醇转化为蜕皮甾类。Ca^{2+} 流在蜕皮甾类生成中的作用很重要。分泌到血淋巴中的蜕皮激素被细胞色素 P450 酶 CYP314A1 转化为外周组织中的活性形式 20E。一旦释放到血淋巴，20E 就会结合 EcR 和 USP 的异二聚体受体复合物。受体结合的 20E 激活早期基因 *BrC*、*E74* 和 *E75* 的转录。早期蛋白参与脂肪体中 *YPP* 编码基因的转录调控，如 Vg。分泌的 YPP 随后在成熟过程中被卵母细胞吸收（仿 Roy 等，2018）。

of *Vg* and other *YPP* genes is essential for transition to the second gonadotropic cycle. HR3 is also critical for turning off target of rapamycin (TOR) signaling and initiating programmed autophagy in the fat body, and activates *βFTZ-F1* for progression to the second gonadotropic cycle.

于过渡到第二个促性腺周期至关重要。HR3 对于关闭 TOR 信号转导靶点和启动脂肪体内的程序性自噬，并为第二个促性腺素周期进程激活 *βFTZ-F1*，也是至关重要的。

16.1.3.3 Interaction of JH and ecdysteroids controls reproduction

JH plays a major role in reproduction of cyclorrhaphous Diptera, but dipterans are too diverse a group to expect JH to play a common role. Adult life history and nutrition determine in large part the way in which JH and the neuroendocrine system exert their control over reproduction. JH Ⅲ the bisepoxide (JHB3) as well as JH Ⅲ and methyl farnesoate have been identified in a number of cyclorrhaphous larval and adult Diptera. In the blowfly, *P. regina*, JH influences mating behavior in both sexes, fat body development, ovary growth and development, and pinocytotic uptake of vitellogenin by oocytes, but JH is synthesized by the CA only at a low level until a protein meal is taken by the flies.

The cecropia moth, *H. cecropia*, initiates Vgs synthesis early in the prepupal stage, and the silkmoth, *B. mori*, begins Vgs synthesis in the early pupae stage. In these two moths JH does not seem to have a role, and 20E stimulates Vgs synthesis. 20E stimulates Vgs synthesis in the gypsy moth, in which Vgs synthesis begins late in the last instar, and experimental treatment with JH inhibits Vgs synthesis. Some moths, including some Pyralid moths, use falling ecdysteroids concentrations to induce Vgs synhesis, and eggs mature before eclosion of adults. In the fall armyworm, *S. frugiperda*, both JH and ecdysone promote Vgs synthesis, but sequestering of Vg by the oocytes is under JH control. Only JH induces vitello-genesis synthesis in the monarch butterfly, *D. plexippus*, the moth, *H. virescens*, and in a number of other lepidopterans that begin vitellogenesis after emergence of the adult. Male transfer JH to female *H. virescens* during mating, and mating itself stimulates the CA in females to synthesize JH Ⅱ (and small amount of JH Ⅰ and JH Ⅲ) and causes inhibition of JH esterase that could potentially destroy JH transferred or synthesized.

Vitellogenesis initiation is an important control point in oogenesis that integrates hormonal cues to match female physiological conditions (i.e., nutrition, mating status, etc.). The major gonadotropic hormones in *D. melanogaster* are juvenile hormone (JH) and the insect steroid hormones ecdysteroids (Fig.16-5; Appendix color graphic Fig.16-5). Ecdysteroids stimulate yolk

16.1.3.3 JH 和蜕皮甾类相互作用对生殖的控制

JH 在双翅目环裂亚目昆虫生殖中起着重要作用，但双翅目昆虫种群如此的多样性以至于不能期待 JH 只起着普通作用。成虫的生活史和营养在很大部分决定了 JH 和神经分泌系统对生殖的控制方式。环氧化保幼激素Ⅲ（JHB3）和 JH Ⅲ及法呢烯酸甲酯一样，在很多双翅目环裂亚目的幼虫和成虫中被鉴定。丽蝇 JH 影响两性交配行为、脂肪体发育、卵巢生长发育和卵母细胞对卵黄原蛋白的胞饮吸收，但丽蝇在取食蛋白粉之前，咽侧体仅低水平合成 JH。

惜古比天蚕在预蛹期早期启动 Vgs 的合成，家蚕在蛹期早期开始合成 Vgs。这两种蛾的 JH 似乎没有作用，20E 刺激 Vgs 的合成。20E 刺激吉普赛蛾的 Vgs 合成，其 Vgs 合成始于末龄后期，用 JH 实验处理抑制 Vgs 的合成。有些蛾，包括一些螟蛾，利用降低蜕皮甾类的浓度诱导 Vgs 的合成，卵在成虫羽化前已经成熟。草地夜蛾 JH 和蜕皮激素共同启动 Vgs 的合成，但卵母细胞螯合 Vgs 则由 JH 控制。黑脉金斑蝶、烟芽夜蛾和数种其他鳞翅目昆虫在成虫期后才开始诱导卵黄的发生，只有 JH 诱导 Vgs 合成。雄性烟蚜夜蛾在交配时将 JH 转移给雌虫，交配本身也刺激雌虫 CA 合成 JH Ⅱ（以及少量 JH Ⅰ和 JH Ⅲ），并导致 JH 酯酶受到抑制，这具有破坏转移或合成的 JH 的潜能。

卵黄发生的起始是卵子发生中的一个重要控制点，它整合了激素信号，以符合雌蝇的生理条件（即营养、交配状态等）。果蝇中的主要促性腺激素是保幼激素（JH）和昆虫甾类激素蜕皮甾类（图 16-5；附彩图 16-5）。蜕皮甾类

protein (YP) synthesis in the fat body, while JH stimulates the synthesis and uptake of YP by the ovary. Thus, JH is essential for the continuing development of vitellogenic follicles past stages 8 and 9. As a female fly completes reproductive maturation, JH levels fall gradually and vitellogenesis ceases. Later, mating once more stimulates the oogenesis that is required to sustain robust egg-laying activity in mated females. In *D. melanogaster*, a male-derived seminal fluid protein called sex peptide (SP) elicits oogenesis progression. SP induces germline stem cells proliferation by stimulating the biosynthesis of ecdysteroids (Zhang et al., 2022).

刺激脂肪体中的卵黄蛋白（YP）合成，而JH刺激卵巢对YP的合成和摄取。因此JH对卵黄形成性滤泡持续发育经过第8阶段和第9阶段是必需的。当雌蝇完成生殖成熟时，JH水平逐渐下降，卵黄形成停止。之后，交配再次刺激卵子发生，这是交配过的雌蝇维持强大产卵活动所必需的。在果蝇中，来自雄性称为性肽（SP）的一种精液蛋白引发卵子发生的进程。SP通过刺激蜕皮甾类的生物合成诱导干细胞增殖（Zhang等，2022）。

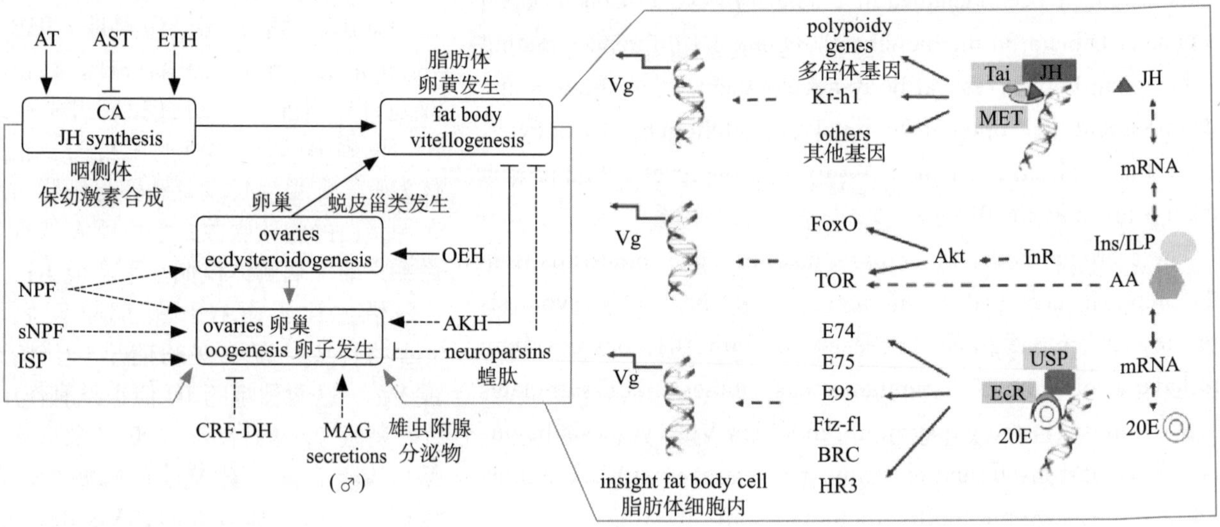

Fig.16-5 Peptide control of insect oogenesis and Vg synthesis in fat bdy. JH receptor complex Met/Tai directly regulates the transcription of target genes encoding Kr-h1 and the polyploidy genes (including DNA replication and cell cycle genes). 20E receptor complex EcR/USP directly activates the expression of 20E-responsive genes.Nutritional signaling is involved in Vg production in the fat body by interplay with JH or 20E. In addition, miRNA participates in JH- or 20E-dominant Vg synthesis in the fat body. These factors directly or indirectly regulate Vg transcription and vitellogenesis. Abbreviations: AT. allatotropin; AST. allatostatin; ETH. ecdysis triggering hormone; JH. juvenile hormone; NPF. neuropeptide F; sNPF. short neuropeptide F; ISP. insulin signalling pathway; CRF-DH. corticotropin-releasing factorlike diuretic hormone; MAG. male accessory gland; OEH. ovary ecdysteroidogenic hormone; AKH. adipokinetic hormone (modified from Lenaerts et al., 2019; Wu et al., 2021).

图16-5 昆虫卵发生的肽类物质控制和脂肪体内Vg的合成。JH受体复合物Met/Tai直接调节编码Kr-h1的靶基因和多倍体基因（包括DNA复制和细胞周期基因）的转录。20E受体复合物EcR/USP直接激活20E应答基因的表达。营养信号通过与JH或20E相互作用参与脂肪体中Vg的生成。此外，miRNA参与脂肪体中JH-或20E显性Vg的合成。这些因子直接或间接调节Vg转录和卵黄发生。AT. 咽侧体促进素；AST. 咽侧体抑制素；ETH. 蜕皮触发激素；JH. 保幼激素；NPF. 神经肽F；sNPF.（短链）神经肽F；ISP. 胰岛素信号通路；CRF-DH. 促肾上腺皮质激素释放因子样—利尿激素；MAG. 雄性附腺；OEH. 卵巢蜕皮甾类生成激素；AKH. 脂动激素（仿Lenaerts等，2019；Wu等，2021）。

16.1.3.3 Peptides control of insect oogenesis

At present, the best documented physiological control mechanism of insect oogenesis is the insulin-related peptide signaling pathway. It regulates different stages of the process and provides a functional link between nutritional status and reproduction. The insulin receptor (InR) is a transmembrane receptor tyrosine kinase, which is capable of activating both the mitogen activated protein kinase (Ras-MAPK) and the phosphatidy-linositol-3-kinase/protein kinase B (PI3K/PKB) pathways. An RNA interference (RNAi) mediated knockdown of InR, TOR or FOXO resulted in impairment of the maturation of the primary oocyte, as well as defective oocyte growth, right after the primary oocytes move to the neck region of the ovariole. Several other peptides, such as neuroparsins and ovary ecdysteroidogenic hormone, allatoregulatory peptides, allatotropins (ATs) and allatostatins (ASTs) , ecdysis triggering hormone (ETH), ecdysiotropic and ecdysiostatic peptides, (Short) neuropeptide F (sNPF), CRF-like diuretic hormone, adipokinetic hormones, and orcokinins etc. have been shown to exert gonado-regulatory activities, their effects on oogenesis could be direct or indirect. Some regulatory peptides, such as the *Drosophila* sex peptide, are being transferred from the male to the female during the mating process (Lenaerts et al.,2019).

The neuropeptide allatostatin-C (AstC)-producing neurons that gate both the initiation of vitellogenesis that occurs posteclosion and its re-initiation post-mating. During sexual maturation, the AstC neurons receive excitatory inputs from sex peptide abdominal ganglion (SAG) neurons. In mature virgin females, high sustained activity of SAG neurons shuts off vitellogenesis *via* continuous activation of the AstC neurons. Upon mating, however, sex peptide inhibits SAG neurons, leading to deactivation of the AstC neurons. As a result, this permits both JH biosynthesis and the progression of vitellogenesis in mated females (Zhang et al., 2022).

16.1.3.4　昆虫卵子发生的肽控制

目前，昆虫卵子发生的生理控制机制最好的记录是胰岛素相关肽信号通路。它调节该过程的不同阶段，并在营养状况和生殖之间提供功能联系。胰岛素受体（InR）是一种跨膜受体酪氨酸激酶，能同时活化丝裂原激活蛋白激酶（Ras-MAPK）和磷脂酰肌醇-3-激酶/蛋白激酶B（PI3K/PKB）通路。RNA干扰（RNAi）介导的InR、TOR或FOXO的下调导致初级卵母细胞的成熟障碍以及在初级卵母细胞转移到卵巢颈区后的卵母细胞生长缺陷。一些其他肽，如神经肽和卵巢蜕皮甾类生成激素、促性腺激素调节肽，即咽侧体促进素（ATs）和咽侧体抑制素（ASTs）、蜕皮触发激素（ETH）、促蜕皮素和促蜕皮激素肽、（短链）神经肽F（sNPF）、CRF样利尿剂激素、脂动激素和亲肌神经肽（Oks）等已被证明具有性腺调节活性，它们对卵子生成的影响可能是直接或间接的。一些调节肽，像果蝇性肽，在交配过程中从雄性转移到雌性（Lenaerts等，2019）。

产生咽侧体抑制素-C（AstC）神经肽的神经元，用于控制羽化后出现的卵黄发生的启动和交配后的重新启动。在性成熟过程中，AstC神经元接受来自腹神经节性肽（SAG）神经元的兴奋性输入。在成熟的未交配过的雌蝇中，SAG神经元的高持续活性通过持续激活AstC神经元从而切断卵黄发生。然而，交配后，性肽抑制SAG神经元，导致AstC神经元失活。因此，这允许JH生物合成和已交配雌性的卵黄发生的进展（Zhang等，2022）。

16.2 卵黄原蛋白和卵黄蛋白

16.2 Vitellogenin and Yolk Proteins

16.2.1 Biochemical characteristic of vitellogenins and yolk proteins

The egg yolk is rich in proteins and lipids. The proteins were sex-specific and found in small amount of the proteins in males of some species, including some lepidopterans, hemipterans, orthopterans and honeybees. In some males, egg proteins can be induced with hormone treatment. Although synthesis of the egg proteins occurs in both fat body and the follicular cells of the ovary in some insects, but major source of the proteins are the fat body in most insects.

The proteins are called vitellogenins (Vgs) while they are being produced by the fat body and during transport to the ovaries by the hemolymph, but after incorporation into developing eggs, the proteins are known as vitellins.

Insect Vgs typically are large glycolipoproteins, from 400-600 kDa, that are composed of large (120-200 kDa) and small (40-60 kDa) subunits. Large insect Vgs contains 7% to 15% lipids, consisting primary of phospholipids and diacylglycerol and apoproteins. Vgs usually exist as dimmers, but monomers are known from the cockroach, *N. cinereae*. Genes controlling synthesis of Vgs have been identified and cloned from a number of insects, including *L. migratoria, A. aegypti*, *A. gambiae* and *D. melanogaster*. The apoproteins of insect and vertebrate Vgs diverged from a superfamily of proteins controlled by genes with an ancient heritage.

Egg proteins of Diptera fall into two classes that split along the lines of the two suborders of Diptera. Lower Diptera in the suborder Nematocera, including mosquitoes and some other dipterans have Vgs similar in structure to those of other insects, i.e. large glycolipoproteins composed of large and small subunits. A major evolutionary shift in gene control of yolk proteins occurred

16.2.1 卵黄原蛋白和卵黄蛋白的生化特征

卵黄富含蛋白质和脂类。其蛋白质是性特异性的，并在某些种包括一些鳞翅目、半翅目、直翅目和蜜蜂的雄虫中被发现数量很少的蛋白质。有些雄蛾经激素处理后可诱导卵蛋白质。在一些昆虫中，尽管卵蛋白质的合成在脂肪体和滤泡细胞完成，但大多数昆虫脂肪体是蛋白质的主要来源。

这些蛋白质在脂肪体产生并通过血液被转运到卵巢的过程中被称为卵黄原蛋白（Vgs），但结合到发育中的卵后，这些蛋白质被称为卵黄磷蛋白。

典型的昆虫卵黄原蛋白是大分子糖脂蛋白，分子质量为 400 ～ 600 kDa，由大亚基（120 ～ 200 kDa）和小亚基（40 ～ 60 kDa）组成。大昆虫 Vgs 含 7% ～ 15% 的脂类，是磷脂和甘油二酯以及载脂蛋白的主要组成成分。Vgs 通常以二聚体形式存在，但是灰色蠊中的是单聚体。控制 Vgs 合成的基因已从几种昆虫中克隆得到，包括蝗虫、埃及伊蚊、冈比亚按蚊和果蝇。昆虫载脂蛋白和脊椎动物 Vgs 是超级蛋白质家族的分支，由远古遗传的基因控制。

双翅目昆虫的卵蛋白沿两个亚目分界线分成两类。低等双翅目在长角亚目，包括蚊虫和一些其他双翅目，具有和其他昆虫相似的 Vgs 结构，即由大小亚基组成的大糖脂蛋白。高等双翅目主要的卵黄蛋白控制基因出现了进化漂移，

in the higher Diptera, and the proteins that go into the yolk are not homologous with the Vgs of other Insects. Consequently, these proteins are not called Vgs, and instead are called yolk proteins. In most insects, vitellogenesis, which is a central event of female reproduction, involves the production and secretion of vitellogenin (Vg) and other yolk protein precursors (YPPs) by the fat body (FB), followed by internalization of YPPs by maturing oocytes through receptor-mediated endocytosis. However, in *Drosophila*, YPPs are also produced by the ovarian follicular epithelium. It has been long established that female insect reproduction is controlled by hormones: neuropeptides, juvenile hormones, and ecdysteroids (Roy et al., 2018).

其进入卵黄的蛋白质与其他昆虫的 Vgs 没有同源性。因此，这些蛋白质不能称为 Vgs，而称之为卵黄蛋白（YPs）。在大多数昆虫中，卵黄发生是雌性生殖首要的事情，它涉及脂肪体（FB）的产生、分泌卵黄蛋白原（Vg）和其他卵黄蛋白前体（YPPs），然后通过受体介导的内吞作用使卵母细胞成熟，内化 YPPs。然而，在果蝇中，YPPs 也由卵巢滤泡上皮产生。长期以来，人们公认雌性昆虫的繁殖受激素控制：神经肽、保幼激素和蜕皮甾类（Roy 等，2018）。

16.2.2 Sequestering of Vgs and YPs by Oocytes

16.2.2.1 Patency of follicular cells

Oocytes take up proteins through channels between the follicular cells. JH acts with membrane receptors to promote Vgs uptake by promoting widening of the intercellular spaces in the follicular epithelium. A Na^+/ K^+ -ATPase is activated and the cells shrink. Phosphatidylinositol and protein kinase C are involved. The opening of spaces between follicular cells is called patency. At the end of vitellogenesis, new junctions between follicle cell channels and protein uptake rapidly falls.

The Vgs and YPs bind to specific receptors at the surface of the oocyte plasma membrane between and at the base of microvilli. The receptor-protein complex tends to sink inward at the oocyte surface, forming a pit, with a clatherin protein coat on the cytoplasmic side. The pits continue to invaginate, close up, and become pinched off as small, coated vesicles (150-190 nm diameter) inside the oocyte. Coated vesicles have been isolated from the ovaries, and the clathrin heavy chains have a M.W. of 180 kDa.

The clathrin coat soon dissociates from the vesicles containing the receptor-protein molecules, and the vesicles are then called endosomes (Fig.16-6). Bound egg proteins dissociate from the receptor, probably as a result of ATP-dependent acidification within the endosome. The clathrin mole- cules and receptors probably recycle to the oocyte surface for reuse.

Endosomes coalesce into a large transitional yolk body, and

16.2.2 卵母细胞对卵黄原蛋白和卵黄蛋白的结合

16.2.2.1 滤泡细胞的开放

卵母细胞通过滤泡细胞间通道吸收蛋白质。JH 与膜上受体结合，通过扩大滤泡细胞间空隙启动 Vgs 的摄取。此时，Na^+/K^+ -ATP 酶被活化，细胞收缩。磷脂酰肌醇和蛋白激酶 C 涉入其中。滤泡细胞间空隙的打开称为开放。卵黄发生的末期，滤泡细胞间的新连接封闭了胞间通道，蛋白质吸收快速下降。

Vgs 和 YPs 与卵母细胞质膜间的表面和基部微绒毛的特异性受体结合。受体 - 蛋白质复合体在卵母细胞表面向内沉积，形成凹陷，质膜一侧形成网格蛋白质包被。继续凹入，凹陷闭合，变成卵母细胞内掐断的包被小泡（直径 150 ～ 190 nm）。包被小泡从卵巢分离，网格蛋白包被的重链分子质量约为 180 kDa。

网格蛋白包被很快与含有受体蛋白分子的小泡解离，此时的小泡称为内体或核内体（图 16-6）。在核内体中，可能是 ATP 依赖性的酸化作用，结合的卵蛋白从受体中解离出来。为了再利用，网格蛋白分子和受体也许会循环到卵细胞表面。

内体合并到一个较大的过渡型卵黄

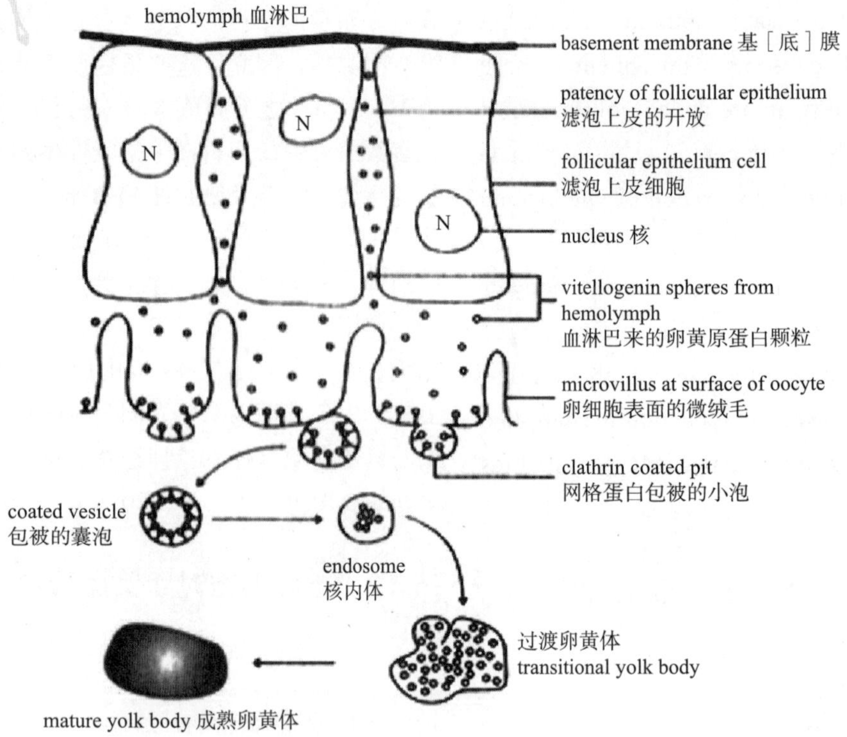

Fig.16-6 The uptake of vitellogenin by a developing oocyte (modified from Raikhel et al., 1992).
图 16-6 发育中的卵母细胞对卵黄原蛋白的摄取（仿 Raikhel 等，1992）。

the egg proteins, now called vitellins (Vns), begin to crystallize. Additional Vns are added to the transitional yolk body until it becomes a mature yolk body. Generally, Vgs and Vns have the same immunelogical properties, chemical composition, and physical properties. Some Vns derived from YPs are sulfated.

体。现在的卵蛋白质称为卵黄磷蛋白（Vns），开始晶体化。其他的 Vns 也加入过渡型卵黄体，直至成为成熟的卵黄体。总体来说，Vgs 和 Vns 具有相同的免疫学特性、化学组成和生理特性。有些从 YPs 衍生而来的 Vns 是硫酸盐。

16.2.2.2 Egg proteins produced by follicular cells

Proteins produced in the ovary, usually by follicular cells lining the follicle, do not enter the hemolymph, but are passed directly into the developing oocytes. In several *Drosophila* species that have been studied, the proportion of the several YPs synthesized in the fat body and in the follicular cells varies with species, but follicular cells are a major source of some YPs. Paravitellogenin is a 70 kDa protein produced in the follicular cells of *H.cecropia* and taken into the oocyte. A similar homologous protein making up to 25% of the egg proteins is produced in the follicular cells of *B. mori*. Proteins produced in the follicular cells of Indian meal moths, *P. interpunctella*, and in several mosquitoes are incorporated into eggs.

16.2.2.2 滤泡细胞产生的卵蛋白

卵巢内产生的蛋白质通常是由滤泡细胞产生的，并不进入血液，而是直接进入发育中的卵母细胞。已经被研究的几个果蝇种的几种 YPs 的大部分是由脂肪体和滤泡细胞合成的，因物种而异，但滤泡细胞是一些 YPs 的主要来源。亚卵黄原蛋白是天蚕蛾滤泡细胞产生并被卵母细胞摄取的 70 kDa 蛋白质。家蚕的滤泡细胞产生的卵蛋白质相似，同源性接近 25%。印度谷斑螟和几种蚊虫滤泡细胞产生的蛋白质都结合进卵。

16.2.2.3 Additional proteins in the egg

Developing eggs often contain varying amounts of proteins that are sequestered from hemolymph in addition to Vgs and YPs. They are not usually considered to be Vgs because they are

16.2.2.3 卵内其他蛋白质

除 Vgs 和 YPs 外，发育中的卵从血淋巴中螯合的不同数量蛋白质。通常，不认为它们是 Vgs，因为其数量少

present in small amounts and/or do not have the general structure of Vgs. Lipophorin, a general lipid transport protein of insects, and microvitellin, a small female-specific protein, are present in eggs in *H. cecropia* and *M. sexta*, respectively. In addition, *M. sexta* eggs contain insecticyanin, a blue biliprotein giving the eggs a pale blue color. A group of storage proteins of about 30 kDa are synthesized during larval stages of both sexes of the commercial silk moth, *B. mori*, and become the dominant hemolymph proteins during larval life and persist into the pupal stage. Substantial quantities are taken into the developing eggs, but they usually not considered being vitellogenins.

和 / 或没有 Vgs 的一般结构特征。载脂蛋白是昆虫的一种脂类运输蛋白；微卵黄磷蛋白质，一种小分子雌性特异蛋白质，分别出现在天蚕蛾和烟草天蛾的卵中。此外，烟草天蛾卵含有虫青素，一种蓝色胆青素使卵具有淡蓝色。两种性别的商品蚕在幼虫阶段都合成一系列贮存蛋白质，分子质量约 30 kDa，在幼虫的生活史中成为主要的血液蛋白质并持续到蛹期。发育中的卵摄取了大量这种蛋白质，但通常也不认为是卵黄原蛋白。

Near the termination of vitellogensis, a thin protein sheet, the vitellin membrane, is secreted at the inner surface of the follicle cells. The vitellin membrane in *D. melanogaster* is composed of numerous proteins ranging from 14-130 kDa that are encoded by a family of genes.

卵黄发生快要结束时会在滤泡细胞内表面上分泌一种蛋白质薄层，称为卵黄膜。果蝇的卵黄膜由多种性质不同的蛋白质组成，分子质量 14 ～ 130 kDa，由一个基因族编码。

16.2.2.4 Chorion

The chorion is composed of a number of sclerotized proteins. It contains no chitin. Sperm, which are released from the spermatheca as the egg passes down the common oviduct, have to enter the egg through a small, usually twisted channel, the micropyle, which passes through the various layers of the chorion. More than one micropyle channel is not uncommon; although most Diptera have only one, *Locusta* has 35 to 43 openings. Follicular epithelial cells secrete chorionic proteins on the outer surface of the vitellin membrane, thus enclosing it on the inside of the chorion. The follicular epithelial cells lay down proteins sequentially, indicative of the sequential expression of a superfamily of genes, to produce a laminar structure. Insect chorion is structured into a number of layers which are conserved in their composition.

Using a Suppression Substractive Hybridization (SSH) library approach, a series of genes were found that are expressed differentially during the postvitellogenic period of *B. germanica*, a hemimtabolan insect with panoistic ovaries, whose reproduction is mainly regulated by juvenile hormone. From the *SSH* library, two genes were especially prominent, *Brownie* and *Citrus*. They were outstanding due to their remarkable abundance, their characteristic pattern of expression and because they had no homologues in insect gene databases. The function of *Brownie* has been recently described as encoding a protein that plays a key role in the formation of the sponge-like body, a complex structure of the eggshell that combines the aeropyle and the micropyle, and which is reminiscent of the *D. melanogaster* egg horns, that permit gas exchange between the egg and the exterior. The most particular

16.2.2.4 卵壳

卵壳是由许多骨质化蛋白质组成的，不含几丁质。当卵经过共通区时，贮精囊释放的精子通过一条小而弯弯曲曲的通道即卵孔，并经过许多层卵壳才能进入卵。多于一个卵孔通道并不罕见；虽然多数双翅目仅有一个，但飞蝗有 35 ～ 43 个开口。滤泡上皮细胞分泌卵壳蛋白于卵黄膜外表面，因而将它密封在卵壳内。滤泡上皮细胞依次分泌蛋白质，表明基因超级家族是有序表达的，从而产生一薄层结构。昆虫的卵壳结构上的层次组成具有保守性。

用抑制底物杂交（SSH）构建基因库的方法发现不完全变态昆虫德国小蠊的无滋式卵巢管在卵黄发生后期一系列基因的差异表达，其生殖主要受保幼激素调控。从 *SSH* 基因库发现两个特别的主基因 *Browine* 和 *Citrus*。它们突出是由于其显著的丰度和表达模式特征，且与其他昆虫基因数据库没有同源性。*Browine* 基因编码的蛋白质在形成海绵样体中起着重要作用。海绵样体在卵壳中组成气孔和卵孔的复杂结构，使人想起果蝇卵的卵角，使卵与外界可进行气体交换。*Citrus* 的最大特点是由一

feature of *Citrus* is that it is formed by a motif which is repeated 33 times. Through RNAi studies, *Citrus* is expressed in all follicular cells in earlier stages of chorion formation, and it is involved in the formation of the exochorion. In the case of *Citrus* absent, the eggs collapse, while in the case of *Brownie* absent, the embryos cannot develop due to gas exchange problems.

The chorion does not contain chitin and, except in a few dipterans, no significant quantity of minerals. Intricate surface sculpturing is characteristic of many insect eggs.

Many eggs have a porus gas-filled meshwork near the inner side. In some cases, this is a plastron, a surface that is not easy to wet. Several channels called aeropyles connect the meshwork or plastron to the external surface of the egg. The function of such hard-to-wet structure is to supply oxygen to the developing embryo if the egg becomes submerged under water for some times or when the natural site for laying the eggs is in wet decaying organic matter, animal manner, fruit, or similar plant tissues.

In wild silk moth, *A. polyphemus*, chorion formation requires about 2 days and more than 100 low molecular weight proteins are secreted. A large gene family controls the secretion of chorine proteins in *A. polyphemus* without gene amplification. By contrast, in *D. melangaster*, only about 20 proteins are secreted under control of a small family of genes. The single copy genes are amplified 20-to 80-fold in the follicular cells about 15 h prior to transcription. These multiple gene copies allow the follicle cells to secrete a large amount of chorionic proteins in short times, and the chorion is completed in about 5 h.

个重复 33 次的结构单元所形成。通过 RNAi 研究，*Citrus* 在卵壳形成早期的所有滤泡细胞中均有表达，在外卵壳形成过程中起着重要作用，缺失该基因引起卵的崩塌。而缺失 *Brownie* 基因，胚胎因气体交换问题不能发育。

卵壳不含几丁质，除极少数双翅目昆虫外，没有明显数量的矿物质。错综复杂的雕纹是许多昆虫卵的特征。

许多卵在靠内侧都有一个充满气体的有孔气室。在有些情况下，这是一个气盾，是不易湿润的表面。数个被称为气孔的通道连接气室或卵外表的花饰。这种难湿润结构的功能是如果卵有时浸泡在水中或产卵的自然场所是潮湿的腐烂有机质、动物粪便、汁液或相似的植物组织时为发育中的胚胎供氧。

野蚕卵壳的形成约需要 2 d，分泌出 100 多种低分子质量的蛋白质。野蚕的卵壳蛋白的分泌是由一个大基因家族所控制的，没有基因的扩增。相反，果蝇仅分泌 20 种蛋白质，由一个小基因族控制。其单基因在转录前约 15 h 在滤泡细胞内扩增了 20 ～ 80 倍。这些多基因拷贝使滤泡细胞在短时内分泌大量卵壳蛋白，卵壳在 5 h 内完成。

16.3 雄性生殖系统

16.3 Male Reproductive System

16.3.1 Composition of male reproductive system

The internal organs of the male reproductive system are the paired testes, the vas deferens, the ejaculatory duct and the accessory glands (Fig.16-7A).

All parts of the system may produce secretions that aid the transfer of sperm to the female. Each testis generally consists of a number of tubes or follicles which spermatozoa are mature. Follicles may vary from 1 to greater than 100 follicles, and may be incompletely separated from each other, such as lepidopteran

16.3.1 雄性生殖系统的组成

雄性生殖系统内部器官由睾丸、输精管、射精管和附腺组成（图 16-7A）。

雄性生殖系统的所有组成部分都能产生分泌物，协助精子转移到雌虫中。总体来说，每个精巢由数条精巢管或精原细胞成熟的生精囊组成。滤泡的数量变化从 1 到多于 100，相互之间有的并不

testis, or the testes may consist of several lobes, each with several follicles.

16.3.2 Spermatogenesis and maturation of sperm

Zones of maturation stages of sperm exist along the length of a typical follicle. The distal part of a testicular tubule is a germarium in which repeated mitotic divisions give rise to undifferentiated, diploid, spermatogonia. In a growth zone (zone Ⅰ), the spermatogonia divided by mitosis into many diploid spermatocytes enclosed within a cyst or capsule of some cells. All spermatocytes within a sac or cyst generally arise from the same spermatogonial cell and their development is synchronized. The spermatocytes may undergo more mitotic divisions; there are 5 to 8 divisions in Acrididae and 7 in *Melanoplus*, but eventually in the "zone of maturation" (zone Ⅱ)meiosis and haploid spermatids are produced. A spermatid has completed its meiotic division, but is an immature sperm (Fig.16-7B). Meiosis in *S. gregaria* male depends on the presence of ecdysteroids. Normally 4 sperm are produced from each spermatocyte. In zone Ⅲ, the region of transformation, the mature sperm develop. Insect spermatozoa tend to be very long (300 μm in *R. prolixus* and 5.08 cm in *D. bifurca*) and have a slender head region, probably as an evolutionally adaptation to the necessity to navigate the micropyle. Usually the mature sperm remain bundled together in Zone Ⅲ .

Many insects contain mature sperm in the late pupal stage, while others may requiie several days as an adult to mature sperm. In *R. prolixus* and grasshoppers, the accessory glands in males are influenced in their development by secretion from the corpora

完全分开，如鳞翅目昆虫的睾丸，由几个睾丸叶组成，每个有几个生精囊。

16.3.2 精子的发生与成熟

精子成熟期的区域沿着生精囊的长度存在。精巢管的顶部是精原区，初级精原细胞经过重复的有丝分裂，产生未分化的、二倍体精原细胞。在成长区（Ⅰ区），精原细胞通过有丝分裂产生许多二倍体精母细胞，包裹在包囊或几个细胞的囊内。一个囊或包囊内的所有精母细胞都来自于一个精原细胞，其发育是同步的。精母细胞可进行更多的有丝分裂；在蝗科中分裂 5 ～ 8 次，蚱蜢中分裂 7 次，最后在“成熟区”（Ⅱ区），经减数分裂产生单倍体精细胞。精细胞完成了其减数分裂，仍是未成熟的精子（图 16-7B）。沙漠蝗雄虫的减数分裂依赖于蜕皮甾类。在正常情况下，一个精母细胞产生 4 个精子。在转化区Ⅲ区发育为成熟精子。昆虫精子很长（吸血蝽 300 μm，小果蝇 5.08 cm），有一细长的头部区域，可能是穿越卵孔所必需的进化适应。通常成熟的精子成束留在Ⅲ区 。

许多昆虫在后蛹期有成熟的精子，而其他昆虫可能需要几天时间才能使精子成熟。吸血蝽和蚱蜢雄性附腺受到咽

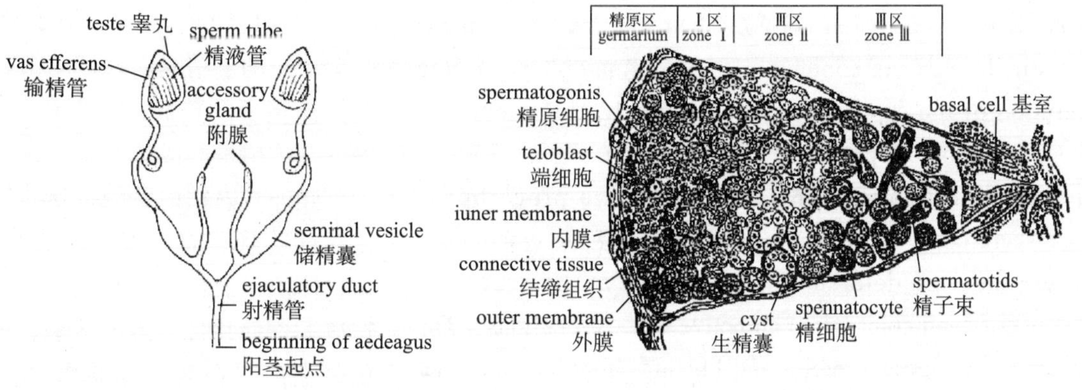

A.the composition of insect male reproductive organ
A. 雄性生殖器官组成

B. section view of testis
B. 睾丸剖面

Fig.16-7 Insect male reproductive organ (modified from Wu Zaide, 1993).

图 16-7 昆虫雄性生殖器官及睾丸剖面（仿吴载德，1993）。

allata. Sperm survival within the gential tract of female insects may be prolonged for weeks, months, or even years. Honeybee queens have been known to lay fertilized eggs after several years (8-9years in one reported case), and queens ants were reported to contain viable sperm after 15 years. Spermatozoa survival in female *R. prolixus* for about a month, and for about 10 weeks in *S. gregaria*.

侧体分泌物的影响。昆虫精子在雌性昆虫生殖道内可存活数周、数月甚至数年。已知蜂后在几年后产下受精卵（报道为 8 ～ 9 年），蚁后据报道可保留精子活力 15 年。雌性吸血蝽精子存活约 1 个月，沙漠蝗约 10 周。

16.3.3 Apyrene and eupyrene sperm of Lepidoptera

Most Lepidoptera produce two types of sperm: apyrene sperm without a nucleus and nucleated eupyrene sperm. Only the latter type can fertilize an egg. The two types of sperm are produced by major differences in the meiotic process. Sperm dichotomy appeared early in the evolution of Lepidoptera, but apyrene sperm are not present in the Micropterigidae, one of the most primitive families of Lepidoptera. Eupyrene sperm are usually packaged into bundles, while apyrene sperm are dissociated as single, but immobile cells. Both types of sperm are incorporated into the spermatophore that is formed in the female bursa copulatrix at the time of mating by secretions from the male. Maturation division (to produce the haploid number of chromsomes) of the eupyrene sperm and development of motility in both types of sperm occur in the spermatophore of some insects. There is a correlationship between decrease of apyrene sperm in the spermatheca and remating patterns, that is the number of apyrene sperm may influence female remating.

Male *B. mori* have an endopeptidase, called initiatorin, in secretion of the posterior segment of the ejaculatory duct that is important in activation of both apyrene and eupyrene sperm and in maturation of the eupyrene sperm. Initiatorin is a serine endoprotease that is active at pH 9.2. Their vigorous movements serve to stir the viscous contents of the spermatophore, aiding the liberation of the eupyrene sperm and facilitating metabolic reactions that promote eupyrene sperm maturation.

In addition, initiatorin converts an inactive procarboxy-peptidase secreted by the ampulla, the region of the ejaculatory duct where the vasa deferentia and ducts from the accessory glands converge and empty, into an active carboxypeptidase. The carboxypeptidase digests proteins and liberates arginine and other amino acids. Arginine is subsequently converted to glutamate, which is metabolized by the sperm to support motility.

16.3.3 鳞翅目昆虫无核和真核精子

多数鳞翅目昆虫产生两种类型的精子：没有细胞核的无核精子和有核的真核精子。只有后者能使卵受精。两类精子是减数分裂过程中的主要差别形成的。精子二分法在鳞翅目进化早期就出现了，但在鳞翅目最原始的科之一的小翅科中却不存在单倍体精子。真核精子常装配成束，而无核精子即解离成单个而不活动的细胞。交配时雄虫的分泌物在雌虫交配囊内形成精囊，两类精子都能掺入精包。有些昆虫的精包内存在真核精子的成熟分裂（产生单倍体数目的染色体）和两类精子游动能力的发育。受精囊内无核精子数量的减少与雌虫的再交配模式有相互关系，即无核精子的数量会影响雌虫的再交配。

雄蚕射精管的后段分泌一种内肽酶，称为熟精内肽酶，对真核精子和无核精了的活化及真核精子的成熟都很重要。熟精内肽酶是一种丝氨酸内切蛋白酶，在 pH 9.2 时活跃。精子充满活力的运动搅动了精囊内黏稠的内容物，帮助真核精子的释放，并促进代谢反应，促进真核精子的成熟。

此外，熟精内肽酶将坛状体释放的无活性羧肽酶前体转换成活性羧肽酶。坛状体位于射精管，是输精管和附腺的管道汇合处并且空闲的区域。羧肽酶消化蛋白质，释放出精氨酸和其他氨基酸。精氨酸随后转换成谷氨酸，代谢后支持精子的运动。

16.4　附腺及其分泌物

16.4　Accessory Glands and Their Products

16.4.1 Female accessory glands

Accessory reproductive glands of female are composed of cement gland, spermathecal gland, egg calyx gland and milk gland, etc (Fig.16-8). From a morphological point of view, the glands of many female insects can be clearly assigned to the one-cell-layer type and multiple-cell-layer type. The glands are mainly of ectodermal origin and form in many insects during the late preadult stage. In *T. commodus*, the gland growth is exclusively caused by a hypertrophy of single gland cells with respective volume gains between quadruple and septuple. According to the stereological results, the volume of mitochondria per cell is marked by an up to fourfold increase during peak differentiation.

Functions of secretions from accessory glands of female: ① produce oviposition pheromone; ② coat and fasten eggs to laying substrates; ③ lubrication and/or egg protection; ④ dissolve spermatophores; ⑤ in *Musca domestica* accessory gland secretions are moved with spermatozoa to the fertilization chamber where they aid micropyle cap removal, allowing fertilization to take place; ⑥ trigger the acrosome reaction when present with micropylar cap substance and, in higher concentrations, cause degradation of spermatozoa; ⑦ interactions with male secretions which influence on aspects of female physiological and behavioral characters.

The behavioral and physiological changes induced in insect upon mating include: ① decrease attractiveness to males; after mating, the female's production of attractive pheromone drops, and she acquires and produce her own an anti-aphrodisiac from her mate; ② decrease receptivity to mating; ③ increase food digestion; ④ elevation of egg laying; ⑤ trigger storage and utilization of sperm; ⑥ decrease life span.

DNA microarray studies have demonstrated that, shortly after mating, transcript abundance of 1,700 genes altered in the female's reproductive tract as well as in other tissues in *Drosophila*. Many

16.4.1　雌性附腺

昆虫雌性附腺有黏腺、受精囊腺、卵萼腺和乳腺等（图 16-8）。许多昆虫雌性生殖腺形态学上分为单细胞层附腺和多细胞层附腺。雌性附腺主要由外胚层发育而来，多数昆虫在成虫早期的后阶段形成。澳洲黑蟋蟀腺体的成长仅仅由单个腺细胞的过度生长而引起，各个腺细胞体积增幅为 4~7 倍。根据体视学研究结果，在分化高峰期，每个细胞的线粒体体积增加了 4 倍。

雌性附腺分泌物的功能：①产生产卵激素；②将卵囊在产卵基质上固定；③润滑和 / 或保护卵；④溶解精包；⑤家蟋蟀附腺分泌物随精子一起进入受精囊，协助移走卵孔盖，完成受精；⑥和卵孔帽物质一起出现时，促进顶体的反应，高浓度时，降解精子；⑦与雄性分泌物相互作用，影响雌虫的生理和行为特征方面。

交配后引起的雌性昆虫行为和生理变化包括：①对其他雄虫的吸引力下降；交配后，雌性的性信息素产物减少，并从伴侣那里获得并生产自己的“抑春素”；②减少再次交配的接纳能力；③增加食物的消化；④提高产卵率；⑤促进精子贮存和利用；⑥缩短了虫体的寿命。

用 DNA 芯片技术检测雌性果蝇生殖腺和其他组织，交配后短时内超过 1 700 个基因的转录丰度发生变化，这些

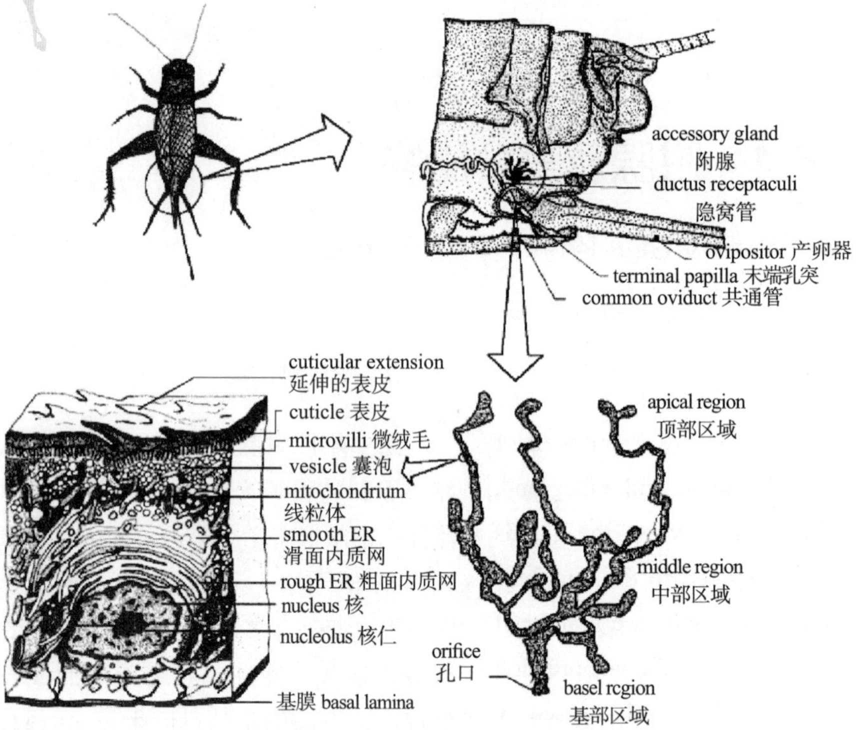

Fig.16-8 Position and shape of the female accessory sex glands in *T. commodus* and simplified ultrastructure of a single gland cell. Bars represent 10 nm, 5 nm, 2 nm, and 10 nm, respectively (modified from Sturm, 2000; Sturm and Pohlhammer, 2000).

图 16-8 雌性蟋蟀附腺的位置和形状以及简化的单个腺细胞的超微结构。横杠分别代表 10 nm、5 nm、2 nm 和 10 nm（仿 Sturm, 2000; Sturm 和 Pohlhammer, 2000）。

of these changes are elicited by sperm and accessory glands proteins (Acps) that males transfer to females. Four individual Acps (ovulin, *Acp*36DE, *Acp*29AB, and *Acp*62F) with unique functions independently affected gene expression in females. Consistent with their early and possibly local action within the female, ovulin and *Acp*36DE caused relatively few gene expression changes in whole bodies of mated females. In contrast, *Acp*29AB and *Acp*62F modulated a large number of transcriptional changes shortly after mating. *Acps* upregulate a large number of genes involved in the female's immune response. In particular, 12 immune response genes whose transcript levels increased in females after receiving *Acps* from their mates, 5 showed twofold or greater differences in transcript levels.

变化中的很多基因是由雄虫交配时转移到雌虫的精子和附腺蛋白（Acps）引起的。4 个具有独特功能的单基因（卵巢内泌素，*Acp*36DE, *Acp*29AB, 和 *Acp*62F）分别影响雌虫基因的表达。其中，卵巢内泌素和 *Acp*36DE 仅影响了少数基因的转录。相反，*Acp*29AB 和 *Acp* 62F 调控了大量基因转录水平的变化。*Acps* 上调大量牵涉到雌虫免疫的基因，其中雌虫的 12 个免疫反应基因转录水平由于交配接收了 *Acps* 而被增加，5 个基因转录水平上调幅度达 2 倍或更大差异。

16.4.2 Male accessory glands

Male accessory glands of *B.tryoni* were found to comprise one pair of mesodermic glands and three pairs of ectodermic glands. The mesodermic accessory glands consist of muscle-lined, binucleate epithelial cells, which are highly microvillated and extrude electron-dense secretions by means of macroapocrine transport into a central lumen. The ectodermic accessory glands consist of muscle-lined epithelial cells which have wide subcuticular cavities, lined with microvilli. The electron-

16.4.2 雄性附腺

澳洲果蝇的雄性附腺可分为 1 对中胚层腺体，3 对外胚层腺体。中胚层附腺由肌肉发达的双核上皮细胞组成，这些细胞具有高度微绒毛，通过巨顶浆分泌挤压将电子密集的分泌物转运进中心管腔。外胚层腺体由肌肉发达的上皮细胞组成，上皮细胞有宽大的表皮下腔，内衬微绒毛。这些腺体分泌的电子透明

transparent secretions from these glands are first extruded into the cavities and then forced out through small pores of the cuticle into the gland lumen. Secretions from the two types of accessory glands then flow into the ejaculatory duct, which is highly muscular, with epithelial cells rich in rough endoplasmic reticulum and lined with a thick, deeply invaginated cuticle. While there are some notable differences, reproductive accessory glands of male flies generally resemble those of the olive fruitfly, *Bactrocera oleae*, and to a lesser extent the Mediterranean fruit fly, *C. capitata*. Accessory glands are lacking in some insects; for example, they are not present in some dipterans, including the housefly and stable fly.

分泌物首先被挤压进入表皮下腔，然后通过表皮的小孔被挤出进入腺腔。之后，两类附腺的分泌物流进射精管。射精管肌肉高度发达，其上皮细胞富含粗面内质网，由深深内陷的厚表皮围成。雄性澳洲果蝇生殖腺的附腺类似于橄榄果蝇，与地中海实蝇存在一些显著差异。但有些昆虫缺乏附腺，如有些双翅目昆虫，包括家蝇和厩螯蝇。

Some of the male secretory products may stimulate contractions in the reproductive tract of female, thus aiding movement of sperm into the spermathecae of the female. Accessory glands of *Drosophila* synthesize more than 85 proteins. One of the proteins, called a “sex peptide,” is passed at mating to the female, which is then much less receptive to remating for 6 to 9 days. Thc peptide also stimulates oviposition.

有些雄性分泌产物能促进雌虫生殖道的收缩，因此协助精子运动到雌虫受精囊。果蝇附腺合成的蛋白质超过 85 种，其中一种蛋白质称为“性肽”，交配时传递给雌虫，此后的 6 ～ 9 d 大幅度减少再交配接受的能力，这个肽也能刺激产卵。

Male accessory gland proteins show evolutionary diversity both within and between species. Following their transfer at mating, the functions of specific Acps include: ① reduce female receptivity; ② stimulate egg production; ③ essential for sperm storage following mating; ④ play a role in sperm competition; ⑤ the protease inhibitory activity of some *Acps*; ⑥ activation of other *Acps* (as *Acp29AB*) share homology with the functional domains of lectins, lipases and cysteine-rich secretory proteins; ⑦ some Acps have antibacterial activity, and one Acp70A stimulates the production of antimicrobial peptides in females following mating.

雄性附腺蛋白质在种内和种间都有较大的进化多样性。通过交配它们的转移，特异性雄性附腺蛋白质（Acps）功能包括：①降低雌虫再交配能力；②刺激卵产量；③交配后贮存精子所必需；④在精子的竞争中起作用；⑤有些 *Acps* 具有蛋白酶抑制活性；⑥有些 *Acps*（如 *Acp29AB*）基因结构上与凝集素、酯酶及富含半胱氨酸分泌蛋白质功能域具有同源性；⑦有些 Acp 具有抗菌活性，Acp70A 能刺激雌虫在交配后产生抗菌物质。

16.4.3 Transfer of sperm

Some male insects transfer packets of sperm to the female reproductive tract through mating. Many insects produce a spermatophore that contains the sperm and is transferred to the female. Accessory glands secrete spermatophorins, proteins that form the spermatophore. Spermatozoa are released from the spermatophore after it is deposited in the bursa copulatrix of the female *Rhodnius*. Mechanical abrasion of the spermatophore or the action of proteolytic enzymes, or both, may play a role in releasing sperm. In a cricket, the loss of protein during spermatophore formation and transfer amounted to 40% of the body weight.

16.4.3 精子的转移

一些雄性昆虫通过交配将精子包转移到雌虫生殖道。许多昆虫都产生有精子的精囊并被转给雌性昆虫。附腺分泌精包蛋白，这些蛋白质形成精包。被放置在雌性吸血蝽的交配囊后，精囊中的精子被释放出来。精囊的机械摩擦或蛋白水解酶的作用或两者都可能在精子释放中起着重要作用。在一种蟋蟀中，精囊形成和转移期间，蛋白质损失量占其体重的 40%。

16.5 昆虫生殖的多样性

16.5 Reproductive Diversity of Insects

Insects have amazing reproductive ability, and different reproductive modes such as egg viviparity, parthenogenesis and larval reproduction. Most insects lay hundreds of eggs. Insects with social and parthenogenetic reproduction are more fertile. If necessary, a bee can lay millions of eggs in her lifetime. If all the offspring of one parthenogenetic aphid survive and continue to reproduce, the total number of aphids can reach about 600 million in half a year. Strong reproductive capability is the basis of population prosperity.

昆虫具有惊人的繁殖能力和不同的生殖方式如卵胎生、孤雌生殖和幼体生殖。大多数昆虫产卵数百粒，社会性与孤雌生殖的昆虫生殖力更强。如果有必要，1 只蜜蜂蜂后一生可产卵数百万粒；1 只孤雌生殖的蚜虫若后代全部成活并继续繁殖，半年后蚜虫总数可达 6 亿只左右。强大的生殖潜能是种群繁荣的基础。

The reproduction strategies of invertebrates, oviparity, ovoviviparity and viviparity, always reflect the relationship between individuals and their surroundings. There is plasticity in the reproductive strategies of sarcosaphagous flies as they adjust to rapidly changing circumstances. The transition from oviparity to ovoviviparity or viviparity involves numerous changes in physiology, morphology and immunology. Oviparity is a reproductive mode in which females oviposit eggs at a relatively early developmental stage, and the place for deposition of the embryo can be different according to patterns of nutrient supply (Yang et al., 2021).

无脊椎动物的繁殖策略有卵生、卵胎生和胎生，往往反映了个体与其周围环境之间的关系。肉蝇在适应迅速变化的环境时，其繁殖策略具有可塑性。从卵生到卵胎生或胎生的过渡涉及生理学、形态学和免疫学的许多变化。卵生是一种生殖模式，其中雌性在相对早期的发育阶段产卵，胚胎的沉积位置可能根据营养供应模式而不同（Yang 等，2021）。

16.5.1 Parthenogenesis

16.5.1 孤雌生殖

Parthenogenesis is defined as the development of individuals from an unfertilized gamete and, among a variety of other reproductive strategies, it is one of the types observed in different phylogenetically unrelated organisms. Thelytoky, the mode where females from unfertilized eggs produce only daughters, is usually referred to as parthenogenesis *sensu stricto*. Besides thelytoky, unfertilized eggs can also develop into males, a mode named arrhenotoky. In organisms with this type of parthenogenesis, females develop from fertilized eggs and have diploid chromosome sets while males are haploid. Similarly, a haplodiploid mode

孤雌生殖被定义为个体从未受精配子发育而来。在各种其他生殖策略中，它是在不同系统发育无关联生物中观察到的类型之一。孤雌产雌，即未受精卵中的雌性只产生雌性后代的模式，通常被称为严格意义上的孤雌生殖。除了孤雌产雌，未受精的卵也可以发育成雄性，这种模式称为孤雌产雄生殖。在这种单性生殖的生物体中，雌性从受精卵发育而来，具有二倍体染色体组，而雄

of reproduction also occurs in species with paternal genome elimination. Another major reproductive mode is deuterotoky where both females and males are produced from unfertilized eggs. The basic system affected by parthenogenetic reproduction is the genome and particularly the set of chromosomes.

The shift from sexual reproduction to parthenogenesis has occurred repeatedly in animals. In the stick insect genus *Timema*, parthenogenesis results in an extreme reduction of heterozygosity and often leads to genetically uniform populations. Less effective positive selection in parthenogenetic species, suggesting that sex is ubiquitous in natural populations because it facilitates fast rates of adaptation. Parthenogenetic species did not show increased transposable element (TE) accumulation, likely because there is little TE activity in the genus.

性是单倍体。同样，在父系基因组消除的物种中也会出现单倍体和二倍体的繁殖模式。另一种主要的繁殖方式是产两性单性生殖，雌雄都是由未受精的卵产生的。受孤雌生殖影响的基本系统是基因组，特别是染色体组。

从有性生殖到单性生殖的转变在动物中不断发生。在竹节虫属中，单性生殖的结果是杂合度的极端降低，并经常导致均匀遗传种群。单性生殖物种的正选择效率较低，表明性别在自然种群中普遍存在，因为它有利于快速适应。孤雌生殖物种没有表现出转座子（TE）积累的增加，可能是因为该属中几乎没有转座子活性。

16.5.1.1 Arrhenotoky

Parthenogenetic mode where females produce only males from unfertilized eggs. Sometimes also referred as androgenesis. Found in taxa with XX/XO sex determining system. There are two mechanisms of diploid arrhenotoky resulting in the XO status of the males: paternal genome elimination, and paternal genome heterochromatinization.

Arrhenotokous and thelytokous strains of the same species often coexist, and, in that case, males coming from bisexual populations can even mate with presumably “asexual” females. Conversely, rare males produced by generally thelytokous strains (which strictly speaking, thus become deuterotokous, producing both male and female offspring parthenogenetically) also can mate with “sexual” females and start new parthenogenetic lineages. These males can therefore spread thelytoky-inducing alleles, exhibiting the so-called contagious parthenogenesis. Coexistence of thelytokous and arrhenotokous strains is especially characteristic of many parasitoid Hymenoptera of the superfamilies Cynipoidea and Chalcidoidea, in which thelytoky is usually caused by symbiotic microorganisms that belong to the genera Wolbachia, Cardinium, and Rickettsia. Endosymbionts are apparently responsible for the origin of one-third of the thelytokous lineages in this order.

16.5.1.1 孤雌产雄

雌性只从未受精的卵中产生雄性的孤雌生殖模式，有时也称为单雄生殖。在具有 XX/XO 性别决定系统的分类群中发现。导致雄性 XO 状态的二倍体孤雌产雄生殖有两种机制：父系基因组消除和父系基因组异染色质化。

孤雌产雄生殖和孤雌产雌品系在同一物种内通常共存，在这种情况下，来自两性群体的雄虫甚至可以与很可能是“无性”的雌虫交配。相反，由通常为孤雌生殖的品系（严格地说，因此成为雌雄同系，即孤雌生殖产生雄性和雌性后代）产生的罕见雄性也可以与“有性”的雌性交配，并开始新的孤雌生殖谱系。因此，这些雄性可以传播产雌孤雌生殖诱导的性等位基因，表现出所谓的传染性孤雌生殖。孤雌产雌和产雄孤雌生殖品系的共存是瘿蜂亚科和小蜂亚科的许多寄生性膜翅目的特征，其中孤雌产雌品系通常由属于沃尔巴氏菌属、卡迪氏菌属和立克次体属的共生微生物引起。内共生菌显然是这个目 1/3 的孤雌产雌品系的起因。

16.5.1.2 Thelytoky

Parthenogenetic mode where females produce only females

16.5.1.2 孤雌产雌

雌性只从未受精的卵中产生雌性的

from unfertilized eggs. There are two forms described: apomixis and automixis (sometimes also referred to as gynogenesis where stimulation by a sperm cell is required to activate the egg into development without fertilization). Apomixis, parthenogenetic mode (thelytoky) where the egg cell is a result of mitosis and offspring consists of true maternal clones; meiosis does not take place during this type of reproduction. Automixis, parthenogenetic mode (thelytoky) where egg cells are produced by meiosis should not be confused with self-fertilization in hermaphrodites. Cyclic parthenogenesis, the alternation of bisexual and parthenogenetic reproduction. Contagious parthenogenesis, a process involving rare functional males produced by a parthenogenetic lineage, which mate with sexually reproducing females resulting in fertile parthenogenetic offspring. Paternal genome elimination, a mode of reproduction where eggs are fertilized, but only the female genome is transmitted to offsprings (sometimes also referred to as pseudo-arrhenotoky or parahaploid). Paternal genome set is eliminated in early embryogenesis (males are somatically haploid) or during spermatogenesis (males are somatically diploid, however, the paternal genome is inactivated by chromatin condensation, also referred to as paternal genome heterochromatinization).

Thysanoptera are haplo-diploid insects. Most common is the arrhenotokous reproduction mode, in which females arise from fertilized diploid eggs, whereas males develop from unfertilized haploid ones. Some species reproduce *via* thelytoky, whereby females produce exclusively female progeny and only very rare males. Furthermore, some species show a reproductive polymorphism, where populations in different geographical regions reproduce *via* arrhenotoky or thelytoky.

Beside genetically based thelytoky, this reproduction mode can also be endosymbiont induced. The recovery of these females from their infection again leads to the development of males. Functionality of these males ranges widely, and this might be associated with sperm structure. Beside the different reproduction modes, and adaptations to their life style, the arrhenotokous species *S. linguis* (Thysanoptera, Tubulifera) and *Echinothrips americanus* (Thysanoptera, Terebrantia) possess typical thysanopteran-like sperm structure. But endosymbiont-cured males from the thelytokous species *Hercinothrips*

孤雌生殖模式。描述了两种形式：无融合生殖和自体融合（有时也称为雌核发育。在这种情况下，需要精子细胞的刺激才能激活卵子进入发育，而无需受精）。无融合生殖、孤雌生殖模式（孤雌产雌），其卵细胞是有丝分裂的结果，后代由真正的母体克隆组成；减数分裂不发生在这种类型的生殖过程中。自体融合，孤雌生殖模式（孤雌产雄）的卵细胞通过有丝分裂产生，不应与雌雄同体的自受精混淆。周期性孤雌生殖，双性生殖和单性生殖的交替。传染性孤雌生殖是一种由孤雌生殖谱系产生的罕见功能性雄性与有性生殖的雌性交配，从而产生可育的孤雌生殖后代的过程。父系基因组消除是一种生殖方式，其中卵受精，但只有雌性基因组传递给后代（有时也称为伪无核生殖或拟单倍体）。父系基因组在早期胚胎发生（雄性为体细胞单倍体）或精子发生中被消除（雄性是体细胞二倍体。然而，父系基因组因染色质浓缩而失活，也称为父系基因组染色体异质化）。

缨翅目昆虫是单 - 双倍体昆虫。最常见的是孤雌产雄生殖的模式，其中雌性从受精的二倍体卵中产生，而雄性从未受精的单倍体卵中发育。有些物种通过孤雌产雌生殖，其中雌性只产生雌性后代，且只有非常罕见的雄性。此外，一些物种表现出生殖多态性，不同地理区域的种群通过产雄孤雌生殖或孤雌产雌进行生殖。

除基于遗传的孤雌产雌外，这种生殖模式也可以是内共生诱导的。这些雌性昆虫从感染中恢复再次导致雄性的发育。这些雄性昆虫的功能范围广泛，这可能与精子结构有关。除不同的生殖模式和对生活方式的适应外，产雄孤雌生殖的物种 *S. linguis*（缨翅目、管尾亚目）和美洲棘蓟马（缨翅目、锥尾亚目）具有典型的缨翅目样精子结构。但是，内共生体矫正的孤雌产雌温室条蓟马

femoralis (Thysanoptera, Terebrantia) possess several malformed spermatozoa and a large amount of secretions in their testes. Spermiophagy seems to be typical. It indicates a highly conserved mechanism of the male developmental pathways (Krueger et al., 2021).

（缨翅目、锥尾亚目）产生的雄性昆虫的睾丸中拥有数个畸形精子和大量分泌物。精子吞噬似乎是典型的。这表明雄性发育途径的高度保守机制（Krueger 等，2021）。

Another way of asexual reproduction in insects is the polyembryony; it occurs when a single egg undergoes fission giving rise to two or more embryos. This sort of reproduction is restricted predominantly to Hymenoptera and few Strepsiptera, and seems to have evolved independently among different groups of animals. In the parasitic wasps, the number of embryos produced can be influenced by the size of the host, varying from 10 to several hundred embryos arising from one egg. Characteristically, the eggs of polyembryonic species are devoid of yolk, and its nutrition is acquired from the host's hemolymph through a specialized enveloping membrane called the trophamnion, a modification of the serosa layer (Rocha et al., 2020).

昆虫无性繁殖的另一种方式是多胚生殖；当一个卵子发生裂变，产生两个或更多的胚胎时，就会发生这种情况。这种生殖主要局限于膜翅目和少数捻翅目，并且似乎在不同的动物群体中独立进化。在寄生蜂中，产生的胚胎数量会受到宿主大小的影响，从一个卵产生 10 个胚胎到几百个胚胎不等。从特征上讲，多胚物种的卵缺乏卵黄，其营养通过一种被称为滋养羊膜，改进的浆膜层的特殊包膜从宿主的血淋巴中获得（Rocha 等，2020）。

16.5.2 Viviparous insect

16.5.2 胎生昆虫

Viviparity is defined as a species laying larva directly, and there is a place in part of the vagina for embryo development. Under viviparity, the embryo can avoid the influence of the external environment, obtain nutrition from the parent and engage in gas exchange. Between oviparity and viviparity, there is an intermediate mode known as ovoviviparity (or oviparous egg retention). Ovoviviparous species oviposit eggs at a relatively mature stage of embryo development, and the eggshell still envelopes the larva during or immediately following oviposition. Ovoviparous species are more similar to viviparous species than to oviparous species because of the irreversible morphological and physiological changes. In spite of this, the ovoviviparous species maybe only provide a place for offspring development (Yang et al., 2021).

胎生被定义为一种直接产幼虫的物种，在阴道的一部分有胚胎发育的地方。胎生时，胚胎可以避免外部环境的影响，从父母那里获得营养，并进行气体交换。在卵生和胎生之间有一种中间模式，称为卵胎生（或卵生的卵子滞留）。卵胎生品种在胚胎发育的相对成熟阶段产卵，在产卵期间或之后，蛋壳仍然包裹幼虫。由于不可逆的形态和生理变化，卵胎生物种比卵生物种更类似于胎生物种。尽管如此，卵胎生物种可能只为后代发育提供了一个场所（Yang 等，2021）。

Most insects hatch from eggs, but others are ovoviviparous or viviparous, i.e eggs hatch within the body of the female, such as some species mayfly in Ephemeroptera, viviparous leaf beetle *Oreina cacaliae* (Coleoptera: Chrysomelidae), and Tsetse flies (Diptera: Glossina spp.) etc. Tsetse flies (*Glossina* spp.) are distinguished by their specialized reproductive biology, defined by adenotrophic viviparity (maternal nourishment of progeny by

大多数昆虫从卵中孵化，但其他昆虫是卵胎生或胎生的，即卵是在昆虫母体内孵化的，如蜉蝣目中的某些种类的蜉蝣、胎生叶甲（鞘翅目：金甲科）和采采蝇（双翅目：舌蝇科）等。采采蝇（舌蝇属）独特的生殖生物学特征是腺营养胎生（母体通过腺分泌物对后代

glandular secretions followed by live birth). This trait has evolved infrequently among insects and requires unique reproductive mechanisms. A key event in *Glossina* reproduction involves the transition between periods of lactation and nonlactation (dry periods). Increased lipolysis, nutrient transfer to the milk gland, and milk-specific protein production characterize lactation, which terminates at the birth of the progeny and is followed by a period of involution. The dry stage coincides with embryogenesis of the progeny, during which lipid reserves accumulate in preparation for the next round of lactation. The obligate bacterial symbiont *Wigglesworthia glossinidia* is critical to tsetse reproduction and likely provides B vitamins required for metabolic processes underlying lactation and/or progeny development (Benoit et al., 2015). Viviparous reproduction is characterized by maternal retention of developing offspring within the reproductive tract during gestation, culminating in live birth. In some cases, a mother will provide nutrition beyond that present in the yolk; this is known as matrotrophic viviparity. While this phenomenon is best associated with mammals, it is observed in insects such as the viviparous cockroach, *Diploptera punctata*. Female *D. punctata* carry developing embryos in the brood sac, a reproductive organ that acts as both a uterus and a placenta by protecting and providing a nutritive secretion to the intrauterine developing progeny (Jennings et al., 2020). Morphological and physiological adaptations required to support unique biology of tsetse flies. These adaptations are driven by demands associated with obligate hematophagy and viviparous reproduction. Obligate viviparity entails intrauterine larval development and the provision of maternal nutrients for the developing larvae. The morphological adaptations associated with viviparous biology include: ① modifications facilitating abdominal distention required during blood feeding and pregnancy; ② abdominal and uterine musculature adaptations for gestation and parturition of developed larvae; ③ reduced ovarian structure and capacity; ④ structural features of the malederived spermatophore optimizing semen/sperm delivery and inhibition of insemination by competing males; ⑤ structural features of the milk gland facilitating nutrient incorporation and transfer into the uterus (Attardo et al., 2020).

进行营养，然后活产）。这种特性在昆虫中很少进化，需要独特的繁殖机制。舌蝇生殖的一个关键事件是哺乳期和非哺乳期（干乳期）之间的过渡。脂肪分解增加、营养物质转移到乳腺和乳汁特异性蛋白质产物增加是泌乳的特征，泌乳在后代出生时终止，随后是一段退化期。干乳期与子代的胚胎发生相一致，在此期间，脂质储备累积，为下一轮哺乳做准备。专性细菌共生魏格沃斯菌对采采蝇的生殖至关重要，并可能提供泌乳和/或后代发育代谢过程所需的B族维生素（Benoit 等，2015）。胎生生殖的特点是母体在妊娠期间将发育中的后代保留在生殖道内，以活胎产为终结。在某些情况下，母体会提供卵黄以外的营养；这被称为滋养胎生。虽然这种现象与哺乳动物最为相似，但在太平洋硕蠊等昆虫中也观察得到。雌性太平洋硕蠊在卵囊中携带发育中的胚胎，卵囊是一种生殖器官，通过保护和提供营养分泌物给宫内发育中的后代，起到子宫和胎盘的作用（Jennings 等，2020）。形态和生理适应是采采蝇独特生物学所必需。这些适应是由与专性造血和胎生生殖相关的需求驱动的。专性胎生需要宫内幼虫发育和为发育中的幼虫提供母体营养。与胎生生物学相关的形态学适应包括：①血液饲养和妊娠期间所需的促进腹胀的改良；②腹部和子宫肌肉组织适应于妊娠和发育幼虫的分娩；③卵巢结构和容量的减少；④雄性精细胞的结构特征优化了精液/精子的输送，抑制了竞争性雄虫的授精；⑤乳腺的结构特征促进了营养物质的吸收和转移到子宫中（Attardo 等，2020）。

Earwigs (Dermaptera) is a small insect order traditionally classified into three subgroups: two epizoic taxa, the *Arixeniina* and *Hemimerina*, and the free-living *Forficulina*. The majority of dermapteran species is, like most insects, oviparous. In contrast, both epizoic groups are viviparous. A novel mechanism of

蠼螋（革翅目）是一个小昆虫目，传统上分为三个亚类：两个体外寄生类群、蝠螋科和鼠螋科以及独立生存的球螋科。与大多数昆虫一样，革翅目

maternal contribution to embryonic development which operates during oogenesis and involves characteristic modification of endoplasmic reticulum cisternae. Conspicuous and apparently inactive paracrystalline stacks of the endoplasmic reticulum are deposited in the oocyte cytoplasm and become activated during early embryonic development. Additionally, in *Hemimerus*, transformed follicular/ ovarian cells (on the mother's side) and an evagination of the dorsal vessel (on the embryo's side) converge to form a cephalic vesicle, structure analogous to a placenta. The cellular architecture of this unusual cephalic placenta points to its participation in an exchange of low molecular weight substances between a mother and developing embryo (Bilinski et al., 2018). In the viviparous dermapteran, *A. esau* (Insecta, Dermaptera), the second (intrauterine) phase of development consists of three physiologically disparate stages: early embryos (before dorsal closure, surrounded by an egg envelope), late embryos (after dorsal closure, surrounded by an egg envelope) and the first instar larvae (after "hatching" from an egg envelope). Early and late embryos float in the fluid filling the uterus, whereas the first instar larvae develop attached to the uterus wall. Tworzydlo et al.(2019) analyses revealed also that in *Arixenia* serial multilobed outgrowths develop on dorso-lateral aspects of all abdominal segments. At the onset of the third developmental stage and after liberation from an egg envelope, these outgrowths (or more precisely their lobes) adhere to the epithelium lining the uterus, forming a series of small contact sites, where the mother and embryo tissues are separated only by a thin, presumably permeable, embryonic cuticle. All these contact sites collectively constitute a dispersed placenta-like organ involved in the nourishment of the embryo. The follicular cells surrounding fully grown oocyte of *Hemimerus* do not degenerate after initiation of embryogenesis, but transform and gradually form the wall of the incubation chamber in which the embryo develops. Amniotic cells of the early embryo remain in direct contact with transformed follicular cells. In the region of contact, short outgrowths of the amniotic cells associate with irregular surface specializations of the transformed follicular cells. Extended "postfertilization" activity of hemimerid follicular cells represents an adaptation to viviparity and matrotrophy in this insect lineage (Bilinski et al., 2020) .

In *A. mellifera*, the worker reproduction is based on apoptosis

物种主要是卵生的。相比之下，两个体外寄生类群都是胎生的。在卵子发生过程中起作用的母体对胚胎发育的一种新机制涉及内质网池的特征性修饰。在卵母细胞质中沉积了显而易见不活跃的内质网副晶体堆叠，并在早期胚胎发育过程中被激活。此外，在鼠螋科，转化的卵泡 / 卵巢细胞（在母体一侧）和背血管外翻（在胚胎一侧）汇聚形成头囊泡，结构类似于胎盘。这种不寻常的头状胎盘的细胞结构表明，它参与了母体和发育中胚胎之间的低分子质量物质交换（Bilinski 等，2018）。胎生革翅目，蠼螋 *A. esau*（昆虫纲，革翅目）发育的第二阶段（宫内）由三个生理上完全不同的阶段组成：早期胚胎（背部闭合前，被卵囊包围）、晚期胚胎（背部关闭后，被卵囊包围）和一龄幼虫（从卵囊"孵化"后）。早期和晚期胚胎漂浮在充满子宫的液体中，而一龄幼虫附着在子宫壁上发育。Tworzydlo 等（2019）的分析揭示了在蠼螋中，其所有腹部节段的背侧发育出一系列多叶突起。在第三个发育阶段开始时，从卵囊中释放出来后，这些突起（或更准确地说，它们的叶）黏附在子宫内膜上皮上，形成一系列小的接触部位，其中母体和胚胎组织仅由一层薄的、可能可渗透的胚胎表皮层隔开。所有这些接触部位共同构成一个分散的胎盘样器官，参与胚胎的营养。在胚胎发生开始后，鼠螋科完全成熟卵母细胞周围的滤泡细胞不会退化，而是转化并逐渐形成胚胎发育的孵化室壁。早期胚胎的羊膜细胞仍与转化的滤泡细胞直接接触。在接触区，羊膜细胞的短突起与转化滤泡细胞的不规则表面特化有关。鼠螋滤泡细胞延长的"受精后"活性代表了该昆虫谱系对胎生和营养的适应（Bilinski 等，2020）。

意大利蜜蜂工蜂的生殖是基于幼虫

of oocytes in the larval and adult phases. Transmission electron microscopy observations indicated that the germinative and somatic cells become fragmented and are engulfed by other cells after the larva molt to the third instar, showing characteristics of apoptosis. The apoptosis of germline and somatic cells in ovarioles becomes more severe in the 4th and 5th instar larvae, causing workers to lose more than 90% of their ovarioles. Although adult workers retain a few functional ovarioles in the ovary, their reproduction is suppressed by the pheromone of the queen. Molecular evidence, including apoptotic gene expression and nuclear morphological features, indicates that the programmed cell death of oocytes and nurse cells is involved in adult worker sterility. In ants, fertilized eggs develop into females, which have the potential to differentiate into queens and workers. A worker possesses a pair of ovaries, which each contain fewer and shorter ovarioles than those of the queen. In their ovarioles, germline stem cells (GSC) undergo division in the germaria, giving rise to normal cysts and oocytes in both castes. In *Drosophila*, the protein encoded by the germline gene *vasa* is specifically restricted to the germ cells, while products of the gap gene *hunchback* (*hb*), a somatic gene, are preferentially expressed in the neuroblasts. Very similar patterns of *hb* expression were also identified in the green peach aphid *M. persicae*, suggesting that germline expression of *hb* is conserved within the Aphididae (Chung et al., 2018). The Vasa (Vas) protein in *A. mellifera* is expressed in GSCs, nurse cells, and early oocytes, with no difference in the Vas pattern between these two castes. In later oogenesis of the queen ovariole, the Vas protein and *nanos* (*nos*) mRNA are localized to the posterior pole of oocytes. These oocytes will become viable oocytes that subsequently develop into embryos. However, the worker oocytes lack posterior-localized Vas and *nos*. These oocytes are destined to become trophic oocytes, the food for the colony. Oocytes destined for reproduction also exist in the worker ovariole, where the posterior localization of Vas and *nos* is observed. However, at later stages, aggregated Vas protein and *nos* mRNA detach from the posterior pole, causing developmental failure of the oocytes (Chung et al., 2022).

Among Aphidinea, only Adelgidae and Phylloxeridae (comprising together about 140 species in the world fauna) are obligately oviparous and even keep the ovipositor in adult females, whereas all other families (with about 5,000 species) demonstrate obligate

期和成年期卵母细胞的凋亡。通过透射电镜观察表明，幼虫蜕皮至3龄后，生殖细胞和体细胞变得碎片化，并被其他细胞吞噬，呈现凋亡特征。在4龄和5龄幼虫中，卵巢中生殖细胞和体细胞的凋亡变得更严重，导致工蜂失去超过90%的卵巢。虽然成年工蜂在卵巢中保留了一些功能性卵巢，但它们的生殖受到蜂后信息素的抑制。通过凋亡基因表达和核形态特征的分子证据表明，卵母细胞和护理细胞的程序性细胞死亡与成年工蜂不育有关。在蚂蚁体内，受精卵发育成雌性，具有分化成蚁后和工蚁的潜力。工蚁拥有一对卵巢，每一对卵巢都比蚁后的卵巢更少、更短。在卵巢中，生殖干细胞（GSC）在生殖原区进行分裂，在两个社会等级中产生正常的包囊和卵母细胞。在果蝇中，种系基因 *vasa* 编码的蛋白质特异性仅限于于生殖细胞，而体细胞缺裂基因 *hb* 的产物则优先在成神经细胞中表达。在桃蚜中也发现了非常相似的 *hb* 表达模式，表明 *hb* 的种系表达在蚜科中是保守的（Chung 等，2018）。Vasa（Vas）蛋白在蜜蜂 GSCs、营养细胞和早期卵母细胞中表达，这两种社会等级之间的 Vas 模式没有差异。在卵母细胞后期卵子发生过程中，Vas 蛋白和 *nanos*（*nos*）mRNA 定位于卵母细胞的后极。这些卵母细胞将成为活卵母细胞，随后发育成胚胎。然而，工蜂卵母细胞缺乏后部定位的 Vas 和 *nos*。这些卵母细胞注定要成为营养卵母细胞，即群体的食物。用于生殖的卵母细胞也存在于工蜂卵巢中，在那里可以观察到 Vas 和 *nos* 的后定位。然而，在后期，聚集的 Vas 蛋白和 *nos* mRNA 从后极分离，导致卵母细胞发育失败（Chung 等，2022）。

在蚜虫中，只有球蚜科和叶蚜科（在世界动物群中总共约140种）是专性卵生的，甚至在成年雌性体内保持产

viviparity (rarely ovoviviparity) in parthenogenetic generations retaining oviparity in the bisexual generation only. The uniqueness of the aphid viviparity lies in the very precocious start of the embryogenesis in the parthenogenetic egg, before the birth of the mother itself. Such eggs are very small, lacking yolk and chorion; the entire embryogenesis occurs inside the vitellarium, and the egg receives nutrition directly from the cells of the follicular epithelium (Gavrilov-Zimin, 2021). Both nymphs and adults of *A. craccivora* attack seedlings, leaves, flowers and pods of *L. sativus*, and cause direct damage to the plant by sucking cell sap. Adult viviparous females lay nymphs on leaves of L. sativus, and during severe infestation, females also lay nymphs on flowers (Mitra et al., 2020). The pea aphid life cycle includes an asexual (viviparous) and a sexual (oviparous) phase.There are striking spatiotemporal differences between viviparous and oviparous development. Viviparous embryogenesis is an order of magnitude faster than oviparous embryogenesis. Viviparous eggs are also substantially smaller at the time of key patterning events in early development, with nearly a fivefold difference in egg length. With the marked fecundity benefits of the asexual/viviparous mode, oviparous development is generally regarded as a seasonal feature (Lin et al., 2022).

Aphids have evolved bacteriocytes or symbiotic host cells that harbor the obligate mutualistic bacterium *Buchnera aphidicola*, which supplies the host with vitamins and essential amino acids, such as arginine and methionine that aphids cannot synthesize or derive insufficiently from their diet, the phloem sap of plants. *B. aphidicola* lives within large aphid cells, called “bacteriocytes.” Bacteriocytes and another type of cell that is small and flattened, called “sheath cells” are grouped into organ-like structures, referred to as “bacteriomes”. Because of the large cell size (approximately 100 μm in diameter) of bacteriocytes and their pivotal role in nutritional symbiosis, it is considered that these cells are highly polyploid and assumed that bacteriocyte polyploidy may be essential for the symbiotic relationship between the aphid and the bacterium. Moreover, aphid bacteriocytes should be highly metabolically active, because they provide *B. aphidicola* with a large amount of non-essential amino acids derived from the phloem sap, such as glutamate and aspartate, and other metabolites that the symbiont can no longer produce,

卵器，而所有其他科（约 5 000 种）在单性生殖世代中表现出专性胎生（极少卵胎生），仅在两性世代保持卵生。蚜虫胎生的独特性在于孤雌生殖卵在母体出生之前就开始胚胎发生。这种卵很小，缺乏卵黄和绒毛膜，整个胚胎发生在卵黄内，卵子直接从滤泡上皮细胞获得营养（Gavrilov-Zimin，2021）。花生蚜的若虫和成虫都会损害山黧豆的幼苗、叶片、花朵和荚果，并通过吮吸细胞汁液对植物造成直接伤害。成年胎生雌虫将若虫产在山黧豆的叶子上，在严重侵染期间，雌虫也将若虫产在花朵上（Mitra 等，2020）。豌豆蚜虫的生命周期包括无性（胎生）和有性（卵生）阶段。胎生和卵生发育之间存在显著的时空差异。胎生胚胎发生比卵生胚胎发生快一个数量级。在早期发育的关键模式事件发生时，胎生卵子也要小得多，卵子长度相差近五倍。由于无性生殖 / 胎生模式具有显著的生殖力效益，卵生发育通常被视为一种季节性特征（Lin 等，2022）。

蚜虫已经进化出细菌细胞或共生宿主细胞，它们是专性共生细菌蚜虫布鲁氏菌属的宿主，它为宿主提供维生素和必需的氨基酸，如精氨酸和蛋氨酸，这些是蚜虫不能合成或从其食物，植物的韧皮部汁液获取不足。蚜虫布鲁氏菌属生活在被称为“细菌细胞”的大蚜虫细胞内。细菌细胞和另一种扁平的小细胞，被称为“鞘细胞”，被归为器官样结构，被称为“怀菌体”。由于细菌细胞的大细胞尺寸（直径约为 100 μm）及其在营养共生中的关键作用，这些细胞被认为是高度多倍体，且细菌细胞多倍体可能是蚜虫和细菌之间共生关系的关键。此外，蚜虫细菌细胞应该具有高度的代谢活性，因为它们为蚜虫提供了大量来自韧皮部汁液的非必需氨基酸，如谷氨酸和天冬氨酸以及共生体由于大量基因丢

owing to the massive gene losses. Therefore, polyploidy may be a key element in the maintenance of the symbiotic system, as has been shown in between leguminous plants and rhizobium bacteria (Nozaki et al., 2022).

The evolution of scale insects seems to show multiple cyclic conversions from oviparous reproduction to ovoviviparous/viviparous reproduction with the emergence of new peculiar adaptations for eggs' protection. Thus, in archaeococcids, the initial facultative ovoviviparity with the formation of loose ovisac ("primitive" genera of Xylococcinae s.l., most of Callipappinnae s.l.) evolves into normal oviparity in their probable descendants (Margarodinae s.s., some Monophlebinae and Ortheziidae) showing different new adaptations, such as laying eggs in a special cavity under the body or in a solid wax sac behind the body. In turn, some divergent Monophlebinae and their descendants (Phenacoleachiidae, Carayonemidae, and "primitive" neococcids) demonstrate again incomplete or complete ovoviviparity putting partly developed embryos inside the marsupium or laying fully developed embryos outside the body (Gavrilov-Zimin, 2021).

16.5.3 Paedogenesis

The two main types of atypical metamorphosis are neometaboly, a deviation of hemimetaboly, where quiescent pupal-like stages precede adult metamorphosis, and metamorphoses leading to paedomorphism, or the retention of juvenile features in adults. Neometaboly was described only in insects that normally undergo hemimetaboly. This includes all members of Thysanoptera, or thrips (>6,000 species), and the Sternorrhyncha lineages Aleyroidoidea, or whiteflies (~1,500 species) and males of Coccomorpha, or scale insects (>8,000 species). In contrast, paedomorphic adults are found in lineages both undergoing hemimetaboly and holometaboly (Vea1 et al., 2021). All female scale insects and the neotenic caste of some termites (Isoptera) belong to normally hemimetabolous groups. In Holometabola, paedomorphic adults are found in the females of the fascinating Strepsiptera or twisted wing insects, but also some 2,500 species of elateroid beetles (Coleoptera) across several families - fireflies Lampyridae being a famous example, and a few species of flies (Cecidomyiidae and Sciaridae).

(Li Wenchu)

失而无法再产生的其他代谢物。因此，多倍体可能是维持共生系统的关键因素，正如豆科植物和根瘤菌细菌之间所显示的那样（Nozaki 等，2022）。

蚧虫的进化似乎表现出从卵生生殖到卵胎生 / 胎生生殖的多重循环转换，并出现了新的卵保护特殊适应。因此，在木珠蚧亚科中，随着松散卵囊的形成，最初的兼性卵胎生（“原始”代木珠蚧亚科，大多数木球球虫科）在其可能的后代（珠蚧亚科，一些单翅目和旌介壳虫科）中演化为正常卵子胎生，显示出不同的新适应，例如在身体下方的特殊腔中或在身体后面的固体蜡囊中产卵。反过来，一些不同的单翅目及其后代（同翅目眼蚧科、蚧科和“原始”新蚧科）再次表现出不完全或完全卵胎生，将部分发育的胚胎放入育儿袋内或将完全发育的胚胎置于体外（Gavrilov-Zimin，2021）。

16.5.3 幼体生殖

非典型变态的两种主要类型是偏离不完全变态的转变态，其中休眠的蛹样阶段先于成年变态以及导致幼体生殖的变态或成虫保留幼虫特征。新变态仅在正常进行不完全变态的昆虫中被发现。这包括刺翅目或蓟马（＞6 000 种）的所有成员以及胸喙亚目谱系的粉虱总科或粉虱（～1 500 种）和蚧亚目的雄性或蚧虫类（＞8 000 种）。相反，在经历不完全变态和完全变态的谱系中都发现有幼体生殖的成虫（Vea1 等，2021）。所有雌性蚧虫类和一些白蚁（等翅目）的新生社会等级都属于正常的不完全变态群。在完全变态物种中，成虫幼体生殖出现在迷人的胸喙亚目或扭翅昆虫的雌性中，也发现有跨越几个科的 2 500 种长角甲虫（鞘翅目）——萤火虫是一个著名的例子和几种苍蝇（瘿蚊科和蚊科）。

（李文楚）

Exercises

1.How to classify the types of female insect ovaries and ovarioles ?

2.How hormones regulate the process of of ovary development and synthesis of egg proteins?

3.What functions of secretions from accessory glands of female insect? Describe the behavioral and physiological changes induced.

4.How insects demonstrate their diversity of reproduction?

思考题

1. 如何对雌性昆虫卵巢和卵巢管进行分类?

2. 激素如何调节卵巢发育和卵蛋白合成的过程?

3. 雌性昆虫副腺分泌物有什么功能? 描述由此引起的行为和生理变化。

4. 昆虫如何展示其繁殖的多样性?

第 17 章 昆虫滞育

Chapter 17 Insect Diapause

Key points

1.Diapause is one of the strategies for insects to cope with adverse environments. Different insects can occur in different developmental stage. It is determined by genetic type. Conversely, stimulation of fitness environmental factors can break diapause and continue development.

2.The general mechanism of diapause regulated by hormones is that sunlight, surviving environmental factors, circadian rhythm, development stage and other factors stimulate insect receptors, and through nerve conduction, it starts the synthesis of neuropeptides such as diapause hormone, insulin-like peptides, juvenile hormone and ecdysone, result in triggering diapause behavior.

3.Surprisingly, none of clock and timing genes directly control diapause, but numerous genes are up-regulated, or down-regulated, or are intermittently expressed during diapause.

4. Insect diapause involves many physiological changes and biochemical reactions, such as sugar metabolism entering the tricarboxylic acid cycle and starting sorbose, polyols, etc. as antifreeze. Through the transmission of various signal molecules, diapause is finally relieved.

知识点

1. 滞育是昆虫应对不良环境的策略之一。不同的昆虫可以在不同的发育阶段发生。这是由遗传类型决定的。相反，适当的环境因素刺激可以解除滞育并继续发育。

2. 滞育受到激素调控的普遍机制是日照、生存环境因素、昼夜节律和发育阶段等刺激昆虫感受器，通过神经传导启动神经肽类物质以及滞育激素、胰岛素样多肽、保幼激素和蜕皮激素等的合成，从而引发滞育行为。

3. 令人惊讶的是，没有一个时钟和定时基因直接控制滞育，但有许多基因在滞育期间上调或下调或间歇性表达。

4. 昆虫滞育牵涉到很多生理变化和生化反应，比如糖的代谢进入三羧酸循环，并启动山梨糖、多元醇等作为抗冻剂。通过各种信号分子传导，最终解除滞育。

Insects have evolved some adaptive mechanisms to respond challenging from the extreme environment. Three of the ways that insect adapt to an unfavorable environment include dormancy, quiescence, and diapause. Dormancy is a broadly encompassing term that has been used to describe any state of suppressed development that is ecologically or evolutionarily meaningful, usually accompanied by suppressed metabolism. So, dormancy encompasses two very different states: quiescence and diapause. Quiescence is driven directly by the dynamics of environmental factors, whereas diapause is determined by a predictive mechanism combining environmental signals and

昆虫进化了一些适应机制，以应对来自极端环境的挑战。昆虫适应不良环境的三个途径包括冬眠、休眠和滞育。冬眠是广义的包含词语，用于描述生态和进化意义上的任何被抑制的发育状态。因此，冬眠包括两种完全不同的状态：休眠和滞育。休眠是由动态环境因素直接驱动的，而滞育是由环境信号和内部

an internal hormonal mechanism.

Quiescence is an immediate response of insect to temporarily unfavorable environmental conditions, such as a few days of unseasonably cold weather in the life of an insect; the insects typically becomes quiet and physiological responses slow down. It is not genetically programmed or regulated, and with the return of favorable environmental conditions, the insect responds immediately with resumption of normal activity.

Diapause is a genetically regulated process which represent an alternative developmental pathway prompted by unique patterns of gene expression that result in the sequestration of nutrient reserves, suppression of metabolism, a halt or slowing of development, and the acquisition of increased tolerance to environmental stress. Diapause is usually classified into two typical phenotypes: low metabolic activity and extreme lifespan extension. Diapause often begins while favorable environmental conditions still exist and does not end with temporary return of favorable conditions. The environmental token cues that cause insects to enter diapause include drought, desiccation, low moisture content of the food, scarcity of food, high temperature, low temperature, critical day length, crowding, maternal diet, maternal age at oviposition, and maternal exposure to certain environmental conditions at particular periods in its life.

Drosophila periodically enter a state of quiescence that meets a series of criteria for sleep: ①this quiescent state is characterized by an increased arousal threshold (decreased responsiveness to sensory stimuli). ② it can be distinguished from coma or anesthesia by its rapid reversibility with a stimulus that is sufficiently strong. ③ the timing of sleep is regulated by the circadian clock, although these two processes can also be separated; flies with mutations in the core clock genes have fragmented sleep across the day, but can have normal overall sleep amounts, and flies with mutations that result in very low total sleep amounts can still show robust circadian activity rhythms. ④ sleep is also homeostatically regulated, such that when flies are deprived of sleep using a mechanical stimulus, they compensate with longer and deeper sleep the following day. This suggests that sleep serves an important restorative function rather than simply reflecting ecologically advantageous periods of inactivity. ⑤flies experience broad changes in neuronal activity during sleep (Dubowy et al., 2017).

激素机制相结合的预期机制决定的。

休眠是昆虫对短暂的不良环境条件的快速反应，如昆虫生活史中少数几天非季节性的寒冷天气；昆虫典型地变得宁静，生理反应缓慢。这种休眠状态不是遗传性的编码或可调节的，一旦回复到有利的环境条件，昆虫立即反应并恢复正常活动。

滞育是遗传调节的过程，代表一种选择性的发育途径，由独特的基因表达模式推动，导致贮存营养的吸收、代谢的抑制、发育的停滞或减速以及获得对环境胁迫耐受的增加。滞育通常分为两种典型的表型：低代谢活性和极端寿命延长。滞育通常在有利的环境条件尚存的时候就开始了，并且不会随着有利条件的短暂回调而终止。造成昆虫进入滞育的环境标记讯号包括干旱、干燥、食物中水分含量低、食物短缺、高温、低温、日长临界、拥挤、母性节食、母体产卵时的年龄以及在母体特定时期暴露在某种环境条件下。

果蝇周期性地进入一种满足一系列睡眠标准的静止状态：①这种静止状态的特征是唤醒阈值增加（对感觉刺激的反应性降低）。②它与昏迷或麻醉的区别在于其具有在足够强的刺激下快速的可恢复性。③睡眠的时间是由昼夜节律调节的，尽管这两个过程也可以分开；核心时钟基因突变的果蝇一整天的睡眠都是零散的，但总睡眠量可以是充足的，而总睡眠量非常低的突变果蝇仍然可以表现出强大的昼夜活动节律。④睡眠也是动态调节的，因此当果蝇被机械刺激剥夺睡眠时，它们会在第二天以更长、更深的睡眠作为补偿。这表明睡眠具有重要的恢复功能，而不仅仅是反映出处于生态有利时期的不活动状态。⑤果蝇在睡眠期间经历了神经元活动的广泛变化（Dubowy 等，2017）。

17.1 滞育的分类与滞育阶段

17.1 Classifications and Phases of Diapause

Species are known to diapause in the embryonic stage, as larvae, as pupae, or as adults. Diapause is obligatory for some insects, typically those with a univoltine life cycle, so that every generation enters diapause during the same point in the life cycle. In most insects, diapause is facultative, and as long as favorable conditions exist, generations continue without diapause, but when unfavorable conditions are eminent, diapause is entered. However, in *N. vitripennis*, maternal diapause induction follows a "switch" pattern: When exposed to short photoperiods, females will produce non-diapausing offspring during the first days of adulthood, before rapidly switching to producing only diapause-destined offspring for the rest of their adult lifespan, which revealed a functional role of the microbiome for nutrient allocation and mobilisation during diapause(Dittmer and Brucker, 2021).

已知昆虫种群在胚胎时期、幼虫期、蛹期或成虫期滞育。滞育在有些昆虫是强制性的，典型的是那些一化性生活史的昆虫，因此，每个世代在生活史的相同时段都会进入滞育。多数昆虫的滞育是任意的，只要存在有利条件，世代继续而不滞育，但当不利条件突出时，即进入滞育。然而，金小蜂母体滞育诱导遵循一种"转换"模式：当暴露于短光周期时，雌性会在成年后的第一天产生非滞育后代，然后在其成年后的剩余寿命内迅速转换为只产生滞育的后代，这揭示了微生物群在滞育期间对养分的分配和动员的功能性作用（Dittmer 和 Brucker, 2021）。

Insects programmed to diapause pass through succes sive phases, including prediapause induction and preparation, initiation of diapause, continuing maintenance, termination of diapause, and postdiapause physiology and behavior.

昆虫程序化的滞育经历连续的阶段，包括滞育前期的诱发和准备期、滞育启动期、连续保持期、滞育终止期和滞育后的生理和行为。

17.1.1 Prediapause: induction and preparation

17.1.1 滞育前期：诱发和准备

The prediapause program typically begins well before environmental conditions become adverse for further development or survival. Enter into diapause is not a sudden, triggered event, but typically a slow process, with changes occurring gradually. Some parasitoids enter diapause when their host enters diapause and when they depend on the host's endocrine system to terminate diapause. Sometimes the presence of a parasitoid in a host can cause the host to fail to enter diapause when it otherwise would have done so.

典型的滞育前期的程序在环境条件变得不利于进一步发育或生存之前开始。进入滞育不是一个突发的触发事件，而是一个典型的缓慢的过程，变化是逐渐发生的。有些寄生虫在其宿主进入滞育时滞育，并依赖宿主的内分泌系统终止滞育。有时，宿主体内出现寄生虫可导致宿主无法进入滞育，否则寄主将进入滞育状态。

During the diapause induction, the metabolism and growth are slowed down intentionally in insects by picking up the photoperiodic clock signals and transmitting them to neurosecretory

在滞育诱发过程中，昆虫的代谢和生长通过获取光周期时钟信号并将其传输到大脑和内分泌器官中的神经分泌细

cells in the brain and endocrine organs. Photoperiod, circadian rhythms, and neuroendocrine systems regulate diapause.

Overwintering insects that diapause in order to survive cold temperatures typically synthesize low molecular weight polyhydric alcohols, such as glycerol, trehalose, and sorbitol, and accumulate energy reserves as lipids that are stored in the fat body as triacylglycerols. The polyols may serve as cryoprotectants, or in some cases, they may have a role in control without feeding or taking water, and accumulation of some reserves before diapause and conservation of resources during diapause are important to survival and to fitness of the niches when diapause is terminated. During the pre-diapause phase, insects perceive the triggering stimulus from the environment but continue to develop. Research on diapause induction in *B.mori* has shown that rearing conditions of 25℃ with a 14h light/10h dark cycle can induce diapause. The external direct development of individual form (morphogenesis) begins to gradually slow and eventually ceases during diapause. Finally, during post-diapause, the insects immediately resume development when advantageous environmental conditions arise (Qi et al., 2015).

Two basic mechanisms by which insects survive subzero temperatures are generally recognized: freeze tolerance, the ability to survive freezing of body fluids, usually only of the extracellular fluids, and freeze avoidance in freeze-intolerant species which are freeze susceptible and die if frozen. Low temperatures can induce numerous potentially lethal problems including metabolic imbalance, membrane malfunction, oxygen stress (hypoxia, ischemia, as well as oxidative stress requiring antioxidants), at temperatures above those where the insect actually freezes, and which are generally exacerbated as temperatures continue to decrease. Consequently, insects that successfully winter at subzero temperature extremes must have adaptations to overcome these pre-freeze problems, as well as those that pertain to actual freezing, both avoidance and tolerance.

The three types of insect ice-binding proteins (IBPs) primarily in the context of their roles in subzero temperature tolerance: ①antifreeze proteins (AFPs) produce significant body fluid thermal hysteresis (TH) and are present in freeze-avoiding species where they function to prevent ice crystal growth and/or ice nucleation. ② recrystallization inhibition proteins (RIPs) produce significantly less TH *in vivo*, and function mainly to inhibit recrystallization in

胞从而有意识地减慢这一过程。光周期、昼夜节律和神经内分泌系统调节滞育。

越冬昆虫滞育是为了在寒冷气温下生存，通常合成低分子质量的多羟基醇，如甘油、海藻糖、山梨醇，并以脂质形式积聚能量，在脂肪体内以三酰基甘油的形式贮存。多元醇可作为冷冻防护剂，或者在某些情况下，它们在不取食或吸收水分时起到控制的作用。滞育前某些贮存物质的积聚和滞育中的资源保存对生存和滞育终止后适应微环境是重要的。在滞育前阶段，昆虫感受到来自环境的触发刺激，但仍在继续发育。对家蚕滞育诱导的研究表明，25 ℃、14 h 光照 /10 h 暗周期的饲养条件可以诱导滞育。个体形态的外部直接发育（形态发生）开始逐渐变慢，并在滞育期间最终停止。最后，在滞育后阶段，当有利的环境条件出现时，昆虫立即恢复发育（Qi 等，2015）。

昆虫在零下温度下生存的两个基本机制得到了普遍认可：抗冻性，在体液，通常只有细胞外液冻结后生存的能力以及在不耐冷冻物种中避免冷冻，这些物种是冷冻敏感的，一旦冷冻就会死亡。低温会引发许多潜在的致命问题，包括温度高于昆虫实际结冰温度时的代谢失衡、膜功能紊乱和氧应激（缺氧、缺血以及需要抗氧化剂的氧化应激），并且随着温度持续的下降，这些问题通常会加剧。因此，在零度以下极端温度下成功过冬的昆虫必须具有适应能力，避免和耐受都可以克服这些预冻问题以及与实际冻结有关的问题。

这三种类型的昆虫冰结合蛋白（IBPs）在低温耐受性方面的主要作用：①抗冻蛋白（AFPs）会产生明显的体液热迟滞（TH），出现在避冻物种中，它们起到阻止冰晶生长和 / 或冰核形成的作用。②重结晶抑制蛋白（RIPs）在体内产生的 TH 显著减少，其功能主要是

freeze-tolerant insects. ③ ice nucleator proteins (INPs) lack TH activity and inhibit supercooling by initiating freezing at higher subzero temperatures than would otherwise be the case (Duman et al.,2020).

抑制耐寒昆虫的重结晶。③ 冰核蛋白（INPs）缺乏 TH 活性，并通过在比其他情况下更高的零下温度开始冷冻来抑制过冷（Duman 等，2020）。

17.1.2 Diapause: initiation and maintenance

17.1.2 滞育：启动和保持

Initiation of diapause defines as that point in time when direct development ceases, as, for example, when the insect molts into a specific diapause stage that has characteristic color or morphological features. Determining that diapause has been initiated requires dissection of insects to detect internal changes in tissue, such as the failure of ovaries to grow and eggs to develop when reproductive diapause occurs. Another characteristic of initiation of diapause is the decrease in metabolic rate as measured by oxygen consumption, but the change in metabolic rate may be gradual in those species that remain mobile in diapause.

Possible molecular mechanism for diapause initiation might: ① transition of metabolism and energy utilization. In addition to a decrease of metabolic activity, metabolic pathways are also changed in diapause-destined pupae at diapause initiation. Anaerobic metabolism pre dominates, and sugars and polylols accumulate in the brain. ② enhancement of stress resistance. The antifreeze agents glycerol and sorbitol as well as HSP, GST, and others are heavily synthesized to protect the insect from rigorous environmental conditions. ③ regulation of cellular development. The cell cycle is arrested, resulting in repression of pupal development toward adulthood. ④ repression of transcription and translation. The upregulation of transcriptional repressors, downregulation of translational activators, and increased protein SUMOylation (small uniquitin-related modifier) result in decreases of both gene transcription and protein translation at diapause initiation (Bao and Xu, 2011).

滞育的启动定义为经过一段时间后发育完全停止的准确时间点，比如，当昆虫蜕皮进入一个特定的滞育期，具有特殊色彩或形态学上的特征。确定滞育已经启动需要解剖昆虫以检测组织的内部变化，如生殖滞育时卵巢无法成长，卵不能发育。另一个滞育启动的特征是通过测定耗氧量确定代谢率的降低，但那些滞育期仍然活动的品种代谢率的变化可能是缓慢的。

滞育起始的可能分子机制：①新陈代谢和能量利用的转变。滞育起始时，除了降低代谢活性，滞育蛹的代谢途径也发生了改变。厌氧代谢占主导地位，在脑中积累糖和多元醇。②增强抗压能力。大量合成防冻剂甘油和山梨醇以及 Hsp、GST 和其他化合物以保护昆虫免受极端环境条件影响。③细胞发育的调节。细胞周期被阻滞，导致向成虫期发育的蛹受到抑制。④转录和翻译的抑制。滞育开始时转录抑制因子上调，翻译活化因子下调，蛋白质的 SUMO 化（小泛素类修饰）增加导致基因转录和蛋白质翻译都减少（Bao 和 Xu, 2011）。

17.1.3 Diapause termination

17.1.3 滞育的终止

Diapausing insects gradually become more sensitive to diapause terminating conditions as diapause intensity decreases, but like the initiation of diapause, this is a gradual process. Termination usually depends on reception of token stimuli concerning the environment. The physiology of termination often commences long before the insect manifests behavior that clearly means diapause is over. Factors that lead to diapause termination usually are environmental conditions, but the transduction mechanisms by which signals are

滞育昆虫在滞育强度降低时对滞育终止条件逐渐变得更为敏感，但正如滞育的启动，这是一个渐变的过程。滞育终止通常依赖于接受与环境有关的标志性的刺激。昆虫生理上的滞育终止通常在显露出明显的滞育结束行为之前很长一段时间就已经开始。导致滞育终止的因素通常是环境条件，但多数昆

converted into physiological and biochemical processes leading to diapause termination and resumption of development in most species are not resolved.

虫通过信号转换到生理和生化过程导致滞育终止并恢复发育的转换机制尚未阐明。

17.2 昆虫滞育的激素调控

17.2 Insect Diapause under Hormone Control

17.2.1 Embryonic diapause

Many temperate zone insects have an egg or embryo diapause. Diapause of *B. mori* is facultative, but a univoltine strain is known in which diapause is obligatory. Whether an embryo is destined to diapause depends on the photoperiod to which the mother was exposed during her development as a larvae. Female silkworm moths that grow up in the early spring under short-day conditions lay eggs that do not diapause, but hatch into the summer generation. A diapause hormone (DH) produced by female growing up under long-day summer conditions is secreted from the subesophageal glands and transferred to her eggs, which then enter diapause about 2 days after being laid. The adaptive benefit for embryos in eggs laid in late summer to diapause is that they would not have time to grow to maturity and pupate before the arrival of winter; thus, they survive the winter in diapause. A period of prolonged cold exposure in the winter followed by a return of springtime temperatures terminates diapause, and the larvae become the spring generation growing under short-day conditions. This spring generation of adults will again lay non-diapausing eggs and the summer generation of adults will lay eggs in which embryos will spend the winter in diapause.

The *B. mori* model is based on embryonic diapause just after the eggs are laid and before the neuroendocrine system of the embryo has developed. Species in which the diapause occurs in older embryos may have different regulatory controls than the maternal one in *B.mori,* such as grasshopper, *M. sanguinipes*. Eggs are laid in late summer or early fall and the embryo usually spends the winter in diapause after completing about 89% of embryonic development. The embryonic diapause may be obligatory or facultative in different parts of its rang, and embryos may enter diapause at different ages.

17.2.1 胚胎滞育

许多温带昆虫经历卵或胚胎滞育。家蚕的滞育是任意的，但一化性品种的滞育却是强制性的。一个胚胎是否预定要滞育，依赖于母体幼虫发育期间所暴露的光周期。在早春短日照条件下长大的雌性蚕蛾产下的卵不滞育，但孵化成夏季世代。雌性家蚕在长日照的夏天成长，产生的滞育激素（DH）由食道下腺分泌并转移到卵，因此，卵在产下后约 2 d 进入滞育。夏季末产下的卵，胚胎进入滞育的适应性优势是它们没有时间成长至发育成熟并在冬季来临时化蛹，这样，它们在滞育中度过了冬天。在冬天长时间的置冷，接着回复春季气温终止了滞育，其幼虫成为在短日照条件下成长的春季世代。春季世代的成虫再产下非滞育卵，夏季世代的成虫将产下胚胎滞育越冬的卵。

家蚕模式是基于产卵后和胚胎内分泌系统发育之前的胚胎滞育。老龄胚胎发生滞育的物种，如蚱蜢，与家蚕的母系调节相比，具有不同的调节控制。卵在夏末或早秋产下，胚胎通常在完成发育的 89% 时于滞育中度过冬天。胚胎的滞育在不同的生活范围内可以是任意的或强制性的，胚胎也可以在不同的时期进入滞育。

17.2.2 Larval diapause

Larval diapause is common among species of Lepidotera, Diptera, Hymenoptera, Coleoptera, Neuroptera, Odonata, Orthoptera, Homoptera, Hemiptera, and Plecoptera. Any instar may diapause, but diapause of last instar is more common. Diapausing larvae may display active movement and feeding. Some diapausing larvae molt "in place," i.e., they molt without growing into a large instar. Those that have stationary molts usually lose weight because they use body reserves without replacement during the diapause. Hormonal regulation of larvae diapause is consistent in many species with a high titer of JH and relative inactivity of the ecdysteroid-producing prothoracic glands or their failure to produce enough ecdysone to counter the level of JH present. Exogenous DH break the larval diapause of *O. fuscidentalis* by increasing the levels of ecdysone in the hemolymph; the hormonal mechanism of DH activates the PG to release ecdysone, and then stimulates the larvae to break the diapause (Suang et al., 2019). There are some species, however, in which there is no apparent role for JH in larval diapause, and a hormonal role, if any, is unknown.

17.2.2 幼虫滞育

幼虫滞育在鳞翅目、双翅目、膜翅目、鞘翅目、脉翅目、蜻蜓目、直翅目、同翅目、半翅目和襀翅目的种内很普遍。任一龄期可滞育，但末龄的滞育更为普遍。滞育幼虫显示出活跃的运动和取食。有些滞育幼虫“适时”的蜕皮，即它们蜕皮后不进入大龄。那些静止蜕皮的昆虫通常会减轻体重，因为它们在滞育中使用了身体的贮备而没有更换。在许多物种内，幼虫滞育的激素调节是一致的，具有高 JH 滴度以及产生蜕皮甾类的前胸腺相对不活跃，或者它们无法产生足够的蜕皮激素来抵消出现的 JH 水平。外源 DH 通过增加竹虎天牛体内血淋巴蜕皮激素的水平来打破幼虫的滞育；激素作用的机制是 DH 激活 PG 释放蜕皮激素，然后刺激幼虫打破滞育（Suang 等，2019）。然而，有些物种幼虫滞育时 JH 没有明显的作用，激素的作用即使有，也尚未可知。

17.2.3 Pupal diapause

Pupal diapause is a very common overwintering strategy. *H. cecropia* is a univoltine species that over-winters as a diapausing pupae, entering diapause shortly after pupation. It does not break diapause in the north-eastern United States until late in the spring or early summer, depending upon the location, when mulberry leaves are available on which females lay eggs and larvae feed. A period of cold exposure was necessary for diapausing *H. cecropia* pupae to break the diapause, and keeping them in a refrigerator at about 5℃ for at least 6 weeks and then returning them to room temperature allowed adult development in a few weeks. Diapause could be prolonged by holding pupae at 5℃ for many months and then return them to room temperature. Brain hormone (PTTH) was not secreted until the brain had been chilled for a certain period of time; without PTTH, the prothorathic glands were not stimulated to secrete ecdysone, and without ecdysone, development of the adult could not occur. Exogenous DH terminates pupal diapause in *H.armigera* by increasing the hemolymph ecdysone titer. JH likely acts on the SG to synthesize DH, and then DH activates the PG to cause ecdysone synthesis.

17.2.3 蛹滞育

蛹滞育是一种非常普遍的越冬策略。天蚕是以蛹越冬一化性物种，化蛹后在很短的时间内进入滞育。在美国北部，它直到春末或初夏才会打破滞育，具体取决于地点，当雌蛾产卵的桑叶充裕时，幼虫即取食。滞育的天蚕要打破滞育必须有一段时间的寒冷。它们被放置在大约 5℃的冰箱至少需要 6 周，然后放回室温使成虫发育数周。将蛹保持在 5℃几个月，然后将它们放回室温可以延长滞育时间。直到脑被冷却一段时间后，脑激素（PTTH）才分泌。没有 PTTH，前胸腺则不被刺激以分泌蜕皮激素；没有蜕皮激素，成虫的发育则不能发生。外源 DH 通过提高血淋巴蜕皮激素滴度终止棉铃虫的蛹滞育。JH 可能作用于 SG 合成 DH，然后 DH 激活 PG 引起蜕皮激素合成。

The flesh flies, *S. crassipalpis* and *S. bullalata*, have a facultative pupal diapause. The flies *S. crassipalpis*, diapause in the pupal stage when exposed to short-day condition (12 hours of light) at 25℃. Ecdysone is absent or at least very low in diapausing pupae, but diapause can be broken by injection of ecdysone, implicating its absence as an important factor in diapause. Expression of EcR remains detectable and is not down-regulated during pupal diapause, but transcripts of the *USP* gene are down-regulated and only begin rising near the end of normal diapause, rising rapidly upon termination of diapause. Availability of USP may be a major factor in diapause termination in *S. crassipalpis*. USP is the partner needed by the EcR protein in order to bind 20E and form a functional transcription factor for genes involved in development.

FMRFamide, a cardioexcitatory tetrapeptide, participate in a variety of physiological processes, including basic life functions, such as feeding, reproduction, diapause, and development in *M.brassicae*, *L.maderae*, and *S. gregaria etc*, probably involved in photoperiodic diapause induction (Hao et al., 2020). The unifying mechanism in insects that diapause as papuae is the lack of sufficient ecdysone to stimulate adult development, but the mechanism that cause the lack of ecdysone may diverse in different species. Some species that have been studied depend on an intact brain only briefly at the beginning of pupac diapause, whereas *H. cecropia* requires that the brain must be chilled for weeks before it becomes competent to secrete PTTH. There is no evidence in *H. cecropia* that JH is involved in diapause, but in diapausing *S. crassipalpis*, JH have a role in initiating, maintaining and terminating pupal diapause.

17.2.4 Adult diapuse

Adult diapause is common in species of Coleoptera, Lepidoptera, Diptera, Heteroptera, Orthoptera, Neuroptera, and Trichoptera. Several hormones are known to be required for reproduction in female insects, but it is generally believed that loss titer or lack of JH is crucial in regulating adult diapause.

Environmental cues and sensory modalities required to accurately interpret environmental condition for both migration and diapause initiation in monarch butterfly (*D. plexippus*) as well as hormonal mechanisms that control monarch diapause development. In preadult diapauses, 20E is most often recognized as the diapause terminator. 20E plays opposing roles in different organisms,

红尾肉蝇和麻蝇有任意的蛹滞育。红尾肉蝇在 25℃短日照条件下（12 h 光照）以蛹期滞育。滞育蛹缺乏蜕皮激素或至少浓度极低，但滞育可被注射蜕皮激素解除，表明缺乏蜕皮激素是滞育的一个重要因素。在滞育中，EcR 保持可检测水平并没有下调，但 *USP* 基因的转录下调了，仅在接近正常滞育末期开始回升，滞育终止时迅速上升。红尾肉蝇可利用的 USP 是滞育终止的主要因素。USP 是 EcR 蛋白质便于结合 20E，并构成参与发育的功能性基因转录因子所需的搭挡。

酰胺化 FMRF 是一种心脏兴奋性四肽，参与多种生理过程，包括基本的生命功能，如甘蓝夜蛾、马德拉蟑螂和沙漠蝗等的取食、繁殖、滞育和发育，这可能与光周期滞育诱导有关（Hao 等，2020）。昆虫以蛹滞育的一致的机制是缺少足够的蜕皮激素以刺激成虫发育，但不同的种导致蜕皮激素缺乏的机制是具有多样性的。有些研究过的种仅在蛹滞育开始时短时间依赖于完整的脑，而天蚕在有能力分泌 PTTH 前需要冷冻脑数周。天蚕滞育没有牵涉到 JH 的迹象，但滞育中的红尾肉蝇，JH 在启动、维持和终止滞育上起作用。

17.2.4 成虫滞育

成虫滞育在鞘翅目、鳞翅目、双翅目、半翅目、直翅目、脉翅目和毛翅目种内常见。已知雌虫的生殖需要几种激素，但大体上认为保幼激素滴度的丧失或缺乏是调节成虫滞育至关重要的。

帝王蝶（*D. plexippus*）的迁徙和滞育起始，如控制滞育发育的激素、机制一样，需要精准诠释具环境因素和感觉模式。在成虫前滞育中，20E 通常被认为是滞育终止因子。20E 在不同的生物

promoting diapause termination and reproduction. The consistent controller of adult reproductive diapause across insects is JH. JH signalling is required for reproductive development in non-migrating monarchs, and exogenous JH analogue application is sufficient to terminate diapause. JH indirectly regulates larval growth rate through controlling ecdysone concentration, versus its timing, whereby influencing insulin/insulin-like signalling (ⅡS) pathway activity, which is a well known controller of larval growth rate.

JH, ecdysone, and ⅡS pathways interact to achieve coherent responses in different contexts. Calcium signalling mediated environmental sensitivity of the diapause timer, and it is speculated that it is a key integrator of environmental condition with downstream hormonal control of diapause. JH signalling changed spontaneously in diapause-inducing conditions, capacitating response to future environmental condition. Although JH signalling is a general marker of postdiapause quiescence and a key timer target, but is not itself the timer. Epigenetic mechanisms are implicated to be the proximate timing mechanism. Ecdysone, JH, and insulin/ insulin-like peptide signalling are major targets of the diapause program used to control response to permissive environmental conditions (Green Ⅱ et al., 2019). JH may also regulate reproductive diapause through a conserved methoprene tolerant (Met), a universal JH receptor-dependent pathway in *G. daurica* (Ma et al., 2021). Juvenile hormone Ⅲ skipped bisepoxide (JHSB3) titers in hemolymph remarkably differ between reproductive males and females of the linden bug *P. apterus*, whereas no JH was detected in diapausing adults of both sexes.

Like in females, ectopic application of JH mimic effectively terminated male diapause through the canonical JH receptor components, Met and Taiman. In contrast to females, long photoperiod induced reproduction even in males with silenced JH reception or in males with removed corpus allatum, the JH-producing gland. JHSB3 was detected in the accessory glands of reproductive males, unexpectedly, even in males without CA. If there is a source of JHSB3 outside CA or a long-term storage of JHSB3 in accessory glands remains to be elucidated. These sex-related idiosyncrasies are further manifested in different dynamics of diapause termination in *P. apterus* by low temperature

体中对促进滞育终止和生殖起着相反的作用。自始至终控制成虫生殖滞育的是JH。非迁徙帝王蝶的生殖发育需要JH信号，应用外源JH类似物足以终止滞育。JH通过控制蜕皮激素浓度和时间来间接调节幼虫的生长速率，从而影响胰岛素/胰岛素样信号（ⅡS）途径的活性，后者是众所周知的幼虫生长速率控制器。

JH、蜕皮激素和ⅡS途径相互作用，在不同的环境中实现一致的反应。钙信号介导滞育计时器的环境敏感性，推测它是环境条件与滞育下游激素控制的关键整合因子。JH信号在滞育诱导条件下自发改变，具有对未来的环境条件产生反应的能力。虽然JH信号是滞育后期休眠的普遍标记和关键的计时器靶点，但它本身并不是计时器。表观遗传学机制被认为是最接近的时间机制。蜕皮激素、用于控制对允许环境条件反应JH和胰岛素/胰岛素样肽信号是滞育程序的主要目标（Green Ⅱ等，2019）。沙葱萤叶甲JH也可能通过保守的methoprene tolerant (Met)，一种通用的JH受体依赖途径来调节生殖滞育（Ma等，2021）。始红蝽的生殖期雄成虫和雌成虫血淋巴中的双环氧化保幼激素（JHSB3）滴度存在显著差异，而在滞育的两性成虫中均未检测到。

与雌性一样，异位应用JH类似物通过典型的JH受体成分，即Met和Taiman，有效终止了雄性滞育。与雌性相比，即使在JH受体沉默的雄虫或去除产生JH的咽侧体的雄虫中，长光周期也能诱导生殖。出人意料的是，JHSB3在雄性生殖附腺中被检测到，即使在没有咽侧体的雄虫也是如此。是否有JHSB3来源于咽侧体之外或JHSB3在生殖附腺中长期贮存仍有待阐明。这些性别关联特质进一步表明低温引起的始红蝽滞育终止的不同动态（Hejníková

(Hejníková et al., 2022). This demonstrated sexual dimorphism of diapause under JH regulation in the hemipteran bug.

等，2022）。这证实了半翅目昆虫保幼激素调节滞育的两性异形。

17.3　滞育与基因表达

17.3　Diapause and Gene Expression

17.3.1 Diapause genes

The ultimate regulation of diapause lies in genes that regulate the hormonal and metabolic changes associated with entry into diapause, diapause maintenance, and diapause termination and return to normal activity. The gene that are directly involved in diapause, however, remain to be identified. A number of clock and timing genes have been identified, including *period, timeless, dClock, cycle, doubletime*, and *vrille*, but none of this is known to directly control diapause.

Numerous genes are up-regulated, or down-regulated, or are intermittently expressed during diapause. More than 300 proteins occurred in brain tissue of non- diapausing *S. crassipalpis*, but only about 180 were detectable in the brain of diapausing pupae. About 40% of genes active in the brain of nondiapausing pupae were silenced during diapause, and about 10% of the genes expressed during diapause are only expressed at that time. Many of the genes down-regulated during diapause are genes one would expect to be silenced because of reduction in metabolic and physical activity. They are likely a consequence of diapause rather than a cause of diapause, and probably do not play a controlling role in diapause. *HSP90* is down-regulated during diapause and up-regulated when diapause is terminated, presumably a consequence of 20E control of the gene *HSP90*. Ecdysteroids up-regulate *HSP90*, and the gene is down-regulated when ecdysteroids are absent. When down-regulated, however, *HSP90* remains responsive to environmental conditions. Because pupal diapause may be, in part at least, a consequence of ecdysteroids deficiency, it is possible that *HSP90* and ecdysteroid are linked in important physiological ways to diapause, possibly with *HSP90* serving in some way in the functioning of the ecdysone receptor/ultraspiracle (EcR/*USP*) dimer that binds 20E and then binds DNA in the nucleus, allowing transcription of many genes.

17.3.1　滞育基因

滞育的最终调节在于基因对进入滞育、维持滞育和滞育终止以及恢复正常活动相关联的激素和代谢变化的调节。然而，直接参与滞育的基因仍有待确定。已鉴定出大量生物钟和计时基因，包括 *period*、*timeless*、*dClock*、*cycle*、*doubletime* 和 *vrille*，但已知没有任何一个直接控制滞育。

许多基因在滞育中是向上或向下调节的，或是间歇表达的。在非滞育的红尾肉蝇脑组织中被发现有多于 300 种蛋白质，但在滞育蛹脑内仅检测到约 180 种。非滞育蛹脑内的活跃基因大约有 40% 在滞育过程中沉默，约 10% 在滞育期表达的基因仅在其时表达。许多在滞育期向下调节的基因是那些预期沉默的基因，因为这些基因降低了代谢和生理活动。与其说它们像滞育的结果，倒不如说它们是滞育的起因，在滞育中也许不起控制作用。*HSP90* 在滞育中是下调的，当滞育终止时上调，据推测是 20E 控制 *HSP90* 的结果。蜕皮甾类上调 *HSP90*，缺乏蜕皮甾类就下调。即使下调，*HSP90* 也仍保持对环境的响应。因为蛹期滞育可能是蜕皮甾类缺乏的结果。可能在滞育的重要生理途径上 *HSP90* 和蜕皮甾类是联动的。可能借助 *HSP90* 某种方式的作用，蜕皮激素受体 / 超气门蛋白（EcR/*USP*）功能二聚体得以与 20E 结合，然后与核 DNA 结合，允许诸多基因转录。

17.3.2 Daily and seasonal biological clock in diapause

Daily (circadian) and seasonal (photoperiod) biological clocks influence daily and seasonal rhythms in insects. Daily repeating circadian rhythms are set by light and dark cues, giving an insect a reference point for a particular time of the 24 h cycle. For many insects, circadian rhythms program such behaviors as pupation, eclosion from eggs and from the pupal stage, and time internal physiological events occur such as pheromone secretion and release. Seasonal activity in many temperate insects is based on monitoring the photoperiod, with insects somehow calculating the duration and accumulation of daily changes in day or night length over a period of time. Daily clock measure the time of day while seasonal clocks monitor the duration of each day and the number of days of a given length, as well as additional environmental factors. Circadian clocks in *D. melanogaster* are heritable and populations can be selected with different circadian rhythms.

In 1984, Jeffrey Hall and Michael Rosbash succeeded in isolating the *period* gene, then went on to discover that PER, the protein encoded by *period*, accumulated during the night and was degraded during the day. For a long time *period* was the only circadian gene known, but rapid progress in the circadian field in the 1990s led to the identification of *timeless* (*tim*), *clock* (*clk*), and *cycle* (*cyc*), which, together with *per*, make up the core transcriptional feedback loop that drives circadian rhythms. The late 1990s and early 2000s saw the gene identifications of a circadian photoreceptor, *cryptochrome* (*cry*), the kinases *double-time* (*dbt*) and *shaggy* (*sgg*), and a transcriptional activator involved in a second feedback loop, *vrille* (*vri*).

They hypothesized that the PER protein blocked the activity of the *period* gene, and reasoned that by an inhibitory feedback loop, PER protein could prevent its own synthesis and thereby regulate its own level in a continuous, cyclic rhythm. When the *period* gene is active, *period* mRNA is made. The mRNA is transported to the cell's cytoplasm and serves as template for the production of PER protein. The PER protein accumulates in the cell's nucleus,

17.3.2 滞育中的日生物钟和季节生物钟

日常生物钟（昼夜节律）和季节生物钟（光周期）影响昆虫日常和季节性的生活节奏。每天重复的昼夜节律由光和黑信号建立，给予昆虫在 24 h 的循环中一个特别的时间参考点。多数昆虫，昼夜节律程序如化蛹行为、从卵中孵化和蛹期羽化以及内部生理环节发生的时间，如信息素的分泌和释放都受到昼夜节律的控制。许多温带昆虫的季节性活动基于对光周期的监测以及昆虫以某种方式精确计算持续时间，并累计一段时间的白天或晚上时长的日变化。就像另外的环境因子一样，日生物钟测定一天时间而季节生物钟即监测每天和给定时长内的天数。果蝇的昼夜节律钟具遗传性，种群能选择不同的昼夜节律。

1984 年，Jeffrey Hall 和 Michael Rosbash 成功地分离出了 *period* 基因，然后发现由 *period* 编码的蛋白质 PER 在夜间积累，在白天降解。在很长一段时间里，*period* 是唯一已知的昼夜节律基因，但在 20 世纪 90 年代，昼夜节律领域研究的快速发展导致了对 *timeless*（*tim*）、*clock*（*Clk*）和 *cycle*（*cyc*）的识别，它们与 *per* 一起构成了驱动昼夜节律的核心转录反馈循环。20 世纪 90 年代末和 21 世纪初，人们弄清楚了一种昼夜节律感光受体基因，即隐花色素（*cry*）、激酶 *double-time*（*dbt*）和 *shaggy*（*sgg*）以及一种参与第二反馈循环的转录激活物 *vrille*（*vri*）等基因。

他们假设 PER 蛋白阻断了 *period* 基因的活性，并推断通过抑制性反馈循环，PER 蛋白可以阻止自身的合成，从而以连续、循环的节奏调节自身水平。当 *period* 基因激活时，*period* mRNA 产生。mRNA 被运输到细胞的细胞质中，并作为产生 PER 蛋白的模板。PER 蛋白在细胞核中积累，在那里 *period* 基因活

where the *period* gene activity is blocked. In 1994, Michael Young discovered a second clock gene, *timeless*, encoding the TIM protein that was required for a normal circadian rhythm. when TIM bound to PER, the two proteins were able to enter the cell nucleus where they blocked *period* gene activity to close the inhibitory feedback loop. He identified yet another gene, *doubletime*, encoding the DBT protein that delayed the accumulation of the PER protein. This provided insight into how an oscillation is adjusted to more closely match a 24-hour cycle.

Circadian clock genes could control winter diapause in insects by regulating photoperiodic diapause induction. Moreover, *tim* instantly affected the diapause induction *via* circadian photoperiodism in *D. melanogaster*. In nondiapause individuals of *P. apterus*, the expression of *per* and *clk* were low, whereas they were expressed at a high level in diapausing types (Fig. 17-1; Appendix color graphic Fig.17-1). The *cycle* plays an important role in ovarian development or diapause, and that it is involved somewhere in the upstream part of the pathway of juvenile hormone and not at the level of the ovaries themselves (Bradshaw et al., 2010). In *M. sexta* larvae, photoperiod and circadian clock interaction leading to limited PTTH release to a definite time of the day–night cycle. Furthermore, PTTH secretion depends on the circadian system in the *R. prolixus*, and *clock* neurons can regulate its rhythmic release and synthesis. PTTH release and photoperiodic termination of pupal diapause in the giant moth, *A. pernyi* are under the control of the circadian clock in the brain. In *S. ocellatella* in the short-day condition during autumn, PER and TIM increased in the brain that leads to a decrease in PTTH titers in the brain and hemolymph, which itself causes decreasing 20E titer in hemolymph that leads to diapause induction. It demonstrated the involvement of circadian clock genes *per* and *tim* in photoperiodic regulation of diapause (Ahmadi et al., 2021).

The receptors for circadian rhythms and for seasonal activity are located in the brain of some insects, in the compound eyes of others, and, in some cases, in both brain and compound eyes. Cryptochrome, a light sensitive pigment encoded by the gene *Cryptochrome*, contains the vitamin riboflavin and a protein. It is mainly involved in sensing circadian light information. Cryptochrome is located in both brain tissue and compound eyes of *D. melanogaster*. In the brain it regulates the morning rhythms

动被阻断。1994 年，Michael Young 发现了第二个时钟基因 *timeless*，编码正常昼夜节律所需的 TIM 蛋白。当 TIM 与 PER 结合时，这两种蛋白质能够进入细胞核，在那里它们阻断了 *period* 基因的活动，以关闭抑制性反馈循环。他还发现了另一个基因 *doubletime*，它编码 DBT 蛋白，延迟了 PER 蛋白的积累。这让我们深入了解了如何调整振荡，使其更接近 24 h 的周期。

生物钟基因可以通过调节光周期滞育诱导来控制昆虫的冬季滞育。此外，果蝇的 *tim* 通过昼夜节律性光影响滞育的诱导。在非滞育始红蝽个体中，*per* 和 *clk* 的表达水平较低，而在滞育个体中，其表达处于高水平（图 17-1；附彩图 17-1）。*cycle* 在卵巢发育或滞育中起着重要作用，它参与了保幼激素途径的上游部分，而不是卵巢本身的水平（Bradshaw 等，2010）。烟草天蛾幼虫的光周期和生物钟的相互作用导致 PTTH 在日夜循环的某个特定时间有限释放。此外，PTTH 的分泌依赖于吸血蝽体内的昼夜节律系统，*clock* 神经元可以调节其节律性释放和合成。PTTH 释放与光周期终止柞蚕蛹的滞育受大脑生物钟控制。甜菜夜蛾在秋季的短日照条件下，大脑中的 PER 和 TIM 增加导致脑和血淋巴中的 PTTH 滴度降低，这本身导致血淋巴中的 20E 滴度降低，从而导致滞育诱导。它证明了生物钟基因 *per* 和 *tim* 参与滞育的光周期调节（Ahmadi 等，2021）。

昼夜节律钟和季节性活动的受体位于一些昆虫的脑内，其他的在复眼内，在有些情况下，脑和复眼都有。隐花色素，一种光敏感色素由隐花色素基因编码，包含维生素、核黄素和蛋白质。其主要参与感觉生理节律的光信息。隐花色素位于果蝇的脑组织和复眼中。在脑内，它调节果蝇的晨早节律，而在复眼

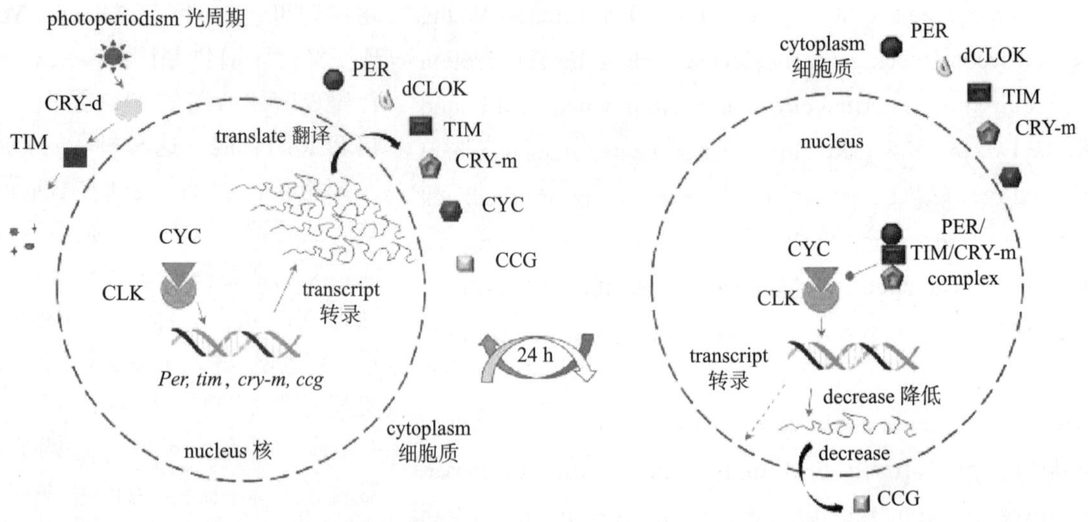

Fig.17-1 Molecular mechanisms of the circadian clock.Positive regulate elements of CYC and CLK form a heterodimer and activate transcription of clock genes *per*, *tim*, and *cry-m* and many other clock-controlled genes (*ccg*). Their protein products are synthesized in the cytoplasm and PER, TIM and CRY-m form a complex. The complex enters the nucleus and acts as a negative element that represses CYC/CLK transcriptional activity. Reduction of *ccg* transcript levels and consequent reduction of CCG protein levels lead to a decrease of repressive regulation of CYC/CLK by PER/TIM/CRY-m, and therefore CYC/CLK-mediated transcription increases again. These phases in which ccg transcription is activated or repressed are repeated in approximately 24 h. CRY-d causes degradation of TIM in a light dependent manner (modified from Numata et al., 2015).

图 17-1 生物钟的分子机制。正向调控成分 CYC 和 CLK 形成异二聚体并激活时钟基因 *per*、*tim* 和 *cry-m* 以及许多其他时钟控制基因（ccg）的转录。它们的蛋白质产物在细胞质中合成后，PER、TIM 和 CRY-m 形成复合物。该复合物进入细胞核，起反向作用抑制 CYC/CLK 的转录活性。ccg 转录水平下降，接着蛋白水平降低导致 PER/TIM/CRY-m 对 CYC/CLK 的抑制性调节减少，因此 CYC/CLK 介导的转录再次增加。ccg 转录激活或抑制的这些阶段在大约 24 h 内重复。CRY-d 以光依赖的方式导致 TIM 降解（仿 Numata 等，2015）。

of *Drosophlia*, while in the compound eyes it controls the evening peak of activity. The chromophore receptor that responds to seasonal changes contains carotinoids or vitamin A, but little is known about how it functions in relation to seasonal changes.

The butterfly, *Polygonia c-album*, both absolute and relative volumes of the olfactory neuropils decrease substantially during the two weeks after dormancy conditions were terminated, despite butterflies being kept at the same conditions as before dormancy conditions were initiated. In addition to this pattern, the relative volumes of calyces and antennal lobes are remarkably smaller, both at eclosion and at two weeks of age, compared with those of females from a non-diapausing generation (Eriksson et al.,2020).

内即控制傍晚的活动高峰。对季节性变化产生反应的发色团受体含类胡萝卜素或维生素 A，但功能上如何与季节变化相关还知之甚少。

尽管黄蛱蝶被保持在与冬眠条件开始前相同的条件下，蝴蝶的嗅觉神经纤维的绝对和相对体积在冬眠条件终止后的两周内大幅度减小。除此之外，在羽化期和两周龄时，萼片和触角的相对体积明显小于非滞育世代的雌性（Eriksson 等，2020）。

17.4 滞育的生物化学与信号转导路径

17.4 Biochemical and Signal Transduction Pathways of Diapause

In diapausing *Chlosyne lacinia* larvae, oxidative metabolism diminished, catalase and glucose 6-phosphate dehydrogenase activities maintenance, total glutathione depleted, and lipid peroxidation reduced. The potential to maintain both the NADPH supply and the control of H_2O_2 metabolism may be crucial at the moment of arousal from diapause toward larval development (Moreira et al., 2012). In the process of glycolysis of diapausing *E. civilis,* as the primary energy metabolism pathway for sugars, *D*-glucose generated glucose 6-phosphate into glycolysis by the action of upregulated hexokinase, which promoted the conversion of pyruvate to acetyl coenzyme A *via* upregulated pyruvate dehydrogenase and thus into the TCA cycle. The upregulation of malate dehydrogenase promoted the production of oxaloacetate. Acetyl CoA and oxaloacetate produced citric acid under the action of the upregulated citrate synthase. Aconitase, isocitrate dehydrogenase, and succinyl-CoA synthetase α subunit were both upregulated (Zhang et al., 2021).

滞育的向日葵毛虫氧化代谢减少，维持过氧化氢酶和 6- 磷酸葡萄糖脱氢酶活性，总谷胱甘肽被消耗，脂质过氧化减少。从滞育到幼虫发育的觉醒时刻，维持 NADPH 供应和控制 H_2O_2 代谢的潜力可能至关重要（Moreira 等，2012）。滞育伞裙追寄蝇在糖酵解过程中，*D*- 葡萄糖作为糖的主要能量代谢途径生成葡萄糖 6- 磷酸，通过上调己糖激酶的作用进入糖酵解，己糖激酶促进丙酮酸的转化，通过丙酮酸脱氢酶上调乙酰辅酶 A，从而进入 TCA 循环。苹果酸脱氢酶上调促进草酰乙酸的产生。在上调的柠檬酸合成酶的作用下乙酰辅酶 A 和草酰乙酸产生柠檬酸。乌头酸酶、异柠檬酸脱氢酶和琥珀酰辅酶 A 合成酶 α 亚基均上调（Zhang 等，2021）。

Diapusing embryos of *B. mori* accumulate sorbitol and glycogen in the eggs and although they may serve as antifreeze compounds as often suggested, more recent evidence suggests that sorbitol may have a much more central role as a controlling factor in maintaining diapause. In addition, diapausing freeze-tolerant insects must withstand the challenges associated with cooling, freezing, and thawing. Many freeze-tolerant insects accumulate cryoprotectants such as polyols, sugars, amino acids, lipids, ice-binding proteins, and transport proteins; and antioxidants, cell adhesion proteins, chaperones, chelators, cytochrome P450s, cytoskeletal protein isoforms, cytoskeletal regulators, disordered proteins (e.g. dehydrins),sirtuins etc. as cytoprotectants (Toxopeus et al., 2018).

滞育中的家蚕胚胎在卵内积累山梨醇和糖原，尽管它们可能像人们常说的那样作为防冻剂，但最新的证据表明，三梨醇可能在保持滞育时，作为控制因子起到更为核心的作用。三梨醇和甘油是利用糖原合成的。此外，滞育抗冻昆虫必须能够承受与降温、冷冻和解冻相关的挑战。许多耐冻昆虫积累冷冻保护剂，如多元醇、糖、氨基酸、脂质、冰结合蛋白和运输蛋白，并以抗氧化剂、细胞黏附蛋白、伴侣、螯合剂、细胞色素 P450、细胞骨架蛋白异构体、细胞骨架调节因子、无序蛋白质（如脱水素）、sirtuins 等作为细胞保护剂（Toxopeus 等，2018）。

An interesting way to break pupal diapause in *S. crassipalpis* and *M. sexta* is to treat diapausing pupae with a few microliters of a nonpolar organic solvent, such as hexane. Within 10 minutes after

一个解除红尾肉蝇和烟草天蛾滞育蛹的有趣方法是用几毫升非极性有机溶

administering hexane, phosphorylation of extracellular receptor-activated kinases(ERKs) occurs in brain tissue and in various peripheral tissues, including epidermis and fat body cells. ERKs and mitogen activated protein kinase (MAPKs) are enzymes that participate in amplification cascade, and, typically, are activated by phosphorylation. They, in turn, phosphorylate other molecules, including members of the MAPKs family. ERKs and MAPKs have been implicated in signal transduction cascades and in termination of diapause in several different insects. Phosphorylated-ERK (p-ERK) in the brain of *S. crassipalpis* leads to secretion of PTTH, which could then activate the prothoracic glands to produce ecdysteroids to initiate development.

ERKs and p38 MAPKs have been implicated in the termination of silkworm embryo diapause. Phosphorylation of ERK in yolk cells increased in diapausing *Bombyx* embryos that were chilled at 5℃ for 45 to 60 days (conditions that break diapause) and then return to 25℃. p-ERK is regulated by MAPK-ERK (MEK) and possibly by p38 MAPK. P-ERK regulates diapause termination by activating transcription of genes controlling enzymes needed to free bound ecdysteroids in the yolk and convert sorbitol to glycogen in *Bombyx*. Another dual-specificity kinase, mycetozoa protein kinase-like (MPKL) promotes photoperiodic diapause induction of *L. migratoria*. It significantly inhibits Akt and FOXO phosphorylation levels in ovaries, and also enhances reactive oxygen species, superoxide dismutase and catalase activities, whereas peroxidase activity is decreased under both photoperiodic regimes (Hao et al., 2019). In *H. armigera*, p-MEK and p-ERK are elevated in the brain of diapause-destined pupae and suppression of MEK/ERK activity terminates diapause progress. Reactive oxygen species (ROS) activate MEK/ERK signalling, causing large-scale phosphorylation of downstream proteins. Terminated diapause progress by 20E injection significantly decreases p-MEK, p-ERK and phospho-ribosomal S6 kinase levels, while phospho-MAPK substrates and ubiquitin-conjugated protein levels increase (Lin et al., 2021).

Research on insect insulin-like peptides (ILP) has shown they play an important role in diapause phenotypes, such as halted reproduction, reserve accumulation, metabolic depression, and enhanced stress-tolerance (Cambron et al., 2021). The transforming growth factor-β (TGF-β) superfamily signaling pathway is quite conservative with versatile functions in animals, such as cell

剂，如正己烷处理滞育蛹。施用正己烷10 min后，脑组织和外周不同的组织包括真皮和脂肪体细胞的胞外受体活化激酶（ERKs）磷酸化。ERKs和分裂素活化蛋白激酶（MAPKs）是参与扩增级联的酶，通常是由磷酸化活化的。它们转而磷酸化其他分子，包括MAPKs族的成员。在几种不同的昆虫中，ERKs和MAPKs牵涉信号转导级联和滞育的终止。红尾肉蝇脑内磷酸化的ERK（p-ERK）引起PTTH的分泌，然后活化前胸腺产生蜕皮甾类从而启动发育。

ERKs和p38 MAPKs涉及家蚕胚胎滞育的终止。家蚕滞育胚胎在5℃冷藏45～60 d（打破滞育的条件），之后返回到25℃，卵黄细胞中的ERK磷酸化增加。p-ERK是由MAPK- ERK (MEK)或可能由p38 MAPK调节。p-ERK通过活化控制酶的基因转录，调节滞育的终止，这些酶是家蚕卵黄内游离结合态蜕皮甾类并将山梨醇转化成糖原所必需的。另一种双特异性激酶，黏菌蛋白激酶样（MPKL）促进了沙漠蝗的光周期滞育诱导。它显著抑制卵巢中的Akt和FOXO磷酸化水平，并增强活性氧类，超氧化物歧化酶和过氧化氢酶的活性，而过氧化物酶活性在两种光周期状态下均降低（Hao等，2019）。棉红铃虫的p-MEK和p-ERK在滞育蛹大脑中升高，抑制MEK/ERK活性则终止滞育进程。活性氧激活MEK/ERK信号传导，导致下游蛋白质大规模磷酸化。注射20E终止滞育进程显著降低了p-MEK、p-ERK和磷酸核糖体S6激酶的水平，而磷酸化MAPK底物和泛素结合蛋白水平增加（Lin等，2021）。

对昆虫胰岛素样肽（ILP）的研究表明，它们在滞育表型中起着重要作用，如生殖停滞、储备积累、代谢抑制和应激耐受增强（Cambron等，2021）。转化生长因子β（TGF-β）超家族信号通路在动物中非常保守并有多功能，如

proliferation, differentiation, morphogenesis, tissue homeostasis and regeneration. The TGF-β superfamily signaling pathway participates in regulating *H. armigera* development or diapause through mitochondrial activity mediated by a cascade of transcription factors. Mitochondrial transcription factor A (TFAM) is well known to regulate mitochondrial gene transcription and mitochondrial DNA replication. Their levels is lower in brains of diapause-destined pupae than in nondiapause pupae, and that low TFAM levels result in low mitochondrial activity, which induces diapause. Transcription factor POU can regulate many gene expressions in insects, such as diapause hormone and *PBAN* gene, hexokinase gene, and dopa decarboxylase gene, and these genes are closely related to insect diapause (Li et al.,2018).

(Li Wenchu)

细胞增殖、分化、形态、发生组织、稳定和再生。TGF-β 超家族信号通路通过转录因子级联介导的线粒体活性参与调节棉铃虫的发育或滞育。众所周知，线粒体转录因子 A（TFAM）调节线粒体基因转录和线粒体 DNA 的复制。它们在滞育蛹大脑中的水平比非滞育蛹的低水平的 TFAM 导致线粒体活性降低，从而导致滞育。转录因子 POU 可以调节昆虫体内许多基因的表达，如滞育激素和 *PBAN* 基因、己糖激酶基因和多巴脱羧酶基因，这些基因与昆虫滞育密切相关（Li 等，2018）。

（李文楚）

Exercise

1.What are the types of insect diapause? What are the differences in the nature of diapause at different stages?

2.What is possible molecular mechanism for diapause initiation?

3.What hormones are involved in the regulation of insect diapause?

4.What are the signal transduction pathways in the process of diapause and diapause termination or breaking?

思考题

1. 昆虫滞育分哪几类？不同阶段发生的滞育的本质有哪些区别？

2. 昆虫滞育起始可能的分子机制是什么？

3. 参与昆虫滞育调控的激素有哪些？

4. 滞育及滞育解除或打破滞育过程中有哪些信号转导途径？

第 18 章 胚胎发育

Chapter 18 Embryonic Development

Key points

1. Insect embryonic development undergoes cleavage, blastoderm formation, formation and differentiation of germ band, segmentation and formation of appendages, organogenesis, blastokinesis and dorsal closure, and embryonic envelop degradation.

2. Insect embryonic development is mainly regulated by maternal genes (or maternal effector genes), zygotic genes (or segmentation genes) and homeotic genes (or selector genes).

3.Both ecdysone and JH are involved in the embryonic development.

知识点

1. 昆虫胚胎发育主要经历合子卵裂、胚盘形成、胚带形成与分化、分节与附肢形成、器官形成、胚动与背合及胚膜退化等过程。

2. 昆虫胚胎发育主要受母体基因（也称为母体效应基因）、合子基因（或体节形成基因）及同源异性基因（或选择基因）的调控。

3. 昆虫蜕皮激素与保幼激素均参与胚胎发育过程。

The male of a sexually mature insect fertilizes the female by injecting sperm into her body through mating. The essence of fertilization is the combination of male and female prokaryotes into zygote. Insect embryonic development refers to the process from the zygotic cleavage to the embryo with internal and external organs ready for incubation. The first division of the zygote marks the beginning of embryonic development of the new individual.

性成熟的昆虫雄虫通过交配将精子注入雌虫体内进行受精。精子使卵受精实质是雄性原核与雌性原核结合为合子。昆虫胚胎发育是指从合子卵裂开始直至发育成内外器官俱全的胚胎并准备孵化的过程。合子的第 1 次分裂标志着新个体胚胎发育的开始。

18.1 昆虫胚胎发育的基本概念

18.1 Basic Concept of Insect Embryogenesis

Embryonic development refers to the process in which multicellular organisms grow from a single cell, including the self-

胚胎发育是指多细胞生物由最初一个单细胞成长起来的过程，包括生物有

construction and self-organization of living organisms. Usually, the zygote first divides and differentiates into the cells with different morphology and functions through a series of processes, and then builds various tissues and organs through cell-cell interactions, finally forming an embryonic organism. The developmental process is the result of the expression of a large number of genes in a certain time, space and order controlled by the developmental genetic program.

机体的自我构建与组织。一般情况下，受精卵首先通过一系列的分裂与分化形成形态和功能不同的细胞，接着通过细胞间的相互作用构建各种组织与器官，最后建成一个胚胎有机体。整个发育过程是大量基因按一定时间、空间和次序表达的结果，而这种次序性由发育遗传程序所控制。

Compared with other higher animals, the insect embryonic development is simpler, shorter in duration, and of more types. The zygote undergoes cleavage to form a large number of daughter nucleus, and then form blastoderm; Then the cells differentiate to germ band; The germ band forms different germ layers by invagination and expansion, and the embryo trapped in the yolk forms embryonic membrane on the ventral surface; The germ layer differentiates, and the embryo segments and forms appendages. With the embryogenesis, the two sides of germ band expand to the dorsal surface around the yolk, and close at the dorsal midline to form a complete embryo (Fig.18-1; Appendix color graphic Fig.18-1). At the same time, the internal organ systems are formed. When the organs mature, the larval body shape appears, and the epidermal pigment synthesizes and deposites. Finally, the embryonic membrane degenerates and the embryonic development completes.

与其他高等动物比，昆虫胚胎发育显得更为简单、历时更短且类型更多。受精卵通过卵裂形成大量子核，继而形成胚盘；胚盘形成后，细胞开始分化形成胚带；胚带通过内陷和扩增形成不同的胚层且此时胚胎陷入卵黄在胚胎腹面生成胚膜；胚层进行分化，同时胚胎开始分节并形成附肢；随着胚胎发育的进行，胚带两侧围绕着卵黄不断向背面扩展，最终胚带两侧在背中线封合，形成一个完整的胚胎（图 18-1；附彩图 18-1）。与此同时，昆虫内部的各个器官系统相继形成。当各器官发育成熟后，幼虫虫体形态开始出现，真皮色素开始合成并积累。幼虫发育完全时，胚膜退化，胚胎发育完成。

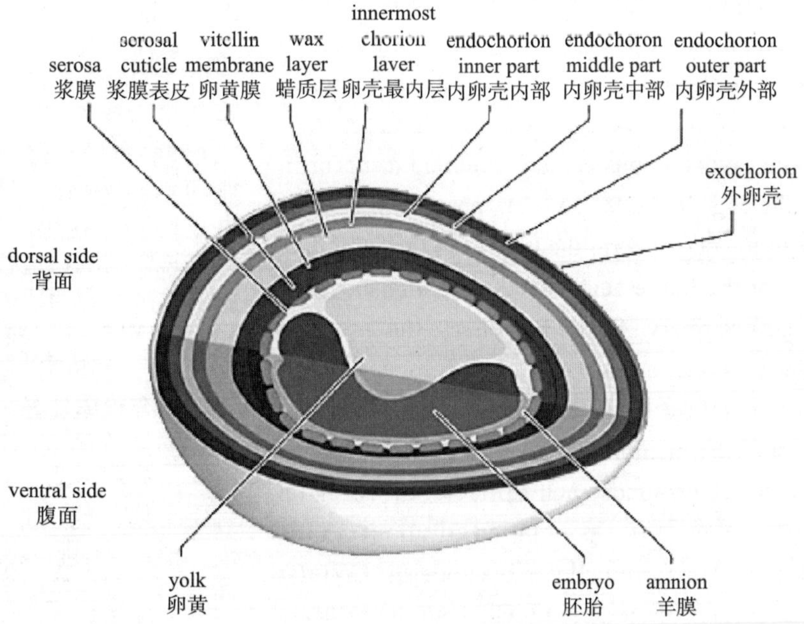

Fig.18-1 A schematic presentation of the radial complexity of egg layers in holometabolous insects. Embryo position related to the yolk varies in hemimetabolous insects (modified from Rezende et al., 2016).

图 18-1 完全变态昆虫卵层径向复杂性。不完全变态昆虫中与卵黄相关的胚胎位置不同（仿 Rezende 等，2016）。

18.2 昆虫胚胎发育

18.2 Process of Insect Embryonic Development

18.2.1 Cleavage and blastoderm

Except for parthenogenesis, insect embryonic development begins after fertilization. The fertilized egg divides first, and insect cleavage includes holoblastic cleavage and superficial cleavage.

Holoblastic cleavage refers to the mode in which all the materials in the fertilized egg split into two daughter cells of roughly the same size, usually seen in the early embryogenesis of Collembola and the embryogenesis of some Hemiptera and Hymenoptera.

Superficial cleavage refers to the unequal division of matter in the fertilized egg, with cell division occurring mainly in the outer layer of the egg. Most insects' cleavage belongs to this category. An interesting feature of this typed cleavage is that the nucleus has been divided, but the cytoplasm does not, resulting in a syncytium containing many nuclei. The zygote nucleus undergoes mitotic division in the center of the egg cytoplasm, but the resulting daughter nucleus or energids, each surrounded by an island of cytoplasm, are not incorporated into new complete cells. Instead, after a series of mitoses, the energids migrate to the egg periphery with their islands of cytoplasm and continue to divide there. In *Drosophila*, The initial seven cleavage divisions are synchronous, but they become more asynchronous as they continue to occur in different regions of the egg; In the 9th division, about five nucleus migrate to posterior of the egg to form the pole cells that will give rise to the germ cells of the future adult; By the 10th divison, they have reached the egg periphery. At the periphery, the energids first form a syncytial blastoderm that lacks an envelope, with all the cleavage nucleus contained in the common cytoplasm of the egg; In the 13th divison, the membrane become trapped between the nucleus and eventally surrounds each of nuclei of syncytial blastoderm, and now consists of a population of about 5,000, creating individual cells that form cellular blastoderm. Cellular blastoderm belongs to surface blastoderm in which all the cells are arranged in a layer around a central yolk. Besides, some daughter nucleus remains in the yolk become vitellophages for providing nutrients to the embryo (Fig.18-2).

18.2.1 卵裂与胚盘形成

除孤雌生殖外，昆虫胚胎发育均需经过卵受精后才能开始。受精卵首先进行卵裂，而昆虫卵裂包括完全卵裂与表面卵裂两种。

完全卵裂是指受精卵内所有物质分裂形成大小大致相同的两个子细胞的卵裂方式，多见于弹尾目昆虫的胚胎发育早期及部分半翅目与膜翅目昆虫的胚胎发育时期。

表面卵裂是指受精卵内物质非均等分裂，且细胞分裂主要在卵的外层进行。大多数昆虫的卵裂属于此类。这一类型卵裂的典型特征是核已经分裂，但细胞质却不分裂，进而形成含有许多细胞核的核胞体。位于卵中央部位的受精卵细胞核进行有丝分裂，但产生的每一个子核（或叫活质体）都被细胞质包围，不能形成新的完整细胞。相反，经过一系列的有丝分裂之后，子核随着包围着它们的细胞质开始迁移到卵周，并在那里继续分裂。在果蝇中，最初的7次卵裂是同步进行的，但随着它们在卵的不同区域进行卵裂，它们开始变得不同步；在第9次分裂周期时，大约有5个细胞核迁移到卵后部形成极细胞，将来发育为成体的生殖细胞；在第10次卵裂后，子核已经迁移到卵的外围。在卵的外围，子核首先形成一个没有膜的合胞体囊胚，所有的细胞核都包含在卵细胞的细胞质中；在第13次卵裂期，卵膜内陷于核之间，并最终包围着每个核，现在由大约5 000个群体组成，把每个细胞核分割成单个细胞，这样就形成了细胞囊胚，即胚盘。细胞囊胚属于表面囊胚，即所有细胞都围绕中央的卵黄排列成一层。同时，在卵裂时还会有部分子核留在卵黄间成为供给胚胎营养的卵黄胞（图18-2）。

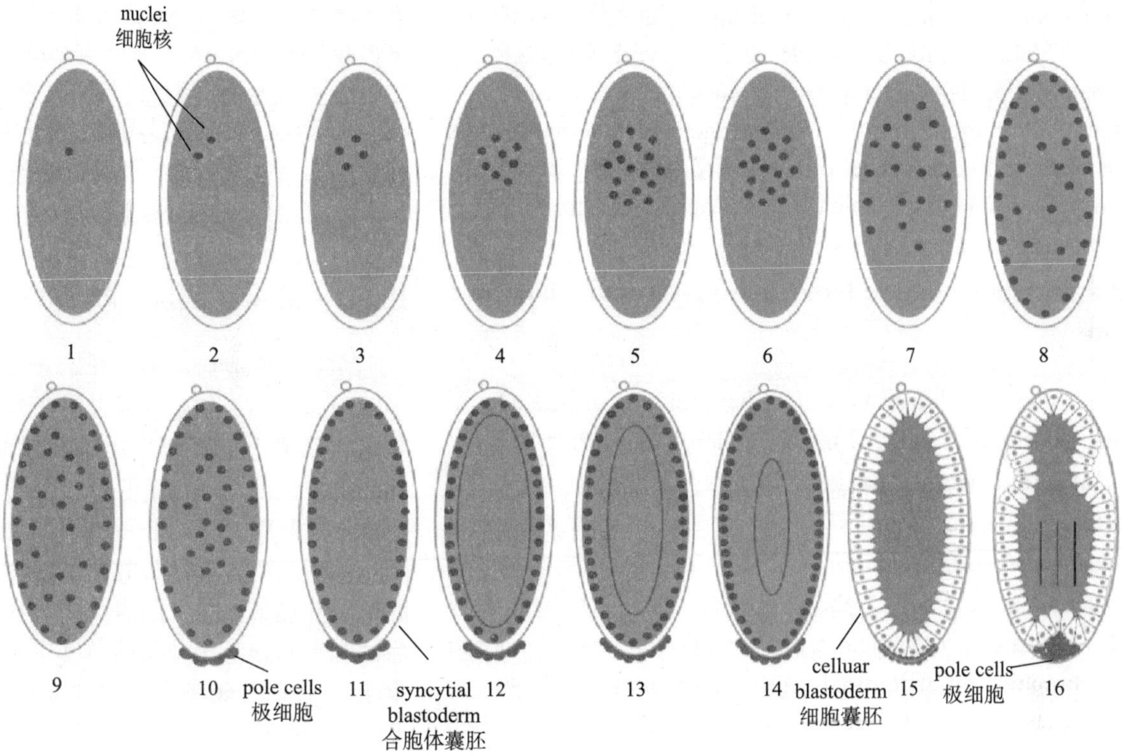

Fig. 18-2 Surface cleavage in *Drosophila* (modified from Gilbert, 2000). The numbers in the figure indicate the number of cleavages.

图 18-2 果蝇的表面卵裂（仿 Gilbert, 2000）。图中数字表示卵裂的次数。

The migration time of the energids to yolk surface in different species is different. In some species, the energids reach the periplasm after 6 divisions, while in some species, the energids migrate into the periplasm after 10 divisions. The cleavage of Collembola changes from holoblastic cleavage to superficial cleavage during 32 daughter nuclear stage.

The cleavage cells have the characteristics of shorter cell cycle and faster division compared with somatic cells. In early cleavage, the zygote genome hardly needs to function, but is maternal-driven and transmitted to all cells through mitosis. In addition, cleavage cells do not need to increase their cytoplasmic volume, but simply distribute the large amount of stored ooplasm to an increasing number of small cells. Therefore, the cell cycle of cleavage is shortened, almost in sync with cell division. *Drosophila* oocyte contains abundant reserves of Cyclin and Cdc25 protein in the cytoplasm before fertilization. Thus, the activity of mitotic promoting factor (MPF) kinase remains constant during the 7th cell cycles, resulting in the rapid cleavage, i.e. DNA replication is

不同种类的昆虫子核迁移至卵黄表面的时间也不同。有些种类在进行 6 次分裂后子核可到达周质，而有些种类在进行 10 次分裂后子核才迁入周质。弹尾目昆虫在 32 个子核期时的卵裂从完全卵裂转变为表面卵裂。

与体细胞相比，卵裂细胞具有细胞周期短与分裂速率快等特点。在早期卵裂时，合子基因组几乎不需要发挥功能，而是在母源性物质的驱动下，通过有丝分裂传递给所有细胞。此外，卵裂细胞的细胞质体积无需增加，只是将事先存储的大量卵质分配到不断增多的小细胞中。因此，卵裂的细胞周期缩短，几乎与细胞分裂同步。果蝇卵在受精前，卵质中含有丰富的细胞周期蛋白质与 Cdc25 蛋白质储备。因此，在前 7 次细胞周期中，有丝分裂促进因子（MPF）激酶活性保持恒定，因而卵裂迅速，即 DNA 复制完成，卵裂也随即完成；从

completed, and cleavage is completed immediately. From the 8th cleavage, Cyclin begins to degrade, and the synthesis of Cyclin using maternal mRNA as template becomes the rate-limiting step after the 8th cleavage. From the 8th cleavage, the maternal mRNA is exhausted, and gene transcription is needed. At this point, cell division shifts from cytoplasmic control to nuclues control.

第 8 次卵裂开始，Cyclin 开始降解，此时以母源性 mRNA 为模板合成的周期蛋白质成为第 8 次分裂以后的限速步骤；到 14 次分裂，母源性 mRNA 消耗殆尽，而基因转录是必需的，此时，细胞分裂从卵细胞质控制转向细胞核控制。

18.2.2 Formation of germ band, germ layer and embryonic envelop

The cells begin to differentiate after the formation of blastoderm. The blastoderm cells located on the abdominal surface become thick, large and tightly packed, forming germ band, while blastoderm cells on the dorsal and lateral surfaces gradually become thinner, forming embryonic envelop. The insects' germ band contains three types: ① the short-germ. The germ band only includes the later protocephalic and mandibular region at first, and the thoracic and abdominal segments gradually grow out of the end of the germ band with the extension of the embryo, which mainly exists in hemimetabolan; ② the long-germ. The germ band includes protocephalic, mandibular, thoracic and abdominal region at the beginning, mainly existing in holometabolan; ③ the intermediate-germ. The germ band initially contains protocephalic, mandibular and thoracic region, which occurs in some Coleoptera.

Continued proliferation of the germ band causes it to penetrate into the interior of the yolk mass, and the presence of morphogenetic gradients governs its continued determination and differentiation. Gradients of morphogenesis such Bicoid protein at the anterior pole and the Nanos protein at the posterior pole activate or inactivate the transcription of specific genes that are associated with anterior or posterior structures.

The germ band is still a single layer of cells at this stage, but it takes an important step when it becomes a double layer by the process of gastrulation. In other animals, gastrulation includes the invagination of the already formed cell balls, but in insects, the invagination is hindered by the large amount of yolk. Instead, the formation of a multicellular layer occurs when some of the cells of germ band migrate inward into the yolk. The reorganization produces a multilayered body consisting of different cell types that are able to interact with each other in different ways. When the cells along the ventral midline of the germ band elongate and migrate upward, a longitudinal ventral furrow is created. Other cells of the germ band soon fill the ventral furrow, with this outer layer of cells remaining as the embryonic ectoderm while the invaginated cells proliferate to form an inner layer of bilayer cells,

18.2.2 胚带、胚层及胚膜形成

在胚盘形成后，细胞开始分化。位于卵腹面的胚盘细胞逐渐增厚变大且排列紧密形成之后发育成胚胎的细胞带，即胚带。背面与侧面的胚盘细胞则逐渐变薄形成胚膜。昆虫胚带可分为 3 种类型：①短胚型，即胚带起初仅包括以后的原头和颚节区域，随着胚胎延长，以后的胸节和腹节区才逐渐从胚带末端长出来，主要出现在不完全变态昆虫中；②长胚型，即胚带从一开始就包括以后的原头、颚节、胸节和腹节区域，主要存在于完全变态发育昆虫中；③中间胚型，胚胎起初包括了以后的原头、颚节和胸节，存在于部分鞘翅目中。

胚带的持续增殖使其渗透到卵黄团的内部，形态决定因子浓度梯度的存在，决定了其持续的决定和分化。普遍认为，形态决定因子的前后极浓度梯度，如前极 Bicoid 蛋白和后极 Nanos 蛋白，可激活或失活与前或后结构相关的特定基因的转录。

胚带在这一阶段仍为单层细胞，但经过原肠形成过程，它成为双层细胞，这是胚带发育的重要一步。在其他动物中，原肠发育包括已形成的细胞球的内陷，但在昆虫中，内陷受到存在大量卵黄的阻碍。相反，当生殖带的一些细胞向内迁移进入卵黄时，就会形成多细胞层。这种重组产生了一个由不同类型的细胞组成的多层体，这些细胞能够以不同的方式相互作用。当细胞沿着生殖带的腹侧中线伸长并向上迁移时，就形成了一条纵向的腹侧沟，即原肠沟。胚带的其他细胞很快就填满了腹侧沟，这一外层的细胞保留为胚胎外胚层，而内陷的细胞增殖形成双层细胞的里层，里

which further differentiates into endoderm and mesoderm (Fig. 18-3; Appendix color graphic Fig.18-3). The migration of mesodermal cells depends on the expression of *heartless* gene that encodes the *Drosophila* fibroblast growth factor receptor. With this initial expression, these cells acquire a different developmental fate, and will form most of the internal organs. A narrow band of cells where two genes *twist* and *snail* initially transcript are destined. *Twist* is a transcriptional activator for mesodermal genes and *snail* is one of its targets, and both of them are responsible for dividing the blastoderm into the ectodermal, endodermal, and mesodermal germ layers.

The embryo is trapped inwards, and both ends of the embryonic envelop gradually extend and heal to the ventral. Therefore, the ventral of the embryo contains two layers, the inner membrane named serosa and the outer membrane named amnion. The intervening cavity is called the amniotic cavity, and it contains amniotic fluid to protect the embryo.

层进一步分化为中带与侧带，前者成为内胚层，后者成为中胚层（图 18-3；附彩图 18-3）。中胚层细胞的迁移依赖于编码果蝇成纤维细胞生长因子受体 *heartless* 基因的表达。随着这一基因的起始表达，这些细胞获得了不同的发育命运，并将形成大多数内部器官。*Twist* 和 *snail* 这两个基因在狭窄的细胞带中的起始转录，决定了该细胞带注定成为中胚层。*Twist* 是中胚层基因转录的激活因子，而 *snail* 是其靶标之一。这两个基因负责将胚层分为外胚层、内胚层和中胚层。

大多数昆虫在胚层形成时的胚胎是向内方陷入的，胚膜两端逐步向腹面延伸并愈合，此时，胚胎腹面包含两层胚膜，内部的胚膜称为浆膜，外部的胚膜称为羊膜。其间的腔称为羊膜腔，腔内充满保护胚胎的羊水。

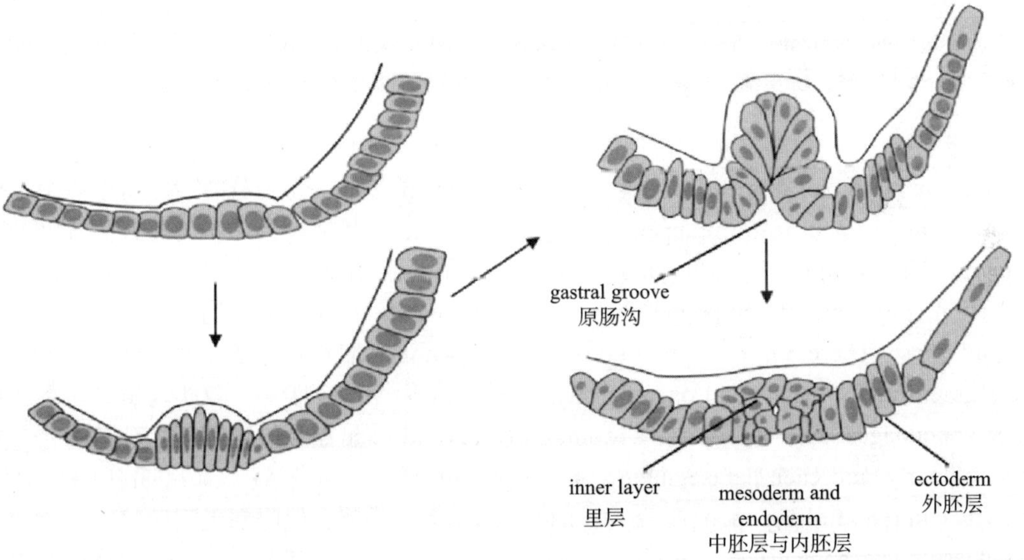

Fig. 18-3 Formation of gastral groove and germ layer (modified from Anderson, 1972).
图 18-3 原肠沟和胚层的形成（仿 Anderson, 1972）。

18.2.3 Embryonic segmentation and appendages formation

The embryo begins to segment and it usually occurs in the mesoderm first and then in the ectoderm. The front end of the wide germ band is the procephalon which is gradually differentiated into lips, mouths, eyes and antennae, and the other narrower part

18.2.3 胚胎分节与附肢形成

胚层形成时，胚胎即开始分节，通常中胚层先分节，之后，外胚层出现横沟。胚带较宽的前端是原头，逐步分化为唇、口、眼和触角等；其余狭窄的部

is protocorm, which gradually differentiates into the jaw, thorax, and abdomen. Most insects segment their bodies backwards from the front, but some beetles segment their bodies forwards from the thorax and backwards. The embryonic body segments of *Drosophila* include Ma, Mx and Lb, which will develop to the head; T1-T3, which will develop into the thorax; A1-A9, which will develop into the abdomen (Fig. 18-4).

分是原躯，逐步分化成颚叶、胸和腹。大多数昆虫原躯的分节从前部往后进行，但一些甲虫分节是从胸往前和后端进行的。果蝇胚胎体节包括将发育成为头部的体节 Ma、Mx 及 Lb，将发育成为胸部的体节 T1 ～ T3，将发育成为腹部的体节 A1 ～ A9（图 18-4）。

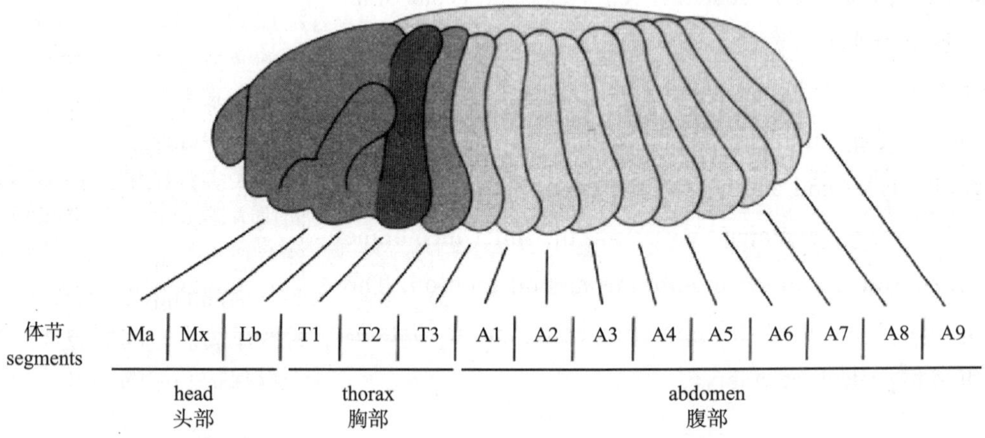

Fig. 18-4 The embryonic segments in *Drosophila*. M, L. Head; T. Thorax; A. Abdomen (Martinz-Arias and Lawrence,1985).
图 18-4 果蝇胚胎体节。M，L. 头部；T. 胸部；A. 腹部（Martinz-Arias 和 Lawrence, 1985）。

Each parasegment has a pair of cystic processes, some of which gradually extend and segment to form appendages. Based on the order of segmentation and appendage, the stage of embryogenesis after segmentation is divided into propopod phase, polypod phase and oligopopod phase. The embryo of propopod phase does not segment or segment unobviously, and only the head and thorax have primary appendages. The embryonic abdomen of polypod phase segment clearly, and each parasegment contains a pair of appendages. The embryo of oligopopod phase segment clearly, but the abdomen doesn't have appendages.

胚胎分节后，每个体节上有一对囊状突起，一些突起逐步延伸并分节形成附肢。按分节与附肢发生次序，胚胎发育在分节后的时期可分为原足期、多足期及寡足期。原足期的胚胎无分节或分节不明显，仅头与胸部有初生附肢；多足期胚胎腹部明显分节且每个体节包含一对附肢；寡足期胚胎具有明显的分节，但腹部无附肢。

The formation of embryonic somite is regulated by the gap gene, pair-rule gene and segment polarity gene. The concentrations of Bicoid, Hunchback and Caudul proteins, the maternal effect factors, determine the transcriptional patterns of the gap genes. The products of the gap genes then regulate the expression of pair-rule genes. When transcripted they further divide the broad gap gene regions into the parasegments. The segment polarity genes guarantee that some repeated structures appear in each segment, establishing the cell fates within each of the parasegment. Subsequent interactions of the gap and pair-rule genes regulate

胚胎体节的形成受缺口基因、成对控制基因、体节极性基因及同源异构型基因的调控。母源效应基因 Bicoid、Hunchback 和 Caudul 蛋白的浓度决定了缺口基因的转录模式。随后，缺口基因的产物调控成对控制基因的表达，当这些基因被转录时，它们进一步将缺口基因所在的区域划分为体节。体节极性基因确保某些重复的结构出现在每个体节中并确定每个体节内的细胞分化。然后，缺口基因和成对控制基因的后续相

the homeotic genes that establish the identity of each segment and its characteristic structures (Fig.18-5; Appendix color graphic Fig.18-5).

互作用，调控了建立每个体节及其特征结构的同源异型基因的表达（图 18-5；附彩图 18-5）。

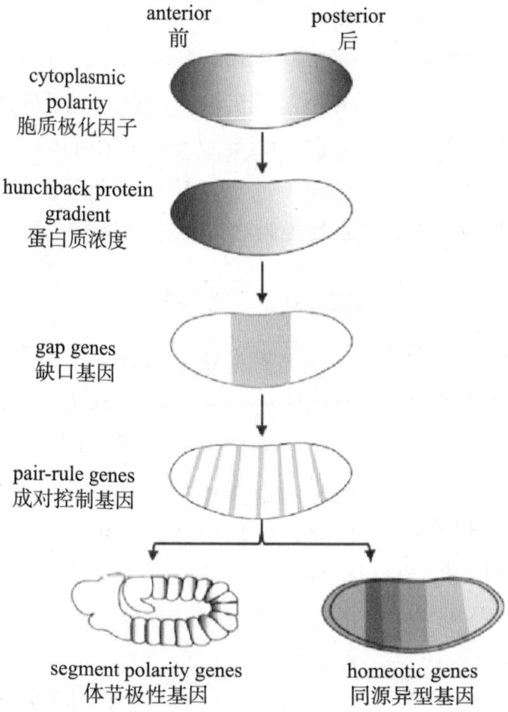

Fig. 18-5 Schematic representation of embryonic segment formation in *Drosophila* (modified from Gilbert, 2010).

图 18-5 果蝇胚胎体节形成示意图（仿 Glibert, 2010）。

18.2.4 Organogenesis

The development of various internal organs occurs after gastrulation and the generation of the new germ layers. Organs are differentiated from germ layers. The ectoderm differentiates into integument, anterior and posterior gut of digestive tract, Malpighian tubule, nervous system, respiratory system, various glands, oenocyte, median oviduct of female and ejaculatory duct of male, etc. The mesoderm differentiates into muscular system, heart and blood cells, fatbody, and partial gonads. The endoderm differentiates to the midgut.

18.2.4.1 Digestive system

The digestive tract is divided into foregut, midgut and hindgut based on its origin and function. When the embryo starts to segment, the ectoderm in the front and the end of the embryo immerses into tubular structures, being called the stomodaeum and proctodaeum which then differentiates into the foregut and hindgut of the digestive tract, respectively. The cells from the endoderm of the anterior and posterior mesenteron rudiments undergo cell division, and finally heal into a midgut that connects stomodaeum and proctodaeum on the dorsal surface. The midgut is adjacent to

18.2.4 器官形成

各内在器官的形成发生在原肠胚和新胚层的产生之后。器官是由不同胚层分化而形成的。外胚层后续分化形成体壁、消化道前肠与后肠、马氏管、神经系统、呼吸系统、多种腺体、绛色细胞、雌虫的中输卵管及雄虫的射精管等；中胚层分化形成肌肉系统、心脏与血细胞、脂肪体及部分生殖腺；内胚层则分化形成中肠。

18.2.4.1 消化系统

消化道按其来源和功能不同可分为前肠、中肠和后肠。在胚胎开始分节时，胚胎前、后端的外胚层分别内陷成管状结构，分别称为口道与肛道，进而分别分化形成消化道的前肠与后肠。源于内胚层的前中肠韧和后中肠韧进行细胞分裂，最后在背面愈合成与口道与肛道相接的中肠。在卵孵化前，中肠与前后肠

foregut and hindgut to form the digestive tract before hatching. Generally, the Malpighian tubule is formed by the invagination of the anterior end of the hindgut, i.e. from the ectoderm.

The formation of digestive system is also regulated by a series of genes. For example, the local expression of several transcription factors, including those encoded by the genes *Sex combs reduced*, *extradenticle* (*Exd*), and *homothorax* (*Htd*), determine where the gut the salivary glands will form primordial cells. The foregut structure formed by the movement of cells depends on the spatial expression of integrins and cell surface adhesion receptors.

18.2.4.2 Tracheal system

The trachea system is the result of invagination of each somite of ectoderm from two sides. The embryo is invaginated in pairs on both sides from the thorax to the abdomen, forming spiracular invagination. With the development of the embryo, the spiracular invagination gradually goes deep into the inner part of the embryo to form the spiracular trachea, and then extends to the dorsal, ventral and anterior-posterior extension. These three bronchus branch continuously to form smaller bronchus and microtrachea distributed in tissues and organs, forming the tracheal system of the larva. At the same time, the spiracular invagination gradually contracts to form an opening which is a valve in the body wall.

The invagination of ectoderm is regulated by transcription factor encoded by *trachealess* in tracheal cells. Branching is guided by the nontracheal cells around invaginating tracheal cells that express the fibroblast growth factor-like ligand encoded by the *branchless*.

18.2.4.3 Nervous system

Before the formation of stomodaeum and proctodaeum, the ectoderm cells depress in the midline of the ventral to form a neural groove, and the ectoderm on both sides of the groove further differentiates into small somatic cells and large neuroblasts. The ectodermal cells become to neuroblasts rather than epidermal cells due to the expression of a group of proneural genes, including *achaete*, *scute*, and *lethal of scute*. The subsequent expression of neurogenic genes prevents any neighboring cells from becoming neuroblasts. The neural ridges result from the proliferation of neuroblasts on both sides of the nerve groove. Three groups of neuroblasts that proliferate in the anterior region ultimately differentiate into the protocerebrum, deutocerebrum, and tritocerebrum of the brain, respectively. Those in other segments give rise to the subesophageal and abdominal ganglia.

相通形成消化道。一般认为，马氏管是后肠前端向体腔进一步内陷而成的，即来自外胚层。

消化系统的分化形成也是受一系列基因调控的。例如，由 *Sex combs reduced*、*extradenticle* (*Exd*) 和 *homothorax* (*Htd*) 等基因编码的转录因子的局部表达，决定了唾液腺在肠道中的何处形成原基细胞；细胞迁移形成前肠，依赖于整合素与细胞表面黏附受体的空间表达。

18.2.4.2 气管系统

气管系统是各体节外胚层从两个侧面内陷的结果。胚胎从胸部至腹部两侧产生成对内陷，形成气门内陷。气门内陷随着发育逐渐深入胚胎内侧形成气门气管，之后分布向背面、腹面及前、后延伸。这三支气管再不断分支形成更小的支气管与微气管分布到组织与器官中，构成幼虫的气管系统。同时，气门内陷逐渐收缩形成开口，即体壁上的气门。

外胚层内陷是由气管细胞中 *trachealess* 基因编码的转录因子所调控的。而分支是由这些内陷的气管细胞周围的非气管细胞中，表达的由 *branchless* 基因编码的成纤维细胞生长因子样配体所引导的。

18.2.4.3 神经系统

在口道与肛道形成之前，外胚层细胞在腹面中线凹陷形成神经沟，沟两侧的外胚层进一步分化出小的体壁细胞与大的神经母细胞。外胚层细胞成为神经母细胞而不是真皮细胞，这是由一组前神经基因 *achaete*、*scute* 及 *lethal of scute* 的表达所决定的。随后的神经源性基因的表达会抑制任何邻近细胞成为神经母细胞。神经母细胞进一步分裂增殖，在神经沟两侧分化为神经脊。胚胎前区增殖的三组神经母细胞最终分化成脑的原脑、中脑和三脑。其他体节的神经母细胞则分别分化形成食管下神经节

Then the internerve part forms the nerve cord. Subsequently, the nervous system then separates from somatic.

Studies on the *Drosophila* central nervous system (CNS) show that the arrangement of the attractor axons in insects' CNS is trapezoidal. Each body segment has two transverse nerve junctions and two longitudinal nerve junctions. The nerve cells at both sides of neural groove extend axon to the midline under the attraction of CNS midline cells, and when they reach the midline, they grow longitudinally. *Drosophila* midline cells are the earliest differentiated cells in the CNS, including a series of glial and neuronal cells, functioning in the organization of the axons of nerve commissure, including anterior commissure and posterior commissure in each body segment.

和腹索神经节。神经间部分形成神经索。随后，神经系统与体壁发生分离。

对果蝇中枢神经系统（CNS）的研究显示，昆虫中枢神经吸引神经轴索的排列方式呈梯形。各体节有两条横向的神经连合与两条纵向的神经联合。神经沟两侧的神经细胞受中枢神经系统中线细胞的吸引分别由两侧向中线伸出轴突，到达中线后，则往纵向生长。果蝇中线细胞是中枢神经系统分化最早的细胞，包括一系列神经胶质细胞与神经元细胞，其作用是组织神经联合的轴突，而每个体节的神经联合包括前联合与后联合。

18.2.4.4 Circulatory system

Most insects have similar dorsal blood vessel formation processes. The cardioblasts, located between the lateral dorsal and ventral part of the coelcomic sac, expand back with the coelomic sac and gradually become crescent shaped and concave facing each other. The paired crescent cells of the mesoderm are fused at the middorsal line to form the heart. The coelomic sac of the head grows backward to form the aorta, connecting to the heart to form the dorsal vessels. The dorsal vascular wall is thin, only slightly protruded in the nucleus, and surrounded by adipocytes and pericardial cells. Blood cells are specialized by dispersed cells in the lateral wall of the coelom sac.

18.2.4.4 循环系统

大多数昆虫的循环系统背血管的形成过程是相似的。位于体腔囊侧背与腹壁相连处的心母细胞随着体腔囊向背面扩展，并逐渐变成新月形且凹面相对。中胚层背合时，成对的新月细胞在背中线处愈合形成心脏。头部的体囊腔向后增长形成大动脉，最后与心脏相连成背血管。背血管的管壁较薄，只有在细胞核处稍微突起，其四周有脂肪体和围心细胞分布。血细胞则是由体腔囊内侧壁的分散细胞特化的。

18.2.4.5 Reproductive system

The germ plasm at the base of the egg differentiates into primordial germ cells during the formation of the blastoderm. The primordial germ cells are located between the germ band and yolk during the formation of amnion, and then gradually move from the posterior pole of the egg to the sides of the hindgut and are divided into two groups. After mesoderm differentiation, the germ cell population is surrounded by a flat layer of mesoderm cells, forming a gential ridge and further differentiating into gonads. The posterior end of the gential ridge differentiates into lateral germ ducts (lateral oviduct and spermaduct), connecting with the middle germ ducts (middle oviduct and ejaculatory ducts) formed by ectoderm invagination, respectively.

18.2.4.5 生殖系统

在形成胚盘时，卵基部的生殖质就分化出了原始生殖细胞。在羊膜形成时，原始生殖细胞位于胚带与卵黄间，之后逐步从卵的后极移到后肠两侧并分成两群。中胚层分化后，生殖细胞群外部被一层扁平的中胚层细胞包围，形成生殖脊并进一步分化成为生殖腺，生殖脊后端则分化为侧生殖管（侧输卵管与输精管），分别与外胚层内陷形成的中牛殖管（中输卵管与射精管）相连。

18.2.4.6 Coelom

During segment differentiation, a cavity is formed in the left and right bands of most somites, namely coelomic sac, and the cavity below the bands in the inner layer is called the epineral sinus. With the continuous expansion of the coelomic sac and the epineral sinus, the left and right coelomic sac is connected with the epineral sinus, and the transverse wall of the coelomic sac of the anterior and posterior somites gradually disappears, leading to the communication between the coelomic sac of each somite.

18.2.4.7 Fatbody and Oenocyte

Fatbody differentiated from mesoderm is appear on both sides of embryo at a certain stage of embryogenesis. After dorsal enclosure, the coelom of the embryo is filled with a large number of fatbody cells. Xu et al. (1998) found that part of fatbody cells in the body cavity of *Chrysopa septempunctata* embryos were deeply stained, and further speculated that fatbody might be an important organ for storing metabolites during embryonic development.

Some layer cells located on the 1st-8th ventral side of embryo gradually separate from the ectoderm and form purple cells under the valve depression at a certain stage of embryogenesis. The nucleus of purple cells have been found to change in cycle during embryogenesis in *Heliothis zea* and *Ephestia kuehniella*, suggesting that purple cells have a secretory function and their secretory activities are related to epidermal formation.

A summary of the contributions of all cell lineages toward adult structure is shown in Fig. 18-6.

18.2.4.6 体腔

在体节分化时，大部分体节的左右侧带内各形成一个空腔，即体腔囊。在里层中带下方出现的空腔称为神经上窦。随着体腔囊与神经上窦的不断膨大，左右体腔囊与神经上窦相连，前后体节体腔囊的横壁也逐渐消失，各体节的体腔囊是相通的。

18.2.4.7 脂肪体与绛色细胞

胚胎发育到一定时期，胚胎两侧出现由体壁中胚层分化而成的脂肪体。背合后，胚胎体腔内充满了大量的脂肪体细胞群。Xu 等（1998）发现大草蛉胚胎后期体腔内的部分脂肪体细胞染色较深，进而推测脂肪体可能是大草蛉胚胎发育贮存代谢产物的重要器官。

胚胎发育到一定时期，位于胚胎 1 ～ 8 腹节侧腹面的一部分胚层细胞逐渐脱离外胚层，在气门凹陷下方形成绛色细胞。在谷实夜蛾和地中海粉斑螟中均发现胚胎发育中的绛色细胞的细胞核会进行循环性变化，表明绛色细胞具有分泌功能且其分泌活动与真皮形成有关。

所有分化为成虫组织和器官的细胞系均在图 18-6 中显示。

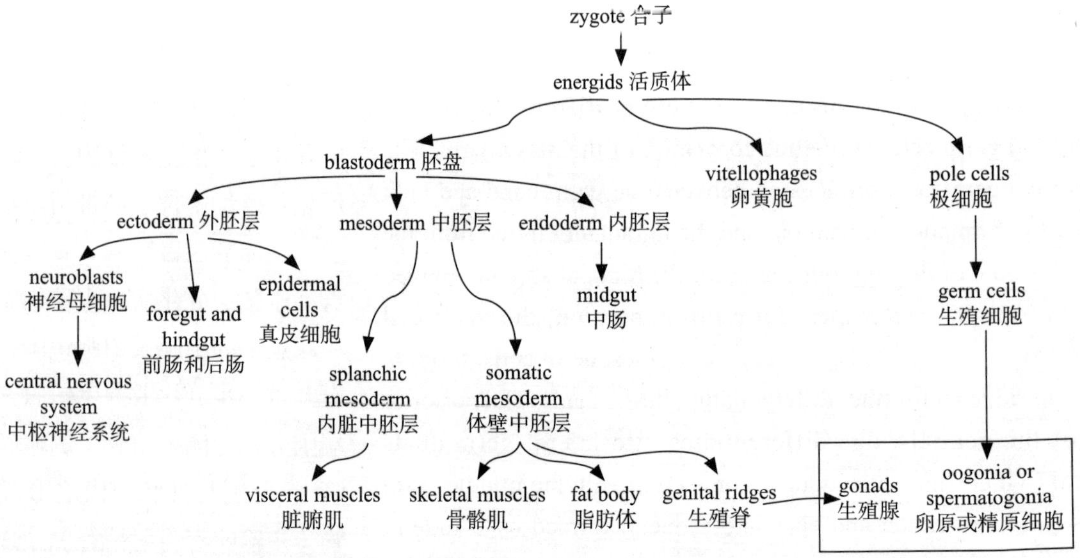

Fig. 18-6 Organ formation during embryonic development in insects (modified from Chapman, 1988).
图 18-6 昆虫胚胎发育过程中的器官形成（仿 Chapman, 1988）。

18.2.5 Blastokinesis, Dorsal enclosure and Disappearance of Embryonic Envelop

18.2.5.1 Blastokinesis

During embryonic development, the position of the embryo on the surface of the egg changes to varying degrees, being called blastokinesis. Blastokinesis allows the embryo to make better use of nutrients within the egg. The amplitude of blastokinesis in short-germ insects is larger than that of insects with long-germ.

18.2.5.2 Dorsal enclosure

With the development of the embryo, both sides of the germ band continue to extend around the yolk to the back. The embryo part gradually envelops the yolk and eventually two sides of the embryo extend to the dorsal midline and close to form a complete embryo, being called dorsal enclosure. In general, insects have an early dorsal enclosure on the abdomen and then dorsal enclosure on the thorax, and usually leave an unclosed hole at the front end of the thorax until hatching. In some insects, such as *Mythimna separata*, the dorsal enclosure occurs in the posterior embryo at first and then forward to the thorax. The completion of the dorsal enclosure marks the formation of the integument and coelom.

18.2.5.3 Disappearance of embryonic envelop

During the middle and late embryogenesis, the serosa and amnion break from their respective heals. Then they are drawn to the back of the embryo at the time of dorsal enclosure, being trapped in yolk as a tubular shape and retreating into dorsal organs. At the end of dorsal enclosure, dorsal organs are dissociated and absorbed by yolk. In a few insects, the serosa or amnion is retained. In Lepidoptera, neither serosa nor amnion is lost, and a small portion of egg yolk is trapped between the two membranes. Some newly hatched larvae often feed on the egg shell as the earliest nutrition.

18.2.6 Formation of integument

The body wall can be formed in two ways. The first is the rupture of amnion from the egg shell to the pouch, which develops a temporary body wall and it gradually degenerates and disappears after the ectoderm is fused to form a true body wall. Examples of this are *Spodoptera exigua* and *C. Septempunctata*. In the second way, the body wall cells differentiated from the ventral center of the embryo

18.2.5 胚动、背合与胚膜消失

18.2.5.1 胚动

在胚胎发育过程中，胚胎在卵表面的位置会发生不同程度的变动，这一现象称为胚动。胚动与胚胎更好地利用卵内的营养有关。短胚带昆虫的胚动幅度较大，而长胚带昆虫的胚动幅度较小。

18.2.5.2 背合

随着胚胎发育的进行，胚带两侧围绕卵黄不断向背面伸展。胚胎部分逐渐包裹卵黄，最终，胚胎的两侧延伸至背中线并闭合，形成一个完整的胚胎，这个过程称为背合。一般来说，昆虫腹部背合较早，然后向胸部背合且常在胸部最前端留一个未背合的小孔，直至孵化前才完全闭合。也有一些昆虫的背合最先出现在胚胎尾部，然后向前至胸部，如黏虫。背合的结束标志着体壁与体腔的形成。

18.2.5.3 胚膜消失

胚胎发育进入中后期时，浆膜和羊膜从各自的愈合处破裂，背合时逐渐被牵引到胚胎背面，多呈管状陷于卵黄中，退缩为背器，背合末期，背器也逐渐解离并被卵黄吸收。有少数昆虫的浆膜或者羊膜会被保留，鳞翅目的浆膜和羊膜均不消失并有一小部分卵黄夹存于两膜之间，一些初孵幼虫常取食卵壳作为最初的营养。

18.2.6 体壁形成

体壁形成一般有 2 种方式。第一种方式是羊膜破裂后沿卵壳向背包起，先于胚胎的外胚层而形成临时性体壁，待外胚层背合形成真正体壁后，临时性体壁逐渐退化并消失，如甜菜夜蛾与大草蛉就属于这种方式。第二种方式是由

divide, proliferate, extend and meet at the middorsal line. Then dorsal enclosure starts from the tail, gradually extends forward, and finally completes the whole dorsal enclosure at the dorsal thorax, indicating the completed development of the body wall. Examples of this are *Mythimna separata* and *Sphaerodema rustius*.

胚体腹面中央分化出的体壁细胞分裂增殖伸展汇合于背中线，背合先由尾部开始，逐渐向前伸展，最后在胸部背面前端完成整个背合，体壁发育完成，如夜蛾科黏虫和半翅目负子蝽就属于这种方式。

18.3 胚胎发育的遗传调控

18.3 Genetic Regulation of Embryonic Development

18.3.1 Genes involved in the regulation of embryogenesis

Molecular regulatory mechanism of insect embryogenesis is mainly derived from the research on *Drosophila*. Generally, *Drosophila* embryonic development is mainly regulated by maternal genes (or maternal effect genes), zygotic genes (or segmentation genes) and homeotic genes (or selector genes) (Fig. 18-7).

Prior to blastoderm formation, the embryonic development, such as the determination of axon, is mainly regulated by maternal genes. More than 30 maternal genes have been identified, and they are divided into four groups: anterior genes, posterior genes, distal genes and dorsal-ventral genes. The anterior-posterior polarity of embryo is mainly affected by the genes of the first three groups, while dorsal-ventral polarity is mainly affected by the genes of dorsal-ventral group. For example, *Bicoid* regulates the embryonic anterior axis. When its transcript is high in the anterior polarity, *Nanos* is activated in the posterior polarity, and then *Drosophila* embryo undergoes normal polar differentiation. The absence of *Bicoid* at the posterior polarity is due to the inhibitory effect of *Nanos*, while the absent expression of *Nanos* at the anterior polarity is due to the inhibitory effect of BCD protein. *Drosophila mago nashi* is a member of the posterior polarity determining gene group and its expressed product is also one of the determining components of gonoplasm. In *mago*⁻ mutant, the maternal determinant *osk* mRNA locates abnormally, resulting in abnormal abdominal segments and inability to form germ cells in the offspring.

18.3.1 调控昆虫胚胎发育的基因

昆虫胚胎发育的分子调控机制主要来源于对果蝇的研究。大体上，果蝇胚胎发育主要受母体基因（也称为母体效应基因）、合子基因（或体节形成基因）及同源异型基因（或选择基因）调控（图 18-7）。

在胚盘形成前，卵的发育，例如胚胎体轴的确定，主要受母体基因调控。已有 30 多种母体基因被鉴定，它们大致可分为前极基因群、后极基因群、末端基因群及背腹基因群这 4 类。胚胎的前后极性主要受前 3 群基因影响，背腹极性主要受背腹群基因影响。例如，*Bicoid* 可调节胚胎的前轴，当该基因的转录在胚胎前极含量高时，后极 *Nanos* 被激活，果蝇胚胎进行正常的极性分化。*Bicoid* 不在后极表达是由于 *Nanos* 的抑制作用，*Nanos* 不能在前极表达是由于 BCD 蛋白的抑制作用。果蝇 *mago nashi* 是后极极性决定基因群成员之一，其表达产物也是生殖质的决定成分之一。在 *mago*⁻ 突变体中，母性决定成分 *osk* mRNA 不能正常定位，可导致产生的后代腹部体节异常且不能形成生殖细胞。

The zygote genes mainly play a role after the formation of the blastoderm, and determine the segments of the body and the polarity of each somite. More than 20 zygotic genes have been identified, and they are divided into gap genes, pair-rule genes and segment polarity genes according to their expression time.

Homeotic genes (*HOX* genes) determine the cell fate and the somite characteristics, and their expression is influenced by some zygote genes and other homeotic genes. The mutation of *HOX* causes the changes in cell fate during development, leading to the appearance of certain segment features at the corresponding positions of different segments, namely homeosis. For example, *Antp* mutation of *Drosophila* results in the transformation of cephalic antennae into a second pair of pereiopoda, and *Bithorax* mutation causes the characteristics of the 3rd thoracic segment to transform into the second one, leading to the generation of another pair of wings.

In addition, nuclear receptor (NR) genes have been found to regulate insect embryonic development. *D. melanogaster* has 18 NR genes and 3 atypical subfamily members. NR gene *Tailless* and *knirps* act as gap genes to regulate segmentation, while *ftz-f1* affects pair-rule segmentation. *Tailless*, *seven-up*, and *eagle* are involved in neuroblast specification in both the brain and ventral nerve cord. *Seven-up* specifies Malpighian tubule primordia and dorsal vessel progenitors, and *Hepatocyte nuclear factor* 4 functions in posterior midgut primodium. *Seven-up* also functions in fat body and oenocyte precursors. EcR/USP drives germ band retraction, dorsal closure and head involution. *DHR3* and *ftz-f1* function in the final phase of trachea development.

合子基因主要在胚盘形成后发挥作用，决定体躯的分节与各体节的极性。已有 20 余个合子基因被鉴定，它们按其在胚盘的表达顺序依次分为缺口基因、成对控制基因及体节极化基因。

同源异型基因（*HOX* 基因）进一步决定各体节的细胞命运决定与体节特征的决定，其表达受某些合子基因与其他同源异型基因的影响。*HOX* 基因突变会引起发育过程中细胞命运发生改变，导致某一部分的体节特征在不同体节的相应位置出现，即同源异型转化。例如，果蝇 *Antp* 突变会导致位于头部的触角生长转化为第二对胸足，*Bithorax* 突变会引起第三胸节特征向第二胸节特征的转变，导致另一对翅的产生。

此外，核受体（NR）基因也被发现调控昆虫胚胎的发育。果蝇包含 18 个典型的核受体基因和 3 个非典型亚家族成员。核受体基因 *Tailless* 和 *knirps* 作为间隙基因调控胚胎分节，而 *ftz-f1* 影响配对法则基因所决定的分节。*Tailless*、*seven-up* 及 *eagle* 有助于大脑和腹神经索的神经母细胞的命运决定。*Seven-up* 参与了马氏管原基和背侧血管祖细胞的命运决定，*Hepatocyte nuclear factor 4* 在中肠后原基中发挥作用。*Seven-up* 也在脂肪体和单核细胞前体中起作用。EcR/USP 驱动了胚带收缩、背合及头对合。*DHR3* 和 *ftz-f1* 在气管发育的最后阶段发挥作用。

18.3.2 Factors involved in the embryogenesis

The phenotype of organisms is the result of the combination of genotype and environment, and insect embryonic development is no exception. External environment factors, including physical factors (light, temperature and humidity), biological factors (parasitic and symbiotic), and cytoplasmic DNA and RNA, affect embryonic development.

18.3.2 影响昆虫胚胎发育的因子

生物的表现型是基因型与环境综合作用的结果，昆虫胚胎发育也不例外。外界环境因子，如光照、温度及湿度等物理因子，寄生与共生物等生物因子及细胞质 DNA 与 RNA 均会影响胚胎发育。

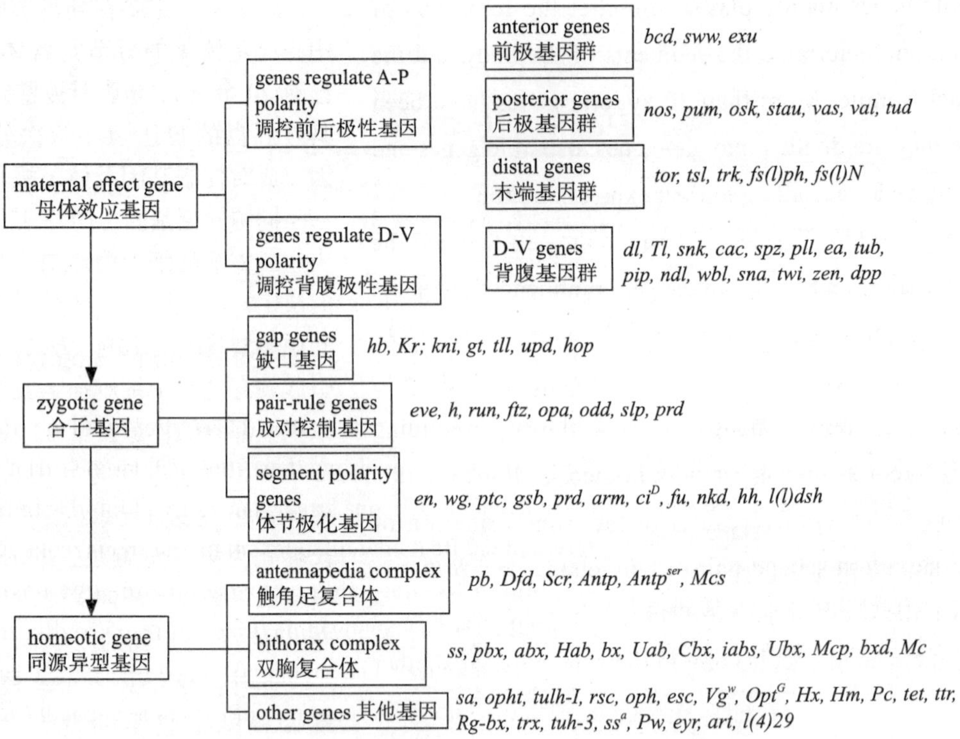

Fig. 18-7 Genes involved in the regulation of embryogenesis.
图 18-7 调控胚胎发育的基因。

18.4 胚胎发育的内分泌学

18.4 Endocrinology of Embryonic Development

Insect embryonic development is also regulated by hormones. Ecdysone is the key regulator of embryonic development. The diapause of *B. mori* eggs is due to the absence of ecdysone. In diapause eggs, the bulk of ecdysone exists as conjugated forms (phosphoric esters), but in nondiapause eggs, free forms coexist with conjugated forms. In *D. melanogaster*, mutant flies of ecdysteroidogenic enzymes exhibit embryonic lethal phenotypes during late embryogenesis.

The presence of ecdysone before the formation of embryonic endocrine organs is the result of their maternal synthesis by ovarian follicle cells during vitellogenesis. They are packaged into the egg largely as inactive phosphorylated conjugates of ecdysone and released from the conjugates as active hormones when they are

昆虫胚胎发育受到激素的调节。蜕皮激素是昆虫胚胎发育的关键调控因子。家蚕虫卵的滞育是由缺乏蜕皮激素所致的。在滞育虫卵中，大部分的蜕皮激素以共轭形式（磷酸酯）存在，但在非滞育虫卵中，蜕皮激素自由形式与共轭形式共存。在果蝇中，蜕皮甾类生成酶突变蝇在发育后期出现致死表型。

在胚胎内分泌器官形成之前，存在的蜕皮激素是母体在卵黄发生过程中，由卵巢滤泡细胞合成的。大量的蜕皮激素主要以不活跃的磷酸化的激素共轭物的形式被贮存在卵黄中，并在需要时作为活性激素从结合物中释放出来。在

required. Among these conjugates, C-22 phosphorylated 2dE (2dE-22P), ecdysone (Ecdysone-22P), and 20E (20E-22P) are considered as storage forms of ecdysone, whereas C-2 and C-3 phosphorylated ecdysone are thought as inactivated metabolites. Ecdysone 22-kinase (Eck) and ecdysone-phosphate phospha-tase (EPPase) have been identified in *B. mori* as enzymes that phosphorylate and dephosphorylate ecdysone, respectively. During embryogenesis, EPPase is expressed and converts inactive ecdysone conjugates into the free form of 2dE and 20E.

When embryonic endocrine organs are formed, the embryo is capable of the hormones synthesis. The ecdysone, 20-hydroxye-cdysone, 26-hydroxy- ecdysone, 20-deoxy-ecdysone, and 20, 26-dihydroxy- ecdysone have been isolated from the insect eggs. Besides, a series of genes related to ecdysonegenesis after organogenesis are detected.

There are usually 2-3 times of cuticle formation during insects' embryogenesis. The embryo of lower wingless insect, *T.domestica*, undergoes two successive cuticle processes, while there are three times of cuticle formation during embryogenesis of exoptera. The embryo of *L. migratoria* produces four distinct cuticles, including the serosal cuticle and three embryonic cuticles, and the last is the cuticle of the 1st instar larva. The levels of ecdysone typically peak with the deposition of thcsc cuticle, suggesting that the esdysone is involved in the formation of cuticle.

The embryonic prothoracic glands are formed after blasto-kinesis and then synthesize the ecdysone. Ecdysone signaling also controls the morphogenesis of the embryo. Ametabolous and hemimetabolous insects have a pronymphal stage that precedes the 1st larval instar, being considered as a continuation of embryonic development after hatching. When enclosed in the embryonic cuticle, the embryo would molt and enter into the prenymphal stage controlled by the peak of ecdysone, while the embryo would molt and enter into the 1st instar larva driven by amother ecdysone peak.

Although the roles of JH in embryogenesis are rarely studied, it is clear that JH is also involved in the embryogenesis. For example, the inhibition of JH synthesis pathway or the mutation of JH receptor Met during the embryogencsis of silkworm leads to the

这些共轭物中，C-22 磷酸化的 2- 脱氢蜕皮酮（2dE-22P），蜕皮酮（Ecdysone-22P）及 20E（20E-22P）被认为是蜕皮激素的贮存形式，而 C-2 和 C-3 磷酸化的蜕皮激素被认为是失活的代谢物。在家蚕卵中，已分离到磷酸化蜕皮激素的蜕皮激素 -22- 激酶（Eck）和去磷酸化的蜕皮激素 - 磷酸磷酸酶（EPPase）。在胚胎发生过程中，EPPase 将非活性的蜕皮激素共轭物转化为自由形式的 2dE 和 20E。

而当胚胎内分泌器官形成时，胚胎具备合成蜕皮激素的能力。在昆虫卵中，分离到了蜕皮激素、20E、26E、20dE 和 20，26E，此外，还检测到了器官发生后一系列蜕皮激素合成相关基因的表达。

昆虫在胚胎期一般有 2 ～ 3 次的几丁质生成过程。低等无翅类昆虫衣鱼（*T. domestica*）的胚胎发育中有 2 次连续的角质层生成过程，外翅类昆虫胚胎期一般有 3 次的角质层生成过程。飞蝗胚胎分泌 4 个角质层，包括 1 个浆膜角质层和 3 个胚胎角质层，最后 1 个即是 1 龄幼虫的角质层。蜕皮激素的水平通常随着这些角质层的沉积而达到峰值，表明蜕皮激素参与了胚胎角质层的形成。

胚胎前胸腺在胚动后形成，然后可合成蜕皮激素。蜕皮激素信号也控制着胚胎的形态发生。无变态和不完全变态昆虫在第一个幼虫龄期之前有一个预若虫期，这被认为是孵化发生后胚胎发育的延续。当胚胎被包裹在胚胎角质层中时，蜕皮激素的一个峰值会导致蜕皮进入预若虫期，而蜕皮激素的另一个峰值则标志着蜕皮进入第一阶段的幼虫。

虽然保幼激素（JH）在胚胎发育过程中的研究较少，但可以肯定的是保幼激素也参与胚胎发育。例如，在家蚕和赤拟谷盗的胚胎发育过程中，保幼激素合成通路受阻或者保幼激素受体 Met

delay of embryonic development, causing the reduction of hatching rate.

(Deng Huimin)

发生突变都会导致胚胎发育延迟，进而引起孵化率的降低。

（邓惠敏）

Exercises

1. What are the main ways of cleavage? In which insects are they found?

2. What organs do the ectoderm, mesoderm and endoderm differentiate into?

3. What is the sequence of gene groups that determine *Drosophila* embryonic development?

思考题

1. 昆虫卵裂方式主要有哪些？分别存在于哪些昆虫中？

2. 外胚层、中胚层及内胚层分别分化成哪些器官？

3. 在果蝇胚胎发育过程中起决定作用的基因群的先后顺序是什么？

参考文献
References

王荫长 . 2001. 昆虫生物化学 . 北京 : 中国农业出版社 .

冯丽春 , 沈卫德 . 2015. 蚕体解剖生理学 . 北京 : 高等教育出版 .

付新华 , Nobuyoshi O., 王余勇 , 等 . 2005. 条背萤的闪光求偶行为 . 昆虫学报 , 2: 227-231.

吴少英 , 段文波 , 李芬 , 等 . 2021. 昆虫钠离子通道的研究进展 . 昆虫学报 , 64(7): 862-874.

段文波 , 王颢 , 李芬 , 等 . 2020. 麦长管蚜电压门控钠离子通道基因克隆及序列分析 . 中国生物防治学报 , 36(2): 307-314.

诸星静次郎 , 1984. 昆虫的生长发育 . 蒋同庆 . 译 . 重庆：重庆出版社 .

Ahmadi F., Mikani A., Moharramipour S. 2021. Induction of diapause by clock proteins period and timeless *via* changes in PTTH and ecdysteroid titer in the sugar beet moth, *Scrobipalpa ocellatella* (Lepidoptera: Gelechiidae). Arch. Insect Biochem. Physiol., 107: e21790.

Alexander D.E. 2018. A century and a half of research on the evolution of insect flight. Arthropod Struct. Dev., 47, 322-327.

Alves-Bezerra M., Gondim K.C. 2012. Triacylglycerol biosynthesis occurs *via* the glycerol-3-phosphate pathway in the insect *Rhodnius prolixus*. Biochim Biophys Acta Mol Cell Biol Lipids., 1821: 1462-1471.

Ameku T., Yoshinari Y., Fukuda R., et al. 2017. Ovarian ecdysteroid biosynthesis and female germline stem cells. Phytopathology, 11:185-193.

Amey, Joanna S., Andrias O., et al. 2015. An evolutionarily-unique heterodimeric voltage-gated cation channel found in Aphids. FEBS Lett., 589(5): 598-607.

Anderson D. T. 1972a. The development of hemimetabolous insects. In: Counce SJ, Waddington CH (eds) Developmental systems: insects, Vol.1. London: Academic Press, 95-163.

Anderson D. T. 1972b.The development of holometabolous insects. In: Counce SJ, Waddington CH (eds) Developmental systems: insects, Vol.1. London: Academic Press, 165-242.

Ando N., Kono P.P. 2020. Toward robotic applications of insect sex pheromone-searching behavior: lessons from the silkmoth. In *Insect Sex Pheromone Research and Beyond*, Entomology Monographs. Springer Nature Singapore Pte Ltd., 287-313.

Ansari S, Troelenberg N.,.Dao V.A., et al. 2018. Double abdomen in a short-germ insect: zygotic control of axis formation revealed in the beetle *Tribolium castaneum*. Proc Natl Acad Sci USA., 116:1819-1824.

Aprelev P., Bruce T.F., Beard C.E., et al. 2019. Nucleation and formation of a primary clot in insect blood. Sci Rep., 9: 3451.

Arakane Y., Noh M.Y., Asano T., et al. 2016.Tyrosine metabolism for insect cuticle pigmentation and sclerotization. In: E. Cohen, B. Moussian (eds.), Extracellular Composite Matrices in Arthropods. Switzerland:

Springer International Publishing., 165-220.

Armstrong C. M. 1981. Sodium channels and gating currents. Physiol Rev., 61(3): 644-683.

Arrese E.L., Soulages J.L. 2010. Insect fat body: energy, metabolism, and regulation. Annu Rev Entomol., 55: 207-225.

Badejo O., Skaldina O., Gilev A., et al. 2020. Benefits of insect colours: a review from social insect studies. Oecologia., 194:27-40.

Baker C.A., Clemens J., Murthy M. 2019.Acoustic pattern recognition and courtship songs: Insights from insects. Annu Rev Neurosci., 8(42): 129-147.

Bao B., Xu W.H. 2011. Identification of gene expression changes associated with the initiation of diapause in the brain of the cotton bollworm, *Helicoverpa armigera*. BMC Genomics., 12:224.

Barsagade D.D., Gharade S.A., Barsagade V.G. 2021. Immunocytochemical localization of leptin hormone in the neurosecretory cells of brain-suboesophageal ganglion complex of tropical tasar silkworm, *Antheraea mylitta* (D.) eco-race *Bhandara*. Cur. Sci., 10, 1611-1615.

Bataillé L., Colombié N., Pelletier A., et al. 2020. Alary muscles and thoracic alary-related muscles are atypical striated muscles involved in maintaining the position of internal organs. Dev., 147(8): dev185645.

Bataillé L., Frendo J.L.,Vincent A. 2015. Hox control of *Drosophila* larval anatomy; The Alary and Thoracic Alary-Related Muscles. Mech Dev., 2: 170-176.

Bechinger B, Gorr S U. 2017. Antimicrobial peptides: mechanisms of action and resistance. J. Dent. Res., 96(3):254-260.

Beetz M.J. 2018. Insect orientation: stay on course with the sun. Current Biol., 28, 933-936.

Beetz M.J., Kraus C. Franzke M., et al. 2022. Flight-induced compass representation in the monarch butterfly heading network. Current Biol., 32, 338–349.

Behr M., Riedel D. 2020. Glycosyl hydrolase genes control respiratory tubes sizes and airway stability. Sci. Rep., 10:13377.

Belles X. 2020. Insect metamorphosis: from natural history to regulation of development and evolution. Cambridge: Academic Press.

Belles X. 2020. Krüppel homolog 1 and E93: The doorkeeper and the key to insect metamorphosis. Arch Insect Biochem Physiol., 103(3):e21609.

Bendena W.G. 2010. Neuropeptide physiology in insects. In Neuropeptide Systems as Targets for Parasite and Pest Control, edited by Geary T.G and Maule A.G. Landes Bioscience and Springer Science+Business Media.

Benoit J.B., Attardo, G.M., Baumann A.A., et al. 2015. Adenotrophic viviparity in tsetse flies: potential for population control and as an insect model for lactation. Annu. Rev. Entomol., 60:351-71.

Bilinski S.M., Sekula M. Tworzydlo W. 2020. Morphogenesis of the ovarian follicular epithelium during initial stages of embryogenesis of the viviparous earwig, *Hemimerus talpoides*. J.Morphol., 281:47-54.

Bilinski S.M., Jaglarz M.K., Halajian A., et al. 2018. Unusual morphological adaptations and processes associated with viviparity in an epizoic dermapteran. PLoS One., 13 (4): e0195647.

Blanco-Obregon M., Katz J., Durrieu L., et al. 2020. Context-specific functions of Notch in *Drosophila* blood cell progenitors. Dev Biol., 462: 101-115.

Blomquist G. J., Tittiger C., Jurenka R. 2020. Cuticular hydrocarbons and pheromones of Arthropods. In: Hydrocarbons, Oils and Lipids: Diversity, Origin, Chemistry and Fate, H. Wilkes (ed.), Handbook of

Hydrocarbon and Lipid Microbiology, Springer Nature Switzerland AG.

Boos A., Distler J., Rudolf H., et al. 2018. A re-inducible gap gene cascade patterns the anterior–posterior axis of insects in a threshold-free fashion. Elife., 7: e41208.

Bradshaw W.E., Holzapfel C.M., 2010. Circadian clock genes, ovarian development and diapause. BMC Biology. 8:115.

Bricker A.B. 2021. Buzz Words: Leave us not honeyless. Lit. Rev. Canada.

Briscoe A. D. 2008. Reconstructing the ancestral butterfly eye: focus on the opsins. J Exp Biol, 211(Pt 11): 1805-1813.

Briscoe A. D., Chittka L. 2001.The evolution of color vision in insects. Annu Rev Entomol., 46: 471-510.

Briscoe J, Small S. 2015. Morphogen rules: design principles of gradient-mediated embryo patterning. Development., 142(23):3996-4009.

Briscoe, A. D., Bybee S.M., Bernard G. D., et al.2010. Positive selection of a duplicated UV-sensitive visual pigment coincides with wing pigment evolution in *Heliconius butterflies*. P Natl Acad Sci USA., 107(8): 3628-3633.

Brito N.F., Moreira N.F., Melo A.C.A. 2016. A look inside odorant-binding proteins in insect chemoreception. J. Insect Physiol., 95: 51–65.

Britt S. G., Feiler R.,Kirschfeld K., et al. 1993. Spectral tuning of rhodopsin and metarhodopsin *in vivo*. Neuron., 11(1): 29-39.

Browne A. O'Donnell M.J. 2018. Mechanisms of calcium sequestration by isolated Malpighian tubules of the house cricket *Acheta domesticus*. Arch. Insect Biochem. Physiol., 97: e21431.

Buehlmann C., Wozniak B., Goulard R., et al. 2020. Mushroom bodies are required for learned visual navigation, but not for innate visual behavior, in Ants. Current Biol. 30, 3438-3443.

Bulet P., Hetru C., Dimarcq J. et al. 1999.Antimicrobial peptides in insects; structure and function. Dev. Comp. Immunol., 23(4/5):329-344.

Bullard B., Pastore A. 2019. Through thick and thin: dual regulation of insect flight muscle and cardiac muscle compared. J. Muscle Res. Cell M., 40:99-110.

Cabrero P., Richmond L., Nitabach M.,et al. 2013. A biogenic amine and a neuropeptide act identically: tyramine signals through calcium in *Drosophila* tubule stellate cells. P Roy Soc B-Biol Sc. 280(1757).

Cambron L.D., Yocum G.D., Yeater KM., et al. 2021. Overwintering conditions impact insulin pathway gene expression in diapausing *Megachile rotundata*. Comp. Biochem.and Physiol., Part A, 256, 110937.

Campos-Ortega J.A., Hartenstein V. 1985.The embyonic development of *Drosophila melanosgaster*. Spinger-Verlag, Berlin.

Cao T.X., Sujkowski A., Cobb T., et al. 2020. The glutamic acid-rich–long C-terminal extension of troponin T has a critical role in insect muscle functions. J. Biol. Chem., 295(12) :3794-3807.

Cao T.X., Jin J.P. 2020. Evolution of flight muscle contractility and energetic efficiency. Front. Physiol., 11,1038.

Cardé R.T. 2021. Navigation along windborne plumes of pheromone and resource-linked odors. Annu. Rev. Entomol., 66:317-36.

Cassau S., Krieger J. 2021. The role of SNMPs in insect olfaction. Cell Tissue Rese, 383:21-33

Cerenius L, Söderhäll K. 2021. Immune properties of invertebrate phenoloxidases. Dev Comp Immunol, 122:104098.

Ceriani M. F., Darlington T. K., Staknis D., et al. 1999. Light-dependent sequestration of TIMELESS by CRYPTOCHROME. Science, 285(5427): 553-556.

Chalisova N. I., Ryzhak G. A., Ivko O. M. 2021. effect of short peptides on the insect nervous system. Biology Bulletin Reviews, 11(6): 597-601.

Chapman R.F., Douglas A.E., Siva-Jothy M.T. 2013. Circulatory system, blood and the immune system. In: The Insects: Structure and Function, ed. SJ Simpson, AE Douglas, Cambridge, UK: Cambridge Univ. Press.

Chari N., Vutukuru S.S., Saicharan R. et al., 2021. Chitinous membranes and analogous material. In Biophysics of Insect Flight, Chari N. et al. (eds.), Springer Series in Biophysics 22, Springer Nature Singapore Pte Ltd.

Chatterjee N., Perrimon N. 2021. What fuels the fly: energy metabolism in *Drosophila* and its application to the study of obesity and diabetes. Sci Adv. 7(24): eabg4336.

Chaturvedi R., Stork T., Yuan C.Y., et al. 2022. Astrocytic GABA transporter controls sleep by modulating GABAergic signaling in *Drosophila* circadian neurons. Curr. Biol. 32, 1895-1908.

Cheatle J.A.M, Pick L. 2017. The function and evolution of nuclear receptors in insect embryonic development. Curr Top Dev Biol, 125:39-70.

Chen Y, Liu Y.Y., Tian H.J., et al. 2019. Distribution of pheromone biosynthesis-activating neuropeptide in the central nervous system of *Plutella xylostella* (Lepidoptera: Plutellidae). J. Eco. *Entom*, 112(6), 2638-2648.

Chen, B.S. Zhang N., Xie S., et al. 2020. Gut bacteria of the silkworm *Bombyx mori* facilitate host resistance against the toxic effects of organophosphate insecticides. Environ. Int., 143, 105886.

Chintapalli R.T.V., Hillyer J.F. 2016. Hemolymph circulation in insect flight appendages: physiology of the wing heart and circulatory flow in the wings of the mosquito *Anopheles gambiae.* J Exp Biol. 219(24): 3945-3951.

Chipman AD. 2020. The evolution of the gene regulatory networks patterning the *Drosophila* blastoderm. Curr Top Dev Biol. 139:297-324.

Choi S., Lim D.S., Chung J. 2015. Feeding and fasting signals converge on the LKB1-SIK3 pathway to regulate lipid metabolism in *Drosophila*. PLoS Genet,11: e1005263.

Chou W.H., Hall K. J., Wilson D. B., et al. 1996. Identification of a novel *Drosophila* opsin reveals specific patterning of the r7 and r8 photoreceptor cells. Neuron, 17(6): 1101-1115.

Chu X., Heinze S., Ian E., et al. 2020. A novel major output target for pheromone-sensitive projection neurons in male moths. Front. Cell. Neurosci, 14:147.

Chung C.Y, Hsiao Y.M., Huang T.Y., et al. 2018. Germline expression of the *hunchback* orthologues in the asexual viviparous aphids: a conserved feature within the Aphididae. Insect Mol.Biol, 27(6), 752-765.

Chung C.Y., Shigenobu S. 2022. Reproductive constraint in the social aphid *Ceratovacuna japonica:* Sterility regulation in the soldier caste of a viviparous insect. Insect Biochem.Mol. Biol, 145, 103756.

Clark E, Peel A.D., Akam M. 2019. Arthropod segmentation. Dev., 146:170480.

Coast G.M, Nachman R.J. 2009. Insect Endocrinology. Introduction. Gen Comp Endocrinol, 162(1):1.

Cohen E., Sawyer J. K., Peterson N.G., et al. 2020. Physiology, development, and disease modeling in the *Drosophila* excretory system. Genetics, 214, 235-264.

Collett M., Chittka L., Collett T.S. 2013. Spatial memory in insect navigation. Curr. Biol., 23, R789-R800.

Contreras H.L., Bradley T.J. 2010. Transitions in insect respiratory patterns are controlled by changes in metabolic rate. J. Insect Physiol., 56: 522-528.

Corrales M., Cocanougher B. T., Kohn A. B., et al.2022. A Single-cell transcriptomic atlas of complete insect nervous systems across multiple life stages. Neural Dev.,17(1): 8.

Costa-Leonardo A.M., Laranjo L.T., Janei V., et al. 2013.The fat body of termites: functions and stored materials. J Insect Physiol., 59(6): 577-87.

Crews ST. 2019. *Drosophila* embryonic CNS development: neurogenesis, gliogenesis, cell fate, and differentiation. Genetics, 213(4):1111-1144.

Cronin T. W., Järvileht o M., Weckström.M., et al. 2000. Tuning of photoreceptor spectral sensitivity in fireflies (Coleoptera: Lampyridae). J Comp Physiol A.,186(1): 1-12.

Csordás G., Gábor E., Honti V. 2021. There and back: the mechanisms of differentiation and transdifferentiation in *Drosophila* blood cells. Dev Biol., 469: 135-143.

Cui H.Y., Zhao Z.W. 2020. Structure and function of neuropeptide F in insects. J.Integr. Agri., 19(6): 1429-1438.

Czaczkes T.J., Grüter C., Ratnieks F.L.W. 2015.Trail pheromones: an integrative view of their role in social insect colony organization. Annu. Rev. Entomol., 60:581-99.

Daimon T., Uchibori M., Nakao H., et al. 2015. Knockout silkworms reveal a dispensable role for juvenile hormones in holometabolous life cycle. Proc Natl Acad Sci U S A., 112: 4226-4235.

Davies S.A., Cabrero P., Povsic M., et al. 2013. Signaling by *Drosophila* capa neuropeptides. Gen. Comp. Endocr., 188: 60-66.

Davis A.K., Clancy K.M., Sasaki T. 2021. How to take an ant's pulse: a procedure for non-destructively monitoring baseline and stimulated heart rate in Formicidae. Ent Exp Appl., 169:807-812.

Davis G.K., Patel N.H. 2002. Short, long, and beyond: molecular and embryological approaches to insect segmentation. Annu Rev Entomol., 47:669-99.

de Azambuja P., Garcia E.S., Ratcliffe N.A. 1991. Aspects of classification of hemiptera hemocytes from six triatomine species. Mem Inst Oswaldo Cruz., 86(1):1-10.

De Loof A. 2008. Ecdysteroids, juvenile hormone and insect neuropeptides: recent successes and remaining major challenges. Gen Comp Endocrinol., 155(1):3-13.

Dickinson M., Farman G., Frye M., et al.2005. Molecular dynamics of cyclically contracting insect flight muscle *in vivo*. Nature., 433:330-333.

Dittmer J., Brucker R.M. 2021. When your host shuts down: larval diapause impacts host-microbiome interactions in *Nasonia vitripennis*. Microbiome. 9:85.

Dong S.Z., Song Q.S. 2012. Bursicon as a potential target for insect control. In: Ishaaya I. et al. (eds.), Advanced Technologies for Managing Insect Pests. Springer Science+Business Media Dordrecht.

Doran C.R., Estevez-Lao T.Y., Hillyer J.F. 2017. Mosquito aging modulates the heart rate and the proportional directionality of heart contractions. J Insect Physiol., 101: 47-56.

Dow J.A.T., Krause S.A. Herzyk P. 2021. Updates on ion and water transport by the Malpighian tubule. Curr Opin Insect Sci., 47:31-37.

Dubrovsky E.B. 2005. Hormonal cross talk in insect development. Trends Endocrinol Metab., 16(1):6-11.

DUBOWY C., SEHGAL A. 2017. Circadian rhythms and sleep in *Drosophila melanogaster*. Genetics., 205, 1373-1397.

Duncan F.D., Hanrahan S. A. 2018. Respiratory patterns in field collected brown locust, *Locustana pardalina*, in the gregarious phase. J. Insect Physiol., 6: 209-216.

Dunlap. J. C. 1999. Molecular bases for circadian clocks. Cell., 96: 271-290.

Galarza J.A. 2021. Comparative transcriptomics of albino and warningly-coloured caterpillars. Ecol.Evol., 11:7507-7517.

Eleftherianos I, Heryanto C., Bassal T., et al. 2021. Haemocyte-mediated immunity in insects: Cells, processes and associated components in the fight against pathogens and parasites. Immunology., 164(3):401-432.

Eleftherianos I, Zhang W, Heryanto C., et al. 2021. Diversity of insect antimicrobial peptides and proteins-A functional perspective: A review. Int. J. Biol. Macromol., 191:277-287.

El-Sherif E, Zhu X, Fu J. 2014. Caudal regulates the spatiotemporal dynamics of pair-rule waves in *Tribolium*. PLoS Genet., 10:e1004677.

Ephrussi A, Lehmann R. 1992. Induction of germ cell formation by oskar. Nature., 358:387-392.

Eriksson M., Janz N., Nylin S., et al. 2020. Structural plasticity of olfactory neuropils in relation to insect diapause. Ecol.Evol., 10:14423-14434.

Eschbach C., Zlatic M. 2020. Useful road maps: studying *Drosophila* larva's central nervous system with the help of connectomics. Curr Opin Neurobiol., 65:129-137.

Ezzati-Tabrizi R., Farrokhi N., Talaei-Hassanloui R., et al. 2013. Insect inducible antimicrobial peptides and their applications. Curr Protein Pept Sci., 14(8):698-710.

Ferguson L, Marlétaz F., Carter J-M., et al. 2017. Ancient expansion of the Hox cluster in Lepidoptera generated four homeobox genes implicated in extra-embryonic tissue formation. PLoS Genet., 10:e1004698.

Feuda, R., Marlétaz F., Bentley M. A., et al. 2016. Conservation, duplication, and divergence of five opsin genes in insect evolution. Genome Biol Evol., 8(3): 579-87.

Fleischer J. Pregitzer P., Breer H. et al. 2018. Access to the odor world: olfactory receptors and their role for signal transduction in insects. Cell. Mol. Life Sci., 75:485-508.

Franke R. R., Sakmar T. P., Graham R. M., et al. 1992. Structure and function in rhodopsin. studies of the interaction between the rhodopsin cytoplasmic domain and transducin. J Biol Chem., 267(21): 14767-14774.

Franz A., Wood W., Martin P.2018. Fat body cells are motile and actively migrate to wounds to drive repair and prevent infection. Dev Cell., 44(4): 460-470.

Fujii T., Banno Y. 2019. Identification of a novel function of the silkworm integument in nitrogen metabolism: Uric acid is synthesized within the epidermal cells in *B. mori*. Insect Biochem. Mol. Biol., 105, 43-50.

Fujinaga D, Gu J., Kawahara H. 2020. Twenty-hydroxyecdysone produced by dephosphorylation and ecdysteroidogenesis regulates early embryonic development in the silkmoth, *Bombyx mori*. Insect Biochem Mol. Biol., 127:103491.

Fuscà D., Kloppenburg P. 2021. Odor processing in the cockroach antennal lobe—the network components. Cell and Tissue Res., 383:59-73.

Gao Y., Liu Y.C., Jia S.Z., et al. 2020. Imaginal disc growth factor maintains cuticle structure and controls melanization in the spot pattern formation of *Bombyx mori*. PLoS Genetics., 16(9): e1008980.

Gavrilov-Zimin I.A. 2021. Egg retention, viviparity and ovoviviparity in Paraneoptera. Comp Cytogen., 15(3): 239-252.

Ghosh A.C., O'Connor M.B. 2014. Systemic Activin signaling independently regulates sugar homeostasis, cellular metabolism, and pH balance in *Drosophila melanogaster*. Proc Natl Acad Sci USA., 111: 5729-5734.

Gibbs A. G., Johnson R. A. 2004. The role of discontinuous gas exchange in insects: the chthonic hypothesis does not hold water. J. Exp. Biol., 207, 3477-3482.

Gilbert L.I, Bollenbacher W.E, Granger N.A. 1980. Insect endocrinology: regulation of endocrine glands, hormone titer, and hormone metabolism. Annu Rev Physiol., 42:493-510.

Gilbert L.I., Rybczynski R., Warren J.T. 2002. Control and biochemical nature of the ecdysteroidogenic pathway. Annu Rev Entomol., 47:883-916.

Gilbert S.F. 2010. Development Biology. 9th ed. Sinauer Associates, Sunderland (MA).

Gilbert S.F. 2000.The origins of anterior-posterior polarity. In: Developmental biology, 6th ed. Sinauer Associates, Sunderland (MA).

Gokhman V.E., Kuznetsova V.G. 2018. Parthenogenesis in Hexapoda: holometabolous insects. J Zool Syst Evol Res., 56: 23-34.

Gołębiowski M., Stepnowski P. 2022. Chemical composition of insect surface waxes: biological functions and analytics. In: B. Buszewski, I. Baranowska (eds.), Handbook of Bioanalytics. Springer Nature Switzerland AG.

Göpfert M. C., Hennig R. M. 2016. Hearing in Insects. Annu. Rev. Entomol., 61:257-76.

Green II D.A., Kronforst M.R. 2019. Monarch butterflies use an environmentally sensitive, internal timer to control overwintering dynamics. Mol Ecol., 28,3642-3655.

Grönke S.D.F., Clarke S., Broughton T.D., et al. 2010. Molecular evolution and functional characterization of *Drosophila* insulin-like peptides. PLoS Genet., 6: e100085.

Gu SH, Chen CH, Lin PL. 2021. Changes in expressions of ecdysteroidogenic enzyme and ecdysteroid signaling genes in relation to *Bombyx* embryonic development. J Exp Zool A Ecol Integr Physiol., 335(5):477-488.

Guo H., Smith D. P. 2022. Time-dependent odorant sensitivity modulation in insects. Insects., 13: 354.

Guo W., Wang. X.H., Ma Z.Y., et al. 2011. CSP and takeout genes modulate the switch between attraction and repulsion during behavioral phase change in the migratory locust. PLoS Genet., 7(2): e1001291.

Ha Y.R., Yeom E., Ryu J., et al. 2017. Three-dimensional structures of the tracheal systems of *Anopheles sinensis* and *Aedes togoi* pupae. Sci. Rep., 7:44490.

Habenstein J., Thamm M., Rössler W. 2021. Neuropeptides as potential modulators of behavioral transitions in the ant *Cataglyphis nodus*. J Comp Neurol., 529:3155-3170.

Hahn T., Tafi E., Von Seggern N., et al. 2022. Purification of chitin from pupal exuviae of the black soldier fly. Waste and Biomass Valorization., 13:1993-2008.

Hamid R., Hajirnisa N., Kushwahab S., et al. 2019. *Drosophila* choline transporter non-canonically regulates pupal eclosion and NMJ integrity through a neuronal subset of mushroom body. Dev. Biol., 446, 80-93.

Handel K, Basal A, Fan X., et al. 2005. *Tribolium castaneum* twist: gastrulation and mesoderm formation in a short-germ beetle. Dev Genes Evol., 215: 13-31.

Handel K, Grünfelder C, Roth S., et al.. 2000. Tribolium embryogenesis: a SEM study of cell shapes and movements from blastoderm to serosal closure. Dev. Genes Evol., 210: 167-179.

Haney E.F, Mansour S. C., Hancock R.E. 2017. Antimicrobial peptides: an introduction. Methods Mol Biol., 1548:3-22.

Hanrahan C. J., Palladino M. J., Ganetzky B., et al. 2000. RNA editing of the *Drosophila* para Na^+ channel transcript. evolutionary conservation and developmental regulation. Genetics, 155(3): 1149-1160.

Hao K., Ullah H., Jarwar A.R., et al. 2020. Functional identification of an FMRFamide-related peptide gene on diapause induction of the migratory locust, *Locusta migratoria* L. Genomics, 112, 1821-1828.

Hao K., Ullah H., Jarwar A.R., et al. 2019. Molecular identification and diapause-related functional characterization of a novel dual-specificity kinase gene, MPKL, in *Locusta migratoria*. FEBS Letters, 593, 3064-3074.

Hao S., Gestrich J.Y., Zhang X., et al. 2021. Neurotransmitters affect larval development by regulating the activity of prothoracicotropic hormone-releasing neurons in *Drosophila melanogaster.* Front. Neurosci, 15, 653858.

Hejníková M., Nouzova M., Ramirez C.E., et al. 2022. Sexual dimorphism of diapause regulation in the hemipteran bug *Pyrrhocoris apterus*. Insect Biochem. Mol. Biol., 142, 103721.

Helfrich-Förster. C. 2018. Sleep in insects. Annu. Rev. Entomol., 63:69-86.

Hellerschmied D, Lehner A, Franicevic N., et al., 2019. Molecular features of the UNC-45 chaperone critical for binding and folding muscle myosin. Nat. Commun., 10:4781.

Henze M. J., Oakley T. H. 2015.The dynamic evolutionary history of pancrustacean eyes and opsins. Integr Comp Biol., 55(5): 830-842.

Hetz S.K., Bradley T.J. 2005. Insects breathe discontinuously to avoid oxygen toxicity. Nature., 433(3): 516-519.

Hillyer J.F., Pass G.. 2020.The insect circulatory system: structure, function, and evolution. Review Annu Rev Entomol., 65:121-143.

Holtof M., Lenaerts C., Cullen D., et al. 2019. Extracellular nutrient digestion and absorption in the insect gut. Cell Tissue Res., 377:397- 414.

Horard B., Loppin B. 2015. Histone storage and deposition in the early *Drosophila* embryo. Chromosoma., 124: 163-175.

Hou L., Guo S., Ding D., et al. 2022. Neuroendocrinal and molecular basis of flight performance in locusts. Cellu. Mol. Life Sci., 79:325.

Hou Q.L., Chen E.H., Jiang H.B., et al. , 2017. Adipokinetic hormone receptor gene identification and its role in triacylglycerol mobilization and sexual behavior in the oriental fruit fly (*Bactrocera dorsalis*). Insect Biochem Mol Biol., 90: 1-13.

Houchmandzadeh B., Wieschaus E., Leibler S. 2002. Establishment of developmental precision and proportions in the early *Drosophila* embryo. Nature., 415(6873):798-802.

Huan Y., Kong Q., Mou H., et al. 2020. Antimicrobial peptides: classification, design, application and research progress in multiple fields. Front. Microbiol., 11:582779.

Imler J. L, Hoffmann J. A. 2002. *Toll* receptors in *Drosophila*: a family of molecules regulating development and immunity. Curr. Top. Microbiol., 270:63.

Imler J. 2014. Overview of *Drosophila* immunity: a historical perspective. Dev. Comp. Immunol.

Imura E., Shimada-Niwa Y., Nishimura T. 2020.The corazonin-PTTH neuronal axis controls systemic body growth by regulating basal ecdysteroid biosynthesis in *Drosophila melanogaster.* Curr. Biol., 30, 2156-2165.

Jaeger J., Manu, Reinitz J. 2012. *Drosophila* blastoderm patterning. Curr Opin Genet Dev., 22(6):533-41.

Jaeger J., Surkova S., Blagov M., et al. 2004. Dynamic control of positional information in the early *Drosophila* embryo. Nature., 430(6997):368-71.

Jaron K.S., Parker D.J., Anselmetti Y., et al. 2022. Convergent consequences of parthenogenesis on stick insect genomes. Sci. Adv., 8: e3842.

Jennings E.C., Korthauer M.W., Hendershot J.M., et al. 2020. Molecular mechanisms underlying milk production and viviparity in the cockroach, *Diploptera punctate*. Insect Biochem Mol Biol., 120: 103333.

Jia Q., Liu Y., Liu H., et al. 2014. Mmp1 and Mmp2 cooperatively induce *Drosophila* fat body cell dissociation with distinct roles. Sci Rep., 4: 7535.

Jindra M., Bellés X. Shinoda T. 2015. Molecular basis of juvenile hormone signaling. Curr Opin Insect Sci., 11:39-46.

Jindra M., Bittova L. 2020. The juvenile hormone receptor as a target of juvenoid "insect growth regulators". Arch Insect Biochem Physiol., 103(3):e21615.

Jindra M, Palli S.R., Riddiford L.M. 2013. The juvenile hormone signaling pathway in insect development. Annu Rev Entomol., 58:181-204.

Jones K.K., Seymour R. S. 2021. Gas exchange and dive behaviour in the diving beetle *Platynectes decempunctatus* (Coleoptera: Dytiscidae). J. Insect Physiol., 133: 104286.

Joseph N.M., Elphick N.Y., Mohammad S., et al. 2018. Altered pheromone biosynthesis is associated with sex-specific changes in life span and behavior in *Drosophila melanogaster*. Mech. Ageing Dev., (176) :1-8.

Josephson, R., Stokes D.M.J.G., 2000. Asynchronous muscle: a primer. J. Exp. Biol., 203(18):2713-2722.

Kaessmann H. O. 2010. Evolution, and phenotypic impact of new genes. Genome Res., 20(10): 1313-1326.

Kaftan F., Kulkarni P., Knaden M., et al. 2020. *Drosophila melanogaster* chemical ecology revisited: 2-D distribution maps of sex pheromones on whole virgin and mated flies by mass spectrometry imaging. BMC Zool., 5:4.

Kaiser A., Hartzendorf, S., Wobschall A., et al. 2010. Modulation of cyclic CO_2 release in response to endogenous changes of metabolism during pupal development of *Zophobas rugipes* (Coleoptera: Tenebrionidae). J. Insect Physiol., 56 :502-512.

Kamiyama T., Niwa R. 2022. Transcriptional regulators of ecdysteroid biosynthetic enzymes and their roles in insect development. Front Physiol., 13:823418.

Kapoor D., Khan A., O'Donnell M.J., et al. 2021. Novel mechanisms of epithelial ion transport: insights from the cryptonephridial system of lepidopteran larvae. Curr Opin Insect Sci., 47:53-61.

Kawasaki H. 2016. Genes of cuticular proteins and their regulation. In: E. Cohen, B. Moussian (eds.),Extracellular Composite Matrices in Arthropods. Springer International Publishing Switzerland.

Khalid MZ, Ahmad S, Ngegba PM., et al. 2021. Role of Endocrine system in the regulation of female insect reproduction. Biol. (Basel)., 10(7):614.

Kiger J.A.J., Natzle J.E., Green M.M. 2007. Hemocytes are essential for wing maturation in *Drosophila melanogaster*. PNAS., 98: 10190-95.

Kim J., Neufeld T.P. 2015. Dietary sugar promotes systemic TOR activation in *Drosophila* through AKH-dependent selective secretion of Dilp3. Nat Commun., 6: 6846.

King, Glenn F. 2007. Modulation of insect cav channels by peptidic spider toxins. Toxicon, 49(4): 513-530.

Klowden M.J. 2013. Physiological systems in insects: circulatory systems. Elsevier Science & Technology.

Knebel D., Rigosi E. 2021. Temporal and structural neural asymmetries in insects. Curr. Opin. Insect Sci., 48:72-78.

Kobayashi Y, Ando H. 1990. Early embryonic development and external features of developing embryos of the caddisfly, *Nemotaulius admorsus* (Trichoptera: Limnephilidae). J. Morphol., 203: 69-85.

Koles K., Nunnari J., Korku C.T., et al. 2012. Mechanism of evenness interrupted (evi)-exosome release at synaptic boutons. J Biol Chem., 287(20): 16820-16834.

Kolosov D., O'Donnell M.J. 2020. Mechanisms and regulation of chloride transport in the Malpighian tubules of the larval cabbage looper *Trichoplusia ni*. Insect Biochem. Mol. Biol., (116): 103263.

Koštál V., Renault D., Rozsypal J. , 2011. Seasonal changes of free amino acids and thermal hysteresis in overwintering heteropteran insect, *Pyrrhocoris apterus*. Comp Biochem Physiol A Mol Integr Physiol., 160(2): 245-251.

Kotwica J., Joachimiak E, Polanska M.A., et al. , 2011. Diurnal rhythm in expression and release of yolk protein in the testis of *spodoptera littoralis* (*Lepidoptera: Noctuidae*). Insect Biochem Mol Biol., 41(4): 264-272.

Koutroumpa F., Monsempès C., Anton S., et al. 2022. Pheromone receptor knock-out affects Pheromone detection and brain structure in a moth. Biomol., 12:341.

Koyama T., Naseem M.T., Kolosov D., et al. 2021. A unique Malpighian tubule architecture in *Tribolium castaneum* informs the evolutionary origins of systemic osmoregulation in beetles. PNAS., 118 (14): e2023314118.

Krenn H. W. 2019. Form and function of insect mouthparts. In Krenn H. W. (ed.) Insect Mouthparts, Zoological Monographs 5. Springer Nature Switzerland AG.

Krueger S., Moritz G. 2021. Sperm ultrastructure in arrhenotokous and thelytokous Thysanoptera. Arth. Struct. Devel., 64: 101084.

Kumar P, Kizhakkedathu J, Straus S. 2018. Antimicrobial peptides: diversity, mechanism of action and strategies to improve the activity and biocompatibility *in vivo*. Biomolecules., 8(1): 4.

Kwon H., Smith R.C. 2019. Chemical depletion of phagocytic immune cells in *Anopheles gambiae* reveals dual role of mosquito hemocytes in anti-plasmodium immunity. PNAS., 116(28): 14119-14128.

Law J. A., Jacobsen S. E. 2010. Establishing, maintaining and modifying dna methylation patterns in plants and animals. Nat Rev Genet., 11(3): 204-220.

Lawley S.D., Reed M. C., Nijhout H F. 2020. Spiracular fluttering increases oxygen uptake. PLoS One., 15(5): e0232450.

Lee J, Kiuchi T, Kawamoto M., et al. 2018. Accumulation of uric acid in the epidermis forms the white integument of *Samia ricini* larvae. PLoS One., 13(10): e0205758.

Leite-Mondin M., DiLegge M.J., Manter D.K., et al. 2021. The gut microbiota composition of *Trichoplusia ni* is altered by diet and may influence its polyphagous behavior. Scientific Reports., 11:5786.

Lemaitre B, Nicolas E., Michaut L., et al. 1996. The dorsoventral regulatory gene cassette spätzle/*Toll*/ cactus controls the potent antifungal response in *Drosophila* adults. Cell, 86(6):973-83.

Lenaerts C., Monjon E., Lommel J. V., et al. 2019. Peptides in insect oogenesis. Curr. Op. Insect Sci., 31:58-64.

Leoni L., Grübel K., Spaethe.J., et al. 2018. Opsin expression patterns coincide with photoreceptor development during pupal development in the honey bee, *Apis mellifera*. BMC Dev Biol, 18:1.

Lerch S., Zuber R., Gehring N., et al. 2020. Resilin matrix distribution, variability and function in *Drosophila*. BMC Biol., 18:195.

Li H.Y., Lin X.W., Geng S. L., et al. 2018. TGF-β and BMP signals regulate insect diapause through Smad1-POU-TFAM pathway. BBA- Mol. Cell Res., 1865 :1239-1249.

Li K, Jia Q.Q, Li S. 2019. Juvenile hormone signaling - a mini review. Insect Sci., 26(4):600-606.

Li R.S., Xiao Y., Li K., et al. 2022. Transcription and post-translational regulation of autophagy in insects. Front Physiol., 13: 825202.

Li S., Yu X., Feng Q.2019. Fat body biology in the last decade. Annu Rev Entomol., 64: 315-333.

Liegertová M., Pergner J., Kozmiková I., et al.2015. Cubozoan genome illuminates functional diversification of opsins and photoreceptor evolution. Sci Rep-Uk., 5(1): 11885.

Lighton J.R. 1996. Discontinuous gas exchange in insects. Annu Rev Entomol., 41:309- 324.

Liman E.R. 2021. A receptor that lets insects sense scents. Nature, 597(2): 37-38.

Lin X.W., Fan Z.Z., Liu Y.H., et al. 2021. High MEK/ERK signalling is a key regulator of diapause maintenance in the cotton bollworm, *Helicoverpa armigera*. Insect Mol., Biol. 30, 508-518.

Lin J. Y., Yang Z., Yang C., et al. 2021. An ionic lock and a hydrophobic zipper mediate the coupling between an insect pheromone receptor BmOR3 and downstream effectors. J. Biol. Chem., 297:101160.

Lopez-Martinez G., Elnitsky M. A., Benoit J. B., et al. 2008. High resistance to oxidative damage in the Antarctic Midge belgica antarctica, and developmentally linked expression of genes encoding superoxide dismutase, catalase and heat shock proteins. Insect Biochem Mol Bio., 38(8): 796-804.

Loughney K., Kreber R, Ganetzky B. 1989. Molecular analysis of the para locus, a sodium channel gene in *drosophila*. Cell, 58(6): 1143-1154.

Lu A., Zhang Q., Zhang J., et al. 2014. Insect prophenoloxidase: the view beyond immunity. Front Physiol., 5:252.

Liu X.J., Zhang J.Z. Zhu K.Y. 2019. Chitin in Arthropods: biosynthesis, modification, and metabolism. In: Q. Yang and T. Fukamizo (eds.), Targeting Chitin-containing Organisms, Advances in Experimental Medicine and Biology, 1142: 174.

Luan Z., Quigley C., Li H.S. 2015. The putative Na^+/Cl^-dependent neurotransmitter/ osmolyte transporter inebriated in the *Drosophila* hindgut is essential for the maintenance of systemic water homeostasis. Sci. Rep., 5: 7993.

Lynch J.A, Brent A.E, Leaf D.S., et al. 2006. Localized maternal orthodenticle patterns anterior and posterior in the long germ wasp Nasonia. Nature, 439:728-732.

Lynch J.A, El-Sherif E, Brown S.J. 2012. Comparisons of the embryonic development of *Drosophila, Nasonia, and Tribolium*. Wiley Interdiscip Rev Dev Biol., 1:16-39.

Lynch J.A, Ozuak O., Khila A., et al. 2011. The phylogenetic origin of oskar coincided with the origin of maternally provisioned germ plasm and pole cells at the base of Holometabora. PLoS Genetics., 7:e1002029.

MacMillan H.A., Nazal B., Wali S., et al., 2018. Anti-diuretic activity of a CAPA neuropeptide can compromise *Drosophila* chill tolerance. J. Expe.l Biol., (221): jeb185884.

Makka T, Seino A, Tomita S., et al. 2022. A possible role of 20- hydroxyecdysone in embryonic development of the silkworm *Bombyx mori*. Arch. Insect Biochem. Physiol., 51:111-120.

Malita A, Rewitz K. 2021. Interorgan communication in the control of metamorphosis. Curr Opin Insect Sci., 43:54-62.

Maljevic S., Lerche H. 2013. Potassium channels: a review of broadening therapeutic possibilities for neurological diseases. J Neurol., 260(9): 2201-2211.

Manniello M D, Moretta A, Salvia R., et al. 2021. Insect antimicrobial peptides: potential weapons to counteract the antibiotic resistance. Cell. Mol. Life Sci., 78(9): 4259-4282.

Marchal E., Vandersmissen H.P, Badisco L., et al. 2010. Control of ecdysteroidogenesis in prothoracic glands of insects: a review. Peptides., 31(3):506-19.

Marco H.G., Katali O.K.H., Gäde G. 2018. Influence of aminergic and peptidergic substances on heart beat frequency in the stick insect *Carausius morosus* (Insect, Phasmatodea). Arch Insect Biochem Physiol., 98:

e21469.

Marin E.C., Büld L., Theiss M., et al. 2020. Connectomics analysis reveals first-, second-, and third-order thermosensory and hygrosensory neurons in the adult *Drosophila* brain. Curr. Biol., 30: 3167-3182,

Martinez-Arias A, Lawrence P.A. 1985. Parasegments and compartments in the *Drosophila* embryo. Nature, 313(6004):639-42.

Matthews, P.G.D.,White C.R. 2013. Reversible brain inactivation induces discontinuous gas exchange in cockroaches. J. Exp. Biol., 216: 2012-2016.

Mattila J., Hietakangas V. 2017. Regulation of carbohydrate energy metabolism in *Drosophila melanogaster.* Genetics., 207(4): 1231-1253.

Medioni C., Astier M., Zmojdzian M., et al. 2008. Genetic control of cell morphogenesis during *Drosophila melanogaster* cardiac tube formation. J Cell Biol., 182: 249-61.

Melcarne C, Lemaitre B, Kurant E. 2019. Phagocytosis in *Drosophila*: From molecules and cellular machinery to physiology. Insect Biochem Mol Biol., 109: 1-12.

Menail H. A., Cormier S.B., Youssef M. B., et al. 2022. Flexible thermal sensitivity of mitochondrial oxygen consumption and substrate oxidation in flying insect species. Front. Physiol., 13: 897174.

Miller T. 1968. Role of cardiac neurons in the cockroach heartbeat. J Insect Physiol., 14(9): 1265-1275.

Mischiati M., Lin H.T., Herold P., et al. 2015. Internal models direct dragonfly interception steering. Nature., 517: 333.

Mitra P , Das S , Barik A . 2020. Leaf waxes from *Lathyrus sativus:* short-range attractant and stimulant for nymph laying in a viviparous insect. Chemoecol., 30(10): 117-129.

Morata G. 1993. Homeotic genes of *Drosophila*. Curr Opin Genet Dev., 3(4):606-14.

Morata G. 2001. How *Drosophila* appendages develop. Nat Rev Mol Cell Biol., 2(2):89-97.

Moreira D.C., Paula D.P., Lima M.H. 2012. Redox metabolism during tropical diapause in a Lepidoptera larva. In: *Living in a Seasonal World.* T. Ruf et al. (eds.), Springer-Verlag Berlin Heidelberg.

Msangosoko K , Bhattacharya R., Ramakrishnan B., et al. 2021. Cellulolytic activity of gut bacteria isolated from the eri silkworm larvae, *Samia ricini*, (Lepidoptera: Saturniidae). Int.J.Trop. Insect Sci., 41:2785-2794.

Muthukrishnan S. Arakane Y., Noh M.Y., et al. 2022. Chitin in insect cuticle. In Insect Cuticle-Chitin, Catecholamine and Chemistry of Complexation. *Ed.* Manickam S. Adv. Insect Physiol. Academic Press, Elsevier Ltd.

Muthukrishnan S. Merzendorfer H., Arakane Y., et al. 2019. Chitin organizing and modifying enzymes and proteins involved in remodeling of the Insect cuticle. In: Q. Yang and T. Fukamizo (eds.), Targeting Chitin-containing Organisms, Adv Exp Med Biol., 1142:83-114.

Nagy L, Riddiford L, Kiguchi K. 1994. Morphogenesis in the early embryo of the Lepidopteran *Bombyx mori.* Dev Biol., 165:137-151.

Nakao H. 2012. Anterior and posterior centers jointly regulate *Bombyx* embryo body segmentation. Dev Biol., 371:293-301.

Nakao H. 2021. Early embryonic development of *Bombyx*. Dev Genes Evol., 231(5/6):95-107.

Namiki S., Kanzaki R. 2019. Morphology and physiology of olfactory neurons in the lateral protocerebrum of the silkmoth *Bombyx mori.* Sci. Rep., 9:16604.

Naruse S, Washidu Y, Miura K. 2020. Methoprene-tolerant is essential for embryonic development of the red flour beetle *Tribolium castaneum*. J Insect Physiol., 121:104017.

Nässel D.R., Wu S.F. 2021. Leucokinins: multifunctional neuropeptides and hormones in insects and other invertebrates. Int. J. Mol. Sci., 22: 1531.

Nathans J., Hogness D. S. 1983. Isolation, sequence analysis, and intron-exon arrangement of the gene encoding bovine rhodopsin. Cell, 34(3): 807-814.

Nation J.L. 2022. Insect physiology and biochemistry. 4th ed, Oxon, UK: CRC Press.

Navizet I., Liu Y. J., Ferré N., et al. 2009. Color-tuning mechanism of firefly investigated by multi-configurational perturbation method. J Am Chem Soc., 132: 706-712.

Nayab S., Aslam M. A., Rahman S U., et al. 2022. A review of antimicrobial peptides: Its function, mode of action and therapeutic potential. Int. J. Pept., Res. Ther. 28(1): 46.

Nguyen H., Donini A. 2010. Larvae of the midge *Chironomus riparius* possess two distinct mechanisms for ionoregulation in response to ion-poor conditions. Am J Physiol Regul Integr Comp Physiol., 299(3): 762-773.

Ni L., Bronk P., Chang E.C., et al. 2013.A gustatory receptor paralogue controls rapid warmth avoidance in *Drosophila*. Nature, 500,29:580-583.

Niederhuber M.J, McKay D.J. 2021. Mechanisms underlying the control of dynamic regulatory element activity and chromatin accessibility during metamorphosis. Curr Opin Insect Sci., 43:21-28.

Niwa Y. S., Niwa R. 2016. Transcriptional regulation of insect steroid hormone biosynthesis and its role in controlling timing of molting and metamorphosis. Dev Growth Differ., 58(1):94-105.

Nozaki T., Shigenobu S.J. 2022. Ploidy dynamics in aphid host cells harboring bacterial symbionts. Sci. Rep., 12:9111.

Numata H., Miyazaki Y. Ikeno T. 2015. Common features in diverse insect clocks. Zool. Lett., 1:10.

Nur Aliah N. A., Ab-Rahim S, Moore H. E., et al.2021. Juvenile hormone: Production, regulation, current application in vector control and its future applications. Trop Biomed., 38(3):254-264.

Oh M., Zhao S., Matveev V., et al. 2012. Neuromodulatory changes in short-term synaptic dynamics may be mediated by two distinct mechanisms of presynaptic calcium entry. J Comput Neurosci., 33(3): 573-585.

Ohde T., Mito T., Niimi T., et al. 2022. A hemimetabolous wing development suggests the wing origin from lateral tergum of a wingless ancestor. Nat. Commun., 13:979.

Ons S. 2017. Neuropeptides in the regulation of *Rhodnius prolixus* physiology. J Insect Physiol., 97:77-92.

O'Tousa, J. E., Baehr W., Martin R. L., et al. 1985. The *Drosophila Ninae* gene encodes an opsin. Cell., 40(4): 839-850.

Ou Q., King-Jones K. 2013. What goes up must come down: transcription factors have their say in making ecdysone pulses. Curr Top Dev Biol., 103:35-71.

Owens G. L., Rennison D. J. 2017. Evolutionary ecology of opsin gene sequence, expression and repertoire. Mol Ecol., 26(5): 1207-1210.

Paluzzi J. P., Vanden Broeck J. 2019. General and comparative endocrinology: special issue on insect neuroendocrinology and neurobiology. Gen Comp Endocrinol., 280:192-193.

Pan X., Connacher R.P., O'Connor M.B. 2021. Control of the insect metamorphic transition by ecdysteroid production and secretion. Curr Opin Insect Sci., 43:11-20.

Pankratz M. J., Jäckle H. 1993. Blastoderm segmentation. In: Bate M, Arias AM (eds) The development of *Drosophila melanogaster*. New York: Cold Spring Harbor Laboratory Press.

Passino K.M., Seeley T.D.,Visscher P.K. 2008. Swarm cognition in honey bees. Behav Ecol Sociobiol., 62:401-414.

Pereira C.A. 2014.Arginine kinase: a potential harmacological target in trypanosomiasis. Infect Disord Drug

Targets., 14(1): 30-36.

Perochon J., Yu Y., C., Aughey G. N., et al. 2021. Dynamic adult tracheal plasticity drives stem cell adaptation to changes in intestinal homeostasis in *Drosophila*. Nature Cell Biol., 23: 485-496.

Pesch Y Y., Riedel D., Patil K. R., et al. 2016. Chitinases and imaginal disc growth factors organize the extracellular matrix formation at barrier tissues in insects. Sci Rep. (6): 18340.

Pfeiffer K., Homberg U. 2014.Organization and functional roles of the central complex in the insect brain. Annu. Rev. Entomol., 59:165-84.

Pick C., Schneuer M., Burmester T., et al. 2010. Ontogeny of hemocyanin in the ovoviviparous cockroach *Blaptica dubia* suggests an embryo-specific role in oxygen supply. J. Insect Physiol., 56: 455-460.

Pogodalla N., Winkler B., Klämbt C. 2022. Glial tiling in the insect nervous system. Front Cell Neurosci., 16:825695.

Ponderville E.N. Puchot J.P. Parvy G., et al. 2020. Hemocyte-targeted gene expression in the female malaria mosquito using the hemolectin promoter from *Drosophila*. Insect Biochem Mol Physiol., 120: 103339.

Ponnusamy M., Karthikeyan C V., Ramanathan B., et al. 2019. Meta-omics in detection of silkworm gut microbiome diversity. In: V. Tripathi et al. (eds.), Microbial Genomics in Sustainable Agroecosystems. Springer Nature Singapore Pte Ltd.

Porter M. L., Blasic J. R., Bok M. J., et al. 2012. Shedding new light on opsin evolution. P Roy Soc B-Biol Sci., 279(1726): 3-14.

Prelic S., Mahadevan V.P.,Venkateswaran V., et al. 2022. Functional interaction between *Drosophila* olfactory sensory neurons and their support cells. Front. Cell Neurosci., 15: 789086.

Qi X.Y., Zhang L.S., Han Y.H., et al. 2015.De novo transcriptome sequencing and analysis of Coccinella septempunctata L. in non-diapause, diapause and diapause terminated states to identify diapause associated genes. BMC Genomics., 16:1086.

Quan Y., Metterville D., Briscoe A., et al. 2007. Insect cryptochromes: gene duplication and loss define diverse ways to construct insect circadian clocks. Mol Biol Evol., 24: 948-955.

Raja J. S. I., Katanayeva N., Katanaev V. L., et al. 2014. Role of G(O/I) subgroup of G proteins in olfactory signaling of *Drosophila melanogaster*. Eur J Neurosci., 39(8): 1245-1255.

Rebers J. E. Riddiford, L. M. 1988. Structure and expression of a *Manduca sexta* larval cuticle gene homologous to Drosophila cuticle genes. J. Mol. Biol., 203: 411-423.

Rebora M., Salerno G., Piersanti S. 2019. Aquatic insect sensilla: morphology and function. In: Aquatic Insects. K. Del-Claro, R. Guillermo (eds.), Springer Nature Switzerland AG: 139-166.

Rezende G. L., Vargas H. C. M., Moussian B., et al. 2016. Composite eggshell matrices: chorionicLayers and sub-chorionic cuticular envelopes. E. Cohen, B. Moussian (eds.), Extracellular Composite Matrices in Arthropods. Switzerland. Springer International Publishing.

Riddiford L.M, Hiruma K., Zhou X., et al. 2003.Insights into the molecular basis of the hormonal control of molting and metamorphosis from *Manduca sexta* and *Drosophila melanogaster*. Insect Biochem Mol Biol., 33(12):1327-38.

Riddiford L.M. 2012. How does juvenile hormone control insect metamorphosis and reproduction. Gen Comp Endocrinol., 179(3):477-84.

Riddiford L.M. 1980. Insect endocrinology: action of hormones at the cellular level. Annu Rev Physiol., 42:511-28.

Ritzmann R., Fox J.L.2022. Insect neuroethology: flight constructs a compass for monarch migration. Current Biol., 32, 72–96.

Roberts, J. A., Catherine V, Helen R., et al.2006. Molecular properties of p2x receptors. Pflugers Arch., 452(5): 486-500.

Rocha I., Roza A., Gonçalves C., et al. 2020. Insect life history. J. Vonk, T. K. Shackelford (eds.), Encyclopedia of Animal Cognition and Behavior. Springer Nature Switzerland AG.

Roma G.C., Bueno O.C., Camargo-Mathias M.I.2010. Morpho-physiological analysis of the insect fat body: a review. Micron., 41(5): 395-401.

Rosales C., Uribe-Querol E., Stockinger H. 2017. Phagocytosis: a fundamental process in immunity. Biomed Res. Int., 90428:18-51.

Rosas D.A., Guilén D.S., Malo E.A., et al. 2019. (S)-2-Heptanol, the alarm pheromone of the stingless bee *Melipona solani* (Hymenoptera, Meliponini). Apidol., 50:277-287.

Rose E., Lee D., Xiao E., et al. 2019. Endocrine regulation of *MFS*2 by *branchless* controls phosphate excretion and stone formation in *Drosophila* renal tubules. Sci. Rep., 9:8798.

Roy S., Saha T.T., Zou Z., et al. 2018. Regulatory pathways controlling female insect reproduction. Annu. Rev. Entomol., 63:489-511.

Salcedo M.K., Socha J.J. 2020. Circulation in insect sings. Integr Comp Biol., 60(5): 1208-1220.

Sander K. 1976. Specification of the basic body pattern in insect embryogenesis. Adv Insect Physiol., 12:125-238.

Santos C.G., Humann F.C., Hartfelder K. 2019. Juvenile hormone signaling in insect oogenesis. Curr Opin Insect Sci., 31:43-48.

SAS Institute. 2002 ."SAS User's Guide"in: Statistics. Cary, NC: SAS Institute..

Saunders D. S. 2010. Controversial aspects of photoperiodism in insects and mites. J Insect Physiol., 56(11) : 1491-1502.

Saunders J., Smith T. 2010. Malnutrition: causes and consequences. Clin Med (Lond), 10(6): 624-627.

Schmid M.R., Anderl I., Vesala L., et al. 2014. Control of *Drosophila* blood cell activation *via Toll* signaling in the fat body. PLoS One., 9(8): e102568.

Scott K. 2018. Gustatory processing in *Drosophila melanogaster.* Annu. Rev. Entomol., 63:15-30.

Seike H., Nagata S. 2022. Different transcriptional levels of Corazonin, Elevenin, and PDF according to the body color of the two spotted cricket, *Gryllus bimaculatus*. Biosci. Biotech. Biochem., 86 (1), 23-30.

Senthilan P. R., Helfrich-Förster C.2016. Rhodopsin 7-the unusual rhodopsin in *Drosophila*. Peer J., 4: 2427.

Senthilkumar R., Srinivasan R. 2019. Sex-specific spatial and temporal gene expressions of pheromone biosynthesis activating neuropeptide (PBAN) and binding proteins (PBP/OBP) in *Spoladea recurvalis*. Sci. Rep., 9:3515.

Shao Y. M., Dong K., Tang Z. H., et al. 2009. Molecular characterization of a sodium channel gene from the silkworm *Bombyx mori*. Insect Biochem Mol Biol., 39(2): 145-151.

Sharkey C. R., Fujimoto M. S., Lord N. P., et al. 2017. Overcoming the loss of blue sensitivity through opsin duplication in the largest animal group, beetles. Sci Rep., 7(1): 8.

Shen R.X., Wang Y.T., Wu J.H., et al.2022. Deltamethrin interacts with *Culex quinquefasciatus* odorant-binding protein: a novel potential resistance mechanism. Parasites & Vectors., 15:2.

Shi P., Huang Z., Chen G., et al. 2006. Effects of azadirachtin on six inorganic cation distributions in

Ostrinia furnacalis (G.). Biol Trace Elem Res., 113(1): 105-112.

Shiotsuki T., Suzuki R., Tsuchiya W., et al. 2023. Characteristics of the takeout protein ce-0330 in the silkworm, Bombyx mori (Lepidoptera: Bombycidae). Appl Entomol Zool., 58:237-244.

Si Y.X., Guo J.M., Liao H. 2022. Functional differentiation of three pheromone binding proteins in *Orthaga achatina* using mixed-type sex pheromones. Pesticide Biochemistry and Physiology., 184:105097.

Sigle L.T., Hillyer J.F.2018a. Mosquito hemocytes associate with circulatory structures that support intracardiac retrograde hemolymph flow. Front Physiol., 9: 1187.

Sigle L.T., Hillyer J.F. 2018b. Easter and draper are involved in the periostial hemocyte immune response in the mosquito *Anopheles gambiae*. Insect Mol Biol., 27(4): 429-438.

Sigle L.T., Hillyer J.F. 2016. Mosquito hemocytes preferentially aggregate and phagocytose pathogens in the periostial regions of the heart that experience the most hemolymph flow. Dev Comp Immunol., 55: 90-101.

Silva V., Muñoz A. P., Volonté M., et al. 2021. Orcokinin neuropeptides regulate reproduction in the fruit fly, *Drosophila melanogaster*. Insect Biochem. Mol. Biol., 139: 103676.

Sokolinskaya E. L., Kolesov D. V., K. Lukyanov A., et al. 2020. Molecular principles of insect chemoreception. Acta Nature., 12 (46): 81-91.

Song W.D., Cheng S., wHong B., et al. 2017. Midgut-derived activin regulates glucagon-like action in the fat body and glycemic control. Cell Metab., 25: 386-399.

Spaethe, J., Briscoe A. D. 2004. Early duplication and functional diversification of the opsin gene family in insects. Mol Biol Evol., 21(8): 1583-1594.

Spindler K. D., Hönl C, Tremmel C.h., et al. 2009. Ecdysteroid hormone action. Cell Mol Life Sci., 66(24):3837-50.

Stathopoulos A., Newcomb S. 2020. Setting up for gastrulation: *D. melanogaster*. Curr Top Dev Biol., 136:3-32.

Strauß J., Stritih-Peljhan N. 2022. Vibration detection in arthropods: Signal transfer, biomechanics and sensory adaptations. Arthropod Struct. Devel., 68: 101167.

Suang S., Hiruma K., Kaneko Y., et al. 2019. Diapause hormone directly stimulates the prothoracic glands of diapause larvae under juvenile hormone regulation in the bamboo borer, *Omphisa fuscidentalis* Hampson. Arch. Insect Biochem. Physiol., 102: e21603.

Suggs J.M., Jones T.H., Murphree S.C., et al. 2016. CCAP and FMRFamide-like peptides accelerate the contraction rate of the antennal accessory pulsatile organs (auxiliary hearts) of mosquitoes. J Exp Biol., 219: 2388-2395.

Sujkowski A., Bazzell B., Carpenter K., et al. 2015. Endurance exercise and selective breeding for longevity extend *Drosophila* healthspan by overlapping mechanisms. Aging., 7: 535-52.

Sun L.L. Liu X.L., Wang Y.N. et al. 2022. Neuronal architecture and functional mapping of the taste center of larval *Helicoverpa armigera* (Lepidoptera: Noctuidae). Insect Sci., 29: 730-748.

Sun R.J., Zhang Y.Y., Xu B.H. 2013. Characterization of the response to ecdysteroid of a novel cuticle protein R&R gene in the honey bee, *Apis cerana* cerana. Comp Biochem Physiol, Part B., 166:73-80.

Surkova S., Golubkova E., Mamon L., et al. 2018. Dynamic maternal gradients and morphogenetic networks in *Drosophila* early embryo. Biosystems, 173: 207-2013.

Tamamouna V., Rahman M. M., Petersson M., et al. 2021. Remodelling of oxygen-transporting tracheoles drives intestinal regeneration and tumorigenesis. in *Drosophila.* Nature Cell Biol., 23: 497-510.

Tan S Y, Dee M K. 2009. Elie Metchnikoff (1845—1916): discoverer of phagocytosis. Singapore Med J.,

50(5): 456-457.

Tang Y.Y., Zhang Y., Zhang Q.L., et al. 2022. Prophenoloxidase-positive tubes derived from the hindguts may be the doorkeeper to detoxify the waste metabolites collected by Malpighian tubules in Lepidoptera insects. Dev. Comp. Immunol., (131),104361.

Terakita A. 2005. The opsins. Genome Biol., 6(3): 213.

Terblanche J.S., Marias, E., Hetz S.K., et al.2008. Control of discontinuous gas exchange in *Samia cynthia:* Effects of atmospheric oxygen, carbon dioxide and moisture. J. Exp. Biol., 211(20): 220949.

Teulier L, Weber J.M, Crevier J., et al. 2016. Proline as a fuel for insect flight: enhancing carbohydrate oxidation in hymenopterans. Proc Biol Sci. 283(1834): 20160333.

The Nobel Prize in Physiology or Medicine 2017, https://www.nobelprize.org/ prizes/medicine /2017/ summary/

Tian L., Ma L., Guo E., et al. 2013. 20-hydroxyecdysone upregulates *Atg* genes to induce autophagy in the *Bombyx* fat body. Autophagy., 9(8): 1167-1182.

Tian Y.N., Jiang C.H., Pan Y., et al. 2021.*Bombyx* neuropeptide G protein-coupled receptor A14 and A15 are two functional G protein-coupled receptors for CCHamide neuropeptides. Insect Biochem. Mol. Biol., 131:103553.

Toprak U. 2020. The role of peptide hormones in insect lipid metabolism. Front Physiol., 11:434.

Toprak U., Hegedus D., Doğan C., et al. 2020. A journey into the world of insect lipid metabolism. Arch Insect Biochem Physiol., 104(2): e21682.

Toxopeus J., Sinclair B. J. 2018. Mechanisms underlying insect freeze tolerance. Biol. Rev., 93: 1891-1914.

Truman JW; Riddiford LM. 2019.The evolution of insect metamorphosis: a developmental and endocrine view. Philos Trans R Soc Lond B Biol Sci., 374(1783):20190070.

Truman JW, Riddiford LM. 2007. The morphostatic actions of juvenile hormone. Insect Biochem Mol Biol., 37(8):761-70.

Truman JW. 2019. The evolution of insect metamorphosis. Curr Biol., 29(23): R1252-R1268.

Tseng-Crank, J., J. A. Pollock, I., Hayashi., et al. 1991. Expression of ion channel genes in *Drosophila.* J Neurogen., 7(4): 229-239.

Tsubouchi A., Yano T., Yokoyama T. K., et al. 2017.Topological and modality-specific representation of somatosensory information in the fly brain. Sci., 358: 615-623.

Tworzydlo, W., Jaglarz, M.K., Pardyak, L., et al. 2019. Evolutionary origin and functioning of pregenital abdominal outgrowths in a viviparous insect, *Arixenia esau.* Sci Rep., 9: 16090.

Ueda A., Wu C. F. 2006. Distinct frequency-dependent regulation of nerve terminal excitability and synaptic transmission by *i-a* and *i-k* potassium channels revealed by *Drosophila* Shaker and Shab mutations. J Neurosci., 26(23): 6238-6248.

Urca T., Gefen E., Ribak G. 2021. Critical PO_2 and insect flight: the role of tracheal volume in the Oogenesis-Flight Syndrome. Comp. Biochem. Physiol., Part A., 254: 110873.

VandenBrooks J.M., Gstrein G., Harmon J., et al. 2018. Supply and demand: how does variation in atmospheric oxygen during development affect insect tracheal and mitochondrial networks? J. Insect Physiol., 106: 217-223.

Vansteensel M. J., Michel S., Meijer J. H., et al. 2008. Organization of cell and tissue circadian pacemakers: a comparison among species. Brain Res Rev., 58(1): 18-47.

Veal I. M.,Minakuchi C. 2021. A typical insects: molecular mechanisms of unusual life history strategies. Curr. Opin. Insect Sci., 43:46-53.

Veenstra J.A. 2021.The neuropeptide SMYamide, a SIFamide paralog, is expressed by salivary gland innervating neurons in the American cockroach and likely functions as a hormone. Peptides, 136: 170466.

Venkiteswaran G., Hasan G. 2009. Intracellular Ca^{2+} signaling and store-operated Ca^{2+} entry are required in *Drosophila* neurons for flight. P Natl Acad Sci USA,106(25): 10326-10331.

Verkhratsky A., Ho M.S., Parpura V. 2019. Evolution of neuroglia. Adv Exp Med Biol., 1175: 15-44.

Vershinina A.O., Kuznetsova V. G. 2016. Parthenogenesis in hexapoda: entognatha and non holometabolous insects. J Zoolog Syst Evol Res., 54(4): 257-268.

Von Hadeln J., Althaus V., Häger L., et al. 2018. Anatomical organization of the cerebrum of the desert locust *Schistocerca gregaria*. Cell and Tissue Res., 374:39-62.

Waléria C.A.F., Dihego O.A., Gustavo F.M., et al. 2013. Histochemistry and ultrastructure of urocytes in the pupae of the stingless bee *Melipona quadrifasciata* (Hymenoptera: Meliponini). Microsc Microanal., 19(6): 1502-1510.

Wang G, Zietz CM, Mudgapalli A., et al. 2022. The evolution of the antimicrobial peptide database over 18 years: Milestones and new features. Protein Sci., 31(1):92-106.

Wang H. L., Liénard M. A., Zhao C. H., et al. 2010. Neofunctionalization in an ancestral insect desaturase lineage led to rare δ6 pheromone signals in the chinese tussah silkworm. Insect Biochem Molec., 40(10): 742-751.

Wang P.Y., Zhao Q.l., Qiu Z.Y., et al., 2019. The silkworm (*Bombyx mori*) neuropeptide orcokinin is involved in the regulation of pigmentation. Insect Biochem. Mol. Biol., 114: 103229.

Wang S.L., Wang W.W., Ma Q., et al. 2019. Elevenin signaling modulates body color through the tyrosine-mediated cuticle melanism pathway. FASEB J., 33(9): 9731-9741.

Wang Y.J., Li S.Y., Zhao J.Y., et al. 2020. Clathrin-dependent endocytosis predominantly mediates protein absorption by fat body from the hemolymph in *Bombyx mori*. Insect Sci., 27(4): 675-686.

Wasserthal L.T. , 2015. Flight-motor-driven respiratory airflow increases tracheal oxygen to nearly atmospheric level in blowflies (*Calliphora vicina*). J Exp Biol., 218: 2201-2210.

Wasserthal L.T., Cloetens P., Fink R.H., et al. 2018. X-ray computed tomography study of the flight-adapted tracheal system in the blowfly *Calliphora vicina*, analyzing the ventilation mechanism and flow-directed valves. J Exp Biol., 221: jeb176024.

Weihrauch D., O'Donnell M.J. 2021. Mechanisms of nitrogen excretion in insects. Cur. Opin. Insect Sci., 47:25-30.

Wicher D., Miazzi F. 2021. Functional properties of insect olfactory receptors: ionotropic receptors and odorant receptors. Cell and Tissue Res., 383:7-19.

Wojda I., Cytryńska M., Zdybicka-Barabas A., et al. 2020. Insect defense proteins and peptides. Subcell Biochem., 94:81-121.

Wu Q, Patočka J, Kuča K. 2018. Insect antimicrobial peptides, a mini review. Toxins (Basel)., 10(11):461.

Wu S., Nomura Y., Du Y., et al.2017. Molecular basis of selective resistance of the bumblebee BiNav 1 sodium channel to tau-fluvalinate. Proc Natl Acad Sci USA., 114(49): 12922-12927.

Wu W.M., Li K., Guo S.Y., et al. 2021. P300/HDAC1 regulates the acetylation/deacetylation and autophagic activities of LC3/Atg8-PE ubiquitin-like system. Cell Death Discov., 7(1): 128.

Wu W.M., Luo M., Li K., et al. 2021. Cholesterol derivatives induce dephosphorylation of the histone deacetylases Rpd3/HDAC1 to upregulate autophagy. Autophagy. 17(2): 512-528.

Wu Z.G., Yang L.B., He Q.G., et al. 2021. Regulatory mechanisms of vitellogenesis in insects. Front. Cell Dev. Biol., 8: 593613.

Wulff J.P., Capriotti N., Ons S.2018. Orcokinins regulate the expression of neuropeptide precursor genes related to ecdysis in the hemimetabolous insect *Rhodnius prolixus*. J Insect Physiol., 108: 31-39.

Xu L , Chen R , Du G., et al. 2016. Anisotropic shrinkage of insect air sacs revealed *in vivo* by X-ray microtomography. Sci. Rep., 6:32380.

Xu W. 2020. How do moth and butterfly taste: molecular basis of gustatory receptors in Lepidoptera. Insect Sci., 27: 1148–1157.

Xu, P., Feuda R., Lu B., et al. 2016. Functional opsin retrogene in nocturnal moth. Mobile Dna-Uk., 7: 18.

Yan H., Jafari S., Pask G., et al. 2020. Evolution, developmental expression and function of odorant receptors in insects. J. Exp. Biol., 223: jeb208215.

Yan S., Wang W.X., Shen J. 2020. Reproductive polyphenism and its advantages in aphids: switching between sexual and asexual reproduction. J. Integr. Agri., 19(6): 1447–1457.

Yan Y., Hillyer J.F. 2019. Complement-like proteins TEP1, TEP3 and TEP4 are positive regulators of periostial hemocyte aggregation in the mosquito *Anopheles gambiae*. Insect Biochem Mol Biol., 107: 1-9.

Yan, S., Zhu J., Zhu W., et al. 2014. The expression of three opsin genes from the compound eye of *Helicoverpa Armigera* (Lepidoptera: Noctuidae) is regulated by a circadian clock, light conditions and nutritional status. PLoS One., 9(10): e111683.

Yang C., Cheng J., Lin J.Y., et al.2022. Sex pheromone receptors of lepidopteran insects. Front. Ecol. Evol., 10, 797287.

Yang L., Qiu L. M., Fang Q., et al. 2021. Cellular and humoral immune interactions between *Drosophila* and its parasitoids. Insect Sci., 28(5):1208-1227.

Yang L., Ren L.P., Shang Y.J., et al. 2021. Progress on reproductive modes of sarcosaphagous flies. J. Asia-Pac. Entomol., 24: 731-738.

Yew J.Y., Dreisewerd K., Luftmann H., et al. 2009. A new male sex-pheromone and novel cuticular cues for chemical communication in *Drosophila*. Curr Biol., 19(15): 1245-1254.

Yi H.Y, Chowdhury M, Huang Y.D., et al. 2014. Insect antimicrobial peptides and their applications. Appl Microbiol Biotechnol., 98(13): 5807-22.

Zandawala M., Hamoudi Z., Lange A.B., et al. 2015. Adipokinetic hormone signalling system in the Chagas disease vector, *Rhodnius prolixus*. Insect Mol Biol., 24: 264-276.

Zhang B., Han H.B., Xu L.B., et al.2021.Transcriptomic analysis of diapause-associated genes in *Exorista civilis* Rondani (Diptera:Tachinidae). Arch. Insect Biochem. Physiol., 107: e21789.

Zhang C., Kim A.J.,Rivera-Perez C., et al. 2022. The insect somatostatin pathway gates vitellogenesis progression during reproductive maturation and the post-mating response. Nat. Commun., 13:969.

Zhang Q.Y, Yan Z.B, Meng Y.M., et al. 2021. Antimicrobial peptides: mechanism of action, activity and clinical potential. Mil Med Res., 8(1):48.

Zhang T.Y., Kang L., Zhang Z.F., et al. 2004. Identification of a POU factor involved in regulating the neuron-specific expression of the gene encoding diapause hormone and pheromone biosynthesis-activating neuropeptide in *Bombyx mori*, Biochem. J., 380: 255-263.

Zhang X, Li S, Liu S. 2022. Juvenile hormone studies in *Drosophila melanogaster.* Front Physiol.,

12:785320.

Zhang, K., Chen M.,Wang H., et al. 2020. Molecular characterization and functional expression of voltage-gated sodium channel variants in *Apolygus lucorum* (MeyerDür). Pest Manag Sci., 76(6): 2095-2104.

Zhao X.F. 2020. G protein-coupled receptors function as cell membrane receptors for the steroid hormone 20-hydroxyecdysone. Cell Commun Signal., 18(1):146.

Zhao Z.L, McBride C. S. 2020. Evolution of olfactory circuits in insects. J Comp Physiol A., 206:353-367

Zheng H., Powell J.E., Steele M.I., et al. 2017. Honeybee gut microbiota promotes host weight gain *via* bacterial metabolism and hormonal signaling. PNAS.,114(18): 4775-4780.

Zheng W., Feng G., Ren D., et al. 1995. Cloning and characterization of a calcium channel alpha 1 subunit from *Drosophila melanogaster* with similarity to the rat brain type D isoform. J Neurosci, 15(2): 1132.

Zheng, X., Sehgal A. ,2008. Probing the relative importance of molecular oscillations in the circadian clock. Genetics, 178(3): 1147-1155.

Zhou Y.T., Li L., Zhou X.R., et al. 2020.Three chemosensory proteins involved in chemoreception of *Oedaleus asiaticus* (Orthopera: Acridoidea). J Chem Ecol., 46:138-149.

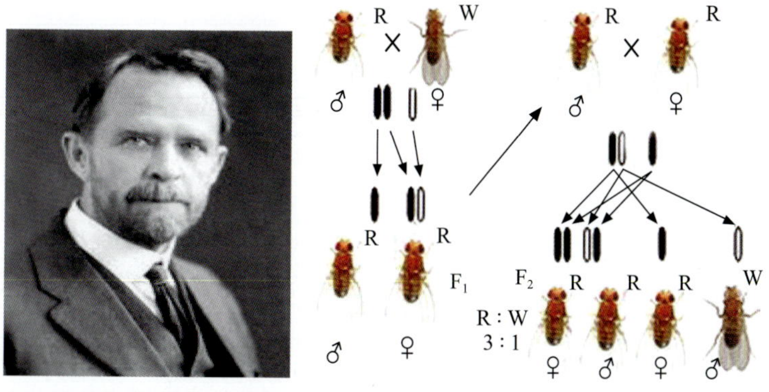

Fig.0-3 Thomas Hunt Morgan and his work on *D.melanogaster* white (eyed) mutant.

图 0-3 摩尔根和他的研究果蝇白眼突变体。

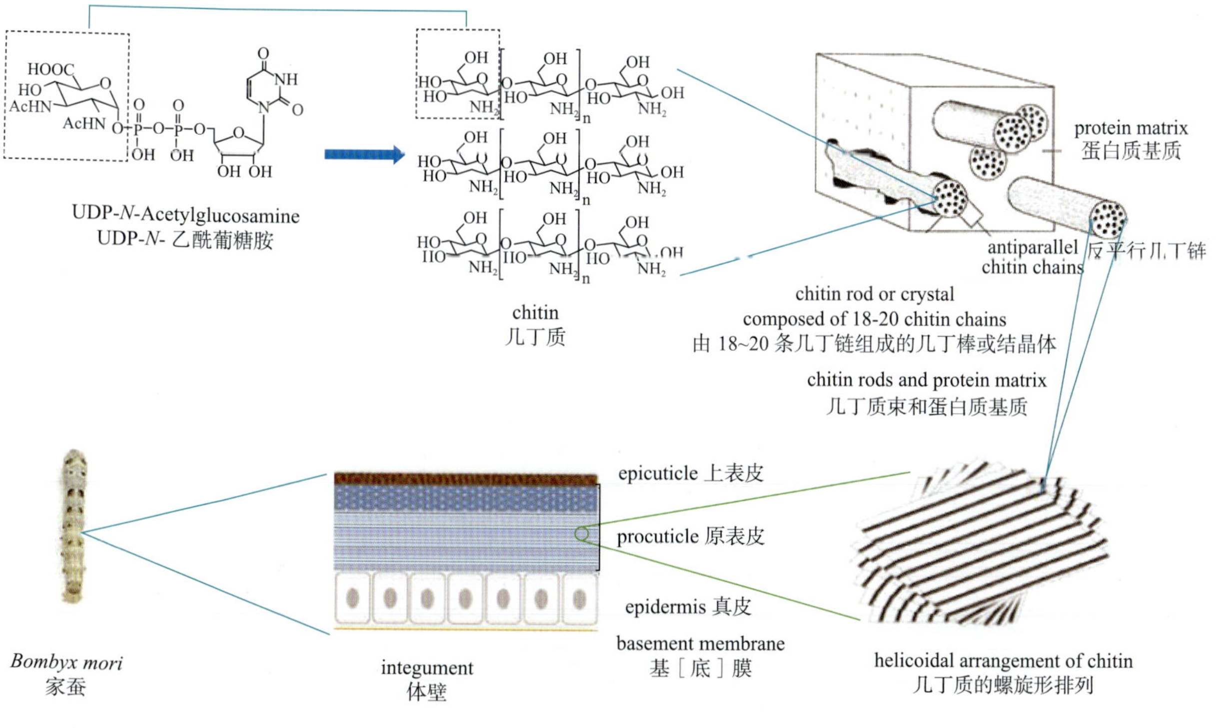

Fig.1-3 Structure of hierarchy of silkworm cuticle. Typically, about 18-20 Chitin chains antiparallel arrangement held together by hydrogen bonds to form chitin rods or crystallites embedded in a protein matrix in cuticle (modified from Liu et al., 2019; Nation,2022).

图 1-3 家蚕表皮的层次结构。通常由 18 ～ 20 根几丁链反平行组成的几丁棒或微晶包埋在表皮的蛋白质基质中（仿 Liu 等，2019 ；Nation，2022）。

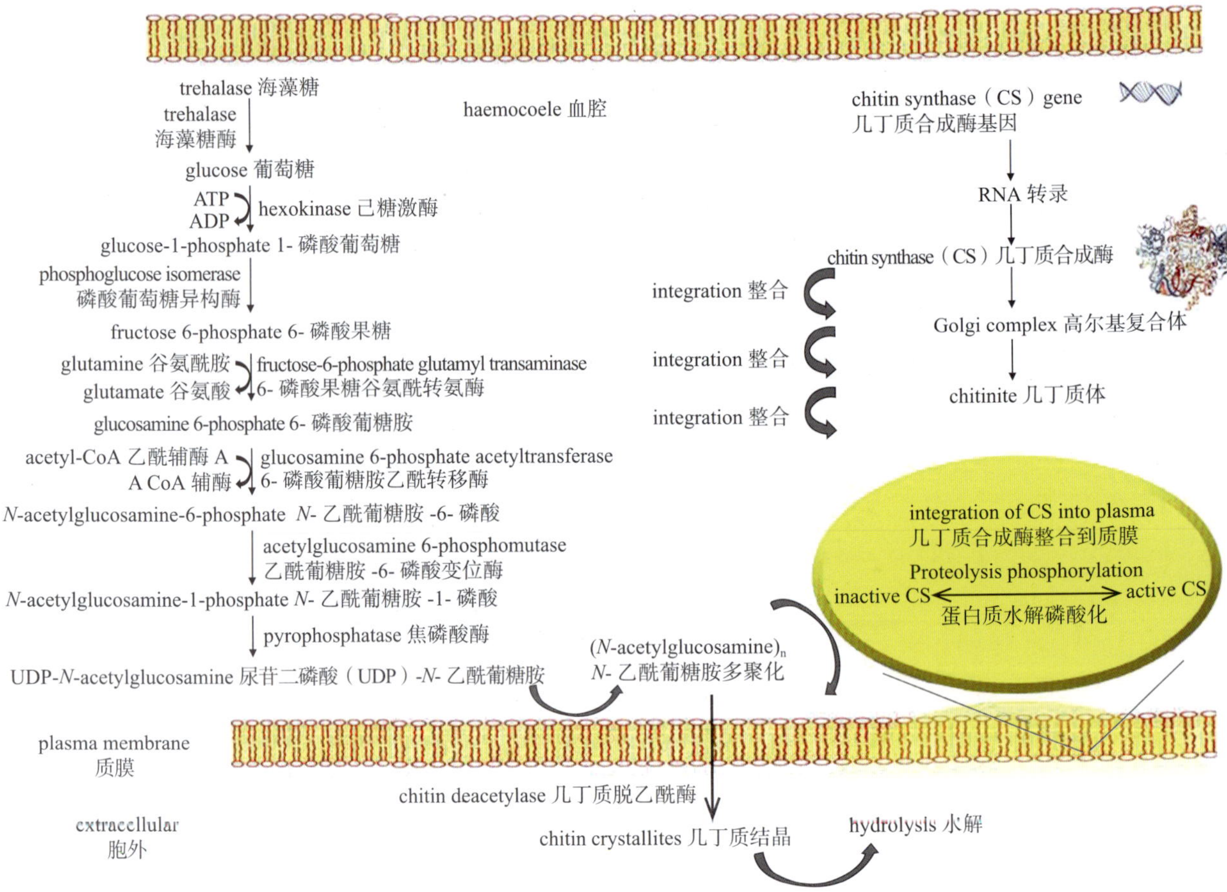

Fig.1-4 Cascade steps in chitin biosynthesis (modified from Nation, 2022).

图 1-4 几丁质合成的级联反应（仿 Nation，2022）。

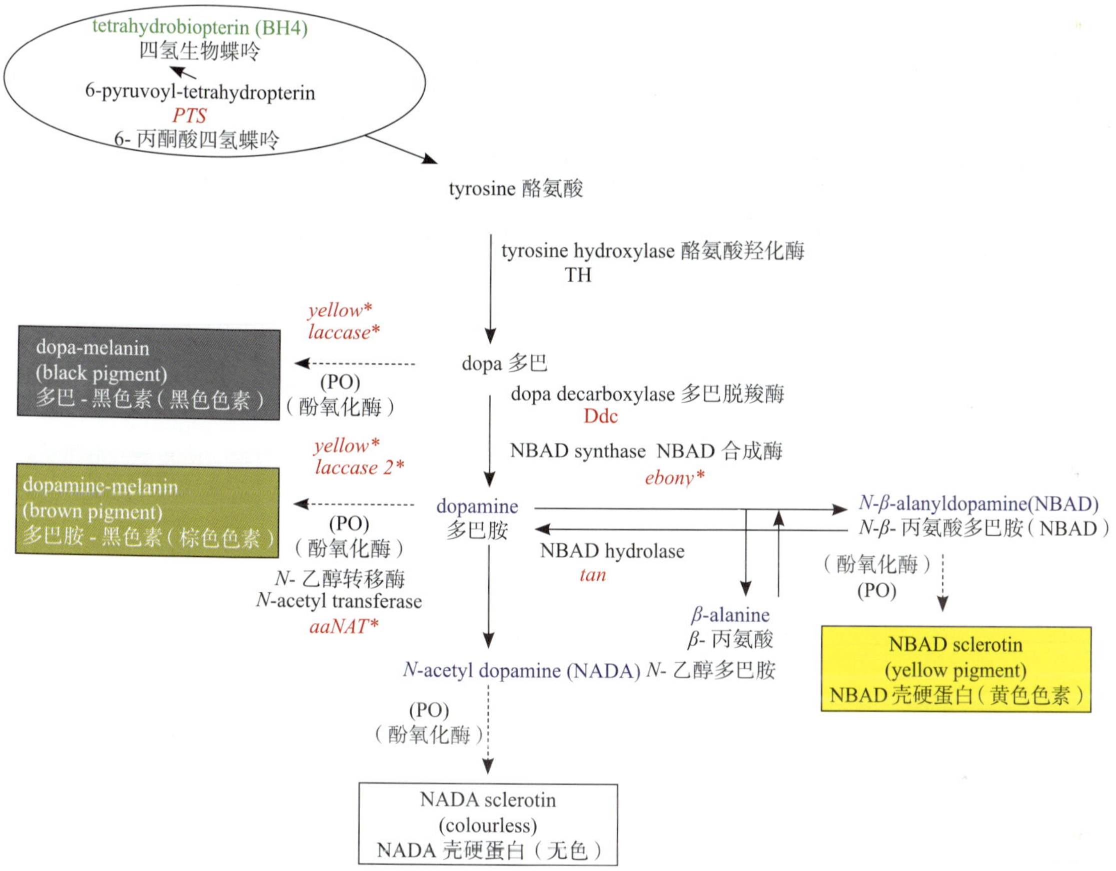

Fig.1-6 Insect body coloration melanin biosynthesis pathway. Green shows the cofactor outside the melanin pathway impacting the hydroxylation of tyrosine, the precursor of melanin biosynthesis. Biochemical phenoloxidase reactions catalyzing pigment conversion are shown in dotted lines, and the final pigments produced are shown in boxes. Asterisk indicates the genes putatively involved in caterpillar albinism (modified from Galarza, 2021).

图 1-6 昆虫体色黑色素生物合成途径。绿色显示黑色素途径外的辅助因子影响酪氨酸的羟基化，酪氨酸是黑色素生物合成的前体。催化色素转化的酚氧化酶生化反应用虚线表示，最终产生的色素用方框表示。星号表明可能与毛虫白化病有关的基因（仿 Galarza，2021）。

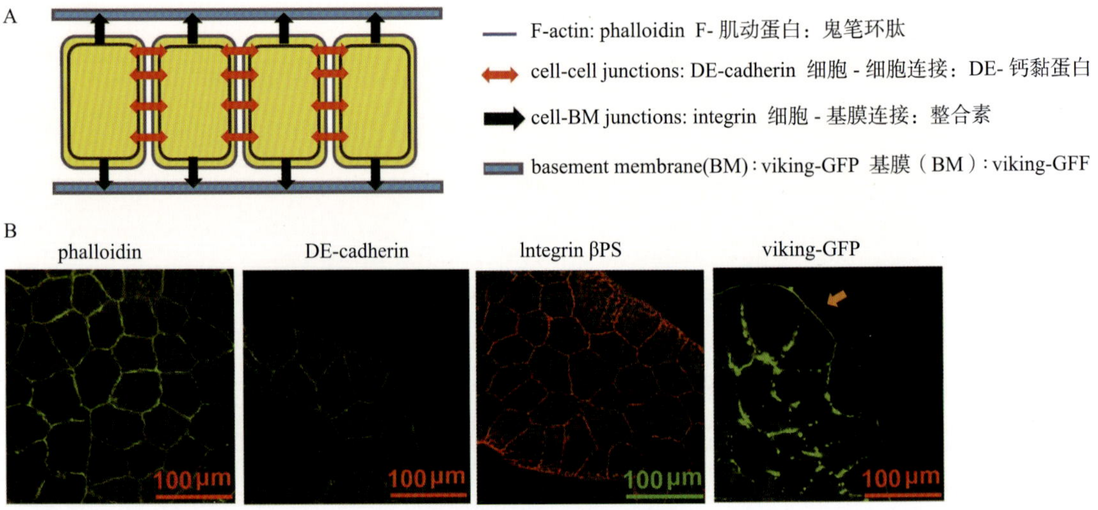

Fig.3-2 A model of the fat body structure and the detection of key molecules of the cell-cell conjugations and basement membrane (BM) junctions. A.Schematic diagram of the different marker molecules localized at cell membrane; B. Integrity of fat body cells stained by phalloidin. Immunofluorescent staining of DE-cadherin and integrin βPS proteins. Fluorescence observation of viking-GFP, viking-GFP: a fluorescently-tagged collagen IV reporter(permitted by Qiangqiang Jia, 2014).

图 3-2 果蝇脂肪体结构示意图及关键连接分子检测。**A.** 脂肪体细胞膜上的不同标志性分子示意图；**B.** 鬼笔环肽染色观察细胞膜的完整性，**DE-** 钙黏蛋白以及整合素的免疫荧光染色及 **viking-GFP** 的荧光标记观察，**viking-GFP** 为荧光标记的胶原蛋白 **IV** 报告蛋白（经 **Qiangqiang Jia** 博士允许，**2014**）。

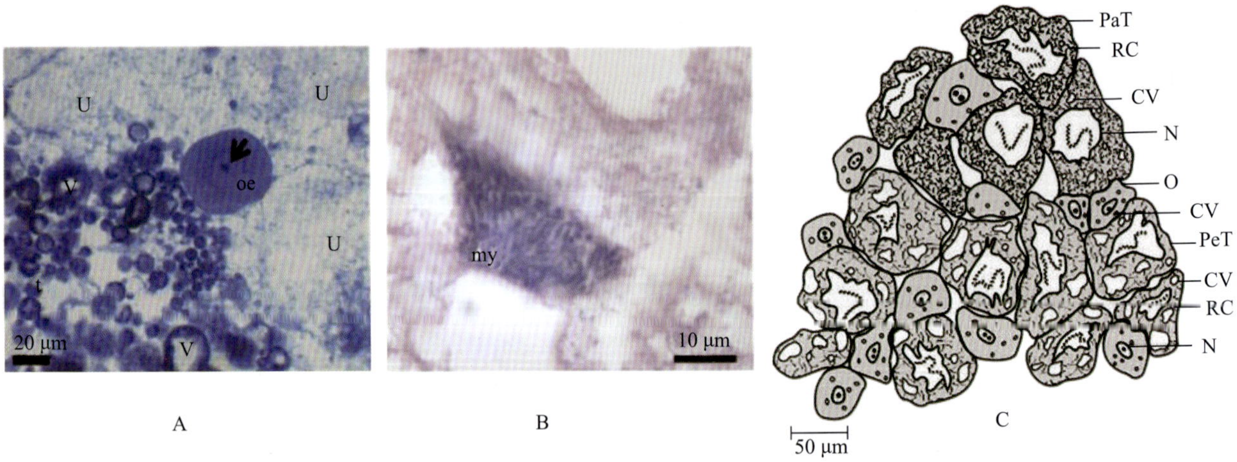

Fig. 3-3 The main types of insect fat body cells: trophocytes, urate cells, oenocytes, and mycetocytes.
A. Histochemical test (mercury bromophenol blue) for protein in the fat body cells of *Melipona quadrifasciata* pupae. Trophocyte (t) with vacuoles (v) positive for protein and oenocyte (oe) with positive cytoplasm and nucleus. u. urocytes (modified from Waléria et al., 2013). B. Detail of a mycetocyte (my) in the fat body of a *Mastotermes darwiniensis* worker. Stain: hematoxylin–eosin (modified from Costa-Leonardo et al., 2013). C. Fat body cells from worker of A. disciger (modified from Roma et al., 2010). PaT. present perivisceral trophocyte; RC. reticular cytoplasm; CV, cytoplasmic vacuole; N, nucleus; O. oenocyte; PeT. perivisceral trophocyte (modified from Waleria et al., 2013).

图 3-3 昆虫脂肪体细胞的主要种类：营养细胞、尿酸盐细胞、绛色细胞和含菌细胞

A. 利用汞溴酚蓝对膜翅目蜜无刺蜂 ***Melipona quadrifasciata*** 蛹期脂肪组织中的蛋白质进行染色观察，**t.** 营养细胞；**u.** 尿酸盐细胞；**oe.** 绛色细胞；**v.** 包含蛋白质的囊泡（仿 **Waléria** 等，**2013**）。**B.** 苏木精 - 伊红染色观察达尔文澳白蚁工蚁脂肪体中的含菌细胞（仿 **Costa-Leonardo** 等，**2013**）。**C.** 切叶蚁（***Acromyrmex disciger***）工蚁脂肪体细胞（仿 **Roma** 等，**2010**）。**PaT.** 腔壁营养细胞；**RC.** 网状细胞质；**CV.** 胞质液泡；**N.** 细胞核；**O.** 绛色细胞；**PeT.** 围脏营养细胞（仿 **Waléria** 等，**2013**）。

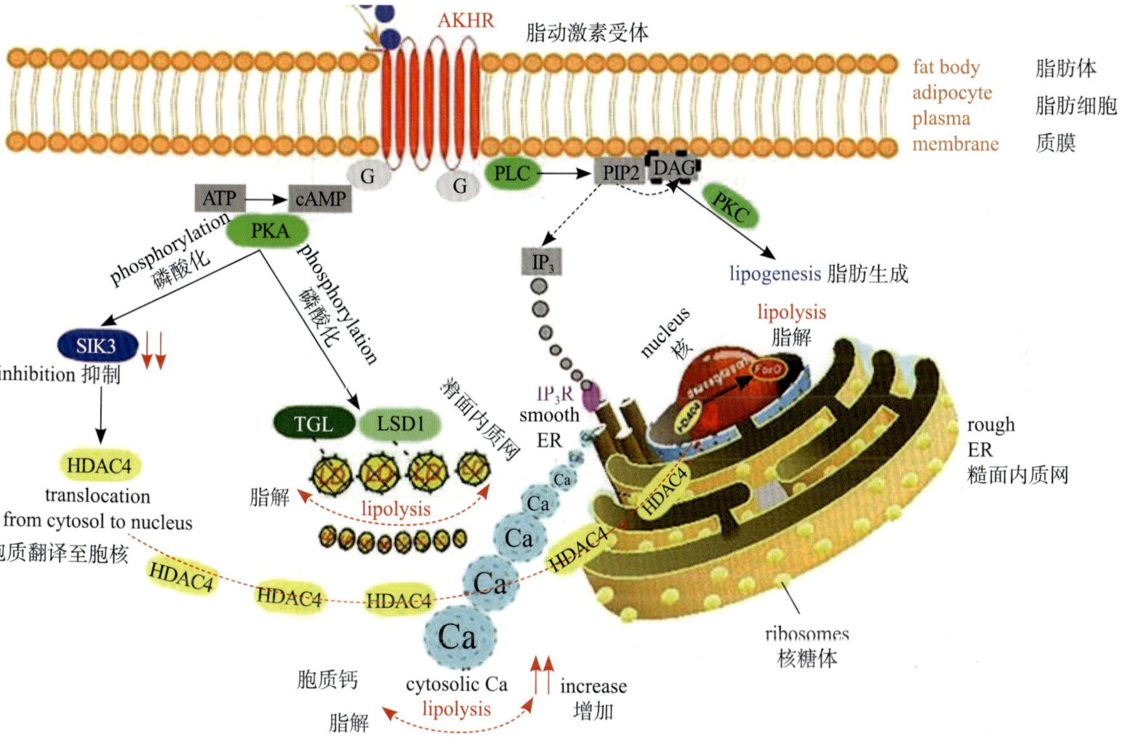

Fig. 3-8 Mode of signaling of adipokinetic hormone (modified from Toprak, 2020). AKH. adipokinetic hormone; AKHR. adipokinetic hormone receptor; Ca. calcium; DAG. diacylglycerol; ER. endoplasmic reticulum; G, g-protein couple; HDAC4. histone deacetylase 4; IP_3. inositol triphosphate; IP_3R. inositol-1,4,5-triphosphate receptor; LD. lipid droplet; LSD1, lipid storage droplet 1 protein; PLC. phospholipase C; PIP2. phosphatidylinositol 4,5-diphosphate; PKA. cAMP-dependent protein kinase; PKC. calcium-dependent protein kinase C; SIK3. salt induced kinase 3; TGL. triglyceride lipase.

图 3-8 脂动激素（AKH）的信号转导（仿 Toprak, 2020）。AKH. 脂动激素；AKHR. 脂动激素受体；Ca. 钙；DAG. 甘油二酯；ER. 内质网；G. G- 蛋白耦联受体；HDAC4. 组蛋白去乙酰化酶 4；IP_3. 三磷酸肌醇；IP_3R. 肌醇 -1，4，5- 三磷酸受体；LD. 脂滴；LSD1. 脂质贮存脂滴 1 蛋白；PLC. 磷脂酶 C；PIP2. 磷脂酰肌醇 4，5- 双磷酸；PKA. 依赖于 cAMP 的蛋白激酶 A；PKC. 钙依赖蛋白激酶 C；SIK3. 盐诱导激酶 3；TGL. 三酰甘油脂肪酶。

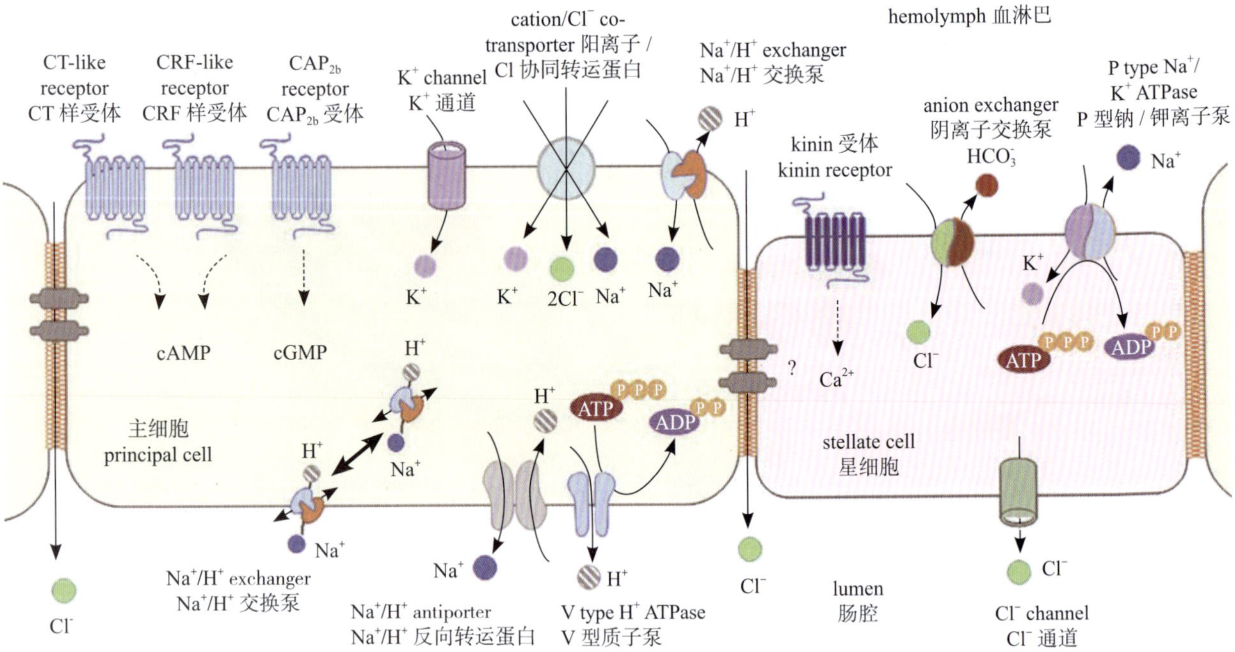

Fig. 4-7 Ion transport mechanism and hormonal regulation in the principal cells and stellate cells of Malpighian tubule in insects (modified from Lu er al., 2011).

图 4-7 昆虫马氏管主细胞和星细胞离子转运机制及激素调控（仿 Lu 等，2011）。

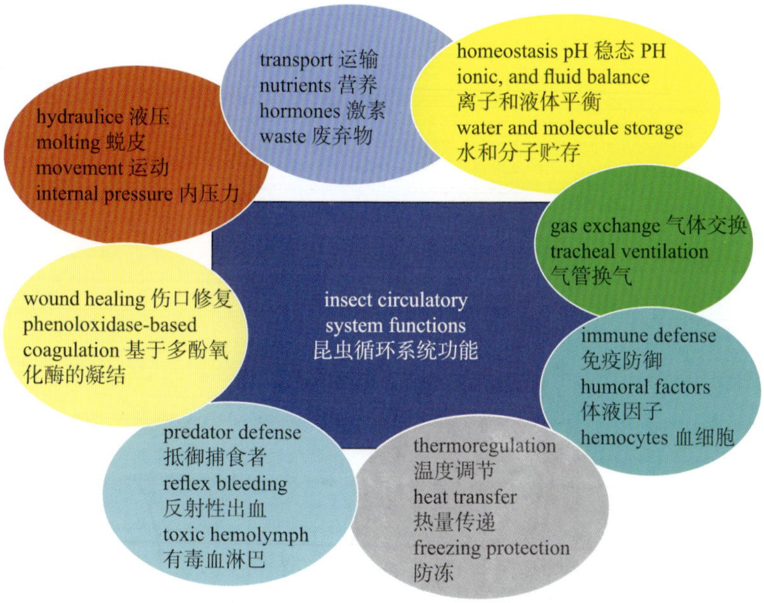

Fig. 6-1 The many functions of the circulatory system of insects (modified from Hillyer and Pass, 2019).

图 6-1 昆虫循环系统的主要功能（仿 Hillyer 和 Pass，2019）。

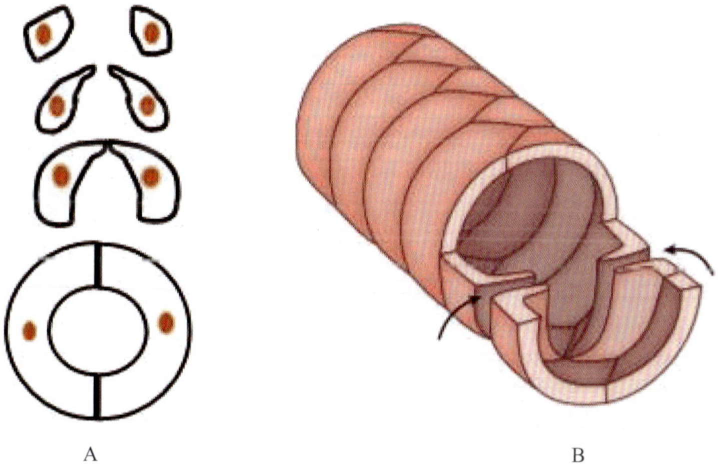

Fig. 6-2 Development and structure of the dorsal vessel. A.Consecutive stages of morphogenesis, from top to bottom; B.The mature dorsal vessel consists of two rows of semicircular cardiomyocytes. The anterior front cut shows the organization of the ostial cells; the two inward directed flaps act together and allows hemolymph to enter the cavity, but hinders outward flow (modified from Hilly and Pass, 2019).

图 6-2 背血管的发育与结构。A. 持续发育过程中背血管的形态，从上至下；B. 成熟的背血管由两排半圆形心肌细胞组成。它们在背侧和腹侧中线的接触点形成螺旋状排列。前部切面显示心门细胞；两个向内的瓣膜共同作用就像一个阀门一样，允许血淋巴进入管腔，但防止其外流（仿 Hilly 和 Pass, 2019）。

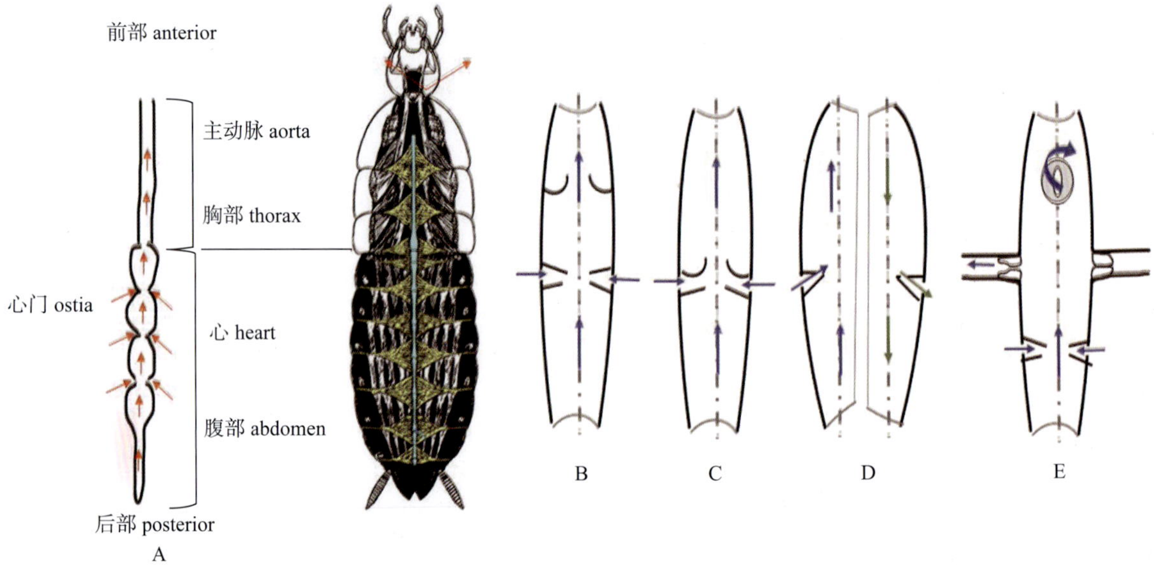

Fig. 6-3 Morphology of the dorsal vessel (heart) and various kinds of ostia in insects. A. Constructure of dorsal vessel; B. Incurrent ostia with paired lips and an intracardiac valve; C. Incurrent ostia with elongate lips that function as a pouch valve; D. Two-way ostia with a single lip. (left) Incurrent flow during the anterograde phase. (right) Excurrent flow during the retrograde phase. E. Incurrent ostia (bottom), excurrent ostia with sphincter-like valves at the base of lateral vessels (middle), and an unpaired excurrent ostium (top). The sphincter-like valve on the left is shown in the open position, whereas the valve on the right is shown in the closed position (modified from Hillyer and Pass, 2019).

图 6-3 昆虫背血管（心脏）的形态及各种心门图。**A.** 背部血管形态与结构；**B.** 长有一对瓣膜结构的流入心门，同时伴有心脏内阀门；**C.** 长有拉长的瓣膜的流入心门，发挥类似于口袋的阀门作用；**D.** 双口单瓣膜的心门，（左）顺行阶段向内的流动，（右）逆行阶段向外流出的流动；**E.** 流入心门（底部），左侧血管的基部具有类似括约肌阀门的流出心门（中央）以及不成对的流出心门（上部）。左侧类似括约肌的阀门显示的是打开的位置，右侧类似括约肌的阀门显示的是关闭的位置（仿 **Hillyer** 和 **Pass, 2019**）。

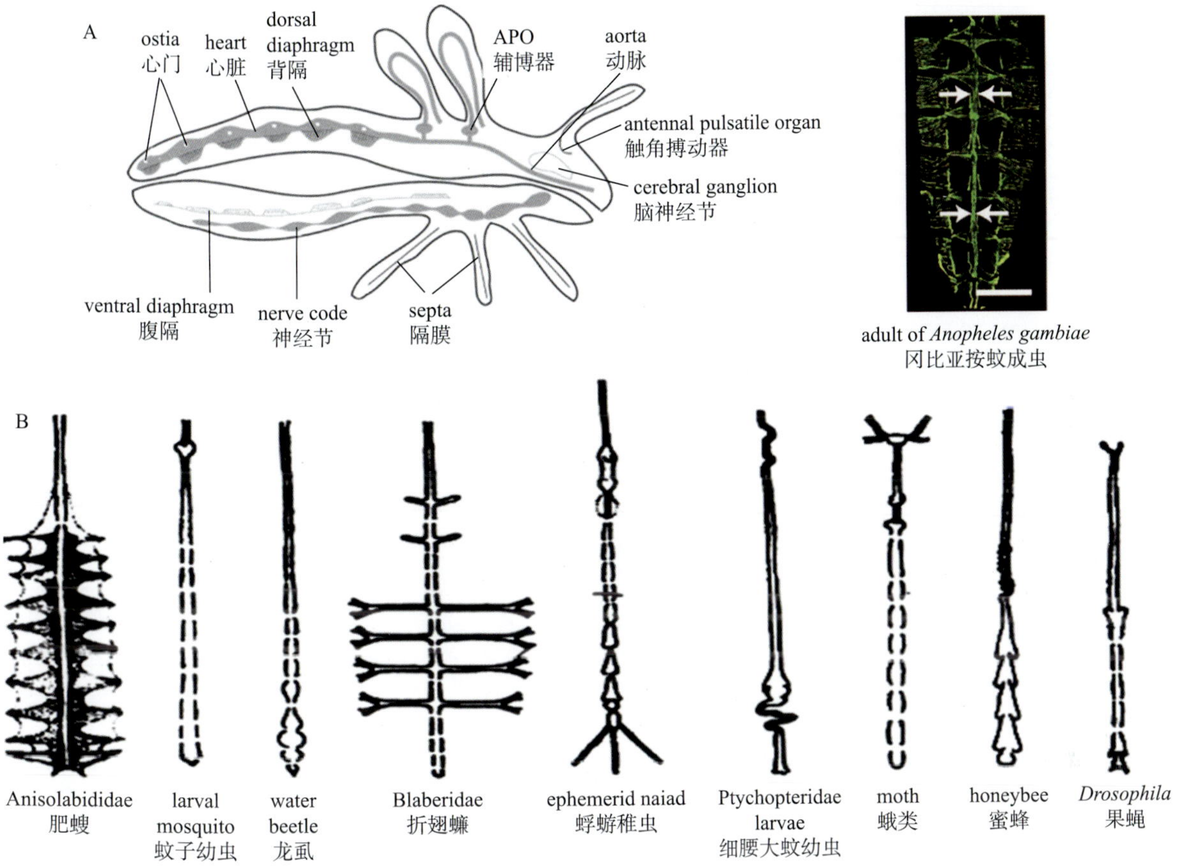

Fig. 6-4 The structure of insect circulatory system. A.The composition of insect hemolymph circulatory system and the routine for circulation as well as the dorsal vessel of adult of *Anopheles gambiae*. B.Variation of the heart and aorta in different insects (modified from other authors).

图 6-4 昆虫血液循环系统结构。**A.** 昆虫血液循环系统组成和血液循环路线以及冈比亚按蚊成虫的背血管图；**B.** 不同昆虫的心脏和动脉的变化（仿其他作者）。

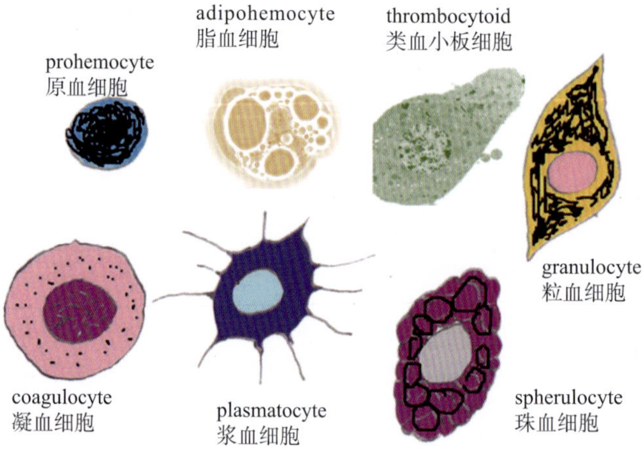

Fig. 6-6 Types of hemocytes found in many insects. Arrows indicate that prohemocytes transform into plasmatocytes, some plasmatocytes may give rise to granulocytes, and granulocytes may transform into sperulocytes (Nation, 2022).

图 6-6 多数昆虫中存在的典型血细胞类型。箭头显示原血细胞转变成浆血细胞，一些浆血细胞可能产生粒血细胞，而粒血细胞可能转变成珠血细胞（Nation, 2022）。

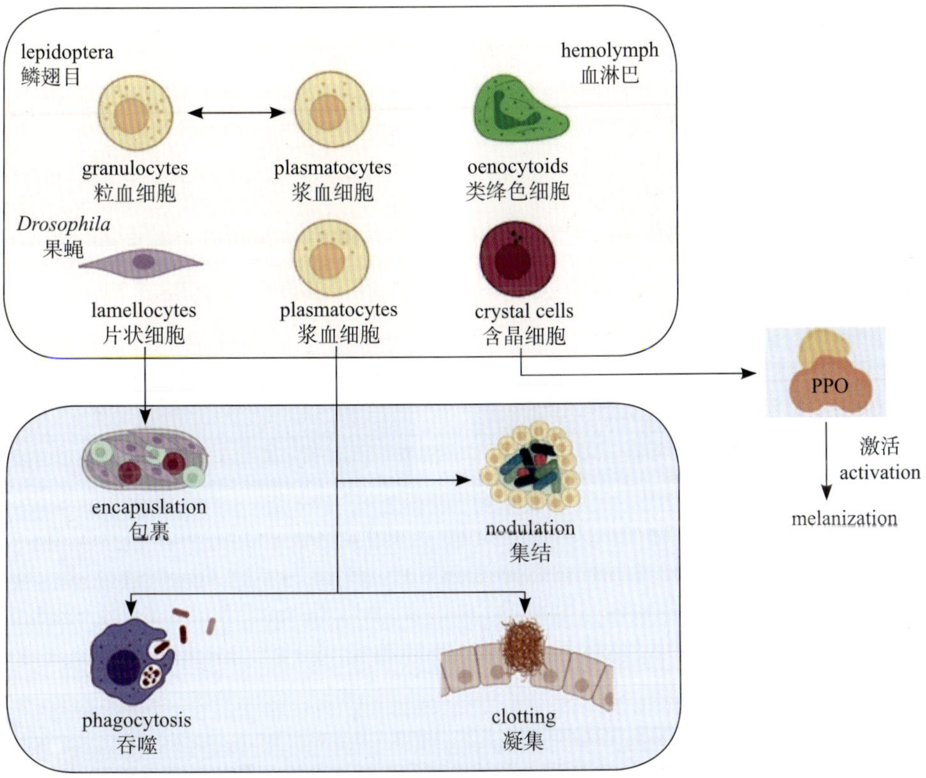

Fig. 7-2 Cellular immunity mediated by hemocytes and melanization (modified from Eleftherianos et al., 2021).

图 7-2 血细胞介导的细胞免疫及黑化反应（仿 Eleftherianos 等，2021 ）。

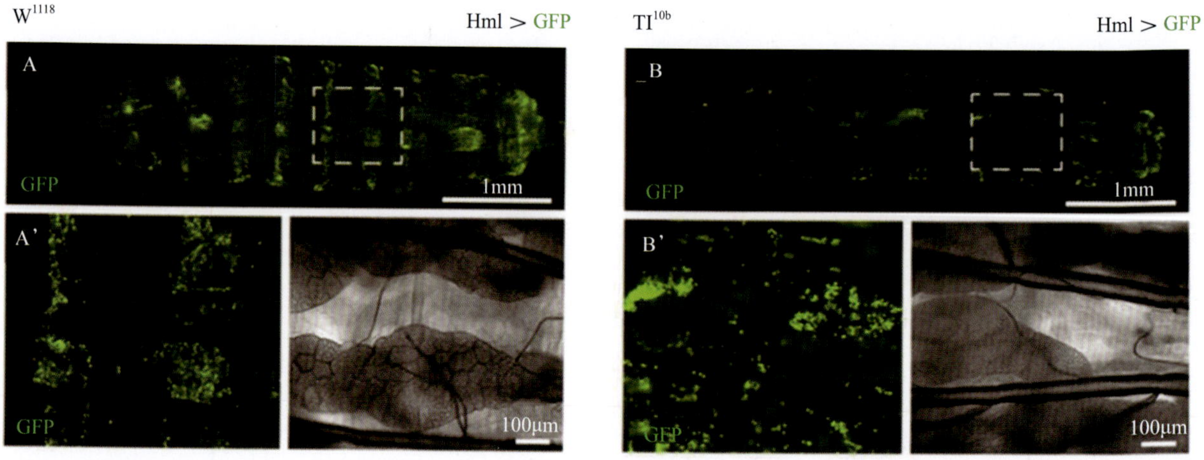

Fig. 7-3 Hemocytes distribution in wild type *Drosophila* larva and a gain-of-function Toll receptor mutant Toll^{10b}. *Hml* (Hemolectin) is a hemocytes-specific driver whose expression is visualized by GFP fluorescence after crossing to the UAS-Gal4. A, A' .Sessile banded pattern was seen in Toll wild-type; B，B' . In Toll^{10b} genetic background, banded pattern was lost. The demarcated areas in A and B are shown enlarged in panels A' and B' , bright-field images to the right (modified from Schmid et al., 2014).

图 7-3 野生型和免疫激活型 Tl^{10b} 果蝇幼虫中的血细胞分布。*Hml*（Hemolectin）. 血细胞特异性的启动子，与 UAS-Gal4 杂交后通过 GFP 荧光显示其表达。A, A' . Toll 野生型幼虫显示固着血细胞呈带状分布；B, B' . Toll^{10b} 遗传背景的果蝇幼虫中带状消失。A' 和 B' 是 A、B 中指示区域的放大图像，右边是白光明场图（仿 Schmid 等，2014）。

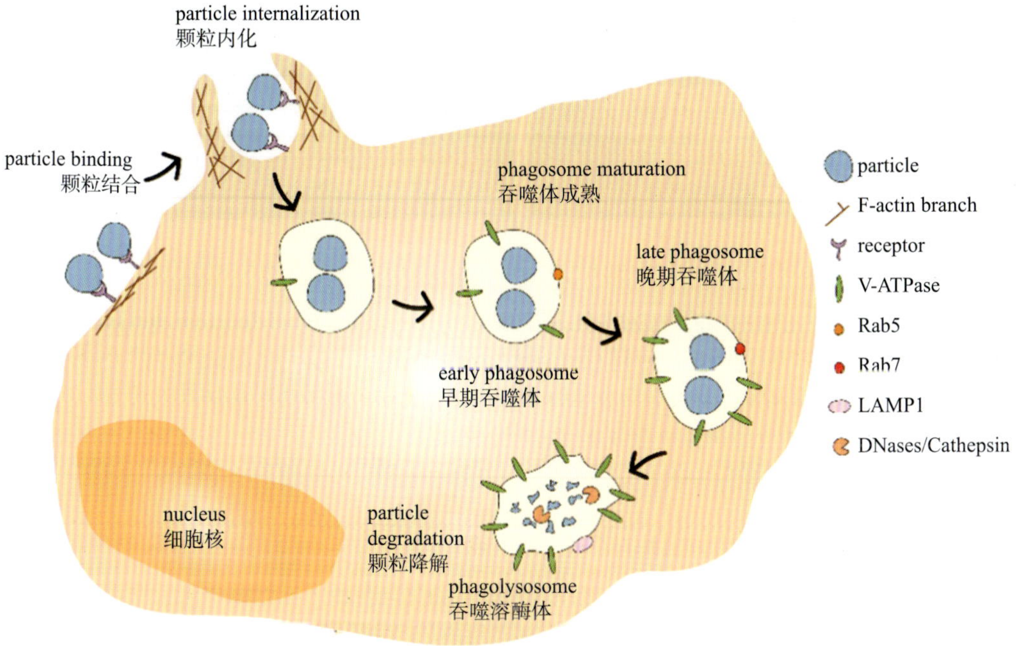

Fig. 7-4 The phagocytosis process in *Drosophila*. particle binding: *Drosophila* specialized phagocytic cell surface receptors recognize and bind target particles; particle internalization; F-actin branches off at the phagocytosis site, forming the phagocytic cup. Actin polymerizes around the granule until phagocyte processes fuse at the edge, producing a newly formed phagosome; maturation of phagosomes: the newly formed phagosomes then undergo a series of fission and fusion events with organelles; particle degradation: In this last step, the phagosome obtains the important components of the final particle destruction step, such as DNase and protease (modified from Melcarne et al., 2019).

图 7-4 果蝇的细胞吞噬过程。颗粒结合：果蝇专业吞噬细胞表面的特异受体识别并结合目标颗粒。颗粒内化；F- 肌动蛋白在吞噬位点形成分支，形成吞噬杯。肌动蛋白在颗粒周围聚合，直到吞噬细胞突起在边缘融合，产生一个新形成的吞噬体。吞噬体成熟：新形成的吞噬体随后与细胞器发生一系列裂变和融合事件。颗粒降解：在这最后一步中，吞噬体获得最终破坏颗粒的重要成分，如 DNase 和蛋白酶（仿 Melcarne 等，2019）。

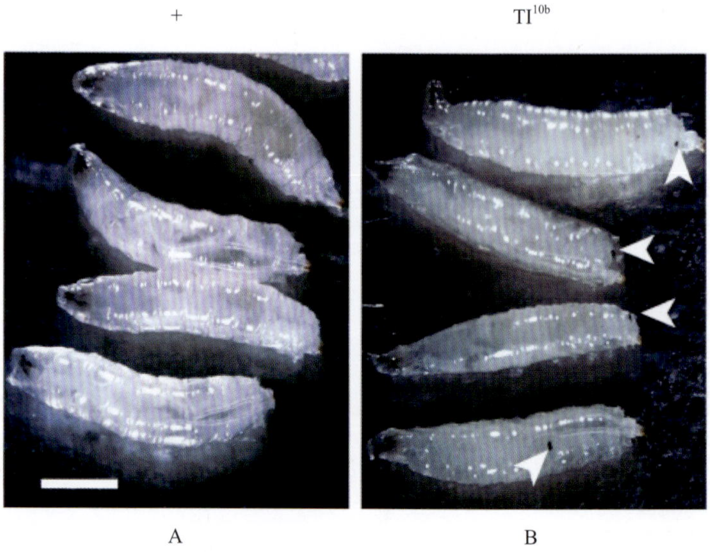

Fig. 7-5 Melanotic nodules in the larvae of *Drosophila* Tl^{10b} mutant. Tl^{10b} is a gain-of-function mutant of Toll. Melanotic nodules are seen black spots in the larvae of *Drosohila* indicated as the arrows. A. +, wild type; B. Tl^{10b} (Schmid et al., 2014).
图 7-5 果蝇 Tl^{10b} 突变体幼虫黑化的结节。Tl^{10b} 为免疫激活型突变，黑化的结节如图中箭头所示 A.+, 野生型；B. Tl^{10b} (Schmid 等，2014)。

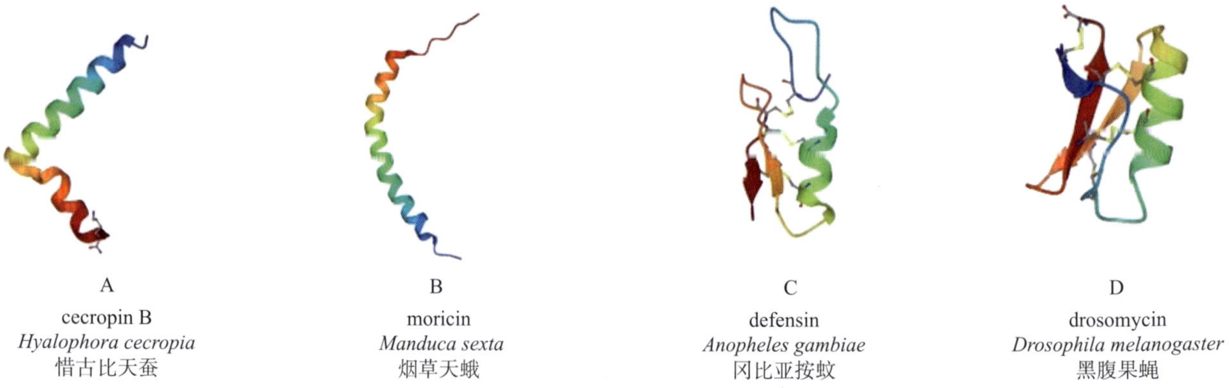

Fig. 7-8 Different structures of AMPs. A.Cecropin adopts two α-helical conformations (PDB : 2IGR); B.Moricin has a typical α-helical conformation (PDB : 2JR8); C.Defensin (PDB : 2NY8); D.Drosomycin (PDB : 1MYN) adopts both alpha-helix and β-sheet conformation.
图 7-8 不同抗菌肽的结构。A. 天蚕素呈现两个 α- 螺旋构象（PDB : 2IGR）；B.Moricin 具有一个典型的 α- 螺旋构象（PDB : 2JR8）；C. 防御素（PDB : 2NY8）；D. 果蝇肽（PDB : 1MYN）采用 α- 螺旋和 β- 折叠混合的构像。

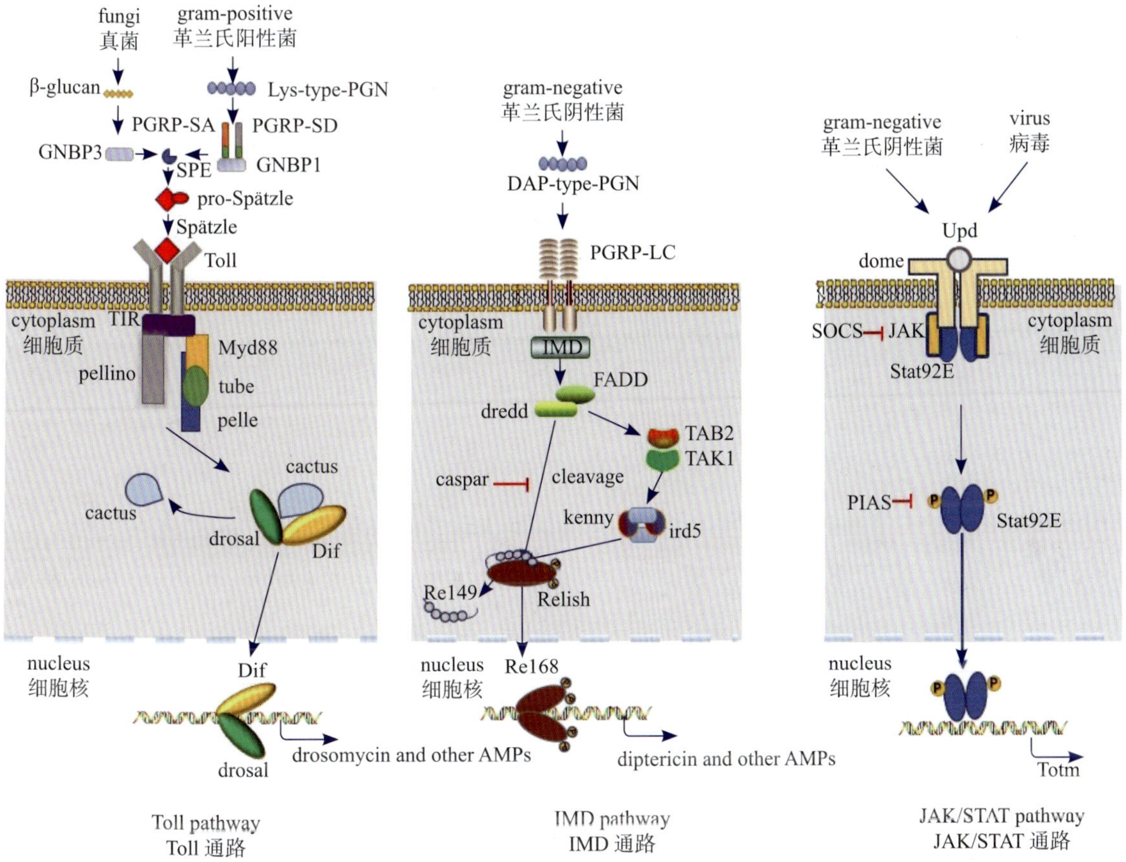

Fig. 7-9 Signaling pathways regulating expression of antimicrobial peptide genes in *Drosophila*. Lys-type-PGN. Lys-tye-peptidoglycan; GNBP3. Gram-negative binding protein 3; GNBP1. Gram-negative binding protcin 1; SPE. spätzle processing enzyme; TIR. Toll/IL-1R homologous region; dmyD88. Drosophila homology of differentiation primary response protein; Tube: adaptor protein containing death domain; Pelle. kinase protein; dif. NF-κB transcription factor; DAP-type-PGN. diaminopimelic acid-tye peptidoglycan; PGRP-LC. peptidoglycan recognition protein LC; dFADD. drosophila Fas-associated with death domain; Dredd. homology of cysteine protease; TAK1. growth factor β-activated kinase; TAB2. TAK1 binding protein; Upd. unpaired cytokines; Stat92E. signal-transducer and activator of transcription protein at 92E.

图 7-9 果蝇调控抗菌肽表达的主要信号传导通路。

Lys-type-PGN. 赖氨酸型肽聚糖；GNBP3. 革兰氏阴性菌结合蛋白 3； GNBP1. 革兰氏阴性菌结合蛋白；SPE. 一种加工 spätzle 的丝氨酸蛋白酶；PGRP-SA, SD. 肽聚糖识别蛋白 SA, SD; TIR. Toll/IL-1 受体同源域； dMyD88. 果蝇骨髓分化初级反应蛋白质同源物 Tube. 含有死亡域接头蛋白； Pelle. 蛋白激酶 cactus. IκB (NF-κB inhibitor protein 抑制蛋白); Dif. NF-κB 转录因子；DAP-type-PGN. diaminopimelic acid-type peptidoglycan; PGRP-LC. 肽聚糖识别蛋白 LC; IMD: immune deficiency protein; dFADD. 含死亡结构与的 Fas 相关蛋白质 dredd. 半胱氨酸蛋白酶同源物；TAK1. 生长因子 *β* 活化酶 1; TAB2. 生长因子 *β* 活化酶 1 结合蛋白；IKK complex. IKK 复合体 (ird 5, key); Relish. NF-κB 转录因子；Upd. 不对称因子、信号转导和转录激活因子 92E.

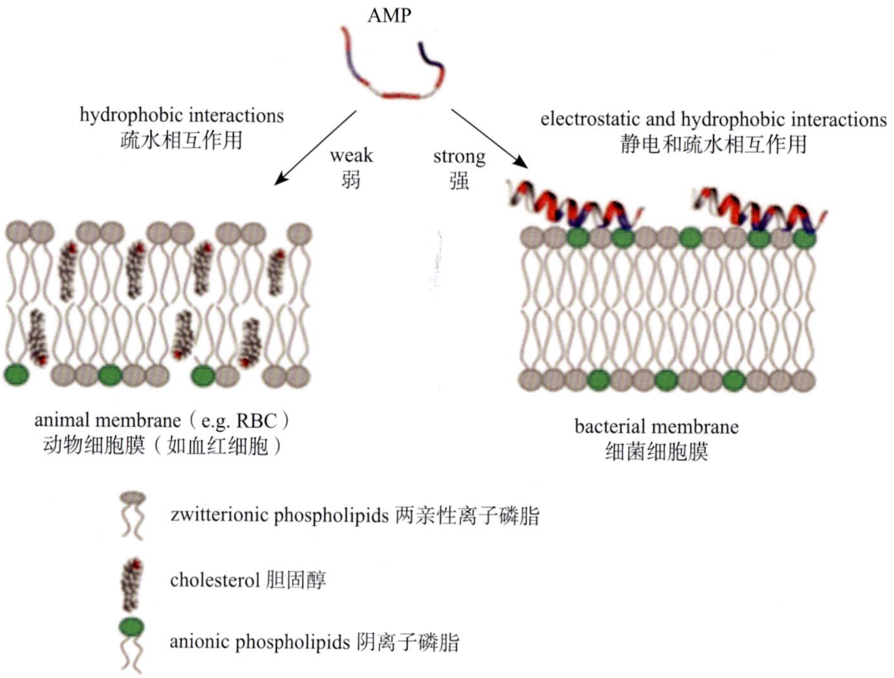

Fig. 7-10 Action mechanism of α helical AMPs and its cytotoxic selectivity (Kumar et al., 2018).
图 7-10 α- 螺旋抗菌肽的杀菌机制及其选择时的细胞毒性（Kumar 等，2018）。

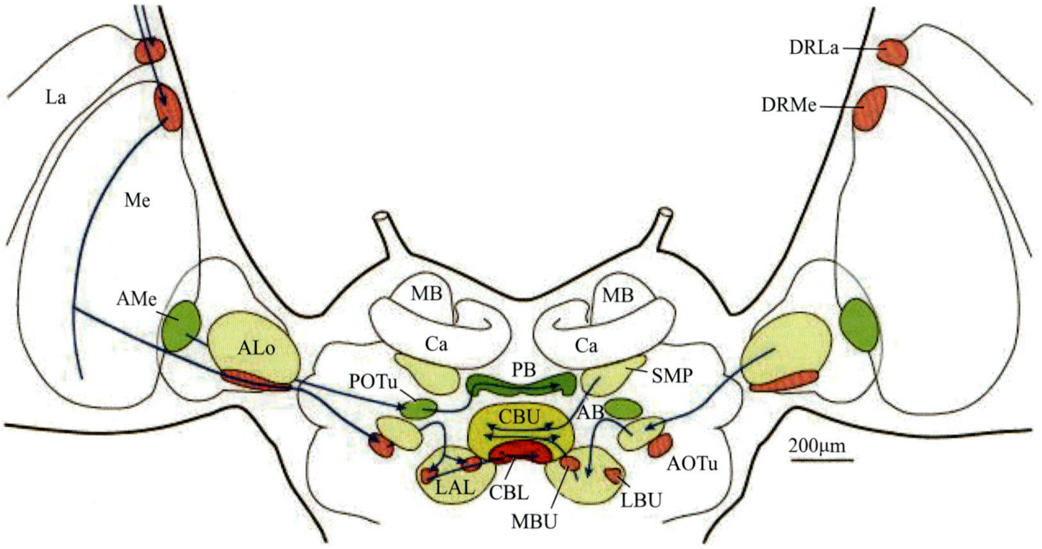

A.illustrating fiber connections of the central complex

A. 显示中心复合体的神经纤维连接

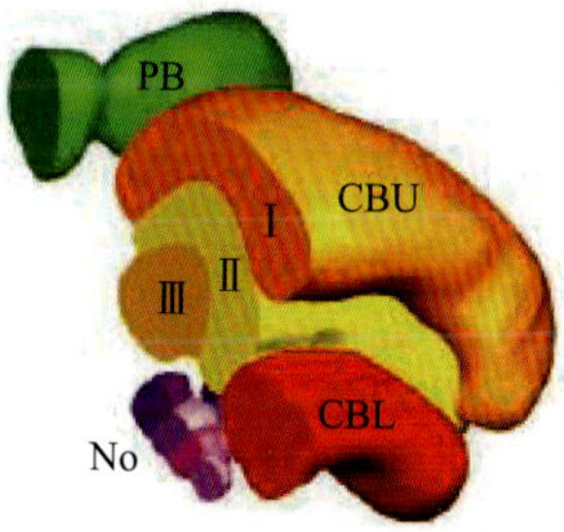

B.Oblique three-dimensional view of the locust central complex

B. 蝗虫中心复合体的倾斜三维视图

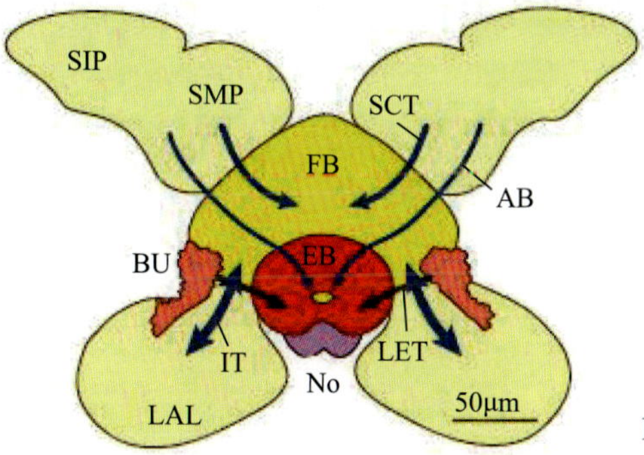

C.Fiber connections of the central body of the blow fly *Lucilia sericata*

C. 丝光绿蝇中心体的神经纤维连接

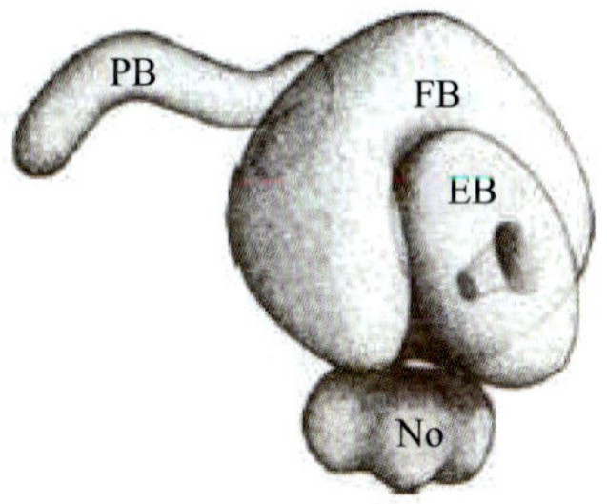

D.Oblique lateral view illustrating the subunits of the central complex of the fruit fly, *Drosophila melanogaster*

D. 侧视图显示果蝇中心复合体的亚单位

Fig. 8-3. Subunits and connections of the insect central complex. A.illustrating fiber connections of the central complex. The anterior polarization pathway connects the dorsal rim area of the lamina and medulla (DRLa, DRMe) *via* the ventral layer of the anterior lobe of the lobula (ALo), the lower unit of the anterior optic tubercle (AOTu), and the medial (MBU) and lateral bulb (LBU) to the lower division of the central body (CBL). A parallel pathway originating in the ALo connects *via* the upper unit of the AOTu and lateral accessory lobe (LAL) to the upper division of the central body (CBU). The superior medial protocerebrum is connected to the CBU *via* the anterior bundles (AB). Finally, projections from the accessory medulla (AMe) extend to the posterior optic tubercle (POTu) and likely target tangential neurons entering the protocerebral bridge (PB). B.Oblique three-dimensional view of the locust central complex. A sagittal cut reveals the positions of the PB, the three layers of the CBU（Ⅰ，Ⅱ，Ⅲ), the CBL, and the four · subunits of the noduli (NO). C.Fiber connections of the central body of the blow fly *Lucilia sericata*. The bulbs (BU) are connected through the lateral ellipsoid tracts (LET) to the ellipsoid body (EB). The isthmus tracts (IT) provide connections between the LALs and the fan-shaped body (FB). The superior intermediate protocerebrum (SIP) and superior medial protocerebrum (SMP) are co1mected to the FB *via* the AB and the superior central-body tract (SCT). D.Oblique latera l view illustrating the subunits of the central complex of the fruit fly, *Drosophila melanogaster* (modified from Von Hadeln et al., 2018).

图 8-3. 昆虫中心复合体的亚单位和连接。A. 显示中心复合体的神经纤维连接。前极化通路通过小叶前叶腹侧层（ALo）、前视突下单元（AOTu）、内侧球（MBU）和外侧球（LBU）连接椎板和髓质的背缘区域（DRLa，DRMe）和中心体下部（CBL）。起源于 ALo 的一条平行通路通过 AOTu 上部单位和外侧副叶（LAL）与中心体上部（CBU）相连。前脑上内侧通过前束（AB）与 CBU 相连。副髓质（AMe）的投射神经元延伸至后视突（POTu），并可能以进入前脑桥（PB）的切向神经元为靶点。B. 蝗虫中心复合体的倾斜三维视图。矢状切面显示了 PB、CBU 的三层（Ⅰ、Ⅱ、Ⅲ）、CDL 和结节（NO）的四个亚单位的位置。C. 丝光绿蝇中心体的神经纤维连接。球茎（BU）通过外侧椭球束（LET）与椭球体（EB）相连。峡部束（IT）在 LAL 和扇形体（FB）之间提供连接。上中间前脑（SIP）和上内侧前脑（SMP）通过 AB 和上中心复合体束（SCT）与 FB 相连。D. 侧视图显示果蝇中心复合体的亚单位（仿 Von Hadeln 等，2018 年）。

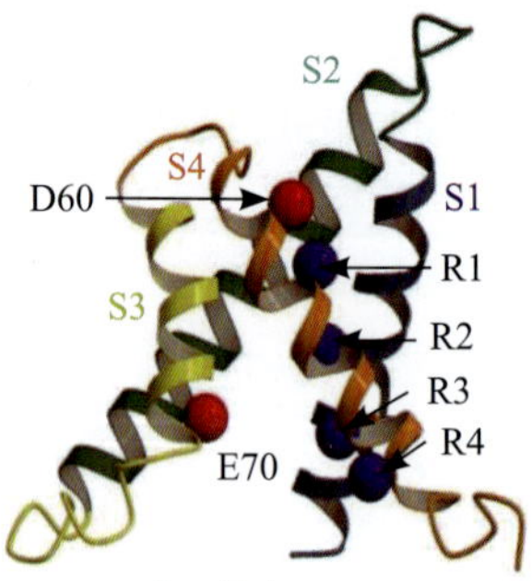

Fig.9-5 Voltage sensing models in both resting and active states. The charged arginine in fragments S1 ～ S4 is R1 ～ R4 in blue, while D60 and E70 are marked in red color. In the resting state, the first arginine gating charge (R1) of the S4 segment is predicted to interact with D60 in the S2 segment. On activation of the voltage sensor, the model predicts that the S4 segment moves outward along a spiral pathway, and the third arginine gating charge (R3) moves into close proximity to D60 in the S2 segment in its activated state. D60 is in an analogous position to negatively charged aspartate or glutamate residues or hydrophilic asparagine residues (modified from DeCaen et al., 2008).

图 9-5 静息和活动状态下的电压感受模型。在 S1 ～ S4 片段中带电精氨酸为 R1 ～ R4，蓝色，D60 和 E70 标记为红色。在静息状态下，预测 S4 片段的第一精氨酸门控电荷（R1）与 S2 片段中的 D60 相互作用。在激活电压感受元件时，该模型预测 S4 片段沿着螺旋路径向外移动，并且第三精氨酸门控电荷（R3）在 S2 片段的激活状态下移动到非常接近 D60。D60 与带负电荷的天冬氨酸或谷氨酸残基或亲水性天冬酰胺残基的位置相似（仿 DeCaen 等，2008）。

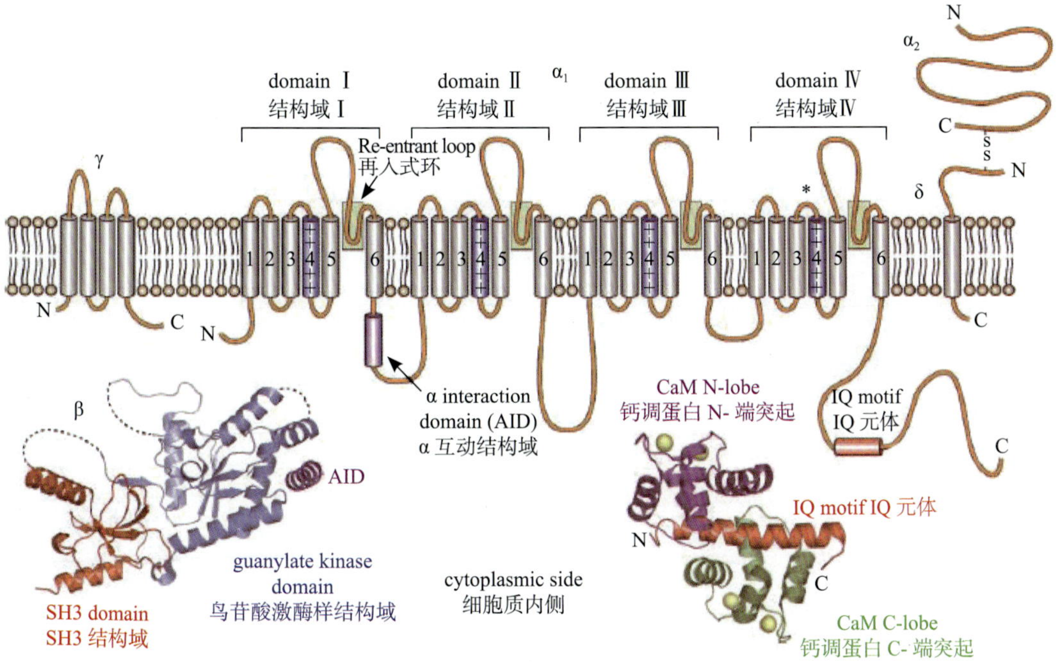

Fig.9-7 Cartoon of the molecular structure and membrane topology of CaV channel subunits. HVA CaV channels usually comprise a single copy of each subunit (α1, α2, δ, β, and γ), whereas LVA CaV channels consist of only the pore-forming α1 subunit. CaM. Calmodulin; the "IQ motif". conserved sequence; AID. α-interaction domain (modified from King, 2007).

图 9-7 电压门控 Ca^{2+}（CaV）通道亚基的分子结构草图和膜的拓扑图。HVA CaV 通道常由单拷贝亚基（α1, α2, δ, β, 和 γ）组成，而 LVA CaV 通道仅由形成通道的 α1 亚基组成。Domain I~IV. 结构域 I~IV；CaM. 钙调蛋白；IQ 元件 . 保守序列；AID. α- 互动结构域（仿 King, 2007）。

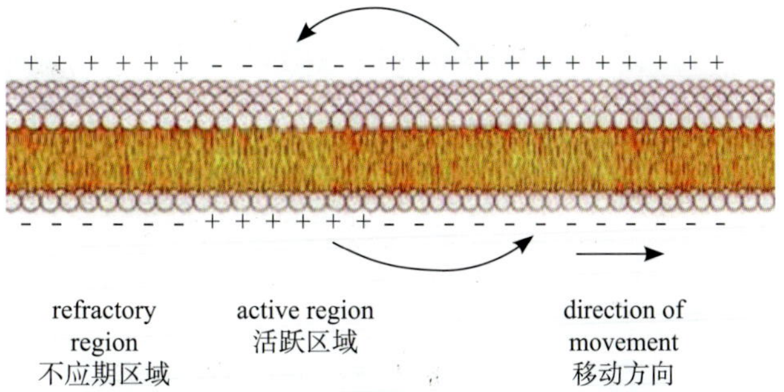

Fig.9-8 Local currents and all illustration of how they function to propagate a spike along the axon. Small currents flowing ahead of the active region open voltage-gated Na^+ channels in the axon. The currents also flow back into the region over which the impulse has just passed, but that region is still partially refractory, and the local currents are not strong enough to open Na^+ channels (modified from Nation, 2022).

图 9-8 局部电流及其如何沿轴突产生峰电位的功能完全图解。流出活跃区域之前的小电流打开轴突电压门控 Na^+ 通道，电流也回流进脉冲刚刚通过的区域，但那个区域仍部分处于不应期，局部电流强度不足以打开 Na^+ 通道（仿 Nation, 2022）。

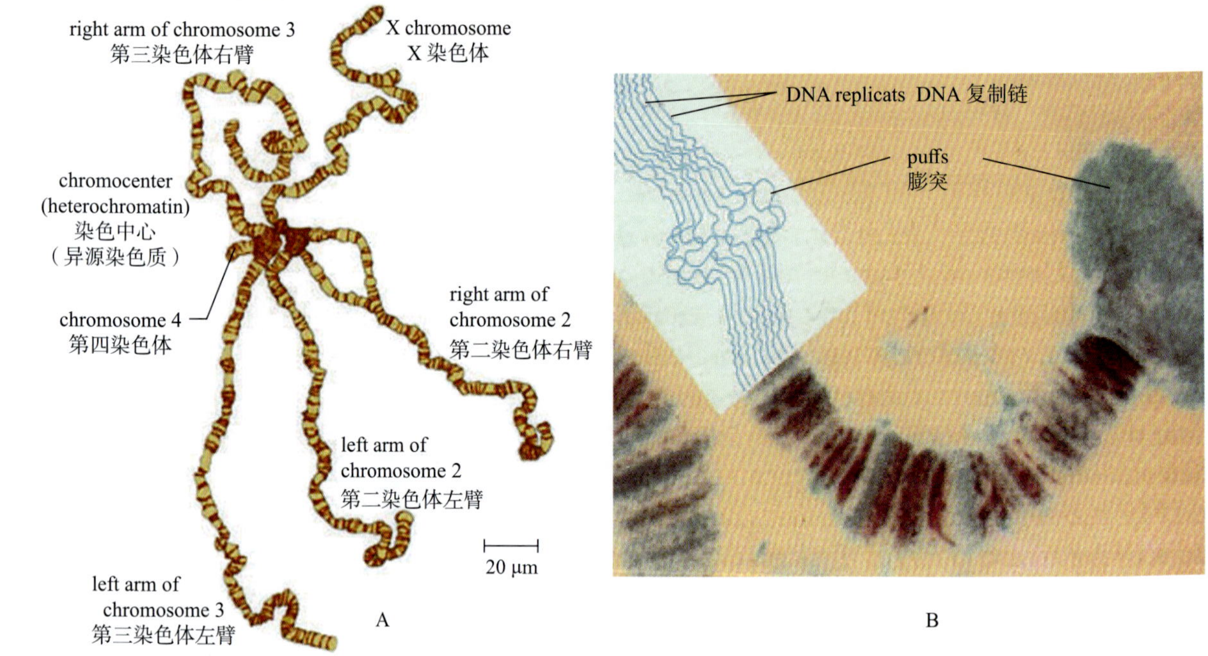

Fig. 10-8 Polytene chromosomes of *D. melanogaster* (modified from other authors). A. The entire set of polytene chromosomes; B. The light micrograph of puffers in polytene chromosome.

图 10-8 果蝇多线染色体（仿其他作者）。A. 果蝇多线染色体全部详情；B. 多线染色体膨突部分的光学显微镜图像。

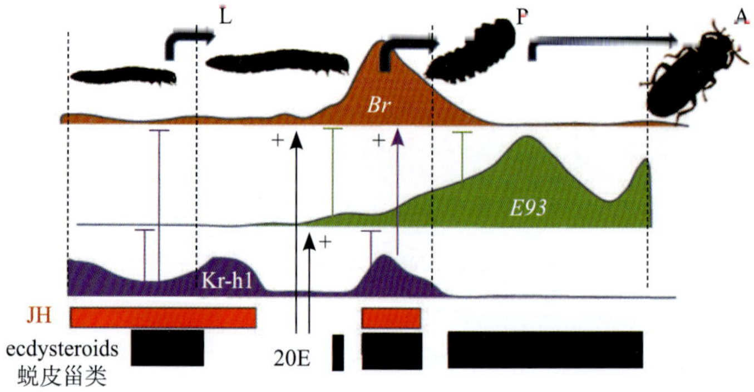

Fig. 10-11 Expression change of *Broad* (*Br*) and *E93* to the normal progression from larva (L) to pupa (P) to adult (A) (modified from Truman et al., 2020).

图 10-11 *Broad* (*Br*) 和 *E93* 在幼虫 - 蛹 - 成虫发育过程的表达变化（仿 Truman et al., 2020）。

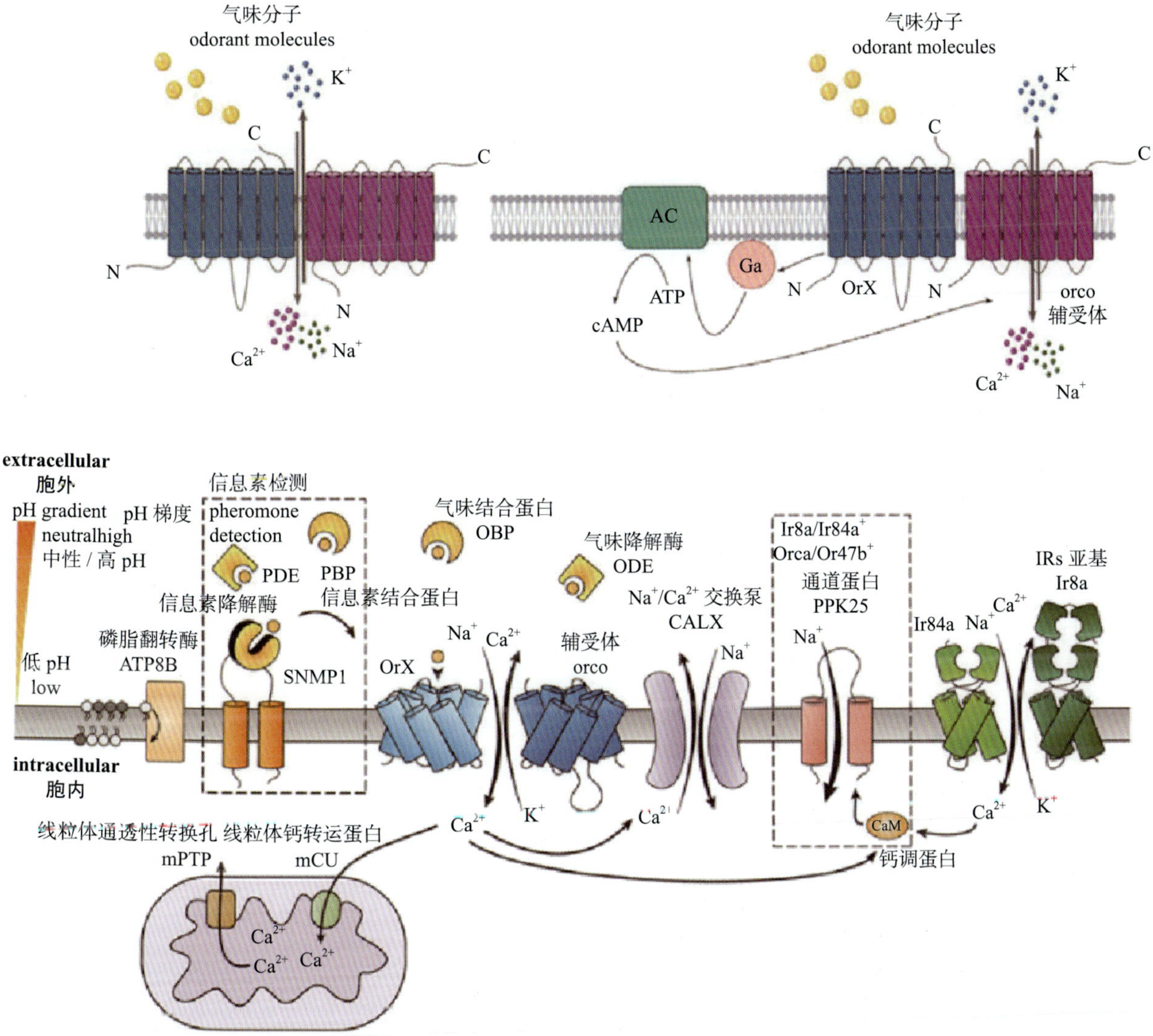

Fig.11-11 Activity modulation of olfactory receptors at the interface between the extracellular sensillum lymph and the cytoplasm (modified from Wicher and Miazzi, 2021). SNMP1. sensory neuron membrane protein 1; DEG/ENaC. degenerin/epithelial sodium.

图 11-11 细胞外感器淋巴和细胞质界面嗅觉受体的活性调节（仿 Wicher 和 Miazzi, 2021）。SNMP1. 感觉神经元膜蛋白 1; DEG/ENaC. 退化蛋白 / 上皮钠。

Fig. 11-13 "Nearby" round dance (modified from Yahya, 2011).

图 11-13 "附近"的圆圈舞（仿 Yahya, 2011）。

Fig.11-14 Tail-wagging long distances dance (modified from Yahya, 2011).

图 11-14 远距离尾部摇摆舞（仿 Yahya, 2011）。

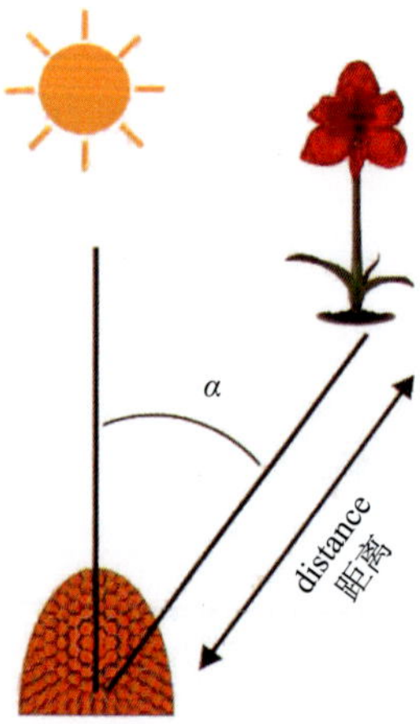

Fig. 11-15 The Tail-Wagging dance tells the other bee information of food distance and direction (modified from Brown, 2011).

图 11-15 工蜂传递给其他蜜蜂关于食物源距离和方向的信息（仿 Brown, 2011）。

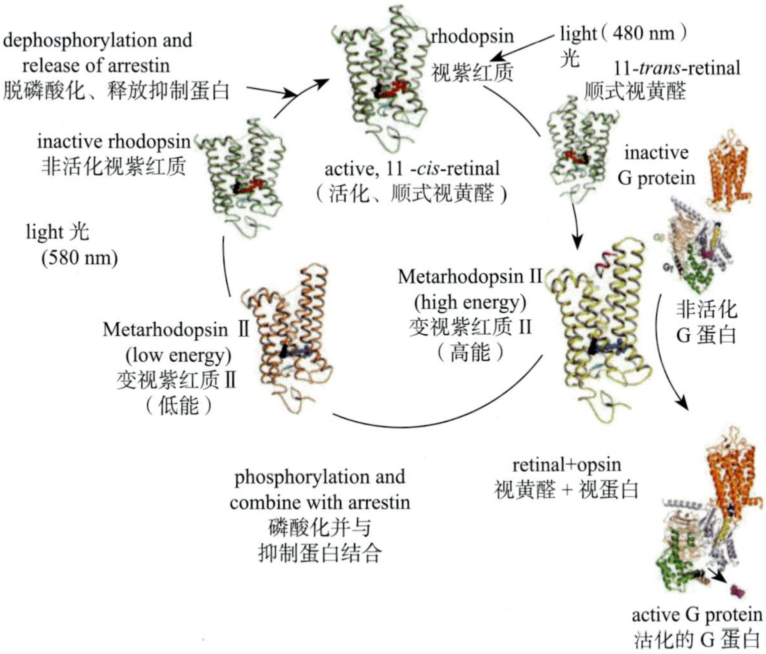

Fig.12-5 Regulation of Visual cascade.

图 12-5 视觉级联的调节。

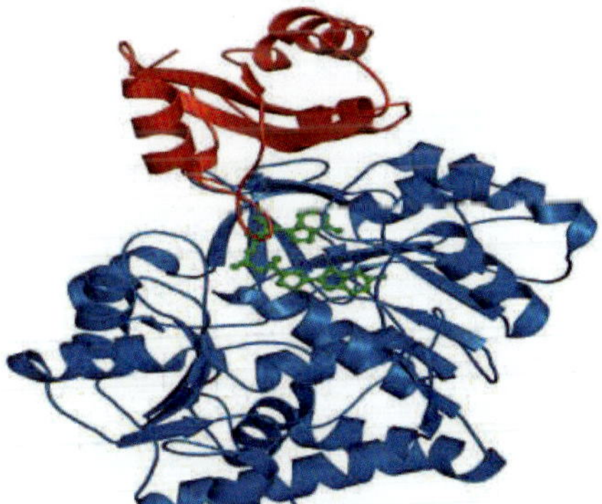

Fig.12-7 Ribbon diagram of the firefly luciferase structure.

图 12-7 荧光素酶的带状结构。

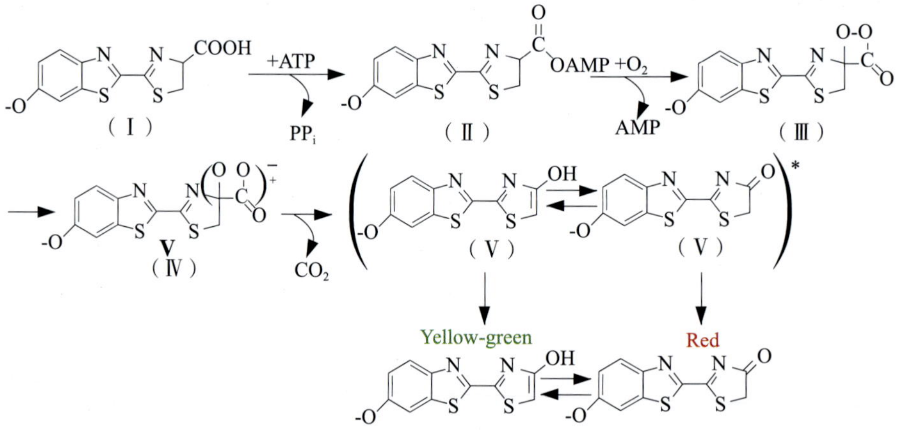

Fig.12-8 Proposed mechanism of the firefly bioluminescence reaction. The carboxylate group of firefly luciferin (I) is activated by reaction with ATP to form the adenylated luciferin (II). A proton is lost, allowing reaction with molecular oxygen to yield (III). Cleavage of the dioxetanone ring (IV) yields the excited state of oxyluciferin (V). (modified from Baldwin, 1996); oxyluciferin (Navizet et al., 2010).

12-8 萤火虫生物发光反应的建议机制。萤火虫荧光素（I）羧基基团与 ATP 反应活化，产生荧光素腺苷酸（II）；失去一个质子，与氧分子反应产生（III）；裂解双氧环杂酮（IV）产生氧化荧光素的激发态（V）（仿 Baldwin, 1996）；氧化荧光素（Navizet 等，2010）。

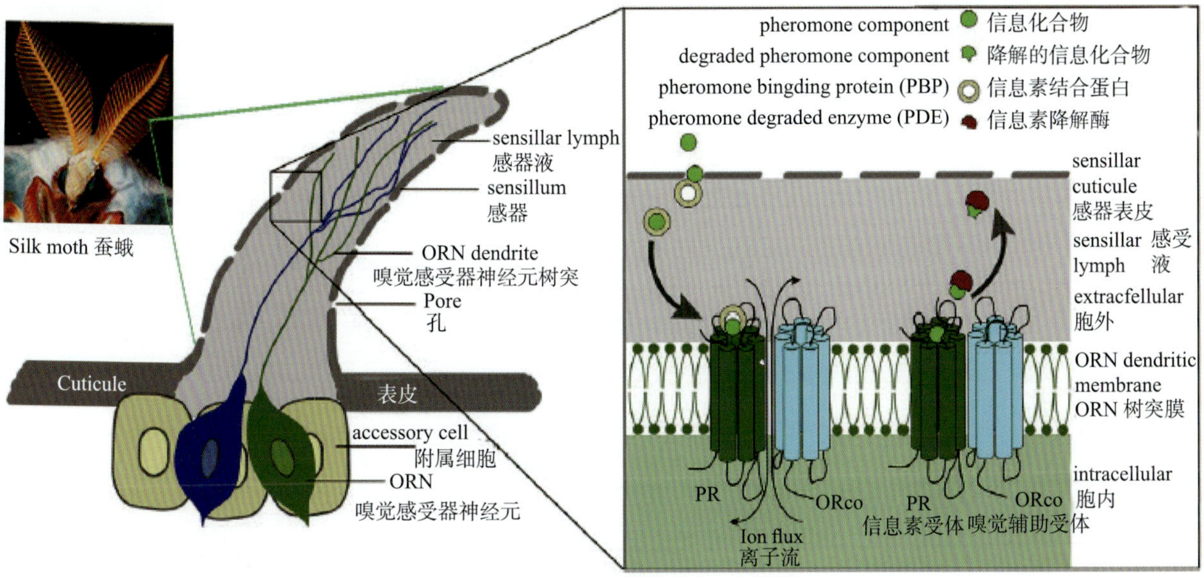

Fig.13-1 The plumose antenna of a male silk moth and the mechanism of receptors. There are thousands of pheromone receptors on the tiny hairs of the antennae (modified from Koutroumpa et al., 2014).

图 13-1 家蚕雄蛾的羽状触角感觉毛及其感受机制。触角小感觉毛上有数以千计的信息素受体（仿 Koutroumpa 等，2014）。

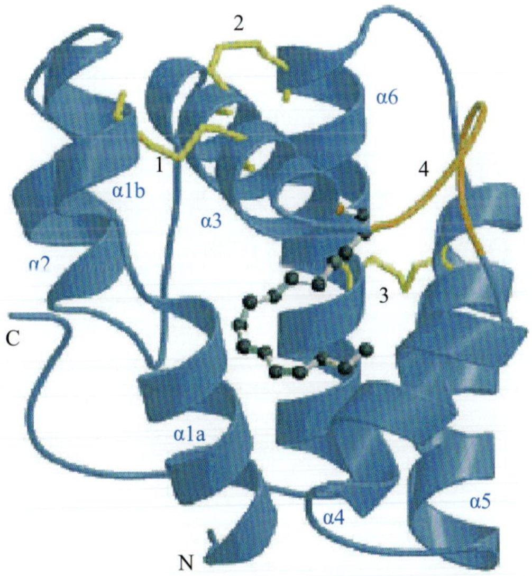

Fig.13-2 Overall structure of the *B. mori* PBP. Disulfide 1-3 bridges are shown between two α-helices, and the loop4 covered the binding pocket. Bombykol is shown in a ball-and-stick representation with double helix (modified from Sandler et al., 2000; Lee et al., 2002).

图 13-2 家蚕 PBP 总体结构。两条螺旋之间有 1～3 双硫键，环肽链 4 覆盖了结合袋区。蚕蛾性诱醇用双螺旋球链表示（仿 Sandler 等，2000; Lee 等，2002）。

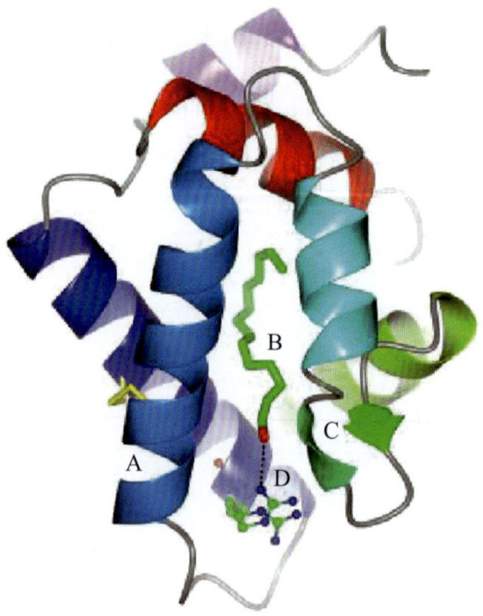

Fig.13-3 The cartoon structure of the silkworm moth GOBPs bound to an analogue of its sex pheromone, shows binding to the arginine amino acid (D) at the entry to the binding pocket (A-C) (modified from Gill, 2009).

图 13-3 蚕蛾 GOBPs 结合到其类似物性信息素的卡通结构，显示在结合袋区（A-C）入口结合到精氨酸（D）（仿 Gill，2009）。

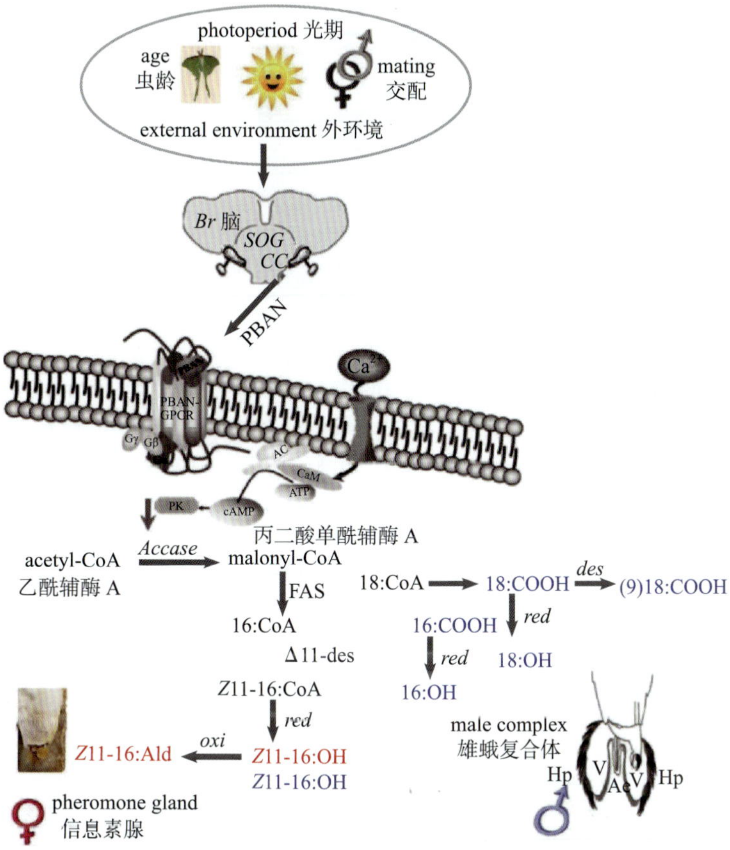

Fig.13-5 A diagrammatic representation of sex pheromone biosynthesis resulting from PBAN release into the hemolymphand its interaction with the PBAN GPCR in female pheromone glands (PG) and male hair pencil- aedaegus-complexes of *H. armigera*. AC. adenylate cyclase; ACCase. acetyl-CoA carboxylase; Br. brain; CaM. calcium-calmodulin; cAMP. cyclic AMP; CC. corpora cardiaca; Δ11-des. Δ11 desaturase; FAS. fatty acids ynthetase; GPCR. G-protein coupled receptor; oxi. oxidase; PBAN. pheromone biosynthesis activating neuropeptide; PK. proteinkinaseA; red. reductase; SOG. subesophageal ganglion. V. valves; HP. hairpencil; Ae. aedaegus(modified from Juernka et al., 2011).

图 13-5 棉铃虫释放 PBAN 进入血液并与雌性信息腺（PG）和雄性笔状阳茎复合体的 PBAN GPCR 相互作用，引起信息素生物合成的图解描述。AC. 腺苷酸环化酶；ACCase. 乙酰辅酶 A 羧化酶；Br. 脑；CaM. 钙 - 钙调蛋白；cAMP. 环腺苷酸；CC. 心侧体；Δ11-des. Δ11 去饱和酶；FAS. 脂肪酸合成酶；GPCR. G- 蛋白耦联受体；Oxi. 氧化酶；PBAN. 信息素生物合成激发神经肽；PK. 蛋白激酶 A；red. 还原酶；SOG. 咽下神经节；V. 瓣膜；HP. 毛鬃；Ae. 阳茎（仿 Jurenka 等，2011）。

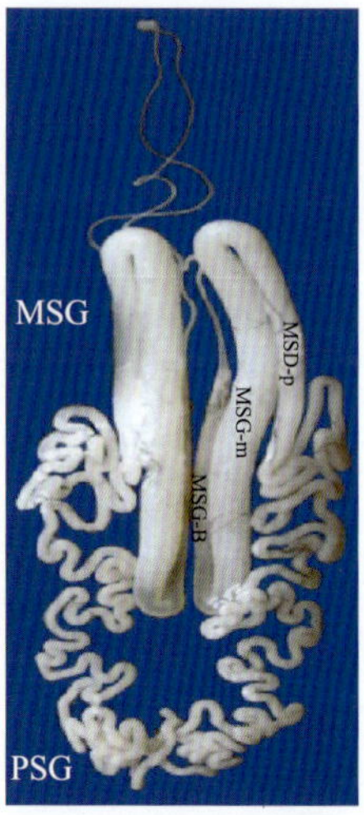

Fig.14-1 Appearance structure of silk gland.

图 14-1 丝腺的外观结构。**fusulus.** 吐丝器；**ASG (anterior silk gland).** 前部丝腺；**MSG (middle silk gland).** 中部丝腺；**PSG (posterior silk gland) .** 后部丝腺；**MSG-a.** 中部丝腺前部；**MSG-m.** 中部丝腺中部；**MSG-p.** 中部丝腺后部。

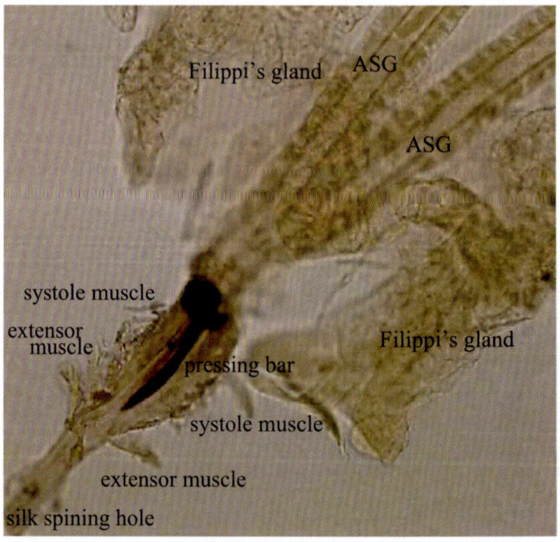

Fig.14-2 Structure of fusulus.

图 14-2 吐丝器的结构。**silk spinning hole.** 吐丝孔；**extensor muscle.** 舒展肌；**systole muscle.** 压缩肌；**pressing bar.** 压杆；**Filippi's gland.** 菲氏腺；**ASG.** 前部丝腺。

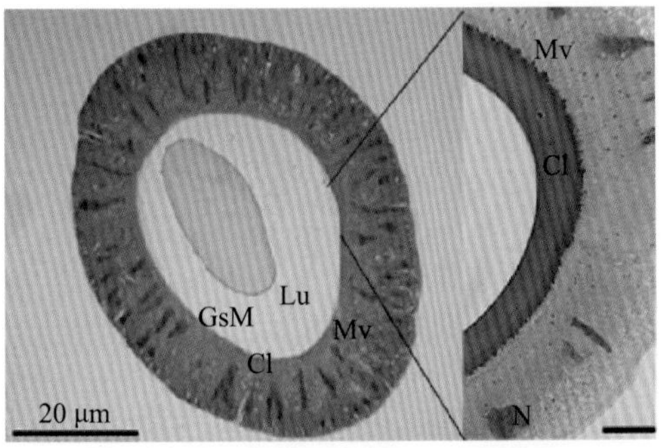

Fig.14-3 Cross sectional structure of silk gland. BsM. basal membrane; CI. cuticular Intima; Lu. lumen; Mv. microvilli; N. nucleus.

图 14-3 丝腺的横截结构。**BsM.** 底膜；**CI.** 表皮质内膜；**Lu.** 丝腺腔；**Mv.** 微绒毛；**N.** 腺细胞核。

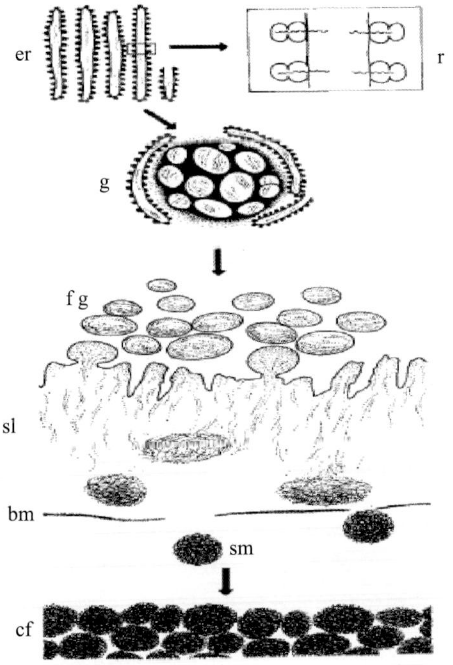

Fig.14-4 Schematic drawing of the process of fibroin secretion in posterior division. bm. Boundary membrane of cuticular intima between the silk layer and the lumen; cf. columnar fibroin accumulated with spherical masses of fibroin fiber in the lumen; er. rough endoplasmic reticulum; fg. fibroin globules; g. Golgi complex; r. ribosome; sl. silk layer; sm. spherical masses of fibroin fiber (Akai, 1983).

图 14-4 后区丝心蛋白分泌过程。**bm.** 丝层和管腔之间的表皮内膜边界膜；**cf.** 柱状丝心蛋白在管腔内堆积成球状丝心蛋白纤维团；**er.** 粗面内质网；**fg.** 纤维蛋白球；**G.** 高尔基复合体；**r.** 核糖体；**sl.** 丝层；**sm.** 丝心蛋白纤维的球形团块（**Akai, 1983**）。

```
B.mori          MKPIFLVLLVATSAYAAPSVTINQY-SDNEIPRDIDDGK--ASSVISRRWDYVDDTDKSIAILNVQEILKDMASQGDYAS   77
B. mandarina    MKPIFLVLLVATSAYAAPSVTINQY-SDNEIPRDIDDGK--ASSVISRAWDYVDDTDKSIAILNVQEILKDMASQGDYAS   77
M.sexta         MKPIVLVLFVAAGAFAAPSVVLNEY-SENELPR--------SNSLTERAFGLIDGGDVNIHILNLQAVLKDL-SERDPRS  70
G. mellonella   MLPFVLVLLVATSALAAPSVVISQD-NINNIAPRVGNGRpISSALIDRAFEIVDGGDTNIYILTIQQILNDLADQPDGLS  79
V. tameamea     MLPIVLVLFVAQSAFAAPSVVGVQFkNFNEVAPVSDNGKlVSSYITDGSIAYIDGGDTPLYAQMIIQSINDMADTSDTDS  80
H. occidentalis --MALSLLIGALLAIQGASFVASSQ---------------ISASLLEETWNLVDQGEVEPFTLLLKKEV--VATGGVY--  59

B.mori          QASAVAQTAGIIAHLSAGIPGDACAAANVINSYTDGVRSGNFAGFRQSLGP-FFGHVGQNLNLINQLVINPGQLRYSVGP  156
B. mandarina    QASAVAQTAGIIAHLSAGIPGDACAAANVINSYTDGVRSGNFAGFRQSLGP-FFGHVGQNLNLINQLVINPGQLRYSVGP  156
M.sexta         QALAFGQTLGVLVDLATGIPGDSCASAALINAYAAGVKSGNFSGFRSVLYN-FVRQLASNLKLIVKLASNPASVRFSTGP  149
G. mellonella   QSLAVTQAVAALGELATGVPGNSCEAAAVIDAYANSVRTGDNSALSIAVAN-YINRLSSNIGLISQLASNPDSLRYSSGP  158
V. tameamea     WANAVIQTIATVGELANGVPGDACEAAALINAYAYSVSTGNSAGLRAALPN-FVQRIAGYVDTIVQLVVNPNGVRYSSGP  159
H. occidentalis ---GLGATLTGVGELAWPRPASGCGHSKLINANVALNDGTLAWGELEDAVDsYAVVLAQAVDNLRILG--LSCIIPAPWP  134

B.mori          AL--GCAGGGRIYDFEAAWDAILASSD----SSFLNEEYCIVKRLYNSRNSQSNNIAAYITAHLLPPVAQVFHQSAGSI--  229
B. mandarina    AL--GCAGGGRIYDFEAAWDAILASSD----SGFLNEEYCIVKRLYNSRNSQSNNIAAYITAHLLPPVAQVFHQSAGSI--  229
M.sexta         SG--GCLGSGRSYQFEAAWDAILSSSS----SSLVHEQYCTAKRLYYAFNSRSNNVGATITALSLPPVAQVLEAAGEST--  222
G. mellonella   AG--NCAGGGRSYQFEAAWDAVLNNANpyQIGLINEEYCAARRLYNAFNSRSNNVGAAITAGAVVAQTQAAQIILPSL--  234
V. tameamea     RG--KCLGGGRSYQFEEAWDGILDKANayQIGLLNEEYCAAKRLYNAFNIRSNNVAAGVTAVTTRPVNQLVQLALPQLaq  237
H. occidentalis TLenSCGDWGRIYDFENSWD-----------LSNVNNGVVCAARRLYTAFGARANNVGAAATSAATDAAITIISDVEDEL--  203

B.mori          TDLLRGVGNGNDATGLVANAQRYIAQAASQVHV     262
B. mandarina    TDLLRGVGNGNDATGLVANAQRYIAQAASQVHV     262
M.sexta         SNLLRAVAGGYDASAAAASANKALLRAVERIQL     255
G. mellonella   VNVLSAVAAGGNVAGAAAQAGQALANAAANVQL     267
V. tameamea     VSFLRVVANGGNPAQAGANAKNALLQSLPKIQY     270
H. occidentalis VSYLEAVLSKSAGPGCKSKQKQLRTLAGSLKAS     253
```

Fig. 14-5 Primary structure of silk L-fibroin of *Bombyx mori, Bombyx mandarina, Manduca sexta, Vanessa tameamea* and the *caddisflies Hesperophylax occidentalis* (Lowercase letters represent nonhomologous amino acid regions）.

图 14-5 家蚕、野蚕、烟草天蛾、夏威夷红蛱蝶和石蛾的丝心蛋白轻链的一级结构比较（小写字母表示非同源氨基酸区域）。

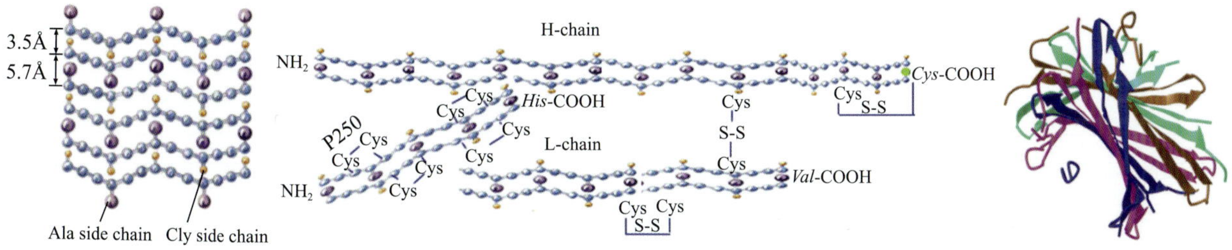

Fig.14-6 Secondary and tertiary structure of fibroin (modified from Tanaka et al., 1999; He et al., 2012).
图 14-6 丝心蛋白的二级和三级结构 (仿 Tanaka 等，1999 ； He 等，2012)。

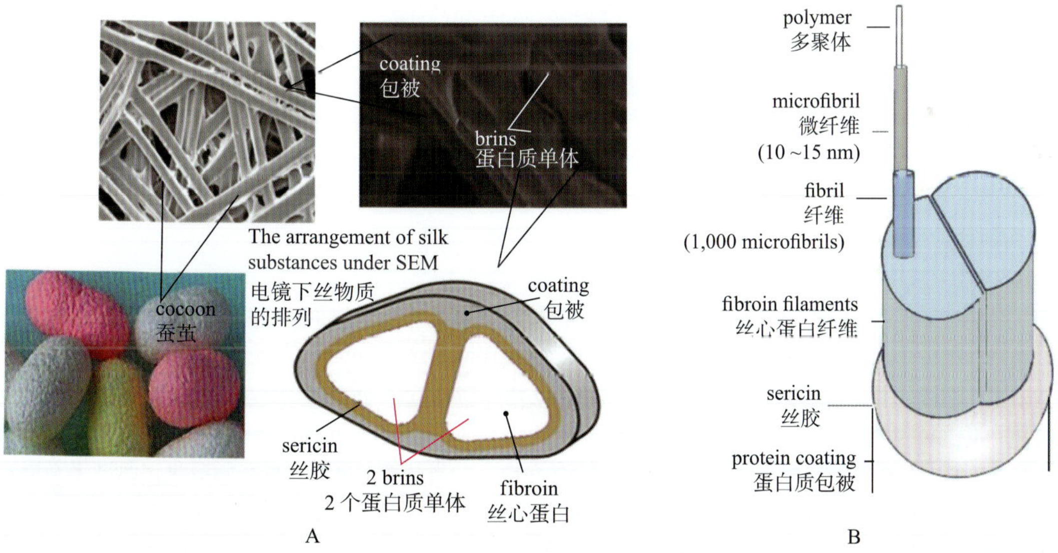

Fig.14-7 Structure of silk thread from macro to micro view. A. The cocoon is composed of silk thread, which consists of coated protein and monofilament. Each monofilament is composed of sericin and fibroin; B. Cross sectional view of monofilament.
图 14-7 丝束宏观到微观结构。A. 蚕茧由丝束组成，丝束有蛋白质包被和单纤丝，每根单纤丝由丝胶和丝心蛋白组成；B. 单纤丝横截面图。

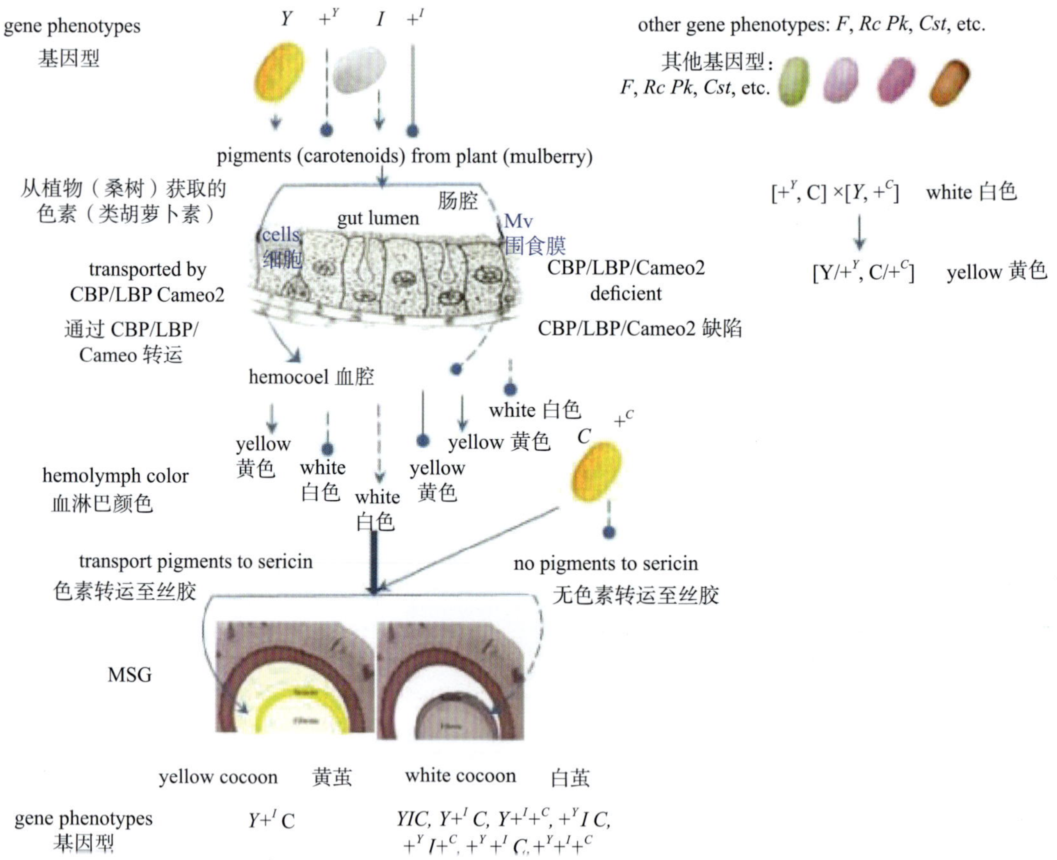

Fig.14-8 Mechanism of cocoon color formation. Silkworm and other insects uptake pigments from mulberry trees or other plants, such as carotenoids, which are absorbed in midgut, and transported into the hemocoel through CBP encoded by *Y* gene to color the silkworm hemolymph, which circulates to the silk gland and infiltrates into the silk gland cells. The pigments accumulated in MSG and incorporated into sericin layer, resulting in color cocoons.

图 14-8 茧色形成机制。家蚕等昆虫取食桑树或其他植物来的色素，如类胡萝卜素等在中肠被吸收，通过 *Y* 基因编码的类胡萝卜素结合蛋白 CBP 转运进入血腔，使蚕血液着色；血液循环至丝腺并渗入丝腺细胞，在中部丝腺累积的色素进入丝胶层，从而产生色茧。

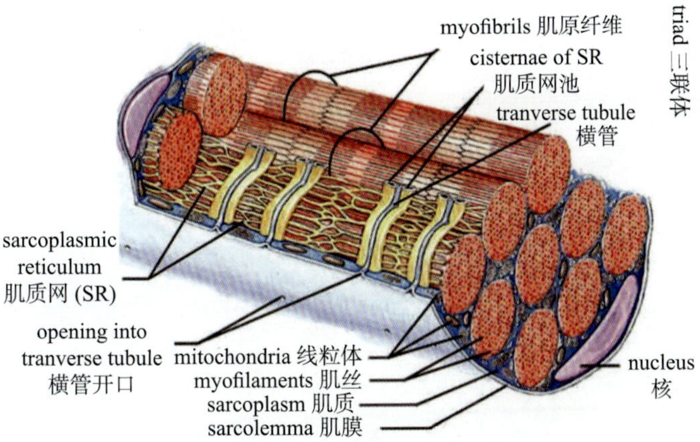

Fig.15-2 Diagrammatic illustration of the sarcoplasmic reticulum (SR) and transverse (T) tubules. Sarcolemma and mitochondria-abundant sarcoplasm enveloped myofibrils.

There are openings into transverse tubules on the surface of the sarcolemma. The linkage of T-tubule intersecting with two SR vesicles (or cisternae) are called triad junction (modified from Mic-Graw Hill Company).

图 15-2 肌质网（SR）和横管（T）概要图解。肌原纤维外围是肌膜和富含大型线粒体的肌质。横管在肌膜表面有开口，与两个肌质网池相连时称为三联体连接（仿 Mic-Graw Hill 公司）。

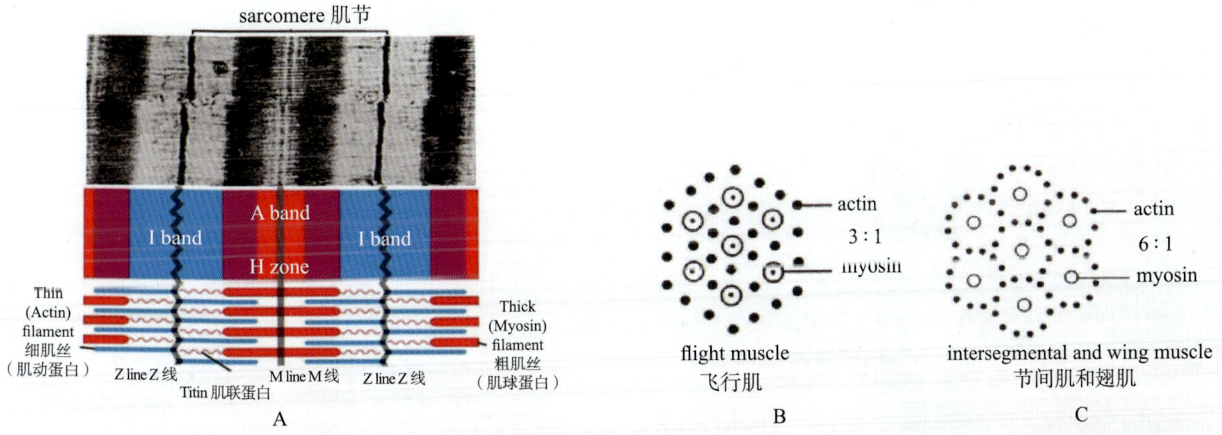

Fig.15-3 Electron microscopic division of sarcomere units and the arrangement of actin and myosin filaments in different muscles. A. Composition of sarcomere; B. In the flight muscle of most insects, producing a 3∶1 ratio of thin to thick filaments; C. An arrangement in which 12 actin filaments are arranged between myosin filaments to give a 6∶1 ratio. (modified from Toselli et al., 1968).

图 15-3 肌节单位的电镜学划分和不同肌肉中的肌动蛋白和肌球蛋白丝的排列。A. 肌节的组成；B. 飞行肌细肌丝和粗肌丝的比例为 3∶1；C. 肌球蛋白丝间 12 根肌动蛋白丝的排列产生 6∶1 的比例（仿 Toselli 等，1968）。

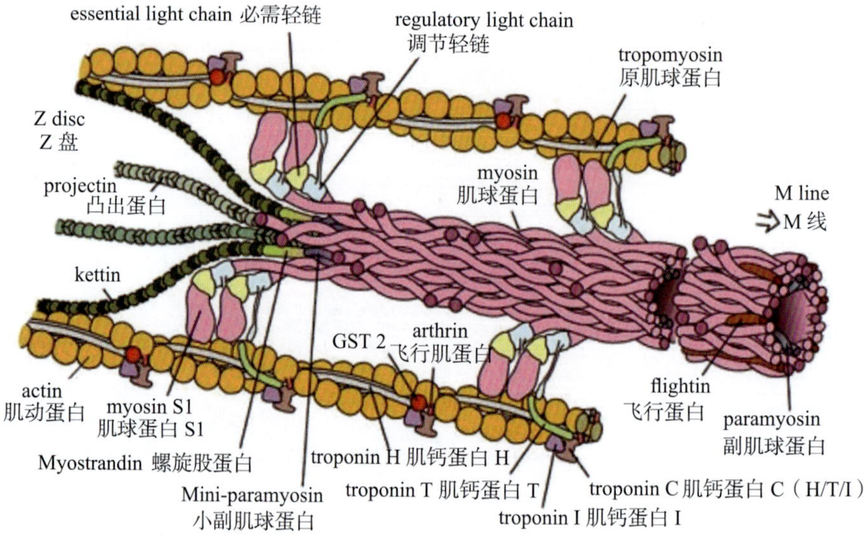

Fig.15-7 The position of varies proteins on muscle thin and thick filament in adult indirect flight muscle (modified from Vigreaux, 2005) .

图 15-7 成虫间接飞行肌各种蛋白质在细肌丝和粗肌纤丝上的位置（仿 Vigreaux, 2005）。

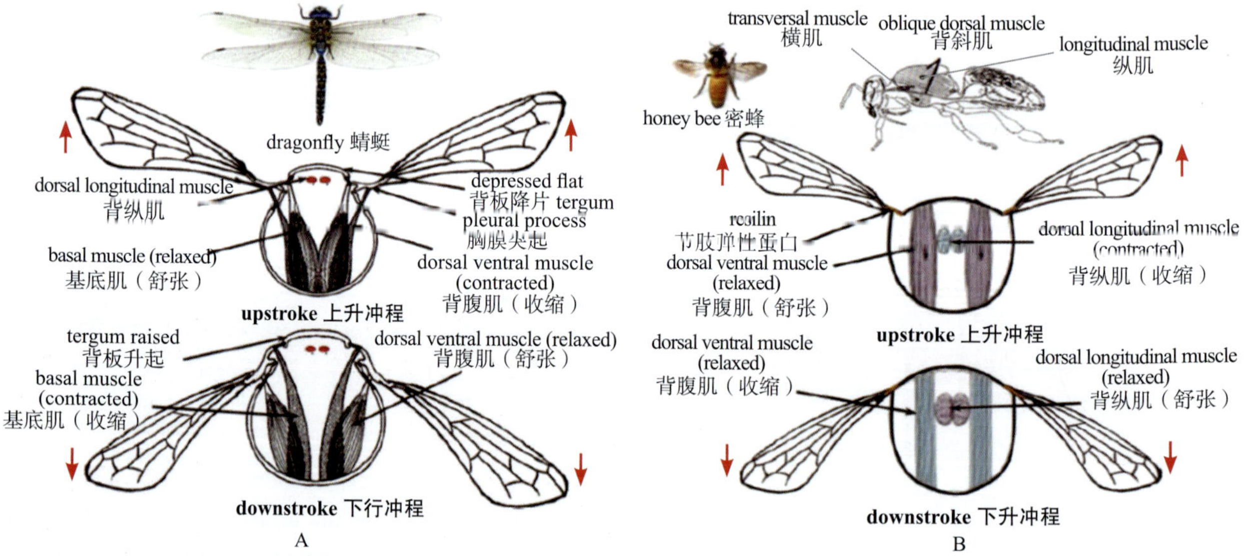

Fig.15-8 The action model of insect direct and indirect muscles during flight. A. The direct muscle of dragonfly; B. Indirect muscle function of bee (modified from Chari et al., 2021).

图 15-8 昆虫间接肌和直接肌在飞行中的作用模式。A. 蜻蜓直接肌的作用；B. 蜜蜂的间接肌的作用（仿 Chari 等，2021）。

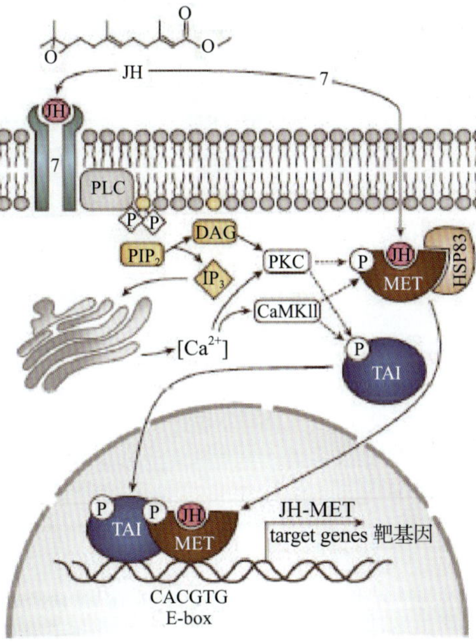

molecular mechanism and signal pathway of JH action
JH 作用的分子机制和信号途径

Fig.16-3 The JH receptor Methoprene tolerant (Met) and its partner Taiman (Tai) are central to JH signaling. In the *Aedes aegypti* mosquito, an additional membrane receptor, working through inositol triphosphate (IP_3), diacylglycerol (DAG), and intracellular calcium ([Ca^{2+}]), activates protein kinase C (PKC) and calcium/calmodulin-dependent protein kinase II (CaMKII), leading to the phosphorylation of Met and Tai. The JH receptor complex binds to E-box elements in the promoter of JH-Met target genes, inducing their expression (modified from Roy et al., 2018).

图 16-3 JH 受体 Methoprene tolerant（Met）及其搭档蛋白 Taiman（Tai）是 JH 信号传导的核心。在埃及伊蚊中，一个额外的膜受体通过三磷酸肌醇（IP_3）、二酰甘油（DAG）和细胞内钙 [（Ca^{2+}）] 发挥作用，激活蛋白激酶 C（PKC）和钙/钙调蛋白依赖性蛋白激酶 II（CaMKII），导致 Met 和 Tai 的磷酸化。JH 受体复合物与 JH-Met 靶基因启动子中的 E-box 元件结合，诱导其表达（仿 Roy 等，2018）。

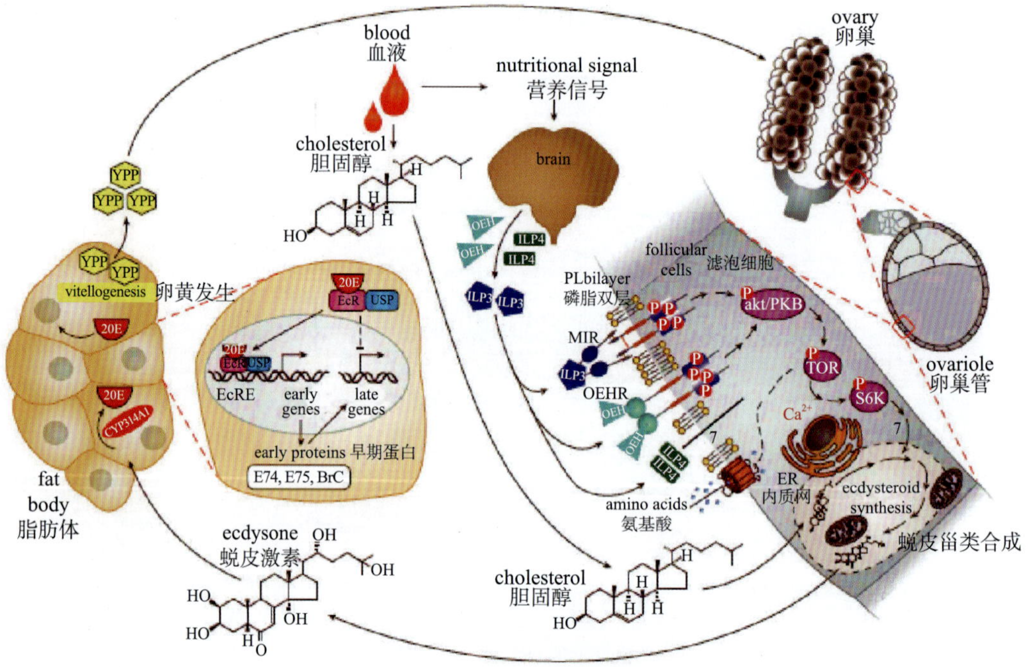

Fig. 16-4 Blood meal–activated nutritional signals trigger the release of the neuropeptide OEH and ILP3 and ILP4 from the brain of *Aedes aegypti* females. OEH binds OEHR; ILP3 binds MIR; and ILP4 binds to an unknown receptor. Phosphorylated OEHR and MIR, along with amino acid signaling, activate insulin and AA/TOR signaling pathways, manifested by phosphorylation of Akt (PKB), TOR, and S6 kinases. These pathways convert cholesterol to ecdysteroids through a multistep process in the ER and mitochondria of the follicle cells. The role of Ca2+ flux in ecdysteroidogenesis is important. Ecdysone secreted into the hemolymph is converted by the cytochrome P450 enzyme CYP314A1 to its active form of 20E in the peripheral tissues. Once released into the hemolymph, 20E binds the heteromeric receptor complex of EcR and USP. Receptor-bound 20E activates the transcription of the early genes, *BrC*, *E74*, and *E75*. The early proteins are involved in the transcriptional regulation of genes encoding for YPPs, such as Vg, in the fat body. The secreted YPPs are subsequently taken up by oocytes during maturation (modified from Roy et al., 2018).

图 16-4 雌性埃及伊蚊血粉激活的营养信号触发大脑释放神经肽 OEH、ILP3 和 ILP4。OEH 结合 OEHR；ILP3 结合 MIR；ILP4 与未知受体结合。磷酸化 OEHR 和 MIR 与氨基酸信号一起激活胰岛素和 AA/TOR 信号通路，表现为 Akt（PKB）、TOR 和 S6 激酶的磷酸化。这些途径通过卵泡细胞内质网和线粒体的多步骤过程将胆固醇转化为蜕皮甾类。Ca2+ 流在蜕皮甾类生成中的作用很重要。分泌到血淋巴中的蜕皮激素被细胞色素 P450 酶 CYP314A1 转化为外周组织中的活性形式 20E。一旦释放到血淋巴，20E 就会结合 EcR 和 USP 的异二聚体受体复合物。受体结合的 20E 激活早期基因 *BrC*、*E74* 和 *E75* 的转录。早期蛋白参与脂肪体中 YPP 编码基因的转录调控，如 Vg。分泌的 YPP 随后在成熟过程中被卵母细胞吸收（仿 Roy 等，2018）。

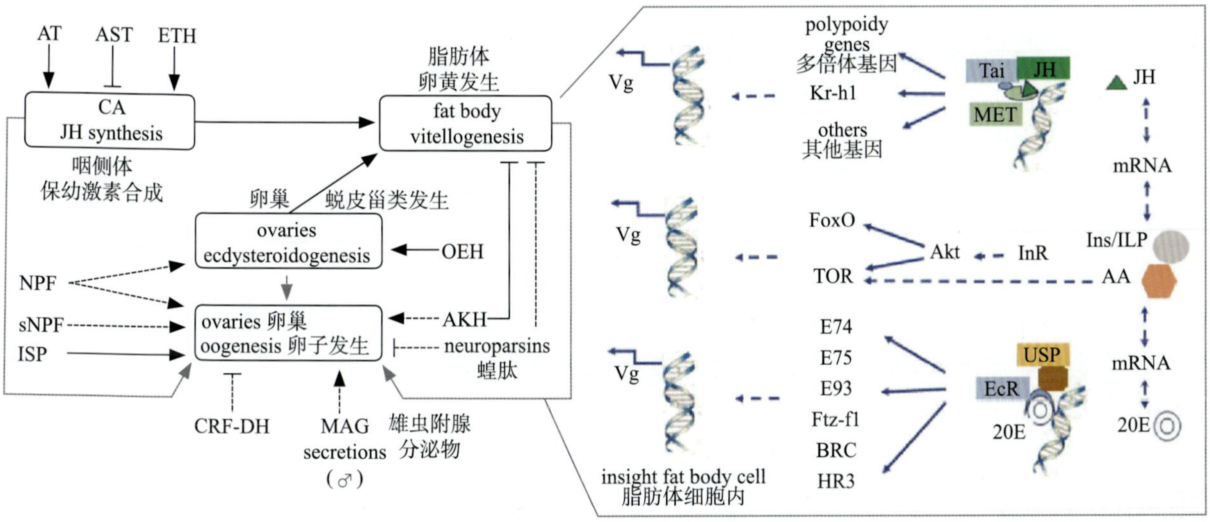

Fig.16-5 Peptide control of insect oogenesis and Vg synthesis in fat bdy. JH receptor complex Met/Tai directly regulates the transcription of target genes encoding Kr-h1 and the polyploidy genes (including DNA replication and cell cycle genes). 20E receptor complex EcR/USP directly activates the expression of 20E-responsive genes.Nutritional signalling is involved in Vg production in the fat body by interplay with JH or 20E. In addition, miRNA participates in JH- or 20E-dominant Vg synthesis in the fat body. These factors directly or indirectly regulate Vg transcription and vitellogenesis. Abbreviations: AT. allatotropin; AST. allatostatin; ETH. ecdysis triggering hormone; JH. juvenile hormone; NPF. neuropeptide F; sNPF. short neuropeptide F; ISP. insulin signalling pathway; CRF-DH. corticotropin-releasing factorlike diuretic hormone; MAG. male accessory gland; OEH. ovary ecdysteroidogenic hormone; AKH. adipokinetic hormone. (modified from Lenaerts, et al., 2019; Wu, et al., 2021).

图 16-5 昆虫卵发生的肽类物质控制和脂肪体内 Vg 的合成。JH 受体复合物 Met/Tai 直接调节编码 Kr-h1 的靶基因和多倍体基因（包括 DNA 复制和细胞周期基因）的转录。20E 受体复合物 EcR/USP 直接激活 20E 应答基因的表达。营养信号通过与 JH 或 20E 相互作用参与脂肪体中 Vg 的生成。此外，miRNA 参与脂肪体中 JH- 或 20E 显性 Vg 的合成。这些因子直接或间接调节 Vg 转录和卵黄发生。AT. 咽侧体促进素；AST. 咽侧体抑制素；ETH. 蜕皮触发激素；JH. 保幼激素；NPF. 神经肽 F；sNPF. 短神经肽 F；ISP. 胰岛素信号通路；CRF-DH. 促肾上腺皮质激素释放因子样—利尿激素；MAG. 雄性附腺；OEH. 卵巢蜕皮甾类生成激素；AKH. 脂动激素（仿 Lenaerts 等，2019；Wu 等，2021）。

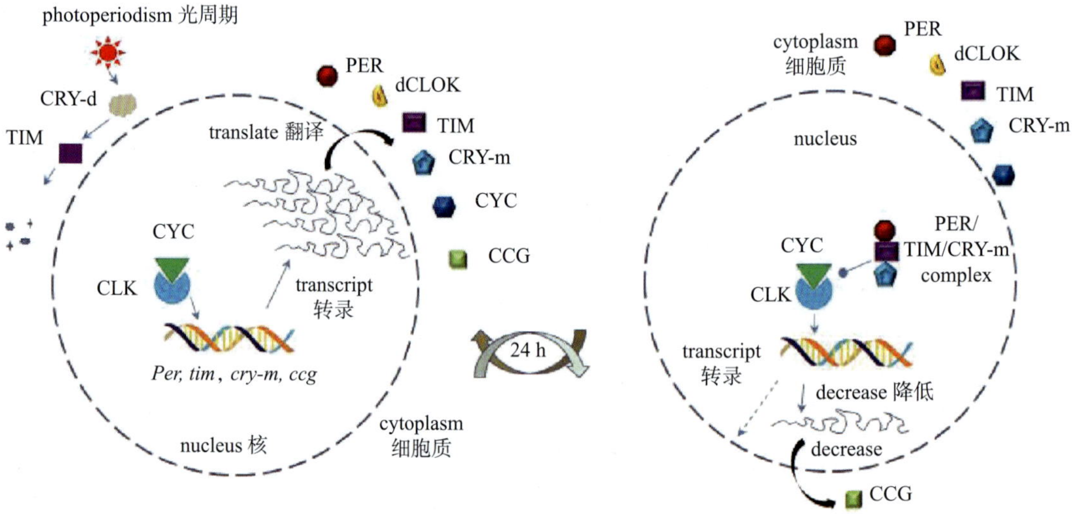

Fig.17-1 Molecular mechanisms of the circadian clock. Positive regulate elements of CYC and CLK form a heterodimer and activate transcription of clock genes *per*, *tim*, and *cry-m* and many other clock-controlled genes (*ccg*). Their protein products are synthesized in the cytoplasm and PER, TIM and CRY-m form a complex. The complex enters the nucleus and acts as a negative element that represses CYC/CLK transcriptional activity. Reduction of *ccg* transcript levels and consequent reduction of CCG protein levels lead to a decrease of repressive regulation of CYC/CLK by PER/TIM/CRY-m, and therefore CYC/CLK-mediated transcription increases again. These phases in which ccg transcription is activated or repressed are repeated in approximately 24 h. CRY-d causes degradation of TIM in a light dependent manner (modified from Numata et al., 2015).

图 17-1 生物钟的分子机制。正向调控成分 CYC 和 CLK 形成异二聚体并激活时钟基因 *per*、*tim* 和 *cry-m*，以及许多其他时钟控制基因（ccg）的转录。它们的蛋白质产物在细胞质中合成后，PER、TIM 和 CRY-m 形成复合物。该复合物进入细胞核，起反向作用抑制 CYC/CLK 的转录活性。ccg 转录水平下降，接着蛋白水平降低导致 PER/TIM/CRY-m 对 CYC/CLK 的抑制性调节减少，因此 CYC/CLK 介导的转录再次增加。ccg 转录激活或抑制的这些阶段在大约 24 h 内重复。CRY-d 以光依赖的方式导致 TIM 降解（仿 Numata 等，2015）。

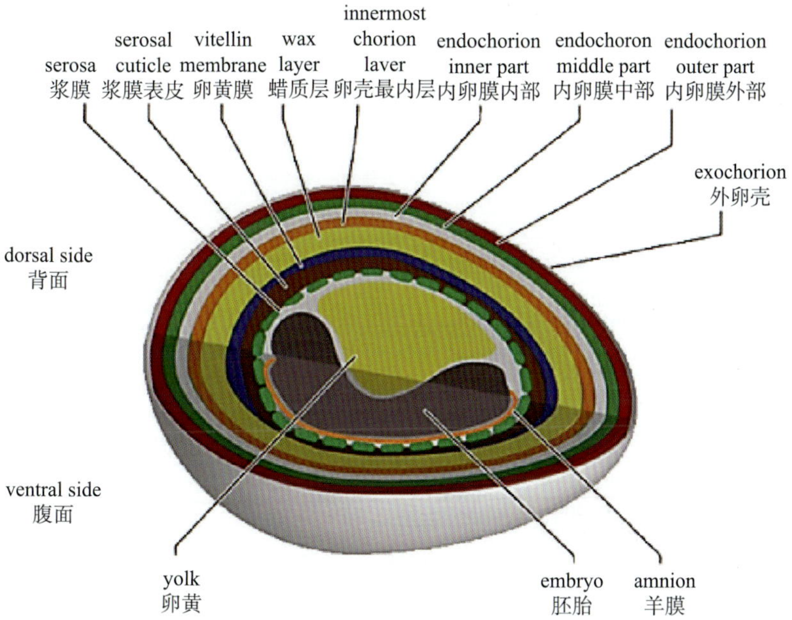

Fig.18-1 A schematic presentation of the radial complexity of a egg layers in holometabolous insects. Embryo position related to the yolk varies in hemimetabolous insects (modified from Rezende et al., 2016).

图 18-1 完全变态昆虫卵层径向复杂性。不完全变态昆虫中与卵黄相关的胚胎位置不同（仿 **Rezende** 等 **, 2016**）。

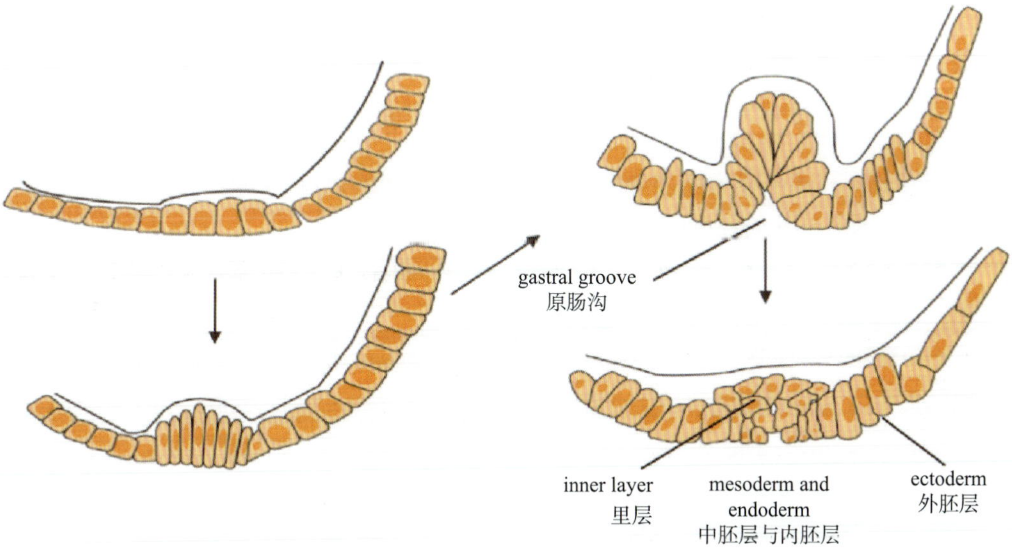

Fig. 18-3 Formation of gastral groove and germ layer (modified from Anderson, 1972).

图 18-3 原肠沟和胚层的形成（仿 **Anderson, 1972**）。

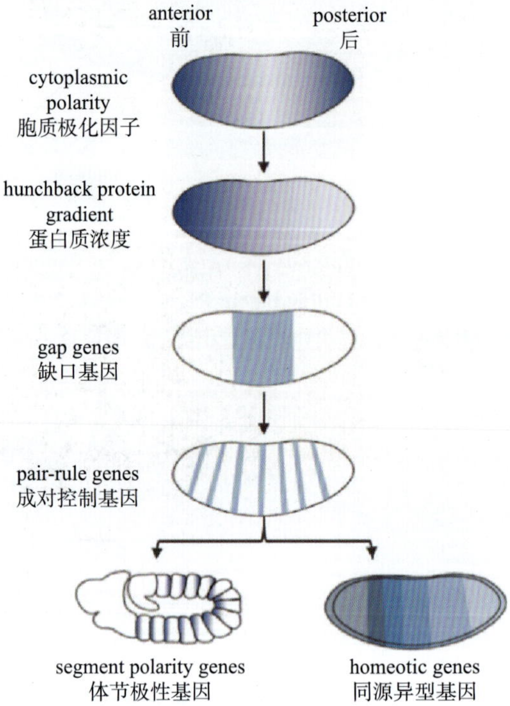

Fig. 18-5 Schematic representation of embryonic segment formation in *Drosophila* (modified from Gilbert, 2010).
图 18-5 果蝇胚胎体节形成示意图（仿 Glibert, 2010）。